Intermittent Rivers and Ephemeral Streams

Ecology and Management

Intermittent Rivers and Ephemeral Streams

Ecology and Management

Edited by

Thibault Datry

Núria Bonada

Andrew Boulton

ACADEMIC PRESS

An imprint of Elsevier

Academic Press is an imprint of Elsevier
125 London Wall, London EC2Y 5AS, United Kingdom
525 B Street, Suite 1800, San Diego, CA 92101-4495, United States
50 Hampshire Street, 5th Floor, Cambridge, MA 02139, United States
The Boulevard, Langford Lane, Kidlington, Oxford OX5 1GB, United Kingdom

Notices
Knowledge and best practice in this field are constantly changing. As new research and experience broaden
our understanding, changes in research methods, professional practices, or medical treatment may become
necessary.

Practitioners and researchers must always rely on their own experience and knowledge in evaluating and using
any information, methods, compounds, or experiments described herein. In using such information or methods
they should be mindful of their own safety and the safety of others, including parties for whom they have a
professional responsibility.

To the fullest extent of the law, neither the Publisher nor the authors, contributors, or editors, assume any
liability for any injury and/or damage to persons or property as a matter of products liability, negligence or
otherwise, or from any use or operation of any methods, products, instructions, or ideas contained in the
material herein.

Library of Congress Cataloging-in-Publication Data
A catalog record for this book is available from the Library of Congress

British Library Cataloguing-in-Publication Data
A catalogue record for this book is available from the British Library

ISBN: 978-0-12-803835-2

For information on all Academic Press publications
visit our website at https://www.elsevier.com/books-and-journals

Working together
to grow libraries in
developing countries

www.elsevier.com • www.bookaid.org

Publisher: Candice Janco
Acquisition Editor: Louisa Hutchins
Editorial Project Manager: Hilary Carr
Production Project Manager: Anitha Sivaraj
Cover Designer: Vicky Pearson Esser

Typeset by SPi Global, India

Contents

Contributors

María Isabel Arce
Institute of Freshwater Ecology and Inland Fisheries (IGB), Berlin, Germany

Brad Autrey
USEPA/ORD/NERL, Cincinnati, OH, United States

Darren S. Baldwin
La Trobe University, Wodonga, VIC, Australia

Susana Bernal
Center for Advanced Studies of Blanes (CEAB-CSIC), Blanes, Spain

Kate S. Boersma
University of San Diego, San Diego, CA, United States

Michael T. Bogan
University of Arizona, Tucson, AZ, United States

Núria Bonada
Universitat de Barcelona (UB), Barcelona, Spain

Nick Bond
Murray-Darling Freshwater Research Centre, La Trobe University, Wodonga, VIC, Australia

Gudrun Bornette
Université de Franche-Comté, Besançon Cedex, France

Andrew J. Boulton
University of New England, Armidale, NSW, Australia

Stephanie M. Carlson
University of California, Berkeley, CA, United States

Eric Chauvet
University of Toulouse, CNRS, Toulouse, France

Edwin T. Chester
Murdoch University, Murdoch, WA, Australia; Deakin University, Warrnambool, VIC, Australia

Ming-Chih Chiu
University of California, Berkeley, CA, United States; National Chung Hsing University, Taichung City, Taiwan

Núria Cid
Universitat de Barcelona (UB), Barcelona, Spain

John Conallin
UNESCO-IHE Institute for Water Education, Delft, Netherlands

Roland Corti
Leibniz-Institute of Freshwater Ecology and Inland Fisheries (IGB), Berlin, Germany; Irstea, UR MALY, centre de Lyon-Villeurbanne, Villeurbanne, France

Katie H. Costigan
University of Louisiana at Lafayette, Lafayette, LA, United States

Clifford N. Dahm
University of New Mexico, Albuquerque, NM, United States

Thibault Datry
Irstea, UR MALY, centre de Lyon-Villeurbanne, Villeurbanne, France

Mélissa De Wilde
Institut Méditerranéen de Biodiversité et d'Ecologie (IMBE), Université d'Avignon et des Pays de Vaucluse, Aix Marseille Université, IUT d'Avignon, Avignon, France

Daniel Escoriza
University of Girona, Girona, Spain

Catherine Febria
University of Canterbury, Christchurch, New Zealand

Debra S. Finn
Missouri State University, Springfield, MO, United States

Ken Fritz
USEPA/ORD/NERL, Cincinnati, OH, United States

Emili García-Berthou
University of Girona, Girona, Spain

Keith Gido
Kansas State University, Manhattan, KS, United States

Rosa Gómez
University of Murcia, Murcia, Spain

Jani Heino
Finnish Environment Institute, Oulu, Finland

Bernard Hugueny
UMR "BOREA" CNRS, DMPA, Museum National d'Histoire Naturelle, Paris, France

Kristin L. Jaeger
USGS Washington Water Science Center, Tacoma, WA, United States

Mark J. Kennard
Griffith University, Nathan, QLD, Australia

Adam Kerezsy
Dr Fish Contracting, Lake Cargelligo, NSW, Australia; Griffith University, Nathan, QLD, Australia

Richard T. Kingsford
UNSW Australia, Sydney, NSW, Australia

Phoebe Koundouri
Athens University of Economics and Business, Athens, Greece; London School of Economics, London, United Kingdom; International Centre for Research on the Environment and the Economy, Athens, Greece

Philip S. Lake
Monash University, Clayton, VIC, Australia

Simone D. Langhans
Leibniz-Institute of Freshwater Ecology and Inland Fisheries (IGB), Berlin, Germany

Catherine Leigh
Griffith University, Nathan, QLD, Australia; Irstea, UR MALY, centre de Lyon-Villeurbanne, Villeurbanne, France; CESAB-FRB, Immeuble Henri Poincaré, Aix-en-Provence, France

David A. Lytle
Missouri State University, Springfield, MO, United States

Maria F. Magalhães
University of Lisbon, Lisbon, Portugal

Eugènia Martí
Center for Advanced Studies of Blanes (CEAB-CSIC), Blanes, Spain

Raphael Mazor
Southern California Coastal Water Research Project, Costa Mesa, CA, United States

Peter A. McHugh
Utah State University, Logan, UT, United States

Angus R. McIntosh
University of Canterbury, Christchurch, New Zealand

Craig A. McLoughlin
Resilient Learning Consultancy, Armidale, NSW, Australia

Katerina Michaelides
University of Bristol, Bristol, United Kingdom; Earth Research Institute, University of California Santa Barbara, Santa Barbara, CA, United States

Marcos Moleón
University of Granada, Granada, Spain

Michael T. Monaghan
Leibniz Institute of Freshwater Ecology and Inland Fisheries (IGB), Berlin, Germany

Juanita Mora-Gómez
University of Girona, Girona, Spain

Ashley L. Murphy
Monash University, Melbourne, VIC, Australia

Vol Norris
Desert Channels Group, Longreach, QLD, Australia

Paul Reich
Arthur Rylah Institute for Environmental Research, Heidelberg, VIC, Australia

Vincent Resh
University of California, Berkeley, CA, United States

Ute Risse-Buhl
Helmholtz Centre for Environmental Research—UFZ, Magdeburg, Germany

Belinda J. Robson
Murdoch University, Murdoch, WA, Australia

Robert J. Rolls
University of Canberra, Bruce, ACT, Australia; Institute for Applied Ecology, University of Canberra, ACT, Australia

Anna M. Romaní
University of Girona, Girona, Spain

Dirk J. Roux
South African National Parks, George, South Africa; Nelson Mandela Metropolitan University, George, South Africa

Albert Ruhi
Arizona State University, Tempe, AZ, United States

Sergi Sabater
University of Girona, Girona, Spain; ICRA, Girona, Spain

María M. Sánchez-Montoya
Leibniz-Institute of Freshwater Ecology and Inland Fisheries (IGB), Berlin, Germany; University of Murcia, Murcia, Spain

José A. Sánchez-Zapata
Miguel Hernández University, Elche, Spain

Eric Sauquet
Irstea, UR MALY, centre de Lyon-Villeurbanne, Villeurbanne, France

Michael Singer
Earth Research Institute, University of California Santa Barbara, Santa Barbara, CA, United States; Cardiff University, Cardiff, United Kingdom

Paul H. Skelton
SAIAB, Grahamstown, South Africa

Ioannis Souliotis
International Centre for Research on the Environment and the Economy, Athens, Greece; Imperial College London, London, United Kingdom

John C. Stella
State University of New York, Syracuse, NY, United States

Alisha L. Steward
Queensland Government, Brisbane, QLD, Australia; Griffith University, Nathan, QLD, Australia

Juliet C. Stromberg
Arizona State University, Tempe, AZ, United States

Rachel Stubbington
Nottingham Trent University, Nottingham, United Kingdom

Nicholas A. Sutfin
Los Alamos National Laboratory, Los Alamos, NM, United States

Pablo A. Tedesco
Université Paul Sabatier Toulouse, Toulouse, France

Xisca Timoner
ICRA, Girona, Spain

Stephen Tooth
Aberystwyth University, Aberystwyth, Wales, United Kingdom

Daniel von Schiller
University of the Basque Country, Bilbao, Spain

Ross Vander Vorste
Virginia Water Resources Research Center, Blacksburg, VA, United States

Markus Weitere
Helmholtz Centre for Environmental Research—UFZ, Magdeburg, Germany

James E. Whitney
Pittsburg State University, Pittsburg, KS, United States

Lydia Zeglin
Kansas State University, Manhattan, KS, United States

Preface

Intermittent rivers and ephemeral streams (IRES) drain over half the world's land surface. Often 'hotspots' of regional biodiversity, especially in arid and semiarid areas, many IRES are also exploited to meet growing human demands for water and other ecosystem services. In the last two decades, research into the ecology of these widespread ecosystems has burgeoned. At the same time, management concerns have intensified because most climate change scenarios predict expansion in the global extent of IRES. Therefore, it is timely to collate the literature on the ecology and management of IRES to see what we know, what we need to know, and how this current and future knowledge might help us to better manage, protect, and restore the diverse types of IRES worldwide.

The idea for this book arose during a meeting of the Intermittent River Biodiversity Analysis and Synthesis (IRBAS, http://irbas.cesab.org) working group in 2013, hosted at the Centre for the synthesis and analysis of biodiversity (CESAB, http://www.cesab.org/index.php/en/) in Aix-en-Provence, France. IRBAS aims to synthesize research on the distribution, ecological features, and biodiversity of IRES to improve our understanding and management of these types of ecosystems. One way to achieve this aim was to invite international experts to write review chapters spanning topics from geomorphology, hydrology, and biogeochemistry to ecology, restoration, and adaptive management in IRES. Elsevier agreed to publish the book, and we were thrilled by the enthusiastic response from the experts we contacted to write various chapters. Some eighty authors from across the world contributed to the 22 chapters of this book and we are grateful to them all. Each chapter was peer-reviewed by at least two referees and we thank these experts for their constructive reviews. Finally, we are indebted to the many photographers who generously provided pictures, the copyright holders who allowed us to reproduce material for the book, and the staff at Elsevier, especially Hilary Carr.

Before closing, some words about the book's content. We urged authors to adopt a global approach, seeking to compare findings from case studies across multiple continents, climates, flow regimes, and land uses to provide a comprehensive perspective of the ecology and management of IRES. We encouraged identification of current knowledge gaps and speculation about future research directions. In particular, we sought insights about how our scientific understanding of IRES might be used to better guide their management. All authors rose admirably to these challenges and this book is the result.

This book is the first to focus entirely on the ecology and management of IRES across the world. We hope you find it interesting, informative, and inspirational. However, there is still much to learn about these dynamic ecosystems and how best to protect their beauty, ecological integrity, and other social values.

Thibault Datry
Núria Bonada
Andrew Boulton

GENERAL INTRODUCTION

1

Thibault Datry*, Núria Bonada[†], Andrew J. Boulton[‡]

Irstea, UR MALY, centre de Lyon-Villeurbanne, Villeurbanne, France[*]
Universitat de Barcelona (UB), Barcelona, Spain[†]
University of New England, Armidale, NSW, Australia[‡]

IN A NUTSHELL

- Intermittent rivers and ephemeral streams (IRES) are flowing waters that cease flow and/or dry at some point along their course
- Multiple natural and anthropogenic processes generate flow intermittence which influences almost every biogeochemical and ecological process in IRES
- IRES are dynamic mosaics of flowing, nonflowing water and terrestrial habitats, harboring distinctive aquatic, semi-aquatic, and terrestrial organisms that use different strategies to persist
- IRES suffer the same impacts of human activities as perennial streams and rivers but are often less valued
- A better understanding of the ecology of IRES will improve their legislative protection, restoration, and management

---

1.1 WHAT ARE INTERMITTENT RIVERS AND EPHEMERAL STREAMS (IRES)?

Flowing waters confined within a channel (except during floods) and moving in one direction are called rivers or streams. Rivers are considered to be larger and deeper than streams, although the distinction is a loose one of common usage rather than one based on fixed sizes and depths. The same common usage applies to describing differences in patterns of flow; "ephemeral" implies a shorter flow duration and lower predictability than "intermittent" but again there are no fixed boundaries. Thus, given the broad association of channel size and flow duration, a stream is more likely to be ephemeral and a river intermittent, prompting the distinction in the title of this book.

Of course, the association of channel size and flow is not this simple in reality (Chapter 2.1) nor is the classification of flow regimes (Chapter 2.2). The scientific literature is peppered with attempts (e.g., Uys and O'Keeffe, 1997; Williams, 2006; Gallart et al., 2012) to assign names to classes of streams and rivers whose flows cease for varying periods with varying predictability (Table 1.1). However, a global consensus remains elusive and probably will continue to do so, especially as many intermittently flowing waterways dry for widely different periods in different years, leading to variation among different categories within a single waterway. Therefore, rather than enter this semantic minefield, we refer to "intermittent rivers and ephemeral streams" and adopt the acronym "IRES" in this book as a shorthand term for all flowing waters that cease flow and/or dry completely at some point along their course.

Intermittent Rivers and Ephemeral Streams. http://dx.doi.org/10.1016/B978-0-12-803835-2.00001-2

Table 1.1 Examples of the diverse definitions of different types of IRES

Term	Definition	References
Ephemeral	Rivers that flow less than 20% of the time	Matthews (1988)
	Rivers that run for short periods after rain has fallen high in their catchments	Day (1990)
	Rivers that flow for less time than they are dry. Flow or flood for short periods of most years in a 5-year period, in response to unpredictable high rainfall events. Support a series of pools in parts of the channel	Uys and O'Keeffe (1997) Stringer and Perkins (2001)
	A channel formed by water during or immediately after precipitation events as indicated by an absence of forest litter and exposure of mineral soil	
	Rivers with dry stream beds in the dry season or even for longer periods	Bonada et al. (2007)
	Rivers/reaches receiving only river runoff because the channel is always above the water table	Larned et al. (2008)
	Rivers which form dry stream beds as the water disappears	Anna et al. (2009)
	Only filled after unpredictable rainfall and runoff. Surface water dries within days to weeks of filling and can support only short-lived aquatic life	Boulton et al. (2014)
Episodic	Streams and rivers—mostly in arid regions—which flow when unpredictable rain has fallen	Bayly and Williams (1973)
	Flow does not necessarily occur every year	King and Tharme (1993)
	Flow that only occurs after rainfall episodes	Davies et al. (1994)
	Highly flashy systems that flow or flood only in response to extreme rainfall events, usually high in their catchments. May not flow in a 5-year period or may flow only once in several years	Uys and O'Keeffe (1997) Williams (2006)
	Dry 9 years out of 10, with rare and very irregular flooding or wet periods which may last for a few months	
	Annual inflow is less than the minimum annual loss in 90% of years. Usually dry but filled after rare and large unpredictable rainfall events. Surface water persists for months to years and often supports longer-lived aquatic life (than ephemeral streams)	Boulton et al. (2014)
Temporary	Rivers with relatively regular, seasonally intermittent discharge	Boulton and Lake (1988)
	Rivers with intermittent discharge	Davies et al. (1994)
	Rivers in which the entire bed dries	Delucchi (1988)
	Flow stops and surface water may disappear along parts of the channel either yearly or during 2 or more years in 5	Uys and O'Keeffe (1997) Williams (2006)
	Natural bodies of water that experience a recurrent dry phase of varying duration	Acuña et al. (2014)
	Waterways that cease to flow at some points in space and time along their course	

Table 1.1 Examples of the diverse definitions of different types of IRES—cont'd

Term	Definition	References
Intermittent	Streams or rivers which flow only seasonally and are otherwise dry	Bayly and Williams (1973)
	Rivers with intermittent discharge	Towns (1985) and Boulton and Lake (1988)
	Rivers that flow for 20%–80% of the year	Matthews (1988)
	Rivers that only dry in parts	Delucchi (1988)
	These rivers cease to flow and may dry along parts of their lengths for a variable period annually, or for 2 or more years in 5. Flow may recommence seasonally, or highly variably, depending on climatic influences and predictability of rainfall in the area. An intermittent river may experience several cycles of flow, no flow, and drying in a single year	Uys and O'Keeffe (1997) Stringer and Perkins (2001)
	Rivers that hold water during wet portions of the year	
	Natural bodies of water that experience a recurrent dry phase of varying duration	Williams (2006)
	Rivers with isolated pools during the dry season that may persist to the end of summer	Bonada et al. (2007)
	Rivers/reaches receiving groundwater when the water table intersects the channel and also might receive runoff	Larned et al. (2008)
	Rivers which form chains of lotic features (isolated pools) as flow disappears but surface water is still present	Anna et al. (2009)
	Alternately wet and dry but less frequently or regularly than seasonal waters. Surface water persists for months to years and often supports longer lived aquatic life (than ephemeral streams)	Boulton et al. (2014)
Seasonal	Flow that occurs for more than half the year, every year, during the same season	King and Tharme (1993)
	Alternately wet and dry every year, according to season. Usually fills and dries predictably and annually. Surface water persists for months, long enough for some plants and animals to complete the aquatic stages of their life cycles	Boulton et al. (2014)
Dryland	The rivers of arid and semiarid regions	Davies et al. (1994)
Interrupted	Rivers that flow for less than 20% of the time	Matthews (1988)
Nonperennial	Streams that dry out completely or for parts of their length during summer	Chester and Robson (2011)
Near permanent	Predictable flooding, though water levels may vary. The annual input of water is greater than the losses in 9 years out of 10	Williams (2006)
	Predictably filled, although water levels may vary. Annual inflow exceeds minimum annual loss in 90% of years. During extreme droughts, these waters may dry. Usually supports diverse aquatic life, much of which cannot tolerate desiccation	Boulton et al. (2014)

The main criteria used to separate them are duration and predictability of flow but these have not been used consistently among authors, leading to further confusion.

We accept that by some definitions (Table 1.1), rivers can be ephemeral and streams intermittent, and that many authors prefer to refer to temporary or nonperennial streams and rivers. We also acknowledge there are many local names such as winterbournes, wadis, arroyos, and ramblas (e.g., Steward et al., 2012) used to describe IRES. This diversity of names is not surprising given the variety of IRES occurring worldwide (Fig. 1.1) and indicates that IRES are often recognized by local residents as unique parts of the landscape that deserve unique names.

Several other terms commonly used in this book to describe aspects of IRES also deserve early definition. The first is "flow cessation," used to refer to the loss of surface flow along the channel, typically evident when riffles dry, resulting in chains of isolated pools (Fig. 1.2) that may or may not be connected by subsurface (hyporheic) flow. Periods of zero flow have important repercussions for biota and abiotic processes that rely on flow and that typify perennial streams and rivers, a key theme repeatedly discussed in the chapters to follow. The second term is "drying," used to refer to the

FIG. 1.1

Different types of IRES from across the world: (a) unnamed karstic stream, West Coast, South Island, New Zealand, (b) Río Seco, Chaco, Bolivia, (c) Asse River, Provence, France, (d) unnamed gravel-bed stream, West Coast, South Island, New Zealand, (e) unnamed stream, Altiplano, Bolivia, (f) Chaki Mayu, Amazonia, Bolivia, (g) Clauge, Jura, France, (h) Calavon, Provence, France, and (i) Río Hozgarganta, Andalucía, Spain.

Photos: Courtesy T. Datry (a–f), B. Launay (g and h), and N. Bonada (i).

FIG. 1.2

Four different hydrological phases in the Calavon River, southeastern France: (a) flowing, (b) nonflowing, (c) dry, and (d) flooding.

Photos: Courtesy B. Launay.

disappearance of surface water in the channel (Fig. 1.2), again with or without hyporheic flow. We prefer this term over "drought," a word that is better reserved for the climatic process of drying and its broader effects (e.g., Lake, 2011) and over "drawdown" which is used in hydrogeology to refer to a decline in the water table. The third term frequently used in this book is "flow intermittence" (sometimes termed "intermittency"), referring to the temporal sequence of flow cessation and resumption that characterizes all IRES. In the chapters to follow, other specific terms may be used but will always include careful definition and explanation.

1.2 CAUSES OF FLOW INTERMITTENCE

Under natural conditions, many different processes generate flow intermittence in IRES (Table 1.2). As these processes often act in concert, disentangling their relative importance is difficult and probably unnecessary. However, some causes of intermittence predominate in different combinations of climate, water source, and channel features (e.g., width, bed permeability). For example, in alpine and arctic regions, flow ceases when water at the surface and in the shallow subsurface zone freezes (Robinson et al., 2016). IRES in the McMurdo Dry Valley in Antarctica (McKnight et al., 1999) flow for only a few months during austral summer before the onset of cooler weather freezes the free water and flow ceases. In contrast, flow intermittence in semiarid and arid regions is largely driven by low precipitation and high rates of direct evaporation and evapotranspiration through plants (Table 1.2). Flowing tracts of water in spatially intermittent IRES such as Sycamore Creek in the Sonoran Desert, Arizona, fluctuate widely in response to the temporal cycles of evapotranspiration and precipitation (Stanley et al., 1997).

In addition to climatic drivers, physical features of the channel, shallow alluvium, and catchment govern flow intermittence in many IRES (Chapter 2.1). Transmission loss is the infiltration of surface water into the streambed as a function of bed porosity and water table depth (Table 1.2). In IRES with coarse sediments (e.g., gravel-bed rivers) and water tables well below the channel (Sophocleous, 2002), transmission losses may be especially high, and these "losing" streams contribute substantially

Table 1.2 Types of IRES classified by predominant form of water loss

IRES type	Processes causing flow intermittence	Examples
Snowmelt and glacial melt water streams	Cessation of melting/ablation, freezing of stream surface and shallow subsurface water	Dry Valley streams, Antarctica; Val Roseg and Macun catchments, Switzerland
Perched and semiperched alluvial rivers	Transmission losses, depletion of bank storage, and floodplain aquifer	Albarine River, France; Selwyn River, New Zealand; Tagliamento River, Italy; Cooper Creek, Australia; Kuiseb River, Namibia; Mojave River, USA; Shashane and Wenlock Rivers, Zimbabwe
Nonperched rivers in arid and semiarid regions	Depletion of surface water and shallow groundwater by direct evaporation and evapotranspiration	Sycamore Creek, USA; Oued Chelif, Algeria
Low-order and headwater streams	Cessation of overland flow, depletion of saturated soil water or hillslope aquifer	Werribee and Lerderderg Rivers, Australia; Riera de Fuirosos, Spain; Maybeso Creek and Stillman Creek watershed, USA
Permafrost streams	Soil freezing, soil water, and wetland recession	Granger Creek and Wolf Creek, Canada
Karstic streams	Transmission loss, cessation of spring discharge, or groundwater inputs	River Lathkill and River Wye, UK; Coulazou River, France; Alme River, Germany
Lake outlets	Lake level drops below outlet elevation	Shadow Lake, Canada

Modified from Larned, S.T., Datry, T., Arscott, D.B., Tockner, K., 2010. Emerging concepts in temporary-river ecology. Freshwater Biology, 55, 717–738.

BOX 1.1 NONNATURAL CAUSES OF FLOW INTERMITTENCE

In addition to the many natural causes of intermittence, flow cessation can result from one or more human activities including alterations of land use, flow regulation, surface and/or groundwater extraction, and reduced precipitation and increased evaporation resulting from climate change (Palmer et al., 2008; Steward et al., 2012). Even large rivers are not immune to intermittence caused by humans. For example, the Colorado River, which flows 2334 km from its headwaters in the Rocky Mountains of Colorado to the Gulf of California, has been transformed from a perennial river to one that has flowed intermittently in most years since the completion of the Glen Canyon Dam in 1963 (Gleick, 2003). Another example of a large river becoming intermittent due to human activities is the Yellow River in China (Changming and Shifeng, 2002; Fu et al., 2004). This 5464-km river, which was perennial before 1972, is now intermittent and connects to the sea only infrequently. Flow intermittence is a direct result of the construction of 12 major dams along the river, coupled with a dramatic increase in water abstraction for direct human consumption and agricultural use. In addition to the barriers presented by the dams, longitudinal hydrological connectivity (Chapter 2.3) is also disrupted by an extended drying segment over 500 km long (Changming and Shifeng, 2002).

to aquifer recharge in many areas (Steward et al., 2012). Downward shifts in water tables in response to seasonal fluctuations and human activities can cause flow intermittence in streams or reaches that are normally "gaining" (Sophocleous, 2002) and may eventually lead to drying along the entire channel (Larned et al., 2008).

The upper reaches of most headwater and low-order streams flow intermittently, largely driven by rainfall (Benstead and Leigh, 2012). Channel flow ceases when overland flow stops and saturated soil water is depleted (Table 1.2); these processes are strongly influenced by vegetation cover, topographic relief, and the extent of bedrock in the upper catchment, along with rainfall. Streams draining lakes typically cease to flow when lake levels fall (Mielko and Woo, 2006) whereas IRES in karstic regions may flow intermittently in response to fluctuating groundwater discharge (Wood et al., 2005) (Table 1.2).

Intermittence also arises nonnaturally, caused by human activities (Box 1.1). In addition to converting perennial streams and rivers into IRES, these activities may impact on naturally IRES by lengthening the duration and extent of the zero-flow periods and altering their frequency and timing (Chapter 5.1). The interactions between natural and anthropogenic intermittence are poorly understood. This is especially true for the complex effects of altered flow regimes on sediment regimes (Chapter 2.1), water physicochemistry (Chapter 3.1), biogeochemical processes (Chapter 3.2), and the biota (Chapters 4.1–4.6) in IRES.

1.3 GLOBAL DISTRIBUTION AND AREAL IMPORTANCE OF IRES

IRES occur on all continents, including Antarctica. They are the predominant water bodies in hyperarid, arid, semiarid, Mediterranean, and dry-subhumid regions, which together represent almost half of the Earth's land surface (Tooth, 2000; Bonada and Resh, 2013). In many parts of the world, most of the channel lengths of natural waterways are intermittent. For example, over 70% of the 3.5 million km of river channels in Australia measured at the 1:250,000 scale are probably intermittent (Sheldon et al., 2010), and in American southwestern states such as Arizona and New Mexico, IRES comprise up to 94% of the total river length (Levick et al., 2008).

Even in temperate and humid areas, IRES are abundant. This is notably true for headwater streams. Headwater streams make up over 70% of the channel length of most river networks and are typically intermittent (Meyer and Wallace, 2001; Fritz et al., 2013; Datry et al., 2014), yet are often ignored when regional estimates are made of flow regimes. For example, Nadeau and Rains (2007) used the National Hydrography Dataset to estimate that 59% of rivers and streams in the United States are intermittent. However, this dataset probably severely underestimated the total extent of IRES because it excluded stream segments less than one mile (1.6 km) in length and was based on 1:100,000 scale topographic maps (Benstead and Leigh, 2012). In addition to headwaters, many large IRES are reported from temperate and humid areas (e.g., Buttle et al., 2012; Snelder et al., 2013). Indeed, IRES appear to be the most common flowing waters on Earth.

1.4 TRENDS IN A CONTEXT OF WATER SCARCITY AND CLIMATE CHANGE

The flow regimes of streams and rivers are changing worldwide, mainly in response to changing climates coupled with rapidly increasing human demands for fresh water (Chapter 5.1). As a result, there is an ongoing global increase in the occurrence and spatial and temporal extent of IRES. This increase is likely to continue, with shifts from perennial to intermittent flow regimes projected by the 2050s for many parts of the world, including northeastern and southwestern Australia, Brazil, California, the Caribbean, southern Africa, West Africa, and around the Mediterranean Basin (Döll and Schmied, 2012). Conversely, some naturally IRES are predicted to become perennial, such as in Siberia and parts of Canada and Alaska due to warmer winters (Döll and Schmied, 2012).

Meanwhile, flows in many naturally IRES are declining and dry periods are getting longer. For example, in the intermittent Albarine River in temperate eastern France, hydrological modeling using various rainfall-runoff models coupled to climate change scenarios suggests that annual flow intermittence will increase on average (considering all Intergovernmental Panel on Climate Change scenarios) by 5% by 2050 (Cipriani et al., 2014). Furthermore, decreased annual precipitation (on average, 6%) and increased summer temperatures (on average, by 2°C) will extend dry periods both spatially and in duration.

The trend of increasing intermittence in IRES is not universal. Controlled releases from dams and weirs; discharge of agricultural, industrial, and urban effluents; and interbasin transfers have all contributed to decreased intermittence in many IRES, turning some of them perennial (Hassan and Egozi, 2001; Steward et al., 2012; Datry et al., 2014). In some water-scarce areas, the baseflow of many formerly intermittent urban rivers is now maintained by wastewater effluent (Luthy et al., 2015). In other areas, seasonal patterns of intermittence have been reversed. For example, where channels are used to carry irrigation flows released from upstream dams, peak flows from dam releases now occur when flow in the channel used to cease whereas instream dams now retain water when the downstream river historically flowed, sometimes causing flow to cease (Barnett and Pierce, 2009).

1.5 ECOLOGICAL FEATURES OF IRES

One distinctive feature of all IRES is that they are tightly coupled aquatic-terrestrial systems that are highly variable in both space and time as a result of flow intermittence and, in many cases, drying. In IRES over time, the wetting-drying cycle generates shifting habitat mosaics of lotic (flowing), lentic (nonflowing pools), and terrestrial (dry) habitats (Datry et al., 2014; Chapters 2.3 and 4.9). All or any combination of these habitats can occur simultaneously and anywhere along the network (Fig. 1.3). As the habitats expand and contract, their spatial arrangement, temporal turnover, and hydrological connectivity constantly shift in response to changes in surface discharge and groundwater level (Larned et al., 2010; Datry et al., 2016a; Fig. 1.3).

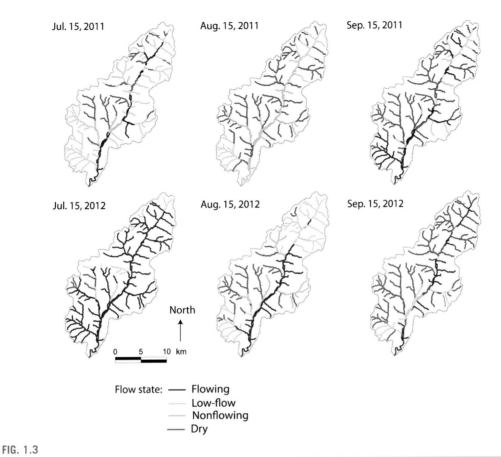

FIG. 1.3

Example of a shifting mosaic of flowing, low-flow, nonflowing, and dry habitats in the Tude River, France.

From Datry, T., Bonada, N., Heino, J., 2016b. Towards understanding metacommunity organisation in highly dynamic systems. Oikos, 125, 149–159.

These habitats are colonized by different communities comprising taxa with preferences for each particular habitat. The regime of intermittent flow, wetting, and drying includes major disturbances for both aquatic and terrestrial organisms of IRES. Aquatic biota must contend with long periods of no flow or no surface water whereas when flow resumes, terrestrial biota must evade or tolerate inundation that may last for months or even years. Different strategies (and combinations of strategies) enable aquatic, semiaquatic, and terrestrial organisms to persist in IRES (Chapters 4.1–4.6). Some organisms withstand the disturbance *in situ* through different physiological or behavioral adaptations, collectively referred to as "resistance" (Chapter 4.8). Species from several aquatic groups can persist for months or years in dry river sediments as cysts, cocoons, or diapausing juveniles or adults (e.g., Boulton et al., 1992; Williams, 2006; Stubbington and Datry, 2013). Conversely, terrestrial organisms can have adaptations, such as the ability to float or to store air, that allow them to escape advancing wetting fronts or to resist submersion (Boumezzough and Musso, 1983; Adis and Junk, 2002; Lambeets et al., 2008).

Other organisms lack resistant stages but instead have the ability to quickly recolonize after disturbance, referred to as "resilience." Examples include various life history traits and behavioral or morphological adaptations that promote high dispersal capacity, extensive use of refuges, short life cycles, and high fecundity (Fritz and Dodds, 2004; Williams, 2006; Datry et al., 2014). Some aquatic species with strong powers of dispersal quickly return to rewetted habitats by drifting from upstream waters or flying in from perennial refuges (Chester and Robson, 2011) whereas others recolonize from the hyporheic zone (Vander Vorste et al., 2016). Thus far, few strategies for resilience of terrestrial communities to inundation have been quantified in IRES, although several semiaquatic taxa avoid inundation by detecting rainfall cues (Lytle, 1999). It has also been suggested that some terrestrial organisms use advancing wetted fronts to colonize downstream terrestrial habitats (Corti and Datry, 2012; Rosado et al., 2014; Chapter 4.8).

Flow intermittence in IRES disrupts hydrological connectivity along longitudinal, lateral, and vertical dimensions (Chapter 2.3), creating a highly dynamic mosaic of habitat patches. These habitat patches are constantly being rearranged at multiple spatial and temporal scales, leading to perpetual fluctuations in the ecology and population or community structure of their residents. The temporal turnover and varying connectivity of these habitat patches in IRES render these ecosystems ideal arenas for studying metapopulations and metacommunities (Box 1.2) and disentangling the intricate ecological interactions among intermittence, biogeochemical processes, and biotic interactions.

BOX 1.2 IRES AS IDEAL ARENAS FOR METASYSTEM APPROACHES

Constant shifts in the spatial arrangement of habitat patches and their temporal turnover and connectivity generate recurrent ecological successions. These, in turn, create highly dynamic metapopulations and metacommunities (Jaeger et al., 2014; Cañedo-Argüelles et al., 2015; Datry et al., 2016a; Chapter 4.9). Recent work (Datry et al., 2016b) indicates that metapopulation and metacommunity patterns reported in perennial rivers are different in IRES, due to the effects of recurrent flow cessation events on population and community processes. For example, the role of regional (dispersal) processes may override that of local (environmental filtering) processes, especially following flow resumption when aquatic populations and communities are recovering (Chapter 4.9).

This could explain the general patterns of nestedness found in IRES. Nested patterns are those in which the species composition of less diverse assemblages (such as might be found in IRES) is a nested subset of more diverse assemblages (such as might be found in perennial streams and rivers). For example, several studies have shown that the number of fish species in IRES decreases as distance to perennial waters increases (e.g., Davey and Kelly, 2007; Colvin et al., 2009), and that many aquatic invertebrate communities are strongly nested along gradients of flow intermittence (Arscott et al., 2010; Datry et al., 2014). Conversely, local factors may predominate during initial phases of drying, and this could induce spatial turnover in species composition at the network scale (Chapter 4.9). Conceptual developments are needed to advance our understanding of metapopulations and metacommunities in dynamic systems such as IRES to guide effective management strategies.

Metasystem approaches are also highly relevant to understand how flow intermittence alters biogeochemical cycles. For example, organic matter dynamics in IRES are more pulsed than in perennial systems (Chapter 3.2); under lentic and dry conditions, the decomposition of leaf litter is strongly reduced or even halted (Corti et al., 2011). Large quantities of leaf litter accumulate on dry streambeds and are then entrained when flow resumes, along with myriads of terrestrial organisms (Corti and Datry, 2012; Rosado et al., 2014). The ecological consequences of these pulses are poorly understood but potentially range from generating anoxic "blackwater" events and altering downstream physicochemistry (Chapter 3.1) through to stimulating food webs downstream (Corti and Datry, 2012; Datry et al., 2014; Bixby et al., 2015; Chapter 4.7). Another major knowledge gap is understanding how the inclusion of IRES would alter current large-scale estimates of how much carbon is processed within river networks (Battin et al., 2009) or how much inland waters contribute to CO_2 outgassing (Raymond et al., 2013).

1.6 LEGISLATION, PROTECTION, RESTORATION, AND MANAGEMENT OF IRES

IRES face the same impacts from human activities (Fig. 1.4) as perennial streams and rivers. Broadly speaking, these impacts are alterations to the timing, volumes and durations of flow, changes to surface and groundwater quality, invasion by exotic species, and changes to the morphology of the channel and its catchment (Cooper et al., 2013; Acuña et al., 2014; Chapter 5.1), and all of them affect ecological processes and native biodiversity (Dudgeon et al., 2006). These effects interact in different ways and are superimposed on the projected effects of climate change to produce complex and often unpredictable alterations of sediment dynamics (Chapter 2.1), flow regime (Chapter 2.2), hydrological connectivity (Chapter 2.3), and water quality (Chapters 3.1 and 3.2). However, scientific research has focused far more on human impacts in perennial waterways than in IRES, and we have much yet to learn about their specific impacts in IRES, especially in the long term (Leigh et al., 2016).

The overwhelming focus on perennial streams and rivers may largely arise because they are more highly valued by the public than IRES (Chapter 5.2). For example, Armstrong et al. (2012), who interviewed landowners living along either perennial rivers or IRES in a small Pennsylvanian catchment, concluded that "Landowner perceptions and attitudes reveal a disproportionate lack of concern toward ephemeral or intermittent streams." The ecosystem services of IRES, especially the more subtle ones of supporting and regulating services (Millennium Ecosystem Assessment, 2005), are currently little appreciated by society (Boulton, 2014), and therefore, it is not surprising that research on the likely effects of human activities has focused on perennial waterways.

Another reason for the focus on perennial waterways may be the difficulty of disentangling the ecological effects of natural flow intermittence from the impacts of particular human activities on aquatic biota. For example, the poor water quality associated with declining concentrations of dissolved oxygen that occurs naturally in drying pools (Chapter 3.1) resembles environmental conditions caused by nutrient enrichment from agricultural runoff or sewage entering a stream, and the effects upon resident aquatic biota are very similar. Standard biological monitoring techniques, using aquatic macroinvertebrates, for example, are seldom able to differentiate changes in lotic assemblage composition caused by organic pollution in IRES from the changes that occur naturally in response to flow cessation or drying (Sheldon, 2005). This has prompted investigations into ecological indices and tools that might be more powerful for assessing the ecological status of IRES (e.g., Mazor et al., 2014; Prat et al., 2014) and improve interpretation of different metrics (Cid et al., 2015) as well as approaches that incorporate IRES when defining reference conditions (Munné and Prat, 2011). Major challenges still exist, such as assessing when is the best time to sample or what metrics are the most reliable for detecting different human impacts on IRES (Chapter 5.1).

Ominously, a rather unfortunate assumption has often been drawn from the similarity of ecological responses by many IRES taxa to flow intermittence and to human impacts on water quality and flow. This assumption is that the biota of IRES are less likely to be adversely affected by human modifications of flow regime and water quality because the residents of IRES are more tolerant of "harsh" conditions and are more resilient or resistant to physical and chemical disturbances (Boulton et al., 2000). Although this may be true in some circumstances, it should not be used as an excuse for not assessing potential impacts of human activities on IRES when conducting environmental impact assessments or for assuming that IRES are somehow "second-class" waterways that can be ignored. The combination of ignorance of the ecosystem services and values of IRES and the assumed tolerance of their biota to

FIG. 1.4

Photos of different impacts in IRES in different parts of the world: (a) an IRES that is now perennial, carrying wastewater effluent (unnamed river, Fes, Morocco), (b) an IRES channelized for a golf course (unnamed river, Málaga, Spain), (c) an artificial IRES created by hydropower generation (Ter River, Llanars, Spain), (d) an urban IRES that is completely channelized (El Duí River, Collioure, France), (e) an IRES with an extensive olive tree plantation in the catchment (unnamed river, Valenzuela, Spain), (f) an IRES invaded by the exotic species *Arundo donax* (Barranc del Carraixet, Valencia, Spain), (g) an IRES heavily used by livestock (Barranc del Carraixet, Valencia, Spain), and (h) an IRES with urbanized river banks and used as a car trail during the dry period (Seco River, Vélez-Málaga, Spain).

Photos: Courtesy N. Bonada (a, b, d, e, h), P. Bonada (c), and N. Cid (f and g).

the impacts of human activities is probably the most serious threat currently jeopardizing the ecological condition of these waterways around the world (Acuña et al., 2014; Boulton, 2014).

Far more research has been done on restoration and conservation of perennial streams and rivers than on intermittently flowing ones (Chapter 5.4). The few areas that have been protected, mostly in national parks, have usually been designed to conserve terrestrial landscapes and seldom adequately protect freshwater taxa (Abellán et al., 2007; Guareschi et al., 2012; Hermoso et al., 2015). In some countries, natural riverine reserves and freshwater protected areas are being implemented but rarely do they specifically target IRES. Furthermore, only a few countries have well-developed legislation that recognizes IRES, assisting their protection and wise management (Chapter 5.3). Finally, there have been very few efforts to specifically restore IRES damaged by human activities (Chapter 5.4), despite the widespread damage to these ecosystems (Leigh et al., 2016). In many cases, alterations of flow regime and water quality in IRES have been severe enough to produce such novel ecosystems that restoration back to some natural or near-natural condition is impossible and "reconciliation ecology" is seen as a more realistic framework for guiding their management (Moyle, 2014). Whatever guiding image is chosen, approaches such as strategic adaptive management (Chapter 5.5) are already improving the integrated management of IRES in some countries, building on the increased scientific understanding and legislative protection in recent years.

1.7 THE STRUCTURE OF THIS BOOK

The ongoing changes in flow patterns and global distribution of IRES raise a number of key questions: What are the effects of such changes on the ecology of IRES? How do these effects vary across different climates, flow durations, and landforms? How should we manage IRES to minimize the negative ecological effects of altered flow regimes yet optimize our use of their ecosystem services, especially access to potable fresh water? Are management strategies developed for perennial streams and rivers directly applicable to IRES? Or do refinements need to be made to acknowledge the centrality of flow intermittence to the ecology of IRES?

To answer these and other questions, we need to understand the physical, chemical, and biological features that influence the ecology of the biotic groups and ecosystem processes that characterize IRES, and the way that organisms in these systems have evolved. We also need to know the impacts that various human activities have upon IRES and the mechanisms (i.e., cause-effect relationships) behind these impacts. Thus, by using our expanding knowledge of the ecology of IRES, we can explore potential management approaches such as flow management, river restoration, and conservation. Using our understanding of the ecology of IRES to inform potential management approaches is a central theme of this book, and the chapters are arranged to follow this logical sequence (Fig. 1.5).

The first section, comprising three chapters (Fig. 1.5), describes the geomorphological and hydrological features of IRES. Chapter 2.1 outlines how different tectonic settings and lithology in different climates interact with the distinctive hydrology of IRES to generate a diverse array of sediment supply processes and channel geomorphology at multiple scales of space and time. The distinctive hydrological features, especially the intermittent and highly variable discharge, that characterize IRES and their flow regimes are discussed more fully in Chapter 2.2. In turn, the diversity of flow regimes in IRES creates spatial trends in longitudinal, lateral, and vertical hydrological connectivity that interact and vary over time (Chapter 2.3). These features and interactions influence almost every physical, chemical, and biological process occurring in IRES at local, catchment, and landscape scales.

Chapter 1: GENERAL INTRODUCTION

FIRST SECTION: PHYSICAL FEATURES AND PROCESSES

Chapter 2.1: Geomorphology and sediment regimes

Chapter 2.2: Flow regimes

Chapter 2.3: Hydrological connectivity

SECOND SECTION: CHEMICAL FEATURES AND PROCESSES

Chapter 3.1: Water physicochemistry

Chapter 3.2: Nutrient and organic matter dynamics

THIRD SECTION: ECOLOGICAL FEATURES AND PROCESSES

THE BIOTA OF INTERMITTENT RIVERS

Chapter 4.1: Prokaryotes, fungi, and protozoans

Chapter 4.2: Algae and vascular plants

Chapter 4.3: Aquatic invertebrates

Chapter 4.4: Terrestrial and semiaquatic invertebrates

Chapter 4.5: Fishes

Chapter 4.6: Amphibians, reptiles, birds, and mammals

ECOLOGICAL AND EVOLUTIONARY PROCESSES

Chapter 4.7: Food webs and trophic interactions

Chapter 4.8: Resistance, resilience, and community recovery

Chapter 4.9: Habitat fragmentation, metapopulation, and metacommunity dynamics

Chapter 4.10: Genetic, evolutionary, and biogeographical processes

FOURTH SECTION: THREATS AND MANAGEMENT

Chapter 5.1: Anthropogenic threats

Chapter 5.2: Ecosystem services, values, and societal perceptions

Chapter 5.3: Governance, legislation, and protection

Chapter 5.4: Restoration

Chapter 5.5: Strategic adaptive management

Chapter 6: CONCLUSIONS

FIG. 1.5

This book's structure showing the main sections and their constituent chapters.

Photos: N. Bonada (all except middle photo) and T. Herrera (middle photo).

The second section, comprising two chapters (Fig. 1.5), deals with the physicochemical and bio-geochemical features of IRES. Spatiotemporal variations in flow regime and hydrological connectivity, particularly during low flows and drying, strongly influence water quality in IRES (Chapter 3.1). This is especially true of nutrient and organic matter dynamics at both reach and catchment scales, and there are also important links with hyporheic and riparian zone ecosystems as described in Chapter 3.2. The major influences of pulsed flows and, in many IRES, drying on physicochemistry and biogeochemical processes are a key theme in both chapters.

The third section (Fig. 1.5) describes the major groups of biota of IRES (six chapters) before re-viewing the main ecological and evolutionary processes (four chapters). The ecology of the major groups is discussed primarily in terms of distribution, ecological roles, and interactions and commences with prokaryotes, fungi and protozoans (Chapter 4.1), and algae and vascular plants (Chapter 4.2). The major groups covered in these first two chapters play fundamental roles in organic matter production and decomposition in all IRES, with rates and fates being largely governed by water availability and flow intermittence. In turn, these major groups underpin the food webs and secondary production of IRES whose components include aquatic (Chapter 4.3) and semiaquatic and terrestrial invertebrates (Chapter 4.4); fishes (Chapter 4.5); and other vertebrates such as amphibians, reptiles, birds, and mam-mals (Chapter 4.6). Common themes throughout all these chapters include each group's ecological responses to rewetting and drying, the various modes of resistance and resilience, and the numerous adaptations to flow intermittence and variable habitat conditions.

Several of these ecological and evolutionary themes are explored in more detail in the next four chapters. Chapter 4.7 focuses on food web dynamics, trophic interactions, and how drying reduces the length of aquatic food chains but may initiate a switch to terrestrial energy pathways. Particularly important is the spatial and temporal fragmentation of habitats that substantially alters the strength of biological interactions such as predator-prey dynamics. This habitat fragmentation also necessitates various strategies for recolonization, resistance, and dispersal (Chapter 4.8) and creates a "dynamic mosaic" of lotic, lentic, and terrestrial habitats in IRES, providing opportunities to test theories about metapopulations, metacommunities, and metaecosystems (Chapter 4.9, Box 1.2). These three levels of system organization are subjected to different evolutionary pressures in IRES resulting in the various species traits and trade-offs described in Chapter 4.10.

The fourth and final section of the book describes the challenges of environmental management in IRES and complements the preceding sections on physical, chemical, and ecological aspects. It comprises five chapters (Fig. 1.5). The many human activities that impact upon IRES are discussed in Chapter 5.1, with particular emphasis on alterations of flow regime, water quality, biota, and ecological processes. Appreciating what is being lost through lack of appropriate management involves under-standing and evaluating the various ecosystem services provided by IRES to help address societal per-ceptions that currently undervalue these assets (Chapter 5.2). Chapter 5.3 describes some of the current legislative frameworks that, in many parts of the world, still fail to fully acknowledge the importance of IRES and their significance to river networks. However, attitudes and legal perspectives are changing, and this chapter also reviews the integration of ecological insights and systems thinking with juris-dictional changes in effective water governance to better protect threatened assets. The fourth chapter in this section explores the restoration of IRES, drawing contrasts, and parallels with the restoration of perennial streams and rivers (Chapter 5.4). Drawing together these threads of mitigating human impacts on IRES against a backdrop of societal ignorance of these ecosystems' values, the legislative limitations in many countries, and the limited understanding of restoration approaches, Chapter 5.5

outlines strategic adaptive management as a way to link science to value-driven objectives. In this way, management strategies both use and inform the physical, chemical, and biological insights presented in the first three sections, and the final chapter of this section highlights the benefits of successful integration of science and management of IRES and their surrounding landscapes.

Principal themes and conclusions from all of the chapters in these four sections are summarized in the book's last chapter (Fig. 1.5). It is important to realize that this research field is rapidly growing, new insights are published frequently, and inevitably some of the new information will challenge ideas and perceptions presented in this book. This is to be expected and is a healthy sign of any fertile discipline. Nonetheless, a global synthesis like this one is crucial to take stock of what is known, collate the currently scattered information on IRES from diverse parts of the world, and provide the first book solely devoted to describing the ecology and management of one of Earth's most common aquatic ecosystems.

ACKNOWLEDGMENTS

We thank P. Bonada, N. Cid, T. Herrera, and B. Launay for permission to reproduce their photos, and Cath Leigh and Rachel Stubbington for valuable comments on an earlier draft of this chapter.

REFERENCES

Abellán, P., Sánchez-Fernández, D., Velasco, J., Millán, A., 2007. Effectiveness of protected area networks in representing freshwater biodiversity: the case of a Mediterranean river basin (south-eastern Spain). Aquat. Conserv. 17, 361–374.

Acuña, V., Datry, T., Marshall, J., Barceló, D., Dahm, C.N., Ginebreda, A., et al., 2014. Why should we care about temporary waterways? Science 343, 1080–1082.

Adis, J., Junk, W.J., 2002. Terrestrial invertebrates inhabiting lowland river floodplains of Central Amazonia and Central Europe: a review. Freshwat. Biol. 47, 711–731.

Anna, A., Yorgos, C., Konstantinos, P., Maria, L., 2009. Do intermittent and ephemeral Mediterranean rivers belong to the same river type? Aquat. Ecol. 43, 465–476.

Armstrong, A., Stedman, R.C., Bishop, J.A., Sullivan, P.J., 2012. What's a stream without water? Disproportionality in headwater regions impacting water quality. Environ. Manag. 50, 849–860.

Arscott, D.B., Larned, S., Scarsbrook, M.R., Lambert, P., 2010. Aquatic invertebrate community structure along an intermittence gradient: Selwyn River, New Zealand. J. North Am. Benthol. Soc. 29, 530–545.

Barnett, T.P., Pierce, D.W., 2009. Sustainable water deliveries from the Colorado River in a changing climate. Proc. Natl. Acad. Sci. U. S. A. 106, 7334–7338.

Battin, T.J., Luyssaert, S., Kaplan, L.A., Aufdenkampe, A.K., Richter, A., Tranvik, L.J., 2009. The boundless carbon cycle. Nat. Geosci. 2, 598–600.

Bayly, I.A.E., Williams, W.D., 1973. Inland Waters and their Ecology. Longman, Melbourne.

Benstead, J.P., Leigh, D.S., 2012. An expanded role for river networks. Nat. Geosci. 5, 678–679.

Bixby, R.J., Cooper, S.D., Gresswell, R.E., Brown, L.E., Dahm, C.N., Dwire, K.A., 2015. Fire effects on aquatic ecosystems: an assessment of the current state of the science. Freshw. Sci. 34, 1340–1350.

Bonada, N., Resh, V.H., 2013. Mediterranean-climate streams and rivers: geographically separated but ecologically comparable freshwater systems. Hydrobiologia 719, 1–29.

Bonada, N., Rieradevall, M., Prat, N., 2007. Macroinvertebrate community structure and biological traits related to flow permanence in a Mediterranean river network. Hydrobiologia 589, 91–106.

Boulton, A.J., 2014. Conservation of ephemeral streams and their ecosystem services: what are we missing? Aquat. Conserv. 24, 733–738.

Boulton, A.J., Lake, P.S., 1988. Australian temporary streams—some ecological characteristics. Verh. Int. Ver. Theor. Angew. Limnol. 23, 1380–1383.

Boulton, A.J., Stanley, E.H., Fisher, S.G., Lake, P.S., 1992. Over-summering strategies of macroinvertebrates in intermittent streams in Australia and Arizona. In: Robarts, R.D., Bothwell, M.L., Robarts, R.D., Bothwell, M.L. (Eds.), Aquatic Ecosystems in Semi-Arid Regions: Implications for Resource Management. NHRI Symposium Series 7, Environment Canada, Saskatoon, pp. 227–237.

Boulton, A.J., Sheldon, F., Thoms, M.C., Stanley, E.H., 2000. Problems and constraints in managing rivers with variable flow regimes. In: Boon, P.J., Davies, B.R., Petts, G.E. (Eds.), Global Perspectives on River Conservation: Science, Policy and Practice. John Wiley and Sons, London, pp. 415–430.

Boulton, A.J., Brock, M.A., Robson, B.J., Ryder, D.S., Chambers, J.M., Davis, J.A., 2014. Australian Freshwater Ecology: Processes and Management, second ed. Wiley-Blackwell, Chichester.

Boumezzough, A., Musso, J.J., 1983. Etude des communautés animales ripicoles du bassin de la rivière Aille (Var-France). I. Aspects biologiques et éco-éthologiques. Ecol. Medit. 9, 31–56.

Buttle, J.M., Boon, S., Peters, D.L., Spence, C., van Meerveld, H.J., Whitfield, P.H., 2012. An overview of temporary stream hydrology in Canada. Can. Water Res. J. 37, 279–310.

Cañedo-Argüelles, M., Boersma, K.S., Bogan, M.T., Olden, J.D., Phillipsen, I., Schriever, T.A., et al., 2015. Dispersal strength determines meta-community structure in a dendritic riverine network. J. Biogeogr. 42, 778–790.

Changming, L., Shifeng, Z., 2002. Drying up of the Yellow River: its impacts and counter-measures. Mitig. Adapt. Strat. Glob. Chang. 7, 203–214.

Chester, E.T., Robson, B.J., 2011. Drought refuges, spatial scale and recolonisation by invertebrates in non-perennial streams. Freshw. Biol. 56, 2094–2104.

Cid, N., Verkaik, I., García-Roger, E.M., Rieradevall, M., Bonada, N., Sánchez-Montoya, M.M., et al., 2015. A biological tool to assess flow connectivity in reference temporary streams from the Mediterranean Basin. Sci. Total Environ. 540, 178–190.

Cipriani, T., Tilmant, F., Branger, F., Sauquet, E., Datry, T., 2014. Impact of climate change on aquatic ecosystems along the Asse River network. In: Daniell, T. (Ed.), Hydrology in a Changing World: Environmental and Human Dimensions. Proceedings of FRIEND-Water 2014, Hanoi, Vietnam, IAHS Publication 363, pp. 463–468.

Colvin, R., Giannico, G.R., Li, J., Boyer, K.L., Gerth, W.J., 2009. Fish use of intermittent watercourses draining agricultural lands in the Upper Willamette River Valley, Oregon. Trans. Am. Fish. Soc. 138, 1302–1313.

Cooper, S.D., Lake, P.S., Sabater, S., Melack, J.M., Sabo, J.L., 2013. The effects of land use changes on streams and rivers in mediterranean climates. Hydrobiologia 719, 383–425.

Corti, R., Datry, T., 2012. Invertebrates and sestonic matter in an advancing wetted front travelling down a dry river bed (Albarine, France). Freshw. Sci. 31, 1187–1201.

Corti, R., Datry, T., Drummond, L., Larned, S.T., 2011. Natural variation in immersion and emersion affects breakdown and invertebrate colonization of leaf litter in a temporary river. Aquat. Sci. 73, 537–550.

Datry, T., Larned, S.T., Tockner, K., 2014. Intermittent rivers: a challenge for freshwater ecology. BioScience 64, 229–235.

Datry, T., Bonada, N., Heino, J., 2016a. Towards understanding metacommunity organisation in highly dynamic systems. Oikos 125, 149–159.

Datry, T., Pella, H., Leigh, C., Bonada, N., Hugueny, B., 2016b. A landscape approach to advance intermittent river ecology. Freshw. Biol. 61, 1200–1213.

Davey, A.J., Kelly, D.J., 2007. Fish community responses to drying disturbances in an intermittent stream: a landscape perspective. Freshw. Biol. 52, 1719–1733.

Davies, B.R., Thoms, M.C., Walker, K.F., O'Keeffe, J.H., Gore, J.A., 1994. Dryland rivers: their ecology, conservation and management. In: Calow, P., Petts, G.E. (Eds.), The Rivers Handbook, vol. 2. Blackwell, Oxford, pp. 484–511.

Day, J.A., 1990. Environmental correlates of aquatic faunal distribution in the Namib Desert. In: Seely, M.K. (Ed.), Namib Ecology: 25 Years of Namib Research. Transvaal Museum, Pretoria, pp. 99–107.

Delucchi, C.M., 1988. Comparison of community structure among streams with different temporal flow regimes. Can. J. Zool. 66, 579–586.

Döll, P., Schmied, H.M., 2012. How is the impact of climate change on river flow regimes related to the impact on mean annual runoff? A global-scale analysis. Environ. Res. Lett. 7, 014037.

Dudgeon, D., Arthington, A.H., Gessner, M.O., Kawabata, Z.I., Knowler, D.J., et al., 2006. Freshwater biodiversity: importance, threats, status and conservation challenges. Biol. Rev. 81, 163–182.

Fritz, K.M., Dodds, W.K., 2004. Resistance and resilience of macroinvertebrate assemblages to drying and flood in a tallgrass prairie stream system. Hydrobiologia 527, 99–112.

Fritz, K.M., Hagenbuch, E., D'Amico, E., Reif, M., Wigington, P.J., Leibowitz, S.G., et al., 2013. Comparing the extent and permanence of headwater streams from two field surveys to values from hydrographic databases and maps. J. Am. Water Resour. Assoc. 49, 867–882.

Fu, G., Chen, S., Liu, C., Shepard, D., 2004. Hydro-climatic trends of the Yellow River basin for the last 50 years. Clim. Change 65, 149–178.

Gallart, F., Prat, N., García-Roger, E.M., Latrón, J., Rieradevall, M., Llorens, P., et al., 2012. A novel approach to analysing the regimes of temporary streams in relation to their controls on the composition and structure of aquatic biota. Hydrol. Earth Syst. Sci. 16, 3165–3182.

Gleick, P.H., 2003. Global freshwater resources: soft-path solutions for the 21st century. Science 302, 1524–1528.

Guareschi, S., Gutiérrez-Canovas, C., Picazo, F., Sánchez-Fernández, D., Abellán, P., Velasco, J., et al., 2012. Aquatic macroinvertebrate biodiversity: patterns and surrogates in mountainous Spanish national parks. Aquat. Conserv. 22, 598–615.

Hassan, M.A., Egozi, R., 2001. Impact of wastewater discharge on the channel morphology of ephemeral streams. Earth Surf. Process. Landf. 26, 1285–1302.

Hermoso, V., Filipe, A.F., Segurado, P., Beja, P., 2015. Effectiveness of a large reserve network in protecting freshwater biodiversity: a test for the Iberian Peninsula. Freshw. Biol. 60, 698–710.

Jaeger, K.L., Olden, J.D., Pelland, N.A., 2014. Climate change poised to threaten hydrologic connectivity and endemic fishes in dryland streams. Proc. Natl. Acad. Sci. U. S. A. 111, 13894–13899.

King, J.M., Tharme, R.E., 1993. Assessment of the instream flow incremental methodology, and initial development of alternative instream flow methodologies for South Africa. Water Research Commission, Pretoria.

Lake, P.S., 2011. Drought and Aquatic Ecosystems: Effects and Responses. Wiley-Blackwell, West Sussex.

Lambeets, K., Vandegehuchte, M.L., Maelfait, J.P., Bonte, D., 2008. Understanding the impact of flooding on trait-displacements and shifts in assemblage structure of predatory arthropods on river banks. J. Anim. Ecol. 77, 1162–1174.

Larned, S.T., Hicks, D.M., Schmidt, J., Davey, A.J.H., Dey, K., Scarsbrook, M., et al., 2008. The Selwyn River of New Zealand: A benchmark system for alluvial plain rivers. River Res. Appl. 24, 1–21.

Larned, S.T., Datry, T., Arscott, D.B., Tockner, K., 2010. Emerging concepts in temporary-river ecology. Freshw. Biol. 55, 717–738.

Leigh, C., Boulton, A.J., Courtwright, J.L., Fritz, K., May, C.L., Walker, R.H., et al., 2016. Ecological research and management of intermittent rivers: an historical review and future directions. Freshw. Biol. 61, 1181–1199.

Levick, L.R., Goodrich, D.C., Hernandez, M., Fonseca, J., Semmens, D.J., Stromberg, J.C., et al., 2008. The Ecological and Hydrological Significance of Ephemeral and Intermittent Streams in the Arid and Semi-Arid American Southwest. US Environmental Protection Agency, Office of Research and Development, Washington, DC.

Luthy, R.G., Sedlak, D.L., Plumlee, M.H., Austin, D., Resh, V.H., 2015. Wastewater-effluent-dominated streams as ecosystem-management tools in a drier climate. Front. Ecol. Environ. 13, 477–485.

Lytle, D.A., 1999. Use of rainfall cues by *Abedus herberti* (Hemiptera: Belostomatidae): a mechanism for avoiding flash floods. J. Insect Behav. 12, 1–12.

Matthews, W.J., 1988. North American prairie streams as systems for ecological study. J. North Am. Benthol. Soc. 7, 387–409.

Mazor, R.D., Stein, E.D., Ode, P.R., Schiff, K., 2014. Integrating intermittent streams into watershed assessments: applicability of an index of biotic integrity. Freshw. Sci. 33, 459–474.

McKnight, D.M., Niyogi, D.K., Alger, A.S., Bomblies, A., Conovitz, P.A., Tate, C.M., 1999. Dry valley streams in Antarctica: ecosystems waiting for water. BioScience 49, 985–995.

Meyer, J.L., Wallace, J.B., 2001. Lost linkages and lotic ecology: rediscovering small streams. In: Press, M.C., Huntly, N.J., Levin, S. (Eds.), Ecology: Achievement and Challenge. Blackwell, Malden, pp. 295–317.

Mielko, C., Woo, M.K., 2006. Snowmelt runoff processes in a headwater lake and its catchment, subarctic Canadian Shield. Hydrol. Process. 20, 987–1000.

Millennium Ecosystem Assessment, 2005. Ecosystems and Human Well-being: Synthesis. World Resources Institute, Washington, DC.

Moyle, P.B., 2014. Novel aquatic ecosystems: the new reality for streams in California and other Mediterranean climate regions. River Res. Appl. 30, 1335–1344.

Munné, A., Prat, N., 2011. Effects of Mediterranean climate annual variability on stream biological quality assessment using macroinvertebrate communities. Ecol. Indic. 11, 651–662.

Nadeau, T.-L., Rains, M.C., 2007. Hydrological connectivity of headwaters to downstream waters: introduction to the featured collection. J. Am. Water Resour. Assoc. 43, 1–4.

Palmer, M.A., Reidy Liermann, C.A., Nilsson, C., Flörke, M., Alcamo, J., Lake, P.S., et al., 2008. Climate change and the world's river basins: anticipating management options. Front. Ecol. Environ. 6, 81–89.

Prat, N., Gallart, F., von Schiller, D., Polesello, S., García-Roger, E.M., Latron, J., et al., 2014. The MIRAGE toolbox: an integrated assessment tool for temporary streams. River Res. Appl. 30, 1318–1334.

Raymond, P.A., Hartmann, J., Lauerwald, R., Sobek, S., McDonald, C., Hoover, M., et al., 2013. Global carbon dioxide emissions from inland waters. Nature 503, 355–359.

Robinson, C.T., Tonolla, D., Imhof, B., Vukelic, R., Uehlinger, U., 2016. Flow intermittency, physico-chemistry and function of headwater streams in an Alpine glacial catchment. Aquat. Sci. 78, 327–341.

Rosado, J., Morais, M., Tockner, K., 2014. Mass dispersal of terrestrial organisms during first flush events in a temporary stream. River Res. Appl. 31, 912–917.

Sheldon, F., 2005. Incorporating natural variability into the assessment of ecological health in Australian dryland rivers. Hydrobiologia 552, 45–56.

Sheldon, F., Bunn, S.E., Hughes, J.M., Arthington, A.H., Balcombe, S.R., Fellows, C.S., 2010. Ecological roles and threats to aquatic refugia in arid landscapes: dryland river waterholes. Mar. Freshw. Res. 61, 885–895.

Snelder, T.H., Datry, T., Lamouroux, N., Larned, S.T., Sauquet, E., Pella, H., et al., 2013. Regionalization of patterns of flow intermittence from gauging station records. Hydrol. Earth Syst. Sci. 17, 2685–2699.

Sophocleous, M., 2002. Interactions between groundwater and surface water: the state of the science. Hydrogeol. J. 10, 52–67.

Stanley, E.H., Fisher, S.G., Grimm, N.B., 1997. Ecosystem expansion and contraction in streams. BioScience 47, 427–435.

Steward, A.L., von Schiller, D., Tockner, K., Marshall, J.C., Bunn, S.E., 2012. When the river runs dry: human and ecological values of dry riverbeds. Front. Ecol. Environ. 10, 202–209.

Stringer, J.W., Perkins, C., 2001. Kentucky Forest Practice Guidelines for Water Quality Management. University of Kentucky Cooperative Extension Service, Lexington, KY.

Stubbington, R., Datry, T., 2013. The macroinvertebrate seedbank promotes community persistence in temporary rivers across climate zones. Freshw. Biol. 58, 1202–1220.

Tooth, S., 2000. Process, form and change in dryland rivers: a review of recent research. Earth Sci. Rev. 51, 67–107.

Towns, D.R., 1985. Limnological characteristics of a South Australian intermittent stream, Brown Hill Creek. Aust. J. Mar. Freshw. Res. 36, 821–837.

Uys, M.C., O'Keeffe, J.H., 1997. Simple words and fuzzy zones: early directions for temporary river research in South Africa. Environ. Manag. 21, 517–531.

Vander Vorste, R., Malard, F., Datry, T., 2016. Is drift the primary process promoting the resilience of river invertebrate communities? A manipulative field experiment in an intermittent alluvial river. Freshw. Biol. 61, 1276–1292.

Williams, D.D., 2006. The Biology of Temporary Waters. Oxford University Press, Oxford.

Wood, P.J., Gunn, J., Smith, H., Abas-Kutty, A., 2005. Flow permanence and macroinvertebrate community diversity within groundwater dominated headwater streams and springs. Hydrobiologia 545, 55–64.

GEOMORPHOLOGY AND SEDIMENT REGIMES OF INTERMITTENT RIVERS AND EPHEMERAL STREAMS

Kristin L. Jaeger*, Nicholas A. Sutfin[†], Stephen Tooth[‡], Katerina Michaelides[§,¶], Michael Singer[¶,**]

USGS Washington Water Science Center, Tacoma, WA, United States[*] Los Alamos National Laboratory, Los Alamos, NM, United States[†] Aberystwyth University, Aberystwyth, Wales, United Kingdom[‡] University of Bristol, Bristol, United Kingdom[§] Earth Research Institute, University of California Santa Barbara, Santa Barbara, CA, United States[¶] Cardiff University, Cardiff, United Kingdom[**]

IN A NUTSHELL

- As intermittent rivers and ephemeral streams (IRES) occur in all climates and across a range of scales and tectonic, lithological, and physiographic settings, generalizing about their geomorphology is challenging
- Valley-floor (i.e., channel and floodplain) processes and forms in IRES are extremely diverse and span a spectrum from distinctly different to overlapping with perennial systems
- Hydrological and sediment regimes are the primary drivers of IRES valley-floor processes and forms
- Variations in valley-floor processes and forms can be analyzed by considering the position along the river network
- Valley-floor morphology in IRES may persist for years to decades or longer, but also can be highly transient, adjusting with every competent flow event

2.1.1 INTRODUCTION

The physical characteristics of a river or stream channel—its size, shape, and dominant substrate—provide the template for physiochemical and biological processes. These physical characteristics are strongly influenced by hydrological and sediment regimes, which collectively encompass the nature of water and sediment delivery to the channel, and the erosional and depositional patterns within and adjacent to the channel. The physical characteristics of intermittent rivers and ephemeral streams (hereafter, IRES) are extremely diverse and span a spectrum from those that resemble their perennial counterparts to those that are distinctly different. This diversity underpins and promotes diverse, and in some respects distinctive, ecologies.

This chapter describes the geomorphology and sediment regimes of IRES occurring across a broad range of climates over scales of space and time that extend from reach-scale, single floods to network-wide, centennial-scale flow sequences. We begin by outlining the main determinants of IRES catchment conditions and how IRES can be analyzed within a catchment-scale framework that identifies and delineates the dominant geomorphological processes driving valley-floor (i.e., channel and floodplain) development. We then describe IRES valley-floor characteristics in different parts of the catchment and

Intermittent Rivers and Ephemeral Streams. http://dx.doi.org/10.1016/B978-0-12-803835-2.00002-4

21

then consider distinct longitudinal trends, the influence of human activities on IRES geomorphology and sediment regimes, and the geomorphological diversity of IRES at the global scale. We conclude by highlighting knowledge gaps and possible future research directions.

DETERMINANTS OF IRES CATCHMENT CONDITIONS

At the broadest scale, climate, geology, and human activities are the fundamental determinants of catchment conditions, and their effect on IRES ecosystems is translated through the catchment by hydrological, sedimentary, and geomorphological processes (Fig. 2.1.1). Catchment conditions, including the topography, soils, vegetation, and river/stream network topology, influence water and sediment delivery to and along the channels, which influences valley-floor morphology and, in turn, ecosystem processes and patterns. These interactions are not unidirectional as ecological processes and patterns can feed back to influence valley-floor morphology, which consequently helps to regulate hydrological and sediment regimes and the longer-term development of catchment conditions.

Various combinations of climate, geology, and human activity (Fig. 2.1.1) contribute to considerable diversity in IRES catchment conditions, hydrological and sediment regimes, and valley-floor forms and processes. For instance, IRES occur across climates ranging from humid to hyperarid. However, the occurrence of IRES in humid regions is generally limited to the upper parts of the catchment

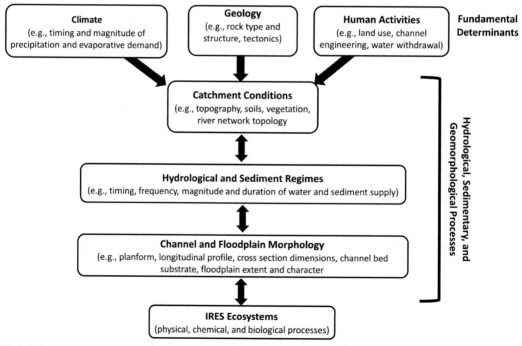

FIG. 2.1.1

Fundamental determinants of catchment conditions and their influence through hydrological, sedimentary, and geomorphological processes on IRES ecosystems. *Single-headed arrows* indicate predominantly unidirectional influences. *Double-headed arrows* indicate feedback interactions.

(e.g., Wohl, 2010) because wetter climates typically provide sufficient surface runoff and groundwater to support perennial systems throughout most of the channel network. In drier climates (e.g., drylands or the seasonal tropics), IRES are more widespread owing to less dependable, more variable surface runoff, and limited groundwater contributions.

Flow regimes of IRES are described in more detail in Chapters 2.2 and 2.3, but merit discussion here as they are a key driver of sediment transfers to and within valley floors, and therefore strongly control channel morphology and development (Fig. 2.1.1). Climates that support extensive IRES tend to be semi-arid and Mediterranean, and are characterized by precipitation or other sources of runoff that are highly variable in space and time (Nicholson, 2011), and by vegetation coverage that is sparse, unevenly distributed, or temporally variable (Tooth, 2013). Regardless of climatic setting, high-intensity convective storms can result in spatially discontinuous and localized flow responses within IRES catchments (Renard and Keppel, 1966; Goodrich et al., 1997; Nicholson, 2011; Chapters 2.2 and 2.3). Even when precipitation or other sources of runoff are more regular and widespread (e.g., climates characterized by seasonal snowmelt or monsoonal rains), the resulting flows can be highly variable and discontinuous as a consequence of transmission losses into unconsolidated alluvium (Graf, 1988; Tooth, 2013; Kampf et al., 2016). In addition, many climates may experience multiyear wet and dry cycles resulting from El Niño-Southern Oscillation (ENSO) phenomena and other global teleconnections. These diverse hydrological characteristics typically translate to long periods of zero flow, irregular floods, and high relative peak flow magnitudes compared to flow in perennial systems (Tooth, 2013).

GEOMORPHOLOGICAL ZONES IN IRES

Within a given climatic and geological setting, many fundamental geomorphological processes that operate in IRES and perennial systems (Box 2.1.1) tend to be influenced by longitudinal position in the catchment. General longitudinal (down-valley) trends in channel morphology and sediment regimes were outlined by Schumm (1977) who described three zones in relation to relative elevation and catchment position (Fig. 2.1.3A). The production zone is dominated by net erosion, which removes sediment from hillslopes and supplies it to the channels. The transfer zone is characterized by down-valley sediment transport; in a stable channel, there is an approximate balance between sediment input and output. The deposition zone is characterized by net sediment accumulation. The process domain concept (Montgomery, 1999) builds on Schumm's conceptual model by outlining how geomorphological processes can influence habitat conditions and ecosystem processes, and identifies regions of a channel network with distinct ecological characteristics that are directly associated with the dominant geomorphological processes. A benefit of Montgomery's (1999) concept is that it allows for longitudinal discontinuities and patch dynamics that occur in many IRES and perennial systems (e.g., Burchsted et al., 2014).

The model presented by Tooth and Nanson (2011), previously developed by Gordon et al. (1992), extends Schumm's (1977) and Montgomery's (1999) conceptual models to dryland river systems and designates four broad geomorphological zones: upland, piedmont, lowland, and floodout (Fig. 2.1.3). In this chapter, we adopt this model as a useful framework for describing IRES and it recurs in discussions of hydrological connectivity in Chapter 2.3. The relative importance, spatial extent, and even character of the four zones vary as a function of climatic, tectonic, and physiographic settings (Section 2.1.7). Therefore, we acknowledge that these zones serve as a very broad generalization and do not capture the full geomorphological diversity of IRES. Nonetheless, characterizing IRES within a context of zones provides a logical, physically meaningful framework for synthesizing information derived from highly variable natural phenomena. In addition, it provides a potential basis for coupling the dominant geomorphology and sediment regimes within each zone to down-valley changes in the characteristic ecological conditions.

BOX 2.1.1 QUANTITATIVE PARAMETERS TO EVALUATE CHANNEL MORPHOLOGY AND SEDIMENT TRANSPORT POTENTIAL

The shape (morphology) of rivers and streams is governed by physical forces studied within the fields of fluvial geomorphology and hydraulic engineering. The ability of flowing water to transport sediment is a function of flow depth (d), velocity (v), discharge (Q), and channel slope (S).

In ungauged rivers and streams, flow velocity (v, in $\mathrm{m\,s^{-1}}$) can be estimated indirectly, using equations such as the Manning formula:

$$v = \left(R^{2/3} S^{1/2} \right) / n \tag{2.1.1}$$

where R is hydraulic radius (channel cross-sectional area (A) divided by the wetted perimeter (P) of the stream bed and banks, but commonly substituted by d in wide, shallow channels) and n is a measure of the roughness of the bed and banks (lower values for smooth bed and banks, higher values for rough bed and banks).

The Manning formula can also be used to estimate discharge (Q, in $\mathrm{m^3\,s^{-1}}$) in ungauged streams by incorporating A into the formula:

$$Q = \left(AR^{2/3} S^{1/2} \right) / n \tag{2.1.2}$$

Many other hydraulic equations encompass these variables, such as those for bed shear stress (τ, in $\mathrm{N\,m^{-2}}$) and unit stream power (ω, in $\mathrm{W\,m^{-2}}$):

$$\tau = \gamma RS \tag{2.1.3}$$

$$\omega = \gamma QS / w \tag{2.1.4}$$

where γ is the specific weight of water and w is channel width. The significance of these equations is their strong association with the potential for geomorphic work (i.e., erosion and sediment transport), which influences many ecological processes and patterns.

Eqs. (2.1.3) and (2.1.4) show that for given values of γ, R, Q, and w, steeper channels (higher S) will be more powerful (higher τ and ω). Relatively steep channels, such as those in the upland and piedmont zones, may have τ and ω values sufficiently competent to erode bedrock and transport coarse-grained sediment, even during relatively moderate floods. Significant bed erosion and rapid sediment transport may lead to channel incision, thereby increasing R, d, and τ, thus creating a positive feedback that continues channel incision. Ultimately, this may lead to the formation of a relatively narrow and deep channel (Figs. 2.1.2A and 2.1.5A and B), depending on the controlling expression of regional climate (Slater and Singer, 2013).

By contrast, less steep channels, such as those in the lowland and floodout zones, may have τ and ω values that are incompetent to consistently transport large volumes even of finer-grained sediment, except perhaps during the largest floods. During low to moderate flood events, some sediment may be deposited within the channel, thereby reducing R, d, and τ. Over time, this may lead to channel aggradation and the formation of a relatively wide and shallow channel (Figs. 2.1.2 and 2.1.5). Arroyos of the American southwest cycle between these contrasting cross-section morphologies.

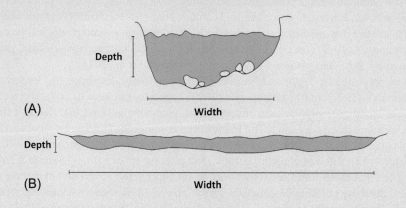

FIG. 2.1.2

Schematic diagram of channel cross-sections showing variability of width and depth. Narrow, deep channels possess a low width:depth ratio (A). Wide, shallow channels possess a larger width:depth ratio (B).

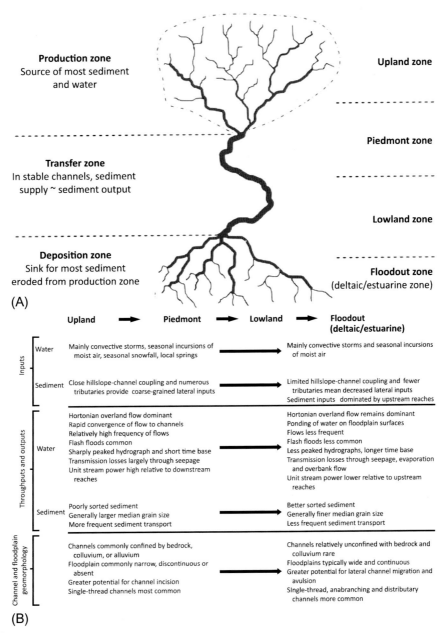

FIG. 2.1.3

Conceptual model showing the four geomorphological zones of an idealized IRES system (A) and the diverse hydrological, sediment regime and geomorphological characteristics across the four zones (B). The four geomorphological zones broadly correspond to the source, transport, and deposition zones of an idealized river, originally proposed by Schumm (1977). Down-valley changes through the zones highlight the tendencies for adjustment of the different characteristics along multiple continua. In any given catchment, down-valley hydrological, sediment regime and geomorphological changes are commonly irregular and nonlinear and may involve step changes or threshold crossings, as influenced by features such as local bedrock outcrop, the spacing of tributary junctions, dams, or artificial abstractions.

Conceptual model (A) is modified from Tooth, S., Nanson, G.C., 2011. Distinctiveness and diversity of arid zone river systems. In Thomas, D.S.G. (Ed.), Arid Zone Geomorphology: Process, Form and Change in Drylands, third ed. Wiley-Blackwell, Chichester, pp. 269–300, with permission.

2.1.2 UPLAND ZONE

In the upland zone, IRES channels are closely coupled with hillslopes, and lateral sediment inputs result from delivery by overland flow and more abrupt, mass failures that may develop into landslides and debris flows (Wohl and Pearthree, 1991). Periodic high-energy flood events may transport large volumes of sediment and erode bedrock. Hence, in the upland zone, IRES systems tend to be primarily erosive and serve as the dominant source of sediment to downstream reaches.

Channel substrate in the upland zone may consist of bedrock, alluvium, or some combination of the two. Bedrock reaches—characterized by a dominance of exposed bedrock in the bed and banks—may possess a relatively thin (typically <10–20 cm thick) veneer of alluvium (Jaeger et al., 2007; Sutfin et al., 2014). Bed sediments in the upland zone tend to be poorly sorted and include a range of grain sizes up to boulders (Fig. 2.1.4). Channels are typically small, steep, and single thread with low width-depth ratios (<20) (Figs. 2.1.2 and 2.1.5A). Floodplains are commonly absent or restricted in width

FIG. 2.1.4

Characteristics of IRES in the upland zone shown in (a) schematic form illustrating key landforms and sedimentary deposits. Examples include this bedrock-incised upland drainage network, Utah, USA (b, view looking up-valley) where the base of the channel is marked by vegetation to the right of the standing figure. Smooth, bare bedrock slopes shed large amounts of overland flow after rainstorms, leading to rapid stage rises in the channel. Other upland examples include this single-thread channel (c, view looking downstream) deeply incised into coarse-grained valley deposits and confined by bedrock outcrop, and this single-thread, step-pool channel (d, view looking downstream), both occurring in Arizona, USA.

Photos: Courtesy S. Tooth (b), N. Sutfin (c), K. Jaeger (d).

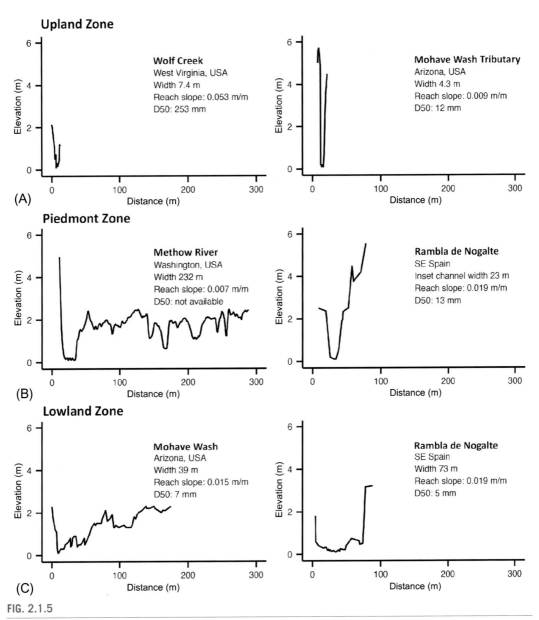

FIG. 2.1.5

Selected IRES cross-sections across the four geomorphological zones. Cross-sections in the (A) upland, (B) piedmont, (C) lowland, and (D) floodout zones are from different rivers. Inset text in (A), (B), and (C) indicates geographic location and cross-section characteristics including channel width, reach slope, and the median grain size (D50). Inset text in (D) indicates river kilometers (rkm), increasing in the downstream direction, and cross-section characteristics. The most upstream cross-section of the Tobiaspruit (D, *right panel*) is from the transition between the lowland and floodout zones, and the most downstream survey is on the floodout itself. Cross-sections are drafted from published (Tooth, 2000; Tooth et al., 2002) and unpublished data (Jaeger, unpublished; Michaelides, unpublished; Sutfin, unpublished), including publicly available LiDAR data (http://pugetsoundlidar.ess.washington.edu).

(Continued)

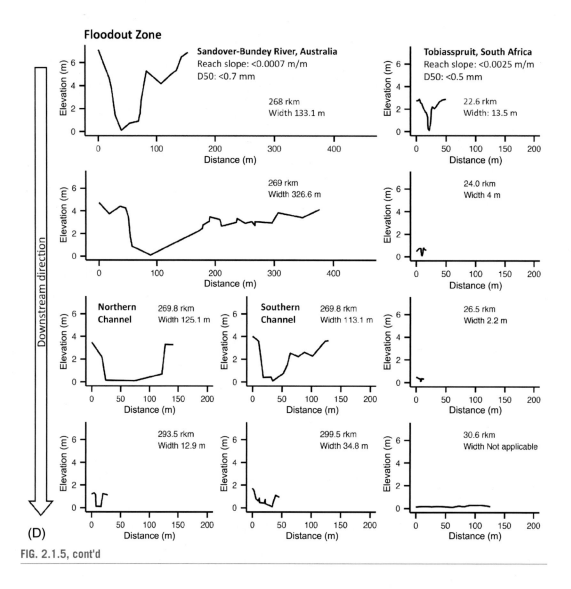

(D)

FIG. 2.1.5, cont'd

and longitudinal extent by outcropping bedrock, colluvium, or older indurated alluvium (Wohl and Pearthree, 1991; Larned et al., 2008; Fig. 2.1.3B).

High-intensity convective precipitation in combination with short overland flowpaths can generate flash floods capable of high rates of coarse-grained sediment transport (Michaelides and Martin, 2012; Lucía et al., 2013). However, very large clasts (>0.5 m diameter) derived from rockfalls, landslides, or debris flows may persist *in situ*, perhaps only slowly abrading while smaller clasts are winnowed from around them (Irwin et al., 2014).

Bedrock exposure and size-selective sediment transport processes may help form the cascade or step-pool morphologies that characterize some IRES in the upland zone (Wohl and Pearthree, 1991; Powell et al., 2012; Billi, 2015; Fig. 2.1.4). Other upland IRES with high sediment inputs, however, may develop comparatively simple, planar-bed channel geometries with only low-relief channel bars, thus facilitating efficient sediment transport (Powell et al., 2012). Where present on the bed, banks, and the floodplain, riparian vegetation can substantially influence in-channel and overbank flow and sediment transport processes (Sandercock et al., 2007; Sandercock and Hooke, 2011; Tooth, 2013). Overbank deposits can be extremely coarse, particularly where infrequent large floods rework debris flows or form levees and boulder berms along channel margins (Wohl and Pearthree, 1991; Macklin et al., 2010). Channel morphology in the upland zone of IRES may appear relatively stable in the short term (years to decades), but sequences of episodic high-energy floods can result in rapid sediment delivery and channel erosion and deposition over longer timescales.

Consistent with other zones in IRES systems, surface water-groundwater interactions are generally limited and highly spatially variable. Except where local springs occur, groundwater is commonly at great depths (>10 m) and disconnected from surface water. As a result, transmission losses can occur through alluvium and/or cracks and fractures in bedrock. Where relatively impermeable bedrock lies at shallow depths, this may promote shallow subsurface flow (Mansell and Hussey, 2005). The general absence of perennial flow and limited groundwater supplies means that aquatic habitats in upland IRES that can serve as important refugia are typically small and transient in time (Bogan and Lytle, 2011; Chapter 4.8) compared to larger features downstream (Bunn et al., 2006). However, localized parts of channel beds can support surface water for extended periods of time during otherwise dry channel conditions (Jaeger and Olden, 2012; Godsey and Kirchner, 2014). For example, stream potholes that are eroded into bedrock outcrops during successive large floods can retain water for several months, providing critical aquatic and terrestrial habitats (Bogan et al., 2014).

2.1.3 PIEDMONT ZONE

In the piedmont zone, valley floors typically are wider relative to the upland zone and channels are commonly flanked by older alluvium, colluvium, and/or bedrock. Hillslopes and channels remain closely coupled, and episodic lateral sediment inputs to steep or moderate-gradient channels are derived from diffusive processes and mass movements on adjacent hillslopes (Box 2.1.2) and from tributaries.

BOX 2.1.2 IMPLICATIONS OF CHANNEL-HILLSLOPE COUPLING FOR SEDIMENT DELIVERY

In many IRES, overland flow is a key contributor to sediment delivery in the piedmont zone. For a given rainstorm, the local sediment supply from hillslopes to channels depends on the characteristics of the adjacent hillslopes and their hydrological response to sequences of spatially variable rainstorms (Michaelides and Martin, 2012). Within a catchment and for a given rainstorm, not all hillslopes will be contributing sediment at the same time. However, over multiple rainfall events, channels will receive sediment intermittently throughout a reach at locations where hillslopes are directly coupled to the channel.

Fig. 2.1.6 demonstrates a plausible scenario of sediment supply in an IRES characterized by substantial fluctuation in valley and channel width but that maintains the typical overarching downstream trend of increasing valley and channel width (Fig. 2.1.6A). The first key aspect is that both mass and grain size of supplied sediment will vary between storms at different locations within a reach (Fig. 2.1.6B and C). A second key aspect is that the overall impact of hillslope sediment supply on the channel is a function of the interaction between channel width and supplied sediment mass and grain size. In particular, an increase in channel width reduces the impact of hillslope sediment supply to the channel because the same mass and size of sediment is spread out over a larger area. This relationship can be articulated for

(Continued)

BOX 2.1.2 IMPLICATIONS OF CHANNEL-HILLSLOPE COUPLING FOR SEDIMENT DELIVERY *(Cont'd)*

discrete locations within a reach by a "hillslope supply index" (HSI) (Michaelides and Singer, 2014), which represents the degree of hillslope impact on the channel as proportional to the mass and the grain size of sediment supplied and inversely proportional to the channel width (Fig. 2.1.6D). As the channel progressively widens downstream and floodplains become more common, the impact of hillslope sediment supply to the channel tends to decline, reflecting a progressive downstream decoupling between hillslope and channels.

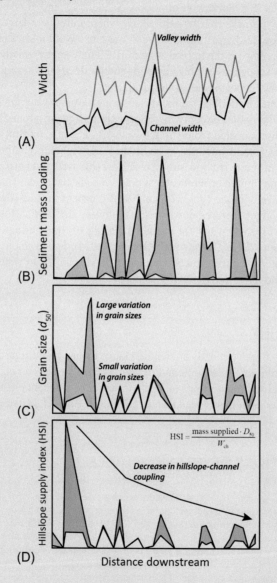

FIG. 2.1.6

Schematic of longitudinal trends in (A) channel and valley-floor width, (B) sediment mass loading from hillslopes to channel, (C) median grain size of sediment supplied from hillslopes to the channel, and (D) the hillslope supply index (HSI) illustrating the decrease in hillslope-channel coupling in the downstream direction as a result of increases in channel width. The gray shading in (B), (C), and (D) represents the range of values based on modeling a range of rainfall events bracketed by the minimum and maximum values. The calculation of the HSI is given within the panel (D).

Bedrock may crop out locally in channel beds, but commonly channel beds are fully alluvial, with a range of sediment sizes up to boulders. Flood events of differing magnitude and frequency sequentially deposit and erode sediment within channels and the wider valley floor, generating a wide variety of fluvial landforms, including floodplains and alluvial terraces (Figs. 2.1.5B and 2.1.7). In general, the characteristics of the geomorphology and sediment regimes of IRES in the piedmont zone reflect the intermediate position between the upland zone of net erosion and the lowland zone of transfer, with sediment being eroded, deposited, and reworked over various spatial and temporal scales (Fig. 2.1.3).

The interplay between episodic hillslope and tributary sediment supply and fluvial reworking in the trunk channel is a key characteristic of IRES in the piedmont zone (Michaelides and Wainwright, 2002; Wainwright et al., 2002; Michaelides and Singer, 2014) and contributes to a wide variety of channel patterns from wide and shallow, single thread, or braided channels (Graf, 1981; Singer and Michaelides, 2014) to narrower and deeper anabranching complexes (Graeme and Dunkerley, 1993; Jacobson et al., 1995; Fig. 2.1.7). The tendency for development of bar forms and bedforms varies

FIG. 2.1.7

Characteristics of IRES in the piedmont zone shown in (a) schematic form illustrating key landforms and sedimentary deposits. Examples include channels on the Northern Plains, central Australia (b, flow from right to left), illustrating confinement by bedrock and by sparsely vegetated older alluvium. Other examples from Arizona, USA, include this single-thread channel (c, view looking downstream), illustrating incision into coarse-grained, indurated Pleistocene alluvium and evidence for coarse-grained sediment transport during high- to moderate-energy floods, and this gravel-bed channel in the Sonoran Desert, Arizona (d, view looking downstream), with abundant boulders and small anabranching side channels.

Photos: Courtesy S. Tooth (b), N. Sutfin (c and d).

widely within and between different channel patterns. Examples include: sand-bed single-thread channels with low-relief channel beds and limited bedforms and bars (Powell et al., 2012); stone-pavement channels created by periglacial freeze-thaw cycles that rotate large clasts to create a relatively flat surface (McKnight et al., 1999); gravel-bed braided channels characterized by development of bars, dunes, and gravel aggregates (Hassan, 2005); and sand- and gravel-bed anabranching channels that display varying bed topographies including dunes and "flow chutes," the latter being influenced by riparian vegetation assemblages (Dunkerley, 2008).

Vegetation dynamics exert strong influences on the morphology and sediment regimes of IRES in the piedmont zone. Vegetation dieback during disturbance (drought or anthropogenic influences) can enhance sediment delivery to the channel (Collins and Bras, 2008), which can lead to partial obstruction of the trunk channel by tributary-junction fans (Larsen et al., 2004) but also to aggradation and alterations to trunk channel morphology. Vegetation recovery may reduce sediment delivery, with subsequent channel incision and terrace formation on the valley margins (Shakesby and Doerr, 2006). Terrace formation may topographically decouple hillslopes from the channels, resulting in locally buffered, heterogeneous patterns of sediment supply (Michaelides and Singer, 2014). Variable patterns of channel, floodplain, and valley-floor vegetation growth mean that overbank flow and sediment transport dynamics can be highly complex (Sandercock et al., 2007; Sandercock and Hooke, 2011). However, as in the upland zone, periodic high-energy floods and relatively short overland and overbank flow paths tend to result in steep lateral hydrological gradients and maintain channel-floodplain hydrological connectivity (Amoros and Burnette, 2002; Michaelides and Chappell, 2009; Trigg et al., 2013; Chapter 2.3).

As in the upland zone, channels in the piedmont zone may be characterized by relative stability over years to decades but then undergo short intervals of abrupt change, depending on the sequence of channel-changing flood events. Runoff and sediment dynamics drive the redistribution of nutrients (Michaelides et al., 2012), which in turn affects primary productivity (Chapter 3.1), plant distribution (Chapter 4.2), and broader ecosystem functioning. Discharges during small to moderate floods may initially increase downstream through the piedmont zone but as a result of limited or no groundwater input, discharge tends to decrease owing to flow transmission losses through the channel bed and banks (Dunkerley and Brown, 1999). Following floods, the retention and persistence of surface water depends largely on local bed topography. Alluvial channels with low-relief channel beds tend to remain largely dry, but where bedrock crops out locally or flow chute morphology has developed, temporary or permanent surface water may be present in topographic lows (Dunkerley and Brown, 1999; Dunkerley, 2008).

2.1.4 LOWLAND ZONE

In the lowland zone, alluvial valleys tend to become wider, thus hillslopes and channels tend to be less closely coupled compared to upland and piedmont zones. This contributes to a wide range of channel-floodplain morphologies, many of which strongly contrast with perennial systems (Fig. 2.1.8). Many IRES in the lowland zone are primarily transfer systems that maintain an approximate balance between sediment input and output, although some systems may be characterized by long-term net deposition.

A key characteristic of IRES in the lowland zone is the typical downstream decreases in flood flow volumes that result from transmission losses into unconsolidated alluvium, losses to overbank flow, and evapotranspiration (Tooth, 2013; Chapter 2.3). In many catchments, tributary inputs are also limited in the lowland zone. These downstream flow decreases are in stark contrast to the downstream flow

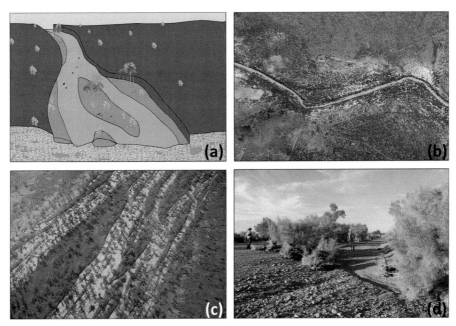

FIG. 2.1.8

Characteristics of IRES in the lowland zone shown in (a) schematic form illustrating key landforms and sedimentary deposits. Examples include a low-sinuosity, single-thread channel in northern South Africa (b, flow from *left to right*), showing a largely vegetation-free channel bed but dense vegetation on the adjacent floodplains. Other examples include anabranching channels in central Australia (c, flow from *upper right to lower left*) showing division and rejoining around vegetated, narrow ridges and wider islands. Vegetated bars along a mixed sand-gravel bed, braided channel in the Sonoran Desert of southwestern Arizona (d) illustrate complex interactions between vegetation and sediment transport processes.

Photos: Courtesy S. Tooth (b and c), N. Sutfin (d).

increases typical of perennial systems and have implications for channel-bed substrate, sediment sorting and transport, and channel and floodplain morphologies. Channel-bed substrates in the lowland zone are strongly influenced by catchment lithology and sediment supply, and so can vary widely, ranging from variable mixtures of alluvium and bedrock (Heritage et al., 1999), through sand and gravel (Reid and Frostick, 1997) to silt and clay (e.g., Knighton and Nanson, 1997).

Some lowland-zone IRES are characterized by low sinuosity, relatively wide and shallow rectangular cross-sections (Leopold et al., 1966; Reid and Frostick, 1997; Singer and Michaelides, 2014; Figs. 2.1.5C and 2.1.8). In these channels, bed and bar forms tend to be poorly developed and seldom persist over decadal to centennial timescales (Hassan, 2005), in part because fluxes of water and sediment can be considered to occur primarily in one dimension (downstream) with negligible cross-stream components (Leopold et al., 1966; Singer and Michaelides, 2014). Other IRES are characterized by more complex channel-floodplain morphologies, such as the extensive anastomosing and anabranching channel systems that occur across wide areas of inland Australia and parts of southern Africa (Heritage et al., 1999; Tooth and McCarthy, 2004; Tooth et al., 2008) or the alluvial gravel-bed rivers in wetter regions in Europe, the US, and New Zealand (Doering et al., 2007; Larned et al., 2008, 2011). In systems with complex channel-floodplain morphologies, various bed and bar forms can develop and persist over centennial to millennial timescales, commonly in association with riparian vegetation patterns (Heritage et al., 1999; Tooth and Nanson, 2000; Tooth et al., 2008). Compared to lowland IRES with rectangular cross-sections, fluxes of water and sediment are influenced by a variety of poorly documented channel-floodplain exchanges. For instance, during extensive flooding in the Channel Country of eastern central Australia, flow in interconnecting anastomosing channels is accompanied by low-energy overbank flows that disperse through networks of extensive floodplain "braid" channels (Knighton and Nanson, 1997), with floodwaters then evaporating or draining back to the anastomosing channels as stage falls.

Studies in central Australia have shown how highly contrasting channel-floodplain morphologies can be found even on closely adjacent IRES, with transitions from wide, shallow single-thread channels to anabranching channels commonly being associated with limited tributary inputs of water and sediment (Tooth and Nanson, 2004). In particular, periodic flows from tributaries into a normally dry trunk channel promote in-channel growth of trees through positive feedbacks between sediment deposition, moisture retention, and vegetation establishment (Tooth and Nanson, 2004). Over time, initially wide, shallow single channels can subdivide into complexes of narrow anabranches divided by tree-lined narrow ridges or wider islands (Fig. 2.1.8). In other settings, marked shifts between complex multi-channel reaches and single-channel reaches also have been associated with local changes in gradient, sediment supply, and influence of riparian vegetation (Larned et al., 2008). Despite these strong contrasts in channel-floodplain morphologies, work in central Australia, Spain, and elsewhere has shown that longitudinal profiles in many lowland-zone IRES are straight (Vogel, 1989; Powell et al., 2012; Singer and Michaelides, 2014), contrasting with the characteristic concave-up profiles of lowland-zone perennial systems (Fig. 2.1.9). The straight longitudinal profile may reflect an oversupply of alluvium that covers the bed surface and is smoothed by incomplete lateral and longitudinal sorting (Singer and Michaelides, 2014). Where bedrock crops out more extensively, however, longitudinal profiles in the lowland zone can be far more irregular, sometimes alternating between moderate-gradient, dominantly bedrock reaches atop resistant outcrop and lower-gradient, mixed bedrock-alluvial or alluvial reaches atop weaker outcrop (Heritage et al., 1999; Tooth and McCarthy, 2004).

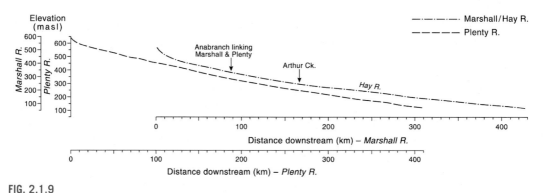

FIG. 2.1.9

Straight longitudinal profiles in the lowland zones of the Plenty and Marshall Rivers, Australia, illustrating the contrast with the concave-up longitudinal profiles typical of many lowland perennial systems.

Reproduced with permission from Tooth, S., Nanson, G.C., 2004. Forms and processes of two highly contrasting rivers in arid central Australia, and the implications for channel-pattern discrimination and prediction. Geol. Soc. Am. Bull. 116, 802–816.

In the lowland zones of IRES, channel-floodplain stability can vary widely. In some IRES, lowland channel-floodplain morphologies can remain relatively stable over centuries or millennia, particularly where the potential for erosion during low-energy floods is restricted by well-developed riparian vegetation (Tooth and Nanson, 2000; Sandercock et al., 2007; Larned et al., 2008). In other settings, sediment aggradation occurs during decadal-scale sequences of small to moderate floods, with extensive channel and floodplain erosion (stripping) and vegetation removal occurring during occasional large floods (Heritage et al., 1999).

Flow transmission losses can be extremely high (e.g., $2.5\,\text{m}^3\,\text{s}^{-1}\,\text{km}^{-1}$, Doering et al., 2007) and contribute to groundwater recharge and in-channel sediment deposition in lowland-zone IRES (Keppel and Renard, 1962; Renard and Keppel, 1966). In many lowland channels, retention of surface water after floods is limited, although exceptions may occur in reaches where bedrock outcrop has resulted in local development of pools (Heritage et al., 1999) or where the confluence of anastomosing channels on extensive muddy floodplains leads to scour and the formation of permanent waterholes (e.g., Knighton and Nanson, 1994, 2000). Although transmission losses are a typical characteristic, groundwater input can contribute to discharge that supports local perennial reaches in the lowland zone (Doering et al., 2007; Larned et al., 2008; Chapter 2.3). Shallow groundwater recharge from channels can sustain water availability to riparian trees that otherwise have limited access to soil moisture (Singer et al., 2014).

2.1.5 FLOODOUT ZONE

For IRES flowing to the coast, there tends to be a gradual transition from the lowland zone to an estuarine or deltaic zone. Across the globe, however, many IRES fail to reach a coastline, instead undergoing various forms of "breakdown" or "failure," and ultimately terminating in inland topographic basins on alluvial plains (Mabbutt, 1977; Tooth, 1999a,b; Billi, 2007) or on the margins of pans or playas (Fisher et al., 2008; Donselaar et al., 2013; Fig. 2.1.10). All these environments are characterized by net deposition, as reflected in a diverse—and, in some instances, distinctive—range of geomorphological features and sedimentary deposits. This section focuses mainly on the geomorphology and sediment regimes of IRES terminating in inland basins (collectively termed the floodout zone), although some characteristics are shared with IRES that end at a coastline.

In most floodout zones, limited or absent hillslope-channel coupling means that flow and substrate characteristics are largely determined by the distance from the upland and piedmont sources of runoff and sediment. In some small IRES, the floodout zone may be within a few tens of kilometers of the uplands, so flow events may occur semiregularly (e.g., after every local convective thunderstorm) and channel-bed sediment may include local cobbles and boulders (Billi, 2007; Craddock et al., 2012). In larger IRES, however, the floodout zone may be located many tens or even hundreds of kilometers from the uplands and the characteristic downstream decreases in discharge described earlier mean that flow events may be very infrequent (perhaps only once every few years or decades), while bed sediment typically is no coarser than granules (Dubief, 1953; Mabbutt, 1977; Tooth, 2000).

Downstream decreases in the magnitude and frequency of discharge can occur in association with downstream gradient decreases caused by changes in lateral confinement and subsequent channel-bed aggradation (Tooth, 1999a) or by lithological/structural factors, such as a change from a harder to a weaker lithology underlying the channel bed (Donselaar et al., 2013; Grenfell et al., 2014). These decreases in discharge and/or gradient mean that unit stream power and sediment transport capacity also decrease. These, in turn, lead to a downstream reduction in the channel size, diversion of an increasing proportion of floodwaters overbank, and sediment deposition (Fig. 2.1.5D). These processes are

FIG. 2.1.10

Characteristics of IRES in the floodout zone shown in (a) schematic form illustrating key landforms and sedimentary deposits. Examples include the lower reaches of channels on the margins of a playa in northwest Iran (b, flow from *left to right*), showing how channel breakdown and the transition to floodouts is accompanied by widespread development of splays and distributary channels. Other examples include this channel terminus on the Northern Plains, central Australia (c, flow *left to right*), showing widespread deposition of sandy sediment in the transition to the floodout, and floodout channels in southern South Africa (d, view looking down-valley) showing evidence for transport of organic material during the occasional floods.

Photos: Courtey S. Tooth (b, c and d).

commonly promoted by the presence of aeolian or bedrock barriers, such as where linear dune formation or fault uplift has occurred across a channel (Bourke and Pickup, 1999; Tooth, 1999a). In floodout zones, channel cross-section morphology can vary widely, from relatively wide and shallow in gravel- and sand-bed channels to narrow and deep in mud-rich channels (Sullivan, 1976; Mabbutt, 1977; Tooth, 1999b; Fig. 2.1.5D). However, channel beds tend to have low relief with limited bed or bar form development owing to the limited sediment transport during the short-lived flow events (e.g., Craddock et al., 2012). Even along channels with well-developed riparian vegetation, channels may be laterally active and subject to bank-line erosion, levee breaching, and periodic avulsion (Tooth, 2005; Billi, 2007). Depending on flood sequencing, channels and floodplains may remain stable for many years or decades and then undergo significant change during short-lived events, resulting in a range of morphological and sedimentary features that include active and inactive distributary channels, paleochannels, splays, waterholes, and various aeolian-fluvial interactions (Tooth, 1999a,b; Billi, 2007).

In some settings, the channel may lose definition and disappear entirely to form a floodout *sensu stricto* (Tooth, 1999a), defined as a site where channelized flow ceases and floodwaters spill across adjacent alluvial surfaces. Floodouts can range widely in scale, reaching up to approximately 1000 km^2 on some larger IRES. The size and shape of floodouts are often strongly influenced by local physiography. In central Australia, floodouts are <500 m wide where rivers terminate between the longitudinal dunes and local bedrock outcrops of the northern Simpson Desert but can reach up to several kilometers wide on the relatively unconfined Northern Plains (Bourke and Pickup, 1999; Tooth, 1999a,b). At the channel terminus, the coarser bedload sediment tends to be deposited in a splay-like form, but low-energy flows and finer suspended sediments may spill across the unchanneled surfaces (Fig. 2.1.10). On terminal floodouts, floodwaters spill across the unchanneled surfaces and eventually dissipate through infiltration or evaporation but on intermediate floodouts at least some floodwaters persist across the unchanneled surfaces and ultimately concentrate into small "reforming channels". Reforming channels commonly develop where the unchanneled floodwaters become constricted by aeolian deposits or bedrock outcrops, or where small tributaries provide additional inflow. These reforming channels either join a larger river or decrease in size downstream before disappearing in another floodout (Tooth, 1999a).

As in the lowland zone, floodwater infiltration in the floodout zone contributes to groundwater recharge (e.g., Morin et al., 2009). Toward the center of some deeper topographic basins, such as those occupied by large playas, groundwater exfiltration may contribute to saturation of near-surface sediments and periodic shallow flooding (Shaw and Bryant, 2011). Overall, the strongly pulsed hydrological and sediment regimes in floodout zones, and the varied morphology of the channels, floodplains, and floodouts result in pronounced down-valley and cross-valley hydrological gradients and a mosaic of ecological patches at various scales. Case studies show that vegetation and other biota commonly respond rapidly to the irregular flood events and the associated material supply (Box 2.1.3).

BOX 2.1.3 MATERIAL TRANSPORT IN IRES SYSTEMS

Most previous investigations of material transport in IRES systems have focused on relatively coarse-grained clastic sediments moving in continuous or near-continuous contact with the channel bed (bedload and saltation load) or finer-grained sediment moving in suspension within the water column (suspended load) although these data sets remain sparse. Less attention has been directed toward other components of material transport, such as the dissolved sediment load and the particulate organic load. In many IRES, concentrations of sediment and certain ions may increase downstream as peak discharges and total flow volumes decrease. Chemical analyses have largely been restricted to inorganic constituents, although some studies have also investigated the nature of organic loads. For example, Jacobson et al. (2000) investigated floodwater chemistry in the ephemeral Kuiseb River, Namibia, by tracking floodwaters along some 200 km of the river's lower reaches. Data from below the confluence of the upper Kuiseb and Gaub rivers (Table 2.1) reveal very high levels of dissolved organic matter but even higher levels of particulate organic matter, derived principally from the riparian vegetation. Some of the dissolved organic matter was derived from leaching of the transported particulate organic matter. In the 2-day January 1994 flood, dissolved inorganic, dissolved organic, and particulate organic matter tended to show an overall downstream increase in response to waning flood volumes (Table 2.1), with all the transported material being deposited in the lower reaches (floodout zone) and no water or matter reaching the Atlantic Ocean.

In a related study focusing on downstream transport of large woody debris (achieved by painting and tracking representative pieces), Jacobson et al. (1999) found that 65% of the painted wood exported from marking sites during the 1994 flood was retained within debris piles associated with in-channel growth of large ana trees (*Faidherbia albida*), and wood retention also peaked in the lower reaches. Debris piles induced deposition of clastic sediment and fine particulate organic matter, and promoted the formation of in-channel islands. Following flood recession, these debris piles and the associated sediments provided moist, organic-rich microhabitats, and these became focal points for decomposition and secondary production. Jacobson et al. (2000) concluded that the lower Kuiseb is a sink for materials transported from upstream, and that the large amounts of labile organic matter provide an important carbon supplement to flood-activated heterotrophic communities among these lower reaches (Chapters 3.2 and 4.7).

(Continued)

BOX 2.1.3 MATERIAL TRANSPORT IN IRES SYSTEMS *(Cont'd)*

Table 2.1 Characteristics of floodwaters along the Kuiseb River during a 2-day flood in January 1994

Station	Catchment area (km²)	Distance (km)	Elevation (m)	Gradient (mm⁻¹)	Discharge (m³ s⁻¹)	Total flood volume (10⁶ m³)	Total susp. solids (g L⁻¹)	Conductivity (µS cm⁻¹)	Dissolved organic matter (g L⁻¹)	Particulate organic matter (g L⁻¹)
Schlesien (upper Kuiseb R)	6520	n/a	760	0.0040	~20	~2	–	–	–	–
Greylingshof (Gaup R tributary)	2490	0	720	0.0055	159	2.75	11.8	302	0.0390	0.78
Confluence (upper Kuiseb and Gaup R)	9500	–	620	0.0035	–	4.75	–	–	–	–
Homeb	–	105	–	–	–	–	30.3	627	0.0557	1.90
Gobabeb	11,700	140	360	0.0030	51	2.3	48.0	703	0.0492	3.24
Rooibank	14,700	197	120	0.0039	<1	0.05	19.7	1035	0.0831	2.36

– indicates data not provided or not available. Distance downstream is the distance downstream from Greylingshof; Rooibank is approximately 18 km from the Atlantic Ocean. Discharge, total suspended solids, conductivity, dissolved organic matter, and particulate organic matter measured during peak discharge at each site. Total suspended solids (Total susp. solids) include organic matter.

Based on data presented in Jacobson, P.J., Jacobson, K.M., Angermeier, P.L., Cherry, D.S., 2000. Variation in material transport and water chemistry along a large ephemeral river in the Namib Desert. Freshw. Biol. 44, 481–491.

2.1.6 DISTINCTIONS IN IRES LONGITUDINAL TRENDS

Considering all four geomorphological zones in down-valley sequence, many IRES appear to possess some distinctive longitudinal trends when compared to perennial systems. These trends are largely related to the distinctive hydrology of IRES, especially the limited and spatially heterogeneous runoff generation, and the tendency for flow transmission losses that leads to downstream decreases in flow volumes and flood peaks.

As in perennial systems, IRES channels initially tend to widen downstream through the upland, piedmont, and lowland zones as catchment area increases. However, two distinctions become evident in some IRES. First, channel width initially may be higher for a given catchment area than it is for perennial streams (Fig. 2.1.11A), at least partly owing to the geomorphological effectiveness of rare extreme

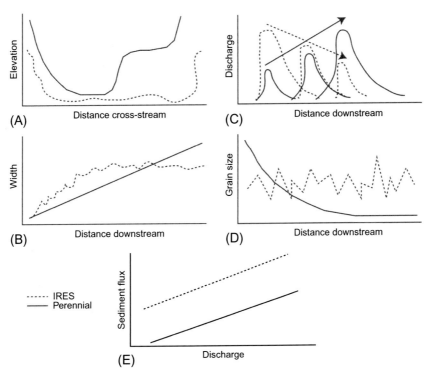

FIG. 2.1.11

Distinctions in IRES longitudinal trends. Channels tend to be wider in IRES (A) compared to perennial systems at a cross-section for a given drainage area. Channel width can be asymptotic to a value of 100–200 m in IRES (B), whereas perennial channel widths continue to increase with increasing drainage area. Discharge magnitudes decrease in the downstream direction in IRES (C) and peak flows are short lived compared to perennial systems. Channel-bed grain size in IRES alternates between coarse- and fine-grained reaches (D) compared to characteristic downstream fining in perennial systems. For a given discharge, sediment flux can be much greater in IRES (E) compared to perennial systems.

events compared to perennial systems (Baker, 1977; Wolman and Gerson, 1978). Second, width has been shown to be asymptotic to a value of 100–200 m for drainage areas ranging from 100 to 1000 km^2 (Fig. 2.1.11B). Wolman and Gerson (1978) attribute this pattern to several possible interrelated factors, notably the widening of channels by extreme flows until they accommodate the largest available discharges, themselves constrained by the higher transmission losses that occur in widened channel beds. In other words, channel widths may be limited by diminishing discharges or muted discharge increases downstream (Fig. 2.1.11C). These trends are not universal, however, and other IRES display different downstream width changes. For example, in some IRES in central Australia, southern Africa, and the Mediterranean, channel widths in the middle to lower reaches alternate downstream between relatively narrow and wide reaches, or are characterized by relatively regular decreases associated with declining discharges (Tooth, 2000, 2013; Fig. 2.1.5).

Downstream decreases in flow volumes and flood peaks also influence sediment transport processes. Some IRES are characterized by fluctuating reaches of coarse and fine sediment with no clear downstream decrease in grain size or any obvious longitudinal transition from gravel to sand, the latter being more typical of perennial streams (Fig. 2.1.11D; Thornes, 1977; Frostick and Reid, 1980; Singer, 2010; Michaelides and Singer, 2014; Singer and Michaelides, 2014). Sediment flux in some IRES channels can be very high compared with their perennial counterparts (Fig. 2.1.11E), a difference attributed to the commonly nonexistent or poor armor-layer development in IRES (Laronne and Reid, 1993). In ephemeral gravel-bed rivers, poor armor-layer development has been attributed to various interrelated factors including abundant sediment supply from sparsely vegetated hillslopes, the substantial particle mixing resulting from scour and fill processes, and the infrequent short-lived floods that minimize the potential for winnowing finer particles from the bed surface (Laronne and Reid, 1993; Laronne et al., 1994; Reid and Laronne, 1995; Reid and Frostick, 1997). Sediment transport data tend to confirm the high rates of bedload or suspended load transport, including evidence for hyperconcentrated washloads in IRES. For instance, measurements from the ephemeral Rio Puerco, a tributary to the Rio Grande River, indicate that it has the fourth highest average annual suspended-sediment concentration for all rivers globally, with the exception of the Yellow River, Asia (Gellis et al., 2004). In addition, bedload transport has been observed to exceed suspended transport in several IRES in Spain (Castillo and Marin, 2011).

2.1.7 INFLUENCE OF HUMAN ACTIVITIES ON IRES MORPHOLOGY AND SEDIMENT REGIMES

In recent decades, there has been increasing awareness of the extent to which a wide range of human activities are affecting IRES (Kingsford et al., 2006; Chapter 5.1). Some regions (e.g., the Mediterranean) have a long history of human activities in and near the riparian zone (Poesen and Hooke, 1997), while in other regions (e.g., the American southwest, Australia, southern Africa), significant human impacts have only occurred during the last few centuries following European exploration and colonization (Graf, 1988; Thoms and Sheldon, 2000; Tooth, 2016). Human activities can affect IRES both indirectly (e.g., through land-use changes that influence hillslope runoff and sediment supply) and directly (e.g., through flow abstraction or various forms of channel and floodplain engineering). Activities may be widespread throughout a catchment, such as grazing which in part has been linked to the extensive incision/depositional behavior of the arroyo cycle in the American southwest (e.g., Cooke and Reeves, 1976).

Other activities may occur in a specific geomorphological zone but nonetheless affect other zones both upstream and downstream. Examples include where dam construction, sediment mining, or vegetation planting in the upland zone alters downstream flow and sediment supply (e.g., Boix-Fayos et al., 2007; Kamp et al., 2013; Chapter 5.1).

Many human-induced changes to the hydrological and sediment regimes of IRES have been associated with profound changes to channel-floodplain geomorphology and ecological processes and patterns. The most visible changes are commonly associated with changes to riparian vegetation assemblages. In many regions drained by IRES, flow regulation through damming and abstraction has fundamentally altered flow regimes, either by reducing hydrological variability (e.g., suppressing peak flows and increasing base flows) or by increasing periods of low and no flow. In many locations, such flow regime alterations have changed the recruitment and survival of many native tree species (e.g., Johnson, 1994, 1997), while commonly promoting colonization by exotic invasive tree species (e.g., Graf, 1978; Everitt, 1979, 1995; Chapter 5.4). In some locations, widespread establishment of vegetation on formerly exposed and transient bars as a consequence of flow regulation has dramatically reduced the cross-sectional area of river channels, with repercussions for other aspects of ecosystem dynamics.

On a smaller scale, some contrasting examples can be found where flow magnitude and frequency have actually increased in IRES owing to agricultural and urban runoff, including wastewater disposal. In Israel and the Palestinian Territories, nutrient-rich wastewater flow from developing urban areas caused rapid shifts from dry ephemeral channels with intermittent floods to vegetated channels with continuous flow (Hassan and Egozi, 2001). In other settings, urban runoff and reclaimed water discharges can contribute to enhanced flow volumes and severe erosion (Chin and Gregory, 2001). For instance, along Las Vegas Wash, Nevada, which drains past one of the United States's fastest growing cities, flash floods in the 1980s and 1990s caused severe erosion along the wash and its associated wetlands, damaging wildlife habitat and threatening homes (Kingsford, 2006).

In many IRES catchments, human impacts have been most widespread and severe in the upland and piedmont zones, but the changes to flow and sediment supply have deprived lower reaches of water, sediment, and nutrient supplies, leading to a variety of changes to channel-floodplain morphology, substrate composition, and ecological habitat (e.g., Casado et al., 2016). These changes can be complex, commonly including channel contraction and stabilization, secondary salinization, widespread vegetation dieback, and ecosystem collapse (Thoms, 1995; Micklin, 2007; Mac Nally et al., 2011; Box 2.1.4). In an era of rapid environmental change and population growth, human pressure on water supplies can only increase, and in coming decades it may be that increasing numbers of perennial river systems start to behave more like IRES. Therefore, study of the geomorphology and sediment regimes of current IRES may provide insights that can help guide management practices in rivers that start to undergo this type of transition (Kingsford, 2006).

2.1.8 DIVERSITY OF IRES AT A GLOBAL SCALE

Although necessarily selective, the coverage of IRES geomorphology and sediment regimes in the previous sections of this chapter illustrates how valley-floor morphology and substrate characteristics vary widely across different climatic and geological settings. A comprehensive assessment of the geomorphological diversity of IRES at a global scale has yet to be attempted (e.g., Tooth and Nanson, 2011),

BOX 2.1.4 DEWATERING THE LOWER COLORADO RIVER AND DELTA

The lower Colorado River and Delta in the United States and Mexico provide good examples of the deleterious effect of human activities on flow and sediment regimes (Mueller and Marsh, 2002). The Colorado River, once the fifth largest river (by discharge) and carrying the largest sediment load in the United States (McDonald and Loeltz, 1976), used to flow onto a vast delta in the Gulf of California. Dams, large-scale diversions, and abstractions over the last 95 years (Carriquiry and Sánchez, 1999) now make the Colorado River one of the most regulated rivers globally (Andrews, 1991).

Historically, the lower Colorado River was a wide (hundreds of meters), braided and shallow (mean depth < 3 m) river. The system supported a highly transient channel, perennial wet meadows, and expansive floodplain forests (reviewed in Mueller and Marsh, 2002). The Delta extended over more than $15,500 \text{ km}^2$ and was characterized by a complex maze of sloughs, wetlands, and oxbows (Mueller and Marsh, 2002).

Today, most flow from the upper and middle reaches of the river only reaches the Delta during major flow events and only about 0.5% of the original sediment load is delivered to the Delta (Carriquiry et al., 2001). As a result, approximately 100 km of the river downstream of the lowermost dam is usually dry, with less than 8% of the wetlands remaining in the Delta (Mueller and Marsh, 2002). In the absence of a significant upstream sediment supply, the Delta is experiencing widespread erosion, with associated habitat loss affecting endemic fish (Carriquiry and Sánchez, 1999). For example, seven of the nine endemic fish in the lower Colorado River are endangered, with several living in small refugial populations (e.g., desert pupfish). Despite active restoration efforts, the losses of water, sediment, and nutrient supplies from the river are difficult to mitigate and the lower Colorado River and Delta are likely to persist in an extremely diminished ecological state.

partly because there is relatively little published, accessible information on the IRES in many regions of Africa, Asia, and South America (notable exceptions include Billi, 2007; Yang and Scuderi, 2010; Irwin et al., 2014). Nonetheless, the conceptual approach adopted in this chapter—namely, characterizing IRES in the context of four geomorphological zones (Fig. 2.1.3)—provides a framework for integrating and synthesizing additional information as it becomes available, as well as a basis for more rigorously comparing the characteristics of IRES within and between different regions (Tooth and Nanson, 2011). As an illustration, a comparison can be made between IRES in the Mediterranean, southern Africa, and Australia. These regions have different physiographies, largely as a function of their contrasting geological setting, especially the degree of tectonic activity. Consequently, the spatial extent and relative proportions of the upland, piedmont, lowland, and floodout zones vary dramatically.

In many Mediterranean catchments, upland and piedmont zones are characterized by high elevation, rugged and tectonically active topography, and are relatively extensive. In comparison, the lowland zone is more restricted in extent and the floodout zone is typically nonexistent, with most deposition instead occurring in deltas, estuaries, or offshore. Consequently, their IRES characteristics are dominated by those commonly associated with relatively high-energy upland and piedmont settings (e.g., bedrock rivers, coarse-grained braided rivers—Fig. 2.1.4, 2.1.5, and 2.1.7).

By contrast, in many Australian catchments, the upland and piedmont zones tend to have lower elevation, and more subdued, tectonically stable topography, and are relatively restricted in spatial extent compared with the extensive lowland and floodout zones. For instance, on the basis of 1:250,000 mapping and an elevation-based definition, more than 90% of the total length of all Australian watercourses can be classified as "lowland rivers," with the majority distributed across the dryland continental interior (Thoms and Sheldon, 2000; Sheldon and Thoms, 2006). Although this broad categorization includes rivers in both the lowland and floodout (deltaic/estuarine) zones, the characteristics associated with lower energy settings are more widespread (e.g., single- and multiple-thread channels, distributary channels, and floodouts—Fig. 2.1.10).

In southern Africa, there is more of a balance between the spatial extent of tectonically quiescent, upland, piedmont, lowland, and floodout (deltaic/estuarine) zones. Consequently, IRES characteristics tend to be associated with a wide variety of energy settings, including various bedrock, alluvial, and mixed bedrock-alluvial styles (e.g., braided, single-thread straight, meandering, anabranching/anastomosing and distributary rivers, and floodouts—Figs. 2.1.4, 2.1.5, 2.1.7, 2.1.8, and 2.1.10).

In these three regions, therefore, the different ecological conditions associated with the channels and floodplains in each of the zones will also vary in spatial extent. However, the precise composition of the biological communities will be a function of regional evolutionary history (Chapter 4.10) and the nature of ecosystem modification by human activities (Chapter 5.1).

2.1.9 SYNTHESIS AND NEW RESEARCH DIRECTIONS

Research on IRES has generally lagged behind that devoted to perennial rivers (Kingsford, 2006), and there are various challenges to improving understanding of the links between hydrology, sediment dynamics, geomorphology, and ecology in IRES. New lines of research are required to extend beyond individual IRES to develop knowledge that will lead to more widely generalizable concepts and theories. Key questions can be posed at a range of temporal and spatial scales, and include: How do flood sequences over years to centuries shape aquatic and riparian community structure, and what are the key feedbacks influencing channel and floodplain development in IRES? How do hillslope vegetation dynamics acting over decades to millennia influence hydrological and sediment fluxes and thus longer term topographic development? Over multimillennial and longer timescales, how does IRES catchment topography develop and what are the dominant drivers?

To address these and related questions, a range of approaches will be required that include a combination of field, flume, and computational model-based enquiry to integrate findings obtained at different spatial and temporal scales. For example, the interlinked hillslope, channel, and floodplain processes must be incorporated in field or flume study designs or in computational model structures. Field and flume studies should extend beyond point and plot scales to help generalize over larger scales, including longer channel reaches or entire catchments. Event-based process models can investigate a particular geomorphological process in isolation within a specific part of a catchment, while landscape evolution models (LEMs) can explore the impacts of external forcing (e.g., climate variability, tectonic activity, human disturbances) on IRES catchment development. Associated vegetation dynamics could be addressed by combining existing ecohydrology models (e.g., Rodriguez-Iturbe et al., 2001; Collins and Bras, 2008) with process models and LEMs.

Finally, recent decades have seen increasing use of a wide range of geochronological techniques (e.g., radiocarbon, luminescence, and cosmogenic radionuclide dating) to investigate geomorphological processes in IRES systems, including the timing and rates of sediment transport and storage, channel and floodplain development, and related vegetation dynamics (e.g., Tooth, 2012). In some cases, geochronology can generate knowledge about IRES dynamics at spatial and temporal scales that is intermediate between short-term field and flume studies and modeling studies of longer-term changes. Given the critical influences of geomorphological processes on ecology, a more comprehensive understanding of IRES dynamics will provide information that can be applied by managers and policy makers.

ACKNOWLEDGMENTS

We thank three reviewers (Allen Gellis and two anonymous referees) and the Handling Editor (Thibault Datry) for valuable comments on an earlier draft.

REFERENCES

Amoros, C., Burnette, G., 2002. Connectivity and biocomplexity in waterbodies of riverine floodplains. Freshw. Biol. 47, 761–776.

Andrews, E.D., 1991. Sediment transport in the Colorado River basin. In: Committee to Review the Glen Canyon Environmental Studies (Ed.), Colorado River Ecology and Dam Management: Proceedings of a Symposium, May 24–25, 1990. National Academy Press, Washington, DC, pp. 54–74.

Baker, V., 1977. Stream-channel response to floods, with examples from central Texas. Geol. Soc. Am. Bull. 88, 1057–1071.

Billi, P., 2007. Morphology and sediment dynamics of ephemeral stream terminal distributary systems in the Kobo Basin (northern Welo, Ethiopia). Geomorphology 85, 98–113.

Billi, P., 2015. Sediment dynamics and morphology of a boulder-bed ephemeral stream. In: Lollino, G., Arattano, M., Rinaldi, M., Giustolisi, O., Marechal, J.-C., Grant, G.E. (Eds.), Engineering Geology for Society and Territory. River Basins, Reservoir Sedimentation and Water Resources, 3. Springer, Dordrecht, pp. 371–375.

Bogan, M.T., Lytle, D.A., 2011. Severe drought drives novel community trajectories in desert stream pools. Freshw. Biol. 56, 2070–2081.

Bogan, M.T., Noriega-Felix, N., Vidal-Aguilar, S.L., Findley, L.T., Lytle, D.A., Gutiérrez-Ruacho, O.G., et al., 2014. Biogeography and conservation of aquatic fauna in spring-fed tropical canyons of the southern Sonoran Desert, Mexico. Biodivers. Conserv. 23, 2705–2748.

Boix-Fayos, C., Barberá, G., López-Bermúdez, F., Castillo, V., 2007. Effects of check dams, reforestation and land-use changes on river channel morphology: case study of the Rogativa catchment (Murcia, Spain). Geomorphology 91, 103–123.

Bourke, M., Pickup, G., 1999. Fluvial form variability in arid central Australia. In: Miller, A.J., Gupta, A. (Eds.), Varieties of Fluvial Form. Wiley, Chichester, pp. 249–271.

Bunn, S.E., Thoms, M.C., Hamilton, S.K., Capon, S.J., 2006. Flow variability in dryland rivers: boom, bust and the bits in between. River Res. Appl. 22, 179–186.

Burchsted, D., Daniels, M., Wohl, E.E., 2014. Introduction to the special issue on discontinuity of fluvial systems. Geomorphology 205, 1–4.

Carriquiry, J., Sánchez, A., 1999. Sedimentation in the Colorado River delta and upper Gulf of California after nearly a century of discharge loss. Mar. Geol. 158, 125–145.

Carriquiry, J.D., Sánchez, A., Camacho-Ibar, V.F., 2001. Sedimentation in the northern Gulf of California after cessation of the Colorado River discharge. Sediment. Geol. 144, 37–62.

Casado, A., Peiry, J.-L., Campo, A.M., 2016. Geomorphic and vegetation changes in a meandering dryland river regulated by a large dam, Sauce Grande River, Argentina. Geomorphology 268, 21–34.

Castillo, L., Marin, M., 2011. Hydrologic properties and sediment transport in ephemeral streams. WIT Trans. Ecol. Environ. 146, 313–324.

Chin, A., Gregory, K.J., 2001. Urbanization and adjustment of ephemeral stream channels. Ann. Assoc. Am. Geogr. 91, 595–608.

Collins, D., Bras, R., 2008. Climatic control of sediment yield in dry lands following climate and land cover change. Water Resour. Res. 44, W10405.

Cooke, R.U., Reeves, R.W., 1976. Arroyos and Environmental Change in the American South-West. Clarendon Press, Oxford.

Craddock, R.A., Howard, A.D., Irwin III, R.P., Tooth, S., Williams, R.M., Chu, P.S., 2012. Drainage network development in the Keanakākoʻi tephra, Kīlauea Volcano, Hawaiʻi: implications for fluvial erosion and valley network formation on early Mars. J. Geophys. Res. Planets 117, E8.

Doering, M., Uehlinger, U., Rotach, A., Schlaepfer, D., Tockner, K., 2007. Ecosystem expansion and contraction dynamics along a large Alpine alluvial corridor (Tagliamento River, Northeast Italy). Earth Surf. Process. Landf. 32, 1693–1704.

Donselaar, M., Gozalo, M.C., Moyano, S., 2013. Avulsion processes at the terminus of low-gradient semi-arid fluvial systems: lessons from the Río Colorado, Altiplano endorheic basin, Bolivia. Sediment. Geol. 283, 1–14.

Dubief, J., 1953. Report on the Superficial Hydrology of the Sahara. General Government of Algeria, Service of Scientific Studies, Birmandreis.

Dunkerley, D., 2008. Flow chutes in Fowlers Creek, arid western New South Wales, Australia: evidence for diversity in the influence of trees on ephemeral channel form and process. Geomorphology 102, 232–241.

Dunkerley, D., Brown, K., 1999. Flow behaviour, suspended sediment transport and transmission losses in a small (sub-bank-full) flow event in an Australian desert stream. Hydrol. Process. 13, 1577–1588.

Everitt, B.L., 1979. Fluvial adjustments to the spread of tamarisk in the Colorado Plateau region: discussion and reply. Geol. Soc. Am. Bull. 90, 1183.

Everitt, B.L., 1995. Hydrologic factors in regeneration of Fremont cottonwood along the Fremont River, Utah. In: Costa, J.E., Miller, A.J., Potter, K.W., Wilcock, P.R. (Eds.), Natural and Anthropogenic Influences in Fluvial Geomorphology, Geophysical Monograph 89. American Geophysical Union, Washington, DC, pp. 197–208.

Fisher, J.A., Krapf, C.B., Lang, S.C., Nichols, G.J., Payenberg, T.H., 2008. Sedimentology and architecture of the Douglas Creek terminal splay, Lake Eyre, central Australia. Sedimentology 55, 1915–1930.

Frostick, L., Reid, I., 1980. Sorting mechanisms in coarse-grained alluvial sediments: fresh evidence from a basalt plateau gravel, Kenya. J. Geol. Soc. London 137, 431–441.

Gellis, A.C., Pavich, M.J., Bierman, P.R., Clapp, E.M., Ellevein, A., Aby, S., 2004. Modern sediment yield compared to geologic rates of sediment production in a semi-arid basin, New Mexico: assessing the human impact. Earth Surf. Process. Landf. 29, 1359–1372.

Godsey, S., Kirchner, J., 2014. Dynamic, discontinuous stream networks: hydrologically driven variations in active drainage density, flowing channels and stream order. Hydrol. Process. 28, 5791–5803.

Goodrich, D.C., Lane, L.J., Shillito, R.M., Milier, S.N., Syed, K.H., Woolhiser, D.A., 1997. Linearity of basin response as a function of scale in a semiarid watershed. Water Resour. Res. 33, 2951–2965.

Gordon, N.D., McMahon, T.A., Finlayson, B.L., 1992. Stream Hydrology: An Introduction for Ecologists. John Wiley and Sons, Chichester.

Graeme, D., Dunkerley, D., 1993. Hydraulic resistance by the river red gum, *Eucalyptus camaldulensis*, in ephemeral desert streams. Aust. Geogr. Stud. 31, 141–154.

Graf, W.L., 1978. Fluvial adjustments to the spread of tamarisk in the Colorado Plateau region. Geol. Soc. Am. Bull. 89, 1491–1501.

Graf, W.L., 1981. Channel instability in a braided, sand bed river. Water Resour. Res. 17, 1087–1094.

Graf, W.L., 1988. Fluvial Processes in Dryland Rivers. Springer-Verlag, New York.

Grenfell, S., Grenfell, M., Rowntree, K., Ellery, W., 2014. Fluvial connectivity and climate: a comparison of channel pattern and process in two climatically contrasting fluvial sedimentary systems in South Africa. Geomorphology 205, 142–154.

Hassan, M.A., 2005. Characteristics of gravel bars in ephemeral streams. J. Sediment. Res. 75, 29–42.

Hassan, M.A., Egozi, R., 2001. Impact of wastewater discharge on the channel morphology of ephemeral streams. Earth Surf. Process. Landf. 26, 1285–1302.

Heritage, G., Van Niekerk, A., Moon, B., 1999. Geomorphology of the Sabie River, South Africa: an incised bedrock-influenced channel. In: Miller, A.J., Gupta, A. (Eds.), Varieties of Fluvial Form. Wiley, Chichester, pp. 53–79.

Irwin III, R.P., Tooth, S., Craddock, R.A., Howard, A.D., de Latour, A.B., 2014. Origin and development of theater-headed valleys in the Atacama Desert, northern Chile: morphological analogs to martian valley networks. Icarus 243, 296–310.

Jacobson, P.J., Jacobson, K.N., Seely, M.K., 1995. Ephemeral Rivers and Their Catchments: Sustaining People and Development in Western Namibia. Desert Research Foundation of Namibia, Windhoek.

Jacobson, P.J., Jacobson, K.M., Angermeier, P.L., Cherry, D.S., 1999. Transport, retention, and ecological significance of woody debris within a large ephemeral river. J. North Am. Benthol. Soc. 18, 429–444.

Jacobson, P.J., Jacobson, K.M., Angermeier, P.L., Cherry, D.S., 2000. Variation in material transport and water chemistry along a large ephemeral river in the Namib Desert. Freshw. Biol. 44, 481–491.

Jaeger, K.L., Olden, J., 2012. Electrical resistance sensor arrays as a means to quantify longitudinal connectivity of rivers. River Res. Appl. 28, 1843–1852.

Jaeger, K.L., Montgomery, D.R., Bolton, S.M., 2007. Channel and perennial flow initiation in headwater streams: management implications of variability in source-area size. Environ. Manag. 40, 775–786.

Johnson, W.C., 1994. Woodland expansions in the Platte River, Nebraska: patterns and causes. Ecol. Monogr. 64, 45–84.

Johnson, W.C., 1997. Equilibrium response of riparian vegetation to flow regulation in the Platte River, Nebraska. Regul. Rivers Res. Manage. 13, 403–415.

Kamp, K.V., Rigge, M., Troelstrup Jr., N.H., Smart, A.J., Wylie, B., 2013. Detecting channel riparian vegetation response to best-management-practices implementation in ephemeral streams with the use of spot high-resolution visible imagery. Rangeland Ecol. Manage. 66, 63–70.

Kampf, S.K., Faulconer, J., Shaw, J.R., Sutfin, N.A., Cooper, D.J., 2016. Rain and channel flow supplements to subsurface water beneath hyper-arid ephemeral stream channels. J. Hydrol. 536, 524–533.

Keppel, R.V., Renard, K., 1962. Transmission losses in ephemeral stream beds. J. Hydraul. Div. ASCE 8, 59–68.

Kingsford, R.T., 2006. Changing desert rivers. In: Kingsford, R.T. (Ed.), Ecology of Desert Rivers. Cambridge University Press, Melbourne, pp. 336–345.

Kingsford, R.T., Lemly, A.D., Thompson, J.R., 2006. Impacts of dams, river management and diversions on desert rivers. In: Kingsford, R.T. (Ed.), Ecology of Desert Rivers. Cambridge University Press, Melbourne, pp. 203–247.

Knighton, A.D., Nanson, G.C., 1994. Waterholes and their significance in the anastomosing channel system of Cooper Creek, Australia. Geomorphology 9, 311–324.

Knighton, A.D., Nanson, G.C., 1997. Distinctiveness, diversity and uniqueness in arid zone river systems. In: Thomas, D.S.G. (Ed.), Arid Zone Geomorphology: Process, Form and Change in Drylands, second ed. John Wiley and Sons, Chichester, pp. 185–203.

Knighton, A.D., Nanson, G.C., 2000. Waterhole form and process in the anastomosing channel system of Cooper Creek, Australia. Geomorphology 35, 101–117.

Larned, S., Hicks, D., Schmidt, J., Davey, A., Dey, K., Scarsbrook, M., et al., 2008. The Selwyn River of New Zealand: a benchmark system for alluvial plain rivers. River Res. Appl. 24, 1–21.

Larned, S.T., Schmidt, J., Datry, T., Konrad, C.P., Dumas, J.K., Diettrich, J.C., 2011. Longitudinal river ecohydrology: flow variation down the lengths of alluvial rivers. Ecohydrology 4, 532–548.

Laronne, J.B., Reid, I., 1993. Very high rates of bedload sediment transport by ephemeral desert rivers. Nature 366, 148–150.

Laronne, J.B., Reid, I., Yitshak, Y., Frostick, L.E., 1994. The non-layering of gravel streambeds under ephemeral flood regimes. J. Hydrol. 159, 353–363.

Larsen, I.J., Schmidt, J.C., Martin, J.A., 2004. Debris-fan reworking during low-magnitude floods in the Green River canyons of the eastern Uinta Mountains, Colorado and Utah. Geology 32, 309–312.

Leopold, L.B., Emmett, W.W., Myrick, R.M., 1966. Channel and hillslope processes in a semiarid area, New Mexico. In: U.S. Geological Survey Professional Paper, 352-G, pp. 193–253.

Lucía, A., Recking, A., Martín-Duque, J.F., Storz-Peretz, Y., Laronne, J.B., 2013. Continuous monitoring of bedload discharge in a small, steep sandy channel. J. Hydrol. 497, 37–50.

Mabbutt, J., 1977. Desert Landforms: An Introduction to Systematic Geomorphology. MIT Press, Massachusetts.

Mac Nally, R., Cunningham, S.C., Baker, P.J., Horner, G.J., Thomson, J.R., 2011. Dynamics of Murray-Darling floodplain forests under multiple stressors: the past, present, and future of an Australian icon. Water Resour. Res. 47, W00G05.

Macklin, M., Tooth, S., Brewer, P., Noble, P., Duller, G., 2010. Holocene flooding and river development in a Mediterranean steepland catchment: the Anapodaris Gorge, south central Crete, Greece. Global Planet. Change 70, 35–52.

Mansell, M., Hussey, S., 2005. An investigation of flows and losses within the alluvial sands of ephemeral rivers in Zimbabwe. J. Hydrol. 314, 192–203.

McDonald, C.C., Loeltz, O.J., 1976. Water resources of lower Colorado River-Salton Sea area as of 1971, summary report. In: U.S. Geological Survey Professional Paper, 486-A, pp. 1–34.

McKnight, D.M., Niyogi, D.K., Alger, A.S., Bomblies, A., Conovitz, P.A., Tate, C.M., 1999. Dry valley streams in Antarctica: ecosystems waiting for water. BioScience 49, 985–995.

Michaelides, K., Chappell, A., 2009. Connectivity as a concept for characterising hydrological behaviour. Hydrol. Process. 23, 517–522.

Michaelides, K., Martin, G.J., 2012. Sediment transport by runoff on debris-mantled dryland hillslopes. J. Geophys. Res. 117, F03014.

Michaelides, K., Singer, M.B., 2014. Impact of coarse sediment supply from hillslopes to the channel in runoff-dominated, dryland fluvial systems. J. Geophys. Res. 119, 1205–1221.

Michaelides, K., Wainwright, J., 2002. Modelling the effects of hillslope–channel coupling on catchment hydrological response. Earth Surf. Process. Landf. 27, 1441–1457.

Michaelides, K., Lister, D., Wainwright, J., Parsons, A.J., 2012. Linking runoff and erosion dynamics to nutrient fluxes in a degrading dryland landscape. J. Geophys. Res. 117, G00N15.

Micklin, P., 2007. The Aral Sea disaster. Annu. Rev. Earth Planet. Sci. 35, 47–72.

Montgomery, D.R., 1999. Process domains and the river continuum. J. Am. Water Resour. Assoc. 35, 397–410.

Morin, E., Grodek, T., Dahan, O., Benito, G., Kulls, C., Jacoby, Y., et al., 2009. Flood routing and alluvial aquifer recharge along the ephemeral arid Kuiseb River, Namibia. J. Hydrol. 368, 262–275.

Mueller, G.A., Marsh, P.C., 2002. Lost, a desert river and its native fishes: a historical perspective of the lower Colorado River: U.S. geological survey information technology report, 2002-0010. U.S. Fish and Wildlife Service, Fort Collins.

Nicholson, S.E., 2011. Dryland Climatology. Cambridge University Press, Cambridge.

Poesen, J., Hooke, J., 1997. Erosion, flooding and channel management in Mediterranean environments of southern Europe. Progr. Phys. Geogr. 21, 157–199.

Powell, D.M., Laronne, J.B., Reid, I., Barzilai, R., 2012. The bed morphology of upland single-thread channels in semi-arid environments: evidence of repeating bedforms and their wider implications for gravel-bed rivers. Earth Surf. Process. Landf. 37, 741–753.

Reid, I., Frostick, L.E., 1997. Channel form, flows and sediments in deserts. In: Thomas, D.S.G. (Ed.), Arid Zone Geomorphology: Processes, Form and Change in Drylands, second ed. Wiley, Chichester, pp. 205–229.

Reid, I., Laronne, J.B., 1995. Bed load sediment transport in an ephemeral stream and comparison with seasonal and perennial counterparts. Water Resour. Res. 31, 773–781.

Renard, K.G., Keppel, R.V., 1966. Hydrographs of ephemeral streams in the Southwest. J. Hydraul. Div. ASCE 92, 33–52.

Rodriguez-Iturbe, I., Porporato, A., Laio, F., Ridolfi, L., 2001. Plants in water-controlled ecosystems: active role in hydrologic processes and response to water stress: I. Scope and general outline. Adv. Water Resour. 24, 695–705.

Sandercock, P., Hooke, J., 2011. Vegetation effects on sediment connectivity and processes in an ephemeral channel in SE Spain. J. Arid Environ. 75, 239–254.

Sandercock, P., Hooke, J., Mant, J., 2007. Vegetation in dryland river channels and its interaction with fluvial processes. Progr. Phys. Geogr. 31, 107–129.

Schumm, S.A., 1977. The Fluvial System. John Wiley and Sons, New York.

Shakesby, R., Doerr, S., 2006. Wildfire as a hydrological and geomorphological agent. Earth Sci. Rev. 74, 269–307.

Shaw, P.A., Bryant, R.G., 2011. Pans, playas and salt lakes. In: Thomas, D.S.G. (Ed.), Arid Zone Geomorphology: Process, Form and Change in Drylands, third ed. Wiley-Blackwell, Chichester, pp. 373–401.

Sheldon, F., Thoms, M., 2006. In-channel geomorphic complexity: the key to the dynamics of organic matter in large dryland rivers? Geomorphology 77, 270–285.

Singer, M.B., 2010. Transient response in longitudinal grain size to reduced gravel supply in a large river. Geophys. Res. Lett. 37, L18403.

Singer, M.B., Michaelides, K., 2014. How is topographic simplicity maintained in ephemeral dryland channels? Geology 42, 1091–1094.

Singer, M.B., Sargeant, C.I., Piégay, H., Riquier, J., Wilson, R.J.S., Evans, C.M., 2014. Floodplain ecohydrology: Climatic, anthropogenic, and local physical controls on partitioning of water sources to riparian trees. Water Resour. Res. 50, 4490–4513.

Slater, L.J., Singer, M.B., 2013. Imprint of climate and climate change in alluvial riverbeds: Continental United States, 1950-2011. Geology 41, 595–598.

Sullivan, M.E., 1976. Drainage Disorganisation in Arid Australia and Its Measurement. Masters in Science Dissertation, University of New South Wales, Sydney.

Sutfin, N.A., Shaw, J.R., Wohl, E.E., Cooper, D.J., 2014. A geomorphic classification of ephemeral channels in a mountainous, arid region, southwestern Arizona, USA. Geomorphology 221, 164–175.

Thoms, M., 1995. The impact of catchment development on a semiarid wetland complex: the Barmah Forest, Australia. IAHS Publications-Series of Proceedings and Reports-Intern Assoc. Hydrol. Sci. 230, 121–130.

Thoms, M., Sheldon, F., 2000. Lowland rivers: an Australian introduction. Regul. Rivers Res. Manage. 16, 375–383.

Thornes, J.B., 1977. Channel changes in ephemeral streams: observations, problems and models. In: Gregory, K.J. (Ed.), River Channel Changes. Wiley, Chichester, pp. 317–335.

Tooth, S., 1999a. Floodouts in central Australia. In: Miller, A.J., Gupta, A. (Eds.), Varieties of Fluvial Form. Wiley, Chichester, pp. 219–247.

Tooth, S., 1999b. Downstream changes in floodplain character on the Northern Plains of arid central Australia. In: Smith, N.D., Rogers, J. (Eds.), Fluvial Sedimentology VI, International Association of Sedimentologists, Special Publication 28. Blackwell, Oxford, pp. 93–112.

Tooth, S., 2000. Downstream changes in dryland river channels: the Northern Plains of arid central Australia. Geomorphology 34, 33–54.

Tooth, S., 2005. Splay formation along the lower reaches of ephemeral rivers on the Northern Plains of arid central Australia. J. Sediment. Res. 75, 636–649.

Tooth, S., 2012. Arid geomorphology: changing perspectives on timescales of change. Progr. Phys. Geogr. 36, 262–284.

Tooth, S., 2013. Dryland fluvial environments: assessing distinctiveness and diversity from a global perspective. In: Shroder, J., Wohl, E.E. (Eds.), Treatise on Geomorphology, Vol. 9, Fluvial Geomorphology. Academic Press, San Diego, pp. 612–644.

Tooth, S., 2016. Changes in fluvial systems during the quaternary. In: Knight, J., Grab, S. (Eds.), Quaternary Environmental Change in Southern Africa: Physical and Human Dimensions. Cambridge University Press, Cambridge, pp. 170–187.

Tooth, S., McCarthy, T., 2004. Anabranching in mixed bedrock-alluvial rivers: the example of the Orange River above Augrabies Falls, Northern Cape Province, South Africa. Geomorphology 57, 235–262.

Tooth, S., Nanson, G.C., 2000. The role of vegetation in the formation of anabranching channels in an ephemeral river, Northern plains, arid central Australia. Hydrol. Process. 14, 3099–3117.

Tooth, S., Nanson, G.C., 2004. Forms and processes of two highly contrasting rivers in arid central Australia, and the implications for channel-pattern discrimination and prediction. Geol. Soc. Am. Bull. 116, 802–816.

Tooth, S., Nanson, G.C., 2011. Distinctiveness and diversity of arid zone river systems. In: Thomas, D.S.G. (Ed.), Arid Zone Geomorphology: Process, Form and Change in Drylands, third ed. Wiley-Blackwell, Chichester, pp. 269–300.

Tooth, S., McCarthy, T.S., Hancox, P.J., Brandt, D., Buckley, K., Nortje, E., et al., 2002. The geomorphology of the Nyl River and floodplain in the semi-arid Northern Province, South Africa. S. Afr. Geogr. J. 84, 226–237.

Tooth, S., Jansen, J.D., Nanson, G.C., Coulthard, T.J., Pietsch, T., 2008. Riparian vegetation and the late Holocene development of an anabranching river: Magela Creek, northern Australia. Geol. Soc. Am. Bull. 120, 1021–1035.

Trigg, M.A., Michaelides, K., Neal, J.C., Bates, P.D., 2013. Surface water connectivity dynamics of a large scale extreme flood. J. Hydrol. 505, 138–149.

Vogel, J.C., 1989. Evidence of past climatic change in the Namib Desert. Palaeogeogr. Palaeoclimatol. Palaeoecol. 70, 355–366.

Wainwright, J., Calvo Cases, A., Puigdefabregas, J., Michaelides, K., 2002. Linking sediment delivery from hillslope to catchment scale. Earth Surf. Process. Landf. 27, 1363–1489.

Wohl, E., 2010. A brief review of the process domain concept and its application to quantifying sediment dynamics in bedrock canyons. Terra Nova 22, 411–416.

Wohl, E., Pearthree, P.P., 1991. Debris flows as geomorphic agents in the Huachuca Mountains of southeastern Arizona. Geomorphology 4, 273–292.

Wolman, M.G., Gerson, R., 1978. Relative scales of time and effectiveness of climate in watershed geomorphology. Earth Surf. Process. 3, 189–208.

Yang, X., Scuderi, L.A., 2010. Hydrological and climatic changes in deserts of China since the late Pleistocene. Quatern. Res. 73, 1–9.

FLOW REGIMES IN INTERMITTENT RIVERS AND EPHEMERAL STREAMS

2.2

Katie H. Costigan*, Mark J. Kennard[†], Catherine Leigh[†,‡,§], Eric Sauquet[‡],
Thibault Datry[‡], Andrew J. Boulton[¶]

University of Louisiana at Lafayette, Lafayette, LA, United States[*] Griffith University, Nathan, QLD, Australia[†]
Irstea, UR MALY, centre de Lyon-Villeurbanne, Villeurbanne, France[‡] CESAB-FRB, Immeuble Henri Poincaré,
Aix-en-Provence, France[§] University of New England, Armidale, NSW, Australia[¶]

IN A NUTSHELL

- Flow regimes (magnitude, frequency, duration, timing, and rates of change in flow events) of all intermittent rivers and ephemeral streams (IRES) are characterized by variable periods of zero flow, caused by processes at different scales and often exacerbated by human activities
- When flow ceases, surface waters in many IRES dry to isolated pools but flow may continue through hyporheic sediments below the streambed. If drying continues, hyporheic flows may also cease
- Intermittence and drying have major consequences for aquatic ecosystem structure and functioning and so it is important to describe flow regimes of IRES. Data from gauging stations can be summarized into hydrological metrics (e.g., variance in frequency, duration, and timing of intermittence) and supplemented by wet-dry mapping, various forms of imagery, and modeling to classify riverine regimes
- These classifications show that IRES are globally abundant and that across much of the world, flow intermittence is increasing through climatic drying and greater rates of water extraction

2.2.1 INTRODUCTION

The flow regime of a river is defined as the temporal variability of its discharge, particularly the quantity, timing, and variability in flow (Poff et al., 1997; Nilsson and Renöfält, 2008; Sauquet et al., 2008). Flow regimes are generally expressed as statistical generalizations of hydrological phenomena (e.g., seasonal runoff patterns, median annual discharge, mean, and variance of peak flows) at a particular location over multiple years or decades (Thoms and Sheldon, 2000). Alternatively, flow regimes may be summarized as probability distribution functions of daily flows (e.g., Doyle et al., 2005) to provide information on average water availability, the extent of discharge fluctuations, and the frequency of high and low flows (Botter et al., 2013). Whatever form is used, information about flow regimes is crucial to most research programs and management strategies in rivers because of the central role played by discharge and its temporal variability in governing geomorphology, water quality, and ecology (Box 2.2.1). Extremes of the flow regime, including frequency, timing, and

BOX 2.2.1 FLOW REGIME FUNDAMENTALS

River flow regimes comprise five broad components: magnitude, frequency, duration, timing, and rate of change of hydrological conditions, as well as the temporal variability and predictability in these components. Magnitude refers to the amount of water at a given time interval and is usually expressed as discharge—the volume of water moving past a fixed point per unit time. Gauging stations typically record river height (stage) at regular time intervals and discharge is estimated from a rating curve that describes the relationship between river height and discharge at the gauging station (Gordon et al., 2004). Magnitude can also be described in relative terms (e.g., the amount of water that links disconnected pools along a river course). Changes in magnitude alter habitat types and availability in all IRES (Chapter 2.3).

Frequency refers to how often a flow above a given magnitude recurs over some specified time interval while duration is the period of time associated with a specific flow condition or event. Frequency and duration can also be described in relative terms; for example, the duration and frequency of a particular flow event that inundates a floodplain in a given area. Among the most important hydrological metrics in IRES are frequency and duration of periods of zero flow (e.g., Snelder et al., 2013) because these have major ecological implications for aquatic and terrestrial biota and ecosystem processes (Jaeger et al., 2014; chapters in Section 4).

Timing refers to when a particular event occurs or its regularity or predictability. Timing is ecologically important when a given flow event is needed at a certain time of the year (e.g., fish spawning) as well as being a useful indicator of the inherent variability of flows. The timing of intermittence can be as ecologically important as frequency and duration of intermittence (Leigh and Datry, 2016). Finally, rate of change (sometimes called "flashiness") refers to how quickly flow changes from one magnitude to another; "flashy" streams have faster rates of change than "stable" streams.

These five components of flow regime can be readily illustrated on a hydrograph (Fig. 2.2.1). Varying the temporal scale of the hydrograph portrays the flow regime components in different ways: over multiple years to show interannual variability and predictability of all five components; over a single year to show durations of zero-flow periods and seasonal timing of floods; and over a single flow pulse to show flood magnitude and rates of change. Across all these temporal scales, these five components of the flow regime govern almost every physical, chemical, and biological aspect of all rivers, including IRES.

Alteration of flow regimes by human activities threatens the ecological integrity of river ecosystems and may result in serious declines in biodiversity and the provision of crucial ecosystem services. This "natural flow regime paradigm", proposed in a seminal paper by Poff et al. (1997), is why river ecologists and water resource managers rely on detailed information about flow regimes: flow regime fundamentals.

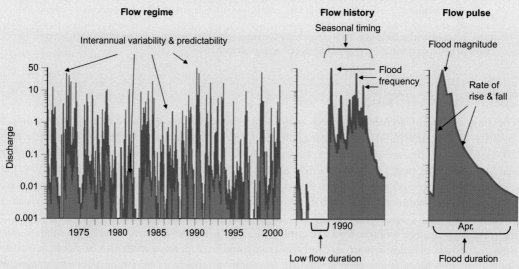

FIG. 2.2.1

Components of the flow regime illustrated at three temporal scales.

Reproduced from Olden, J.D., Kennard, M.J., Pusey, B.J., 2012. A framework for hydrologic classification with a review of methodologies and applications in ecohydrology. Ecohydrology 5, 503–518, with permission.

duration of periods of flow cessation, are particularly important drivers in river ecosystems at multiple spatiotemporal scales (Walker et al., 1995; Larned et al., 2010b), yet their full effects remain poorly understood (Leigh et al., 2016).

In all intermittent rivers and ephemeral streams (hereafter, IRES), the flow regime is defined by periods of flow cessation at some time at one or more points along their network (Chapter 1). These periods vary widely, depending on regional and local factors (Bhamjee and Lindsay, 2011; McDonough et al., 2011; Costigan et al., 2016), and alter the extent, timing, and duration of hydrological connectivity in the longitudinal, lateral, and vertical dimensions of IRES (Chapter 2.3). All river networks include intermittent segments which may be concentrated in one part of the network (e.g., headwaters, lower reaches) or interspersed throughout the entire network (Lake, 2003; Turner and Richter, 2011; Chapter 2.3). In humid and temperate regions, IRES are often confined to the headwaters but in many arid regions, IRES account for most of the stream length and occur throughout the network (Caruso and Haynes, 2011).

As in perennial rivers, the flow regime of IRES governs channel form and sediment dynamics (Godsey and Kirchner, 2014; Chapter 2.1). Similarly, temporal fluctuations in water quality in many IRES correlate with variations in flow regime components (Chapters 3.1 and 3.2), especially timing and duration of intermittence (Sheldon and Fellows, 2010; Leigh, 2013; Costelloe and Russell, 2014). Although most research exploring associations between flow regimes and riverine ecology has focused on perennial rivers (Leigh et al., 2016), similar principles apply in IRES (Leigh and Sheldon, 2008; Datry et al., 2014; chapters in Section 4 of this book). For example, alteration of the natural flow regime (Poff et al., 1997, Box 2.2.1) typically reduces biodiversity of native aquatic species and impairs most riverine ecological processes (Bunn and Arthington, 2002; Rolls et al., 2012). These ecological impacts can be associated with specific aspects of the flow regimes of IRES, particularly changes to the timing, duration, and magnitudes of streamflow and intermittence (Leigh et al., 2012; Datry et al., 2016b; Leigh and Datry, 2016).

Another aspect to consider is that when flow ceases, surface water can persist in remnant pools that may be connected by hyporheic flow below the streambed (Boulton et al., 1992; Anna et al., 2009). In other cases, surface and hyporheic water quickly disappears (Datry, 2012; Vander Vorste et al., 2016a). The effects on biota and ecological processes differ depending on whether surface water remains or not (e.g., Anna et al., 2009; Corti et al., 2011). Traditional approaches to measuring flow regime (e.g., flow gauging) often cannot distinguish these two situations, promoting searches for alternative methods to describe IRES flow regimes. Finally, understanding the ecological effects of altered flow regimes in IRES helps water resource managers to protect and restore key components of flow regimes in IRES using tools such as environmental watering (Chapter 5.4) and "cease-to-pump" rules (Reinfelds et al., 2004).

This chapter reviews our hydrological knowledge about IRES flow regimes to provide context for understanding how intermittence creates temporal sequences of aquatic environments that fluctuate between flowing and nonflowing states, sometimes including complete loss of surface water. After summarizing factors that control the natural flow regimes of IRES, we outline various approaches such as wet-dry mapping, hydrological metrics, and modeling that have been used to describe IRES flow regimes. Case studies from different parts of the world illustrate the application and interpretation of these various approaches. We conclude by discussing knowledge gaps and likely directions for research into the flow regimes of IRES, especially how they might advance our understanding of the effects of altered flow regimes on geomorphology, water quality, biota, and ecological processes in IRES so that we might manage these systems more wisely.

2.2.2 CONTROLS ON THE NATURAL FLOW REGIME OF IRES

The natural flow regime refers to the characteristic pattern of a river's flow magnitude, duration, timing, frequency, and rate of change (Box 2.2.1). In IRES, intermittence is the defining feature of the natural flow regime and generates complex inter- and intraannual expansions and contractions of hydrological connectivity throughout the river network (Stanley et al., 1997; Costigan et al., 2016; Chapter 2.3). Flows in IRES may range from hours following precipitation events through to multiple months or even years. Much of the research on IRES flow regimes has focused on identifying physical controls of intermittence (e.g., Fritz et al., 2008) but many other environmental factors also appear relevant. These factors are reviewed by Costigan et al. (2016) across broad spatiotemporal scales, particularly the way that meteorological, geological, and land cover characteristics control the tendency of flow regimes toward intermittence (Fig. 2.2.2).

As with all rivers, the flow regime of any IRES reflects climatic and catchment processes that mediate how water enters and moves through the river network, interacts with local channel properties along the way. Flows occur when inputs such as rainfall or snowmelt exceed losses from infiltration and evapotranspiration or when water is released from existing storage (Godsey and Kirchner, 2014; Costigan et al., 2015). The interaction between catchment structure (e.g., topography, geology, vegetation) and climate is the primary control of all flow regimes (Jencso et al., 2009; Nippgen et al., 2011; Snelder et al., 2013) and occurs across spatiotemporal scales ranging from entire watersheds to microhabitats (<1 m) and from millennia to hours (Poff et al., 1997; Fig. 2.2.2). Some controls on streamflow can operate at small spatiotemporal scales (e.g., duration and intensity of local rainfall; Fleckenstein et al., 2006; Costigan et al., 2015) so it is important to look at the full range of spatiotemporal scales when understanding controls on intermittence (Costigan et al., 2016).

There have been very few investigations of the multiple mechanisms that control intermittence (Godsey and Kirchner, 2014), despite the long-recognized need for this understanding (Gomi et al., 2002). This paucity of research identifying how hydrological processes interact with geomorphic, climatic, and biological ones may arise from the challenges of integrating different research fields that adopt discipline-specific methods, philosophies, and spatiotemporal extents (e.g., Benda et al., 2002). Therefore, much of our understanding of the controls of stream flow in IRES has been inferred from perennial rivers (Costigan et al., 2016). There is clearly a need to focus on IRES, especially those subject to human activities that alter intermittence (Chapter 5.1). To guide this work, it is useful to review the methods currently used to characterize flow regimes in IRES.

2.2.3 METHODS TO CHARACTERIZE FLOW REGIMES OF IRES

A major challenge in characterizing IRES flow regimes is acquiring hydrological data because many IRES are ungauged, unmapped, or inaccurately depicted on topographic maps (Hansen, 2001; Caruso and Haynes, 2011). Precise spatial distributions of IRES are difficult to obtain (Benstead and Leigh, 2012; Datry et al., 2014), especially for intermittent headwaters routinely underestimated by traditional mapping techniques (Meyer and Wallace, 2001) and seldom gauged.

Where river networks are gauged, stations are usually sparsely distributed and tend to be on perennial rivers close to urban areas or places of economic interest. Headwater streams and other IRES are severely underrepresented (De Girolamo et al., 2015; Eng et al., 2016), leading to underestimation of

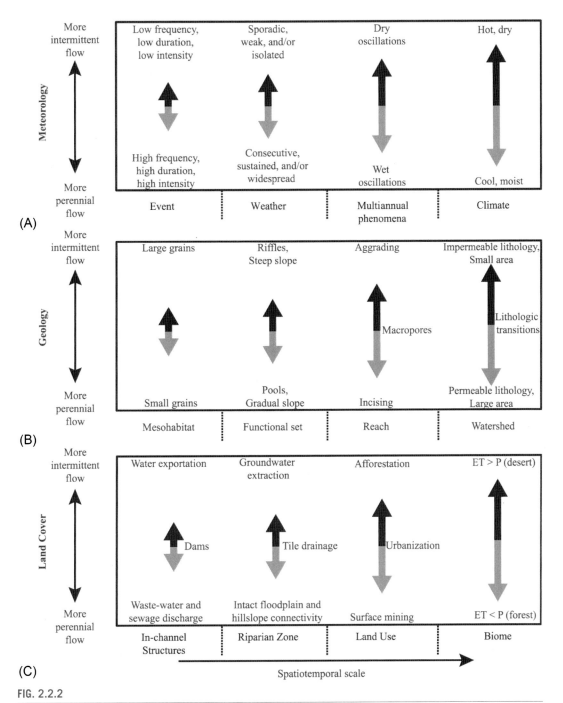

FIG. 2.2.2

Spatiotemporal hierarchy (*x*-axis) of meteorological (A), geological (B), and land-cover (C) controls on flow regime, with factors promoting either intermittence (*upper black arrows*) or perennial flow (*lower gray arrows*) along the *y*-axes. Alternative controls orthose with unknown directions of influence are presented near the origins of the arrows.

Reproduced from Costigan, K.H., Jaeger, K.L., Goss, C.W., Fritz, K.M., Goebel, P.C., 2016. Understanding controls on flow permanence in intermittent rivers to aid ecological research: integrating meteorology, geology and land cover. Ecohydrology doi: 10.1002/eco.1712, with permission.

the regional extent and distribution of flow intermittence (Snelder et al., 2013). Extrapolation of gauge data to characterize patterns across the entire network is perilous because the point measurements at the gauging stations seldom represent spatiotemporal variability in ungauged locations, particularly in IRES in arid and semiarid areas. Added to this is the notoriously variable quality and quantity of discharge data from most gauging stations (Kennard et al., 2010b). Finally, to adequately assess the inherently greater variability of hydrological data from most IRES, long time series (>15–30 years) of streamflow data are desirable (Kennard et al., 2010b; Snelder et al., 2013).

Therefore, alternative approaches to using gauged data to characterize IRES flow regimes are increasing in popularity and technical quality (Datry et al., 2016a; Gallart et al., 2016). These approaches include wet-dry mapping and recent advances in various imaging and modeling techniques. Data from these approaches can be used to supplement traditional sources such as gauged discharge data and often provide valuable extra information such as the duration of isolated pool phases which are of great ecological significance to IRES biota (Boulton, 1989; Rossouw et al., 2005).

WET/DRY MAPPING

On-ground mapping to track temporal changes in the distribution of surface water in IRES networks (e.g., Stanley et al., 1997) has provided useful insights into altered hydrological connectivity (Chapter 2.3) and spatial variability in flow regimes along IRES. Although wet/dry mapping generates valuable data, it demands considerable time in the field, especially when the mapping is conducted at frequent intervals and over large areas. To address this intense demand on field time, citizen scientists have been recruited to assist in wet/dry mapping of IRES (Turner and Richter, 2011; Datry et al., 2016a; Fig. 2.2.3).

Properly trained citizen scientists have proven to be as effective as experts for tasks such as mapping the extent of reaches with flow, isolated pools, and dry channels along IRES (Turner and Richter,

FIG. 2.2.3

Field surveys of flow regime and water presence in IRES demand considerable time and personnel, two issues that can be addressed by involving citizen scientists in the work.

Photo: Courtesy A. Boulton.

2011), and these studies are especially useful for the inexpensive collection of long-term data. One example is the 12 years of wet/dry mapping carried out along an 80-km segment of the San Pedro River (Arizona, USA) described by Turner and Richter (2011). A second example is the visual mapping of some 4000 km of wet and dry river reaches of over 20 IRES networks in Western France by trained volunteers every 15 days during summer that has been done since 2009 (Datry et al., 2016a). This has yielded catchment-scale maps illustrating temporal changes in flow regime at multiple points along different IRES and changes in surface-water connectivity that lead to habitat fragmentation (Chapter 4.9).

Data collected by citizen scientists help compensate for the limited number of gauging stations along many IRES and have improved our ability to characterize IRES flow regimes at specific locations in the network. Coupled with data from existing gauging stations or groundwater monitoring bores nearby, wet/dry mapping can be used to describe, model, and predict the typically complex and dynamic spatiotemporal flow patterns in IRES (Chapter 2.3). Wet-dry mapping is particularly valuable when qualitative states of IRES flow regimes are described such as the six "critical steps" in a drying sequence (Boulton, 2003), the five "hydrological conditions" (Fritz et al., 2006), or the six "aquatic states" (Gallart et al., 2012) simplified to three by Gallart et al. (2016) and used by Datry et al. (2016a). With appropriate training and coordination in river mapping, citizen scientists could monitor real-time flow patterns and qualitative states of IRES flow regimes in nearby IRES, aided by the increasing availability of smartphones with cameras, GPS devices, and high-speed internet connections.

IMAGERY: FROM SATELLITES TO SITE CAMERAS

Various remote sensing techniques have been used to measure proxies for discharge (e.g., river height or width) from which to estimate flow regimes. These techniques include air- and space-borne measurements of surface velocity; radar altimeters to measure surface-water elevations; space-borne measurements of gravity fluctuations to estimate river discharge; and measurements of wetted areas and bank heights to estimate flow volumes, hydrological persistence, and flood extents (e.g., Puckridge et al., 2000; Costa et al., 2013; Gleason and Smith, 2014). Remotely sensed data provide effective proxies for flows in large rivers and their floodplains but are unsuitable for small channels such those of headwater streams because of resolution issues and the density of vegetation cover (Benstead and Leigh, 2012). As most headwater streams are intermittent, this limits the application of satellite imagery for mapping many IRES. Although techniques are being increasingly refined for integrating very-high-spatial-resolution multispectral remote sensing with updated algorithms to map ephemeral stream networks (e.g., Hamada et al., 2016), imagery using light aircraft, drones, and mounted cameras is more practical in smaller streams.

Repeated imagery (e.g., time-lapse photography) can be used to characterize temporal variation in the presence or absence of surface water in sections of smaller IRES, enabling derivation of flow regimes as changes in stream level over time (Puckridge et al., 2000). Fixed cameras, often accompanied by stream gauging instrumentation, can be deployed for long periods and programmed to take photographs at specified intervals (Fig. 2.2.4a-c) or when initiated by movement or flow sensors. In larger rivers, the combined use of multitemporal satellite data with flow gauge and groundwater level data can provide an efficient tool to characterize and model flow regimes (Costa et al., 2013). In smaller rivers, repeated photographs at a single location (Fig. 2.2.4d and e) are visually powerful ways to portray the flow regime of IRES at different times of the year. These repeated images can then be downloaded on websites to reveal long-term variations in stream flow and channel morphology (e.g., http://www.konza.k-state.edu/keep/sci_adventures/streams/oxbow/oxbow_gallery.htm;

FIG. 2.2.4

(a-c) Time-lapse images of a reach in Salome Creek, Arizona, where the stream dries during the day but water reappears at night, providing a resource for local wildlife. Repeated camera images from consistent photo points, such as these paired images (d, e) from Spring Creek, Tennessee, are a powerful way to illustrate seasonal variation in flow in IRES.

Photos: Courtesy M. Smart (a-c), A. Doll (d), W. Curtis (e).

http://datawarehouse.wrl.unsw.edu.au/maulescreek/eastlynne/). Recent developments in airborne technology and remote camera image processing represent promising advances toward better mapping and understanding of IRES and their flow regimes.

FIELD LOGGERS AND FLOW SURROGATES

Various types of data loggers are commonly used to characterize IRES flow regimes when flow gauges are absent. These loggers measure flow surrogates such as electrical conductivity (e.g., Gungle, 2007; Chapin et al., 2014), water temperature (e.g., Constantz et al., 2001; Gungle, 2007), and water level and/or the presence-absence of water (e.g., Costelloe et al., 2005; Bhamjee and Lindsay, 2011; Vander Vorste et al., 2016b). Time series data from the loggers can be used to track the movement of wetting and drying fronts (Bhamjee and Lindsay, 2011) and the persistence of surface waters in different reaches (Vander Vorste et al., 2016b). These data can then be translated into hydrological metrics to assess and compare flow regimes (*Hydrological Metrics* section of this chapter). For example, time series data of water presence and absence provided by water-state loggers generate point estimates of the duration and frequency of drying that can then be matched with ecological changes in IRES (e.g., Jaeger et al., 2014; Vander Vorste et al., 2016b).

Thanks to their relatively low prices (Chapin et al., 2014), arrays of multiple loggers can be placed within and across IRES networks to spatially describe flow and water presence-absence regimes and connectivity in space and time (Jaeger and Olden, 2012; Vander Vorste et al., 2016b). These arrays are especially useful if the study areas are remote and frequent visitation is impractical. However, there are several limitations to using loggers. For example, in-stream and riparian loggers may be washed out or buried during flood events and subject to vandalism. Therefore, regular maintenance and data downloads are required, particularly where flow regimes are flashy or risks of vandalism are high. Most sensors are unable to differentiate flowing or standing waters. Sometimes detecting the presence of water is difficult; if moist sediment builds up on the sensor probes, the sensors interpret that as wet conditions. Regular site visits are required to ground truth the sensor datasets.

HYDROLOGICAL METRICS

Hydrological metrics are indices or statistics computed from multiyear time series of discharge data and used to characterize statistical properties of various components of riverine flow regimes. They have been widely used to evaluate spatial variation among flow regimes (e.g., Puckridge et al., 1998; McMahon et al., 2007), classify rivers (e.g., Hughes and James, 1989; Poff, 1996), and explore associations between flow regimes and biotic structure or ecological processes (e.g., Kennard et al., 2007; Leigh, 2013). Despite the array of hydrological metrics available to characterize river flow regimes (reviewed by Olden et al., 2012), relatively few are employed to characterize intermittent flows. The principal ones are number of zero-flow days, number of zero-flow periods, average duration of zero-flow periods, percent flow permanence in a year, and percent flow permanence by season (e.g., Snelder et al., 2013; Jaeger et al., 2014; Costigan et al., 2015; Schriever et al., 2015).

In an example illustrating the capacity of hydrological metrics to discriminate different flow regimes, including those of IRES, Kennard et al. (2010a) used 120 metrics to classify discharge data from 830 stream gauges across Australia. Eight of the 12 classes of rivers in this study were classified as intermittent: 4 classes of rivers that rarely ceased flow, 3 that regularly ceased flow, and 1 that was extremely intermittent. The hydrological metrics that split perennial rivers from IRES included the number of zero-flow days, baseflow contributions, and magnitude of daily runoff (Fig. 2.2.5), indicating that not just flow intermittence but other (nonzero) flow conditions are important in characterizing IRES flow regimes. There was often greater variance (expressed by the larger boxes and longer "whiskers") of these hydrological metrics within the eight classes of IRES compared to the four perennial classes (Fig. 2.2.5); this pattern of greater variance for IRES was also evident for many of the other metrics (Kennard et al., 2010a).

Suites of hydrological metrics have also been used to identify distinct groups of flow regimes of IRES in other parts of the world. Using 10 hydrological metrics to classify long-term (15–58 years) daily streamflow data from 806 relatively undisturbed streams in the United States, Poff (1996) discerned three classes of IRES from seven perennial-stream classes. Again, number of zero-flow days was a key metric, along with higher unpredictability and coefficients of variation for flows measured at different temporal

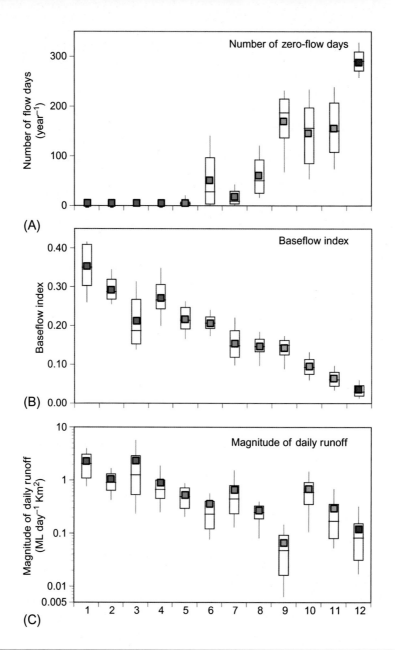

FIG. 2.2.5

Box plots of number of zero-flow days (A), baseflow index (B), and magnitude of daily runoff (C) in 12 classes from a classification of flow regimes of Australian rivers: 1, stable baseflow; 2, stable winter baseflow; 3, stable summer baseflow; 4, unpredictable baseflow; 5, unpredictable winter rarely intermittent; 6, predictable winter intermittent; 7, unpredictable intermittent; 8, unpredictable winter intermittent; 9, predictable winter highly intermittent; 10, predictable summer highly intermittent; 11, unpredictable summer highly intermittent; 12, variable summer extremely intermittent. Rivers in classes 1–4 (*blue boxes*) have primarily perennial flow; those in classes 5–8 rarely cease flow; those in classes 9–11 regularly cease flow; and class 12 rivers are extremely intermittent. *Colored symbols* are means; *horizontal lines at the top, middle,* and *bottom* of each *open box* represent the 75th, 50th (median), and 25th percentiles, respectively, and *vertical bars* (whiskers) represent 90th and 10th percentiles.

Modified from Kennard, M.J., Pusey, B.J., Olden, J.D., Mackay, S.J., Stein, J.L., Marsh, N., 2010a. Classification of natural flow regimes in Australia to support environmental flow management. Freshw. Biol. 55, 171–193; Kennard, M.J., Mackay, S.J., Pusey, B.J., Olden, J.D., Marsh, N., 2010b. Quantifying uncertainty in estimation of hydrologic metrics for ecohydrological studies. River Res. Appl. 26, 137–156, with permission.

scales (days to years). Worldwide, studies of regional flow regimes have used number or frequency of zero-flow days as the primary hydrological metric to define or discriminate classes of intermittence (e.g., South Africa—Smakhtin and Toulouse, 1998; France—Snelder et al., 2013; European Mediterranean region—Gallart et al., 2012; United States—Reynolds et al., 2015; Eng et al., 2016). Consequently, it is logical to focus analyses and comparisons of the flow regimes of IRES on hydrological metrics related to zero flows, arrayed in Table 2.2.1 according to the five main components of frequency, duration, timing and seasonality, and rate of change in flow events before and after periods of zero flows.

Table 2.2.1 Candidate hydrological metrics used to characterize zero-flow conditions in IRES flow regimes

Hydrological metric	Definition
Frequency of flow events—zero-flow conditions	
Zero-flow spell count[a]	Mean number of annual, seasonal, or monthly occurrences during which the magnitude of flow remains at or below some threshold defined as zero flow
CV zero-flow spell count[a]	Coefficient of annual, seasonal, or monthly occurrences during which the magnitude of flow remains at or below some threshold defined as zero flow
Duration of flow events—zero-flow conditions	
Zero-flow spell duration[a]	Mean duration of annual, seasonal, or monthly occurrences during which the magnitude of flow remains at or below some threshold defined as zero flow
CV zero-flow spell duration[a]	Coefficient of variation in duration of annual, seasonal, or monthly occurrences during which the magnitude of flow remains at or below some threshold defined as zero flow
Number of zero-flow days	Mean annual number of days having a magnitude of flow at or below some threshold defined as zero flow
CV number of zero-flow days	Coefficient of variation in annual number of days having a magnitude of flow at or below some threshold defined as zero flow
Timing and seasonality of flow events—zero-flow conditions	
Julian date of annual zero flow	The mean Julian date of the 1-day annual zero flow over all years
CV Julian date of annual zero flow	Coefficient of variation in Julian date of the 1-day annual zero flow over all years
Six-month seasonal predictability of zero-flow periods	Multiannual frequencies of zero-flow months for the contiguous 6 wetter months of the year divided by the multiannual frequencies of zero-flow months for the remaining 6 drier months. Wet and dry 6-month periods are those with fewer and more zero-flow frequencies, respectively
Predictability (P) of zero-flow days[b]	Colwell's (1974) predictability (P) of zero-flow days
Seasonality (M/P) of zero-flow days[b]	Colwell's (1974) seasonality (M/P) of zero-flow days
Rate of change in flow events—before/after zero-flow spell[a]	
Rise rate	Mean rate of increases in flow magnitude (rising limb of hydrograph) over a given time period
CV rise rate	Coefficient of variation in rate of increases in flow magnitude over a given time period

Continued

Table 2.2.1 Candidate hydrological metrics used to characterize zero-flow conditions in IRES flow regimes—cont'd

Hydrological metric	Definition
Fall rate	Mean rate of decreases in flow magnitude (falling limb of hydrograph) over a given time period
CV fall rate	Coefficient of variation in rate of decreases in flow magnitude over a given time period
Number of reversals	Number of increases then decreases in flow magnitude over a given time period
CV reversals	Coefficient of variation in number of increases then decreases in flow magnitude over a given time period

CV, *coefficient of variation.*

[a] *Independence criteria for zero-flow spell frequency and duration need to be specified (e.g., at least 7 days between spells).*
[b] *Colwell's (1974) predictability (P) of flow is composed of two independent, additive components: constancy (C—a measure of temporal invariance) and contingency (M—a measure of periodicity).*

Compiled or adapted from Olden, J.D., Poff, N.L., 2003. Redundancy and the choice of hydrologic indices for characterizing streamflow regimes. River Res. Appl. 19, 101–121; Kennard, M.J., Mackay, S.J., Pusey, B.J., Olden, J.D., Marsh, N., 2010b. Quantifying uncertainty in estimation of hydrologic metrics for ecohydrological studies. River Res. Appl. 26, 137–156; and Gallart, F., Prat, N., García-Roger, E.M., Latrón, J., Rieradevall, M., Llorens, P., et al., 2012. A novel approach to analysing the regimes of temporary streams in relation to their controls on the composition and structure of aquatic biota. Hydrol. Earth System Sci. 16, 3165–3182 and supplemented with some proposed metrics associated with rates of rise and fall in flow.

The frequency and duration of zero-flow periods are particularly useful in characterizing intermittency (Poff, 1996; Knighton and Nanson, 2001) and remain the most common indices used in describing intermittence (e.g., Larned et al., 2010b; Costigan et al., 2015; Reynolds et al., 2015). These two metrics are also particularly relevant because of their direct translation to habitat availability and persistence, so they are increasingly reported in ecological studies of IRES (Jaeger et al., 2014; Schriever et al., 2015; Chapter 4.9).

MODELING

When flow gauges have only limited data for a location of interest, modeling approaches can be used to estimate data from which metrics can be calculated. Such models are also employed to simulate flows under changing environmental conditions, such as global change and water extraction for human uses (Chapter 5.1). Hydrologists often distinguish physical or deterministic models that typically predict time series of simulated daily flows (and from which metrics can be calculated) from statistical models used to directly estimate flow metrics (e.g., Carlisle et al., 2010).

However, most models for deriving flow regimes have been developed for perennial rivers and there remain serious limitations in accurately simulating hydrological extremes, including low-flow events that may lead to zero flows. There are various reasons for these difficulties in simulating zero-flow conditions. For example, most rainfall-runoff models concentrate on representing surface water processes (e.g., Pilgrim et al., 1988) and fail to adequately capture the effects of surface-groundwater interactions or local geological peculiarities such as karstic areas on river flow regime. Poor performances also result from complex nonlinear processes in runoff generation in low-yielding basins (Ye et al., 1997), from inadequate representation of spatial variation in the catchment characteristics that determine runoff or storage (Costelloe et al., 2005) and/or from an unsuitable formulation of the recession phase that does not allow streams to completely dry (Ivkovic et al., 2014).

To address some of these limitations, Ivkovic et al. (2014) developed a modified version of the IHACRES (Identification of Hydrographs And Components from Rainfall, Evaporation, and Streamflow data) model which is a hybrid conceptual-metric model that uses metric approaches to reduce inherent parameter uncertainty while simultaneously attempting to represent more detail of the internal processes than typical metric models (Croke and Jakeman, 2007). This modified version includes explicit representation of groundwater storage and a linear relation between storage and baseflow that replaces the classical slow transfer function. Ivkovic et al. (2014) applied this modified version to modeling the flow regime in an ephemeral river system in New South Wales, Australia, and found that it successfully simulated the transitions between baseflows and zero-flow periods within the ephemeral stream.

A second example of hydrological modeling of flow regimes in IRES is the work by Cipriani et al. (2014) who applied a postprocessing technique, based on a quantile-mapping approach (e.g., Snover et al., 2003), to the outputs of two rainfall-runoff models to simulate zero-flow events. The postprocessing technique is a convenient way to simulate zero-flow events without modifying the structures of the models. Application of this approach to a river in southeastern France that ceased to flow due to losses toward the underlying aquifer yielded a good match between the simulated and bias-corrected discharge time series from two rainfall-runoff models.

A third modeling approach is the Soil Water Assessment Tool (SWAT, Gassman et al., 2007), a public-domain program that has become one of the most widely used catchment-scale hydrological transport models in the world. The hydrological component of the model accounts for precipitation, evapotranspiration, surface runoff, infiltration, lateral flow, and percolation and has been successful in modeling intermittence in IRES (Chahinian et al., 2011; De Girolamo et al., 2016; Tzoraki et al., 2016), with a >85% match between modeling projections and real-world conditions. The SWAT model has also been effective in projecting IRES flow regimes under natural, current, and future conditions (Jaeger et al., 2014; Tzoraki et al., 2016). Finally, random forest modeling approaches use models that are first trained on gauges with environmental data and then extrapolated to areas without gauges. These models rely upon conditional interference trees that assess relationships between environmental variables and flow metrics. Random forest modeling explained up to 83%, but typically 45%–50%, of the variance in the streamflow metrics of ungauged portions of watersheds within the Upper Colorado River basin (Reynolds et al., 2015; Reynolds and Shafroth, 2016).

However, severe limitations persist in using models to derive IRES flow regimes. All models, whether deterministic, statistical, or a combination, are constrained by the amount and quality of available training data and only perform as well as our understanding of processes allows. In most cases, we lack adequate understanding of the different processes driving intermittence in different IRES (Smakhtin, 2001; Snelder et al., 2013; Section 2.2.2) and seldom have adequate data to train models to accurately predict flow regimes of any river type, especially IRES. Major advances in modeling flow regimes in IRES are needed (Section 2.2.5), supplemented with better datasets from IRES around the world.

2.2.4 DESCRIBING AND CLASSIFYING FLOW REGIMES OF IRES: CASE STUDIES

To illustrate applications and results of methods described in the previous section, we present several case studies of their use in IRES. In particular, we wanted to illustrate challenges and approaches to obtaining suitable flow-regime data, various ways of illustrating flow-regime data, and how these data can be used to draw hydrological conclusions about IRES in different parts of the world.

As mentioned in Section 2.2.3, Kennard et al. (2010a) used multiple hydrological metrics to classify flow regimes in Australian rivers but there were some major challenges in obtaining an adequate dataset. As the study was conducted at a continental scale, the strategy for data collection needed to maximize the number and spatial coverage of gauges for analysis while ensuring they had comparable data quality and quantity. The initial dataset comprised mean daily discharge data for 2686 gauges with >10 years of records. Data from these 2686 gauges were then screened to provide a subset that (i) had few or no hydrological modifications caused by human activities; (ii) spanned at least 15 years from Jan. 1, 1965 to Dec. 31, 2000, preferably extending throughout a common year (i.e., 1980); and (iii) comprised continuous mean daily discharge data where possible. The first criterion reflected the study's aim to classify "natural" flow regimes. The second criterion was based on analyses of data adequacy by Kennard et al. (2010b) which also stipulated the need for the discharge records to lie within a discrete temporal window sharing >50% overlap among records. The third criterion also related to data adequacy by seeking the most complete records possible; where fewer than 30 days of data were missing, the missing records were imputed using linear interpolation or regression.

This screening process yielded a final dataset of 830 gauges covering most of the climatic, and hence potential flow regime, types across Australia (Kennard et al., 2010a) and sharing comparable lengths and timing of discharge records (Kennard et al., 2010b). Hydrological classification was undertaken using Bayesian mixture modeling which models the observed distribution of data as a mixture of a finite number of component distributions to determine the number of distributions, their parameters, and object memberships (Webb et al., 2007). The approach is fully probabilistic and uncertainty can be explicitly represented in terms of data specification (in this case, hydrological metric estimation uncertainty), class specification, and the final classification chosen. The most likely classification from the Bayesian analysis generated 12 classes of distinct flow regime types, eight of which featured intermittence. The geographical distribution of these eight classes varied in spatial cohesion (Fig. 2.2.6); the two classes described as "predictable winter highly intermittent" and "variable summer highly intermittent" were particularly scattered across the continent whereas other classes such as "predictable winter intermittent" had most of their representative gauges in the same part of the continent. In addition to geographic factors, climatic and several topographic factors were also generally strong discriminators of flow-regime classes (Kennard et al., 2010a). Major differences among the various classes of IRES related to their degree of intermittence (as number and frequency of zero-flow days) and the predictability and seasonal timing of flow.

Another study that classified rivers by flow regime is that by Snelder et al. (2013) of rivers in France and is noteworthy because it focused specifically on intermittence. Again, there was an extensive data screening process to derive a suitable dataset. The initial dataset comprised records of daily mean flow from over 3800 gauging stations across France. The first screening step removed poor quality data and records from gauges whose flows were likely to be modified by reservoirs, diversions, or significant abstractions upstream. The second screening step was to select data from the remaining stations that spanned the 35-year period from 1975 to 2009. Finally, these records were scanned for gaps longer than 20 days; where these gaps were found, the year of record was removed to yield a final dataset of 628 stations with 23–35 years of record.

Daily flow data from each gauging station were used to produce two sets of time series describing flow intermittence in each calendar year of record: the frequency of zero-flow periods (consecutive days of zero flow) and the total number of zero-flow days. Two further hydrological metrics were generated from these for each station: the mean annual frequency of zero-flow periods and the mean

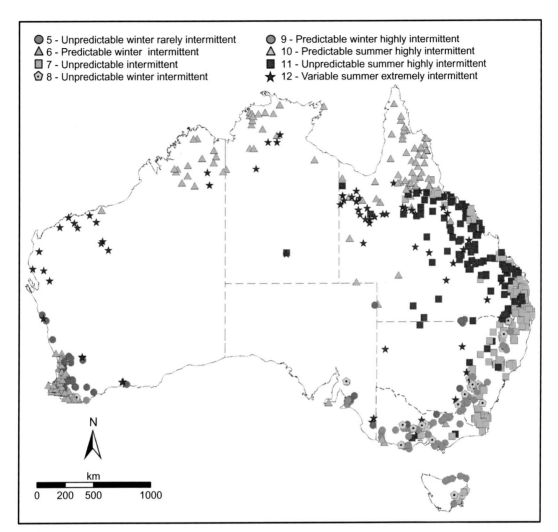

5 - Unpredictable winter rarely intermittent
6 - Predictable winter intermittent
7 - Unpredictable intermittent
8 - Unpredictable winter intermittent
9 - Predictable winter highly intermittent
10 - Predictable summer highly intermittent
11 - Unpredictable summer highly intermittent
12 - Variable summer extremely intermittent

FIG. 2.2.6

Distribution of Australian rivers in eight classes of IRES (5–12, see Fig. 2.2.5) based on the analyses by Kennard et al. (2010a).

Modified from Kennard, M.J., Pusey, B.J., Olden, J.D., Mackay, S.J., Stein, J.L., Marsh, N., 2010a. Classification of natural flow regimes in Australia to support environmental flow management. Freshw. Biol. 55, 171–193; Kennard, M.J., Mackay, S.J., Pusey, B.J., Olden, J.D., Marsh, N., 2010b. Quantifying uncertainty in estimation of hydrologic metrics for ecohydrological studies. River Res. Appl. 26, 137–156, with permission.

duration of zero-flow periods (Snelder et al., 2013). These metrics were then used to classify the flow regimes from the 628 gauging stations into four classes: a single perennial-stream class whose regimes lacked zero-flow days, and three subclasses of IRES whose boundaries were based on a scatterplot of mean annual frequency of zero-flow periods versus mean duration of zero-flow periods (Fig. 2.2.7a).

FIG. 2.2.7

(a) Three classes (colored symbols) of IRES in continental France, grouped by their mean duration and mean annual frequency of zero-flow periods derived from 23 to 35 years of record between 1975 and 2009. Examples of the classes are la Clauge, rarely intermittent (b); la Vis, intermittent (c); and les Valsaintes, highly ephemeral (d).

Plot modified from Snelder, T.H., Datry, T., Lamouroux, N., Larned, S.T., Sauquet, E., Pella, H., et al., 2013. Regionalization of patterns of flow intermittence from gauging station records. Hydrol. Earth Syst. Sci. 17, 2685–2699, with permission. Photos: Courtesy B. Launay.

The three subclasses of IRES (examples in Fig. 2.2.7b–d) can be described as rarely intermittent with infrequent brief zero-flow periods, intermittent with infrequent long zero-flow periods, or highly ephemeral with frequent long zero-flow periods. However, it is clear from Fig. 2.2.7a that these subclasses are imposed upon what is, in reality, a spectrum and is a salient reminder that all classifications of flow regimes impose boundaries on continua.

Snelder et al. (2013) then explored the match between the flow-regime classifications and various environmental characteristics of the catchments of the 628 gauging stations. There were significant associations of intermittence with regional-scale climate patterns and catchment area, shape, and slope. However, these associations were only moderate, interpreted by Snelder et al. (2013) to arise because intermittence is also controlled by processes such as groundwater table fluctuations and seepage through permeable channels that operate at subcatchment scales (Section 2.2.2). This substantial fine-scale heterogeneity was also evident in the limited spatial synchronization of zero flows (Snelder et al., 2013) and is a major constraint when applying flow-regime classifications to finer-scale ecological processes for ecohydrological analyses (e.g., Leigh et al., 2016). Nonetheless, regional classifications and derivative predictive models of patterns in flow intermittence are valuable sources of information when setting environmental flows, designing regional biomonitoring programs, and predicting broad-scale changes in flow regimes likely to arise from the effects of climate change and altered patterns of human exploitation of water resources (Section 2.2.4).

Potential effects of climate change on flow regimes in IRES are attracting particular attention because of the implications for river-dependent ecosystems and availability of potable water in arid and semiarid regions. Runoff and streamflow (especially minimum flows) may begin to decrease in the future, with numbers of zero-flow days increasing in several regions (e.g., McKerchar and Schmidt, 2007; Das et al., 2011; Döll and Schmied, 2012; Jaeger et al., 2014). However, there is substantial uncertainty about how climate change will affect the frequency, occurrence, and temporal dynamics of IRES flow regimes. Not all regions of the world are expected to get drier under climate change, and there is some evidence that IRES flow regimes are fairly resilient to climate change (e.g., Botter et al., 2013; Eng et al., 2016).

In arid and semiarid western North America, climate change scenarios predict longer and drier summers with decreases in mean annual streamflow, late summer precipitation, and late summer streamflow (Seager et al., 2013). To assess the likely effects of these predicted changes on flow regimes of perennial and intermittent streams in the region, especially hydrological thresholds associated with shifts from perennial to intermittent streamflow, Reynolds et al. (2015) analyzed historic flow records from streams in the upper Colorado River Basin. The hydrological metrics of mean number of zero-flow days per year and the percent of months in the flow record that had no flow for the entire month were used to distinguish categories of flow intermittence. Stream reaches were classed as "strongly intermittent" when >5% of months over the period of record had zero flows and there were >20 zero-flow days per year on average, "weakly intermittent" when 0%–5% of months had zero flows and the number of zero-flow days averaged 1–19 days per year, and "perennial" when both the percent of zero-flow months and the average number of zero-flow days were zero.

To assess relationships between climate and flow intermittence, Reynolds et al. (2015) analyzed how the Palmer Drought Severity Index (PSDI) and annual variability in temperature and precipitation influenced the number of zero-flow days for the two intermittent stream types. Although none of the individual precipitation or temperature variables had significant explanatory power, PDSI was a significant predictor of the number of zero-flow days in both the intermittent streamflow classes. Reynolds et al. (2015) then modeled the relationship between minimum daily flow and the predictor variables of mean

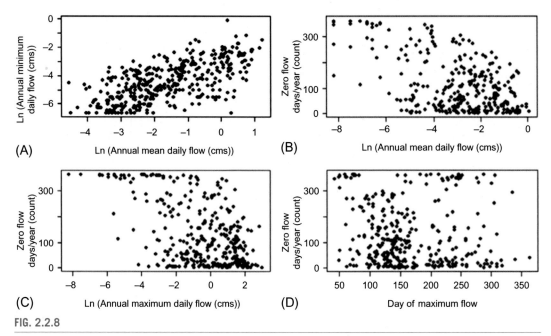

FIG. 2.2.8

Relationships between annual daily flow and annual minimum daily flow (A) and between zero-flow days per year and annual daily flow (B), maximum daily flow (C), and timing of maximum flow (D) in IRES of the upper Colorado River Basin. Each point represents a year's flow data from a given stream reach.

Modified from Reynolds, L.V., Shafroth, P.B., Poff, N.L., 2015. Modeled intermittency risk for small streams in the Upper Colorado River Basin under climate change. J. Hydrol. 523, 768–780, with permission.

daily flow magnitude, maximum daily magnitude, and annual timing of maximum flow for strongly and weakly intermittent streams. Annual mean daily flows were positively associated with annual minimum daily flows (Fig. 2.2.8A). There was a negative relationship between the number of zero-flow days per year and annual mean daily flow (Fig. 2.2.8B) but not maximum daily flow or its timing (Fig. 2.2.8C and D). These findings suggest that under a drier future climate, decreased mean flows will be associated with reduced minimum flows and increased number of zero-flow days, and match the predictions by Jaeger et al. (2014) that the frequency of zero-flow days would increase and flowing portions of the stream would decrease in the Verde River basin in Arizona under future climate scenarios.

A broader-scale analysis of potential climate change effects on the flow regimes of IRES across North America focused on the sensitivity of several hydrological metrics to historical variability in climate (Eng et al., 2016). To eliminate streams that seldom dry, only streams with a minimum annual average over the entire period of record of at least 15 days of zero flow per year were used. Further data screening sought only streams minimally impacted by direct human modifications and that had a minimum length of record of 10 years between 1950 and 2012, resulting in a final dataset of 256 streams. Daily flow records were used to calculate annual time series of three metrics: number of zero-flow days, average of the central 50% range of flows, and average of the largest 10% of flows. The 256 IRES were classified based on the seasonality of zero-flow periods into five different groups as fall, fall-to-winter, nonseasonal, summer, and summer-to-winter IRES (Fig. 2.2.9). Not surprisingly, many

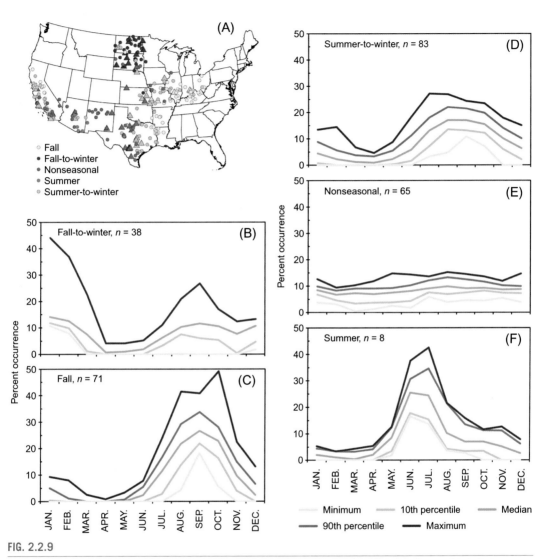

FIG. 2.2.9

(A) Locations of US Geological Survey stream gauges on 265 IRES with 10–47 years of record (*circles*) and >48 years of record (*triangles*). *Colors* represent five broad patterns of timing when the zero-flow values occurred: fall (*yellow*), fall-to-winter (*red*), nonseasonal (*blue*), summer (*orange*), and summer-to-winter (*green*). Percent occurrence of zero flows across all gauges in each class between 1950 and 2012 for (B) fall-to-winter, (C) fall, (D) summer-to-winter, (E) nonseasonal, and (F) summer IRES.

Reproduced from Eng, K., Wolock, D. M., Dettinger, M.D., 2016. Sensitivity of intermittent streams to climate variations in the USA. River Res. Appl. 32, 885–895, with permission.

of the representatives of these five groups of IRES flow regimes are spatially segregated although the extent of this segregation varies (Fig. 2.2.9A); IRES with a fall-to-winter flow regime are more tightly grouped regionally than fall and summer-to-winter subclasses. One of the most interesting conclusions from this study is that future climatic changes may alter different components of the flow regimes of IRES, many of which are crucial ecological cues (Chapters 4.3, 4.5, and 4.6). This study also highlights the value of using multiple analytical approaches to flow-regime analysis, including analyses of changes to seasonal hydrographs for different hydrological metrics.

The final case study in this section is perhaps the most ambitious one, seeking to document the spatial distribution of IRES across much of the globe as well as the year-to-year variability of zero-flow events. Undertaken as part of the international research project "Intermittent River Biodiversity Analysis and Synthesis" (IRBAS, http://irbas.cesab.org), daily discharge data from more than 6069 gauging stations with little or no significant human influence on flow were collected from France, Spain, Australia, and the conterminous United States. Datasets comprised those with at least 30 years of records (spanning 1970–2013) and missing <5% of their data.

IRES were considered to be those averaging at least 5 zero-flow days per year, yielding a final dataset of 525 stations (Fig. 2.2.10). The proportion of IRES ranged from 8% (France) to 65% (Australia), with rivers in Australia and Spain by far the most subject to flow intermittence. Across all four countries, almost 60% of IRES experienced zero flow for at least 10% of their record length. The flow regimes of the IRES were described by two climate descriptors and 19 hydrological metrics, including 11 metrics related to the different facets of intermittence. These included the mean frequency of days without flow per year, the mean total number of zero-flow days per year, and the mean timing of zero-flow events (θ) and its dispersion. At the global scale, the most relevant explanatory factor of intermittence was an aridity index; not unexpectedly, the more arid the climate, the higher the probability of a river being intermittent.

All of these case studies have substantially advanced our knowledge about the regional and global prevalence and diversity of IRES as well as the major challenges of describing their flow regimes at multiple spatial and temporal scales using various techniques and metrics. In the context of this book, these case studies and their various approaches inform the reader about hydrological aspects of flow regimes, especially intermittence, in IRES and provide the technical background for understanding how components of the flow regime (Box 2.2.1) influence physicochemical (Section 3) and ecological (Section 4) features of IRES as well as their management (Section 5). However, there is still considerable scope for further hydrological research and refinements of our perspectives about flow regimes and ecosystem processes in IRES.

2.2.5 CONCLUSIONS: RESEARCH NEEDS AND FUTURE PERSPECTIVES

The current Anthropocene era of dramatic changes to local and global hydrological cycles (Vörösmarty and Sahagian, 2000; Sterling et al., 2013) is one where IRES are increasingly prevalent across much of the globe. Many of these additional IRES may be considered "novel" ecosystems (*sensu* Hobbs et al., 2014) with altered flow regimes of increased intermittence (Fig. 2.2.11) artificially created by burgeoning demands for fresh water (Chapter 5.1). Consequently, it is now imperative that we appropriately

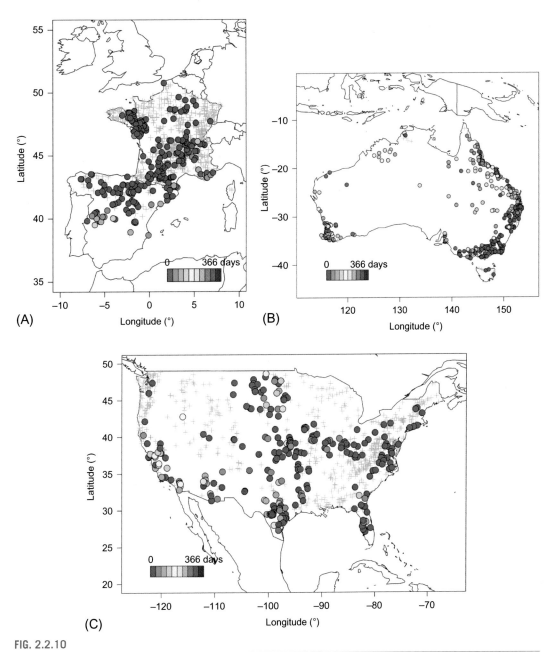

FIG. 2.2.10

Distribution of perennial rivers (*crosses*) and IRES (*colored circles*) in France and Spain (A), Australia (B), and the conterminous United States (C), expressing intermittence as the mean frequency of zero-flow days per year in equal 36-day intervals (i.e., *purple*=<36, *blue*=36–71, etc.).

FIG. 2.2.11

Human activities modify flow regimes to create "novel" ecosystems with increased or reduced intermittence, as seen here in the Pukaki River (a) looking downstream from the spillway of the Pukaki Dam (b), South Island, New Zealand.

Photos: Courtesy A. Boulton.

protect or manage the flow regimes of natural and modified IRES to support them as functional ecosystems (Arthington et al., 2014), potentially requiring strategies such as "reconciliation ecology" to restore severely impaired IRES (Moyle, 2014). Such protection and management often involves implementing environmental flows to supply water in the quantities and flow regimes needed to sustain natural or human-modified IRES (Acreman et al., 2014). This requires knowledge of historical, current, and likely future flow regimes of IRES, including frequency, timing, and duration of intermittence and the implications for hyporheic flow and surface water availability.

One likely future research initiative will be a far greater focus on IRES whose flow regimes have been so modified that they are either "novel" ecosystems with no current analogs or are "hybrid" combinations that retain some original characteristics but have novel elements (Hobbs et al., 2014). The most significant research gap here is understanding the relationship between components of the flow regime (e.g., duration or frequency of zero-flow periods) and the occurrence and persistence of pools and hyporheic flow in IRES and their ecosystem services (Chapter 5.2), biota (Chapters 4.1–4.6), and ecological processes (Chapters 4.7–4.9). Without this information at catchment or broader scales, restoration or protection strategies involving environmental flows (e.g., Seaman et al., 2016) are severely hampered (Chapter 5.4).

Another future initiative takes advantage of the rapid development of new technologies for gathering remotely sensed data, including time series information about the presence of surface water across the landscape, along with wet-dry mapping by citizen scientists (Section 2.2.3). With the growing realization of the increasing extent of intermittence in many parts of the world (Chapter 1), there is a need to better document "hot spots" of change in flow regimes, groundwater-surface water interactions, and catchment-level responses to hydrological impacts of human activities and climate change in IRES (Section 2.2.4). These technologies include real-time tracking of water persistence and quality in critical refugia during dry periods (King et al., 2015) so that remediation strategies can be timely and effective. The perspective that IRES may provide crucial "early warning" indicators of landscape-scale changes in flow regimes and hydrological cycles (Palmer et al., 2008; Brooks, 2009) is likely to evolve into one that uses technological advances in mapping and flow monitoring to identify which aspects of IRES are especially sensitive, where nonlinear responses may occur in hydrological and ecological indicators to changes in flow regime, and what drivers are responsible for these changes in different IRES.

We know very little about the importance of antecedent conditions in IRES as a determinant of subsequent flow regimes and ecohydrological responses to intermittence. We are unaware of any hydrological metrics that specifically describe the rate of change in discharge before and after zero-flow periods (e.g., how fast is the change between flowing and nonflowing, and between nonflowing and flowing, periods?) yet such metrics (proposed in Table 2.2.1) are easy enough to calculate given the existence and wide use of the general rate-of-change metrics in the literature (Olden and Poff, 2003) and software such as RAP (River Analysis Package, Marsh et al., 2003). The challenge will be to decide how long before and after the zero-flow period, and which flow events, should be included in the calculations. Information across a broad range of IRES on the persistence of pools after flow cessation is scant (but see Hamilton et al., 2005) and more use could be made of citizen science (Datry et al., 2016a,b) and interviews and aerial photography (Gallart et al., 2016).

From an ecological perspective, rates of change around zero-flow periods may govern physical, chemical, and biological responses that occur during and after these events. For example, rapid resumption of flow may be associated with mass displacement of organic and inorganic materials (Corti and Datry, 2012; Rosado et al., 2015), potentially leading to poor water quality and fish kills during the dry-wet transition periods (Datry et al., 2014; King et al., 2015). Rapid loss of flow and surface water may be challenging aquatic biota more than slow loss, especially if escape mechanisms require flow (e.g., drift to downstream perennial reaches). The hydrological conditions to which biota and their habitats are exposed prior to each flow event probably determine ecological responses to zero-flow periods in IRES but currently we lack adequate empirical data from which to derive reliable generalizations.

Central to all these research needs is more detailed knowledge of the flow regimes of IRES from the short-term local scales of hourly changes in a drying pool or a riffle where flow has recently ceased through to the long-term broad scales of interannual changes across multiple catchments in landscapes presently subject to, for example, regional warming. The flow regime of all IRES is characterized by intermittence; in the next chapter, we explore how this characteristic feature affects spatial patterns of hydrological connectivity in longitudinal, lateral, and vertical dimensions. The interaction of these three spatial dimensions of connectivity with the temporal dimension of the flow regime (particularly intermittence) governs virtually all physical, chemical, and biological processes in IRES yet many hydrological aspects remain elusive and poorly understood.

ACKNOWLEDGMENTS

We thank Walter Dodds, Bartosz Grudzinski, Keith Gido, Joshuah Perkin, Amy Doll, Will Curtis, Justin Murdock, B. Launay, and Meghan Smart for providing advice and images for this chapter; we regret not being able to use all the images but appreciate the efforts to supply them. We also thank Francesc Gallart and Ted Grantham for constructive comments that improved an earlier draft of this chapter.

REFERENCES

Acreman, M., Arthington, A.H., Colloff, M.J., Couch, C., Crossman, N.D., Dyer, F., et al., 2014. Environmental flows for natural, hybrid, and novel riverine ecosystems in a changing world. Front. Ecol. Environ. 12, 466–473.

Anna, A., Yorgos, C., Konstantinos, P., Maria, L., 2009. Do intermittent and ephemeral Mediterranean rivers belong to the same river type? Aquat. Ecol. 43, 465–476.

Arthington, A.H., Bernardo, J.M., Ilhéu, M., 2014. Temporary rivers: linking ecohydrology, ecological quality and reconciliation ecology. River Res. Appl. 30, 1209–1215.

Benda, L.E., Poff, N.L., Tague, C., Palmer, M.A., Pizzuto, J., Cooper, S.D. et al., 2002. How to avoid train wrecks when using science in environmental problem solving. BioScience 52, 1127–1136.

Benstead, J.P., Leigh, D.S., 2012. An expanded role for river networks. Nat. Geosci. 5, 678–679.

Bhamjee, R., Lindsay, J.B., 2011. Ephemeral stream sensor design using state loggers. Hydrol. Earth Syst. Sci. 15, 1009–1021.

Botter, G., Basso, S., Rodriguez-Iturbe, I., Rinaldo, A., 2013. Resilience of river flow regimes. Proc. Natl. Acad. Sci. U.S.A. 110, 12925–12930.

Boulton, A.J., 1989. Over-summering refuges of aquatic macroinvertebrates in two intermittent streams in central Victoria. Trans. R. Soc. South Aust. 113, 23–34.

Boulton, A.J., 2003. Parallels and contrasts in the effects of drought on stream macroinvertebrate assemblages. Freshw. Biol. 48, 1173–1185.

Boulton, A.J., Valett, H.M., Fisher, S.G., 1992. Spatial distribution and taxonomic composition of the hyporheos of several Sonoran Desert streams. Arch. Hydrobiol. 125, 37–61.

Brooks, R.T., 2009. Potential impacts of global climate change on the hydrology and ecology of ephemeral freshwater systems of the forests of the northeastern United States. Clim. Change 95, 469–483.

Bunn, S.E., Arthington, A.H., 2002. Basic principles and consequences of altered hydrological regimes for aquatic biodiversity. Environ. Manag. 30, 492–507.

Carlisle, D.M., Falcone, J., Wolock, D.M., Meador, M.R., Norris, R.H., 2010. Predicting the natural flow regime: models for assessing hydrological alteration in streams. River Res. Appl. 26, 118–136.

Caruso, B.S., Haynes, J., 2011. Biophysical-regulatory classification and profiling of streams across management units and ecoregions. J. Am. Water Resour. Assoc. 47, 386–407.

Chahinian, N., Tournoud, M.-G., Perrin, J.-L., Picot, B., 2011. Flow and nutrient transport in intermittent rivers: a modelling case study on the Vène River using SWAT 2005. Hydrol. Sci. J. 56, 268–287.

Chapin, T.P., Todd, A.S., Zeigler, M.P., 2014. Robust, low-cost data loggers for stream temperature, flow intermittency, and relative conductivity monitoring. Water Resour. Res. 50, 6542–6548.

Cipriani, T., Tilmant, F., Branger, F., Sauquet, E., Datry, T., 2014. Impact of climate change on aquatic ecosystems along the Asse River Network. In: FRIEND-Water 2014, Hydrology in a Changing World: Environmental and Human Dimensions, February 2014, Hanoi, Vietnam IAHS Publication 36X, pp. 463–468.

Constantz, J., Stonestorm, D., Stewart, A.E., Niswonger, R., Smith, T.R., 2001. Analysis of streambed temperatures in ephemeral channels to determine streamflow frequency and duration. Water Resour. Res. 37, 317–328.

Corti, R., Datry, T., 2012. Invertebrates and sestonic matter in an advancing wetted front travelling down a dry river bed (Albarine, France). Freshw. Sci. 31, 1187–1201.

Corti, R., Datry, T., Drummond, L., Larned, S.T., 2011. Natural variation in immersion and emersion affects breakdown and invertebrate colonization of leaf litter in a temporary river. Aquat. Sci. 73, 537–550.

Costa, A.C., Foerster, S., Araújo, J.C., Bronstert, A., 2013. Analysis of channel transmission losses in a dryland river reach in north-eastern Brazil using streamflow series, groundwater level series and multi-temporal satellite data. Hydrol. Process. 27, 1046–1060.

Costelloe, J.F., Russell, K.L., 2014. Identifying conservation priorities for aquatic refugia in an arid zone, ephemeral catchment: a hydrological approach. Ecohydrology 7, 1534–1544.

Costelloe, J.F., Grayson, R.B., McMahon, T.A., 2005. Modelling stream flow for use in ecological studies in a large, arid zone river, central Australia. Hydrol. Process. 19, 1165–1183.

Costigan, K.H., Daniels, M.D., Dodds, W.K., 2015. Fundamental spatial and temporal disconnections in the hydrology of an intermittent prairie headwater network. J. Hydrol. 522, 305–316.

Costigan, K.H., Jaeger, K.L., Goss, C.W., Fritz, K.M., Goebel, P.C., 2016. Understanding controls on flow permanence in intermittent rivers to aid ecological research: integrating meteorology, geology and land cover. Ecohydrology 9, 1141–1153. http://dx.doi.org/10.1002/eco.1712.

Croke, B.F.W., Jakeman, A.J., 2007. Use of the IHACRES rainfall-runoff model in arid and semi-arid regions. In: Wheater, H., Sorooshian, S., Sharma, K.D. (Eds.), Hydrological Modelling in Arid and Semi-Arid Areas. International Hydrology Series, Cambridge University Press, Cambridge, pp. 41–48.

Das, T., Pierce, D.W., Cayan, D.R., Vano, J.A., Lettenmaier, D.P., 2011. The importance of warm season warming to western U.S. streamflow changes. Geophys. Res. Lett. 38, L23403.

Datry, T., 2012. Benthic and hyporheic invertebrate assemblages along a flow intermittence gradient: effects of duration of dry events. Freshw. Biol. 57, 563–574.

Datry, T., Larned, S.T., Tockner, K., 2014. Intermittent rivers: a challenge for freshwater ecology. BioScience 64, 229–235.

Datry, T., Pella, H., Leigh, C., Bonada, N., Hugueny, B., 2016a. A landscape approach to advance intermittent river ecology. Freshw. Biol. 61, 1200–1213.

Datry, T., Moya, N., Zubieta, J., Oberdorff, T., 2016b. Determinants of local and regional communities in intermittent and perennial headwaters of the Bolivian Amazon. Freshw. Biol. 61, 1335–1349.

De Girolamo, A.M., Lo Porto, A., Pappagallo, G., Tzoraki, O., Gallart, F., 2015. The hydrological status concept: application at a temporary river (Candelaro, Italy). River Res. Appl. 31, 892–903.

De Girolamo, A.M., Barca, E., Ielpo, P., Rulli, M.C., 2016. Characterising the hydrological regime of an ungauged temporary river system: a case study. Environ. Sci. Pollut. Res. http://dx.doi.org/10.1007/s11356-016-7169-0.

Döll, P., Schmied, H.M., 2012. How is the impact of climate change on river flow regimes related to the impact on mean annual runoff? A global-scale analysis. Environ. Res. Lett. 7, 014037.

Doyle, M.W., Stanley, E.H., Strayer, D.L., Jacobson, R.B., Schmidt, J.C., 2005. Effective discharge analysis of ecological processes in streams. Water Resour. Res. 41, W11411.

Eng, K., Wolock, D.M., Dettinger, M.D., 2016. Sensitivity of intermittent streams to climate variations in the USA. River Res. Appl. 32, 885–895.

Fleckenstein, J.H., Niswonger, R.G., Fogg, G.E., 2006. River–aquifer interactions, geologic heterogeneity, and low-flow management. Groundwater 44, 837–852.

Fritz, K.M., Johnson, B.R., Walters, D.M., 2006. Field Operations Manual for Assessing the Hydrologic Permanence and Ecological Condition of Headwater Streams. US Environmental Protection Agency, Washington, DC. EPA/600/ R-06/126 (accessed 3.11.16), http://www.epa.gov/eerd/methods/headwater.html.

Fritz, K.M., Johnson, B.R., Walters, D.M., 2008. Physical indicators of hydrologic permanence in forested headwater streams. J. North Am. Benthol. Soc. 27, 690–704.

Gallart, F., Prat, N., García-Roger, E.M., Latrón, J., Rieradevall, M., Llorens, P., et al., 2012. A novel approach to analysing the regimes of temporary streams in relation to their controls on the composition and structure of aquatic biota. Hydrol. Earth Syst. Sci. 16, 3165–3182.

Gallart, F., Llorens, P., Latron, J., Cid, N., Rieradevall, M., Prat, N., 2016. Validating alternative methodologies to estimate the regime of temporary rivers when flow data are unavailable. Sci. Total Environ. 565, 1001–1010.

Gassman, P., Reyes, M., Green, C., Arnold, J., 2007. The soil and water assessment tool: historical development, applications, and future research directions. Trans. Am. Soc. Agric. Biol. Eng. 50, 1211–1250.

Gleason, C.J., Smith, L.C., 2014. Toward global mapping of river discharge using satellite images and at-many-stations hydraulic geometry. Proc. Natl. Acad. Sci. U.S.A. 111, 4788–4791.

Godsey, S.E., Kirchner, J.W., 2014. Dynamic, discontinuous stream networks: hydrologically driven variations in active drainage density, flowing channels and stream order. Hydrol. Process. 28, 5791–5803.

Gomi, T., Sidle, R.C., Richardson, J.S., 2002. Understanding processes and downstream linkages of headwater systems. BioScience 52, 905–916.

Gordon, N., McMahon, T.A., Finlayson, B.L., Gippel, C.J., Nathan, R.J., 2004. Stream Hydrology: An Introduction for Ecologists, second ed. Wiley, Chichester.

Gungle, B., 2007. Timing and duration of flow in ephemeral streams of the Sierra Vista subwatershed of the Upper San Pedro Basin, Cochise County, southeastern Arizona: U.S. Geological Survey Scientific Investigations Report 2005-519. http://pubs.usgs.gov/sir/2005/5190/. (accessed 28.09.16) Version 2.0.

Hamada, Y., O'Connor, B.L., Orr, A.B., Wuthrich, K.K., 2016. Mapping ephemeral stream networks in desert environments using very-high-spatial-resolution multispectral remote sensing. J. Arid Environ. 130, 40–48.

Hamilton, S.K., Bunn, S.E., Thoms, M.C., Marshall, J.C., 2005. Persistence of aquatic refugia between flow pulses in a dryland river system (Cooper Creek, Australia). Limnol. Oceanogr. 50, 743–754.

Hansen, W.F., 2001. Identifying stream types and management implications. For. Ecol. Manage. 143, 39–46.

Hobbs, R.J., Higgs, E., Hall, C.M., Bridgewater, P., Chapin, F.S., Ellis, E.C., et al., 2014. Managing the whole landscape: historical, hybrid, and novel ecosystems. Front. Ecol. Environ. 12, 557–564.

Hughes, J.M.R., James, B., 1989. A hydrological regionalization of streams in Victoria, Australia, with implications for stream ecology. Aust. J. Mar. Freshw. Res. 40, 303–326.

Ivkovic, K.M., Croke, B.F.W., Kelly, R.A., 2014. Overcoming the challenges of using a rainfall–runoff model to estimate the impacts of groundwater extraction on low flows in an ephemeral stream. Hydrol. Res. 45, 58–72.

Jaeger, K.L., Olden, J.D., 2012. Electrical resistance sensor arrays as a means to quantify longitudinal connectivity of rivers. River Res. Appl. 28, 1843–1852.

Jaeger, K.L., Olden, J.D., Pelland, N.A., 2014. Climate change poised to threaten hydrologic connectivity and endemic fishes in dryland streams. Proc. Natl. Acad. Sci. U.S.A. 111, 13894–13899.

Jencso, K.G., McGlynn, B.L., Gooseff, M.N., Wondzell, S.M., Bencala, K.E., Marshall, L.A., 2009. Hydrologic connectivity between landscapes and streams: transferring reach- and plot-scale understanding to the catchment scale. Water Resour. Res. 45, W04428.

Kennard, M.J., Olden, J.D., Arthington, A.H., Pusey, B.J., Poff, N.L., 2007. Multi-scale effects of flow regime, habitat, and their interaction on fish assemblage structure in eastern Australia. Can. J. Fish. Aquat. Sci. 64, 1346–1359.

Kennard, M.J., Pusey, B.J., Olden, J.D., Mackay, S.J., Stein, J.L., Marsh, N., 2010a. Classification of natural flow regimes in Australia to support environmental flow management. Freshw. Biol. 55, 171–193.

Kennard, M.J., Mackay, S.J., Pusey, B.J., Olden, J.D., Marsh, N., 2010b. Quantifying uncertainty in estimation of hydrologic metrics for ecohydrological studies. River Res. Appl. 26, 137–156.

King, A.J., Townsend, S., Douglas, M.M., Kennard, M.J., 2015. Implications of water extraction on the low-flow hydrology and ecology of tropical savannah rivers: an appraisal for northern Australia. Freshw. Sci. 34, 741–758.

Knighton, A.D., Nanson, G.C., 2001. An event-based approach to the hydrology of arid zone rivers in the channel country of Australia. J. Hydrol. 254, 102–123.

Lake, P.S., 2003. Ecological effects of perturbation by drought in flowing waters. Freshw. Biol. 48, 1161–1172.

Larned, S.T., Datry, T., Arscott, D.B., Tockner, K., 2010b. Emerging concepts in temporary-river ecology. Freshw. Biol. 55, 717–738.

Leigh, C., 2013. Dry-season changes in macroinvertebrate assemblages of highly seasonal rivers: responses to low flow, no flow and antecedent hydrology. Hydrobiologia 703, 95–112.

Leigh, C., Datry, T., 2016. Drying as a primary hydrological determinant of biodiversity in river systems: a broad-scale analysis. Ecography. http://dx.doi.org/10.1111/ecog.02230.

Leigh, C., Sheldon, F., 2008. Hydrological changes and ecological impacts associated with water resource development in large floodplain rivers in the Australian tropics. River Res. Appl. 24, 1251–1270.

Leigh, C., Stewart-Koster, B., Sheldon, F., Burford, M.A., 2012. Understanding multiple ecological responses to anthropogenic disturbance: rivers and potential flow regime change. Ecol. Appl. 22, 250–263.

Leigh, C., Boulton, A.J., Courtwright, J.L., Fritz, K., May, C.L., Walker, R.H., et al., 2016. Ecological research and management of intermittent rivers: an historical review and future directions. Freshw. Biol. 61, 1181–1199.

Marsh, N., Stewardson, M.J., Kennard, M., 2003. River Analysis Package (RAP). Cooperative Research Centre for Catchment Hydrology, Melbourne. <http://www.toolkit.net.au/tools/RAP> accessed 28.09.16.

McDonough, O.T., Hosen, J.D., Palmer, M.A., 2011. Temporary streams: the hydrology, geography, and ecology of non-perennially flowing waters. In: Elliot, S.E., Martin, L.E. (Eds.), River Ecosystems: Dynamics. Management and Conservation. Nova Science Publishers, New York, pp. 259–290.

McKerchar, A.I., Schmidt, J., 2007. Decreases in low flows in the lower Selwyn River? J. Hydrol. (New Zealand) 46, 63–72.

McMahon, T.A., Vogel, R.M., Peel, M.C., Pegram, G.G.S., 2007. Global streamflows—Part 1: characteristics of annual streamflows. J. Hydrol. 347, 243–259.

Meyer, J.L., Wallace, J.B., 2001. Lost linkages and lotic ecology: rediscovering small streams. In: Press, M.C., Huntly, N., Levin, S. (Eds.), Ecology: Achievement and Challenge. Blackwell Scientific Press, New York, pp. 295–317.

Moyle, P.B., 2014. Novel aquatic ecosystems: the new reality for streams in California and other Mediterranean climate regions. River Res. Appl. 30, 1335–1344.

Nilsson, C., Renöfält, B.M., 2008. Linking flow regime and water quality in rivers: a challenge to adaptive catchment management. Ecol. Soc. 13. Article 18.

Nippgen, F., McGlynn, B.L., Marshall, L.A., Emanuel, R.E., 2011. Landscape structure and climate influences on hydrologic response. Water Resour. Res. 47, W12528.

Olden, J.D., Poff, N.L., 2003. Redundancy and the choice of hydrologic indices for characterizing streamflow regimes. River Res. Appl. 19, 101–121.

Olden, J.D., Kennard, M.J., Pusey, B.J., 2012. A framework for hydrologic classification with a review of methodologies and applications in ecohydrology. Ecohydrology 5, 503–518.

Palmer, M.A., Reidy Liermann, C.A., Nilsson, C., Flörke, M., Alcamo, J., Lake, P.S., et al., 2008. Climate change and the world's river basins: anticipating management options. Front. Ecol. Environ. 6, 81–89.

Pilgrim, D.H., Chapman, T.G., Doran, D.G., 1988. Problems of rainfall-runoff modelling in arid and semiarid regions. Hydrol. Sci. J. 33, 379–400.

Poff, N.L., 1996. A hydrogeography of unregulated streams in the United States and an examination of scale-dependence in some hydrological descriptors. Freshw. Biol. 36, 71–91.

Poff, N.L., Allan, J.D., Bain, M.B., Karr, J.R., Prestegaard, K.L., Richter, B.D., et al., 1997. The natural flow regime: a paradigm for river conservation and restoration. BioScience 47, 769–784.

Puckridge, J.T., Sheldon, F., Walker, K.F., Boulton, A.J., 1998. Flow variability and the ecology of large rivers. Mar. Freshw. Res. 49, 55–72.

Puckridge, J.T., Walker, K.F., Costelloe, J.F., 2000. Hydrological persistence and the ecology of dryland rivers. Regul. Riv. Res. Manage. 16, 385–402.

Reinfelds, I., Haeusler, T., Brooks, A.J., Williams, S., 2004. Refinement of the wetted perimeter breakpoint method for setting cease-to-pump or minimum environmental flows. River Res. Appl. 20, 671–685.

Reynolds, L.V., Shafroth, P.B., 2016. Modeled streamflow metrics on small, ungaged stream reaches in the Upper Colorado River Basin. U.S. Geol. Surv. Data Ser. 974, 1–11.

Reynolds, L.V., Shafroth, P.B., Poff, N.L., 2015. Modeled intermittency risk for small streams in the Upper Colorado River Basin under climate change. J. Hydrol. 523, 768–780.

Rolls, R.J., Leigh, C., Sheldon, F., 2012. Mechanistic effects of low-flow hydrology on riverine ecosystems: ecological principles and consequences of alteration. Freshw. Sci. 31, 1163–1186.

Rosado, J., Morais, M., Tockner, K., 2015. Mass dispersal of terrestrial organisms during first flush events in a temporary stream. River Res. Appl. 31, 912–917.

Rossouw, L., Avenant, M.F., Seaman, M.T., King, J.M., Barker, C.H., du Preez, P.J., et al., 2005. Environmental water requirements in non-perennial systems: WRC report no: 1414/1/05. Water Research Commission, Pretoria, South Africa. http://www.wrc.org.za/KnowledgeHubDocuments/ResearchReports/1414.pdf. accessed 3.11.16.

Sauquet, E., Gottschalk, L., Krasovskaïa, I., 2008. Estimating mean monthly runoff at ungauged locations: an application to France. Hydrol. Res. 39, 403–423.

Schriever, T.A., Bogan, M.T., Boersma, K.S., Cañedo-Argüelles, M., Jaeger, K.L., Olden, J.D., et al., 2015. Hydrology shapes taxonomic and functional structure of desert stream invertebrate communities. Freshw. Sci. 34, 399–409.

Seager, R., Ting, M., Li, C., Naik, N., Cook, B., Nakamura, J., et al., 2013. Projections of declining surface-water availability for the southwestern United States. Nat. Clim. Change 3, 482–486.

Seaman, M., Watson, M., Avenant, M., King, J., Joubert, A., Barker, C., et al., 2016. DRIFT-ARID: a method for assessing environmental water requirements (EWRs) for non-perennial rivers. Water SA 42, 356–367.

Sheldon, F., Fellows, C.S., 2010. Water quality in two Australian dryland rivers: spatial and temporal variability and the role of flow. Mar. Freshw. Res. 61, 864–874.

Smakhtin, V.U., 2001. Low flow hydrology: a review. J. Hydrol. 240, 147–186.

Smakhtin, V.Y., Toulouse, M., 1998. Relationships between low-flow characteristics of South African streams. Water SA 24, 107–112.

Snelder, T.H., Datry, T., Lamouroux, N., Larned, S.T., Sauquet, E., Pella, H., et al., 2013. Regionalization of patterns of flow intermittence from gauging station records. Hydrol. Earth Syst. Sci. 17, 2685–2699.

Snover, A.K., Hamlet, A.F., Lettenmaier, D.P., 2003. Climate-change scenarios for water planning studies. Bull. Am. Meteorol. Soc. 84, 1513–1518.

Stanley, E.H., Fisher, S.G., Grimm, N.B., 1997. Ecosystem expansion and contraction in streams. BioScience 47, 427–435.

Sterling, S.M., Ducharne, A., Polcher, J., 2013. The impact of global land-cover change on the terrestrial water cycle. Nat. Clim. Change 3, 385–390.

Thoms, M.C., Sheldon, F., 2000. Water resource development and hydrological change in a large dryland river: the Barwon–Darling River, Australia. J. Hydrol. 228, 10–21.

Turner, D.S., Richter, H.E., 2011. Wet/dry mapping: using citizen scientists to monitor the extent of perennial surface flow in dryland regions. Environ. Manag. 47, 497–505.

Tzoraki, O., De Girolamo, A.M., Gamvroudis, C., Skoulikidis, N., 2016. Assessing the flow alteration of temporary streams under current conditions and changing climate by soil and water assessment tool model. Int. J. River Basin Manage. 14, 9–18.

Vander Vorste, R., Malard, F., Datry, T., 2016a. Is drift the primary process promoting the resilience of river invertebrate communities? A manipulative field experiment in an intermittent alluvial river. Freshw. Biol. 61, 1276–1292.

Vander Vorste, R., Corti, R., Sagouis, A., Datry, T., 2016b. Invertebrate communities in gravel-bed, braided rivers are highly resilient to flow intermittence. Freshw. Sci. 35, 164–177.

Vörösmarty, C.J., Sahagian, D., 2000. Anthropogenic disturbance of the terrestrial water cycle. BioScience 50, 753–765.

Walker, K.F., Sheldon, F., Puckridge, J.T., 1995. A perspective on dryland river ecosystems. Regul. Riv. Res. Manage. 11, 85–104.

Webb, J.A., Bond, N.R., Wealands, S.R., Mac Nally, R., Quinn, G.P., Vesk, P.A., et al., 2007. Bayesian clustering with AutoClass explicitly recognises uncertainties in landscape classification. Ecography 30, 526–536.

Ye, W., Bates, B.C., Viney, N.R., Sivapalan, M., Jakeman, A.J., 1997. Performance of conceptual rainfall-runoff models in low-yielding catchments. Water Resour. Res. 33, 153–166.

FURTHER READING

Larned, S.T., Arscott, D.B., Schmidt, J., Diettrich, J.C., 2010a. A framework for analysing longitudinal and temporal variation in river flow and developing flow-ecology relationships. J. Am. Water Resour. Assoc. 46, 541–553.

Larned, S.T., Schmidt, J., Datry, T., Konrad, C.P., Dumas, J.K., Diettrich, J.C., 2011. Longitudinal river ecohydrology: flow variation down the lengths of alluvial rivers. Ecohydrology 4, 532–548.

HYDROLOGICAL CONNECTIVITY IN INTERMITTENT RIVERS AND EPHEMERAL STREAMS

Andrew J. Boulton*, Robert J. Rolls[†], Kristin L. Jaeger[‡], Thibault Datry[§]

University of New England, Armidale, NSW, Australia[*] University of Canberra, Bruce, ACT, Australia[†] USGS Washington Water Science Center, Tacoma, WA, United States[‡] Irstea, UR MALY, centre de Lyon-Villeurbanne, Villeurbanne, France[§]

IN A NUTSHELL

- Hydrological connectivity extends and interacts over longitudinal, lateral, and vertical dimensions in intermittent rivers and ephemeral streams (IRES), varying more than in perennially flowing systems
- In IRES, hydrological connectivity is primarily controlled by interactions between channel features (e.g., size, shape, substrate composition) and flow regime (e.g., duration and timing of streamflow), often modified by human activities
- Flow intermittence disrupts hydrological connectivity at multiple scales, creating nonlinear and patchy "discontinua" that strongly influence physical, chemical, and ecological patterns and processes in space and time
- A proposed conceptual framework explores three aspects of hydrological connectivity and potential effects of intermittence

2.3.1 INTRODUCTION

In all streams and rivers, hydrological connectivity, defined as the "water-mediated transfer of matter, energy, or organisms within and/or between elements of the hydrologic cycle" (Pringle, 2003: 2685), has three spatial dimensions—longitudinal, lateral, and vertical—that interact along a fourth dimension, time (Ward, 1989). In intermittent rivers and ephemeral streams (hereafter, IRES), cessation of surface flow disrupts hydrological connectivity in one or more spatial dimensions, with repercussions for most physical, chemical, and biological processes (Fig. 2.3.1). Longitudinally, cessation of surface flow halts downstream transport of sediments, other materials, and biota (Hooke, 2003; Rolls et al., 2012), and usually heralds the onset of drying of shallow channel sections, especially riffles. Laterally, aquatic habitats on the floodplain and along the riparian zone that were hydrologically linked to the main channel during overbank flows or through bank storage become isolated when water levels decline (Fig. 2.3.1), interrupting the two-way transfer of energy, sediments, and various organisms (Nakano and Murakami, 2001; Paetzold et al., 2006). Vertically, most of the exchange of water between the surface channel and the shallow saturated sediments below (i.e., hyporheic zone, White, 1993) ceases when surface flow stops, impairing processes such as oxygenation of the hyporheic zone by

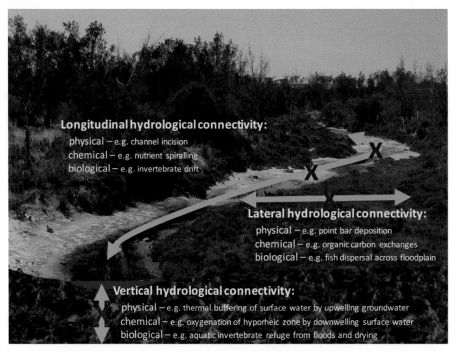

FIG. 2.3.1

Flow cessation and drying in IRES interrupt physical, chemical, and biological processes that rely on hydrological connectivity along three spatial dimensions—longitudinal, lateral, and vertical—portrayed as *blue lines*. The *double-headed arrows* imply many processes operate in both directions, including downstream to upstream (e.g., fish migration). Interruption of hydrological connectivity is indicated by *red crosses*.

Photo: Courtesy A. Boulton.

downwelling water or the flux of nutrients to the surface water in upwelling areas (Datry and Larned, 2008; Boulton et al., 2010, Fig. 2.3.1).

Thus, flow and hydrological connectivity are tightly linked, often creating complex hydrological patterns along all three spatial dimensions. In this chapter, "flow" refers to the net movement of water in a riverscape; where necessary, surface flow is explicitly distinguished from subsurface flow because IRES are defined by intermittence of surface flow even though flow may continue in the sediments below and alongside the channel. In IRES, hydrological connectivity in all three spatial dimensions is disrupted by intermittence interacting with geomorphology (Chapter 2.1) and flow regime (Chapter 2.2).

In IRES, hydrological connectivity spans the full spatial spectrum from the entire stream system (*sensu* Frissell et al., 1986, $>10^3$ m), exemplified by longitudinal connectivity from headwaters to the river terminus, to within microhabitat systems ($<10^{-1}$ m, Frissell et al., 1986) where hydrological connectivity along a few millimeters of interstitial flowpaths can generate sharply contrasting conditions of redox potential (Hendricks and White, 1991). Although hydrological connectivity also occurs in perennially flowing streams, the key difference in IRES is that streamflow intermittence dramatically affects hydrological connectivity in all three dimensions and promotes its dynamic nature in space and time. This spatiotemporal dynamism in hydrological connectivity exerts powerful influences on the biogeochemistry, biota, and ecosystem processes in all IRES, providing unique research opportunities to exploit such variability in connectivity to advance our understanding of general stream ecology. (e.g., Datry et al., 2014; Jaeger et al., 2014; chapters in Sections 3 and 4).

In this chapter, we start by reviewing the factors driving hydrological connectivity at multiple spatial scales and the implications of these drivers for different IRES. We then describe patterns of hydrological connectivity occurring along each of the three spatial dimensions and how they are affected by geomorphological features (Chapter 2.1) and flow regime (Chapter 2.2), especially intermittence. A preliminary conceptual framework, drawing on ecological connectivity theory (Beger et al., 2010), is proposed as a basis for generating hypotheses that predict physical, chemical, and biological responses to intermittence along interacting hydrological dimensions. We conclude by synthesizing the key aspects of hydrological connectivity in IRES by way of an introduction to the next three sections of this book.

2.3.2 WHAT GOVERNS HYDROLOGICAL CONNECTIVITY IN IRES?

Hydrological connectivity along longitudinal, lateral, and vertical dimensions in IRES is largely governed by interactions between the multiple components of flow regime (magnitude, frequency, duration, timing, and rate of change of flow, Chapter 2.2) and fluvial geomorphology (e.g., channel and floodplain shape, size, gradient, sediment composition, and location along the network). Many of these components are modified by catchment land use and other human activities (Fig. 2.3.2). Other important drivers of hydrological connectivity at a landscape scale include climatic conditions such as timing and amounts of precipitation and evaporation, hydrogeological features governing gains and losses of groundwater along the channel, tectonic activity (e.g., fault lines, volcanism), and underlying lithology and geology, including geological transitions along the channel (Fig. 2.3.2; Graf, 1988; Tooth and Nanson, 2011). At finer scales, patterns of hydrological connectivity in IRES may be influenced by biological processes (Fig. 2.3.2) such as microbial activity and benthic algae altering substrate permeability (Mendoza-Lera and Mutz, 2013; Hartwig and Borchardt, 2015) or the presence of vegetation within the channel modifying surface flows and bed topography (Sandercock et al., 2007; Sandercock and Hooke, 2011).

In IRES, hydrological connectivity in all three dimensions can change markedly over time. Natural streams and rivers are spatially dynamic ecosystems that undergo cycles of expansion and contraction of their aquatic boundaries in response to variations in flow regime (Allan and Castillo, 2007; Larned et al., 2011). However, as IRES have especially variable flow regimes (Chapter 2.2), these cycles of expansion and contraction often include fragmentation of surface water hydrological connectivity during drying (Stanley et al., 1997; Jaeger and Olden, 2012). The fluctuations of expansion, contraction, and fragmentation generate dynamic mosaics of habitat patches (Larned et al., 2010; Datry et al., 2014; Chapter 4.9) with their own environmental conditions, biota, and ecological functions.

Along all three spatial dimensions, the changing cycles of expansion, contraction, and fragmentation in IRES produce a hydrological "discontinuum" (*sensu* Poole, 2002). Thus, hydrological patterns along these dimensions are seldom gradual trends over distance but instead exist as sequences or networks of diverse patches with varying streamflow, presence of water, and hydrological connectivity, especially in highly intermittent IRES. This is consistent with the patch dynamics perspective (Pickett and White, 1985) that has proved successful for describing ecohydrological patterns at multiple scales in river ecosystems (Fausch et al., 2002; Winemiller et al., 2010). These ecohydrological patterns are controlled by the pervasive gravitational movement of water downstream and vertically which imposes

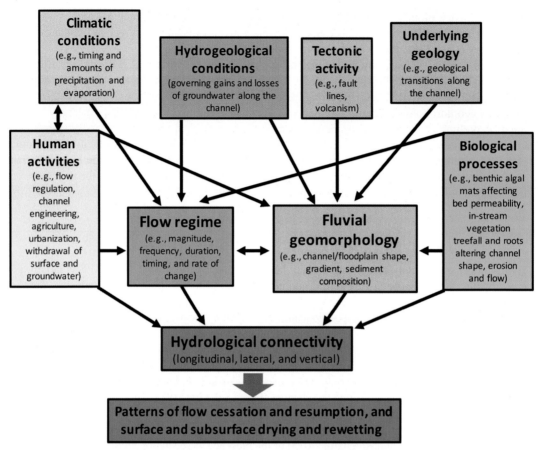

FIG. 2.3.2

Interactions among factors affecting hydrological connectivity in longitudinal, lateral, and vertical dimensions in IRES.

a "catchment hierarchy" (Townsend, 1996), acknowledging that environmental features at a given location along the river network typically reflect both local and upstream hydrological conditions.

Location along the river network also governs the importance of other drivers of hydrological connectivity at that point (Fig. 2.3.2). For example, in the upper reaches of many IRES, precipitation is a strong driver of in-channel flow but this coupling weakens in lowland reaches where groundwater inputs increase (Tooth and Nanson, 2011). Where headwaters of IRES arise in hills and mountains, channel shape is largely bedrock-controlled and strongly influenced by local geology compared with downstream reaches where alluvial sediments and groundwater exchanges are more prevalent (Chapter 2.1; Stanford and Ward, 1993; Tonina and Buffington, 2009). However, all portions of the river network are subject to various human activities that affect hydrological connectivity along longitudinal, lateral, and vertical dimensions (Fig. 2.3.2), either directly such as through water extraction or indirectly from altered flow regimes and/or channel and floodplain geomorphology (Costigan et al., 2016).

These activities include modifications to topography, vegetation cover, runoff, and water quality through agriculture, mining and urbanization, changes to channels by in-stream works (e.g., weirs, dams), and withdrawals of surface water and groundwater (Chapter 5.1).

2.3.3 HYDROLOGICAL CONNECTIVITY, INTERMITTENCE, AND SURFACE WATER DRYING AND REWETTING IN IRES

The characteristic hydrological feature of all IRES is intermittence (Chapter 1), often followed by drying of part or all of the channel (Fig. 2.3.3). Rewetting usually coincides with the onset of sustained flow although in some IRES, there may be several "false starts" where surface water is present or pools

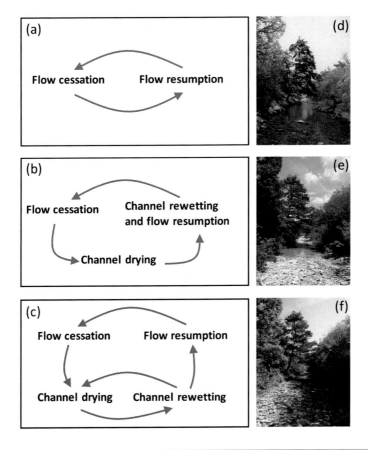

FIG. 2.3.3

Three common sequences of hydrological transitions governing hydrological connectivity in IRES. Surface-flow cessation either involves no channel drying (a), channel drying followed by rewetting and immediate flow resumption (b), or several "false starts" of drying and rewetting before surface flow resumes in earnest (c). Snapshots of this sequence in a French intermittent stream illustrate the flowing (d), drying (e), and dry (f) stages.

Photos: Courtesy B. Launay.

fill and then dry for several weeks or even months before streamflow begins in earnest (e.g., Boulton and Lake, 1990; Bhamjee et al., 2016). Drying in most IRES is a prolonged and gradual "ramp" disturbance (Lake, 2003) whereas rewetting is usually faster, especially when it coincides with resumption of surface flow (Fig. 2.3.3). This hysteresis in channel wetting and drying occurs in nearly all IRES, even those in in temperate regions (e.g., Day, 1983).

Another difference between patterns of drying and rewetting is their directionality along the river channel. In most IRES, rewetting extends in a downstream direction, either as surface runoff down the channel (Fisher et al., 1982; Jacobson et al., 2000) or inputs from rejected recharge springs and seeps (Shaw, 2016) arising from unconfined aquifers that initiate expression of surface water into the channel (Doering et al., 2007; Larned et al., 2011), subject to regional geology and geomorphology (Section 2.3.4, Eltahir and Yeh, 1999). In contrast, drying typically occurs in both upstream and downstream directions, particularly in "losing" streams where the water table lies below the channel (Mansell and Hussey, 2005). Again, this pattern of drying is controlled by regional geomorphology (Larned et al., 2011) and groundwater levels relative to the stream channel (Shaw, 2016) as well as exposure to solar radiation and evaporation.

Rewetting and drying patterns in IRES occur across multiple spatial extents (Stanley et al., 1997; Figure 4 in Costigan et al., 2016). At the scale of tributaries and individual river segments, localized precipitation can lead to rewetting and surface flow resumption in some channels whereas adjacent ones stay dry (e.g., Figure 2 in Datry et al., 2014). This asynchrony in rewetting can be highly unpredictable in arid and semiarid regions where brief storms cause flash floods to surge down channels in ephemeral streams that dry within days or weeks (Fig. 2.3.4). In these areas, surface runoff is rapidly initiated, Hortonian overland flow is the dominant source of water, and

FIG. 2.3.4

Flash flood down the channel of an intermittent stream in the Flinders Ranges, South Australia. This channel was completely dry only three hours before the photograph was taken, and the flood originated from a brief but heavy rainstorm in the headwaters.

Photo: Courtesy A. Boulton.

the hydrographs are sharply peaked (Tooth and Nanson, 2011). Further, such spatial variability can vary among events, seasons, and years (Costigan et al., 2015), contributing to geomorphological discontinua (Section 2.3.4) and habitat heterogeneity in IRES at broad scales. There are also broad-scale differences in patterns and rates of drying, reflecting spatial differences in transmission losses along channels (e.g., seepage, evaporation) and local climatic conditions. Human activities extracting surface water, groundwater, or both also influence drying patterns; in some parts of the world (e.g., Australia, Reinfelds et al., 2004), extraction from IRES is governed by regulations such as cease-to-pump rules to minimize ecological harm from rapid anthropogenic water drawdown or extended periods of intermittence.

At stream-reach and finer spatial extents, patterns of drying and rewetting are often asynchronous within IRES, largely governed by channel topography because depressions fill earlier and hold water longer than surrounding areas (Woodward et al., 2015). These fine-scale patterns of drying and wetting create heterogeneous mosaics of habitat patches of varying permanence, with implications for physicochemical (Chapters 3.1 and 3.2) and ecological processes (Chapters 4.7–4.9). This fine-scale heterogeneity also has an important temporal component governed by flow regime operating at the broader scales of the stream segment or river network. For example, rewetting in IRES may coincide with bed-moving floods that erode new pool basins in the channel while filling old ones and depositing sediments across bars and floodplains (Hooke, 2016). These floods flush mats of desiccated algae and leaf litter downstream (Corti and Datry, 2012), redistributing habitats while resetting succession trajectories for benthic algae, aquatic invertebrates, and other organisms (Fisher et al., 1982; Rosado et al., 2015). Again, we see evidence of the catchment hierarchy (Townsend, 1996) but with fine-scale patterns of drying and rewetting nested within broader-scale patterns of surface flow resumption and cessation in IRES.

Ultimately, patterns and variability in hydrological connectivity in IRES arise from spatial and temporal differences in the water budget. Put simply, flow and presence of surface water are a function of water inputs (e.g., precipitation, groundwater), storage (within, below, and alongside channels and floodplains), and outputs (e.g., seepage, evaporation) at different scales (Gordon et al., 2004; Godsey and Kirchner, 2014). In all IRES, across spatial extents from the entire river network to the smallest microhabitat, drying occurs when water losses exceed gains whereas rewetting occurs when gains exceed losses of water. Different patterns of drying and rewetting at multiple spatial scales in IRES create constantly changing hydrological connectivity in all three dimensions, modified by multiple drivers that include channel geomorphology and location within the riverscape.

2.3.4 LONGITUDINAL HYDROLOGICAL CONNECTIVITY IN IRES

In all rivers, gravity drives unidirectional flow of water down the physical gradients of channels. This typically results in longitudinal trends in geomorphology (e.g., channel width, median substrate particle size), hydrology (e.g., median current velocity, discharge), water chemistry (e.g., dissolved oxygen, total dissolved solids), and, ultimately, ecology (e.g., distribution of aquatic plants and animals) (Allan and Castillo, 2007; Boulton et al., 2014).

Broadly speaking, longitudinal geomorphological processes in IRES resemble those typical of perennial fluvial systems although there are some distinct differences as described in Chapter 2.1. In an idealized perennial fluvial system (Schumm, 1977), sediment production predominates in the upland zone, sediment transport in the transfer zone, and sediment deposition in the lower floodout zone which either enters an inland water body, peters out in uncoordinated channels, or exits into an estuary or directly to the coast. The trend of surface expression of longitudinal hydrological connectivity in an idealized perennial system is for progressively larger volumes of water down the network in response to increasing drainage area and channel capacity. Except during overbank events, longitudinal hydrological dimensions of surface flow and volume are constrained by channel capacity and configuration (Gordon et al., 2004). However, with changes in slope, geology, and geomorphology, differences in channel form in the different zones also influence the relative amounts of water exchanging laterally onto floodplains (Fig. 2.3.5A) and vertically into the sediments below and beside the channel (Fig. 2.3.5B, Stanford and Ward, 1993).

At any point in time and space, channel morphology reflects the interaction of hydraulic forces, substrate resistance, and sediment supply (Schumm, 1977). In many IRES, these interactions are discontinuous because most of the channel adjustment and sediment transport occurs during brief floods followed by long periods of very low or zero surface flow (Wohl, 2014). Crossing thresholds associated with flood magnitude or sediment accumulation leads to sharp changes in channel behavior (e.g., Patton and Schumm, 1981; Gómez-Gutiérrez et al., 2009). These geomorphological discontinua and thresholds mean that longitudinal hydrological connectivity and flow intermittence in IRES are not necessarily a predictable function of channel profile, topography, and bed composition (Tooth and Nanson, 2000; Wohl, 2014). Instead, flow cessation, drying, and rewetting may occur along different sections of the channel according to other drivers such as local patterns of alluvial porosity and permeability, precipitation and evaporation, and groundwater gains and losses (Larned et al., 2011; Costigan et al., 2016). This diversity of drivers, acting in various combinations across the broad geographic spread of IRES globally (Chapter 1), creates an equally diverse array of spatial patterns of intermittence, drying, and rewetting.

To help classify this diversity at the network scale, Lake (2003) described three broad patterns of drying along streams affected by drought (Box 2.3.1). To these three patterns, Datry (2016) added two more: complete drying and point drying. Complete drying is when entire river networks remain dry for weeks or months, only flowing briefly before drying again within days or even hours. This pattern is common in ephemeral streams that lie above the water table, exemplified by many streams in arid (Stanley et al., 1997) and karstic (Meyer and Meyer, 2000) environments as well as in temperate IRES with fluctuating groundwater tables (Winter et al., 1998). Point drying is also common (Godsey and Kirchner, 2014; Datry et al., 2016) and is when multiple isolated sections along the channel dry, interspersed by variable lengths of perennial surface water (Chapter 4.9).

Variations in longitudinal hydrological connectivity generated by these patterns of drying in IRES arise from the interplay between discharge generated by different water gains (e.g., precipitation, groundwater) and transmission losses (Fig. 2.3.6). Transmission losses arise from (i) infiltration and recharge to channel and/or floodplain sediments; (ii) evapotranspiration; and (iii) ponding in terminal storages, such as ephemeral pools, channels, and other wetlands (Knighton and Nanson, 1994). The relative magnitudes of these three losses are governed by catchment area, channel gradient, flow regime, and climatic aridity as well as the geomorphological variables discussed earlier. For example, in many IRES draining small- to mid-sized catchments with low or moderate channel gradients and coarse sediments, infiltration and recharge to shallow aquifers is the major transmission loss (e.g., Hughes and Sami, 1992; Constantz et al., 2002). In contrast, the Orange River, South Africa, largely underlain by bedrock, flows through an extremely hot region and evapotranspiration is the major transmission loss (McKenzie and Craig, 2001).

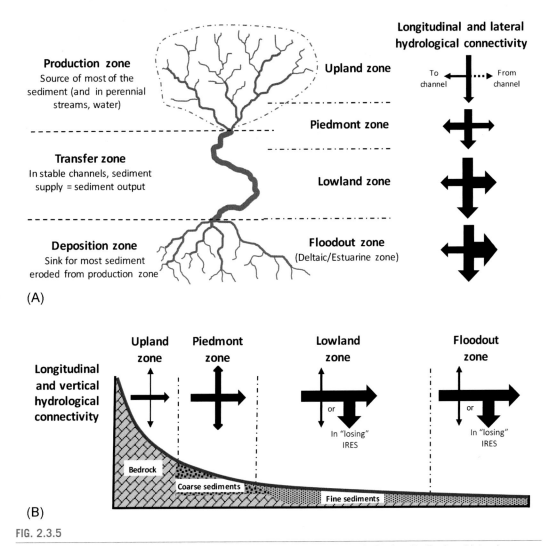

FIG. 2.3.5

Plan (A) and transverse (B) views of geomorphological zones and their relative longitudinal and lateral hydrological connectivity (*black arrows*; thickness represents relative magnitudes of flow; *broken line* indicates minimal flow) in an idealized single-thread river network. In lowland and floodout zones of "losing" IRES (Fig. 2.3.10), transmission losses of surface water (e.g., infiltration and recharge) are often greater, represented by the large down-pointing *arrows* to the right.

The converse to drying is rewetting. Patterns of rewetting are seldom simply a reversal of drying patterns (Fig. 2.3.3) because the causes of the two events differ. Longitudinally, rewetting in many IRES is triggered by increased precipitation and runoff in the headwaters, sometimes intense enough to cause flash floods that swiftly transform downstream channels from completely dry to flowing strongly (Jacobson et al., 2000; Cohen and Laronne, 2005). Alternatively, the increase in catchment precipitation and runoff can be slower, recharging alluvial aquifers and causing more gradual rewetting

BOX 2.3.1 THREE COMMON PATTERNS OF LONGITUDINAL DRYING

At least three broad longitudinal patterns of drying occur in perennial streams during drought (Lake, 2003) and also apply to many IRES. The first is where streams arise from permanent sources (e.g., springs, headwater lakes) and the upper reaches retain perennial flow whereas the mid and lower reaches dry either completely or to isolated pools (Fig. 2.3.6A). This form is common in IRES whose headwaters arise in mountain ranges or well-watered uplands before draining into lowland channels that extend out into more arid areas (Boulton and Williams, 1996; Jaeger and Olden, 2012). Other situations where this pattern arises are when the river course reaches a highly permeable alluvial plain where infiltration losses exceed inflows (Datry, 2012) and in many karstic systems (Meyer and Meyer, 2000).

The second is where the headwaters dry out but flow persists in the lower reaches (Fig. 2.3.6B). In most of these cases, surface water in piedmont and lowland zones is sustained by groundwater inputs, facilitated by larger and deeper channels downstream (Lake, 2003). Datry et al. (2016) describe IRES in Bolivia where a fall in groundwater causes headwater drying yet lowland reaches remain perennial owing to their deeper and less permeable channels. The third common pattern is when the mid-reaches dry while the headwaters and lower reaches retain flow or permanent water (Fig. 2.3.6C). In the spatially intermittent Selwyn River, New Zealand, the upper reaches sustain perennial flow for about three kilometers and although "losing", are supplied by hillslope runoff which peters out before a 45-km segment in the mid-reaches with highly variable rates of drying (Arscott et al., 2010). The next 8 km of the channel intersect confined alluvial aquifers at or near the surface, the river becomes "gaining," and perennial flow is reestablished, this time primarily fed by groundwater instead of runoff. Other examples of this mid-reach drying pattern include IRES in northern Victoria, Australia (Lake, 2003) and Sycamore Creek in Arizona, USA (Stanley et al., 1997).

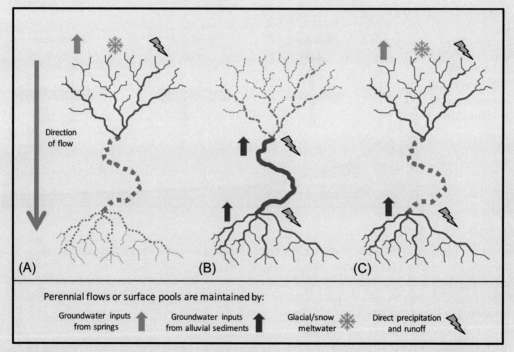

FIG. 2.3.6

Three broad patterns of drying (*dotted brown lines*) occur in IRES and drought-affected perennial streams: (A) drying in mid- and lower reaches, (B) drying in upper reaches, and (C) drying in mid-reaches.

FIG. 2.3.7

Floodwaters spread from the channel of Narran River, Australia, (in 2008) across vast inland floodplains, triggering "boom" periods of biological productivity.

Photo: Courtesy P. Terrill.

driven by a combination of runoff, streamflow, and groundwater inputs (Stanley et al., 1997; Tockner et al., 2000). Initial floods often suspend large amounts of organic material that have accumulated in the channel during the dry phase and carry them downstream, with implications for downstream water quality (e.g., pulses of anoxic water, Hladyz et al., 2011; Chapter 3.2).

Most rewetting events in inland Australian IRES with vast floodplains ($>10,000 \, km^2$) are relatively short-lived and transmission losses are high (e.g., 70%–98% in the Diamantina River, Costelloe et al., 2003). However, occasionally heavy rain will initiate huge slow-moving floods that take many weeks or even months to reach terminal wetlands such as Lake Eyre, hundreds of kilometers downstream. As these floods move down along the channel, they spread laterally across immense floodplains (Fig. 2.3.7) whose clays swell and restrict infiltration once their crack capacity is exceeded (Knighton and Nanson, 1994). These longitudinal and lateral rewetting events trigger extraordinary "booms" in ecological productivity eventually followed by "bust" periods when drying occurs (Bunn et al., 2006; Leigh et al., 2010).

2.3.5 LATERAL HYDROLOGICAL CONNECTIVITY IN IRES

At the river network scale, the size and predominant directionality of the lateral hydrological dimensions in IRES, as in perennial rivers, are primarily a function of location along the river network (Fig. 2.3.5), correlated with differences in channel topography, form, and substrate composition among the main geomorphological zones (Box 2.3.2). The generalities that follow acknowledge that substantial finer-scale variation exists along IRES, especially in how channel and floodplain morphology are altered by geology, vegetation patterns, and human activities, and that this longitudinal variation inevitably affects patterns of lateral hydrological connectivity.

Dimensions of lateral hydrological connectivity in upland zones are typically shorter than in piedmont, lowland, and floodout zones because of the differences in channel cross-section, gradient, and extent of floodplain development (Box 2.3.2). Floodplains in lowland and floodout zones are usually well developed (Fig. 2.3.8C) and may be so vast that during extreme floods, lateral hydrological connectivity of surface water extends far from the main channel or channels (Fig. 2.3.7), transporting sediments, organic matter, and aquatic biota for long distances (Bunn et al., 2006; Muehlbauer et al., 2014). Further, in contrast to upland and piedmont zones, the directionality of hydrological connectivity in lowland and floodout zones can be complex, resulting in variable rates and locations of the delivery of water and materials to adjacent riparian and other terrestrial ecosystems. Surface water floods in lowland and floodout zones of IRES have attracted much attention from ecologists interested in how organisms exploit transient lateral hydrological connectivity to colonize new habitats (e.g., fish, Kerezsy et al., 2013; Chapter 4.5), emerge from resting stages in sediments (e.g., zooplankton, Jenkins and Boulton, 2007; Chapter 4.8), and redistribute energy and organic material in complex aquatic-terrestrial subsidies (Bunn et al., 2003; Chapter 4.7).

The spatial extent, flow rates, and residence time of surface water spreading laterally across the floodplains of IRES are strongly controlled by the topography, substrate composition, and vegetation cover of the floodplain because these govern runoff and rates of evapotranspiration and infiltration (Hamilton, 2010). Although lateral hydrological connectivity in many IRES is complex, there are several broad trends associated with the different zones along the river network. One trend is that the extent and residence time of surface floodwaters in piedmont zones are typically shorter than on the floodplains of lowland and floodout zones. This is because in piedmont zones, compared to lowland and floodout zones, travel distances of water from the channel onto the floodplain are often shorter

BOX 2.3.2 CHANNEL FORM AND SUBSTRATE COMPOSITION IN GEOMORPHOLOGICAL ZONES ALONG IRES

Channels in the upland zone are usually narrow V- or U-shaped single-thread or sometimes multi-threaded channels, often overlying bedrock, with a low width-to-depth ratio, steep banks, and limited if any floodplain (Fig. 2.3.8A; Sutfin et al., 2014). Consequently, lateral hydrological connections in the upland zones of IRES are short because of the relatively short travel distances down hillslopes to the channel. Surface flow along these connections is usually toward the channel (Fig. 2.3.5A), and rewetting along this dimension occurs in response to precipitation, overland flow, and subsurface interflow (Lake et al., 2006). During overbank flows, these lateral inputs are briefly reversed away from the channel (Rassam et al., 2006).

In the piedmont zone, channels start to broaden, increase in width-to-depth ratio, and become more rectangular in cross-section while the banks are less steep and may include narrow floodplains (Fig. 2.3.8B, Singer and Michaelides, 2014). Channels are single or multi-thread but can vary widely (Sutfin et al., 2014) and gradients are usually less steep than in the upland zone. Bed sediments typically comprise poorly sorted mixtures of coarse sediments (e.g., cobbles, pebbles) with large voids among the particles. Lateral hydrological connections frequently include subsurface flowpaths in this zone (Fig. 2.3.5) with much faster transmission rates than in lowland and floodout zones where the sediments are finer.

In the lowland and floodout zones, single-thread channels are very broad with high width-to-depth ratios and may be "under-fit" with surface flow usually occupying only a small part of the channel (Fig. 2.3.8C). Commonly, these two low-gradient zones are drained by multi-thread braided and anastomosing channels, varying in stability and geomorphic complexity (Chapter 2.1). Bed sediments are typically fine, dominated by sands and clays although sometimes larger particles and even outcropping bedrock may occur (Tooth and Nanson, 2000).

BOX 2.3.2 CHANNEL FORM AND SUBSTRATE COMPOSITION IN GEOMORPHOLOGICAL ZONES ALONG IRES *(Cont'd)*

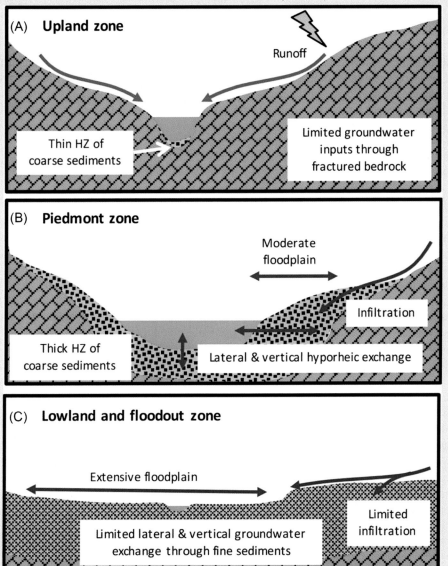

FIG. 2.3.8

In many IRES, the typical channel cross-section form, extent of floodplain development, and substrate composition vary among zones: (A) in upland zones, steep-banked V- or U-shaped channels overlie bedrock and have minimal floodplain development; (B) in piedmont zones, shallow-banked rectangle-shaped channels overlie coarse, poorly sorted alluvium and have moderate floodplain development; (C) in lowland and floodout zones, very shallow-banked broad channels overlie fine, well-sorted alluvium and have extensive floodplains. Only single-thread channels are shown here but there may be multiple channels (see text). *HZ*, hyporheic zone. Geomorphological zones illustrated in Fig. 2.3.5.

(Box 2.3.2), evapotranspiration demands are lower, and lateral and vertical infiltration rates are greater because of this zone's coarser sediments (Tonina and Buffington, 2009; Fig. 2.3.8).

Another trend is that when infiltration capacity is exceeded, flow velocities of surface water on flood-plains of piedmont zones are generally faster than on the floodplains of lowland and floodout zones. These faster flows are caused by the steeper slopes; in lowland and floodout zones, floodplain topographic gradients may be extremely low (e.g., 0.0005–0.002, Tooth, 2005) and floodwaters move more sluggishly, further slowed by the drag resistance of inundated vegetation. Implications of this difference among geomorphological zones in flood velocities include contrasting rates of lateral erosion and deposition (Chapter 2.1), altered water residence times, and different hydraulic conditions for aquatic biota. Other IRES with well-developed floodplains are braided rivers, and again, major differences in their lateral hydrological connectivity arise from the interplay of gradient, substrate porosity, and groundwater table. This interplay creates complex spatial patterns where almost-perennial channels adjoin highly intermittent ones (e.g., Selwyn River, Arscott et al., 2010). During floods, water extends laterally across these braids, temporarily reducing spatial variability in water quality across different channels (Thomaz et al., 2007).

Finally, lateral dimensions of hydrological connectivity arising from overbank flows in IRES are strongly controlled by the type of flood and the shape of its hydrograph, associated with location along the river network. Flash floods, as their name implies, are short-lived events (Fig. 2.3.9) that travel swiftly as roiling pulses of water and suspended material that can substantially modify channel form and bed composition in many IRES (Cohen and Laronne, 2005), causing dramatic but only transient increases in lateral connectivity. Flash floods in IRES are more typical of upland and piedmont zones than lowland and floodout zones where rainfall runoff in the catchment travels longer distances on average (dissipating the water's energy) and floodwaters are able to spread laterally in the broader channels and across the wider floodplains downstream.

However, not all floods in IRES are simple one-off events, prompting Knighton and Nanson (2001) to classify floods as single, multiple, or compound according to the number of flood peaks and the shapes of the rising and falling limbs of the flood hydrograph (Fig. 2.3.9; Table 2.3.1). These different types of floods have different implications for longitudinal, lateral, and vertical hydrological connectivity (Table 2.3.1), especially the frequency and duration of intermittence. For example, although a single-peak flood may initiate surface flow in a previously dry channel at a given location along a river, the duration of subsequent flow may be shorter than after a multiple or compound flood with the same peak discharge as the single-peak event (Leigh et al., 2010). Further, the timing and sequence of individual flood events as well as the combination of different types will have variable effects on patterns of intermittence and, consequently, longitudinal, lateral, and vertical hydrological connectivity at multiple scales. Flood timing and sequence affect ecological processes and biota in many IRES (Arscott et al., 2010; Leigh and Datry, 2017).

2.3.6 VERTICAL HYDROLOGICAL CONNECTIVITY IN IRES

Apart from evaporation and direct rainfall, vertical hydrological connectivity in IRES arises from water exchanging between the surface (channel and inundated floodplain) and the subsurface. This hydrological dimension governs surface-flow intermittence in piedmont, lowland, and floodout zones of IRES where groundwater gains and losses dominate the water budget (Fig. 2.3.5). Conversely, in streams with fine-grained sediments, infiltrated water is stored in the vadose zone and depleted by evapotranspiration. Intermittence in streams with limited vertical hydrological connectivity is largely

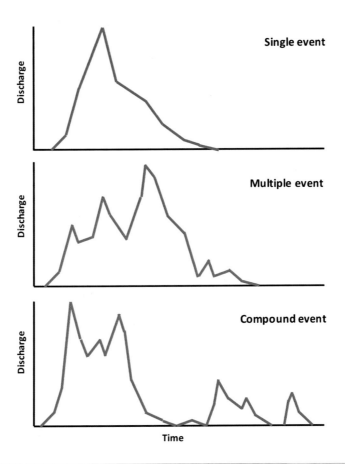

FIG. 2.3.9

Hydrographs of three different types of flood events classified by Knighton and Nanson (2001) according to the number of flood peaks and the shapes of the rising and falling limbs. Implications of these flood events for longitudinal, lateral, and vertical hydrological connectivity are presented in Table 2.3.1.

determined by the spatial variability of rainfall-runoff than by groundwater inputs, and spatial patchiness of rainfall may even lead to independent streamflow events in different reaches of the same network, sometimes never connecting.

When streamflow is lost from a channel to groundwater or the subsurface, the stream is described as "losing" or influent (Fig. 2.3.10A) whereas groundwater enters the channel in "gaining" or effluent streams (Winter et al., 1998; Fig. 2.3.10B). Along its length, a channel can alternate between gaining and losing in response to temporal fluctuations in water table elevation as well as longitudinal variations in geology, topography, substrate composition, and other factors (Cardenas, 2009). Therefore, at a given location in a river network, fluctuations in the elevation of the water table (e.g., in response to precipitation or evapotranspiration) can reverse the direction of vertical flow. The same can happen when high surface flows pass down a river (Malard et al., 2002), causing a transient change in bank storage that recharges the channel after the flows have passed.

Two key points must be made. The first is that the spatial patterns and cycles of surface flow cessation, drying, rewetting, and flow resumption largely result from the dynamics of gaining and losing exchanges in many IRES as a function of the water table level relative to the channel. Although surface runoff usually triggers rewetting and streamflow in upland zones, rising groundwater levels in response to recharge by precipitation often initiate rewetting and surface flow resumption in piedmont and possibly lowland zones of IRES as their channels become gaining, especially in IRES with shallow water tables or in extremely permeable catchments (e.g., karsts, Meyer and Meyer, 2000). When water tables fall below the channel (Fig. 2.3.10A), surface flows and water levels decline; prolonged losing conditions can result in cessation of surface flow and, ultimately, surface and subsurface drying (Boulton, 2003; Larned et al., 2011). So pivotal to flow intermittence is this alternation of gaining and losing conditions, it is used by many hydrologists (Gordon et al., 2004) to distinguish ephemeral streams as

Table 2.3.1 Floods in IRES can be classified according to the number of flood peaks and the shapes of the rising and falling limbs of the flood hydrograph (Fig. 2.3.9; Knighton and Nanson, 2001), and these have implications for hydrological connectivity in longitudinal, lateral, and vertical dimensions

Event and characteristic features (Knighton and Nanson, 2001)	Implications for each hydrological dimension (examples)		
	Longitudinal	Lateral	Vertical
Single event			
One well-defined peak discharge to which a progressive rise from zero is followed by a progressive fall to zero (Fig. 2.3.9A). Often arising from isolated rainfall event as flash flood	Rewet channel sediments and fill channel pools may initiate brief periods of flow. As flash floods may erode sediments and transport leaf litter down channel, often as a pulse of material	Unless an especially large event, only limited effects on surface lateral hydrological connectivity for a short distance onto the floodplain and water does not persist. Very limited recharge through floodplain sediments	In piedmont zones, rewet channel hyporheic sediments reestablish subsurface flowpaths. After bed-moving flash floods, hydraulic conductivity may be high in reworked sediments and new gravel bars
Multiple event			
More than one peak discharge, with a progressive increase in the magnitude of peaks on the rising limb toward a well-defined maximum, followed by a progressive decrease in the magnitude of peaks on the falling limb as flow returns to zero (Fig. 2.3.9B). Often arising from sequences of rainfall events or combined effect of lagged tributary inflows	Rewet channel sediments and fill channel pools initiate sustained flow. Progressive increases in peaks may erode sediments in different parts of channel, promoting in-stream geomorphological complexity. Particulate organic matter (e.g., leaf litter accumulated during dry periods in the channel and on the banks) is carried downstream in multiple pulses, declining after the maximum peak	Progressively greater areas of riparian zone and floodplain in lowland and floodout zones are inundated laterally as peaks approach the maximum. If subsequent peaks occur soon after the maximum, their additive effect further lengthens lateral hydrological connectivity. Pulsed extensions to lateral hydrological dimensions have geomorphological and ecological implications	Hyporheic flowpaths are reestablished in piedmont zones; longer flows promote complex biogeochemical gradients along subsurface flowpaths. Pulsed events alter advective patterns and water quality of downwelling water. Infiltration and recharge of floodplain sediments are substantial and may subsequently return water to the channel to prolong postflood flow

Table 2.3.1 Floods in IRES can be classified according to the number of flood peaks and the shapes of the rising and falling limbs of the flood hydrograph (Fig. 2.3.9; Knighton and Nanson, 2001), and these have implications for hydrological connectivity in longitudinal, lateral, and vertical dimensions—cont'd

Event and characteristic features (Knighton and Nanson, 2001)	Implications for each hydrological dimension (examples)		
	Longitudinal	Lateral	Vertical
Compound event			
More than one peak discharge, with nonprogressive rises and falls before or after the maximum discharge (Fig. 2.3.9C). Often arising from sequences of rainfall events or combined effect of lagged tributary inflows	As in multiple events, rewet channel sediments and fill channel pools initiate sustained flow periods. However, pulsed scour and transport of sediments and particulate organic matter are more idiosyncratic, potentially promoting greater microhabitat diversity than for multiple events	Unlike multiple events, inundation of riparian zone and floodplain in lowland and floodout zones may not increase progressively in spatial extent, depending on the shape of the hydrograph. Different lateral hydrological extensions may arise after each event. These extensions are still pulsed but less predictably	As in multiple events, hyporheic flowpaths and associated complex biogeochemical gradients are reestablished in piedmont zones because of longer flows. Pulsed events alter advective patterns and water quality of downwelling water but less predictably. Substantial infiltration and recharge of floodplain sediments may subsequently return water to the channel to prolong postflood flows

channels always lying above the water table and constantly losing, intermittent rivers that fluctuate between gaining and losing, and perennial systems that are predominantly gaining (Fig. 2.3.10C).

The second key point is that the direction and water source of pathways of vertical hydrological connectivity are largely dictated by whether conditions are gaining or losing. Trends in physical and chemical conditions along a vertical flowpath in a losing section of a stream reach where the source water is exposed to the surface differ greatly from those along a vertical flowpath below a gaining reach where the source water is groundwater. For example, along a vertical flowpath of water downwelling from the surface into the hyporheic zone, concentrations of dissolved oxygen decline while concentrations of reduced forms of nutrients usually increase (Stanley and Boulton, 1995; Capderrey et al., 2013). Where nutrient-enriched alluvial groundwater upwells back into the surface stream, there may be local patches of enhanced primary productivity where benthic algae flourish, benefiting from the increased concentrations of nutrients (Fig. 2.3.11).

Much of our understanding of the functional significance to surface stream ecosystems of hyporheic biogeochemical processes and vertical hydrological connectivity arose from pioneering work by Grimm and Fisher (1984) in Sycamore Creek, an Arizonan IRES. Between floods, the surface stream is shallow (<20 cm) and meanders down a broad channel (Fisher et al., 2004), with most of the stream's discharge occurring along subsurface flowpaths in the alluvial gravels below the stream or in the adjacent parafluvial zone. Along these subsurface flowpaths occur sequences of biogeochemical processes (Fig. 2.3.12A) governed by factors such as redox potential, availability and lability of organic matter and rates of microbial activity (e.g., Holmes et al., 1994; Jones et al., 1995).

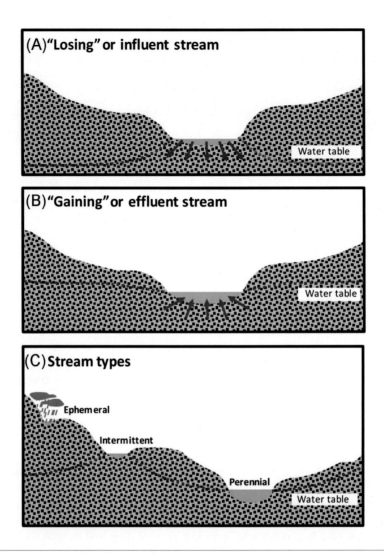

FIG. 2.3.10

"Losing" or influent streams lose groundwater from their channel (A) whereas groundwater enters the channel in "gaining" or effluent streams (B). In a landscape context (C), ephemeral streams are viewed by many hydrologists as channels always lying above the water table and constantly losing, intermittent streams as fluctuating between gaining and losing, and perennial streams as predominantly gaining.

As a result of this sequence of biogeochemical processes, the chemical composition of upwelling water is largely determined by the length of the subsurface flowpath. For example, water upwelling at the end of long subsurface flowpaths, such as those that might occur between widely isolated pools during drying (Fig. 2.3.12B), would be expected to be rich in reduced forms of nitrogen whereas at the end of short subsurface flowpaths (Fig. 2.3.12C), more oxidized forms such as nitrate would dominate.

FIG. 2.3.11

Upwelling nutrient-rich water at the end of a subsurface flowpath promotes algal growth (mainly filamentous *Spirogyra* spp.) in intermittent Brachina Creek, Flinders Ranges, South Australia.

Photo: Courtesy A. Boulton.

The length of subsurface flowpaths, the dominant biogeochemical processes, and the water chemistry of upwelling water all vary temporally in Sycamore Creek in response to floods, surface flow cessation, and drying (Jones et al., 1995; Fisher et al., 2004). There is also substantial fine-scale patchiness and variations in solute-specific processing lengths along the multiple flowpaths, adding to the biogeochemical heterogeneity in these systems (Lewis et al., 2007).

Since these seminal studies in Sycamore Creek, similar trends in concentrations of nutrients and other materials along vertical flowpaths have been identified in IRES elsewhere (Boulton and Williams, 1996; Gómez et al., 2009). Likewise, these hydrological trends have been shown to be strongly associated with spatial trends in microbial activity and invertebrate distribution in various IRES (e.g., Claret and Boulton, 2003; Capderrey et al., 2013; Larned and Datry, 2013). Although much of the empirical work on vertical hydrological connectivity in IRES has been done at the reach scale, the additive and interactive effects of subsurface flowpaths and their collective influence on physicochemical heterogeneity at broad scales have also attracted attention, with particular focus on the influence of flooding and intermittence on the spatial configuration and lengths of subsurface and surface flowpaths (e.g., Fisher et al., 1998; Malard et al., 2002). For example, the combined effects of upwelling water from a sequence of several short flowpaths are likely to generate physicochemical conditions very different from those downstream of a single long flowpath (Malard et al., 2002; Capderrey et al., 2013). Network-scale spatial configurations of hydrological connectivity in all three dimensions constantly vary in many IRES owing to the prevailing cycles of surface flow intermittence.

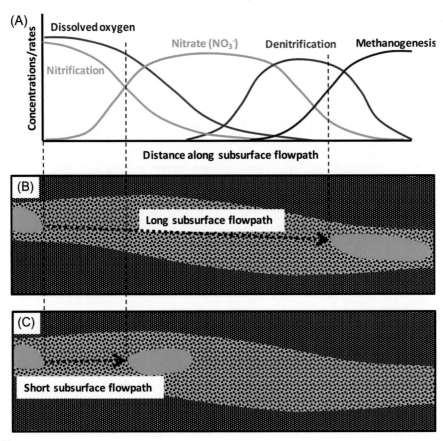

FIG. 2.3.12

Along subsurface flowpaths occur sequences of biogeochemical processes, presented as hypothetical trends (A). Consequently, water chemistry of upwelling water at the ends of a long flowpath (plan-view, B) is likely to differ from that at the end of a short flowpath (plan-view, C), illustrated by the vertical *broken lines* between the top panel and lower two panels.

Trends in (A) are redrawn from Malard, F., Tockner, K., Dole-Olivier, M.-J., Ward, J.V., 2002. A landscape perspective of surface-subsurface hydrological exchanges in river corridors. Freshw. Biol. 47, 621–640, Figure 4, with permission.

2.3.7 A PRELIMINARY CONCEPTUAL FRAMEWORK FOR EXPLORING INTERMITTENCE, CONNECTIVITY, AND INTERACTING HYDROLOGICAL DIMENSIONS IN IRES

Inevitably, disruption of connectivity along one hydrological dimension during intermittence has repercussions for hydrological patterns along the other two dimensions, creating a constantly changing mosaic of interacting flowpaths with different directions and lengths in three dimensions (Poole et al., 2006, 2008). Currently, these three-dimensional hydrological interactions in IRES pose substantial knowledge gaps, ranging from technical ones (Jaeger and Olden, 2012; Bhamjee et al., 2016) through to conceptual gaps such as which aspects of these hydrological dimensions are the most relevant for predicting biogeochemical and ecological responses to different patterns of intermittence in IRES.

To help address this conceptual knowledge gap, we introduce a preliminary framework that draws on theory about the types of interfaces and ecological connectivity. Our logic in adopting this perspective is that many of the implications of intermittence in IRES relate directly to water's role as an agent of connectivity (Sponseller et al., 2013). Examples range from nutrient spiraling and organic matter transport (Chapter 3.2) to the transfer of genetic material among metapopulations (Chapter 4.9). Our framework aims to help researchers refine their research questions about specific aspects of hydrological connectivity when exploring biogeochemical and ecological effects of intermittence in IRES. For example, one hydrological aspect seldom explicitly stated in discussions of the effects of intermittence is the distinction between the presence of water and the presence of flow yet this distinction is crucial because it dictates whether advection is likely to play a role along a given hydrological dimension. In IRES, the situation is further complicated because at times, flowing and nonflowing conditions cooccur.

Thus, the first criterion of the framework (Table 2.3.2) is whether the hydrological connectivity is mediated by flowing water, and therefore influenced by features associated with spatial and temporal patterns of the flow regime (Poff et al., 1997; Chapter 2.2). In the absence of flow, hydrological connectivity in all three dimensions relies on the presence of surface water, subsurface water, or both. However, lacking transport by flowing water, gradients in concentrations of dissolved and suspended

Table 2.3.2 Preliminary conceptual framework for hypothesizing effects of intermittence on hydrological connectivity and geomorphological, biogeochemical, and ecological processes in different geomorphological zones in IRES

Criterion	Upland	Piedmont	Lowland/Floodout
	Hypothesized differences in processes and effects of intermittence in different geomorphological zones in IRES (examples)		
1. Hydrological connectivity mediated by flow?			
Yes	Promotes erosion of channel and transport of sediments and organic matter	Underpins advective exchange of nutrients between surface and hyporheic water	Carries fine sediments and organic matter from upstream out to floodplain and downstream
No	Increases likelihood that drying will fragment longitudinal hydrological connectivity into isolated pools with implications for water quality	Loss of advective exchange but infiltration of surface water may still recharge alluvial aquifers	Allows silt to settle on inundated floodplain which may reduce turbidity, increasing aquatic photosynthesis; allows weak swimmers (e.g., zooplankton) to move upstream
2. Relative width of the interface between realms (e.g., terrestrial-aquatic; river-groundwater)?			
Narrow	Common in this zone and underpins tight hydrological coupling between precipitation and streamflow, leading to steep rising limb of flood hydrograph and rapid rises in water level after heavy rain	Rare in this zone	Rare laterally but common where fine sediments limit hyporheic zone and exchange of river and groundwater, promoting reduced conditions (with biogeochemical implications) in shallow groundwater and soil water

Continued

Table 2.3.2 Preliminary conceptual framework for hypothesizing effects of intermittence on hydrological connectivity and geomorphological, biogeochemical, and ecological processes in different geomorphological zones in IRES—cont'd

Criterion	Hypothesized differences in processes and effects of intermittence in different geomorphological zones in IRES (examples)		
	Upland	Piedmont	Lowland/Floodout
Broad	Rare in this zone	Well-developed hyporheic zone creates broad interface between river and groundwater with long subsurface flowpaths and associated trends in biogeochemistry influencing nutrient concentrations in upwelling water	Extensive floodplains with long lateral hydrological connections that span the broad aquatic-terrestrial interface with implications for fish movement and boom-bust cycles of productivity
3. Connectivity along hydrological dimensions constrained or diffuse?			
Constrained	Well-defined longitudinal hydrological connections mediate transport of sediments and organic materials	Subsurface flowpaths in hyporheic zone mediate transport of dissolved materials	Rare in this zone (except in downwelling termini of some IRES)
Diffuse	Invertebrate drift and fish movement along longitudinal hydrological connections; variation in pool persistence may influence terrestrial oviposition by flying aquatic insects	Across lateral, vertical, and longitudinal hydrological dimensions, active movements of biota may be highly variable and unpredictable	Variation in movement of fish and other aquatic biota (including waterbirds) across inundated floodplains, influencing functional connectivity of metapopulations

materials are governed by drivers associated with standing waters (e.g., stratification, Boulton et al., 2014). Spatially, the implications of this criterion vary among geomorphological zones (Table 2.3.2). For example, in the upland zone, flow-mediated processes along longitudinal and lateral hydrological dimensions erode sediment and transport leaf litter from vegetated upper reaches to deposit them in piedmont and lowland/floodout zones. Hyporheic exchange in the piedmont zone relies heavily on advection (Malard et al., 2002), generating subsurface flowpaths and biogeochemical gradients in all three dimensions simultaneously. Temporally, the criterion is especially relevant to IRES because it explicitly acknowledges that even when flow ceases, hydrological connectivity can persist. Again, this temporal component varies among the geomorphological zones. For example, hydrological connectivity in high-gradient upland zones with limited subsurface flows is likely to be more prone to fragmentation by drying than low-gradient piedmont or lowland zones with substantial groundwater inputs. However, in long pools or stagnant floodwaters of lower gradient zones, the lack of surface flow may allow weak swimmers to move upstream or laterally out on the floodplain (Sheldon et al., 2010; Table 2.3.2).

The second criterion is drawn from ecological connectivity theory, particularly the "taxonomy" proposed by Beger et al. (2010). One of their couplets distinguishes connectivity across narrow versus broad interfaces where interfaces are defined as entities where elements of two or more realms (e.g., aquatic versus terrestrial) mix. Narrow interfaces are where realms adjoin with relatively little spatial

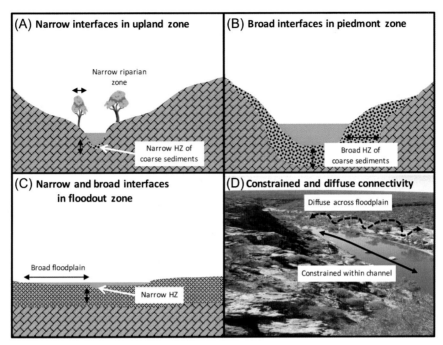

FIG. 2.3.13

Examples in IRES of narrow (A) and broad (B) interfaces (*double-headed black arrows*) and both occurring simultaneously (C), and of constrained and diffuse connectivity (*double-headed black arrows* indicate movement can be in either direction) (D).

separation (Fig. 2.3.13A), such as along riverbanks in the upland zone, whereas broad interfaces have a wide region of mixing between realms (Fig. 2.3.13B), such as the hyporheic zone in the piedmont zone or floodplains in the lowland floodout zone. In IRES, hydrological connectivity across narrow interfaces is likely to be over short distances and transient whereas across broad interfaces, hydrological connectivity will be longer and more persistent.

This criterion's implications relate to distance and water residence time. For example, in upland zones of IRES, tight hydrological coupling between precipitation and streamflow arises from short lateral hydrological connections across narrow riparian zones and steep banks, accelerating rates of water level rise that might drown terrestrial invertebrates (Table 2.3.2; Chapter 4.4). Conversely, the duration and magnitude of "boom" periods of plants, fish, and aquatic invertebrates occupying the broad interfaces represented by floodplains of IRES (Chapter 4.3) rely heavily on the long hydrological connections extending laterally and persisting for months to years after high flows. Narrow and broad interfaces can cooccur in the same river segment, usually in different dimensions (Fig. 2.3.13C), which is why this criterion must be explicitly acknowledged when exploring potential interactions among the three hydrological dimensions.

The third criterion, constrained versus diffuse connectivity (Beger et al., 2010), distinguishes structural and functional connectivity. Throughout this chapter, we have discussed structural connectivity, a physical attribute of the landscape based on features such as the configuration of patches and their linkages (LaPoint et al., 2015). In contrast, functional connectivity is organism oriented where biological (e.g., shared genetic composition, Chapter 4.10) or ecological features are used to determine whether patches in the landscape are perceived as connected by an organism (Pe'er et al., 2011). In IRES, structural hydrological connectivity may influence functional connectivity even if water itself is not a physical vector (e.g., through modifying movement patterns of flying aquatic insects in catchments of IRES (Chapter 4.3)).

In constrained connections, the link is well defined and spatially stable or subject to only minor movements (at least over timescales of months or years) whereas diffuse connections vary widely in space and time and are difficult to delineate (Fig. 2.3.13D). Often, these diffuse connections are mediated by the movement of biota such as fish or active invertebrates (Gallardo et al., 2009; Kerezsy et al., 2013) that create connections between, for example, feeding and breeding areas via routes that are variable, not necessarily water mediated and not always predictable. These connections in IRES are relevant even where these routes leave the water, such as when aquatic insects deposit their eggs terrestrially yet near pools that persist after surface flow has ceased (Table 2.3.2). Again, the two types of connectivity can coincide in the same river segment (Fig. 2.3.13D) and can alternate in response to changing flow regimes.

Although there may be other criteria to add to this framework, the three described here are not mutually exclusive and, in various combinations, characterize most forms of hydrological connectivity in IRES. The criteria by Beger et al. (2010) were proposed as a basis for conservation planning across freshwater, terrestrial, and marine realms, and the framework in Table 2.3.2 could be used similarly when managing the effects of human activities on hydrological connectivity. Many anthropogenic threats to IRES (Chapter 5.1) include disruption of hydrological connectivity (e.g., weirs and dams, extraction of water leading to increased aquatic fragmentation). Another benefit of the framework lies in encouraging researchers to define specific features of hydrological connectivity when describing associations of biogeochemical or ecological processes with aspects of intermittence such as rate of drying or surface flow cessation along one or more hydrological dimensions in IRES.

Finally, the framework can be used to derive hypotheses about how interacting hydrological processes in different dimensions and geomorphological zones along IRES might affect biogeochemical or ecological processes. For example, changes in hydrological connectivity along longitudinal, lateral, and vertical subsurface flowpaths in response to changes in breadth of interface potentially lead to differences in chemical composition of upwelling water in piedmont compared with lowland zones. Other examples are hypothesized in Table 2.3.2. The challenge of unraveling the complexity of these associations, especially in how hydrological connectivity in IRES relates to functional connectivity, is aided by a conceptual framework such as the preliminary one proposed here.

2.3.8 CONCLUSIONS

Hydrological connectivity occurs longitudinally, laterally, and vertically in all flowing waters but is more variable in IRES because of flow intermittence. Spatial and temporal variations in hydrological connectivity are primarily driven by flow regime and catchment geomorphology, often modified by human activities. These variations govern spatial and temporal patterns of surface flow intermittence and, in many cases, drying and rewetting of surface and subsurface water. Concurrent interactions and discontinuities along all three hydrological dimensions produce complex mosaics of physicochemical patches at different spatial scales whose boundaries fluctuate temporally in response to the pulsed flow regime. Hydrological connectivity in IRES seldom exists as gradual trends in, for example, flows or spatial extent of surface water along the three hydrological dimensions. Instead, they are evident as spatially variable discontinua, according well with current conceptual models of river ecosystems that emphasize nonlinearities in ecological patterns (Poole et al., 2008; Sponseller et al., 2013).

Associations between hydrological connectivity among the three spatial dimensions and the four broad geomorphological zones of IRES (Chapter 2.1) can be explored using three criteria proposed in a preliminary conceptual framework to help generate hypotheses about how hydrological connectivity influences biogeochemical and ecological processes at multiple scales. Researchers are now taking advantage of emerging tools to better quantify spatial and temporal patterns along all three hydrological dimensions at increasingly finer resolution over landscape-level spatial scales. These tools include individual or combined application sensors (e.g., electrical resistance, temperature, camera) along with isotope and other environmental tracers (Costigan et al., 2016) and remote sensing data (Sando and Blasch, 2015) and will substantially improve our understanding of the timing, frequency, magnitude, and rate of exchange along these dimensions. In doing so, we will gain a clearer understanding of how the dynamics of hydrological connectivity in IRES interact with intermittence to influence the biogeochemical and ecological patterns and processes described in the next two sections of this book.

ACKNOWLEDGMENTS

We thank Peter Terrill and Bertrand Launay for giving us permission to use their excellent photographs and we thank three reviewers (Margaret Zimmer, Justin Costelloe, and Chris Konrad) and the Handling Editor (Núria Bonada) for valuable comments on an earlier draft.

REFERENCES

Allan, J.D., Castillo, M.M., 2007. Stream Ecology: Structure and Function of Running Waters, second ed. Springer, Dordrecht, The Netherlands.

Arscott, D.B., Larned, S., Scarsbrook, M.R., Lambert, P., 2010. Aquatic invertebrate community structure along an intermittence gradient: Selwyn River, New Zealand. J. North Am. Benthol. Soc. 29, 530–545.

Beger, M., Grantham, H.S., Pressey, R.L., Wilson, K.A., Peterson, E.L., Dorfman, D., et al., 2010. Conservation planning for connectivity across marine, freshwater, and terrestrial realms. Biol. Conserv. 143, 565–575.

Bhamjee, R., Lindsay, J.B., Cockburn, J., 2016. Monitoring ephemeral headwater streams: a paired-sensor approach. Hydrol. Process. 30, 888–898.

Boulton, A.J., 2003. Parallels and contrasts in the effect of drought on stream macroinvertebrate assemblages. Freshw. Biol. 48, 1173–1185.

Boulton, A.J., Lake, P.S., 1990. The ecology of two intermittent streams in Victoria, Australia. I. Multivariate analyses of physicochemical features. Freshw. Biol. 24, 123–141.

Boulton, A.J., Williams, W.D., 1996. Aquatic biota. In: Twidale, C.R., Tyler, M.J., Davies, M. (Eds.), The Natural History of the Flinders Ranges. Royal Society of South Australia, Adelaide, pp. 102–112.

Boulton, A.J., Datry, T., Kasahara, T., Mutz, M., Stanford, J.A., 2010. Ecology and management of the hyporheic zone: stream–groundwater interactions of running waters and their floodplains. J. North Am. Bentholog. Soc. 29, 26–40.

Boulton, A.J., Brock, M.A., Robson, B.J., Ryder, D.S., Chambers, J.M., Davis, J.A., 2014. Australian Freshwater Ecology: Processes and Management. Wiley-Blackwell, Chichester.

Bunn, S.E., Davies, P.M., Winning, M., 2003. Sources of organic carbon supporting the food web of an arid zone floodplain river. Freshw. Biol. 48, 619–635.

Bunn, S.E., Thoms, M.C., Hamilton, S.K., Capon, S.J., 2006. Flow variability in dryland rivers: boom, bust and the bits in between. River Res. Appl. 22, 179–186.

Capderrey, C., Datry, T., Foulquier, A., Claret, C., Malard, F., 2013. Invertebrate distribution across nested geomorphic features in braided-river landscapes. Freshw. Sci. 32, 1188–1204.

Cardenas, M.B., 2009. Stream-aquifer interactions and hyporheic exchange in gaining and losing sinuous streams. Water Resour. Res. 45, W06429.

Claret, C., Boulton, A.J., 2003. Diel variation in surface and subsurface microbial activity along a gradient of drying in an Australian sand-bed stream. Freshw. Biol. 48, 1739–1755.

Cohen, H., Laronne, J.B., 2005. High rates of sediment transport by flashfloods in the Southern Judean Desert, Israel. Hydrol. Process. 19, 1687–1702.

Constantz, J., Stewart, A.E., Niswonger, R., Sarma, L., 2002. Analysis of temperature profiles for investigating stream losses beneath ephemeral channels. Water Resour. Res. 38, 52-1–52-13.

Corti, R., Datry, T., 2012. Invertebrates and sestonic matter in an advancing wetted front travelling down a dry river bed (Albarine, France). Freshw. Sci. 31, 1187–1201.

Costelloe, J.F., Grayson, R.B., Argent, R.M., McMahon, T.A., 2003. Modelling the flow regime of an arid zone floodplain river, Diamantina River, Australia. Environ. Model. Softw. 18, 693–703.

Costigan, K.H., Daniels, M.D., Dodds, W.K., 2015. Fundamental spatial and temporal disconnections in the hydrology of an intermittent prairie headwater network. J. Hydrol. 522, 305–316.

Costigan, K.H., Jaeger, K.L., Goss, C.W., Fritz, K.M., Goebel, P.C., 2016. Understanding controls on flow permanence in intermittent rivers to aid ecological research: integrating meteorology, geology and land cover. Ecohydrology 9, 1141–1153.

Datry, T., 2012. Benthic and hyporheic invertebrate assemblages along a flow intermittence gradient: effects of duration of dry events. Freshw. Biol. 57, 563–574.

Datry, T., 2016. Ecological effects of flow intermittence in gravel bed rivers. In: Tsutsumi, D., Laronne, J. (Eds.), Gravel-Bed Rivers: Processes and Disasters. Wiley-Blackwell, Chichester.

Datry, T., Larned, S.T., 2008. River flow controls ecological processes and invertebrate assemblages in subsurface flowpaths of an ephemeral river reach. Can. J. Fish. Aquat. Sci. 65, 1532–1544.

Datry, T., Larned, S.T., Tockner, K., 2014. Intermittent rivers: a challenge for freshwater ecology. BioScience 64, 229–235.

Datry, T., Pella, H., Leigh, C., Bonada, N., Hugueny, B., 2016. A landscape approach to advance intermittent river ecology. Freshw. Biol. 61, 1200–1213.

Day, D.G., 1983. Drainage density variability and drainage basin outputs. J. Hydrol. 22, 3–17.

Doering, M., Uehlinger, U., Rotach, A., Schlaepfer, D.R., Tockner, K., 2007. Ecosystem expansion and contraction dynamics along a large Alpine alluvial corridor (Tagliamento River, Northeast Italy). Earth Surf. Process. Landf. 32, 1693–1704.

Eltahir, E.A., Yeh, P.J.F., 1999. On the asymmetric response of aquifer water level to floods and droughts in Illinois. Water Resour. Res. 35, 1199–1217.

Fausch, K.D., Torgersen, C.E., Baxter, C.V., Li, H.W., 2002. Landscapes to riverscapes: bridging the gap between research and conservation of stream fishes. BioScience 52, 483–498.

Fisher, S.G., Gray, L.J., Grimm, N.B., Busch, D.E., 1982. Temporal succession in a desert stream ecosystem following flash flooding. Ecol. Monogr. 52, 93–110.

Fisher, S.G., Grimm, N.B., Martí, E., Gomez, R., 1998. Hierarchy, spatial configuration, and nutrient cycling in a desert stream. Aust. J. Ecol. 23, 41–52.

Fisher, S.G., Sponseller, R.A., Heffernan, J.B., 2004. Horizons in stream biogeochemistry: flowpaths to progress. Ecology 85, 2369–2379.

Frissell, C.A., Liss, W.J., Warren, C.E., Hurley, M.D., 1986. A hierarchical framework for stream habitat classification: viewing streams in a watershed context. Environ. Manag. 10, 199–214.

Gallardo, B., Gascón, S., González-Sanchis, M., Cabezas, A., Comín, F.A., 2009. Modelling the response of floodplain aquatic assemblages across the lateral hydrological connectivity gradient. Mar. Freshw. Res. 60, 924–935.

Godsey, S.E., Kirchner, J.W., 2014. Dynamic, discontinuous stream networks: hydrologically driven variations in active drainage density, flowing channels and stream order. Hydrol. Process. 28, 5791–5803.

Gómez, R., García, V., Vidal-Abarca, R., Suárez, L., 2009. Effect of intermittency on N spatial variability in an arid Mediterranean stream. J. North Am. Benthol. Soc. 28, 572–583.

Gómez-Gutiérrez, Á., Schnabel, S., Lavado-Contador, J.F., 2009. Gully erosion, land use and topographical thresholds during the last 60 years in a small rangeland catchment in SW Spain. Land Degrad. Develop. 20, 535–550.

Gordon, N., McMahon, T.A., Finlayson, B.L., Gippel, C.J., Nathan, R.J., 2004. Stream Hydrology: An Introduction for Ecologists, second ed. Wiley, Chichester.

Graf, W.L., 1988. Fluvial Processes in Dryland Rivers. Springer-Verlag, Berlin.

Grimm, N.B., Fisher, S.G., 1984. Exchange between interstitial and surface water: implications for stream metabolism and nutrient cycling. Hydrobiologia 111, 219–228.

Hamilton, S.K., 2010. Floodplains. In: Likens, G.E. (Ed.), River Ecosystem Ecology: A Global Perspective. Elsevier, Amsterdam, pp. 190–198.

Hartwig, M., Borchardt, D., 2015. Alteration of key hyporheic functions through biological and physical clogging along a nutrient and fine-sediment gradient. Ecohydrology 8, 961–975.

Hendricks, S.P., White, D.S., 1991. Physicochemical patterns within a hyporheic zone of a northern Michigan river, with comments on surface water patterns. Can. J. Fish. Aquat. Sci. 48, 1645–1654.

Hladyz, S., Watkins, S.C., Whitworth, K.L., Baldwin, D.S., 2011. Flows and hypoxic blackwater events in managed ephemeral river channels. J. Hydrol. 401, 117–125.

Holmes, R.M., Fisher, S.G., Grimm, N.B., 1994. Parafluvial nitrogen dynamics in a desert stream ecosystem. J. North Am. Benthol. Soc. 13, 468–478.

Hooke, J., 2003. Coarse sediment connectivity in river channel systems: a conceptual framework and methodology. Geomorphology 56, 79–94.

Hooke, J.M., 2016. Geomorphological impacts of an extreme flood in SE Spain. Geomorphology 263, 19–38.

Hughes, D.A., Sami, K., 1992. Transmission losses to alluvium and associated moisture dynamics in a semiarid ephemeral channel system in southern Africa. Hydrol. Process. 6, 45–53.

Jacobson, P.J., Jacobson, K.M., Angermeier, P.L., Cherry, D.S., 2000. Variation in material transport and water chemistry along a large ephemeral river in the Namib Desert. Freshw. Biol. 44, 481–491.

Jaeger, K.L., Olden, J.D., 2012. Electrical resistance sensor arrays as a means to quantify longitudinal connectivity of rivers. River Res. Appl. 28, 1843–1852.

Jaeger, K.L., Olden, J.D., Pelland, N.A., 2014. Climate change poised to threaten hydrologic connectivity and endemic fishes in dryland streams. Proc. Natl. Acad. Sci. U. S. A. 111, 13894–13899.

Jenkins, K.M., Boulton, A.J., 2007. Detecting impacts and setting restoration targets in arid-zone rivers: aquatic microinvertebrate responses to reduced floodplain inundation. J. Appl. Ecol. 44, 823–832.

Jones, J.B., Fisher, S.G., Grimm, N.B., 1995. Nitrification in the hyporheic zone of a desert stream ecosystem. J. North Am. Benthol. Soc. 14, 249–258.

Kerezsy, A., Balcombe, S.R., Tischler, M., Arthington, A.H., 2013. Fish movement strategies in an ephemeral river in the Simpson Desert, Australia. Austral Ecol. 38, 798–808.

Knighton, A.D., Nanson, G.C., 1994. Flow transmission along an arid zone anastomosing river, Cooper Creek, Australia. Hydrol. Process. 8, 137–154.

Knighton, A.D., Nanson, G.C., 2001. An event-based approach to the hydrology of arid zone rivers in the Channel Country of Australia. J. Hydrol. 254, 102–123.

Lake, P.S., 2003. Ecological effects of perturbation by drought in flowing waters. Freshw. Biol. 48, 1161–1172.

Lake, S., Bond, N., Reich, P., 2006. Floods down rivers: from damaging to replenishing forces. Adv. Ecol. Res. 39, 41–62.

LaPoint, S., Balkenhol, N., Hale, J., Sadler, J., van der Ree, R., 2015. Ecological connectivity research in urban areas. Funct. Ecol. 29, 868–878.

Larned, S.T., Datry, T., 2013. Flow variability and longitudinal patterns in parafluvial water chemistry, aquatic invertebrates and microbial activity. Freshw. Biol. 58, 2126–2143.

Larned, S.T., Datry, T., Arscott, D.B., Tockner, K., 2010. Emerging concepts in temporary-river ecology. Freshw. Biol. 55, 717–738.

Larned, S.T., Schmidt, J., Datry, T., Konrad, C.P., Dumas, J.L., Diettrich, J.C., 2011. Longitudinal ecohydrology: flow variation down the lengths of alluvial rivers. Ecohydrology 4, 532–548.

Leigh, C., Datry, T., 2017. Drying as a primary hydrological determinant of biodiversity in river systems: a broad-scale analysis. Ecography. 40, 487–499.

Leigh, C., Sheldon, F., Kingsford, R.T., Arthington, A.H., 2010. Sequential floods drive 'booms' and wetland persistence in dryland rivers: a synthesis. Mar. Freshw. Res. 61, 896–908.

Lewis, D.B., Grimm, N.B., Harms, T.K., Schade, J.D., 2007. Subsystems, flowpaths, and the spatial variability of nitrogen in a fluvial ecosystem. Landsc. Ecol. 22, 911–924.

Malard, F., Tockner, K., Dole-Olivier, M.-J., Ward, J.V., 2002. A landscape perspective of surface-subsurface hydrological exchanges in river corridors. Freshw. Biol. 47, 621–640.

Mansell, M.G., Hussey, S.W., 2005. An investigation of flows and losses within the alluvial sands of ephemeral rivers in Zimbabwe. J. Hydrol. 314, 192–203.

McKenzie, R.S., Craig, A.R., 2001. Evaluation of river losses from the Orange River using hydraulic modelling. J. Hydrol. 241, 62–69.

Mendoza-Lera, C., Mutz, M., 2013. Microbial activity and sediment disturbance modulate the vertical water flux in sandy sediments. Freshw. Sci. 32, 26–35.

Meyer, A., Meyer, E.I., 2000. Discharge regime and the effect of drying on macroinvertebrate communities in a temporary karst stream in East Westphalia (Germany). Aquat. Sci. 62, 216–231.

Muehlbauer, J.D., Collins, S.F., Doyle, M.W., Tockner, K., 2014. How wide is a stream? Spatial extent of the potential "stream signature" in terrestrial food webs using meta-analysis. Ecology 95, 44–55.

Nakano, S., Murakami, M., 2001. Reciprocal subsidies: dynamic interdependence between terrestrial and aquatic food webs. Proc. Natl. Acad. Sci. U.S.A. 98, 166–170.

Paetzold, A., Bernet, J.F., Tockner, K., 2006. Consumer-specific responses to riverine subsidy pulses in a riparian arthropod assemblage. Freshw. Biol. 51, 1103–1115.

Patton, P.C., Schumm, S.A., 1981. Ephemeral-stream processes: implications for studies of quaternary valley fills. Quatern. Res. 15, 24–43.

Pe'er, G., Henle, K., Dislich, C., Frank, K., 2011. Breaking functional connectivity into components: a novel approach using an individual based model, and first outcomes. PLoS One 6, e22355.

Pickett, S.T.A., White, P.S. (Eds.), 1985. The Ecology of Natural Disturbance and Patch Dynamics. Academic Press, San Diego.

Poff, N.L., Allan, J.D., Bain, M.B., Karr, J.R., Prestegaard, K.L., Richter, B.D., et al., 1997. The natural flow regime: a paradigm for river conservation and restoration. BioScience 47, 769–784.

Poole, G.C., 2002. Fluvial landscape ecology: addressing uniqueness within the river discontinuum. Freshw. Biol. 47, 641–660.

Poole, G.C., Stanford, J.A., Running, S.W., Frissell, C.A., 2006. Multiscale geomorphic drivers of groundwater flow paths: subsurface hydrologic dynamics and hyporheic habitat diversity. J. North Am. Benthol. Soc. 25, 288–303.

Poole, G.C., O'Daniel, S.J., Jones, K.L., Woessner, W.W., Bernhardt, E.S., Helton, A.M., et al., 2008. Hydrologic spiralling: the role of multiple interactive flow paths in stream ecosystems. River Res. Appl. 24, 1018–1031.

Pringle, C., 2003. What is hydrologic connectivity and why is it ecologically important? Hydrol. Process. 17, 2685–2689.

Rassam, D.W., Fellows, C., DeHayr, R., Hunter, H., Bloesch, P., 2006. The hydrology of riparian buffer zones: two case studies in an ephemeral and a perennial stream. J. Hydrol. 325, 308–324.

Reinfelds, I., Haeusler, T., Brooks, A.J., Williams, S., 2004. Refinement of the wetted perimeter breakpoint method for setting cease-to-pump or minimum environmental flows. River Res. Appl. 20, 671–685.

Rolls, R.J., Leigh, C., Sheldon, F., 2012. Mechanistic effects of low-flow hydrology on riverine ecosystems: ecological principles and consequences of alteration. Freshw. Sci. 31, 1163–1186.

Rosado, J., Morais, M., Tockner, K., 2015. Mass dispersal of terrestrial organisms during first flush events in a temporary stream. River Res. Appl. 31, 912–917.

Sandercock, P., Hooke, J., 2011. Vegetation effects on sediment connectivity and processes in an ephemeral channel in SE Spain. J. Arid Environ. 75, 239–254.

Sandercock, P., Hooke, J., Mant, J., 2007. Vegetation in dryland river channels and its interaction with fluvial processes. Prog. Phys. Geogr. 31, 107–129.

Sando, R., Blasch, K.W., 2015. Predicting alpine headwater stream intermittency: a case study in the northern Rocky Mountains. Ecohydrol. Hydrobiol. 15, 68–80.

Schumm, S.A., 1977. The Fluvial System. Wiley, New York.

Shaw, S.B., 2016. Investigating the linkage between streamflow recession rates and channel network contraction in a mesoscale catchment in New York state. Hydrol. Process. 30, 479–492.

Sheldon, F., Bunn, S.E., Hughes, J.M., Arthington, A.H., Balcombe, S.R., Fellows, C.S., 2010. Ecological roles and threats to aquatic refugia in arid landscapes: Dryland river waterholes. Mar. Freshw. Res. 61, 885–895.

Singer, M.B., Michaelides, K., 2014. How is topographic simplicity maintained in ephemeral dryland channels? Geology 42, 1091–1094.

Sponseller, R.A., Heffernan, J.B., Fisher, S.G., 2013. On the multiple ecological roles of water in river networks. Ecosphere 4, 17.

Stanford, J.A., Ward, J.V., 1993. An ecosystem perspective of alluvial rivers: connectivity and the hyporheic corridor. J. North Am. Benthol. Soc. 12, 48–60.

Stanley, E.H., Boulton, A.J., 1995. Hyporheic processes during flooding and drying in a Sonoran Desert stream. Arch. Hydrobiol. 134, 1–26.

Stanley, E.H., Fisher, S.G., Grimm, N.B., 1997. Ecosystem expansion and contraction in streams. BioScience 47, 427–435.

Sutfin, N.A., Shaw, J., Wohl, E.E., Cooper, D., 2014. A geomorphic classification of ephemeral channels in a mountainous, arid region, southwestern, Arizona, USA. Geomorphology 221, 164–175.

Thomaz, S., Bini, L., Bozelli, R., 2007. Floods increase similarity among aquatic habitats in river-floodplain systems. Hydrobiologia 579, 1–13.

Tockner, K., Malard, F., Ward, J.V., 2000. An extension of the flood pulse concept. Hydrol. Process. 14, 2861–2883.

Tonina, D., Buffington, J.M., 2009. Hyporheic exchange in mountain rivers. I: mechanics and environmental effects. Geogr. Compass. 3, 1063–1086.

Tooth, S., 2005. Splay formation along the lower reaches of ephemeral rivers on the northern plains of arid central Australia. J. Sediment. Res. 75, 636–649.

Tooth, S., Nanson, G.C., 2000. Equilibrium and nonequilibrium conditions in dryland rivers. Phys. Geogr. 21, 183–211.

Tooth, S., Nanson, G.C., 2011. Distinctiveness and diversity of arid zone river systems. In: Thomas, D.S.G. (Ed.), Arid Zone Geomorphology: Process, Form and Change in Drylands, third ed. Wiley-Blackwell, Chichester, pp. 269–300.

Townsend, C.R., 1996. Concepts in river ecology: pattern and process in the catchment hierarchy. Arch. Hydrobiol. 113, 3–21.

Ward, J.V., 1989. The four-dimensional nature of lotic ecosystems. J. North Am. Benthol. Soc. 8, 2–8.

White, D.S., 1993. Perspectives on defining and delineating hyporheic zones. J. North Am. Benthol. Soc. 12, 61–69.

Winemiller, K.O., Flecker, A.S., Hoeinghaus, D.J., 2010. Patch dynamics and environmental heterogeneity in lotic ecosystems. J. North Am. Benthol. Soc. 29, 84–99.

Winter, T.C., Harvey, J.W., Franke, O.L., Alley, W.M., 1998. Ground Water and Surface Water—A Single Resource. United States Geological Survey Circular 1139, Denver, Colorado.

Wohl, E., 2014. Dryland channel networks: resiliency, thresholds, and management metrics. Geol. Soc. Am. Rev. Eng. Geol. 22, 147–158.

Woodward, K.B., Fellows, C.S., Mitrovic, S.M., Sheldon, F., 2015. Patterns and bioavailability of soil nutrients and carbon across a gradient of inundation frequencies in a lowland river channel, Murray–Darling Basin, Australia. Agric. Ecosyst. Environ. 205, 1–8.

FURTHER READING

Datry, T., Scarsbrook, M., Larned, S., Fenwick, G., 2008. Lateral and longitudinal patterns within the stygoscape of an alluvial river corridor. Fundam. Appl. Limnol. 171, 335–347.

Montgomery, D.R., 1999. Process domains and the river continuum. J. Am. Water Resour. Assoc. 35, 397–410.

WATER PHYSICOCHEMISTRY IN INTERMITTENT RIVERS AND EPHEMERAL STREAMS

3.1

Rosa Gómez*, María Isabel Arce[†], Darren S. Baldwin[‡], Clifford N. Dahm[§]

University of Murcia, Murcia, Spain* Institute of Freshwater Ecology and Inland Fisheries (IGB), Berlin, Germany[†] La Trobe University, Wodonga, VIC, Australia[‡] University of New Mexico, Albuquerque, NM, United States[§]

IN A NUTSHELL

- Water physicochemistry in intermittent rivers and ephemeral streams (IRES) is highly modulated by their surface flow intermittence and the disruption of hydrological connectivity.
- Flow cessation in IRES determines both the commonly noted high solute concentrations and particular vulnerability to pollution.
- Compared with perennial streams and rivers, the dramatic alteration of hydrological connectivity by intermittence in IRES leads to their spatial and temporal variability in water physicochemistry.
- Drying and flooding largely drive the temporal variability in IRES physicochemistry.
- From an ecosystem management perspective, water quality standards should be adapted to the hydrological uniqueness of IRES.

3.1.1 INTRODUCTION

Life in streams and rivers is conditioned by physicochemical parameters such as dissolved oxygen (DO), light, temperature, pH, and salinity that influence both the biotic structure (species richness and diversity) and ecosystem functioning (metabolism and biogeochemical processes). This is true for all aquatic ecosystems, including intermittent rivers and ephemeral streams (IRES). Variability of physicochemical parameters at different spatial and temporal scales is a global feature of all fluvial ecosystems. However, the pattern of spatial and temporal changes in water physicochemistry is substantially different in IRES from their perennial counterparts, as the latter are seldom if ever affected by surface flow cessation.

Streamflow cessation, drying and rewetting, and episodic surface flows in IRES alter hydrological connectivity along longitudinal, lateral, and vertical dimensions (Chapter 2.3), contributing to physicochemical variation at different spatial and temporal scales (e.g., Dent and Grimm, 1999; Holloway and Dahlgren, 2001; Robinson et al., 2016). During periods of low flow prior to flow cessation, interaction of surface water with streambed sediments is important because of the large benthic surface area-to-volume ratio in IRES. Biogeochemical processes at the streambed–surface water interface have a proportionally greater impact on the physicochemistry of surface waters (e.g., Holmes et al., 1994;

Jones et al., 1995; Stanley and Boulton, 1995). However, drying and rewetting are probably the main processes occurring in IRES but absent in perennial rivers that modify their physicochemical environment (Stanley et al., 1997; Fisher et al., 1998). During these periods, lower volumes of water also limit IRES capacity to mitigate anthropogenic inputs through dilution (e.g., Brooks et al., 2006).

The effect of the wetting–drying cycle on key physicochemical parameters (e.g., temperature, dissolved oxygen, salinity, turbidity, pH) in IRES will depend on a number of key local (IRES-specific), interrelated, and often interacting variables (Fig. 3.1.1) including:

- Substrate type (bedrock, sand or silt, organic-rich or organic-poor) of the channel.
- Groundwater interactions, including hyporheic flows.
- Whether pools form after flow cessation; if so, pool morphology (e.g., length, width, depth, and orientation to the prevailing winds).

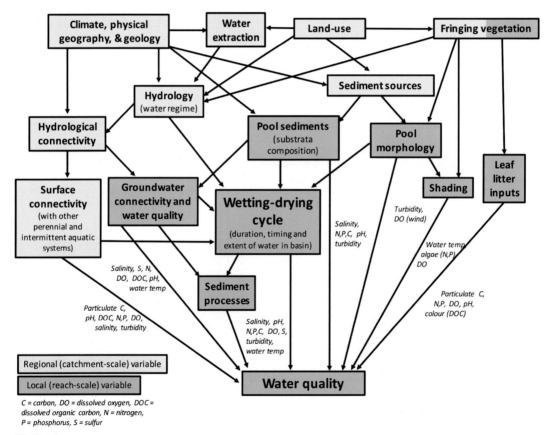

FIG. 3.1.1

Physicochemical parameters are influenced by a number of regional and local (IRES-specific) scale variables. These variables are interrelated and often interacting, many of them with the wetting–drying cycle, a key element shaping the water quality in IRES.

Adapted with permission from an original by A. Boulton for the Australian Water Quality Guidelines: Guidance Document for Assessing and Managing Water Quality in Temporary Waters.

- Water quality in the IRES at the time flow ceases.
- Extent and type of fringing vegetation (specifically shading and the amount and quality of litter fall entering the streambed or pool).

These variables are influenced by climate, hydrology, geology, geomorphology, and land use in the surrounding catchment (Fig. 3.1.1). Hence, changes in water quality during these wetting–drying cycles will differ in different IRES, even among reaches in the same IRES (e.g., Baldwin et al., 2005; Zimmer et al., 2013; Al-Qudah et al., 2015).

In this chapter, we first explore how key physicochemical parameters vary spatially and temporally, using examples of IRES from various regional settings. We then detail the factors that influence these parameters in IRES during a generalized drying and rewetting cycle. Although we mention some factors affecting nutrients and organic matter in IRES in this chapter, they are specifically addressed in Chapter 3.2.

3.1.2 SPATIAL VARIABILITY OF PHYSICOCHEMISTRY IN IRES

Differences in overall water chemistry across streams and rivers are determined by climate, geology, and vegetation in the catchment (Hynes, 1975). Water chemistry along streams integrates many "signals" (hydrological inputs, outputs, and storage and transformation of matter and energy) coming from the hillslope and zones along the channel (e.g., hyporheic, parafluvial and riparian zones, Stanford and Ward, 1993; Fisher et al., 1998) and from within the channel (autogenic processes such as biological uptake and release, sedimentation and resuspension of particles, e.g., Fisher et al., 1998; Tockner et al., 1999). Because the distribution and intensity of these signals change along lotic systems, it is expected that surface water physicochemistry will be spatially variable in all streams and rivers. Following, we describe the factors determining the spatial variability of water physicochemistry in IRES along the three hydrological dimensions discussed in Chapter 2.3.

THE LONGITUDINAL DIMENSION

In perennial rivers, physicochemical variables are expected to change along a continuous gradient from headwaters to mouth. As the River Continuum Concept postulates (Vannote et al., 1980), this gradient of physical and chemical conditions elicits a series of responses resulting in a continuum of biotic adjustments and consistent patterns of loading, transport, utilization, and storage of organic matter down the channel. This generalized conceptual framework for integrating predictable and observable biological features is based on the existence of flow continuity along the longitudinal dimension of lotic systems. However, in IRES, such surface flow connectivity is periodically lost. Hence, generalizations derived from the river continuum concept do not necessarily apply. Change in surface flow connectivity has important consequences for the transport of material and energy downstream, and increases the spatial variability of water physicochemistry along the entire IRES network (e.g., Gomi et al., 2002; Zimmer et al., 2013; Robinson et al., 2016).

At the reach scale, fluctuating low flows in IRES influence water physicochemistry by creating a distinctive mosaic of local environmental conditions (e.g., water residence time, organic matter content, sediment–water interactions, the structure and activity of biological communities) near the streambed. Consequently, biogeochemical process rates and reactions should have a patchy distribution (*sensu* McClain et al., 2003). In addition, fluctuating flows also affect the composition and canopy cover of riparian vegetation (Chapter 4.2).

Moreover, because stream flow and advection rates (i.e., the movement of material dissolved or suspended in the water) are lower, opportunities for uptake and sorption are more commonplace in IRES reaches (shorter processing lengths) than in perennial reaches (e.g., Fisher et al., 1998; Fig. 3.1.2). Solute concentrations (e.g., major anions and cations, nitrogen [N], phosphorus [P], dissolved organic carbon [DOC]) are likely to be spatially heterogeneous at spatial scales from a few meters to several kilometers, and commonly increase as surface flows diminish (e.g., Gómez et al., 2009; Fellman et al., 2011; Vázquez et al., 2011; Fig. 3.1.3). Even when perennial reaches alternate with intermittent reaches (Chapter 2.3), downstream water quality can be strongly modified (Box 3.1.1).

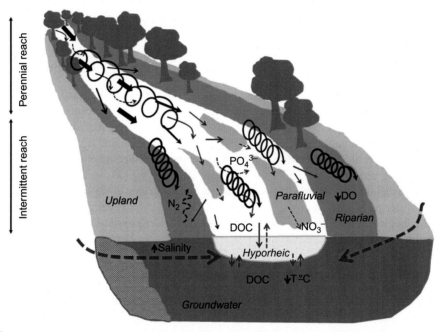

FIG. 3.1.2

Conceptual model showing how differences in flow, advection, and processing length between perennial and intermittent reaches determine spatial heterogeneity of physicochemistry in IRES. Black and red arrows distinguish flow paths in perennial and intermittent reaches, respectively. Larger arrows indicate dominant flow paths. For intermittent reaches, low flows and previous drying create a distinctive mosaic of local environmental conditions (e.g., organic matter content, water residence time, structure and activity of biological communities). Simultaneously, the effects of interactions between subsystems (e.g., surface waters, hyporheic, parafluvial and riparian zones) also are present in perennial reaches and are reinforced in intermittent reaches. Consequently, biogeochemical process rates and reactions have a patchy distribution. In addition, because stream flow and advection are lower, opportunities for uptake and sorption are greater in intermittent (shorter processing lengths) than in perennial reaches. As smaller and mostly closed material cycles are in specific zones or reaches, higher spatial heterogeneity in water chemistry can be expected (Lewis et al., 2006). For convenience, abbreviations indicating biogeochemical processes are only shown in the intermittent reach as the same processes occur within the perennial reach. DO: dissolved oxygen; DOC: dissolved organic carbon; T: temperature; N_2: nitrogen.

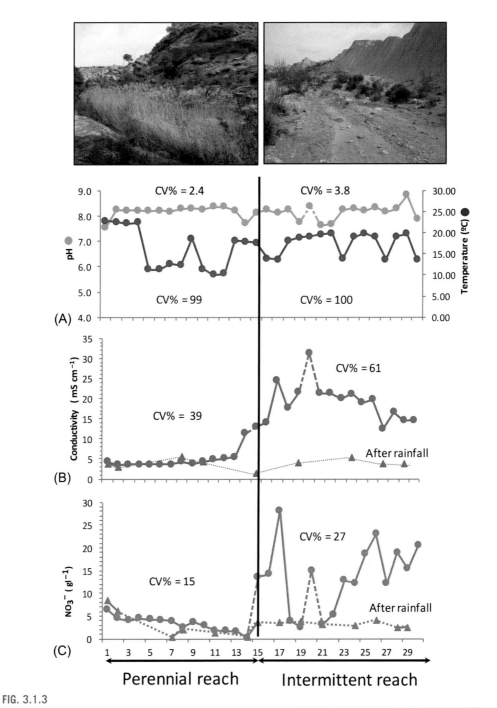

FIG. 3.1.3

Spatial variability of pH and water temperature (A), conductivity (B), and nitrate concentrations (C) along the Chicamo Stream (southeast Spain). Plots show spatial variation in physicochemical conditions along both reaches at different sampling points (1–29) during baseflow (end of March). Discontinuous lines indicate discontinuous surface flow between sampling points. Dotted lines (panels B and C) show data following a spring rain event (May).

Data from García-García (2005). Photos: Courtesy R. Gómez.

BOX 3.1.1 THE EFFECT OF INTERMITTENT REACHES ON STREAM CHEMISTRY

An example of the effect of intermittent reaches on stream surface water quality along the longitudinal dimension is the Chicamo Stream (Murcia, southeast Spain, Gómez et al., 2009). Chicamo Stream is a spatially intermittent river within the Segura River Basin, with a perennial upper reach (5.8 km) followed by an intermittent reach (7.7 km). No tributaries exist along this stream. Differences between reaches in surface water velocity and discharge during baseflow are around 14-fold and 60-fold lower in the intermittent reach, respectively. In the intermittent reach, the low discharge and the streambed topography result in discontinuous surface flow that only becomes continuous after rainfall.

Water conductivity and pH typically remain almost constant along the perennial reach during baseflow, whereas nitrate (NO_3^-) concentration decreases gradually in response to biotic uptake and retention (Fig. 3.1.3). However, in the intermittent reach, these parameters vary widely with spatial coefficients of variation (CV%) exceeding those measured in the perennial reach (García-García, 2005). Conversely, as seen in other lotic systems (e.g., Brown and Hannah, 2008), stream water temperature in Chicamo Stream is highly heterogeneous in both reaches (CV = 99% and 100% for the perennial and intermittent reaches, respectively). In this stream, variations in water temperature are hypothesized to be mainly influenced by the heterogeneous distribution of riparian vegetation in the perennial reach and by differences in water advection in the intermittent reach. Increased hydrological connectivity and advection also influence physicochemistry at certain times of the year. In particular, both the magnitude and spatial variability of conductivity and NO_3^- concentrations decline sharply after spring rains.

In IRES during rewetting, water chemistry evolves as stream runoff is generated and transported. This evolution arises through the interaction of sediments, groundwater, and precipitation or meltwater. The effect of sediment weathering on surface water chemistry (such as Cl^-, SO_4^{2-}, and Na^+ concentration) after storms seems to be related to their concentration in sediments, whereas trace elements are more closely related with the precipitation chemistry (e.g., Elsenbeer et al., 1994; Al-Qudah et al., 2015). If other water sources such as groundwater seepage or meltwaters are present, the final composition of surface waters is difficult to predict (e.g., Robinson et al., 2016). Because of the variability in geomorphology, lithology, hydrology, and land use across IRES, even in the same network, the contribution of sporadically flowing watercourses to downgradient water quality will be heterogeneous (Zimmer et al., 2013). The overall effect of such spatial variability on IRES water quality will depend on the network configuration, specifically the extent and location of intermittent and ephemeral reaches (Chapter 2.3) relative to perennial reaches in the network.

THE VERTICAL DIMENSION

Although we cannot leave out the connections that exist between the water column and the atmosphere via evaporation, typically the vertical hydrological connections in lotic systems are defined by the exchange of water, materials, and energy between the surface water and groundwater that occurs in streams and rivers with permeable substrates (Chapter 2.3). Two vectors connect surface water and groundwater: (i) surface water recharging groundwater aquifers (surface–hyporheic exchanges) and (ii) groundwater flow from the catchment through riparian zones to the active channel (Dahm et al., 1998).

The hyporheic zone (the region of saturated sediments below the water surface where surface waters and groundwaters mix) is an active ecotone. As surface water flows downstream, vertical hydrological exchange between the hyporheic zone and the surface water column occurs at hyporheic upwelling and downwelling locations in response to variations in discharge, bed topography, and porosity

(Boulton et al., 1998). Contributions of the hyporheic zone to surface stream chemistry are determined by the type and rate of subsurface biogeochemical processes taking place and the proportion of subsurface flow emerging at the surface (e.g., Valett et al., 1990; Jones et al., 1995). Compared to perennial streams, such contributions can be greater in IRES, especially when surface flow diminishes (e.g., Grimm and Fisher, 1984; Dahm et al., 2003) or during specific seasons of flow in temperate and polar catchments where weathering contributions from hyporheic zones affect catchment ionic budgets (e.g., Gooseff et al. 2002; McKnight et al., 2004).

Surface water temperature is influenced by hyporheic upwelling zones but not necessarily in the same way in different rivers or river reaches. Some studies report that hyporheic water is cooler in summer and warmer in winter than surface waters (Hynes, 1983; White et al., 1987). Conversely, in Sycamore Creek, a Sonoran Desert intermittent stream, the opposite occurs. In this stream, subsurface temperatures likely exceed surface temperatures in summer, accelerating carbon and nutrient cycling (Valett et al., 1990). The strong diurnal temperature variability of surface waters compared to subsurface waters is a major difference in these desert streams.

Biogeochemical processes occurring in the hyporheic zone are mediated by DO levels that are regulated by the balance between supply from surface or groundwater zones and rates of oxygen consumption during organic matter decomposition (Jones et al., 1995). Interstitial oxygen is high in regions of infiltration (downwelling zones) and generally reduced at upwelling zones after subsurface water flows through the hyporheic zone (Valett et al., 1990) and organic matter is processed in the sediments. Under these circumstances, denitrification (that leads to NO_3^- loss as N_2); sulfate, iron, and manganese reduction; and methanogenesis can occur in the interstitial waters (e.g., Harvey and Fuller, 1998). Under aerobic conditions, nitrification (that leads to NO_3^- production) usually takes place, and hyporheic sediments change from a NO_3^- sink to a NO_3^- source to surface waters (e.g., Triska et al., 1989; Stanley and Boulton, 1995). In streams with coarse sediments, increased surface water NO_3^- concentrations in hyporheic upwelling zones are well documented (e.g., Valett et al., 1990; Triska et al., 1993). In Sycamore Creek, these upwelling zones cause high spatial variation in nutrient concentrations and water conductivity over both long (10 km) and short spatial scales (30 m) (Dent et al., 2001), with coefficients of variation (CV) ranging between 104%–145% and 69%–161%, respectively, during low discharge periods (Dent and Grimm, 1999).

Like hyporheic flows, groundwater discharge can influence surface water temperature, DO, and nutrients and be an important source of DOC in IRES (Vervier et al., 1993; Dahm et al., 2003; Bernal et al., 2006). The magnitude of groundwater inflows is linked to channel morphology, floodplain and streambed sediment features, and catchment lithology. Consequently, physicochemical responses are spatially variable at both large and small spatial scales (e.g., Dahm et al., 1998; Butturini et al., 2003; Zimmer et al., 2013). One example of groundwater effects on surface chemistry at small spatial scales occurs in an unimpacted IRES in the Iberian southeast (Murcia, Spain) where N concentrations changed from 0.31 to 2.02 mg l^{-1} NO_3^- and from 0.24 to 1.32 mg l^{-1} NH_4^+ along a 20-m reach (Gómez, 1995). These distinctive variations in N concentrations were attributed to local groundwater discharge and lateral seepage from chemically reduced sediments, respectively.

Groundwater can also influence surface water chemistry through dilution or by increasing water salinity if waters have previously been in contact with water-soluble minerals, as typically occurs in many saline intermittent Mediterranean streams (Millán et al., 2011) or of cyclic marine origin, as typically occurs in southern Australia (Herczeg et al., 2001). In addition, the differential chemical composition of groundwater inflows (local, intermediate, and regional flows) is especially relevant for interpreting spatial variability of surface-water ionic composition (e.g., Dahm et al., 2003).

THE LATERAL DIMENSION

Along the lateral dimension, water physicochemistry in streams and rivers is affected by the influence of parafluvial (the part of the active channel lateral to the surface stream including gravel and sand bars) and riparian zones (e.g., Lewis et al., 2006). In lotic ecosystems, interactions between these two zones and the stream channel differ longitudinally and are strongly influenced by the distribution of depositional zones—especially in middle and lower reaches, where gravel and sand bars typically appear and there is a broader and more structured riparian zone (Holmes et al., 1994; Fisher et al., 1998).

The parafluvial zone is a biogeochemically active space that can affect DOC, nutrients, and oxygen supply to surface waters in a similar way to exchanges between surface waters and the hyporheic zone (Vervier et al., 1993; Holmes et al., 1994, 1996; Claret et al., 1997, Fig. 3.1.4c). In addition, the geomorphic structure of a specific stream reach (number, size and shape of sand or gravel bars) may influence downstream water physicochemistry (Fisher et al., 1998; Chapter 2.3).

Riparian zones are also highly heterogeneous. Organic matter, redox conditions, soil texture, and plant species composition all influence surface water chemistry (e.g., Harms and Grimm, 2008). Elevated hydrological retention and transient storage of water in riparian zones enhance the transformation and retention of solutes. Therefore, major changes, mainly in N, P, and DOC concentrations, are expected in those reaches where surface waters interact with riparian zones (Chapter 3.2). Such interactions have implications for maintaining high water quality standards in streams and rivers (e.g., Peterjohn and Correll, 1984). Besides nutrient uptake, riparian vegetation can control the chemistry of surface waters through indirect mechanisms. For example, the presence of riparian shrubs growing in parafluvial zones can increase nitrate loss from subsurface waters by creating favorable conditions for denitrification and increasing microbial activity, which in turn reduces DOC concentrations in the water (Schade et al., 2001; Vázquez et al., 2007).

In addition to natural variation along longitudinal, vertical, and lateral dimensions, human impacts can contribute to spatial variability of physicochemistry in IRES (Chapter 5.1). This can be either directly (e.g., sewage inputs and water from different sources; Fig. 3.1.4d) or indirectly (e.g., modification of sediment regime, riparian zone alteration, water abstraction; Fig. 3.1.4f). Although almost all lotic ecosystems are subject to human impacts, IRES are especially vulnerable to such human impacts (Chapter 5.1).

3.1.3 TEMPORAL VARIABILITY OF WATER PHYSICOCHEMISTRY IN IRES

Water physicochemistry in almost all aquatic ecosystems changes through time as a result of the daily, seasonal, annual, and interannual variation of major natural environmental drivers such as light, climatic conditions, and biotic processes as well as human activities (e.g., forest clearance, sewage inputs, water diversions). However, temporal variability in IRES physicochemistry is especially influenced by their drying and rewetting cycles (e.g., von Schiller et al., 2011; Arce et al., 2014, Al-Qudah et al., 2015).

FIG. 3.1.4

See legend on next page.

DAILY VARIABILITY

Water temperature, DO, pH, and nutrient concentration are the main major parameters subject to diurnal changes in streams and rivers (Wetzel, 2001). This is because they are, for the most part, controlled by the activity of organisms. However, the reduced glacial inputs in glacier-fed IRES during the night in late spring and summer also lead to diel variation in nutrient concentration (Robinson et al., 2016).

Typically, over a 24-hour cycle, DO concentrations tend to increase during the day and decrease during night. These changes are based on relative differences in rates of photosynthesis and respiration between day and night (Fig. 3.1.4e). There is also a concomitant but opposite change in dissolved CO_2 (pCO_2), which in turn will decrease pH through liberation of protons when carbonic acid forms. Coupled to changes in light and air temperature, water temperature also fluctuates daily in aquatic ecosystems. Given the ease with which shallow waters heat and cool, IRES may experience daily amplitudes of water temperature much larger than in perennial systems and more similar to air conditions (Williams, 2006). Likewise, because of their low discharge, daily changes in water temperature and DO in IRES are tightly linked, especially during periods of flow cessation. Daily changes in water temperature also drive variation in water salinity by evapoconcentration, especially in saline IRES (Gómez, personal observation).

An example of the extraordinary diel variability of physicochemical conditions in IRES is found in Chicamo Stream (southeast Spain). Vidal-Abarca et al. (2002) observed that ranges of diurnal variation (averaged for the four seasons of the year) in water temperature, pH, DO, and salinity in this highly productive intermittent stream exceeded seasonal ranges. Daily water temperature, DO, pH, and water salinity ranged between 9.6 and 34°C, 1.6 and 20 mg l^{-1}, 6.5 and 9.2, and 1 and 4 g l^{-1}, respectively.

A key element that may exacerbate the high daily variability in physicochemical conditions in many IRES is the lack of well-developed riparian vegetation (Chapter 4.2) that would otherwise limit solar radiation. This is common to a wide range of IRES, including low-order prairie streams (Hill and Gardner, 1987), arid and Mediterranean streams (Gasith and Resh, 1999), and desert streams (Stanley et al., 1997). The high level of irradiation reaching the surface of IRES in arid zones typically increases the daily maximum water temperatures (e.g., Rutherford et al., 2004). Such a scenario is prone to conditions of oxygen supersaturation (>100%), especially at midday. To take an extreme example,

FIG. 3.1.4

(a) Subsurface flow through fine organic-rich sediments with low interstitial oxygen in Chicamo Stream (southeast Spain) leads to the release of soluble PO_4^{3-} and Fe^{2+} to surface waters. Where iron is oxidized at the surface (orange precipitates), available P is rapidly assimilated by algae in this P-limited intermittent stream. (b) In N-limited Sycamore Creek (Arizona), nitrification occurs as subsurface water flows through the parafluvial zone. Thus, coarse sand-bar sediments increase nitrate availability in the water column and benthic primary production at the sand-bar edges. (c) Sediment heterogeneity at the reach scale can promote patchy sediment redox conditions with consequences for nutrient and metal availability (Chicamo Stream). (d) Iron precipitation in Brilka Creek (New South Wales, Australia) caused by anthropogenic pumping of saline groundwater into the stream channel. (e) Bubble formation in a highly productive reach during midday. (f) Water abstraction for agriculture diminishes surface flow in lowland reaches of Chicamo Stream. (g) Because of the low dilution capacity of IRES, even small inputs of allochthonous organic matter like this horse dropping in Chicamo Stream has a disproportional effect on biotic communities.

Photos: Courtesy R. Gómez (a, b, c, e, f and g); D. Baldwin (d).

the daily range of percent oxygen saturation in an isolated pool in Chicamo Stream was 21-224% in June, with steady supersaturation during most of the day (Vidal-Abarca et al., 2002). Stocks of organic matter in sediments and links between surface and subsurface compartments during drying are other factors strongly coupled with daily fluctuations in water chemistry and temperature in IRES (see later).

The activity of aquatic organisms can also drive daily changes in nutrient concentrations, especially of limiting nutrients (Chapter 3.2). For example, in N-limited Sycamore Creek, daily fluctuations of total inorganic N in surface waters have been attributed to the intense assimilatory demand of benthic algae (Grimm et al., 1981). Similarly, high-frequency hydrological and biogeochemical monitoring of the Koliaris River (Crete) during flooding showed that the diurnal variability in pH, DO, and nutrients was depressed but variability began to increase once biota started to recolonize the streambed after flooding (Moraetis et al., 2010).

The range of daily variations in water physicochemistry in IRES typically changes over the year because of the alternation between expansion, contraction, and fragmentation periods. The degree of daily fluctuations of major physicochemical parameters in IRES is, at some point, directly proportional to the flow conditions, with wider and narrower fluctuations during lower and higher flow phases, respectively (Fig. 3.1.5). A good example of such a trend is Pardiela Stream, a southeast Portugal intermittent river where the daily range of water temperature, DO, and N and P concentrations increased from winter to summer, reflecting the transition from winter lotic to summer lentic conditions (Lillebø et al., 2007).

SEASONAL VARIABILITY

Within an annual cycle, water physicochemistry for all lotic ecosystems is naturally variable because of seasonal changes in environmental conditions, discharge, temperature, and biological activity (Wetzel, 2001). Nonetheless, periodic drying and rewetting events make these seasonal changes especially dramatic in IRES (Chapter 2.2) with notable effects on surface water physicochemistry (Fisher et al., 1982). Even in ephemeral streams, the effect of episodic stream runoff after rainstorm events or meltwater flows accentuates the seasonal and event-scale variation in water physicochemistry at the watershed scale (e.g., Elsenbeer et al., 1994; Holloway and Dahlgren, 2001; Gooseff et al. 2002).

Seasonal variability along the longitudinal dimension

Seasonally, high-discharge floods connect rivers and river-reaches within their watershed and cause ecosystem expansion whereas drying leads to ecosystem contractions and fragmentation (Stanley et al., 1997; Robinson et al., 2016). This hydrological setting confers IRES with large annual variability in water physicochemistry. Usually, drying may be more seasonally predictable than flooding, which often depends on episodic and often unpredictable rainfall events (e.g., Bunn et al., 2006). With expansion of surface flow, increased concentrations in SO_4^{2-}, Na^+, Ca^{2+}, SiO_2, K^+, Cl^-, DOC, and N are common in IRES after rainfall events (Mulholland 1993; Elsenbeer et al., 1994; Al-Qudah et al., 2015) and meltwater flows (Gooseff et al. 2002; McKnight et al., 2004; Robinson et al., 2016). In contrast, rainfall events commonly dilute the naturally high levels of Na^+, Cl^-, SO_4^{2-}, and K^+ in saline IRES (García-García, 2005).

Floods also have the potential to mobilize and transport large amounts of particulate and dissolved materials from soils and upstream areas. Depending on their intensity, timing, and spatial extent, floods may contribute substantially to the production, processing, and transport of organic matter downstream

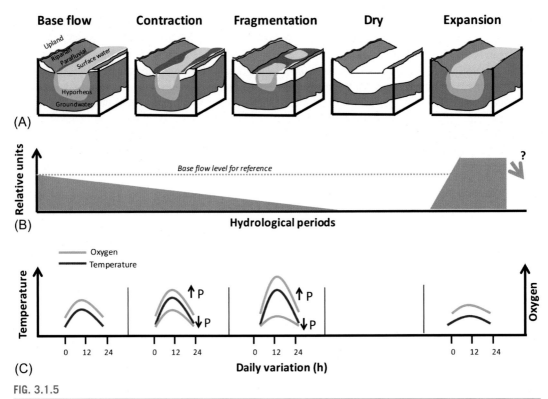

FIG. 3.1.5

(A) Hydrological connectivity between subsystems (upland, riparian zone, surface water, hyporheic zone, and groundwater) in an IRES or IRES reach at different hydrological periods. (B) Temporal variation in discharge and connectivity between subsystems at different hydrological periods (relative to baseflow). Stream connectivity during baseflow becomes differentially disrupted as streams contract and fragment, caused by the cessation of upstream–downstream longitudinal links and the weakening of lateral and vertical links. During severe drought, entire streams or stream reaches may lose all surface water and hydrological connectivity can be completely lost (dry phase). Conversely, floods either during baseflow or following dry periods amplify hydrological connectivity and reconnect exchange. (C) Magnitude and daily variability of DO and temperature. Temporal trends of both parameters from baseflow to expansion assume the seasonal transition from spring–summer–autumn. Daily changes in DO assume net primary production (P) during the day and respiration during the night. Vertical black arrows indicate DO trends in systems with high and low net primary production.

(Jacobson et al., 2000; Obermann et al., 2009; Robinson et al., 2016). For example, rainfall events triggered water fronts that transported accumulated detritus and nutrients from upstream reaches along formerly dry channels in the Albarine River, France (Datry et al., 2014). Concentrations of solutes and organic matter at the leading edge of these fronts can exceed baseflow concentrations by orders of magnitude and produce spikes in solute concentrations during these events in IRES (e.g., Obermann et al., 2009; Skoulikidis and Amaxidis, 2009). Mobilization of large amounts of dissolved and particulate organic matter can cause downstream water hypoxia by promoting respiration and mineralization processes (Hladyz et al., 2011; Corti and Datry, 2012; Dahm et al., 2015).

During drying, some physicochemical parameters may increase dramatically compared with baseflow levels, especially in open-canopied streams. Low flow promotes higher water temperature and evapoconcentration leads to an increase in water salinity and some nutrient species, particularly reduced solute forms (e.g., Zale et al., 1989; Acuña et al., 2005; von Schiller et al., 2011). Declining hydrological connectivity during drying in IRES increases the significance of local processes (e.g., water residence time, sediment redox, biological activity, Fig. 3.1.4g) that influence surface water physicochemistry (e.g., Holmes et al., 1996; Vázquez et al., 2011). Thus, along the hydrological sequence of flow contraction–fragmentation–expansion, the regulation of IRES physicochemistry is expected to alternate temporally between mostly in-stream (local conditions) and external drivers (those operating at watershed scale) as the longitudinal connectivity decreases or increases, respectively (Fig. 3.1.6).

Finally, seasonal changes also exert a considerable effect on the spatial variability of physicochemical conditions in IRES, with considerable implications when designing water quality monitoring programs. Longitudinal spatial variability in IRES is expected to rise during contraction and fragmentation as discharge drops and spatial patchiness increases. Conversely, after rainfall when discharge increases and the stream expands relative to the drying phase, spatial variability tends to decline (e.g., Dent and Grimm, 1999; Acuña et al., 2005; Lewis et al., 2006; von Schiller et al., 2011). An extensive study

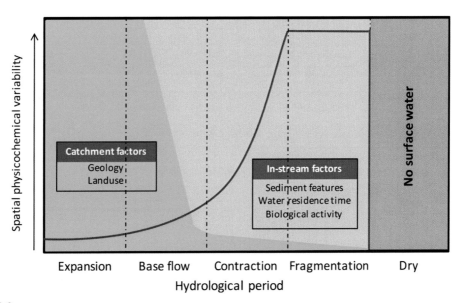

FIG. 3.1.6

Conceptual model of hydrochemical variability of surface water along the longitudinal dimension (solid red line) in IRES through differing hydrological periods (modified from Vidal-Abarca et al., 2004). During floods and stream expansion, large-scale factors govern surface water characteristics and the heterogeneity of water chemistry from upstream to downstream is low. As discharge recedes and drying progresses, local in-stream factors gain importance in modulating water chemistry. Longitudinal heterogeneity in water chemistry peaks when the stream fragments.

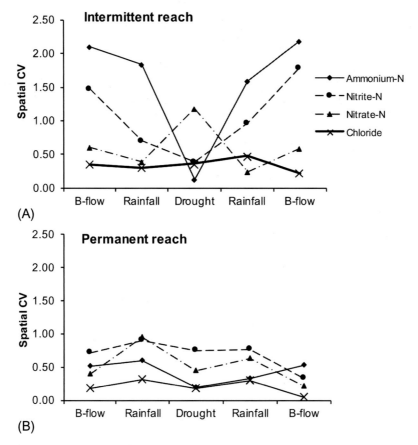

FIG. 3.1.7

Spatial coefficients of variation (CV) for solute concentrations during different hydrological conditions in intermittent (A) and permanent (B) reaches of Chicamo Stream (data from Gómez et al., 2009). B-Flow: baseflow.

throughout Chicamo stream demonstrated the effects of flow intermittence on spatial and temporal variability in water chemistry (Fig. 3.1.7, Gómez et al., 2009).

Seasonal variability along the vertical dimension

Seasonal shifts in vertical interactions (Chapter 2.3) profoundly affect water physicochemistry in IRES. Changes in vertical water exchange alter DO levels in the hyporheic zone that, in turn, influence the dominance of subsurface aerobic and anaerobic biogeochemical processes (Stanley and Boulton, 1995). This has concomitant effects on concentrations of redox-sensitive solutes (e.g., Dahm et al., 1998).

On the other hand, the interaction between meltwater from snow and ice with the hyporheic zone has the effect of flushing out weathered materials and providing inorganic solutes to the surface (Chapter 2.3), especially in IRES of glacial watersheds (e.g., McMurdo Dry Valleys in Antarctica, Gooseff et al. 2002; McKnight et al., 2004). Similarly, changes in the inflow and nature of groundwater can substantially modify surface-water concentrations of inorganic solutes and DOC during droughts

(Dahm et al., 2003). IRES can also be a source of cold subsurface discharge to downstream waters during the dry season, potentially affecting biota (Ebersole et al., 2015).

Seasonal variability along the lateral dimension

Throughout the seasons, the surface area and spatial configuration of the parafluvial zone (number, length, and shape of sand and/or gravel bars) can fluctuate markedly in response to seasonal hydrological changes (e.g., Fisher et al., 1998). For example, the percentage of active channel represented by the parafluvial zone in Sycamore Creek varies from zero to greater than 75% depending on hydrological conditions (Holmes et al., 1994). During floods, surface waters extend across the entire active channel, inundating the parafluvial zone. As discharge decreases, parafluvial sediments progressively emerge and their degree of interaction with surface water changes throughout time.

Coupled with these changes in spatial configuration, seasonal variations in interstitial water chemistry and parafluvial sediment also occur that, in turn, influence surface water physicochemistry in different ways. As an example in Sycamore Creek, Holmes et al. (1994) observed the highest decrease in DO as water moved through gravel bars in summer as a consequence of accelerated metabolism. In parallel, they also observed the largest increases in NO_3^-. However, as flow decreases, DO can be depleted whereas DOC availability in interstitial waters increases, and then anaerobic processes (e.g., denitrification) predominate. The growth of riparian vegetation on gravel bars as drying progresses reinforces this effect (Schade et al., 2001).

Another key component of lateral connectivity is the riparian zone (Lewis et al., 2006; Chapter 4.2). Linkages among uplands, riparian zones, and streams differ in IRES depending on the climatic zone and hydrological conditions. Flood events raise water tables, reactivate biogeochemical processes in the newly inundated riparian soils, and increase organic matter availability (Baker et al., 2000; Bernal et al., 2013). Reestablishment of hydrological connectivity caused by flooding increased N removal in the riparian zone during the summer monsoon season in the spatially intermittent San Pedro River in Arizona (Harms and Grimm, 2008). Conversely, other studies have found that rising water tables promoted significant inputs of N and P (Butturini et al., 2003; Vázquez et al. 2007) and DOC (Baker et al. 2000; Heffernan and Sponseller, 2004) to the stream water from newly inundated riparian soils.

Riparian zones remain active biogeochemical hot spots (*sensu* McClain et al., 2003) during periods of baseflow and low connectivity, especially at the stream–riparian edge. Further stream contraction and drying lower water tables in the riparian zone. At these times, the stream channel and riparian zone are decoupled, and effects on water physicochemistry of IRES are minimal. This situation persists until floods reconnect the two subsystems (Schade et al., 2002).

INTERANNUAL VARIABILITY IN PHYSICOCHEMISTRY: SOME FACTORS INVOLVED

Climate variability plays a major role in the duration and spatial extent of zero flow in most IRES. Hydrological variability throughout the world is affected by major global climatic factors such as the El Niño Southern Oscillation (ENSO) phenomenon (Molles and Dahm, 1990). For example, extremes of the ENSO (El Niño and La Niña) phenomenon produce climatic teleconnections that increase or decrease precipitation in various parts of the world. A specific example is the tendency for enhanced precipitation in the southern United States during El Niño years and for reduced precipitation in these regions during La Niña years. Duration and spatial extent of drying in IRES on interannual timescales in these regions of the United States are linked to these larger scale climate drivers; similar patterns are found in many regions of the Earth. Thus, climate variability and its causes are critical drivers to both the hydrology and physicochemistry of IRES worldwide.

Studies on the influence of interannual climate variability on IRES physicochemistry are limited (Butturini et al., 2008). Dahm and Molles (1992) and Molles et al. (1992) studied the intermittent Rio Salado in New Mexico, USA, that flows only during the summer monsoonal thunderstorm season. Flow during summer ranged from 7 to 70 days over a 38-year record. Interannual variability in flood flows was strongly linked to antecedent moisture. High-magnitude summer flows were associated with dry antecedent conditions whereas little or no surface flows in summer followed wet antecedent conditions. Hypothesized mechanisms causing this inverse flow phenomenon include: (1) increased vegetation cover and reduced runoff due to greater infiltration and evapotranspiration in wet years and (2) the development of hydrophobic soils during prolonged dry periods leading to increased surface runoff when summer thunderstorms occurred. The effects of stronger surface flows after dry periods in arid and semiarid catchments on the physicochemical characteristics of the runoff deserve further research attention. Similarly, the effect of hydrological fluctuations and biogeochemical processes occurring during the interplay between drought-flood extremes should be explored in more detail.

Interannual flow variability in IRES is also influenced by disturbance within catchments. For example, catastrophic forest fires increased the water yield from precipitation events into IRES flowing into perennial streams and rivers in the Jemez Mountains of north-central New Mexico, and the impacts on physicochemistry were striking (Dahm et al., 2015; Reale et al., 2015; Sherson et al., 2015). Major forest fires caused sags in DO and pH, large increases in turbidity, and elevated nutrient concentrations (N and P) in IRES and downstream perennial streams and rivers. The interaction of disturbance and interannual variability in flow can produce large changes in the physicochemistry of IRES.

TEMPORAL VARIABILITY AT LONGER SCALES: IRES IN THE CONTEXT OF GLOBAL CHANGE

What happens when surface flows return to IRES after long periods without surface flows? Molles et al. (1998) and Valett et al. (2005) studied the effects of returning surface flows to a disconnected floodplain of the Rio Grande, New Mexico, after four decades of no flow. Flows were applied in late spring as a flood pulse. Very high rates of metabolism quickly removed DO from the flood waters, and anaerobic pathways (e.g., denitrification, metal reduction, and sulfate reduction) of decomposition became prevalent when flow was reinitiated. Connecting IRES with floodplains rich in organic carbon can trigger high rates of biogeochemical processes. These bursts of metabolism can draw down DO concentrations and produce chemically reduced solutes within the surface waters.

Another example of physicochemical responses to large-scale inundation of a floodplain after a long dry period occurred in the southern Murray-Darling Basin at the end of the Millennium drought (1996–2010). This led to major water quality changes for the river ecosystems in the spring and summer of 2010–2011 (Hladyz et al., 2011; Whitworth et al., 2012). Flood waters inundated forested and agricultural floodplains, producing hypoxic blackwater events (DO $< 2\,mg\,l^{-1}$) and high DOC concentrations over 2000 km of river channel for up to six months. Such hypoxic blackwater events can be exacerbated by higher temperatures and human regulation of flows. Across much of the world, global change and longer warmer periods with droughts are predicted in the coming decades, and the impacts of these changes on IRES physicochemistry deserve further attention.

3.1.4 CHANGES IN WATER PHYSICOCHEMISTRY DURING DRYING AND REWETTING

Although many interrelated and often interacting variables determine the effect of the wetting–drying cycle on key physicochemical parameters in IRES (Fig. 3.1.1), it is possible to draw several generalizations. The effect of local and regional-scale variables on water physicochemistry can be conceptualized as two distinct phases following flow cessation. The first phase is the development of discontinuous flow followed by water recession and, typically in runs, formation of remnant pools (Stanley et al., 1997). The second phase is the exposure to air and subsequent drying of streambed sediments. Obviously, it is impossible to measure physicochemistry when there is no water present. However, sediment exposure and subsequent drying will impact water quality in the IRES once flow resumes during rewetting.

Pool persistence in IRES will depend on the morphology of the pool, atmospheric conditions, and the presence of hyporheic and/or groundwater interactions. Channel morphology will determine the size, depth, and distance between pools. In some IRES or their reaches, pools do not form at all; rather, the channel simply dries out (Fig. 3.1.8). Because isolated pools often form in IRES, factors affecting water physicochemistry in remnant pools and then rewetted sediments are described in detail. Many of the variables and processes affecting the physicochemistry in pools during drying and rewetting resemble those in other shallow waters.

FIG. 3.1.8

Not all IRES or IRES reaches form remnant pools during drying (a, Chicamo Stream, Murcia, Spain) and therefore this phase may not be relevant for all systems. Because of streambed topography, pools are sometimes nonexistent and the reach simply dries out (b, Rambla de la Parra, Murcia, Spain).

Photos: Courtesy R. Gómez.

PHYSICOCHEMISTRY IN REMNANT POOLS

Temperature

The temperature regime within residual pools in IRES is influenced by many factors including ambient air temperature, solar irradiation, wind speed and direction, pool depth, and whether there is interaction with groundwater. Water temperature will be strongly influenced by ambient air temperature and the amount of solar irradiance reaching the pool surface (Kelly and Kohler, 1996). In temperate climates, flow generally ceases in the warmer months, which coincides with the highest levels of solar irradiance so that water temperatures at these sites tend to increase during the drying phase. This is not necessarily the case in the wet/dry tropics, where flow cessation may actually occur in the cooler months (Townsend et al., 2011). The influence of solar irradiation on the water temperature regime in a residual pool can be tempered by shading either by riparian vegetation (Rutherford et al., 2004) or landscape features such as adjacent mountains (Hrachowitz et al., 2010).

Pool depth strongly influences the temperature regime. Shallow pools tend to heat up faster than deeper pools. Also, because shallow pools have less thermal mass than deeper ones, diurnal variation in them would be expected to be greater than in deeper pools—especially at depth (e.g., Turner and Erskine, 2005). Deep (ca. 4 m) residual pools in ephemeral rivers have been shown to thermally stratify, with stratification lasting in excess of 100 days (Baldwin and Wallace, 2009). Cooler water in deeper zones, sometimes influenced by cooler groundwater (e.g., White et al., 1987), may provide a thermal refuge for heat-stressed organisms (e.g., Nielsen et al., 1994).

Dissolved oxygen

Concentrations of DO in a residual pool largely depend on the overall temperature of the pool, whether stratification has occurred, whether the pool can support and sustain substantial levels of primary productivity, and how much allochthonous carbon enters the pool (Chapter 3.2). The amount of oxygen dissolved in water directly depends on water temperature because of saturation kinetics; the warmer the water, the less oxygen can dissolve in it (e.g., Hladyz et al., 2011). Temperature can also indirectly influence DO levels through thermal stratification and effects on the rates of both primary productivity and microbial activity (e.g., Baldwin and Wallace, 2009). Following stratification, if the pool contains organic-rich sediments, sediment oxygen demand can deplete DO. If anoxia (zero oxygen) occurs in deep pools, this leads to the mobilization of nutrients, particularly N and P, from the sediments to the overlying water column (Baldwin and Williams, 2007; Chapter 3.2). If the thermocline breaks down because of wind action (Turner and Erskine, 2005) or a small unseasonal inflow into the pool (Baldwin and Wallace, 2009), the anoxic nutrient-rich water mixes with the overlying water, immediately causing a decline in the overall DO in the pool (e.g., Acuña et al., 2005; Baldwin and Wallace, 2009; von Schiller et al., 2011).

After flow ceases, rates of primary production can increase in pools (e.g., Townsend et al., 2011; Chapter 3.2), in turn leading to substantial modification of diurnal DO concentrations—rising during the day into early evening and then declining overnight (e.g Zale et al., 1989). Primary production is favored by the warmer temperatures associated with flow cessation (described earlier), but the increase in plant and/or algal biomass also requires a source of bioavailable nutrients. Fine organic-rich sediments are more likely to be a source of nutrients to the overlying water column than bedrock, cobbles, or sand. Nutrients can also enter remnant pools through groundwater exchange or anthropogenic sources such as stock access and sewage disposal.

Another modifier of DO in remnant pools is allochthonous carbon input from riparian litter (Zale et al., 1989; Hladyz et al., 2011; Chapter 3.2). DOC leached from leaves falling into the pool can be metabolized by bacteria, in the process utilizing oxygen. Consequently, DO levels in the pool begin to decline. Generally, the impact will be greatest when most of the litter enters the pool during the warmer months as the rate of microbial metabolism and hence potential to reduce oxygen concentrations is temperature dependent (Whitworth and Baldwin, 2012).

Salinity

The salinity within a residual pool will depend on water quality in the pool at the time of disconnection, its volume, and whether groundwater interaction occurs. In the absence of groundwater, the salinity in the pool will increase during the drying phase principally through evapoconcentration. For example, Hamilton et al. (2005) used both conservative ion tracer concentrations and oxygen and hydrogen isotope fractionation to show that evaporative water loss controlled water levels in 15 residual pools in Cooper Creek, central Australia. If the pool does interact with groundwater, the salinity in the pool will, at some point, reflect the salinity in the groundwater (Costelloe et al., 2007).

Turbidity

Turbidity is an important factor affecting DO in the water column through its effect on rates of photosynthesis. Turbidity in residual pools is governed by the nature of the sediments, pool morphology relative to prevailing winds (e.g., Kreiling et al., 2007; Chapter 2.1), the presence of organisms capable of bioturbation (Fletcher et al., 1985; King et al., 1997; Braig and Johnson, 2003), and whether algal blooms occur. Based on the principles of sedimentology, the turbidity in residual pools with coarse substrate (bedrock, cobbles, or sand) should remain low through the drying phase (Mulder and Alexander, 2001). Conversely, if finer substrate is present, it can be suspended through wind action or bioturbation. Wind-driven turbidity will be more prevalent in shallower water than deeper water, and therefore turbidity should increase as water depth declines with evaporation.

pH

The pH in residual pools will be influenced by the inherent buffering capacity in the water column and sediments, whether there is an increase in primary productivity and, in certain circumstances, if the pool interacts with groundwater. Buffering capacity of both the water column and the sediments is strongly influenced by catchment geology and hence will most likely be spatially variable. Diel fluctuations of primary productivity also affect pool pH.

Extremely low pH levels (pH < 4 and as low as 1.8) have been observed in many IRES and other wetlands during the drying phase (Lamers et al., 2001; Hall et al., 2006; Lamontagne et al., 2006). Investigations have shown that these waterbodies have been exposed to saline groundwater containing elevated levels of sulfate. Sulfate reduction in the sediments led to the accumulation of large stores of reactive metal sulfides. While the sediments were submerged, the sulfidic material remained inert but as the water levels in these systems fell, the sulfidic material began to oxidize to oxides, producing acid in the process (Baldwin and Fraser, 2009; Fig. 3.1.4d).

PHYSICOCHEMISTRY DURING COMPLETE DRYING AND THE REWETTING FRONT

Changes to the substrate following complete drying of IRES can have impacts on water quality following flow resumption (e.g., Baldwin and Mitchell, 2000; Skoulikidis and Amaxidis 2009; Gómez et al. 2012; Al-Qudah et al., 2015; Merbt et al., 2016). Probably the most important processes are (i) oxidation and subsequent crystallization of sediments on exposure to air followed by desiccation,

BOX 3.1.2 SEDIMENT CHARACTERISTICS AND AUTOGENIC SPATIAL HETEROGENEITY IN IRES DURING DRYING

In Chicamo Stream, the sediments are characterized by high iron content that generally limits dissolved P availability through its adsorption to iron oxides and oxyhydroxides (Gómez et al., 1994). However, subsurface waters flowing through organic-rich fine sediments can cause anoxia in parafluvial zones, leading to higher levels of dissolved ferrous iron (Fe^{2+}), reduced manganese (Mn^{2+}), ammonium (NH_4^+), and phosphate (PO_4^{3-}). In those areas where parafluvial water enters the surface, different biogeochemical processes occur. The higher availability of soluble P in the water column increases primary production and Fe^{2+} converts rapidly to ferric iron (Fe^{3+}) in the presence of oxygen and then precipitates as iron oxyhydroxides (Fig. 3.1.4a).

In addition, the increase in algal standing crop at groundwater upwelling zones reinforces the spatial heterogeneity in stream physicochemistry by affecting water velocity, temperature, pH, DO, DOC concentrations, and nutrient uptake downstream. This process, called autogenic spatial heterogeneity, has also been described in other IRES (e.g., Fisher et al., 1998; Henry and Fisher, 2003). Conversely, nitrification along parafluvial zones in Sycamore Creek, a nitrogen-limited Sonoran Desert stream, is a source of NO_3^- (Chapter 2.3) and thus gravel bars not only can increase its availability in the water column, but also influence benthic primary production (Fig. 3.1.4b). However, changes in sediment features over successional time may turn gravel bars into a NO_3^- sink (Holmes et al., 1994, 1998).

(ii) the effect of desiccation on soil/sediment micro-organisms, and (iii) the buildup of organic matter and/or salt in the bed of the watercourse.

As many fine and organic-rich sediments are anoxic below the top few millimeters when covered by water, any reduced mineral phases will be oxidized when exposed to air. As described earlier, if reduced sulfur species (e.g., metal sulfides, elemental sulfur) are present in the sediments, exposure to air promotes acid production and a fall in sediment pH. On rewetting, this acid can be mobilized, leading to acidic water in the wetting front (Baldwin and Fraser, 2009).

During desiccation, mineral phases begin to convert to more crystalline forms. In the case of iron (Box 3.1.2), amorphous iron oxyhydroxides are converted during desiccation to more crystalline phases such as goethite and hematite. As the iron oxides become more crystalline, their affinity for P declines and upon rewetting, will release a pulse of bioavailable P to the water column (Attygalla et al., 2016).

Extreme desiccation of sediments can lead to a process called the 'Birch effect' with the first rewetting flush (Chapter 3.2), where there is a pulse of N and C released from the sediments, coupled with a sharp increase in sediment respiration (Baldwin and Mitchell, 2000; Mikha et al., 2005). There is still conjecture on the underlying drivers of this phenomenon, but current thinking is that the pulse of C and N comes from organic osmolytes produced by bacteria (e.g., Lado-Monserrat et al., 2014; Chapter 3.2).

During the dry phase, organic matter from different sources can also accumulate in the channel of the watercourse. Some comes from aquatic organisms, although some of this organic matter is also a subsidy (*sensu* Polis et al., 1997) for terrestrial food webs (Steward et al., 2012; Chapter 3.2). Amphibious and terrestrial plants often colonize drying watercourses following exposure of the sediments (e.g., Baldwin et al., 2013). Finally, litter from fringing vegetation, imported by wind or sporadic rain, can accumulate in the dry bed. On reflooding, the C and other nutrients stored in the streambed are mobilized (e.g., Skoulikidis and Amaxidis, 2009; Al-Qudah et al., 2015). Consumption of DOC released from accumulated litter by aquatic bacteria can rapidly reduce DO, resulting in hypoxia or anoxia in the wetted front (Hladyz et al., 2011). Taken together and depending on the circumstances, the wetted front following inundation of a dry watercourse has the potential to be low in DO, saline, nutrient-rich, and/or acidic.

3.1.5 CONCLUSIONS

Streamflow cessation, drying and rewetting, and episodic surface flows in IRES alter hydrological connectivity along longitudinal, lateral, and vertical dimensions, contributing to variation in key physicochemical variables (temperature, light, salinity, pH, dissolved oxygen, and turbidity) at spatial scales from meters to kilometers along, below, and laterally from the channel occurring over various temporal scales (daily, seasonal, and supraseasonal). This spatial and temporal variability in water physicochemistry created by flow intermittence must be considered when designing water quality monitoring programs and be incorporated into IRES management plans. Similarly, it must be recognized that regulation of IRES physicochemistry is expected to alternate temporally between mostly in-stream (interacting local biotic and abiotic conditions) and external drivers (those operating at watershed scale) as the longitudinal connectivity (Chapter 2.3) decreases or increases, respectively.

Flow cessation in IRES often elevates solute concentrations, increasing vulnerability to direct and indirect human impacts. Flooding of previously dry channels typically leads to an export of solute-rich water downstream. However, sediment condition in IRES during drying will influence the water quality of the wetted front following streambed inundation. Because changes in the physicochemistry are particularly important in defining ecological conditions in IRES, water quality standards in ecosystem management should be adapted to the hydrological uniqueness of IRES.

REFERENCES

Acuña, V., Muñoz, I., Giorgi, A., Omella, M., Sabater, F., Sabater, S., 2005. Drought and post drought recovery cycles in an intermittent Mediterranean stream: structural and functional aspects. J. N. Am. Benthol. Soc. 24, 919–933.

Al-Qudah, O.M., Walton, J.C., Woocay, A., 2015. Chemistry and evolution of desert ephemeral stream runoff. J. Arid Environ. 122, 169–179.

Arce, M.I, Sánchez-Montoya, M.M., Gómez, R., 2015. Nitrogen processing following experimental sediment rewetting in isolated pools in an agricultural stream of a semiarid region. Ecol. Eng. 77, 233–241.

Arce, M.I., Sánchez-Montoya, M.M., Vidal-Abarca, M.R., Suárez, M.L., Gómez, R., 2014. Implications of flow intermittency on sediment nitrogen availability and processing rates in a Mediterranean headwater stream. Aquat. Sci. 76, 173–186.

Attygalla, N.W., Baldwin, D.S., Silvester, E., Kappen, P., Whitworth, K.L., 2016. The severity of sediment desiccation affects the adsorption characteristics and speciation of phosphorus. Evnviron. Sci. Process. Impacts 18, 64–71.

Baker, M.A., Valett, H.M., Dahm, C.N., 2000. Organic carbon supply and metabolism in a shallow groundwater ecosystem. Ecology 81, 3133–3148.

Baldwin, D.S., Fraser, M., 2009. Rehabilitation options for inland waterways impacted by sulfidic sediments—a synthesis. J. Environ. Manag. 91, 311–319.

Baldwin, D.S., Mitchell, A.M., 2000. The effects of drying and re-flooding on the sediment and soil nutrient dynamics of lowland river-floodplain systems: a synthesis. River Res. Appl. 16, 457–467.

Baldwin, D.S., Rees, G.N., Mitchell, A.M., Watson, G., 2005. Spatial and temporal variability of nitrogen dynamics in an upland stream before and after drought. Mar. Freshw. Res. 56, 457–464.

Baldwin, D.S., Rees, G.N., Wilson, J.S., Colloff, M.J., Whitworth, K.L., Pitman, T.L., et al., 2013. Provisioning of bioavailable carbon between the wet and dry phases in a semi-arid floodplain. Oecologia 172, 539–550.

Baldwin, D.S., Wallace, T.A., 2009. Biogeochemistry. In: Overton, I.C., Colloff, M.J., Doody, T.M., Henderson, B., Cuddy, S.M. (Eds.), Ecological Outcomes of Flow Regimes in the Murray-Darling Basin. CSIRO, Canberra, pp. 47–57.

Baldwin, D.S., Williams, J., 2007. Differential release of nitrogen and phosphorus from anoxic sediments. Chem. Ecol. 23, 243–249.

Bernal, S., Butturini, A., Sabater, F., 2006. Inferring nitrate sources through end member mixing analysis in an intermittent Mediterranean stream. Biogeochemistry 81, 269–289.

Bernal, S., von Schiller, D., Sabater, F., Martí, E., 2013. Hydrological extremes modulate nutrient dynamics in mediterranean climate streams across different spatial scales. Hydrobiologia 719, 21–42.

Boulton, A.J., Findlay, S., Marmonier, P., Stanley, E.H., Valett, M., 1998. The functional significance of the hyporheic zone in streams and rivers. Annu. Rev. Ecol. Syst. 29, 59–81.

Braig, E.C., Johnson, D.L., 2003. Impact of black bullhead (*Ameiurus melas*) on turbidity in a diked wetland. Hydrobiologia 490, 11–21.

Brooks, B.W., Riley, T.M., Taylor, R.D., 2006. Water quality of effluent-dominated ecosystems: ecotoxicological, hydrological, and management considerations. Hydrobiologia 556, 365–379.

Brown, L.E., Hannah, D.M., 2008. Spatial heterogeneity of water temperature across an alpine river basin. Hydrol. Process. 22, 954–967.

Bunn, S.E., Thoms, M.C., Hamilton, S.K., Capon, S.J., 2006. Flow variability in dryland rivers: boom, bust and the bits in between. River Res. Appl. 22, 179–186.

Butturini, A., Bernal, S., Nin, E., Hellin, C., Rivero, L., Sabater, F., et al., 2003. Influences of the stream groundwater hydrology on nitrate concentration in unsaturated riparian area bounded by an intermittent Mediterranean stream. Water Resour. Res. 39, 1–13.

Butturini, A., Alvarez, M., Bernal, S., Vazquez, E., Sabater, F., 2008. Diversity and temporal sequences of forms of DOC and NO_3-discharge responses in an intermittent stream: Predictable or random succession? J. Geophys. Res. Biogeosci. 113, G03016.

Claret, C., Marmonier, P., Boissier, J., Fontville, D., Blanc, P., 1997. Nutrient transfer between parafluvial interstitial water and river water. Influence of gravel bar heterogeneity. Freshw. Biol. 37, 657–670.

Corti, R., Datry, T., 2012. Invertebrates and sestonic matter in an advancing wetted front travelling down a dry river bed (Albarine, France). Freshwater Sci. 31, 1187–1201.

Costelloe, J.F., Western, A.W., Irvine, E.C., 2007. Development of hypersaline groundwater in alluvial aquifers of ephemeral rivers. In: Oxley, L., Kulasiri, D. (Eds.), MODSIM 2007: International Congress on Modelling and Simulation. Modelling and Simulation Society of Australia and New Zealand, Christchurch, pp. 1416–1422.

Dahm, C.N., Molles Jr., M.C., 1992. Streams in semiarid regions as sensitive indicators of global climate change. In: Firth, P., Fisher, S.G. (Eds.), Global Warming and Freshwater Ecosystems. Springer, New York, pp. 250–260.

Dahm, C.N., Grimm, N.B., Marmonier, P., Valett, H.M., Vervier, P., 1998. Nutrient dynamics at the interface between surface waters and groundwaters. Freshw. Biol. 40, 427–451.

Dahm, C.N., Baker, M.A., Moore, D.I., Thibault, J.R., 2003. Coupled biogeochemical and hydrological responses of streams and rivers to drought. Freshw. Biol. 48, 1219–1231.

Dahm, C.N., Candelaria-Ley, R.I., Reale, C.S., Reale, J.K., Van Horn, D.J., 2015. Extreme water quality degradation following a catastrophic forest fire. Freshw. Biol. 60, 2584–2599.

Datry, T., Larned, S.T., Tockner, K., 2014. Intermittent rivers: A challenge for freshwater ecology. Bioscience 64, 229–235.

Dent, C.L., Grimm, N.B., 1999. Spatial heterogeneity of stream water nutrient concentrations over successional time. Ecology 80, 2283–2298.

Dent, C.L., Grimm, N.B., Fisher, S.G., 2001. Multiscale effects of surface–subsurface exchange on stream water nutrient concentrations. J. N. Am. Benthol. Soc. 20, 162–181.

Elsenbeer, H., West, A., Bonell, M., 1994. Hydrological pathways and stormflow hydrochemistry at South Creek, northeast Queensland. J. Hydrol. 162, 1–21.

Ebersole, J.L., Wigington, P.J., Leibowitz, S.G., Comeleo, R.L., Van Sickle, J., 2015. Predicting the occurrence of cold-water patches at intermittent and ephemeral tributary confluences with warm rivers. Freshwater Sci. 34, 111–124.

Fellman, J.B., Dogramaci, S., Skrzypek, G., Dodson, W., Grierson, P.F., 2011. Hydrologic control of dissolved organic matter biogeochemistry in pools of a subtropical dryland river. Water Resour. Res. 47, W06501.

Fisher, S.G., Gray, L.J., Grimm, N.B., Busch, D.E., 1982. Temporal succession in a desert stream ecosystem following flash flooding. Ecol. Monogr. 52, 93–110.

Fisher, S.G., Grimm, N.B., Martí, E., Gómez, R., 1998. Hierarchy, spatial configuration, and nutrient cycling in a desert stream. J. N. Am. Benthol. Soc. 23, 41–52.

Fletcher, A.R., Morison, A.K., Hume, D.J., 1985. Effects of carp, *Cyprinus carpio* L., on communities of aquatic vegetation and turbidity of waterbodies in the lower Goulburn River basin. Aust. J. Mar. Freshwat. Res. 36, 311–327.

García-García, V., 2005. Análisis de la variabilidad espacio-temporal del nitrógeno en un río temporal mediterráneo. Masters Thesis, University of Murcia, Spain.

Gasith, A., Resh, V.H., 1999. Streams in Mediterranean climate regions: abiotic influences and biotic responses to predictable seasonal events. Annu. Rev. Ecol. Evol. Syst. 30, 51–81.

Gómez, R., 1995. Función de los humedales en la dinámica de nutrientes (N y P) de una cuenca de características áridas: experiencias en el sureste ibérico. Doctoral Thesis, University of Murcia, Spain.

Gómez, R., Arce, M.I., Sánchez, J., Sánchez-Montoya, M.M., 2012. The effects of drying on sediment nitrogen content in a Mediterranean intermittent stream: a microcosms study. Hydrobiologia 679, 43–59.

Gómez, R., García, V., Vidal-Abarca, R., Suárez, L., 2009. Effect of intermittency on N spatial variability in an arid Mediterranean stream. J. N. Am. Benthol. Soc. 28, 572–583.

Gómez, R., Vidal-Abarca, M.R., Suárez, M.L., 1994. Bioavailability of phosphorus in Ajauque Stream wetland (SE Spain). Verh. Int. Ver. Limnol. 25, 1357–1360.

Gomi, T., Sidle, R.C., Richardson, J.S., 2002. Understanding processes and downstream linkages of headwater systems. Bioscience 52, 905–916.

Gooseff, M.N., McKnight, D.M., Lyons, W.B., Blum, A.E., 2002. Weathering reactions and hyporheic exchange controls on stream water chemistry in a glacial meltwater stream in the McMurdo Dry Valleys. Water Resour. Res. 38, 1–17.

Grimm, N.B., Fisher, S.G., 1984. Exchange between interstitial and surface water: implications for stream metabolism and nutrient cycling. Hydrobiologia 111, 219–228.

Grimm, N.B., Fisher, S.G., Minckley, W.L., 1981. Nitrogen and phosphorus dynamics in hot desert streams of Southwestern USA. Hydrobiologia 83, 303–312.

Hall, K.C., Baldwin, D.S., Rees, G.N., Richardson, A.J., 2006. Distribution of inland wetlands with sulfidic sediments in the Murray-Darling Basin, Australia. Sci. Total Environ. 370, 235–244.

Hamilton, S.K., Bunn, S.E., Thoms, M.C., Marshall, J.C., 2005. Persistence of aquatic refugia between flow pulses in a dryland river system (Cooper Creek, Australia). Limnol. Oceanogr. 50, 743–754.

Harms, T.K., Grimm, N.B., 2008. Hot spots and hot moments of carbon and nitrogen dynamics in a semiarid riparian zone. J. Geophys. Res. 113, 1–14.

Harvey, J.W., Fuller, C.C., 1998. Effect of enhanced manganese oxidation in the hyporheic zone on basin-scale geochemical mass balance. Water Resour. Res. 34, 623–636.

Heffernan, J.B., Sponseller, R.A., 2004. Nutrient mobilization and processing in Sonoran desert riparian soils following artificial re-wetting. Biogeochemistry 70, 117–134.

Henry, J.C., Fisher, S.G., 2003. Spatial segregation on periphyton communities in a desert stream: causes and consequences for N cycling. J. N. Am. Benthol. Soc. 22, 511–527.

Herczeg, A.L., Dogramaci, S.S., Leaney, F.W.J., 2001. Origin of dissolved salts in a large, semi-arid groundwater system: Murray Basin, Australia. Mar. Freshw. Res. 52, 41–52.

Hill, B.H., Gardner, T.J., 1987. Benthic metabolism in a perennial and an intermittent Texas prairie stream. Southwest. Nat. 32, 305–311.

Hladyz, S., Watkins, S.C., Whitworth, K.L., Baldwin, D.S., 2011. Flows and hypoxic blackwater events in managed ephemeral river channels. J. Hydrol. 401, 117–125.

Holmes, R.M., Fisher, S.G., Grimm, N.B., 1994. Parafluvial nitrogen dynamics in a desert stream ecosystem. J. N. Am. Benthol. Soc. 13, 468–478.

Holmes, R.M., Jones, J.B., Fisher, S.G., Grimm, N.B., 1996. Denitrification in a nitrogen-limited stream ecosystem. Biogeochemistry 33, 125–146.

Holmes, R.M., Fisher, S.G., Grimm, N.B., Harper, B.J., 1998. The impact of flash floods on microbial distribution and biogeochemistry in the parafluvial zone of a desert stream. Freshw. Biol. 40, 641–654.

Holloway, J.M., Dahlgren, R.A., 2001. Seasonal and event-scale variations in solute chemistry for four Sierra Nevada catchments. J. Hydrol. 250, 106–121.

Hrachowitz, M., Soulsby, C., Imholt, C., Malcolm, I.A., Tetzlaff, D., 2010. Thermal regimes in a large upland salmon river: A simple model to identify the influence of landscape controls and climate change on maximum temperatures. Hydrol. Process. 24, 3374–3391.

Hynes, H.B.N., 1975. The stream and its valley. Verh. Int. Ver. Limnol. 19, 1–15.

Hynes, H.B.N., 1983. Groundwater and stream ecology. Hydrobiologia 100, 93–99.

Jacobson, P.J., Jacobson, K.M., Angermeier, P.L., Cherry, D.S., 2000. Variation in material transport and water chemistry along a large ephemeral river in the Namib Desert. Freshw. Biol. 44, 481–491.

Jones, J.B., Fisher, S.G., Grimm, N.B., 1995. Vertical hydrological exchange and ecosystem metabolism in a Sonoran desert stream. Ecology 76, 942–952.

Kelly, A.M., Kohler, C.C., 1996. Climate site and pond design. In: Egna, H.S., Boyd, C.E. (Eds.), Dynamics of Pond Aquaculture. CRC Press, Boca Raton, FL, pp. 109–134.

King, A.J., Robertson, A.I., Healey, M.R., 1997. Experimental manipulations of the biomass of introduced carp (*Cyprinus carpio*) in billabongs. 1. Impacts on water-column properties. Mar. Freshw. Res. 48, 435–443.

Kreiling, R.M., Yin, Y., Gerber, D.T., 2007. Abiotic influences on the biomass of *Vallisneria americana* Michx. in the upper Mississippi River. River Res. Appl. 23, 343–349.

Lado-Monserrat, L., Lull, C., Bautista, I., Lidon, A., Herrera, R., 2014. Soil moisture increment as a controlling variable of the "Birch effect". Interactions with the pre-wetting soil moisture and litter addition. Plant Soil 379, 21–34.

Lamers, L.P.M., Ten Dolle, G.E., Van den Berg, S.T.G., Van Delft, S.P.J., Roelofs, J.G.M., 2001. Differential responses of freshwater wetland soils to sulphate pollution. Biogeochemistry 55, 87–102.

Lamontagne, S., Hicks, W.S., Fitzpatrick, R.W., Rogers, S., 2006. Sulfidic materials in dryland river wetlands. Mar. Freshw. Res. 57, 775–788.

Lewis, D.B., Schade, J.D., Huth, A.K., Grimm, N.B., 2006. The spatial structure of variability in a semi-arid, fluvial ecosystems. Ecosystems 9, 386–397.

Lillebø, A.I., Morais, M., Guilherme, P., Fonseca, R., Serafim, A., Neves, R., 2007. Nutrient dynamics in Mediterranean temporary streams: a case study in Pardiela catchment (Degebe River, Portugal). Limnol.-Ecol. Manag. Inland Waters 37, 337–348.

McClain, M.E., Boyer, E.W., Dent, C.L., Gergel, S.E., Grimm, N.B., Groffman, P.M., et al., 2003. Biogeochemical hot spots and hot moments at the interface of terrestrial and aquatic ecosystems. Ecosystems 6, 301–312.

McKnight, D.M., Runkel, R.L., Tate, C.M., Duff, J.H., Moorhead, D.L., 2004. Inorganic N and P dynamics of Antarctic glacial meltwater streams as controlled by hyporheic exchange and benthic autotrophic communities. J. N. Am. Benthol. Soc. 23, 171–188.

Merbt, S.N., Proia, L., Prosser, J.I., Martí, E., Casamayor, E.O., von Schiller, D., 2016. Stream drying drives microbial ammonia oxidation and first-flush nitrate export. Ecology 97, 2192–2198.

Mikha, M.M., Rice, C.W., Milliken, G.A., 2005. Carbon and nitrogen mineralization as affected by drying and wetting cycles. Soil Biol. Biochem. 37, 339–347.

Millán, A., Velasco, J., Gutiérrez-Cánovas, C., Arribas, P., Picazo, F., Sánchez-Fernandez, D., et al., 2011. Mediterranean saline streams in southeast Spain: What do we know? J. Arid Environ. 75, 1352–1359.

Molles Jr., M.C., Dahm, C.N., 1990. A perspective on El Niño and La Niña: global implications for stream ecology. J. N. Am. Benthol. Soc. 9, 68–76.

Molles Jr., M.C., Dahm, C.N., Crocker, M.T., 1992. Climatic variability and streams and rivers in semi-arid regions. In: Robarts, R.D., Bothwell, M.L. (Eds.), Aquatic Ecosystems in Semi-arid Regions: Implications for Resource Management. N.H.R.I. Symposium Series, 7. Environment Canada, Canada, pp. 197–202.

Molles Jr., M.C., Crawford, C.S., Ellis, L.M., Valett, H.M., Dahm, C.N., 1998. Managed floods: restoration of riparian forest ecosystem structure and function along the Middle Rio Grande. Bioscience 48, 749–756.

Moraetis, D., Efstathiou, D., Stamati, F., Tzoraki, O., Nikolaidis, N.P., Schnoor, J.L., et al., 2010. High-frequency monitoring for the identification of hydrological and bio-geochemical processes in a Mediterranean river basin. J. Hydrol. 389, 127–136.

Mulder, T., Alexander, J., 2001. The physical character of subaqueous sedimentary density flows and their deposits. Sedimentology 48, 269–299.

Mulholland, P.J., 1993. Hydrometric and stream chemistry evidence of three storm flowpaths in Walker Branch Watershed. J. Hydrol. 151, 291–316.

Nielsen, J.L., Lisle, T.E., Ozaki, V., 1994. Thermally stratified pools and their use by steelhead in northern California streams. Trans. Am. Fish. Soc. 123, 613–626.

Obermann, M., Rosenwinkel, K.H., Tournoud, M.G., 2009. Investigation of first flushes in a medium-sized Mediterranean catchment. J. Hydrol. 373, 405–415.

Peterjohn, W.T., Correll, D.L., 1984. Nutrient dynamics in an agricultural watershed: observations on the role of a riparian forest. Ecology 65, 1466–1475.

Polis, G.A., Anderson, W.B., Holt, R.D., 1997. Toward an integration of landscape and food web ecology: the dynamics of spatially subsidized food webs. Annu. Rev. Ecol. Syst. 28, 289–316.

Reale, J.K., Van Horn, D.J., Condon, K.E., Dahm, C.N., 2015. The effects of catastrophic wildfire on water quality along a river continuum. Freshwater Sci. 34, 1426–1442.

Robinson, C.T., Tonolla, D., Imhof, B., Vukelic, R., Uehlinger, U., 2016. Flow intermittency, physico-chemistry and function of headwater streams in an Alpine glacial catchment. Aquat. Sci. 78, 327–341.

Rutherford, J.C., Marsh, N.A., Davies, P.M., Bunn, S.E., 2004. Effects of patchy shade on stream water temperature: how quickly do small streams heat and cool? Mar. Freshw. Res. 55, 737–748.

Schade, J., Fisher, S.G., Grimm, N.B., Seddon, J.A., 2001. The influence of a riparian shrub on nitrogen cycling in a Sonoran Desert stream. Ecology 82, 3363–3376.

Schade, J.D., Martí, E., Welter, J.R., Fisher, S.G., Grimm, N.B., 2002. Sources of nitrogen to the riparian zone of a desert stream: implications for riparian vegetation and nitrogen retention. Ecosystems 5, 68–79.

Sherson, L.R., Van Horn, D.J., Gomez-Velez, J.D., Crossey, L.J., Dahm, C.N., 2015. Nutrient dynamics in an alpine headwater stream: use of continuous water quality sensors to examine responses to wildfire and precipitation events. Hydrol. Process. 29, 3193–3207.

Skoulikidis, N.T., Amaxidis, Y., 2009. Origin and dynamics of dissolved and particulate nutrients in a minimally disturbed Mediterranean river with intermittent flow. J. Hydrol. 373, 218–229.

Stanford, J.A., Ward, J.V., 1993. An ecosystems perspective of alluvial rivers: connectivity and the hyporheic corridor. J. N. Am. Benthol. Soc. 12, 48–60.

Stanley, E.H., Boulton, A.J., 1995. Hyporheic processes during flooding and drying in a Sonoran Desert stream. I: Hydrologic and chemical dynamics. Arch. Hydrobiol. 134, 1–26.

Stanley, E.H., Fisher, S.G., Grimm, N.B., 1997. Ecosystem expansion and contraction in streams. Bioscience 47, 427–435.

Steward, A.L., von Schiller, D., Tockner, K., Marshall, J.C., Bunn, S.E., 2012. When the river runs dry: Human and ecological values of dry riverbeds. Front. Ecol. Environ. 10, 202–209.

Tockner, K., Pennetzdofer, D., Reiner, N., Schiemer, F., Ward, J.V., 1999. Hydrological connectivity and the exchange of organic matter and nutrients in a dynamic river-floodplain system (Danube, Austria). Freshw. Biol. 41, 521–535.

Townsend, S.A., Webster, I.T., Schult, J.H., 2011. Metabolism in a groundwater-fed river system in the Australian wet/dry tropics: Tight coupling of photosynthesis and respiration. J. N. Am. Benthol. Soc. 30, 603–620.

Triska, F.J., Kennedy, V.C., Avanzino, R.J., Zellweger, G.W., Bencala, K.E., 1989. Retention and transport of nutrients in a third-order stream in northwestern California: hyporheic processes. Ecology 70, 1893–1905.

Triska, F.J., Duff, J.H., Avanzino, R.J., 1993. Patterns of hydrological exchange and nutrient transformation in the hyporheic zone of a gravel-bottom stream: examining terrestrial-aquatic linkages. Freshw. Biol. 29, 259–274.

Turner, L., Erskine, W.D., 2005. Variability in the development, persistence and breakdown of thermal, oxygen and salt stratification on regulated rivers of southeastern Australia River. Res. Appl. 21, 151–168.

Valett, H.M., Fisher, S.G., Stanley, E.H., 1990. Physical and chemical characteristics of the hyporheic zone of a Sonoran Desert stream. J. N. Am. Benthol. Soc. 9, 201–215.

Valett, H.M., Baker, M.A., Morrice, J.A., Crawford, C.S., Molles Jr., M.C., Dahm, C.N., et al., 2005. Biogeochemical and metabolic responses to the flood pulse in a semiarid floodplain. Ecology 86, 220–234.

Vannote, R.L., Minshall, G.W., Cummins, K.W., Sedell, J.R., Cushing, E., 1980. The river continuum concept. Can. J. Fish. Aquat. Sci. 37, 130–137.

Vázquez, E., Amalfitano, S., Fazi, S., Butturini, A., 2011. Dissolved organic matter composition in a fragmented Mediterranean fluvial system under severe drought conditions. Biogeochemistry 102, 59–72.

Vázquez, E., Romaní, A.M., Sabater, F., Butturini, A., 2007. Effects of the dry-wet hydrological shift on dissolved organic carbon dynamics and fate across stream-riparian interface in a Mediterranean catchment. Ecosystems 10, 239–251.

Vervier, P., Dobson, M., Pinay, G., 1993. Role of interaction zones between surface and groundwaters in DOC transport and processing: consideration for river restoration. Freshw. Biol. 29, 275–284.

Vidal-Abarca, M.R., Gómez, R., Suárez, M.L., 2004. Los ríos de las regiones semiáridas. Rev. Ecosistemas 13, 16–28.

Vidal-Abarca, R., Suárez, M.L., Gómez, R., Moreno, J.L., Guerrero, C., 2002. Diel variations in physical and chemical parameters in a semi-arid stream in Spain (Chicamo Stream). Verh. Int. Ver. Limnol. 28, 1111–1115.

Von Schiller, D., Acuña, V., Graeber, D., Martí, E., Ribot, M., Sabater, S., et al., 2011. Contraction, fragmentation and expansion dynamics determine nutrient availability in a Mediterranean forest stream. Aquat. Sci. 73, 485–497.

Wetzel, R.G., 2001. Limnology: Lake and River Ecosystems, third ed. Academic, San Diego, CA.

White, D.S., Elzinga, C.H., Hendricks, S.O., 1987. Temperature patterns within the hyporheic zone on a northern Michigan river. J. N. Am. Benthol. Soc. 6, 85–91.

Whitworth, K.L., Baldwin, D.S., Kerr, J.L., 2012. Drought, floods and water quality: drivers of a severe hypoxic blackwater event in a major river system (the southern Murray–Darling Basin, Australia). J. Hydrol. 450, 190–198.

Williams, D.D., 2006. The Biology of Temporary Waters. Oxford University Press, Oxford.

Zale, A.V., Leslie Jr., D.M., Fisher, W.L., Merrifield, S.G., 1989. The physicochemistry, flora, and fauna of intermittent prairie streams: a review of the literature: Biological Report 89. U.S. Department of the Interior, Fish and Wildlife Service, Washington, DC.

Zimmer, M.A., Bailey, S.W., McGuire, K.J., Bullen, T.D., 2013. Fine scale variations of surface water chemistry in an ephemeral to perennial drainage network. Hydrol. Process. 27, 3438–3451.

FURTHER READING

Acuña, V., Giorgi, A., Muñoz, I., Uehlinger, U.R.S., Sabater, S., 2004. Flow extremes and benthic organic matter shape the metabolism of a headwater Mediterranean stream. Freshw. Biol. 49, 960–971.

Kerr, J.L., Baldwin, D.S., Whitworth, K.L., 2013. Options for managing hypoxic blackwater events in river systems: a review. J. Environ. Manag. 114, 139–147.

Warren, C.R., 2014. Response of osmolytes in soil to drying and rewetting. Soil Biol. Biochem. 70, 22–32.

Whiteside, B.G., McNatt, M., 1972. Fish species diversity in relation to stream order and physicochemical conditions in the Plum Creek drainage basin. Am. Midl. Nat. 88, 90–101.

NUTRIENT AND ORGANIC MATTER DYNAMICS IN INTERMITTENT RIVERS AND EPHEMERAL STREAMS

3.2

Daniel von Schiller*, Susana Bernal[†], Clifford N. Dahm[‡], Eugènia Martí[†]

University of the Basque Country, Bilbao, Spain* Center for Advanced Studies of Blanes (CEAB-CSIC), Blanes, Spain[†] University of New Mexico, Albuquerque, NM, United States[‡]

IN A NUTSHELL

- Flow regimes in intermittent rivers and ephemeral streams (IRES) confer a unique 'biogeochemical heartbeat' with high temporal and spatial variation in nutrient and organic matter dynamics.
- Intermittence and the dry, wet, and transitional phases strongly influence nutrient and organic matter inputs, in-stream processing, and downstream transport.
- A comprehensive understanding of the biogeochemistry of IRES must consider the coupling between stream surface and subsurface subsystems as well as stream-catchment linkages.
- Future biogeochemical research in IRES should expand the geographical coverage, seek a more mechanistic understanding of processes, investigate interactions with human stressors, and develop additional modeling and upscaling tools.

3.2.1 THE 'BIOGEOCHEMICAL HEARTBEAT' OF IRES

Streams and rivers transport a myriad of compounds in dissolved and particulate forms (Meybeck, 1982; Howarth et al., 1996). Among these, nutrients and organic matter compounds are biologically important. Key nutrients such as nitrogen (N) and phosphorus (P) can limit productivity in stream ecosystems (Elser et al., 2007), although in oversupply they can lead to significant water quality issues (Conley et al., 2009). Organic matter (OM), the biological material in the process of decaying or decomposing, is a key source of carbon and energy in stream ecosystems (Tank et al., 2010). Nutrient and OM dynamics in streams refer to the input, removal, transformation, production, and export of these compounds into, within, and out of the ecosystem (Mulholland and Webster, 2010; Tank et al., 2010). Among other ecological roles, nutrients and OM constitute the basis of stream food webs; therefore, the study of their dynamics is a fundamental step toward a better understanding of ecosystem structure and functioning.

The dynamics of nutrients and OM have been widely investigated in perennial rivers and streams (Allan and Castillo, 2007; Mulholland and Webster, 2010; Tank et al., 2010) but this knowledge is more constrained for intermittent rivers and ephemeral streams (IRES) (Steward et al., 2012; Bernal et al., 2013; Datry et al., 2014). In many aspects, the controlling factors and mechanisms driving dynamics of nutrients and OM are similar between IRES and their perennial counterparts. However, the particular hydrological regimes of IRES, which include intermittence (Chapter 2.2); the alternation of wet and dry phases; and the highly variable lateral, vertical, and longitudinal hydrological connections among the stream's subsystems and between the stream and the surrounding terrestrial environment (Chapter 2.3) exert a strong influence on nutrient and OM dynamics (Bunn et al., 2006; Bernal et al., 2013). The hydrological phases that characterize IRES are those that precede and follow stream drying (Fig. 3.2.1). In general, flow gradually decreases (contraction), followed by flow cessation and the formation of isolated pools (fragmentation), before the surface stream completely dries up (dry or desiccation phase). After the dry phase, flow resumes (expansion) with the decrease in evapotranspiration and the occurrence of rainfall (e.g., temperate, arid, and semiarid IRES) or inputs of meltwater (e.g., glacial IRES). However, this hydrological pattern is not the norm in all types of IRES, where some phases may not occur or transitions between phases may follow a different chronology (Larned et al., 2010).

As a result of their highly dynamic and extreme hydrology, IRES show a unique 'biogeochemical heartbeat' with pulsed temporal and spatial variations in nutrient and OM inputs, in-stream processing, and downstream transport. Accordingly, IRES can be viewed as tightly coupled aquatic-terrestrial ecosystems that function as 'punctuated biogeochemical reactors' that store, process, and transport nutrients and OM in response to spatiotemporal fluctuations in drying and rewetting (Larned et al., 2010; Jacobson and Jacobson, 2013; Datry et al., 2014). Although these patterns may differ considerably among different types of IRES, changes in the sources and in-stream processes during dry, wet, and transitional phases need to be considered to fully understand nutrient and OM dynamics in these ecosystems.

Moreover, the extent and direction of hydrological interactions between the surface stream and hyporheic, parafluvial, and riparian zones are highly dynamic over time in IRES due to the extreme range in hydrological conditions in these ecosystems (Chapter 2.3). Therefore, a comprehensive understanding of IRES biogeochemistry should go beyond the sole consideration of water chemistry in the surface stream channel by considering the interactions with subsurface and lateral compartments, and the linkages between terrestrial ecosystems and IRES networks at the catchment scale (Jones and Mulholland, 2000). One implication of this view of IRES is that large-scale estimates of nutrient and OM fluxes in river networks may be biased if these ecosystems are not properly considered (Acuña and Tockner, 2010; Datry et al., 2014; von Schiller et al., 2014). Furthermore, strategies of effective river monitoring and conservation are challenged if the unique biogeochemical characteristics of IRES are not integrated into management programs, especially considering the expected increase in the geographical extent of IRES in many regions as a consequence of global change (Leigh et al., 2016).

This chapter reviews existing knowledge on the dynamics of N, P, and OM (both dissolved and particulate forms) in IRES under the different hydrological phases that precede and follow stream drying (i.e., contraction, fragmentation, desiccation, and expansion). We use an ecosystem perspective that accounts for in-stream processes, interactions with subsurface and lateral compartments (i.e., hyporheic, parafluvial, and riparian zones), and linkages between terrestrial ecosystems and IRES networks at the catchment scale. Our focus is on research from flowing waters, although studies from other ecosystems such as wetlands, lakes, and reservoirs are also used to illustrate some particular aspects. We conclude with a discussion of knowledge gaps and research needs.

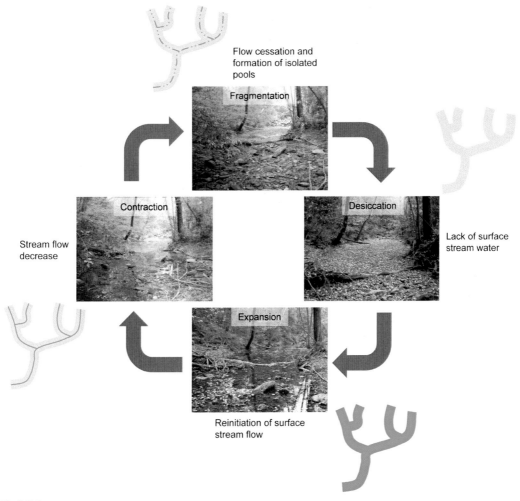

Flow cessation and formation of isolated pools

Fragmentation

Contraction

Desiccation

Stream flow decrease

Lack of surface stream water

Expansion

Reinitiation of surface stream flow

FIG. 3.2.1

Typical hydrological phases that drive the high temporal and spatial variation in nutrient and OM dynamics characteristic of IRES. Representative pictures from a reach in a Mediterranean intermittent stream (Catalonia, Spain) are shown with idealized schemes of surface water presence (blue lines) along a river section during four hydrological phases.

Photos: Courtesy D. von Schiller.

3.2.2 NUTRIENT AND OM DYNAMICS ACROSS HYDROLOGICAL PHASES IN IRES

CONTRACTION

During the contraction phase, the flow and velocity of surface water decrease and the stream becomes hydrologically disconnected from the adjacent catchment (Stanley et al., 1997; Costigan et al., 2015).

Low water availability reduces groundwater inputs from the adjacent catchment into the stream, and discharge in the surface stream channel tends to decrease along reaches which become hydrologically losing. The decrease in surface water discharge is primarily explained by the increase in the extent of downwelling sites along the reaches (Dahm et al., 2003). Streams tend to lose water toward the aquifer, thereby enhancing hydrological retention and water residence time (Valett et al., 1996; Harvey et al., 2003).

Hydrological disconnection occurs laterally and longitudinally but also vertically. Consequently, relevance of subsurface flow increases relative to surface flow (Stanley and Boulton, 1995; Dahm et al., 1998; Acuña et al., 2004). These hydrological conditions favor water movement from stream channels into riparian zones (Marti et al., 2000; Butturini et al., 2003), which is further enhanced by high riparian tree evapotranspiration during summer (Lupon et al., 2016). In high-latitude and alpine IRES, hydrological contraction reaches its maximum in winter when glaciers freeze (McKnight et al., 1999; Robinson et al., 2015). If streams are fed exclusively by snowmelt, the hydrological connectivity between terrestrial and aquatic ecosystems stops with glacier freezing (Robinson and Matthaei, 2007; Robinson et al., 2015). However, if streams are fed by groundwater or lakes, contraction slows down and the transfer of solutes from uplands to streams persists, as observed in the Roseg River in the Swiss Alps (Malard et al., 2000).

Hydrological contraction exerts a strong influence on stream nutrient and OM dynamics in IRES, increasing the overall importance of in-stream nutrient and OM sources relative to terrestrial sources (Bernal et al., 2013). Hydrological contraction reduces downstream export (McKnight et al., 1999; Jacobson et al., 2000; Dahm et al., 2003). In addition, due to the losing nature of the streams (Chapter 2.3), biogeochemistry in the surface water can influence nutrient and OM processing, and finally their dynamics in hyporheic, parafluvial, and riparian zones (Jones and Mulholland, 2000; Dahm et al., 2003). Ultimately, these surface-to-subsurface hydrological interactions increase the water residence time along stream reaches, thus increasing the potential for nutrient and OM retention and transformation through biogeochemical processes (Harvey et al., 2003; Lautz and Siegel, 2007).

In this sense, the spatial configuration and sizes of hyporheic, parafluvial, and riparian zones relative to the stream channel can also affect the amount and form of transported nutrients in the stream because they dictate the rate of surface–subsurface interactions and the length of subsurface flowpaths (Fisher et al., 1998; Dent et al., 2001). Water residence time along the subsurface flowpaths is also an important factor determining the spatial heterogeneity of nutrients and OM in hyporheic and parafluvial zones (Dent et al., 2007; Lewis et al., 2007). The consequences of this subsurface biogeochemical processing capacity during hydrological contraction can resonate at the catchment scale. For instance, Bernal and Sabater (2012) found that nitrate export in a Mediterranean stream decreased by 90% along a 3-km losing reach, attributed to biological uptake at the stream–riparian interface. Schade et al. (2005) also reported substantial declines in inorganic N concentration in a desert stream, which were related to biological uptake by riparian trees during periods of hydrological contraction.

Over the contraction phase, concentrations of some nutrients in the surface water can either decrease or increase, while others may show no significant changes. A decrease in nitrate concentration due to increased relevance of the denitrification pathway is a common pattern, while other N forms (e.g., ammonium and dissolved organic N) tend to remain stable (Stanley and Boulton, 1995; Dahm et al., 1998; von Schiller et al., 2011). For P, results are more contradictory. Some studies show decreases in dissolved P concentration over time (Finn et al., 2009) whereas others show increases (von Schiller et al., 2011) during hydrological contraction.

In subsurface water, the inputs of dissolved oxygen and OM from surface water at downwelling locations can drive the rates of the different biogeochemical processes associated with N and P cycling along subsurface flowpaths (Baker et al., 2000; Zarnetske et al., 2011). In this sense, the spatial distribution of dissolved oxygen may explain differences in the dominant biogeochemical processes among subsurface subsystems (Lewis et al., 2007). For instance, nitrification is an important process in well-oxygenated parafluvial zones (Holmes et al., 1994; Claret et al., 1997) and at downwelling sites in hyporheic zones (Triska et al., 1990; Edwardson et al., 2003). In contrast, denitrification and methanogenesis are more relevant at the edges with hypoxic conditions between the stream channel and the riparian zone (Jones et al., 1995b). Overall, reach-scale nutrient retention efficiency is predicted to significantly increase during hydrological contraction by enhancing the contact time between water and biologically active surfaces (Fisher et al., 1998). In fact, studies that have measured nutrient spiraling in IRES generally report an increase in nutrient retention efficiency during the periods of low flow just before drying (Marti et al., 1997; von Schiller et al., 2008).

The dynamics of OM are also affected by hydrological contraction, although results are contradictory. Studies in a semiarid intermittent stream in Spain reported no changes in total DOM concentration in the surface water but decreases in the proportion of high molecular weight substances (e.g., polysaccharides) during contraction (Ylla et al., 2010; von Schiller et al., 2015). In contrast, decreases in DOM concentrations due to hydrological disconnection have been reported for other semiarid IRES (Freeman et al., 1994; Dahm et al., 2003). These differences among IRES have been hypothesized to depend on the metabolic balance of autotrophy versus heterotrophy (von Schiller et al., 2015).

Low flows, such as those occurring during hydrological contraction, increase the retention and accumulation of both fine and coarse benthic POM, such as leaves and small woody debris (Dewson et al., 2007). Decomposition of POM can be reduced in the slowly flowing stream channel due to a decrease in microbial activity and/or a decrease in the abundance of leaf litter decomposers (Rolls et al., 2012). For instance, a laboratory experiment showed how leaf litter breakdown by shredders became significantly lower as experimental drought stress increased (Leberfinger et al., 2010). Moreover, POM may accumulate on dry parts of the contracting channel, where decomposition is significantly slowed down (Langhans and Tockner, 2006; Abril et al., 2016). Nonetheless, relatively high rates of POM decomposition may be maintained in the moister subsurface zone (Solagaistua et al., 2015). Overall, these observations suggest that hydrological contraction may reduce energy inputs to streams despite increased OM availability on the streambed.

FRAGMENTATION

In most IRES, hydrological contraction ends with the longitudinal disconnection of surface water between upper and lower reaches (Malard et al., 1999; Doering et al., 2007; Fazi et al., 2013). The cessation of flow typically leads to surface fragmentation, creating a heterogeneous landscape of isolated pools within a matrix of dry streambeds. Some isolated pools may last for several weeks, months, or even years, while others may only last for a few days, depending on the presence of groundwater inputs (Fellman et al., 2011; von Schiller et al., 2011). In isolated pools, OM accumulation, high respiration rates, lack of water renewal, and weak reaeration rates commonly create hypoxic or anaerobic environments (Boulton and Lake, 1990; Acuña et al., 2005; Lillebø et al., 2007; von Schiller et al., 2011; Fig. 3.2.2). Microaerophilic or anaerobic processes dominate under these circumstances. As a result, substantial changes in the concentration and forms of nutrients and DOM occur.

FIG. 3.2.2

Organic matter accumulation in an isolated pool with hypoxic conditions during the fragmentation phase (a) and on the dry bed during the desiccation phase (b) in a Mediterranean intermittent stream (Catalonia, Spain).

Photos: Courtesy D. von Schiller.

Furthermore, isolated pools in IRES act to some extent as individual ecosystems with differing magnitudes of biogeochemical processes. Consequently, ecosystem fragmentation tends to increase the spatial heterogeneity of both nutrient and DOM concentrations along stream sections (Dent and Grimm, 1999; Gómez et al., 2009; Fellman et al., 2011; Vazquez et al., 2011; von Schiller et al., 2011). This pattern may differ in glacial streams, where increased dominance of groundwater sources and decreased diversity of water sources and subsurface flowpaths can lead to low spatial heterogeneity among isolated pools (Malard et al., 2000; Tockner et al., 2002). Overall, spatial heterogeneity along IRES networks during hydrological fragmentation is driven by several factors, including in-stream metabolism and succession (Dent and Grimm, 1999), nutrients and OM accumulation, redistribution and cycling (Jacobson et al., 2000; Welter et al., 2005), and the relative contribution of groundwater sources (Dent and Grimm, 1999, Malard et al., 2000b).

In the surface water of isolated pools, the concentration of nitrate (an oxidized form of dissolved inorganic N) decreases due to denitrification favored by the hypoxic conditions (Kemp and Dodds, 2002; von Schiller et al., 2011). Conversely, the concentration of ammonium (a reduced form of dissolved inorganic N) increases. Rapid mineralization of leachates from accumulated organic detritus is often considered the main source of ammonium (Acuña et al., 2005; von Schiller et al., 2011), and large inputs of riparian leaf litter to the stream during the drying phase, sometimes driven by water stress of riparian trees, may enhance this process (Boulton, 1991; Acuña et al., 2007). Dissimilatory nitrate reduction to ammonium has been suggested as an additional source of ammonium in isolated pools with low redox potential (Arce et al., 2015), while other processes may also contribute, such as desorption from sediments (Baldwin and Mitchell, 2000), inhibition of nitrification (Baldwin and Mitchell, 2000), photodegradation of dissolved OM (Larson et al., 2007), microbial-biofilm cell lysis (Humphries and Baldwin, 2003), and evaporation (McLaughlin, 2008). In the case of P, anaerobic conditions may favor the release of phosphate from sediments to the water column through changing pH and metal reduction (Baldwin et al., 2000). Major changes in the concentrations of N and P in the subsurface water of isolated pools have also been observed (Lillebø et al., 2007).

With regards to DOM, the few available studies have found no considerable changes in surface water DOM concentration during hydrological fragmentation (Vazquez et al., 2011; von Schiller et al., 2015). However, some changes in DOM composition have been observed. For instance, DOM becomes less aromatic with time since pool isolation, suggesting an increase in the contribution of in-stream algal and microbial processes to the total DOM pool with hydrological fragmentation (Vazquez et al., 2011; von Schiller et al., 2015). Similarly, increased concentrations of low-molecular potentially labile DOM can occur in older isolated pools (Fellman et al., 2011; von Schiller et al., 2015). Furthermore, abrupt changes in DOM quantity and composition have been reported at the end of the fragmentation phase, just before the stream completely dries up at the surface. These most likely result from abrupt microbial cell lysis and/or DOM exudation under stress conditions (e.g., high temperature, low DO, low pH) (von Schiller et al., 2015). Nonetheless, leaching and degradation of the accumulated leaf litter from riparian trees seem to represent the major DOM source in isolated pools of forested intermittent IRES (Casas-Ruiz et al., 2016).

Autochthonous DOM sources may be more relevant in IRES with less canopy cover (e.g., desert streams), where algal DOM sources may increase as a result of lower leaf inputs and higher light availability (Dahm et al., 2003; Fellman et al., 2011). Inputs of alluvial groundwater to isolated pools can partly affect the quantity and composition of DOM through dilution and moderating evapoconcentration (Siebers et al., 2016). Although POM tends to decompose slowly compared to flowing waters because of low microbial and shredder decomposing activity (Pattee et al., 2001; Schlief and Mutz, 2011; Abril et al., 2016), it can be a relevant source of DOM in isolated pools (Casas-Ruiz et al., 2016). As a result, high respiration rates and carbon dioxide emissions are expected in isolated pools. Moreover, extremely low oxygen conditions at later stages of isolation may enhance methane production. Gómez-Gener et al. (2015) reported high concentrations of dissolved carbon dioxide and methane in isolated pools along an intermittent river in Spain, indicating respiration and methanogenesis, respectively, and accumulation of dissolved gases. However, measured emission rates for both gases were relatively low, mainly because of the low turbulence limiting gas exchange and the lack of methane ebullition (Gómez-Gener et al., 2015).

DRYING (DESICCATION)

In most IRES, fragmentation typically evolves toward the surface stream drying—a phase sometimes termed 'desiccation.' At early stages of desiccation, subsurface water persists, enhancing vertical differences in biogeochemical processes (Timoner et al., 2012; Merbt et al., 2016). Subsurface water becomes disconnected from surface water and the length of subsurface flowpaths in hyporheic and parafluvial zones increases (Dahm et al., 2003). Under these conditions, nutrient and DOM dynamics in the stream are restricted to changes along subsurface flowpaths, and local characteristics of streambed sediments become important determinants of the spatial heterogeneity of subsurface chemistry (Dahm et al., 2003; Schade et al., 2005). This is illustrated in Fig. 3.2.3, which shows that spatial heterogeneity in dissolved oxygen and the proportion of nitrate and nitrite relative to total dissolved inorganic N in the subsurface water is higher when the surface stream channel is dry. In addition, these results suggest that under dry conditions, local characteristics of streambed sediments can significantly influence the spatial heterogeneity in physicochemical variables, resulting in localized 'hot spots' of low oxygen and N. These conditions can prevail in some IRES for extended periods of the year, and nutrient and OM concentrations in subsurface water can become highly variable even at small spatial scales.

In the dry stream sediments, low water availability reduces overall microbial activity through direct physiological effects, reduced diffusion of soluble substrates, and lowered microbial mobility (Humphries and Baldwin, 2003; Amalfitano et al., 2008). In parallel, abiotic processes such as photodegradation, physical disruption, or the precipitation of solutes through evaporation become more important (McLaughlin, 2008; Borken and Matzner, 2009; Dieter et al., 2011). Nonetheless, the studies that have investigated nutrient and OM dynamics during dry phases indicate that dry streambeds are not as inactive as they may seem (Tzoraki et al., 2007; McIntyre et al., 2009b; Timoner et al., 2012). Stream biofilms associated with dry sediments can process organic carbon (Zoppini and Marxsen, 2011; Timoner et al., 2012; Pohlon et al., 2013) and carbon dioxide can be released from these systems when they are dry (Gallo et al., 2013; von Schiller et al., 2014; Gómez-Gener et al., 2016).

Heterotrophic processes associated with exoenzymatic activities are more resistant to stream desiccation than autotrophic processes, especially in subsurface sediments (Zoppini and Marxsen, 2011; Timoner et al., 2012; Acuña et al., 2015). When sediments of IRES dry, they come into direct contact with the air, thus creating an oxygenated environment that favors aerobic transformation processes (Mitchell and Baldwin, 1999; Baldwin and Mitchell, 2000). Moreover, the high mortality

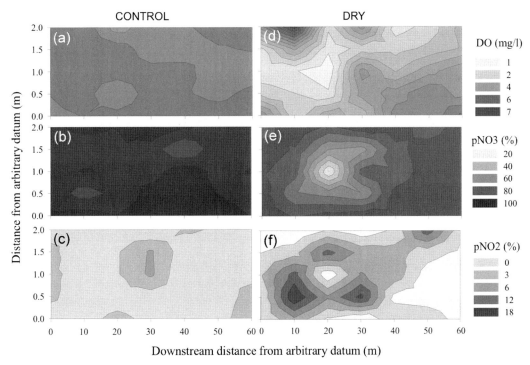

FIG. 3.2.3

Surface plots for dissolved oxygen (DO) and the proportion of nitrate and nitrite relative to total dissolved inorganic N in subsurface water at a stream-reach with surface water (control conditions, a–c) and at a reach section with no surface flow (dry conditions, d–f). Data were obtained from a regular grid of 35 piezometers (five across the stream channel at seven transects along the stream) arrayed across a 2-m wide, 60-m long reach in a Mediterranean stream (Catalonia, Spain). Piezometers were inserted 25 cm into surface sediment. All samples were collected as a snapshot in Oct. 2011. Plots show higher spatial variation in subsurface DO and relative proportions of nitrate and nitrite when surface flow was absent (d–f).

of animals, plants, and microbes during sediment drying lowers immobilization and releases large amounts of nutrients and OM (Amalfitano et al., 2008; Zoppini and Marxsen, 2011). Overall, as a consequence of drying, the spatial heterogeneity of microbial processes tends to increase within stream channels in response to the greater variability in biogeochemical conditions (Claret and Boulton, 2003).

As in soils, the nutrient content and biogeochemical processing in the streambed sediments of dry IRES are strongly affected by desiccation. Despite the relatively low microbial activity, several studies have shown that sediment drying enhances nitrification and restricts denitrification to anaerobic microsites, leading to the accumulation of inorganic N mainly in the form of nitrate (Gómez et al., 2012; Arce et al., 2015; Merbt et al., 2016). In contrast, other studies have reported a significant reduction in both processes (Austin and Strauss, 2011) or that nitrification and denitrification were not significantly affected by sediment desiccation (Mitchell and Baldwin, 1999). However, a common limitation of most of these studies (but see Merbt et al., 2016) is that they used experimentally rewetted sediments to measure N processing rates. Therefore, results from these studies may not be representative of *in situ* dry conditions, but instead indicate the potential for recovery of these processes once stream flow resumes.

The P content and its biogeochemical processing in sediments are also strongly altered by drying. Field studies in different types of aquatic ecosystems have generally shown that drying of sediments results in a decrease in P binding capacity (Twinch, 1987; Qiu and McComb, 1994; Baldwin, 1996; Watts, 2000; de Vicente et al., 2010), whereas some studies found an increase (Jacoby et al., 1982) or no clear effect (Baldwin et al., 2000). The effect of desiccation on P speciation has been less explored, and results seem to be less consistent. For instance, Kerr et al. (2010) showed a shift toward more bioavailable P compounds in river and lake sediments on drying, whereas de Vicente et al. (2010) reported a shift toward less bioavailable P in lake sediments as a consequence of desiccation. In a recent study with floodplain wetland sediments under controlled laboratory conditions, Attygalla et al. (2016) reported shifts in P speciation toward more reactive P pools, a decline in P adsorption capacity through increases in crystallinity of iron phases, an increase in the equilibrium P concentration, and a decline in microbial P with drying. These results support previous studies indicating that P sequestration and release may be at least partially mediated by bacteria (Mitchell and Baldwin, 1998). In another laboratory experiment, drying reduced P adsorption and the fraction of stable P, stimulated mineralization of organic P, and increased the proportions of labile and reductant-soluble P forms in lake sediments (Dieter et al., 2015). Overall, results from these studies reveal marked inconsistencies in the response of P dynamics to water loss; future research should explore the reasons for this inconsistent response among experiments and sites.

In IRES with well-developed riparian vegetation, the drop in the groundwater table can give rise to massive leaf abscission and accumulation of leaf litter on the stream bed during the desiccation phase (Acuña et al., 2007; Fig. 3.2.2). The decomposition of accumulated POM slows down dramatically in the dry channels of IRES (Pattee et al., 2001; Datry et al., 2011; Mariluan et al., 2015; Abril et al., 2016). Biotic decomposing activity declines due to the scarcity of macroinvertebrate decomposers and the low activity of microbial decomposers under dry conditions (Langhans and Tockner, 2006; Steward et al., 2012). Under these conditions, abiotic processes such as photodegradation become relevant for POM decomposition (Abril et al., 2016), especially in the dry beds of arid and semiarid IRES that often have an open canopy cover (Steward et al., 2012).

The few studies on microbial communities associated with decaying POM in dry IRES indicate that fungi are generally more resistant to desiccation than bacteria, probably because their hyphal development facilitates the search for water and nutrients (Langhans and Tockner, 2006; Abril et al., 2016). On the other hand, OM in the dry sediments of IRES is predicted to be subject to similar effects of desiccation as those observed in soils (Borken and Matzner, 2009). However, results from available studies show that the dry sediments of IRES can differ significantly in OM content and composition from adjacent terrestrial soils, probably as a result of differences in OM inputs, sediment texture, and hydrological transport (Jacobson and Jacobson, 2013; Gómez-Gener et al., 2016). Together, these results suggest that the variability in POM composition and the associated microbial communities (Chapter 4.1) during the desiccation phase can affect both POM decomposition and DOM release upon rewetting (Datry et al., 2011; Dieter et al., 2011; Dieter et al., 2013; Abril et al., 2016).

The "terrestrialization" of IRES channels, marked by the presence of terrestrial vegetation (Fig. 3.2.4), can also strongly influence nutrient and OM dynamics in these ecosystems. This process

FIG. 3.2.4

Dry bed colonized by terrestrial vegetation at early (a) and late (b) stages of terrestrialization during the desiccation phase (Catalonia, Spain).

Photos: Courtesy E. Martí.

is particularly relevant in arid and semiarid landscapes, where IRES channels often contain more vegetation than the surrounding terrestrial ecosystems because of the residual moisture in the stream sediments (Steward et al., 2012). Terrestrialization may substantially influence sediment processes through changes in OM availability and the redox potential and pH of sediments (Kleeberg and Heidenreich, 2004). For instance, the presence of shrubs on dry stream beds has been shown to favor denitrification along subsurface flowpaths because the root system of the shrubs provides an additional source of OM (Schade et al., 2001).

The maintenance of sediment moisture, which depends on factors such as temperature, rainfall, and evaporation, can be a key driver of biogeochemical processes occurring in the dry beds of IRES during the desiccation phase (Larned et al., 2010; Steward et al., 2012). Periodic rainfall events that rewet sediment surfaces without creating a flow front can be particularly important. A 'pulse-reserve' effect, such as that seen in desert soils following rain (Noy-Meir, 1973; Reynolds et al., 2004), is expected to occur in the dry beds of IRES; however, this topic has been widely overlooked in IRES research. Rainfall-driven rewetting of dry sediments could result in a pulse of OM mineralization and release of nutrients. This effect, also known as the 'Birch effect' (Birch, 1958), is attributed to the mineralization of previously unavailable, easily decomposable organic substrates (Wilson and Baldwin, 2008; Borken and Matzner, 2009). The frequency and intensity of rainfall events may thus regulate the magnitude of the Birch effect in IRES, with consequences for nutrient and OM dynamics at larger spatial and temporal scales (Lado-Monserrat et al., 2014).

EXPANSION

Resumption of surface flow after the dry phase in arid and semiarid IRES occurs due to decreases in evapotranspiration and the occurrence of heavy rainfall that recharges groundwater reservoirs (Stanley et al., 1997; Butturini et al., 2003; Jacobson and Jacobson, 2013). Overland flow, regulated by rainfall intensity and the infiltration capacity of the soil, contributes to hydrological connectivity and stream flow resumption (Belnap et al., 2005). In contrast, warming temperatures and sudden snowmelt typically initiate flow in glacial IRES (McKnight et al., 1999; Robinson et al., 2015). As a result, a first flow event takes place, which can differ in its intensity from gradual to abrupt (Fig. 3.2.5). In any case, while stream discharge increases, lateral, vertical, and longitudinal hydrological connections along the stream channel get reestablished (Chapter 2.3). In contrast to patterns observed during drying, high hydrological connectivity during the expansion phase increases the relative importance of allochthonous terrestrial nutrient and OM sources with respect to those from in-stream origin (Bernal et al., 2013). In parallel, many in-stream microbial processes and metabolic pathways tend to recover very quickly after surface rewetting (Dodds et al., 2004; Amalfitano et al., 2008; Sabater et al., 2016), resulting in a pulse of nutrient and OM mineralization (i.e., the Birch effect) (Wilson and Baldwin, 2008). Nevertheless, the magnitude of allochthonous and autochthonous sources contributing to stream nutrient and OM exports remains unclear and may differ greatly among IRES, depending on factors such as the type and timing of the rewetting event as well as bioclimatic and geomorphological conditions.

Large amounts of particulate and dissolved materials, including nutrients and OM, are mobilized downstream during these first flow pulses (Fisher and Minckley, 1978; Tzoraki et al., 2007; Obermann et al., 2009; von Schiller et al., 2011; Fig. 3.2.5). Stream nutrient and DOM concentrations can increase several times above baseflow concentrations. However, except in IRES characterized by intense pulse events that recede rapidly, the contribution of these events to annual loads is generally limited because solute concentrations return quickly to low levels and runoff coefficients are small compared to those found later in the season (Butturini et al., 2002; Bernal et al., 2005).

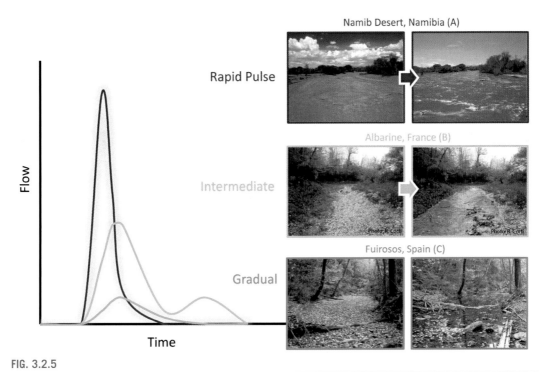

FIG. 3.2.5

Rapid, intermediate, and gradual first flow events transporting high amounts of dissolved and particulate materials in different IRES, illustrated by examples from Africa (A) and Europe (B and C).

Photos: Courtesy (A) P. J. Jacobson, (B) R. Corti, and (C) D. von Schiller.

In most IRES, dissolved nutrients and DOM originate primarily from near-stream zones through the elevation of the groundwater table (Butturini et al., 2003; Vázquez et al., 2007) and from upstream reaches through the leaching and transport of detritus accumulated during the dry phase (Humphries and Baldwin, 2003; Ylla et al., 2010). Release of nutrients and DOM from previously dry stream sediments may also constitute an important source (Baldwin et al., 2005; McIntyre et al., 2009b; Merbt et al., 2016). The timing, duration, and frequency of drying can strongly influence the concentrations and species of nutrients and DOM released from detritus and sediments after rewetting (Baldwin, 1999; Dieter et al., 2013; Abril et al., 2016). For example, Wilson and Baldwin (2008) showed that dried sediments from reservoirs that undergo frequent drying and rewetting cycles released lower amounts of ammonium upon rewetting than those that were dry for longer periods. Therefore, the severity of the preceding dry phase has implications for nutrient and OM dynamics upon flow reconnection.

In high-latitude and alpine IRES, glaciers and reconnected groundwaters represent the most relevant sources of nutrients and OM upon rewetting (McKnight et al., 1999; Robinson et al., 2015). In these ecosystems, spring snowmelt initiates the expansion of the fluvial network and represents a substantial fraction of the annual flux of water, nutrients, and DOM exported downstream (Tockner et al., 2002). These glacial inputs represent sources of water and associated materials that reach the main channel via surface flowpaths and supply large amounts of nutrients and OM originating from atmospheric deposition (Tockner et al., 2002). The biogeochemical signatures of glacial waters are quite distinct between each other and from those of deep and shallow groundwater, contributing to increased spatial heterogeneity of OM across the fluvial network (Malard et al., 2000; Tockner et al., 2002). Extreme examples of glacier-fed IRES are those in the Dry Valleys of Antarctica, which flow only for a few weeks during the austral summer (McKnight et al., 1999). These streams usually receive no lateral inflow or terrestrial runoff, except during exceptionally warm years (Lyons et al., 2005).

The most abundant N species during the initial flow pulse is either nitrate or ammonium (Obermann et al., 2009; Skoulikidis and Amaxidis, 2009; von Schiller et al., 2011), although DON concentrations can also be high (Baldwin et al., 2005; von Schiller et al., 2011). The main instream source of inorganic N is most likely mineralization of accumulated detrital OM as well as that from nitrification in streambeds during the dry phase (Baldwin et al., 2005; Woodward et al., 2015; Merbt et al., 2016). Desorption and redissolution of salts from sediments may constitute an additional in-stream N source (Baldwin and Mitchell, 2000). The predominance of one or the other N species may strongly depend on the relative importance of mineralization, nitrification, and denitrification rates in stream sediments and riparian soils during the antecedent dry period (Bernal et al., 2005). Along with the changes in nutrient concentrations and forms, many in-stream N transformation processes such as denitrification and nitrification are rapidly activated upon rewetting (McIntyre et al., 2009a; Austin and Strauss, 2011; Arce et al., 2015). This generally translates into rapid recovery of the whole-reach N retention efficiency after the first flow events (Marti et al., 1997).

The peak in P concentration during surface flow recovery tends to be low compared to that of N, probably as a result of dilution and the higher relative release of N over P from sediments and soils (Tzoraki et al., 2007; Skoulikidis and Amaxidis, 2009; von Schiller et al., 2011). However, a high proportion of P may be in particulate form (Tzoraki et al., 2007; Skoulikidis and Amaxidis, 2009), and, as described earlier, the proportion of bioavailable P may be higher as a result of sediment drying during the preceding dry phase (Kerr et al., 2010; Attygalla et al., 2016).

DOM concentrations in the surface stream increase upon rewetting in a range of IRES (Jacobson et al., 2000; Bernal et al., 2005; Inamdar et al., 2011; Catalán et al., 2013). This DOM is typically characterized by a high proportion of high molecular weight molecules (Romaní et al., 2006; Vázquez et al., 2007; Inamdar et al., 2011; von Schiller et al., 2015), most likely associated with an increase in the relative amount of polysaccharides at rewetting (Ylla et al., 2010). Studies in semiarid IRES in Spain also report increases of aromatic and humic DOM of terrestrial origin upon rewetting (Catalán et al., 2013; von Schiller et al., 2015), although the concentrations of small highly biodegradable nitrogenous organic compounds may also increase (Vázquez et al., 2015). An increase in protein-like DOM during the first flow event after drought was found in an intermittent stream in Maryland, USA (Inamdar et al., 2011). The reason for the differences in DOM composition observed among IRES may be related to differences in the amount and type of OM in sediments and soils, and its decomposition during the preceding dry phase. Therefore, the timing of flow resumption with respect to leaf fall may be an important factor determining

FIG. 3.2.6

Ephemeral channel with anoxic conditions just after a blackwater event (north-central Victoria, Australia).

Photo: Courtesy D. Baldwin.

the composition of the first flow DOM in IRES, especially those in forested regions (von Schiller et al., 2015). In high altitude and latitude IRES, where drying–rewetting dynamics are different, glaciers may represent a source of highly diverse and bioavailable DOM upon melting (Singer et al., 2012).

One area of concern related to increases in surface water DOM concentrations after rewetting from dry conditions in IRES is the potential occurrence of hypoxic blackwater events (Fig. 3.2.6, Valett et al., 2005; Hladyz et al., 2011; Whitworth et al., 2012). These events are characterized by high levels of DOM, the metabolism of which lowers dissolved oxygen in the water column which can kill fish, crustaceans, and other animals. Furthermore, these events could be 'hot moments' of carbon dioxide emissions to the atmosphere that remain widely unexplored (Raymond et al., 2013).

During first flow events in IRES, POM can also contribute substantially to total OM loads (Corti and Datry, 2012). This seems to be especially relevant in ephemeral desert streams, where riparian vegetation can deliver large amounts of OM that accumulates during the long dry periods in the channel, yielding high POM (and DOM) when surface flow resumes (Jacobson and Jacobson, 2013). Terrestrial plants growing on streambeds during the dry phase could also be an important source of POM when flow resumes in this type of IRES. All this material is transported downstream during first pulses and subsequent floods, and redistributed and sorted by size and weight along the channel as water energy decreases (Jacobson et al., 2000; Jacobson and Jacobson, 2013).

Decomposition of POM deployed on the sediments of streams after rewetting resumes relatively quickly (Pattee et al., 2001; Schlief and Mutz, 2011), strongly affected by preconditioning during the antecedent fragmentation and dry phases (Langhans and Tockner, 2006; Bruder et al., 2011; Dieter et al., 2011).

3.2.3 KNOWLEDGE GAPS AND RESEARCH OPPORTUNITIES

Research in IRES is blooming, and much knowledge on the biogeochemistry of IRES has been gained during recent years. Yet, as evidenced throughout this chapter, there is still much to be learned about nutrient and OM dynamics in these ecosystems. We suggest four main research lines that should be developed on this topic.

First, we need to expand biogeochemical research across a wider geographical extent to increase our understanding of nutrient and OM dynamics in a greater variety of IRES. To date, most biogeochemical studies of IRES have been performed in arid, semiarid, and temperate regions of North America, Europe, and Australia (Larned et al., 2010; Datry et al., 2014). There is still limited knowledge on the biogeochemistry of IRES in arid and semiarid streams from Africa, Asia, and South America; glacial IRES from Alpine or high-latitude regions; and tropical IRES from around the world. Global initiatives such as the "1000 Intermittent Rivers Project" (Datry et al., 2016a) represent excellent opportunities to cover these less studied geographical areas. In addition, there is a general bias toward the study of seasonally intermittent streams, while ephemeral streams that flow more unpredictably have been much less studied (Jacobson and Jacobson, 2013). Moreover, little effort has been put into the study of the biogeochemistry of 'artificial IRES' formed when perennial streams become intermittent due to human interventions (Steward et al., 2012). Available knowledge suggests there are notable differences in nutrient and OM dynamics among all these different IRES types; this hypothesis deserves further research.

Second, we should seek a more mechanistic understanding of nutrient and OM dynamics in IRES. To date, biogeochemical research in IRES has focused mostly on describing changes in structural variables over time and space. More effort should be put into directly measuring the mechanisms driving those changes both under natural conditions and using manipulative laboratory and field experiments (Acuña et al., 2015). Among the most interesting but least known mechanisms are nutrient and OM transformation pathways in dry streambeds of IRES as well as their contribution to gas emissions and downstream export during first flow pulses. For instance, recent evidence suggests that the dry sediments of IRES are biogeochemically active and able to emit large quantities of carbon dioxide to the atmosphere (Gallo et al., 2013; von Schiller et al., 2014; Gómez-Gener et al., 2016). However, the sources of these emissions and how they are affected by drying history are unknown. Furthermore, upon rewetting, many microbial activities are reactivated, large quantities of organic materials are mobilized, and strong changes in redox conditions and pH occur. These first flow events could thus be 'hot moments' of carbon dioxide outgassing that have been scarcely explored (Amalfitano et al., 2008; Datry et al., 2014).

Third, we need to understand better the interactions between flow intermittence and human stressors on nutrient and OM dynamics. Many IRES suffer severe human degradation, often because they are not sufficiently recognized (Steward et al., 2012; Acuña et al., 2014; Leigh et al., 2016). Studies on the effect of human stressors, such as water abstraction and contaminants on biogeochemical processes during dry–wet cycles in IRES are only now beginning (Corcoll et al., 2015; Pesce et al., 2016) (Box 3.2.1 and Fig. 3.2.7).

Fourth, we need to improve and further develop modeling and upscaling of biogeochemical processes in IRES networks. To do this, there is an urgent need to develop large-scale maps of the spatial and temporal variation of the hydrology of IRES networks, including nonflow periods (Costigan et al., 2015; Datry et al., 2016b), as well as common datasets and the maintenance of long-term water quality monitoring programs in IRES (Mosley, 2015). Upscaling and modeling will allow us to address the implications of hydrological fragmentation and drying at river network and catchment scales (Fuller et al., 2015). Although longitudinal discontinuities potentially affect processes at the community, ecosystem, and landscape levels, this topic has still not received much attention in the stream biogeochemical literature (Fisher et al., 2004). Moreover, upscaling and modeling exercises will help elucidate the contribution of internal versus external sources to first flow pulses as well as the contribution of IRES to whole catchment and global elemental balances (Acuña et al., 2014; Datry et al., 2014; von Schiller et al., 2014). Finally, the development of coupled hydro-biogeochemical models could serve to predict the responses of stream networks and their resilience to the expected increase in the extent of IRES under global change in many regions worldwide.

BOX 3.2.1 HUMAN INFLUENCES ON NUTRIENT AND OM DYNAMICS IN IRES

Streams and rivers worldwide are strongly influenced by human activities, and IRES are no exception. This influence is especially severe in IRES in populated semiarid regions, where water resources are scarce (Chapter 5.1). The tremendous demand for water together with the need to control floods has led to the proliferation of water diversions and infrastructure such as dams and impoundments throughout these regions. For instance, in the Sacramento River Basin (California, USA), there are over 400 dams that can store 100% of the river's mean annual discharge (Grantham et al., 2013). Dams typically disrupt longitudinal and lateral hydrological connections, flatten stream hydrographs, and smooth the seasonality of discharge (Magdaleno and Fernández, 2011). Consequently, the length and intensity of hydrological phases in IRES are altered, and thus the nutrient and OM dynamics associated with them.

In extreme cases, formerly intermittent sections may be transformed into perennial ones or, more commonly, perennial stream sections may be turned into intermittent ones (Fig. 3.2.7a). Runoff from agricultural and urban landscapes contributes nutrients and OM to receiving streams, decreases dissolved oxygen concentrations, and alters stream metabolism and in-stream nutrient cycling (Bernot et al., 2006; von Schiller et al., 2008). This can be especially marked in IRES during the fragmentation and desiccation phases when in-stream dilution capacity is least. One example of this paradigm of human influences on the low dilution capacity of IRES is evident in urban areas where IRES become hydrologically perennial downstream from the inputs from wastewater treatment plant effluents (Marti et al., 2004; Brooks et al., 2006) (Fig. 3.2.7b), completely changing the characteristic hydrological regime of IRES. In addition, these inputs dictate the concentrations, in-stream processing, and export of nutrients and OM in the receiving IRES (Merseburger et al., 2005). The benefits of exploiting water resources from IRES bring with them high environmental costs which are difficult to manage and which compromise the long-term sustainability of these fragile and essential ecosystems. Nonetheless, IRES remain poorly recognized in most river management policies (Chapter 5.3); as a result, they are rarely considered in river health monitoring and assessment programs.

(Continued)

BOX 2.1.2 HUMAN INFLUENCES ON NUTRIENT AND OM DYNAMICS IN IRES *(Cont'd)*

FIG. 3.2.7

Human activities with direct influences on nutrient and OM dynamics in streams: (a) exposure of dry sediments in a perennial stream reach (Basque Country, Spain) caused by water abstraction for energy production and (b) temperate intermittent stream (Basque Country, Spain) dominated by industrial waste water effluent discharge in summer.

Photos: Courtesy A. Elosegi.

REFERENCES

Abril, M., Muñoz, I., Menéndez, M., 2016. Heterogeneity in leaf litter decomposition in a temporary Mediterranean stream during flow fragmentation. Sci. Total Environ. 553, 330–339.

Acuña, V., Tockner, K., 2010. The effects of alterations in temperature and flow regime on organic carbon dynamics in Mediterranean river networks. Glob. Chang. Biol. 16, 2638–2650.

Acuña, V., Giorgi, A., Munoz, I., Uehlinger, U., Sabater, S., 2004. Flow extremes and benthic organic matter shape the metabolism of a headwater Mediterranean stream. Freshw. Biol. 49, 960–971.

Acuña, V., Muñoz, I., Giorgi, A., Omella, M., Sabater, F., Sabater, S., 2005. Drought and postdrought recovery cycles in an intermittent Mediterranean stream: structural and functional aspects. J. N. Am. Benthol. Soc. 24, 919–933.

Acuña, V., Giorgi, A., Muñoz, I., Sabater, F., Sabater, S., 2007. Meteorological and riparian influences on organic matter dynamics in a forested Mediterranean stream. J. N. Am. Benthol. Soc. 26, 54–69.

Acuña, V., Datry, T., Marshall, J., Barceló, D., Dahm, C.N., Ginebreda, A., et al., 2014. Why should we care about temporary waterways? Science 343, 1080–1081.

Acuña, V., Casellas, M., Corcoll, N., Timoner, X., Sabater, S., 2015. Increasing extent of periods of no flow in intermittent waterways promotes heterotrophy. Freshw. Biol. 60, 1810–1823.

Allan, J.D., Castillo, M.M., 2007. Stream Ecology: Structure and Function of Running Waters, second ed. Springer, Netherlands.

Amalfitano, S., Fazi, S., Zoppini, A., Caracciolo, A.B., Grenni, P., Puddu, A., 2008. Responses of benthic bacteria to experimental drying in sediments from Mediterranean temporary rivers. Microb. Ecol. 55, 270–279.

Arce, M.I., del Mar Sánchez-Montoya, M., Gómez, R., 2015. Nitrogen processing following experimental sediment rewetting in isolated pools in an agricultural stream of a semiarid region. Ecol. Eng. 77, 233–241.

Attygalla, N.W., Baldwin, D.S., Silvester, E., Kappen, P., Whitworth, K.L., 2016. The severity of sediment desiccation affects the adsorption characteristics and speciation of phosphorus. Evniron. Sci. Process. Impacts 18, 64–71.

Austin, B.J., Strauss, E.A., 2011. Nitrification and denitrification response to varying periods of desiccation and inundation in a western Kansas stream. Hydrobiologia 658, 183–195.

Baker, M.A., Valett, H.M., Dahm, C.N., 2000. Organic carbon supply and metabolism in a shallow groundwater ecosystem. Ecology 81, 3133–3148.

Baldwin, D.S., 1996. Effects of exposure to air and subsequent drying on the phosphate sorption characteristics of sediments from a eutrophic reservoir. Limnol. Oceanogr. 41, 1725–1732.

Baldwin, D.S., 1999. Dissolved organic matter and phosphorus leached from fresh and "terrestrially" aged river red gum leaves: implications for assessing river–floodplain interactions. Freshw. Biol. 41, 675–685.

Baldwin, D.S., Mitchell, A.M., 2000. The effects of drying and re-flooding on the sediment and soil nutrient dynamics of lowland river–floodplain systems: a synthesis. Regul. Rivers Res. Manag. 16, 457–467.

Baldwin, D.S., Mitchell, A.M., Rees, G.N., 2000. The effects of in situ drying on sediment-phosphate interactions in sediments from an old wetland. Hydrobiologia 431, 3–12.

Baldwin, D.S., Rees, G.N., Mitchell, A.M., Watson, G., 2005. Spatial and temporal variability of nitrogen dynamics in an upland stream before and after a drought. Mar. Freshw. Res. 56, 457–464.

Belnap, J., Welter, J.R., Grimm, N.B., Barger, N., Ludwig, J.A., 2005. Linkages between microbial and hydrologic processes in arid and semiarid watersheds. Ecology 86, 298–307.

Bernal, S., Sabater, F., 2012. Changes in discharge and solute dynamics between hillslope and valley-bottom intermittent streams. Hydrol. Earth Syst. Sci. 16, 1595–1605.

Bernal, S., Butturini, A., Sabater, F., 2005. Seasonal variations of dissolved nitrogen and DOC:DON ratios in an intermittent Mediterranean stream. Biogeochemistry 75, 351–372.

Bernal, S., von Schiller, D., Sabater, F., Martí, E., 2013. Hydrological extremes modulate nutrient dynamics in mediterranean climate streams across different spatial scales. Hydrobiologia 719, 31–42.

Bernot, M.J., Tank, J.L., Royer, T.V., David, M.B., 2006. Nutrient uptake in streams draining agricultural catchments of the midwestern United States. Freshw. Biol. 51, 499–509.

Birch, H.F., 1958. The effect of soil drying on humus decomposition and nitrogen availability. Plant Soil 10, 9–31.

Borken, W., Matzner, E., 2009. Reappraisal of drying and wetting effects on C and N mineralization and fluxes in soils. Glob. Chang. Biol. 15, 808–824.

Boulton, A.J., 1991. Eucalypt leaf decomposition in an intermittent stream in south-eastern Australia. Hydrobiologia 211, 123–136.

Boulton, A.J., Lake, P.S., 1990. The ecology of two intermittent streams in Victoria, Australia. I. Multivariate analyses of physicochemical features. Freshw. Biol. 24, 123–141.

Brooks, B.W., Riley, T.M., Taylor, R.D., 2006. Water quality of effluent-dominated ecosystems: ecotoxicological, hydrological, and management considerations. Hydrobiologia 556, 365–379.

Bruder, A., Chauvet, E., Gessner, M.O., 2011. Litter diversity, fungal decomposers and litter decomposition under simulated stream intermittency. Funct. Ecol. 25, 1269–1277.

Bunn, S.E., Thoms, M.C., Hamilton, S.K., Capon, S.J., 2006. Flow variability in dryland rivers: boom, bust and the bits in between. River Res. Appl. 22, 179–186.

Butturini, A., Bernal, S., Sabater, S., Sabater, F., 2002. The influence of riparian-hyporheic zone on the hydrological responses in an intermittent stream. Hydrol. Earth Syst. Sci. 6, 515–526.

Butturini, A., Bernal, S., Hellin, C., 2003. Influences of the stream groundwater hydrology on nitrate concentration in unsaturated riparian area bounded by an intermittent Mediterranean stream. Water Resour. Res. 39, 1110.

Casas-Ruiz, J.P., Tittel, J., von Schiller, D., Catalán, N., Obrador, B., Gómez-Gener, L., et al., 2016. Drought-induced discontinuities in the source and degradation of dissolved organic matter in a Mediterranean river. Biogeochemistry 127, 125–139.

Catalán, N., Obrador, B., Alomar, C., Pretus, J.L., 2013. Seasonality and landscape factors drive dissolved organic matter properties in Mediterranean ephemeral washes. Biogeochemistry 112, 261–274.

Claret, C., Boulton, A.J., 2003. Diel variation in surface and subsurface microbial activity along a gradient of drying in an Australian sand-bed stream. Freshw. Biol. 48, 1739–1755.

Claret, C., Marmonier, P., Boissier, J.-M., Fontvieille, D., Blanc, P., 1997. Nutrient transfer between parafluvial interstitial water and river water: influence of gravel bar heterogeneity. Freshw. Biol. 37, 657–670.

Conley, D.J., Paerl, H.W., Howarth, R.W., Boesch, D.F., Seitzinger, S.P., Havens, K.E., et al., 2009. Controlling eutrophication: nitrogen and phosphorus. Science 323, 1014–1015.

Corcoll, N., Casellas, M., Huerta, B., Guasch, H., Acuña, V., Rodríguez-Mozaz, S., et al., 2015. Effects of flow intermittency and pharmaceutical exposure on the structure and metabolism of stream biofilms. Sci. Total Environ. 503, 159–170.

Corti, R., Datry, T., 2012. Invertebrates and sestonic matter in an advancing wetted front travelling down a dry river bed (Albarine, France). Freshw. Biol. 31, 1187–1201.

Costigan, K.H., Daniels, M.D., Dodds, W.K., 2015. Fundamental spatial and temporal disconnections in the hydrology of an intermittent prairie headwater network. J. Hydrol. 522, 305–316.

Dahm, C.N., Grimm, N.B., Marmonier, P., Valett, H.M., Vervier, P., 1998. Nutrient dynamics at the interface between surface waters and groundwaters. Freshw. Biol. 40, 427–451.

Dahm, C.N., Baker, M.A., Moore, D.I., Thibault, J.R., 2003. Coupled biogeochemical and hydrological responses of streams and rivers to drought. Freshw. Biol. 48, 1219–1231.

Datry, T., Corti, R., Claret, C., Philippe, M., 2011. Flow intermittence controls leaf litter breakdown in a French temporary alluvial river: the "drying memory". Aquat. Sci. 73, 471–483.

Datry, T., Larned, S.T., Tockner, K., 2014. Intermittent rivers: a challenge for freshwater ecology. Bioscience 64, 229–235.

Datry, T., Corti, R., Foulquier, A., von Schiller, D., Tockner, K., 2016a. One for all, all for one: a global river research network. Eos 97. http://dx.doi.org/10.1029/2016EO053587.

Datry, T., Pella, H., Leigh, C., Bonada, N., Hugueny, B., 2016b. A landscape approach to advance intermittent river ecology. Freshw. Biol. 61, 1200–1213.

de Vicente, I., Andersen, F.Ø., Hansen, H.C.B., Cruz-Pizarro, L., Jensen, H.S., 2010. Water level fluctuations may decrease phosphate adsorption capacity of the sediment in oligotrophic high mountain lakes. Hydrobiologia 651, 253–264.

Dent, C.L., Grimm, N.B., 1999. Spatial heterogeneity of stream water nutrient concentrations over successional time. Ecology 80, 2283–2298.

Dent, C.L., Grimm, N.B., Fisher, S.G., 2001. Multiscale effects of surface-subsurface exchange on stream water nutrient concentrations. J. N. Am. Benthol. Soc. 20, 162–181.

Dent, C.L., Grimm, N.B., Martí, E., Edmonds, J.W., Henry, J.C., Welter, J.R., 2007. Variability in surface-subsurface hydrologic interactions and implications for nutrient retention in an arid-land stream. J. Geophys. Res. 112, G04004.

Dewson, Z.S., James, A.B.W., Death, R.G., 2007. Stream ecosystem functioning under reduced flow conditions. Ecol. Appl. 17, 1797–1808.

Dieter, D., von Schiller, D., García-Roger, E.M., Sánchez-Montoya, M.M., Gómez, R., et al., 2011. Preconditioning effects of intermittent stream flow on leaf litter decomposition. Aquat. Sci. 73, 599–609.

Dieter, D., Frindte, K., Krüger, A., Wurzbacher, C., 2013. Preconditioning of leaves by solar radiation and anoxia affects microbial colonisation and rate of leaf mass loss in an intermittent stream. Freshw. Biol. 58, 1918–1931.

Dieter, D., Herzog, C., Hupfer, M., 2015. Effects of drying on phosphorus uptake in re-flooded lake sediments. Environ. Sci. Pollut. Res. 22, 17065–17081.

Dodds, W.K., Gido, K., Whiles, M.R., Fritz, K.M., Matthews, W.J., 2004. Life on the edge: the ecology of Great Plains prairie streams. Bioscience 54, 205–216.

Doering, M., Uehlinger, U., Rotach, A., Schlaepfer, D.R., Tockner, K., 2007. Ecosystem expansion and contraction dynamics along a large Alpine alluvial corridor (Tagliamento River, Northeast Italy). Earth Surf. Process. Landf. 32, 1693–1704.

Edwardson, K.J., Bowden, W.B., Dahm, C., Morrice, J., 2003. The hydraulic characteristics and geochemistry of hyporheic and parafluvial zones in Arctic tundra streams, north slope, Alaska. Adv. Water Resour. 26, 907–923.

Elser, J.J., Bracken, M.E.S., Cleland, E.E., Gruner, D.S., Harpole, W.S., Hillebrand, H., et al., 2007. Global analysis of nitrogen and phosphorus limitation of primary producers in freshwater, marine and terrestrial ecosystems. Ecol. Lett. 10, 1135–1142.

Fazi, S., Vázquez, E., Casamayor, E.O., Amalfitano, S., Butturini, A., 2013. Stream hydrological fragmentation drives bacterioplankton community composition. PLoS One 8, e64109.

Fellman, J.B., Dogramaci, S., Skrzypek, G., Dodson, W., Grierson, P.F., 2011. Hydrologic control of dissolved organic matter biogeochemistry in pools of a subtropical dryland river. Water Resour. Res. 47, W06501.

Finn, M.A., Boulton, A.J., Chessman, B.C., 2009. Ecological responses to artificial drought in two Australian rivers with differing water extraction. Fundam. Appl. Limnol. 175, 231–248.

Fisher, S.G., Minckley, W.L., 1978. Chemical characteristics of a desert stream in flash flood. J. Arid Environ. 1, 25–33.

Fisher, S.G., Grimm, N.B., Marti, E., Gomez, R., 1998. Hierarchy, spatial configuration, and nutrient cycling in a desert stream. Aust. J. Ecol. 23, 41–52.

Fisher, S.G., Sponseller, R.A., Heffernan, J.B., 2004. Horizons in stream biogeochemistry: flowpaths to progress. Ecology 85, 2369–2379.

Freeman, C., Gresswell, R., Guasch, H., Hudson, J., Lock, M.A., Reynolds, B., et al., 1994. The role of drought in the impact of climatic change on the microbiota of peatland streams. Freshw. Biol. 32, 223–230.

Fuller, M.R., Doyle, M.W., Strayer, D.L., 2015. Causes and consequences of habitat fragmentation in river networks. Ann. N. Y. Acad. Sci. 1355, 31–51.

Gallo, E.L., Lohse, K.A., Ferlin, C.M., Meixner, T., Brooks, P.D., 2013. Physical and biological controls on trace gas fluxes in semi-arid urban ephemeral waterways. Biogeochemistry 121, 189–207.

Gómez, R., García, V., Vidal-Abarca, R., Suárez, L., 2009. Effect of intermittency on N spatial variability in an arid Mediterranean stream. J. N. Am. Benthol. Soc. 28, 572–583.

Gómez, R., Arce, M.I., Sánchez, J.J., del Mar Sánchez-Montoya, M., 2012. The effects of drying on sediment nitrogen content in a Mediterranean intermittent stream: a microcosms study. Hydrobiologia 679, 43–59.

Gómez-Gener, L., Obrador, B., von Schiller, D., Marcé, R., Casas-Ruiz, J.P., Proia, L., et al., 2015. Hot spots for carbon emissions from Mediterranean fluvial networks during summer drought. Biogeochemistry 125, 409–426.

Gómez-Gener, L., Obrador, B., Marcé, R., Acuña, V., Catalán, N., Casas-Ruiz, J.P., et al., 2016. When water vanishes: magnitude and regulation of carbon dioxide emissions from dry temporary streams. Ecosystems 19, 710. http://dx.doi.org/10.1007/s10021-016-9963-4.

Grantham, T.E., Figueroa, R., Prat, N., 2013. Water management in mediterranean river basins: a comparison of management frameworks, physical impacts, and ecological responses. Hydrobiologia 719, 451–482.

Harvey, J.W., Conklin, M.H., Koelsch, R.S., 2003. Predicting changes in hydrologic retention in an evolving semi-arid alluvial stream. Adv. Water Resour. 26, 939–950.

Hladyz, S., Watkins, S.C., Whitworth, K.L., Baldwin, D.S., 2011. Flows and hypoxic blackwater events in managed ephemeral river channels. J. Hydrol. 401, 117–125.

Holmes, R.M., Fisher, S.G., Grimm, N.B., 1994. Parafluvial nitrogen dynamics in a desert stream ecosystem. J. N. Am. Benthol. Soc. 13, 468–478.

Howarth, R.W., Billen, G., Swaney, D., Townsend, A., Jaworski, N., Lajtha, K., et al., 1996. Regional nitrogen budgets and riverine N & P fluxes for the drainages to the North Atlantic Ocean: natural and human influences. In: Howarth, R.W. (Ed.), Nitrogen Cycling in the North Atlantic Ocean and Its Watersheds. Springer, New York, pp. 75–139.

Humphries, P., Baldwin, D.S., 2003. Drought and aquatic ecosystems: an introduction. Freshw. Biol. 48, 1141–1146.

Inamdar, S., Singh, S., Dutta, S., Levia, D., Mitchell, M., Scott, D., et al., 2011. Fluorescence characteristics and sources of dissolved organic matter for stream water during storm events in a forested mid-Atlantic watershed. J. Geophys. Res. 116, G03043.

Jacobson, P.J., Jacobson, K.M., 2013. Hydrologic controls of physical and ecological processes in Namib Desert ephemeral rivers: implications for conservation and management. J. Arid Environ. 93, 80–93.

Jacobson, P.J., Jacobson, K.M., Angermeier, P.L., Cherry, D.S., 2000. Variation in material transport and water chemistry along a large ephemeral river in the Namib Desert. Freshw. Biol. 44, 481–491.

Jacoby, J.M., Lynch, D.D., Welch, E.B., Perkins, M.A., 1982. Internal phosphorus loading in a shallow eutrophic lake. Water Res. 16, 911–919.

Jones Jr., J.B., Holmes, R.M., Fisher, S.G., Grimm, N.B., Greene, D.M., 1995. Methanogenesis in Arizona, USA dryland streams. Biogeochemistry 31, 155–173.

Jones, J.B., Mulholland, P.J., 2000. Streams and Ground Waters. Academic Press, San Diego, CA.

Kemp, M.J., Dodds, W.K., 2002. The influence of ammonium, nitrate, and dissolved oxygen concentrations on uptake, nitrification, and denitrification rates associated with prairie stream substrata. Limnol. Oceanogr. 47, 1380–1393.

Kerr, J.G., Burford, M., Olley, J., Udy, J., 2010. The effects of drying on phosphorus sorption and speciation in subtropical river sediments. Mar. Freshw. Res. 61, 928–935.

Kleeberg, A., Heidenreich, M., 2004. Release of nitrogen and phosphorus from macrophyte stands of summer dried out sediments of a eutrophic reservoir. Arch. Hydrobiol. 159, 115–136.

Lado-Monserrat, L., Lull, C., Bautista, I., Lidón, A., Herrera, R., 2014. Soil moisture increment as a controlling variable of the "Birch effect". Interactions with the pre-wetting soil moisture and litter addition. Plant Soil 379, 21–34.

Langhans, S.D., Tockner, K., 2006. The role of timing, duration, and frequency of inundation in controlling leaf litter decomposition in a river-floodplain ecosystem (Tagliamento, northeastern Italy). Oecologia 147, 501–509.

Larned, S.T., Datry, T., Arscott, D.B., Tockner, K., 2010. Emerging concepts in temporary-river ecology. Freshw. Biol. 55, 717–738.

Larson, J.H., Frost, P.C., Lodge, D.M., Lamberti, G.A., 2007. Photodegradation of dissolved organic matter in forested streams of the northern Great Lakes region. J. N. Am. Benthol. Soc. 26, 416–425.

Lautz, L.K., Siegel, D.I., 2007. The effect of transient storage on nitrate uptake lengths in streams: an inter-site comparison. Hydrol. Process. 21, 3533–3548.

Leberfinger, K., Bohman, I., Herrmann, J., 2010. Drought impact on stream detritivores: experimental effects on leaf litter breakdown and life cycles. Hydrobiologia 652, 247–254.

Leigh, C., Boulton, A.J., Courtwright, J.L., Fritz, K., May, C.L., Walker, R.H., et al., 2016. Ecological research and management of intermittent rivers: an historical review and future directions. Freshw. Biol. 61, 1181–1199.

Lewis, D.B., Grimm, N.B., Harms, T.K., Schade, J.D., 2007. Subsystems, flowpaths, and the spatial variability of nitrogen in a fluvial ecosystem. Landsc. Ecol. 22, 911–924.

Lillebø, A.I., Morais, M., Guilherme, P., Fonseca, R., Serafim, A., Neves, R., 2007. Nutrient dynamics in Mediterranean temporary streams: a case study in Pardiela catchment (Degebe River, Portugal). Limnol.-Ecol. Manag. Inland Waters 37, 337–348.

Lupon, A., Sabater, F., Miñarro, A., Bernal, S., 2016. Contribution of pulses of soil nitrogen mineralization and nitrification to soil nitrogen availability in three Mediterranean forests. Eur. J. Soil Sci. 67, 303–313.

Lyons, W.B., Welch, K.A., Carey, A.E., Doran, P.T., Wall, D.H., Virginia, R.A., et al., 2005. Groundwater seeps in Taylor Valley Antarctica: an example of a subsurface melt event. Ann. Glaciol. 40, 200–206.

Magdaleno, F., Fernández, J.A., 2011. Hydromorphological alteration of a large Mediterranean river: relative role of high and low flows on the evolution of riparian forests and channel morphology. River Res. Appl. 27, 374–387.

Malard, F., Tockner, K., Ward, J.V., 1999. Shifting dominance of subcatchment water sources and flow paths in a glacial floodplain, Val Roseg, Switzerland. Arct. Antarct. Alp. Res. 31, 135–150.

Malard, F., Tockner, K., Ward, J.V., 2000. Physico-chemical heterogeneity in a glacial riverscape. Landsc. Ecol. 15, 679–695.

Mariluan, G.D., Villanueva, V.D., Albariño, R.J., 2015. Leaf litter breakdown and benthic invertebrate colonization affected by seasonal drought in headwater lotic systems of Andean Patagonia. Hydrobiologia 760, 171–187.

Marti, E., Grimm, N.B., Fisher, S.G., 1997. Pre- and post-flood retention efficiency of nitrogen in a Sonoran Desert stream. J. N. Am. Benthol. Soc. 16, 805–819.

Marti, E., Fisher, S.G., Schade, J.D., Grimm, N.B., 2000. Flood frequency and stream-riparian linkages in arid lands. In: Jones, J.B., Mulholland, P.J. (Eds.), Streams and Ground Waters. Academic Press, San Diego, pp. 111–136.

Marti, E., Aumatell, J., Godé, L., Poch, M., Sabater, F., 2004. Nutrient retention efficiency in streams receiving inputs from wastewater treatment plants. J. Environ. Qual. 33, 285–293.

McIntyre, R.E.S., Adams, M.A., Ford, D.J., Grierson, P.F., 2009a. Rewetting and litter addition influence mineralisation and microbial communities in soils from a semi-arid intermittent stream. Soil Biol. Biochem. 41, 92–101.

McIntyre, R.E.S., Adams, M.A., Grierson, P.F., 2009b. Nitrogen mineralization potential in rewetted soils from a semi-arid stream landscape, north-west Australia. J. Arid Environ. 73, 48–54.

McKnight, D.M., Niyogi, D.K., Alger, A.S., Bomblies, A., Conovitz, P.A., Tate, C.M., 1999. Dry valley streams in Antarctica: ecosystems waiting for water. Bioscience 49, 985–995.

McLaughlin, C., 2008. Evaporation as a nutrient retention mechanism at Sycamore Creek, Arizona. Hydrobiologia 603, 241–252.

Merbt, S.N., Proia, L., Prosser, J.I., Martí, E., Casamayor, E.O., von Schiller, D., 2016. Stream drying drives microbial ammonia oxidation and first flush nitrate export. Ecology 97, 2192–2198.

Merseburger, G.C., Martí, E., Sabater, F., 2005. Net changes in nutrient concentrations below a point source input in two streams draining catchments with contrasting land uses. Sci. Total Environ. 347, 217–229.

Meybeck, M., 1982. Carbon, nitrogen, and phosphorus transport by world rivers. Am. J. Sci. 282, 401–450.

Mitchell, A., Baldwin, D.S., 1998. Effects of desiccation/oxidation on the potential for bacterially mediated P release from sediments. Limnol. Oceanogr. 43, 481–487.

Mitchell, A.M., Baldwin, D.S., 1999. The effects of sediment desiccation on the potential for nitrification, denitrification, and methanogenesis in an Australian reservoir. Hydrobiologia 392, 3–11.

Mosley, L.M., 2015. Drought impacts on the water quality of freshwater systems; review and integration. Earth Sci. Rev. 140, 203–214.

Mulholland, P.J., Webster, J.R., 2010. Nutrient dynamics in streams and the role of J-NABS. J. N. Am. Benthol. Soc. 29, 100–117.

Noy-Meir, I., 1973. Desert ecosystems: environment and producers. Annu. Rev. Ecol. Syst. 4, 25–51.

Obermann, M., Rosenwinkel, K.-H., Tournoud, M.-G., 2009. Investigation of first flushes in a medium-sized mediterranean catchment. J. Hydrol. 373, 405–415.

Pattee, E., Maamri, A., Chergui, H., 2001. Leaf litter processing and its agents in a temporary Moroccan river. Verh. Int. Ver. Theor. Angew. Limnol. 27, 3054–3057.

Pesce, S., Zoghlami, O., Margoum, C., Artigas, J., Chaumot, A., Foulquier, A., 2016. Combined effects of drought and the fungicide tebuconazole on aquatic leaf litter decomposition. Aquat. Toxicol. 173, 120–131.

Pohlon, E., Fandino, A.O., Marxsen, J., 2013. Bacterial community composition and extracellular enzyme activity in temperate streambed sediment during drying and rewetting. PLoS One 8, e83365.

Qiu, S., McComb, A.J., 1994. Effects of oxygen concentration on phosphorus release from reflooded air-dried wetland sediments. Mar. Freshw. Res. 45, 1319–1328.

Raymond, P.A., Hartmann, J., Lauerwald, R., Sobek, S., McDonald, C., Hoover, M., et al., 2013. Global carbon dioxide emissions from inland waters. Nature 503, 355–359.

Reynolds, J.F., Kemp, P.R., Ogle, K., Fernández, R.J., 2004. Modifying the `pulse-reserve' paradigm for deserts of North America: precipitation pulses, soil water, and plant responses. Oecologia 141, 194–210.

Robinson, C.T., Matthaei, S., 2007. Hydrological heterogeneity of an alpine stream—lake network in Switzerland. Hydrol. Process. 21, 3146–3154.

Robinson, C.T., Tonolla, D., Imhof, B., Vukelic, R., Uehlinger, U., 2015. Flow intermittency, physico-chemistry and function of headwater streams in an Alpine glacial catchment. Aquat. Sci. 78, 327–341.

Rolls, R.J., Leigh, C., Sheldon, F., 2012. Mechanistic effects of low-flow hydrology on riverine ecosystems: ecological principles and consequences of alteration. Freshwater Sci. 31, 1163–1186.

Romaní, A.M., Vázquez, E., Butturini, A., 2006. Microbial availability and size fractionation of dissolved organic carbon after drought in an intermittent stream: biogeochemical link across the stream-riparian interface. Microb. Ecol. 52, 501–512.

Sabater, S., Timoner, X., Borrego, C., Acuña, V., 2016. Stream biofilm responses to flow intermittency: from cells to ecosystems. Front. Environ. Sci. 4, 14.

Schade, J.D., Fisher, S.G., Grimm, N.B., Seddon, J.A., 2001. The influence of a riparian shrub on nitrogen cycling in a Sonoran Desert stream. Ecology 82, 3363–3376.

Schade, J.D., Welter, J.R., Marti, E., Grimm, N.B., 2005. Hydrologic exchange and N uptake by riparian vegetation in an arid-land stream. J. N. Am. Benthol. Soc. 24, 19–28.

Schlief, J., Mutz, M., 2011. Leaf decay processes during and after a supra-seasonal hydrological drought in a temperate lowland stream. Int. Rev. Hydrobiol. 96, 633–655.

Siebers, A.R., Pettit, N.E., Skrzypek, G., Fellman, J.B., Dogramaci, S., Grierson, P.F., 2016. Alluvial ground water influences dissolved organic matter biogeochemistry of pools within intermittent dryland streams. Freshw. Biol. 61, 1228–1241.

Singer, G.A., Fasching, C., Wilhelm, L., Niggemann, J., Steier, P., Dittmar, T., et al., 2012. Biogeochemically diverse organic matter in Alpine glaciers and its downstream fate. Nat. Geosci. 5, 710–714.

Skoulikidis, N., Amaxidis, Y., 2009. Origin and dynamics of dissolved and particulate nutrients in a minimally disturbed Mediterranean river with intermittent flow. J. Hydrol. 373, 218–229.

Solagaistua, L., Arroita, M., Aristi, I., Larrañaga, A., Elosegi, A., 2015. Changes in discharge affect more surface than subsurface breakdown of organic matter in a mountain stream. Mar. Freshw. Res. 67, 1826–1834.

Stanley, E.H., Boulton, A.J., 1995. Hyporheic processes during flooding and drying in a Sonoran Desert stream. I: hydrologic and chemical dynamics. Arch. Hydrobiol. 134, 1–26.

Stanley, E.H., Fisher, S.G., Grimm, N.B., 1997. Ecosystem expansion and contraction in streams. Bioscience 47, 427–435.

Steward, A.L., von Schiller, D., Tockner, K., Marshall, J.C., Bunn, S.E., 2012. When the river runs dry: human and ecological values of dry riverbeds. Front. Ecol. Environ. 10, 202–209.

Tank, J.L., Rosi-Marshall, E.J., Griffiths, N.A., Entrekin, S.A., Stephen, M.L., 2010. A review of allochthonous organic matter dynamics and metabolism in streams. J. N. Am. Benthol. Soc. 29, 118–146.

Timoner, X., Acuña, V., von Schiller, D., Sabater, S., 2012. Functional responses of stream biofilms to flow cessation, desiccation and rewetting. Freshw. Biol. 57, 1565–1578.

Tockner, K., Malard, F., Uehlinger, U., Ward, J.V., 2002. Nutrients and organic matter in a glacial river-floodplain system (Val Roseg, Switzerland). Limnol. Oceanogr. 47, 266–277.

Triska, F.J., Duff, J.H., Avanzino, R.J., 1990. Influence of exchange flow between the channel and hyporheic zone on nitrate production in a small mountain stream. Can. J. Fish. Aquat. Sci. 47, 2099–2111.

Twinch, A.J., 1987. Phosphate exchange characteristics of wet and dried sediment samples from a hypertrophic reservoir: implications for the measurements of sediment phosphorus status. Water Res. 21, 1225–1230.

Tzoraki, O., Nikolaidis, N.P., Amaxidis, Y., Skoulikidis, N.T., 2007. In-stream biogeochemical processes of a temporary river. Environ. Sci. Technol. 41, 1225–1231.

Valett, H.M., Morrice, J.A., Dahm, C.N., Campana, M.E., 1996. Parent lithology, surface-groundwater exchange, and nitrate retention in headwater streams. Limnol. Oceanogr. 41, 333–345.

Valett, H.M., Baker, M.A., Morrice, J.A., Crawford, C.S., Molles Jr., M.C., Dahm, C.N., et al., 2005. Biogeochemical and metabolic responses to the flood pulse in a semiarid floodplain. Ecology 86, 220–234.

Vazquez, E., Amalfitano, S., Fazi, S., Butturini, A., 2011. Dissolved organic matter composition in a fragmented Mediterranean fluvial system under severe drought conditions. Biogeochemistry 102, 59–72.

Vázquez, E., Ejarque, E., Ylla, I., Romani, A.M., Butturini, A., 2015. Impact of drying/rewetting cycles on the bioavailability of dissolved organic matter molecular-weight fractions in a Mediterranean stream. Freshwater Sci. 34, 263–275.

Vázquez, E., Romaní, A.M., Sabater, F., Butturini, A., 2007. Effects of the dry-wet hydrological shift on dissolved organic carbon dynamics and fate across stream-riparian interface in a Mediterranean catchment. Ecosystems 10, 239–251.

von Schiller, D., Martí, E., Riera, J.L., Ribot, M., Marks, J.C., Sabater, F., 2008. Influence of land use on stream ecosystem function in a Mediterranean catchment. Freshw. Biol. 53, 2600–2612.

von Schiller, D., Acuña, V., Graeber, D., Martí, E., Ribot, M., Sabater, S., et al., 2011. Contraction, fragmentation and expansion dynamics determine nutrient availability in a Mediterranean forest stream. Aquat. Sci. 73, 485–497.

von Schiller, D., Marcé, R., Obrador, B., Gómez, L., Casas, J.P., Acuña, V., et al., 2014. Carbon dioxide emissions from dry watercourses. Inland Waters 4, 377–382.

von Schiller, D., Graeber, D., Ribot, M., Timoner, X., Acuña, V., Martí, E., et al., 2015. Hydrological transitions drive dissolved organic matter quantity and composition in a temporary Mediterranean stream. Biogeochemistry 123, 429–446.

Watts, C.J., 2000. Seasonal phosphorus release from exposed, re-inundated littoral sediments of two Australian reservoirs. Hydrobiologia 431, 27–39.

Welter, J.R., Fisher, S.G., Grimm, N.B., 2005. Nitrogen transport and retention in an arid land watershed: influence of storm characteristics on terrestrial—aquatic linkages. Biogeochemistry 76, 421–440.

Whitworth, K.L., Baldwin, D.S., Kerr, J.L., 2012. Drought, floods and water quality: drivers of a severe hypoxic blackwater event in a major river system (the southern Murray–Darling Basin, Australia). J. Hydrol. 450–451, 190–198.

Wilson, J.S., Baldwin, D.S., 2008. Exploring the "Birch effect" in reservoir sediments: influence of inundation history on aerobic nutrient release. Chem. Ecol. 24, 379–386.

Woodward, K.B., Fellows, C.S., Mitrovic, S.M., Sheldon, F., 2015. Patterns and bioavailability of soil nutrients and carbon across a gradient of inundation frequencies in a lowland river channel, Murray–Darling Basin, Australia. Agric. Ecosyst. Environ. 205, 1–8.

Ylla, I., Sanpera-Calbet, I., Vázquez, E., Romaní, A.M., Muñoz, I., Butturini, A., et al., 2010. Organic matter availability during pre-and post-drought periods in a Mediterranean stream. Hydrobiologia 657, 217–232.

Zarnetske, J.P., Haggerty, R., Wondzell, S.M., Baker, M.A., 2011. Dynamics of nitrate production and removal as a function of residence time in the hyporheic zone. J. Geophys. Res. 116, G01025.

Zoppini, A., Marxsen, J., 2011. Importance of extracellular enzymes for biogeochemical processes in temporary river sediments during fluctuating dry–wet conditions. In: Shukla, G., Varma, A. (Eds.), Soil Enzymology SE-6, Soil Biology. Springer, Berlin, pp. 103–117.

THE BIOTA OF INTERMITTENT RIVERS AND EPHEMERAL STREAMS: PROKARYOTES, FUNGI, AND PROTOZOANS

**Anna M. Romaní*, Eric Chauvet[†], Catherine Febria[‡], Juanita Mora-Gómez*,
Ute Risse-Buhl[§], Xisca Timoner**, Markus Weitere[§], Lydia Zeglin[¶]**

University of Girona, Girona, Spain University of Toulouse, CNRS, Toulouse, France[†]
University of Canterbury, Christchurch, New Zealand[‡] Helmholtz Centre for Environmental
Research—UFZ, Magdeburg, Germany[§] Kansas State University, Manhattan, KS, United States[¶] ICRA, Girona, Spain***

IN A NUTSHELL

- Microbial diversity and function in intermittent rivers and ephemeral streams (IRES) are tightly linked to specific habitat availability and hydrological phases
- The intensity and frequency of different phases (especially drying and rewetting) affect community composition and key functions, mainly linked to biogeochemical processes
- Resistance and resilience strategies are distinct among microorganisms and highly dependent on different types of refuges
- Microbial food webs in IRES can differ between hydrological phases, affecting key ecosystem functions and higher trophic levels

4.1.1 ROLE AND RELEVANCE OF MICROBES IN IRES

Microbial communities represent a critical link between the terrestrial environment and freshwater food webs. Microorganisms (microbes) broadly include bacteria, archaea, protozoa (i.e., ciliates, heterotrophic flagellates, amoebas), and fungi as well as photosynthetic algae and cyanobacteria. These highly variable communities are found throughout the water column, on submerged surfaces such as rocks, sediments, leaves, and wood, and in interstitial water of benthic sediments (e.g., Aumen et al., 1983; Findlay et al., 1993; Lock, 1993; Artigas et al., 2009). Microbes grow and develop in benthic habitats, forming biofilms which are "hot spots" of riverine microbial metabolism (Pusch et al., 1998; Battin et al., 2016). In benthic habitats, microbial food web interactions are usually complex; protozoans may feed on bacteria, algae, other protozoans, and detritus, and bacteria, archaea, and fungi may compete for available organic matter (OM) (e.g., Romaní et al., 2006a; Risse-Buhl et al., 2012; Wey et al., 2012). However, over time and usually in downstream reaches and large rivers, a diverse and active microbial community also develops within the water column (e.g., Weitere and Arndt, 2003).

The growth and development of microbes in river ecosystems are highly sensitive to changes in environmental conditions, including changes in available resources such as dissolved and particulate OM (DOM and POM) and inorganic nutrients, and also changes in physical and chemical conditions (temperature, oxygen concentration, pH, water chemistry, flow velocity) affecting their metabolism. Seasonality in physicochemical conditions and OM availability typifies many perennial rivers and is linked to hydrology and the phenology of riparian vegetation (Allan and Castillo, 2007). In IRES, physicochemical conditions also vary seasonally, but fluctuations are usually more extreme and unpredictable than in perennial systems (Gasith and Resh, 1999; Chapters 3.1 and 3.2).

These hydrological fluctuations can either provide beneficial conditions for microbial growth and development or lead to extremes in certain physical and chemical conditions where microbes cannot develop or even survive (Table 4.1.1). Fluctuating and extreme conditions in stream microhabitats are

Table 4.1.1 Environmental factors, many associated with fluctuating hydrology that may significantly affect microorganisms in IRES

Environmental factor	Extremes and conditions in IRES	Effects on microorganisms	References
Temperature	Extremely high in shrinking pools of stagnant water during drying phase	Increase metabolism, interact with other harsh environmental conditions	Ylla et al. (2010) Mora-Gómez et al. (2015, 2016)
Oxygen	Low oxygen values in pools, especially with accumulated organic matter; may reach anoxia	Inhibit decomposition Shift to anaerobic metabolism	Medeiros et al. (2009)
Dissolved organic matter (DOM)	Accumulation of dissolved humic compounds in pools, especially in forested catchments	Inhibit or modify microbial activity and community composition	Fazi et al. (2013)
	High DOM availability in rewetting episodes after drought	Enhance heterotrophic metabolism	Romaní et al. (2006b)
Inorganic nutrients	High NH_4 concentrations during drying period, high NO_3 concentrations during rewetting period	Possible cause for shifting community composition of microbes	Tzoraki et al. (2007)
Flow velocity	High flow velocity during floods	Shear stress, abrasion, loss (transport) of benthic microbial biomass	Zoppini et al. (2010)
	Stagnant water in pools during drying phase	Limit diffusion of nutrients	
Light	Intense UV radiation during drought	Inhibit microbial metabolism Photodegradation of plant/algal material	Dieter et al. (2013) Austin and Vivanco (2006)

Selected references are included as examples.

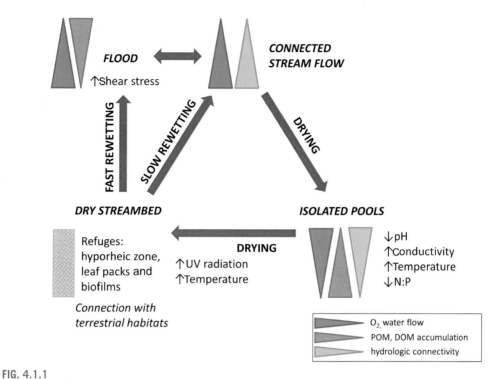

FIG. 4.1.1

Hydrological phases and associated environmental and physical-chemical factors affecting microbial communities in IRES.

linked to the different hydrological phases in IRES (Fig. 4.1.1, Chapters 2.2 and 2.3). During drying, surface-water longitudinal connectivity is gradually lost (Chapter 2.3) and pools form. In these pools, there are extremes of low oxygen and high temperature and conductivity (Chapter 3.1) that collectively affect DOM characteristics and determine changes in microbial community structure and function (Table 4.1.1). High temperatures in pools with large accumulations of POM such as leaf litter determine the decrease in oxygen and pH due to respiration. During the dry-streamed phase, the microbiology and geochemistry of aquatic sediments share striking similarities to surface soils (Burns et al., 2013). During this phase, very high solar irradiation can be critical for microbes, especially when unshaded by riparian vegetation. Rewetting and flow resumption leads to dramatic changes in environmental conditions, affecting microbes either negatively (e.g., shear stress) or positively (e.g., increasing DOM availability, Table 4.1.1). Microbial dynamics often occur in pulses in concert with the rising and falling limb of hydrographs and changes in associated environmental drivers (usually linked to POM and DOM).

Microbes play key roles in riverine ecosystem function, being involved in most biogeochemical processes linked to carbon and nutrient cycling as well as particle retention (Leff et al., 2016). Many microbial processes depend on OM availability, and the quantity and/or quality of OM varies with river compartment (e.g., water column, sediment, coarse and fine POM, and rock surface substrata) and hydrological variability in IRES (Chapter 3.2). Thus, any modification of the microbial environment, community composition, and functioning may affect crucial ecosystem functions. At the same time, although IRES are subject to flow intermittence, many of them periodically reconnect to a perennial fluvial network (Chapter 2.3), and processes in one IRES reach or segment often have repercussions downstream.

In this chapter, we review heterotrophic microbial communities—from prokaryotes to fungi and protozoans—in IRES, linking them with hydrological phases, microhabitats, and various ecosystem functions. Specific strategies that determine differential capacities for resistance and resilience of these groups are highlighted. After an overview of the dynamics of microbial functioning and diversity in IRES, we conclude with a synthesis of concepts and propose some promising hypotheses and avenues for future research.

4.1.2 DIVERSITY OF PROKARYOTES IN IRES
FACTORS CONTROLLING PROKARYOTIC COMMUNITIES IN IRES

Prokaryotes are ubiquitous (Box 4.1.1) but also highly sensitive to changing environmental conditions; the most relevant ones that may affect them in IRES are changes in OM availability (quantity and quality), dissolved oxygen, and temperature (e.g., Fazi et al., 2013; Febria et al., 2015) which are mainly driven by hydrology. Several studies describe the effect of hydrology on prokaryote communities in IRES and highlight the relevance of river biogeochemistry, especially OM availability. For example, the high temporal variation in bacterial community composition in the river water column of one IRES was related to hydrological variability (flooding episodes) and concurrent biogeochemical changes (Portillo et al., 2012). In a Mediterranean IRES, DOM quality was the main driver—among the biogeochemical parameters—of bacterial community composition changes between the base-flow and low-flow periods (Freixa et al., 2016).

The key role of hydrology on prokaryote communities in IRES is further supported by a subset of data summarized by Zeglin (2015) (Fig. 4.1.2). In this evaluation, "hydrology" effects were broadly

BOX 4.1.1 PROKARYOTES IN FRESHWATERS

Prokaryotes—bacteria and archaea—are found in nearly all environments (terrestrial, aquatic, and aerial, including extreme habitats) and often form biofilms. They are ancient life forms, showing very diverse physiologies and metabolic pathways which contrast with their morphological uniformity. Although invisible to the naked eye (most of them are less than 1 µm in size), prokaryotes are essential because they are responsible for key biogeochemical transformations and, at the same time, represent a large portion of global genetic diversity (Whitman et al., 1998). Bacteria and archaea are also important in oxic-anoxic interfaces, being involved in carbon, nitrogen, and sulfur metabolism (Brune et al., 2000).

The dominant phylum of aquatic bacteria is the Proteobacteria. In biofilms, heterotrophic bacteria from the phyla Actinobacteria, Bacteroidetes, and Firmicutes are also typical and, when light is available, Cyanobacteria grow (e.g., Zeglin, 2015). The initial discovery of archaea was from harsh environmental conditions such as salt-saturated lakes, high-temperature terrestrial springs, and deep sea vents. Since then, this group has been found to be ubiquitous (Auguet et al., 2010). Bacteria and archaea have been detected coexisting in stream biofilms (Battin et al., 2001; Herrmann et al., 2011) and on decaying leaves (Manerkar et al., 2008).

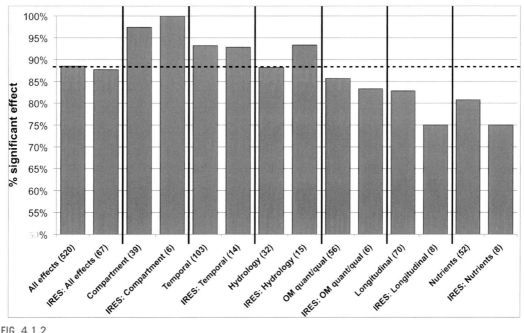

FIG. 4.1.2

Distribution of significant effects of categories of spatiotemporal variation on stream microbial diversity, with data from IRES highlighted and averaged separately. The number of studies for each category is in brackets; the dashed line indicates the percentage of significant effects for all comparisons combined. Some categorical factors included in "all effects" are not compared separately (metals, temperature, and land use) due to <2 studies from IRES. 'OM quant/qual' refers to 'Quantity/quality of organic matter'.

Adapted from Zeglin, L.H., 2015. Stream microbial diversity responds to environmental changes: review and synthesis of existing research. Front. Microbiol. 6, 454.

categorized to include any differences in stream discharge, flow regime, drying, rewetting, or flooding. In addition to hydrology, the most commonly significant factors for prokaryote communities in IRES were among compartment and temporal variation whereas longitudinal and nutrient-driven variation was least commonly significant (Fig. 4.1.2). The nonsignificant effect of OM quantity/quality in this evaluation was probably due to the small sample of studies ($n=6$). Different compartments (water column, sediment, coarse and fine POM, and rock surface substrata) present starkly contrasting environments in terms of OM availability and quality, and there is strong evidence for among-compartment selection leading to dominance by different microbial heterotrophic taxa (Fazi et al., 2005; Gao et al., 2005; Zeglin, 2015).

The heterogeneity of bacterial communities on different Mediterranean stream biofilm substrata was reviewed by Romaní et al. (2013) who found that grain size and OM content strongly determined bacterial community composition. This implies that substrates with greater water-holding capacity (i.e., greater surface area and OM content) might constitute refugia protecting microbial cells from seasonal drying. In addition, specific local conditions in IRES (e.g., conductivity, redox condition, OM quality, temperature) may have selective effects on community composition (e.g., Souza et al., 2006; Zeglin et al., 2011; Joelsson et al., 2013).

Given the variable hydrology of IRES, studies on temporal variation, including drying, rewetting, and flooding effects on prokaryote diversity in them are common, often revealing significant effects on community structure. Predrying conditions and soil moisture appear responsible for changes in bacterial community composition (Febria et al., 2015). Dry sediments, moist sediments, and sediments underlying stagnant pools and flowing waters often have contrasting microbial community composition and activity (Fazi et al., 2005, 2013; Rees et al., 2006; Zeglin et al., 2011; Febria et al., 2012; Pohlon et al., 2013). During the drying (nonflowing) phase, bacterial richness and diversity may decrease, and only a few desiccation-tolerant bacterial taxa dominate, including terrestrial or airborne microbes (Rees et al., 2006; Febria et al., 2012; Timoner et al., 2014a).

High-flow conditions also modulate microbial community structure and function (Lyautey et al., 2005; Zoppini et al., 2010). In desert streams, brief rains may restore needed humidity for some microorganisms and may substantially change the bacterial community composition (Abed et al., 2011). Yet, flow status does not always have marked effects on stream microbial communities (Marxsen et al., 2010; Febria et al., 2015), possibly as a result of populations persisting in isolated favorable microhabitats within the stream reach (Section 4.1.5) or of community turnover at timescales beyond those within the studies.

PROKARYOTIC DIVERSITY IN IRES VS PERENNIAL RIVERS AND STREAMS

IRES bacterial communities diverge from those in streams with more stable hydrology. Although sequence-library data (based on 16S rRNA) are limited, Zeglin (2015) reported highest relative abundances in gamma-Proteobacteria from IRES sediments, while IRES water-column communities had the lowest beta-Proteobacteria representation compared to all other streams (Fig. 4.1.3). Sequence-library data from a headwater stream in the mid-Atlantic United States also show distinctive bacterial communities at intermittent versus perennial sites (Fig. 4.1.4). This comparative network analysis suggested that perennial waterways had central microbial associations with generalist methanogenic bacteria whereas IRES had microbial associations dominated by Nitrospirales, or N-fixers known to have flexible metabolisms highly suited for changing redox conditions (Febria et al., 2015; Koch et al., 2015). Actinobacteria and Firmicutes might be more dominant in IRES (Fazi et al., 2008; Timoner et al., 2014a) since these phyla have several rRNA operons in the genomes (seven and three on average, respectively), which is considered key to the competitive success during periods of maximum desiccation (Klappenbach et al., 2000). Also, both phyla have a gram-positive cell-wall type and Firmicutes includes many endospore-forming genera, which may also favor surviving desiccation in IRES.

Riverine ecosystems may harbor a diverse archaeal community but there are few data on archaea from rivers and fewer from IRES. Although we cannot directly compare archaeal diversity between IRES and perennial rivers, we predict a key role of this microbial group in IRES because many archaea resist harsh environmental conditions (Gao et al., 2005; Box 4.1.1). For instance, archaea occur in significant numbers in arid soils, microbial mats, and freshwater sediments exposed to desiccation (Rothrock and Garcia-Pichel, 2005; Soule et al., 2009; Conrad et al., 2014).

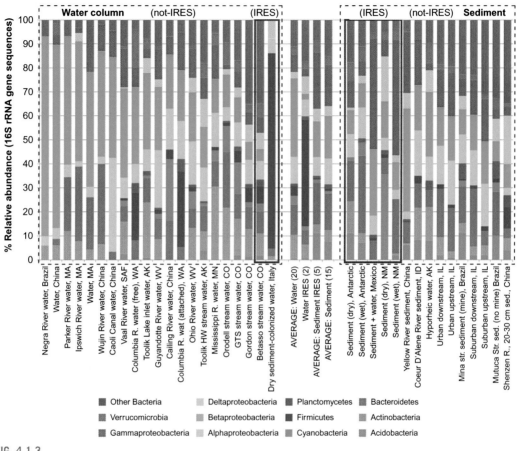

FIG. 4.1.3

Relative abundance of major bacterial phyla and subphyla (based on 16S rRNA gene-sequence libraries) for surface waters and sediments, with data from IRES highlighted and averaged separately. Water-column data are sorted by Beta-Proteobacteria relative abundance (high to low, left to right); sediment data are sorted by "other Bacteria" relative abundance (within IRES, high to low, left to right; and within perennial rivers and streams, low to high, left to right). Beta-Proteobacteria and "other Bacteria" are the most abundant groups within the water column and sediment compartments, respectively, among all streams on average.

Adapted from Zeglin, L.H., 2015. Stream microbial diversity responds to environmental changes: review and synthesis of existing research. Front. Microbiol. 6, 454.

4.1.3 DIVERSITY OF FUNGI IN IRES
FACTORS CONTROLLING FUNGAL COMMUNITIES IN IRES

Fungal communities and their role in decomposing OM are influenced by environmental variables such as temperature, pH, concentrations of oxygen and dissolved nutrients, physical abrasion, and hydromorphological parameters (Webster and Benfield, 1986; Young et al., 2008; Tank et al., 2010).

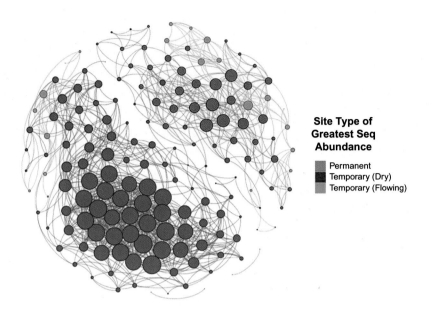

Site Type of Greatest Seq Abundance

■ Permanent
■ Temporary (Dry)
■ Temporary (Flowing)

FIG. 4.1.4

Microbial associations differ between permanent and intermittent streams. A network analysis reveals 167 distinct bacterial operational taxonomic units (OTUs; each represented as an individual circle) connected to one another with lines. Nodes (*circles*) represent individual OTUs and edges represent significant Spearman correlations ($\rho > 0.75$ and $P < 0.05$). Node size (i.e., size of the circle) is determined by the weight of the edges connected to each node. The resulting network clusters the OTUs into two groups: one associated with permanent streams and a second associated with intermittent streams. Bacterial communities from the permanent streams showed high centrality among OTUs that were almost entirely members of the Proteobacteria phylum with only one OTU belonging to Bacteroidetes.

Modified from Febria, C.M., Hosen, J.D., Crump, B.C., Palmer, M.A., Williams, D.D., 2015. Microbial responses to changes in flow status in temporary headwater streams: a cross-system comparison. Front. Microbiol. 6, 522.

In IRES, fungi are potentially exposed to extreme seasonal variation principally associated with the drying phase. Consequently, temperature variation and water-quality changes during summer drying appear to be the principal drivers of fungal community composition (Dang et al., 2009; Mora-Gómez et al., 2016; Box 4.1.2). Conditions in isolated pools during streambed drying have opposing effects on fungal activity: low pH and oxygen are adverse while high conductivity and temperature favor fungal development (Medeiros et al., 2009; Mora-Gómez et al., 2015; Canhoto et al., 2016). Before being exposed to drying, leaf-associated fungal communities are often exposed to increased temperatures in dwindling pools which may stimulate fungal development and favor adapted species (Canhoto et al., 2016). During the drying phase, leaf-associated fungi are strongly affected by desiccation, their development being faster and more extended when drying is less severe (e.g., Bruder et al., 2011).

The drying period may be fatal for fungi except when mycelia and propagules are protected from complete desiccation (e.g., in humid refuges, Section 4.1.5). Aquatic hyphomycetes show exemplary adaptations to flowing waters, including conidial shape (mostly tetracladiate, branched or elongated), production of mucilage, and rapid germination (Read et al., 1992). Conversely, their thin-walled and delicate spores and mycelia make them vulnerable to potentially fatal desiccation (Bärlocher, 2009).

BOX 4.1.2 FRESHWATER FUNGI

Aquatic fungi are quite diverse, although less so than terrestrial fungi, with more than 3000 reported species, including from marine habitats (Bärlocher and Boddy, 2016). Five out of the seven phyla in the Kingdom Fungi have freshwater representatives: Chytridiomycota, Cryptomycota, Blastocladiomycota (which contains only one order), Ascomycota, and Basidiomycota (Hibbett et al., 2007; Jones et al., 2014; Wurzbacher et al., 2015). Most freshwater taxa belong to Ascomycota, including mitosporic taxa, and Chytridiomycota (Shearer et al., 2007). Basidiomycota and fungal-like organisms (except Oomycota, Saprolegniales) are much less common in aquatic than in terrestrial environments.

While several taxa are parasites of or cause diseases to various organisms, some freshwater fungi are saprotrophs and are involved in the decomposition and mineralization of organic matter. Aquatic hyphomycetes, also known as Ingoldians, are anamorphic fungi that, based on phylogenetic studies, are presumed to belong mainly to Ascomycota (Shearer et al., 2007). This group includes more than 320 species (Descals, 2005; Fig. 4.1.5). Their roles as early colonists and decomposers of leaf litter in streams are well documented (Bärlocher, 2015).

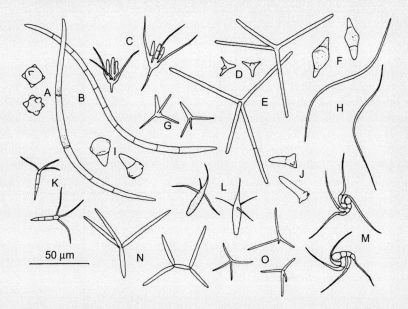

FIG. 4.1.5

Some conidia of aquatic hyphomycetes: (A) *Goniopila monticola*, (B) *Anguillospora longissima*, (C) *Tetracladium setigerum*, (D) *Heliscella stellata*, (E) *Lemonniera aquatica*, (F) *Tumularia aquatica*, (G) *Clavatospora longibrachiata*, (H) *Flagellospora curvula*, (I) *Tumularia tuberculata*, (J) *Heliscus lugdunensis*, (K): *Alatospora flagellata*, (L) *Jaculispora submersa*, (M) *Gyoerffyella rotula*, (N) *Articulospora tetracladia*, and (O) *Alatospora acuminata*. A. acuminata, A. longissima, L. aquatica, and F. curvula, appear to be particularly common in IRES.

From Chauvet, E., 1990. Hyphomycètes aquatiques du Sud-Ouest de la France. Gaussenia 6, 3–31.

Spores in particular are fragile and quickly lose their ability to germinate, even under benign conditions (Sridhar and Bärlocher, 1994). However, aquatic hyphomycetes have been regularly reported from various environments outside running waters, including terrestrial niches such as stream banks, dew, canopy waters, and tree holes, implying some potential for survival in conditions that are not permanently aquatic (Chauvet et al., 2016).

In addition, low precipitation in summer produces water stress on riparian vegetation leading to early and prolonged leaf abscission (Chapter 4.2). As fungal development in freshwater systems is highly linked to the quantity and quality of available decaying plant material, such drying effects on riparian vegetation can affect fungal community composition and their capabilities for degrading OM. For instance, the fungal community is directly affected via abiotic processes such as solar radiation and photodegradation effects on OM during the terrestrial phase (Austin and Vivanco, 2006; Gallo et al., 2006; Uselman et al., 2011). Leaf litter quality might be changed by direct UV radiation in the dry streambed (Dieter et al., 2011), which can have some legacy effects once flow resumes, affecting decomposition rates and fungal community composition (Dieter et al., 2013; Mora-Gómez, 2014).

Finally, streambed composition may affect fungal diversity. Sandy clay loam in intermittent streamlets had richer communities than sandy loam, apparently due to a better filterability of filiform than branched conidia (Ghate and Sridhar, 2015).

FUNGAL DIVERSITY IN IRES VS PERENNIAL RIVERS AND STREAMS

Fungal community composition in IRES has rarely been compared with that of perennial rivers although it is expected that fungal communities in IRES have traits that better adapt them to IRES environmental conditions. For example, the use of humid refugia, such as the hyporheic zone, may select for species with filiform spores and thus able to disperse in interstitial waters (e.g., *Flagellospora curvula*) against those with compact (e.g., *Heliscus lugdunensis*) or branched/tetraradiate morphologies (e.g., *Lemonniera aquatica*, Fig. 4.1.5; Cornut et al., 2014). Also, in contrast to aquatic hyphomycetes, the development of aeroaquatic mitosporic fungi depends on periodic drying of their habitat (Shearer et al., 2007), which may argue for adaptation to IRES. Nevertheless, they have rarely been reported in IRES, probably due to the lack of appropriate studies. Chytrids, which may survive periods of desiccation and occupy soil environments as long as a periodic film of surface water allows the dissemination of their zoospores (Shearer et al., 2007), exhibit potential adaptations to IRES. However, a number of chytrid taxa are substrate- or host-specific which restricts their development in constrained environments.

There is limited evidence allowing us to answer basic questions about differences in richness and diversity of fungal communities between IRES and perennial rivers. Maamri et al. (2001) found higher numbers of sporulating species in a perennial stream than in an IRES in Morocco, although similar species dominated in both systems. Using molecular techniques, Foulquier et al. (2015) did not find differences in fungal community composition and response to emersion frequency between an intermittent and a perennial reach in a French river. However, this study evaluated two nearby reaches in the same IRES and they could have been connected during wet season. A general comparison of fungal communities between IRES and perennial rivers does not exist in the literature, perhaps because discrepancies in methodologies and environmental differences among regions and rivers can affect the number of species found in each study. Notwithstanding, we compiled data from IRES from the temperate region (Table 4.1.2), which indicate some 10–27 different fungal taxa may occur, among which a few species of aquatic hyphomycete tend to dominate (Fig. 4.1.5).

Table 4.1.2 Number of taxa and dominant fungal species found in different IRES

Location	Number of taxa	Leaf species	Dominant fungal species	References
Eastern France	27 (OTUs)	*Alnus glutinosa*		Foulquier et al. (2015)
Pyrenees, France		*Alnus glutinosa, Quercus ilex*	*Flagellospora curvula, Lemonniera terrestris, L. aquatica*	Bruder et al. (2011)
Southwestern India	18 species (sporulating)	(sediments)	*Anguillospora longissima, Cylindrocarpon* sp., *Flagellospora curvula*	Ghate and Sridhar (2015)
Northern Spain	10–25 (OTUs)	*Populus nigra*		Mora-Gómez (2014)
Eastern Morocco	19 species (sporulating)	*Salix pedicellata*	*Alatospora acuminata, Anguillospora longissima, Lemonniera aquatica, Tetracladium marchalianum, Dactylella submersa*	Maamri et al. (2001)
Northern Spain	11 species (sporulating)	*Platanus acerifolia, Populus nigra*	*Clavariopsis aquatica, Alatospora acuminata, Lemonniera* sp.	Artigas et al. (2008)

OTUs refer to Operational Taxonomic Units based on molecular analysis of fungal community.

4.1.4 DIVERSITY OF PROTOZOANS IN IRES
FACTORS CONTROLLING PROTOZOAN COMMUNITIES IN IRES

Protozoans can cope with changing environmental conditions during contraction and fragmentation of IRES because many species tolerate changes in salinity, oxygen, and temperature (Fenchel et al., 1989; Norf et al., 2007; Finlay and Esteban, 2009). This is supported by the high diversity of protozoan assemblages, which provides a pool of genotypes well adapted to different environmental conditions. Furthermore, protozoans also occur in soil systems (e.g., Coûteaux and Darbyshire, 1998; Domonell et al., 2013) and can thus survive in sediments and biofilms of IRES during the dry phase if pore water is still available. Even though community structure of protozoan assemblages differs between contrasting habitats, most protozoan groups are found in a range of different habitats and the overlap in species between habitats is considerable (Box 4.1.3, compare Weitere and Arndt (2003) as example for aquatic

BOX 4.1.3 PROTOZOA IN FRESHWATERS

Protozoans are small eukaryotic, heterotrophic, and unicellular organisms ranging from 1 μm to more than 1 mm that usually (but not exclusively) feed on particles including bacteria, algae, other protozoans, small metazoans, and detritus. They are widely distributed in almost all habitats worldwide (e.g., Finlay, 2002). In streams and rivers, they occur in all compartments, including the water column, sediment, biofilms on both organic and mineral surfaces, and the hyporheic zone (e.g., Franco et al., 1998; Cleven, 2004, Plebani et al., 2015) where they play a key role in microbial food webs. Due to their high reproduction rates and dispersal rates from neighboring habitats (e.g., Wey et al., 2009), they are able to rapidly colonize new habitats. Common protozoans found in rivers and streams include flagellates, ciliates, naked amoeba, and testate amoeba (Fig. 4.1.6).

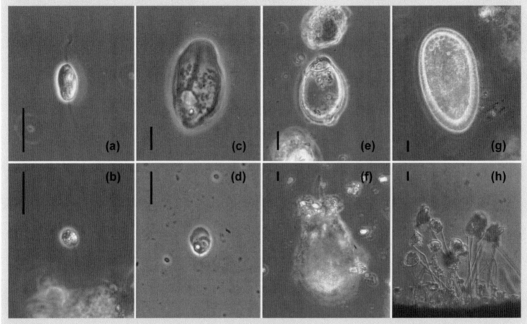

FIG. 4.1.6

Images of different protozoan groups typical of stream biofilms: (a) mobile flagellate, Kinetoplastea, (b) mobile flagellate, Ochromonadales, (c) naked striate amoeba, (d) naked fan-shaped amoeba, Discosea, Vannellida, (e) filose testate amoeba, Euglyphida, *Corythion* sp., (f) lobose testate amoeba, Difflugina, *Difflugia* sp., (g) mobile filter-feeding ciliate, Hymenostomatia, *Frontonia* sp., and (h) sessile filter-feeding ciliate, Peritrichia, *Vorticella* sp. Scale = 10 μm.

Photos: Courtesy U. Risse-Buhl.

protozoans with Domonell et al. (2013) as example for terrestrial protozoans). Nevertheless, the main physical factor in IRES affecting protozoa is drying since their occurrence depends on the availability of at least some free water.

Changes in temperature and biogeochemical conditions are especially relevant for protozoan development in IRES. Increases in water temperature affect all protozoan physiological rates such as respiration, growth, and feeding (e.g., Müller and Geller, 1993; Kathol et al., 2009). However, the effect of increasing temperature on population density depends on resource availability, and effects on protozoan community structure are often small (Norf and Weitere, 2010). Fluctuating temperatures reduce the importance of competition in population control and promotes more even abundances of protozoan species (Eddison and Ollason, 1978).

Changing biogeochemical conditions (e.g., concentrations of DOM, $N-NH_4^+$, $N-NO_3^-$, salts) in IRES also may alter protozoan community structure. Amoebas tolerate a broad range of ion concentrations (Hauer and Rogerson, 2005) while slow changes and some preconditioning ensure the survival of ciliate species to habitats with increasing ion concentrations (Smurov and Fokin, 1999; Salvadó et al., 2001). Interestingly, aquatic flagellates seem to be more tolerant to changes in salt concentrations than their soil counterparts (Ekelund, 2005). During rewetting of IRES, increasing concentrations of easily degradable OM is expected to stimulate activity and production of the protozoan community (e.g., Christensen et al., 1992).

In the case of oxygen, the microscopic size of protozoans and thus the short diffusion distances imply that oxygen concentrations are less critical than for metazoans (Hausmann et al., 2003). Some species adapt their metabolism under hypoxia (Raugi et al., 1975) whereas other protozoans living in hypoxic or anaerobic conditions harbor cytoplasmic, single-celled autotrophs or methanogenic bacteria that either provide oxygen derived from photosynthesis or metabolize hydrogen.

PROTOZOAN DIVERSITY IN IRES VS. PERENNIAL RIVERS AND STREAMS

The diversity and importance of protozoans in IRES have rarely been analyzed. However, attributes of protozoans that are important for tolerating and surviving the dynamic changes in IRES have been studied in other freshwater (e.g., eutrophic streams and rivers, splash zones) and terrestrial habitats such as soils. For instance, flow intermittence and the concurrent accumulation of leaf litter may affect ciliate abundances and community composition as seen in an intermittent pond where small bacterivore species associated with nutrient- or OM-enriched conditions dominated the ciliate community (Andrushchyshyn et al., 2006).

On the other hand, the temporal occurrence and frequency of disturbance (e.g., drying and rewetting) affects species richness and structure of protozoan communities. In a Mediterranean IRES, leaf litter was colonized by a high diversity of protozoan species, triggered by hydrological variability that promoted rapid colonization during stable wet phases but caused a drift of the fauna and subsequent colonization after a flood (Gaudes et al., 2009). In microcosms simulating rain pools, periodic disturbance by drying at 5-day intervals interrupted succession, preventing development of "late successional" species such as large predatory ciliates (McGrady-Steed and Morin, 1996). The authors concluded that the variation in hydroperiod rather than the frequency of disturbance contributes to enhanced species diversity. These studies suggest that changing environmental factors in IRES act as filters on the occurrence and biodiversity of the protozoan community.

4.1.5 RESISTANCE AND RESILIENCE OF MICROBES IN IRES

Bacteria, archaea, fungi, and protozoans have evolved several strategies to resist hydrological stress and concurrent changes in environmental conditions. These strategies vary depending on the compartment where they develop and the specific hydrological phase (Fig. 4.1.7).

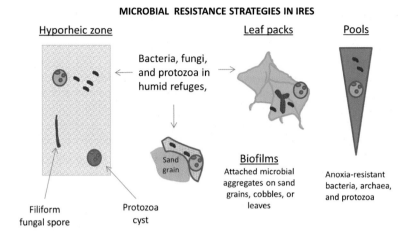

FIG. 4.1.7

Summary of the main microbial resistance strategies in IRES to cope with the harsh conditions at different hydrological phases. Typical refuges in distinct microhabitats and life stage adaptations are shown.

REFUGIAL HABITATS

Some river habitats and microbial life strategies are especially useful for survival of drying and rewetting periods. At the habitat scale, the most important compartments providing refuges for microbes are the hyporheic zone and leaf packs (Ylla et al., 2010; Febria et al., 2012). Both types of compartments retain water and provide aquatic microhabitats. The role of the hyporheic zone as a refuge for bacteria was suggested by Timoner et al. (2014a) in a Mediterranean IRES where the bacterial communities on sand experienced only small changes due to flow intermittence in contrast to abrupt changes in communities on cobbles (Fig. 4.1.8). Bacterial communities in IRES during dry periods are subsets of communities present during inundation, especially in sandy compartments, and thus resilience is insured by the "seed bank" community that resists drying during nonflowing periods (Zeglin et al., 2011).

In the hyporheic zone, large fluctuations in oxic-anoxic conditions can limit the persistence of some bacterial species but many archaea species may persist. For instance, methanogens survive oxygen and desiccation stress in aerated soils (Angel et al., 2012), and Conrad et al. (2014) observed that desiccation deactivates methane production which resumes immediately after reflooding. Ammonia-oxidizing archaea also occur in soils and sediments exposed to drying-rewetting stress (Herrmann et al., 2011; Thion and Prosser, 2014).

The hyporheic zone is also a potential source of fungal propagules when surface flow ceases although conditions for fungal development in this habitat are not optimal. Biomass and diversity of aquatic hyphomycetes are much lower in the hyporheic zone than in benthic compartments (Bärlocher et al., 2006; Cornut et al., 2010; Sudheep and Sridhar, 2012). These patterns are consistent with the lower rates of leaf litter decomposition reported from this habitat, mainly due to the lower dissolved oxygen concentrations, even though some aquatic hyphomycetes show adaptations to these stresses (Field and Webster, 1983; Medeiros et al., 2009).

During the drying phase, protozoans can also retreat into the hyporheic zone. At the microscale, capillary and hygroscopic water adhering to inorganic or OM particles is sufficient to provide a niche

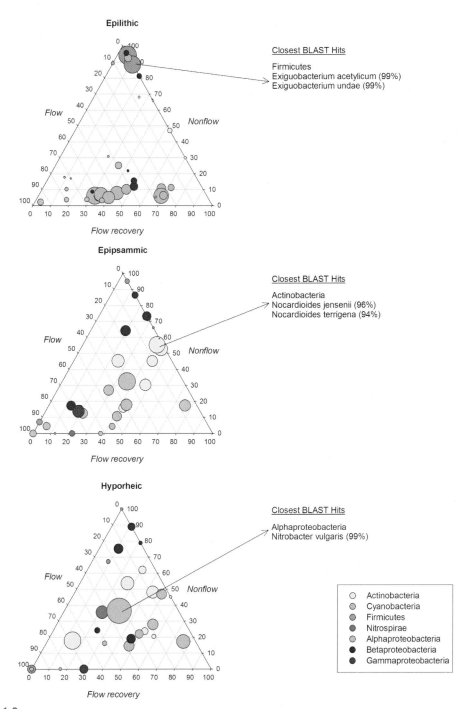

FIG. 4.1.8

Ternary plots showing the distribution of the 43 most common Operational Taxonomic Units (OTUs) (>500 members) associated with each phase (flow, nonflow, flow recovery) in each biofilm compartment (epilithic, epipsammic, and hyporheic) from a Mediterranean IRES. The bubble's position in the triangle indicates the relative abundance of each OTU among the three phases and the size of the bubble represents the relative abundance of each OTU.

Data modified from Timoner, X., Borrego, C.M., Acuña, V., Sabater, S., 2014a. The dynamics of biofilm bacterial communities is driven by flow wax and wane in a temporary stream. Limnol. Oceanogr. 59, 2057–2067.

for active protozoan cells. The ability to persist under these conditions is size dependent; nano-sized protozoans (2–20 µm, especially flagellates and amoebas) have an advantage over microsized protozoans. Also, the evolution of different motility strategies allows the exploitation of bacterial populations that grow in the capillaries and on surfaces of sediment or soil grains (Wang et al., 2005).

Leaf packs also serve as refuges for IRES microbes, especially fungi. Mycelia embedded in the leaf matrix within leaf accumulations exposed to air on the streambed may survive longer due to the leaf packs' water-holding capacity. Maamri et al. (1998, 2001) reported that leaf-associated aquatic fungi in Mediterranean IRES may survive drying-rewetting cycles, with drying phases lasting up to several months. In addition, water films on litter surfaces can transport spores, contributing to fungal dissemination downward and potentially upward (Chauvet et al., 2016). Flow intermittence may represent a challenge for aquatic fungi similar to survival in terrestrial habitats. A major limitation, however, is the patchy distribution of leaf packs on streambed compared to the continuous layer in forest floor, reducing the water-holding capacity and confining spore inoculation to the limited number of adjacent leaves. Nevertheless, leaf-associated microbial decomposers appear highly resilient to intermittence compared to macroinvertebrate decomposers (Datry et al., 2011).

Remnant pools also serve as refuges for microbes. However, conditions, such as low oxygen content and/or accumulation of leaf leachates, can limit the persistence of some species. In the isolated pools of a Mediterranean IRES, the bacterioplankton community changed and declined in diversity according to the shifting redox regime and the quality of the DOM (Fazi et al., 2013). Fungi are sensitive to the harsh conditions in pools, while bacteria seem to be more resistant (Mora-Gómez et al., 2016).

During extended dry periods, which are more characteristic of desert streams, aquatic microbial habitats may be restricted to deep groundwater flowpaths (Dahm et al., 2003), only appearing at the surface near rare perennial springs fed by these waters. The geochemistry of these refuge waters, which often accumulate high concentrations of solutes (Grimm et al., 1997), may select for microbial salt tolerance (Souza et al., 2006; Zeglin et al., 2011) and this halotolerance could contribute to microbial resilience (Northcott et al., 2009).

MICROBIAL LIFE STRATEGIES

Microbial life strategies are also relevant for their survival in IRES. The heterogeneous assemblage within biofilm is advantageous for microbial communities because its three dimensions and consistency protect against desiccation and are a "hot spot" for nutrients and OM trapped within the biofilm matrix (Lock, 1993; Battin et al., 2016). Within the biofilm, prokaryotes feed on exudates from living algae and on dried and decaying algae. The extracellular polymeric matrix is crucial for the maintenance of humidity within the biofilm microenvironment during drying (Roberson and Firestone, 1992) and, at the same time, living tightly attached to a surface and within a biofilm structure helps tolerate shear forces during rapid rewetting by floods. A similar life strategy is the development of microbial mats, typical in extreme conditions such as drought and/or high salinity (e.g., polar desert streams, Stanish et al., 2013). Microbial mats are formed by cyanobacteria and heterotrophic bacteria (usually related to sulfur metabolism; Dupraz and Visscher, 2005), by trapping and binding sediment, precipitating carbonate to create layered structures similar to stromatolites. They occur where grazing pressure is low (Bottjer et al., 1996), as may happen under extreme drought conditions. These strategies resemble the encrusting form of life named "biological soil crust" dominating terrestrial arid environments (Ferrenberg et al., 2015).

Other microbial life strategies relevant to surviving in IRES are related to specific life phases, including dormant and spore-forming cells (Chapter 4.8). Several representatives of all protozoan taxonomic groups form resistant resting cysts (e.g., Verni and Rosati, 2011). Cyst formation is stimulated by worsening conditions such as starvation, high population density, increasing salinity and temperature, and presence of toxic substances (e.g., Dallai et al., 1987; Ekelund et al., 2002). The cysts are extremely resistant against environmental stress including high temperatures and prolonged desiccation. The cell retreats and is protected by the cyst wall, which is mainly composed of chitin. Some species also use cellulose or silica plates for the cyst's wall. During encystment, metabolic processes slow down and the cell ceases all activities. Once beneficial growth conditions are restored, excystment usually occurs rapidly (Verni and Rosati, 2011). Consequently, the process of encystment and excystment enables rapid colonization in the rewetting phase, similar to recolonization from refuges or from cysts imported by wind and animals. Species richness and abundance of protozoans increase rapidly within a few days (McGrady-Steed and Morin, 1996). As soon as water levels rise and isolated pools reconnect, protozoans in upstream pools can be dispersed to downstream pools by flow.

Similarly, some aquatic hyphomycetes have a dual life cycle, with a terrestrial phase as an endophyte, surviving as mycelia or dormant structures associated with litter on the soil (Selosse et al., 2008; Bärlocher, 2009), completing their saprotrophic aquatic phase. The whole forest canopy (especially riparian tree roots) contains various niches harboring aquatic hyphomycetes which may serve as a reservoir of propagules for IRES (Chauvet et al., 2016). Aquatic hyphomycetes have consistently been reported from surrounding terrestrial environments such as the forest floor, with fungal diversity decreasing with the distance from the stream (Bandoni, 1972; Sridhar and Bärlocher, 1993).

4.1.6 MICROBIAL DIVERSITY AND ECOSYSTEM FUNCTIONING IN IRES

ROLES OF MICROBES IN RIVER ECOSYSTEM FUNCTIONING

Microbes in rivers feed on available OM, utilizing dissolved molecules from flowing or interstitial water or from particulate material. Since available OM, including living or dead algae, macrophytes, bryophytes and terrestrial plants, leaves, wood, and DOM, is usually nutrient-poor with high molar ratios of carbon to nitrogen (C:N) or phosphorus (C:P), microbes in streams and rivers are often nutrient-limited and actively retain nutrients in inorganic dissolved forms of N and P (Aumen et al., 1983; Adams et al., 2015; Manning et al., 2015). Activity of prokaryotes is mainly relevant in benthic compartments (sediment and biofilms) where reduction and oxidation processes occur, driving nutrient cycling and OM decomposition (Leff et al., 2016). In contrast, due to their rapid establishment on leaves and their enzymatic capabilities, aquatic hyphomycetes convert a considerable amount of leaf carbon into CO_2, dissolved and fine particulate organic carbon, and mycelial carbon in streams (Gessner et al., 2007).

Prokaryotes and fungi are not only important OM decomposers but also contribute to the transfer of energy and nutrients to higher trophic levels in stream food webs. Fungal and bacterial biomass accrual on leaves tends to increase leaf N content, and the enzymatic maceration of leaves by microbes results in smaller and less refractory plant polymers, both processes making leaf material more palatable to invertebrates (Bärlocher and Sridhar, 2014; Chapter 4.7). On the other hand, the growth and development of microbial biofilms in benthic compartments (sand, cobbles, hyporheic zone), thanks to their use of available carbon sources, feed higher trophic levels such as small metazoans (Risse-Buhl et al., 2012).

Protozoans mediate significant functions related to their predatory activity. They are the dominant consumers of bacteria and can control densities of both suspended and surface-associated (biofilm) bacteria in streams and rivers (e.g., Berninger et al., 1991; Norf et al., 2009). As they act as trophic links between bacteria and higher trophic levels, they are relevant players in the mineralization of bacterial biomass (Augspurger et al., 2008; Norf et al., 2009). Protozoans can also control densities of both suspended and surface-associated algae (e.g., Kathol et al., 2011; Kanavillil and Kurissery, 2013) as well as stimulate leaf litter decomposition (e.g., Ribblett et al., 2005).

MICROBIAL FUNCTIONING IN IRES AND LINKS WITH MICROBIAL COMMUNITY STRUCTURE

The spatial and temporal distribution of microbial communities in IRES, their resistance strategies, and their resilience capacities influence food web interactions and ecosystem functioning. During the different hydrological phases, some microbially mediated processes are enhanced while others are inhibited. As found for diversity, fluctuating flow, especially intermittence, is the strongest factor governing microbial functioning in IRES.

Streambed desiccation reduces bacterial and fungal biomass and activity (Tzoraki et al., 2007; Amalfitano et al., 2008; Chauvet et al., 2016). Conditions during drying, such as prolonged anoxia, may enhance the activity of certain microbial groups such as archaea and many protozoans which are thought to be better suited to drying than fungi and several bacterial taxa. As pools become disconnected and oxygen becomes depleted, anoxia and the accumulation of OM might select for anaerobic communities such as methanogenic archaea and sulfate-reducing bacteria responsible for producing reduced gases such as methane and hydrogen sulfide as by-products of OM mineralization (Briée et al., 2007).

Analysis, by means of extracellular enzymes, of microbial OM decomposition in IRES indicates shifts in degradation capabilities by different microbial heterotrophs at different hydrological phases. During drying in Mediterranean IRES, peptidase activity revealed heavy use of organic-N compounds in stream biofilms while phosphatase activity remained in the sediment (Romaní et al., 2013). At a larger scale, the heterotrophic microbial community in the surface sediments mainly relied on organic-N compounds while degrading accumulated algal biomass (Freixa et al., 2016). This study also revealed greater heterotrophic functional diversity (related to the use of carbon compounds) during drying than during base-flow periods because of the greater patchiness and spatial heterogeneity within the river environment.

In the dry streamed, carbon processing is maintained to some degree (Zoppini and Marxsen, 2011), evident by CO_2 emissions (Gómez-Gener et al., 2016). Similarly, in another Mediterranean IRES, enzyme activity and bacterial density in the dry streamed declined but still persisted at significant levels (Timoner et al., 2012). During drying, including pool phases and when the streamed is dry, a reduction in prokaryote diversity is expected (Section 4.1.2) and this may reduce the capacity to maintain multiple functions (e.g., Peter et al., 2011). However, studies on microbial functional-diversity relationships in IRES are scarce.

During drying, leaf litter decomposition is inhibited (Schlief and Mutz, 2011; Mora-Gómez et al., 2015). In isolated pools, low oxygen and accumulation of phenolic compounds from litter leachates inhibit microbial activity (Schlief and Mutz, 2007; Medeiros et al., 2009). When the bed is dry, there is an associated deceleration of microbial processes (fungal biomass accrual, sporulation rate, and microbial respiration). Both intensity and total duration, but not frequency of drying events, affect fungal biomass and litter decomposition in streams (Langhans and Tockner, 2006; Bruder et al., 2011;

Foulquier et al., 2015). Mora-Gómez et al. (2016) found that summer drying conditions inhibited lignocellulolytic enzyme activities and modified fungal community structure in a Mediterranean IRES, suggesting a link between structure and function of fungal decomposers. However, most aquatic hyphomycete species exhibit a wide range of enzymatic capacities and so there is redundancy among species in leaf-associated assemblages. In a microcosm experiment, Gonçalves et al. (2016) found higher fungal richness was related to greater leaf decomposition, and this relationship was maintained after drying. Leaf-associated bacterial and fungal community composition affected respiration but this interacted with humidity and N-availability more than with temperature (Matulich and Martiny, 2015). Species identities may have a greater effect on leaf decomposition rate than diversity *per se* (Krauss et al., 2011) although the predictability of decomposition rate tends to increase with increasing species richness (Dang et al., 2005).

Declines in leaf litter decomposition rates during drying are further affected by the reduced contribution of macroinvertebrates (Schlief and Mutz, 2009), although protozoans can partly replace their roles. This is due to protozoans coping better with unfavorable conditions during the drying phase in IRES and recovering much faster than the macrofauna. Microcosm experiments showed that oxygen concentrations marginally affected microbially (bacteria, fungi, and protozoa) mediated leaf litter processing, and that the presence of protozoans increased leaf-associated respiration (Risse-Buhl et al., 2015). Furthermore, as protozoans can remain active when the streambed is dry (Coûteaux and Darbyshire, 1998), protozoan-mediated functions probably continue, at least partly, during all hydrological phases in IRES.

Upon rewetting, microbial functions rapidly resume. Extracellular enzyme activities in IRES are quickly recovered (Zoppini and Marxsen, 2011; Pohlon et al. 2013) and the use of polysaccharides is enhanced, linked to the increase in available DOM in the flowing water (Romaní et al., 2006b; Ylla et al., 2010). In a Mediterranean IRES, enzymes recovered instantly, peaking in activity only 3 hours after rewetting (Romaní and Sabater, 1997), and in an arid-zone stream, rehydration caused a significant increase in functional diversity (Timoner et al., 2014b). These observations might be linked to the observed resuscitation strategies of some microbial groups that become activated in few hours of rewetting, contributing to a significant peak in carbon dioxide (Placella et al., 2012). Leaf-associated fungal activity also recovers rapidly after immersion (Maamri et al., 1998, 2001; Langhans and Tockner, 2006; Datry et al., 2011). The fungal communities and microbial enzymatic performance observed after rewetting reflect modification in leaf quality that occurred during the previous dry phase (Mora-Gómez, 2014).

Embracing the variability in structure and function over space and time in IRES, we hypothesize a sequence of changes over the hydrological cycle (Fig. 4.1.9) associated with dominance by different microbial groups during different phases. A key attribute of IRES driving microbial communities is the highly variable nature of microhabitats created during various phases of the hydrological cycle. We predict that this mosaic of microhabitats in IRES leads to patchy distributions in microbial taxa and changes in ecosystem functions over space and time. Testing whether specific functions occur in "hot spots" and "hot moments" (Chapter 3.2) will improve our understanding of the drivers of microbial variability and inform efforts to scale up fine-scale measurements of microbially mediated processes to ecosystem-scale responses. For example, "waves" of hot spots (Fig. 4.1.9) may have implications for higher taxa and downstream functions but more information is needed to assess these hypotheses.

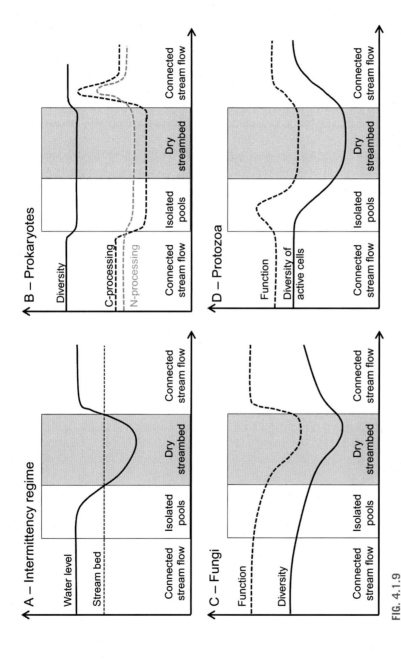

FIG. 4.1.9

Hypothetical model of microbial dynamics in IRES, based on research on several systems. Changes in the water level as a result of the intermittency regime (A) are associated with changes in prokaryote diversity and functioning, distinguishing C and N processing (B), changes in fungal diversity and function such as leaf litter decomposition (C), and changes in protozoan diversity and function such as grazing activity of protozoans and their control of bacterial populations (D). Diversity refers to that of active cells (although the overall diversity including spores would probably not change much).

4.1.7 FUTURE CHALLENGES

IRES, especially headwaters, are well known for their microbially mediated contributions to nutrient cycling in downstream reaches (Meyer et al., 2007), with many implications for ecosystem and global-scale processes as well as management and restoration. Improved understanding of the relationships between IRES structure and function is critical if the challenges posed by issues (e.g., greenhouse gas emissions and pollution, including eutrophication) are to be managed effectively. Comprehensive studies of microbial community structure and function in IRES could guide predictions about ecosystem processes (Graham et al., 2016), answering questions such as whether changes in diversity result in similar shifts in functions and/or if there are keystone taxa in IRES.

Exploring microbial structure-function relationships in IRES is becoming increasingly feasible with the decreasing costs and wider availability of molecular approaches. Of the microbial taxa, bacterial communities are currently the most widely investigated but the lack of consistent methodologies to measure either structural or functional aspects of their communities limits the potential for widespread comparisons. Opportunities for future studies may include the characterization of multiple microbial taxa in a single study, over multiple hydrological phases across a range of microhabitats or compartments. Despite the paucity of data, the occurrence and relevance of archaea in IRES streambed sediments should not be underestimated given their potential roles in stream biogeochemistry. Moreover, the contribution of protozoans to the range of possible microbially mediated ecosystem processes remains underexplored. We surmise that in IRES, there are increases in protozoan functions (e.g., grazing activity and control of bacterial populations, acceleration of carbon flow, and contribution to nutrient recycling) compared to those by higher taxa when conditions become unfavorable for the macrofauna.

These future challenges also have management implications. For example, an increased prevalence of IRES globally may lead to more degraded ecosystem functions. If so, this further reinforces the need to improve our understanding of microbial communities, the controls on the functions they carry out, and how they influence other taxa levels and processes in connected ecosystems.

REFERENCES

Abed, R.M.M., Al Kindi, S., Schramm, A., Barry, M.J., 2011. Short-term effects of flooding on bacterial community structure and nitrogenase activity in microbial mats from a desert stream. Aquat. Microb. Ecol. 63, 245–254.

Adams, H.E., Crump, B.C., Kling, G.W., 2015. Isolating the effects of storm events on arctic aquatic bacteria: temperature, nutrients, and community composition as controls on bacterial productivity. Front. Microbiol. 6, 250.

Allan, D., Castillo, M.D., 2007. Stream Ecology: Structure and Function of Running Waters. Springer, Dordrecht.

Amalfitano, S., Fazi, S., Zoppini, A., Caracciolo, A.B., Grenni, P., Puddu, A., 2008. Responses of benthic bacteria to experimental drying in sediments from mediterranean temporary rivers. Microb. Ecol. 55, 270–279.

Andrushchyshyn, O.P., Magnusson, A.K., Williams, D.D., 2006. Responses of intermittent pond ciliate populations and communities to in situ bottom-up and top-down manipulations. Aquat. Microb. Ecol. 42, 293–310.

Angel, R., Claus, P., Conrad, R., 2012. Methanogenic archaea are globally ubiquitous in aerated soils and become active under wet anoxic conditions. ISME J. 6, 847–862.

Artigas, J., Romaní, A.M., Sabater, S., 2008. Effect of nutrients on the sporulation and diversity of aquatic hyphomycete communities in stream substrata. Aquat. Bot. 88, 32–38.

Artigas, J., Romaní, A.M., Gaudes, A., Muñoz, I., Sabater, S., 2009. Organic matter availability structures microbial biomass and activity in a Mediterranean stream. Freshw. Biol. 54, 2025–2036.

Augspurger, C., Gleixner, G., Kramer, C., Küsel, K., 2008. Tracking carbon flow in a 2-week-old and 6-week-old stream biofilm food web. Limnol. Oceanogr. 53, 642–650.

Auguet, J.-C., Barberan, A., Casamayor, E.O., 2010. Global ecological patterns in uncultured Archaea. ISME J. 4, 182–190.

Aumen, N.G., Bottomley, P.J., Ward, G.M., Gregory, S.V., 1983. Microbial decomposition of wood in streams: distribution of microflora and factors affecting [^{14}C]lignocellulose mineralization. Appl. Environ. Microbiol. 46, 1409–1416.

Austin, A.T., Vivanco, L., 2006. Plant litter decomposition in a semi-arid ecosystem controlled by photodegradation. Nature 442, 555–558.

Bandoni, R.J., 1972. Terrestrial occurrence of some aquatic hyphomycetes. Can. J. Bot. 50, 2283–2288.

Bärlocher, F., 2009. Reproduction and dispersal in aquatic hyphomycetes. Mycoscience 50, 3–8.

Bärlocher, F., 2015. Aquatic hyphomycetes in a changing environment. Fungal Ecol. 19, 14–27.

Bärlocher, F., Boddy, L., 2016. Aquatic fungal ecology–how does it differ from terrestrial? Fungal Ecol. 19, 5–13.

Bärlocher, F., Sridhar, K.R., 2014. Association of animals and fungi in leaf decomposition. In: Jones, E.B.G., Hyde, K.D., Pang, K.-L. (Eds.), Freshwater Fungi and Fungal-Like Organisms. De Gruyter, Berlin, pp. 413–441.

Bärlocher, F., Nikolcheva, L.G., Wilson, K.P., Williams, D.D., 2006. Fungi in the hyporheic zone of a springbrook. Microb. Ecol. 52, 708–715.

Battin, T.J., Wille, A., Sattler, B., Psenner, R., 2001. Phylogenetic and functional heterogeneity of sediment biofilms along environmental gradients in a glacial stream. Appl. Environ. Microbiol. 67, 799–807.

Battin, T.J., Besemer, K., Bengtsson, M.M., Romani, A.M., Packmann, A.I., 2016. The ecology and biogeochemistry of stream biofilms. Nat. Rev. Microbiol. 14, 251–263.

Berninger, U.G., Finlay, B.J., Kuupo-Leinikki, P., 1991. Protozoan control of bacterial abundances in freshwater. Limnol. Oceanogr. 36, 139–147.

Bottjer, D.J., Schubert, J.K., Droser, M.L., 1996. Comparative evolutionary palaeoecology: assessing the changing ecology of the past. Geol. Soc. Lond. Spec. Publ. 102, 1–13.

Briée, C., Moreira, D., López-García, P., 2007. Archaeal and bacterial community composition of sediment and plankton from a suboxic freshwater pond. Res. Microbiol. 158, 213–227.

Bruder, A., Chauvet, E., Gessner, M.O., 2011. Litter diversity, fungal decomposers and litter decomposition under simulated stream intermittency. Funct. Ecol. 25, 1269–1277.

Brune, A., Frenzel, P., Cypionka, H., 2000. Life at the oxic-anoxic interface: microbial activities and adaptations. FEMS Microbiol. Rev. 24, 691–710.

Burns, R.G., DeForest, J.L., Marxsen, J., Sinsabaugh, R.L., Stromberger, M.E., Wallenstein, M.D., et al., 2013. Soil enzymes in a changing environment: current knowledge and future directions. Soil Biol. Biochem. 58, 216–234.

Canhoto, C., Gonçalves, A., Bärlocher, F., 2016. Biology and ecological functions of aquatic hyphomycetes in a warming climate. Fungal Ecol. 19, 201–218.

Chauvet, E., Cornut, J., Sridhar, K.R., Selosse, M.A., Bärlocher, F., 2016. Beyond the water column: aquatic hyphomycetes outside their preferred habitat. Fungal Ecol. 19, 112–127.

Christensen, S., Griffiths, B., Ekelund, F., Rønn, R., 1992. Huge increase in bacterivores on freshly killed barley roots. FEMS Microbiol. Ecol. 86, 303–310.

Cleven, E.-J., 2004. Seasonal and spatial distribution of ciliates in the sandy hyporheic zone of a lowland stream. Eur. J. Protistol. 40, 71–84.

Conrad, R., Ji, Y., Noll, M., Klose, M., Claus, P., Enrich-Prast, A., 2014. Response of the methanogenic microbial communities in Amazonian oxbow lake sediments to desiccation stress. Environ. Microbiol. 16, 1682–1694.

Cornut, J., Elger, A., Lambrigot, D., Marmonier, P., Chauvet, E., 2010. Early stages of leaf decomposition are mediated by aquatic fungi in the hyporheic zone of woodland streams. Freshw. Biol. 55, 2541–2556.

Cornut, J., Chauvet, E., Mermillod-Blondin, F., Assemat, F., Elger, A., 2014. Aquatic hyphomycete species are screened by the hyporheic zone of woodland streams. Appl. Environ. Microbiol. 80, 1949–1960.

Coûteaux, M.M., Darbyshire, J.F., 1998. Functional diversity amongst soil protozoa. Appl. Soil Ecol. 10, 229–237.

Dahm, C.N., Baker, M.A., Moore, D.I., Thibault, J.R., 2003. Coupled biogeochemical and hydrological responses of streams and rivers to drought. Freshw. Biol. 48, 1219–1231.

Dallai, R., Miceli, C., Luporini, P., 1987. *Euplotes rariseta* Curds et al. (Ciliophora, Hypotrichida) from the Somalian coast: description and preliminary observations on cyst induction and ultrastructure. Monitore Zoologico Italiano, Supplemento 22, 263–280.

Dang, C.K., Chauvet, E., Gessner, M.O., 2005. Magnitude and variability of process rates in fungal diversity-litter decomposition relationships. Ecol. Lett. 8, 1129–1137.

Dang, C.K., Schindler, M., Chauvet, E., Gessner, M.O., 2009. Temperature oscillation coupled with fungal community shifts can modulate warming effects on litter decomposition. Ecology 90, 122–131.

Datry, T., Corti, R., Claret, C., Philippe, M., 2011. Flow intermittence controls leaf litter breakdown in a French temporary alluvial river: the "drying memory". Aquat. Sci. 73, 471–483.

Descals, E., 2005. Diagnostic characters of propagules of Ingoldian fungi. Mycol. Res. 109, 545–555.

Dieter, D., von Schiller, D., García-Roger, E.M., Sánchez-Montoya, M.M., Gómez, R., Mora-Gómez, J., et al., 2011. Preconditioning effects of intermittent stream flow on leaf litter decomposition. Aquat. Sci. 73, 599–609.

Dieter, D., Frindte, K., Krüger, A., Wurzbacher, C., 2013. Preconditioning of leaves by solar radiation and anoxia affects microbial colonisation and rate of leaf mass loss in an intermittent stream. Freshw. Biol. 58, 1918–1931.

Domonell, A., Brabender, M., Nitsche, F., Bonkowski, M., Arndt, H., 2013. Community structure of cultivable protists in different grassland and forest soils of Thuringia. Pedobiologia 56, 1–7.

Dupraz, C., Visscher, P.T., 2005. Microbial lithification in marine stromatolites and hypersaline mats. Trends Microbiol. 13, 429–438.

Eddison, J.C., Ollason, R.O., 1978. Diversity in constant and fluctuating environments. Nature 275, 309–310.

Ekelund, F., 2005. Tolerance of soil flagellates to increased NaCl levels. J. Eukaryot. Microbiol. 49, 324–328.

Ekelund, F., Frederiksen, H.B., Rønn, R., 2002. Population dynamics of active and total ciliate populations in arable soil amended with wheat. Appl. Environ. Microbiol. 68, 1096–1101.

Fazi, S., Amalfitano, S., Pernthaler, J., Puddu, A., 2005. Bacterial communities associated with benthic organic matter in headwater stream microhabitats. Environ. Microbiol. 7, 1633–1640.

Fazi, S., Amalfitano, S., Piccini, C., Zoppini, A., Puddu, A., Pernthaler, J., 2008. Colonization of overlaying water by bacteria from dry river sediments. Environ. Microbiol. 10, 2760–2772.

Fazi, S., Vazquez, E., Casamayor, E.O., Amalfitano, S., Butturini, A., 2013. Stream hydrological fragmentation drives bacterioplankton community composition. PLoS One 8, e64109.

Febria, C.M., Beddoes, P., Fulthorpe, R.R., Williams, D.D., 2012. Bacterial community dynamics in the hyporheic zone of an intermittent stream. ISME J. 6, 1078–1088.

Febria, C.M., Hosen, J.D., Crump, B.C., Palmer, M.A., Williams, D.D., 2015. Microbial responses to changes in flow status in temporary headwater streams: a cross-system comparison. Front. Microbiol. 6, 522.

Fenchel, T., Finlay, B.J., Gianní, A., 1989. Microaerophily in ciliates: responses of an *Euplotes* sp. to oxygen tension. Arch. Protistenkd. 137, 317–330.

Ferrenberg, S., Reed, S.C., Belnap, J., 2015. Climate change and physical disturbance cause similar community shifts in biological soil crusts. Proc. Natl. Acad. Sci. U. S. A. 112, 12116–12121.

Field, J.I., Webster, J., 1983. Anaerobic survival of aquatic fungi. Trans. Br. Mycol. Soc. 81, 365–369.

Findlay, S., Strayer, D., Goumbala, C., Gould, K., 1993. Metabolism of streamwater dissolved organic carbon in the shallow hyporheic zone. Limnol. Oceanogr. 38, 1493–1499.

Finlay, B.J., 2002. Global dispersal of free-living microbial eukaryote species. Science 296, 1061–1063.

Finlay, B.J., Esteban, G.F., 2009. Oxygen sensing drives predictable migrations in a microbial community. Environ. Microbiol. 11, 81–85.

Foulquier, A., Artigas, J., Pesce, S., Datry, T., 2015. Drying responses of microbial litter decomposition and associated fungal and bacterial communities are not affected by emersion frequency. Freshw. Sci. 34, 1233–1244.

Franco, C., Esteban, G.F., Tellez, C., 1998. Colonization and succession of ciliated protozoa associated with submerged leaves in a river. Limnologica 28, 275–283.

Freixa, A., Ejarque, E., Crognale, S., Amalfitano, S., Fazi, S., Butturini, A., 2016. Sediment microbial communities rely on different dissolved organic matter sources along a Mediterranean river continuum. Limnol. Oceanogr. 61, 1389–1405.

Gallo, M.E., Sinsabaugh, R.L., Cabaniss, S.E., 2006. The role of ultraviolet radiation in litter decomposition in arid ecosystems. Appl. Soil Ecol. 34, 82–91.

Gao, X.Q., Olapade, O.A., Leff, L.G., 2005. Comparison of benthic bacterial community composition in nine streams. Aquat. Microb. Ecol. 40, 51–60.

Gasith, A., Resh, V.H., 1999. Streams in Mediterranean climate regions: abiotic influences and biotic responses to predictable seasonal events. Annu. Rev. Ecol. Syst. 30, 51–81.

Gaudes, A., Artigas, J., Romaní, A.M., Sabater, S., Munoz, I., 2009. Contribution of microbial and invertebrate communities to leaf litter colonization in a Mediterranean stream. J. N. Am. Benthol. Soc. 28, 34–43.

Gessner, M.O., Gulis, V., Kuehn, K.A., Chauvet, E., Suberkropp, K., 2007. Fungal decomposers of plant litter in aquatic ecosystems. In: Kubikak, C., Druzhinina, I. (Eds.), The Mycota—Environmental and Microbial Relationships, vol. IV. Springer, Berlin, pp. 301–324.

Ghate, S.D., Sridhar, K.R., 2015. Diversity of aquatic hyphomycetes in streambed sediments of temporary streamlets of Southwest India. Fungal Ecol. 14, 53–61.

Gómez-Gener, L., Obrador, B., Marcé, R., Acuña, V., Catalán, N., Casas-Ruiz, J.P., et al., 2016. When water vanishes: magnitude and regulation of carbon dioxide emissions from dry temporary streams. Ecosystems 19, 710.

Gonçalves, A.L., Lírio, A.V., Graça, M.A.S., Canhoto, C., 2016. Fungal species diversity affects leaf decomposition after drought. Int. Rev. Hydrobiol. 101, 78–86.

Graham, E.B., Knelman, J.E., Schindlbacher, A., Siciliano, S., Breulmann, M., Yannarell, A., et al., 2016. Microbes as engines of ecosystem function: when does community structure enhance predictions of ecosystem processes? Front. Microbiol. 7, 214.

Grimm, N.B., Chacon, A., Dahm, C.N., Hostetler, S.W., Lind, O.T., Starkweather, P.L., et al., 1997. Sensitivity of aquatic ecosystems to climatic and anthropogenic changes: the Basin and Range, American Southwest and Mexico. Hydrol. Process. 11, 1023–1041.

Hauer, G., Rogerson, A., 2005. Remarkable salinity tolerance of seven species of naked amoebae (Gymnamoebae). Hydrobiologia 549, 33–42.

Hausmann, K., Hülsmann, N., Radek, R., 2003. Protistology. E. Schweizerbart'sche Verlagsbuchhandlung, Stuttgart.

Herrmann, M., Scheibe, A., Avrahami, S., Küsel, K., 2011. Ammonium availability affects the ratio of ammonia-oxidizing bacteria to ammonia-oxidizing archaea in simulated creek ecosystems. Appl. Environ. Microbiol. 77, 1896–1899.

Hibbett, D.S., Binder, M., Bischoff, J.F., Blackwell, M., Cannon, P.F., Eriksson, O.E., et al., 2007. A higher-level phylogenetic classification of the Fungi. Mycol. Res. 111, 509–547.

Joelsson, J.P., Fridjonsdottir, H., Vilhelmsson, O., 2013. Bioprospecting a glacial river in Iceland for bacterial biopolymer degraders. Cold Reg. Sci. Technol. 96, 86–95.

Jones, E.B.G., Hyde, K.D., Pang, K.-L., 2014. Introduction. In: Jones, E.B.G., Hyde, K.D., Pang, K.-L. (Eds.), Freshwater Fungi and Fungal-like Organisms. De Gruyter, Berlin, pp. 1–22.

Kanavillil, N., Kurissery, S., 2013. Dynamics of grazing protozoa follow that of microalgae in natural biofilm communities. Hydrobiologia 718, 93–107.

Kathol, M., Norf, H., Arndt, H., Weitere, M., 2009. Effects of temperature increase on the grazing of planktonic bacteria by biofilm-dwelling consumers. Aquat. Microb. Ecol. 55, 65–79.

Kathol, M., Fischer, H., Weitere, M., 2011. Contribution of biofilm-dwelling consumers to pelagic–benthic coupling in a large river. Freshw. Biol. 56, 1160–1172.

Klappenbach, J.A., Dunbar, J.M., Schmidt, T.M., 2000. rRNA operon copy number reflects ecological strategies of bacteria. Appl. Environ. Microbiol. 66, 1328–1333.

Koch, H., Lücker, S., Albertsen, M., Kitzinger, K., Herbold, C., Spieck, E., et al., 2015. Expanded metabolic versatility of ubiquitous nitrite-oxidizing bacteria from the genus *Nitrospira*. Proc. Natl. Acad. Sci. U. S. A. 112, 11371–11376.

Krauss, G.-J., Solé, M., Krauss, G., Schlosser, D., Wesenberg, D., Bärlocher, F., 2011. Fungi in freshwaters: ecology, physiology and biochemical potential. FEMS Microbiol. Rev. 35, 1–32.

Langhans, S.D., Tockner, K., 2006. The role of timing, duration, and frequency of inundation in controlling leaf litter decomposition in a river-floodplain ecosystem (Tagliamento, northeastern Italy). Oecologia 147, 501–509.

Leff, L., Van Gray, J.B., Martí, E., Merbt, S.N., Romaní, A.M., 2016. Aquatic biofilms and biogeochemical processes. In: Romaní, A.M., Guasch, H., Balaguer, M.D. (Eds.), Aquatic Biofilms: Ecology, Water Quality and Wastewater Treatment. Caister Academic Press, Poole, pp. 89–108.

Lock, M.A., 1993. Attached microbial communities in rivers. In: Ford, T.E. (Ed.), Aquatic Microbiology: An Ecological Approach. Blackwell Scientific, Oxford, pp. 113–138.

Lyautey, E., Jackson, C.R., Cayrou, J., Rols, J.L., Garabetian, F., 2005. Bacterial community succession in natural river biofilm assemblages. Microb. Ecol. 50, 589–601.

Maamri, A., Chauvet, E., Chergui, H., Gourbière, F., Pattee, E., 1998. Microbial dynamics on decaying leaves in a temporary Moroccan river. I—Fungi. Arch. Hydrobiol. 144, 41–59.

Maamri, A., Bärlocher, F., Pattee, E., Chergui, H., 2001. Fungal and bacterial colonisation of *Salix pedicellata* leaves decaying in permanent and intermittent streams in Eastern Morocco. Int. Rev. Hydrobiol. 86, 337–348.

Manerkar, M.A., Seena, S., Bärlocher, F., 2008. Q-RT-PCR for assessing archaea, bacteria, and fungi during leaf decomposition in a stream. Microb. Ecol. 56, 467.

Manning, D.W.P., Rosemond, A.D., Kominoski, J.S., Gulis, V., Benstead, J.P., Maerz, J.C., 2015. Detrital stoichiometry as a critical nexus for the effects of streamwater nutrients on leaf litter breakdown rates. Ecology 96, 2214–2224.

Marxsen, J., Zoppini, A., Wilczek, S., 2010. Microbial communities in streambed sediments recovering from desiccation. FEMS Microbiol. Ecol. 71, 374–386.

Matulich, K.L., Martiny, J.B.H., 2015. Microbial composition alters the response of litter decomposition to environmental change. Ecology 96, 154–163.

McGrady-Steed, J., Morin, P.J., 1996. Disturbance and the species composition of rain pool microbial communities. Oikos 76, 93–102.

Medeiros, A.O., Pascoal, C., Graça, M.A.S., 2009. Diversity and activity of aquatic fungi under low oxygen conditions. Freshw. Biol. 54, 142–149.

Meyer, J.L., Strayer, D.L., Wallace, J.B., Eggert, S.L., Helfman, G.S., Leonard, N.E., 2007. The contribution of headwater streams to biodiversity in river networks. J. Am. Water Resour. Assoc. 43, 86–103.

Mora-Gómez, J., 2014. Leaf Litter Decomposition in Mediterranean Streams: Microbial Processes and Responses to Drought Under Current Global Change Scenario. University of Girona, Girona. PhD Thesis.

Mora-Gómez, J., Elosegi, A., Mas-Marti, E., Romaní, A.M., 2015. Factors controlling seasonality in leaf litter breakdown for a Mediterranean stream. Freshw. Sci. 34, 1245–1258.

Mora-Gómez, J., Elosegi, A., Duarte, S., Cássio, F., Pascoal, C., Romaní, A.M., 2016. Differences in the sensitivity of fungi and bacteria to season and invertebrates affect leaf litter decomposition in a Mediterranean stream. FEMS Microbiol. Ecol. 92. http://dx.doi.org/10.1093/femsec/fiw121.

Müller, H., Geller, W., 1993. Maximum growth rates of aquatic ciliated protozoa: dependence on body size and temperature reconsidered. Arch. Hydrobiol. 126, 315–327.

Norf, H., Weitere, M., 2010. Resource quantity and seasonal background alter warming effects on communities of biofilm ciliates. FEMS Microbiol. Ecol. 74, 361–370.

Norf, H., Arndt, H., Weitere, M., 2007. Impact of local temperature increase on the early development of biofilm-associated ciliate communities. Oecologia 151, 341–350.

Norf, H., Arndt, H., Weitere, M., 2009. Responses of biofilm-dwelling ciliate communities to planktonic and benthic resource enrichment. Microb. Ecol. 57, 687–700.

Northcott, M.L., Gooseff, M.N., Barrett, J.E., Zeglin, L.H., Takacs-Vesbach, C.D., Humphrey, J., 2009. Hydrologic characteristics of lake- and stream-side riparian wetted margins in the McMurdo Dry Valleys, Antarctica. Hydrol. Process. 23, 1255–1267.

Peter, H., Ylla, I., Gudasz, C., Romaní, A.M., Sabater, S., Tranvik, L., 2011. Multifunctionality and diversity in bacterial biofilms. PLoS One 6, e23225.

Placella, S.A., Brodie, E.-L., Firestone, M.K., 2012. Rainfall-induced carbon dioxide pulses result from sequential resuscitation of phylogenetically clustered microbial groups. Proc. Natl. Acad. Sci. U. S. A. 109, 10931–10936.

Plebani, M., Fussmann, K.E., Hansen, D.M., O'Gorman, E.J., Stewart, R.I.A., Woodward, G., et al., 2015. Substratum-dependent responses of ciliate assemblages to temperature: a natural experiment in Icelandic streams. Freshw. Biol. 60, 1561–1570.

Pohlon, E., Ochoa Fandino, A., Marxsen, J., 2013. Bacterial community composition and extracellular enzyme activity in temperate streambed sediment during drying and rewetting. PLoS One 8, e83365.

Portillo, M.C., Anderson, S.P., Fierer, N., 2012. Temporal variability in the diversity and composition of stream bacterioplankton communities. Environ. Microbiol. 14, 2417–2428.

Pusch, M., Fiebig, D., Brettar, I., Eisenmann, H., Ellis, B.K., Kaplan, L.A., et al., 1998. The role of micro-organisms in the ecological connectivity of running waters. Freshw. Biol. 40, 453–495.

Raugi, G.J., Liang, T., Blum, J.J., 1975. Effect of oxygen on regulation of intermediate metabolism in *Tetrahymena*. J. Biol. Chem. 250, 445–460.

Read, S., Moss, S., Jones, E., 1992. Attachment and germination of conidia. In: Bärlocher, F. (Ed.), The Ecology of Aquatic Hyphomycetes. Springer-Verlag, Berlin, pp. 135–151.

Rees, G.N., Watson, G.O., Baldwin, D.S., Mitchell, A.M., 2006. Variability in sediment microbial communities in a semipermanent stream: impact of drought. J. N. Am. Benthol. Soc. 25, 370–378.

Ribblett, S.G., Palmer, M.A., Coats, D.W., 2005. The importance of bacterivorous protists in the decomposition of stream leaf litter. Freshw. Biol. 50, 516–526.

Risse-Buhl, U., Trefzger, N.M., Seifert, A.-G., Schönborn, W., Gleixner, G., Küsel, K., 2012. Tracking the authochthonous carbon flow in stream biofilm food webs. FEMS Microbiol. Ecol. 79, 118–131.

Risse-Buhl, U., Schlief, J., Mutz, M., 2015. Phagotrophic protists are a key component of microbial communities processing leaf litter under contrasting oxic conditions. Freshw. Biol. 60, 2310–2322.

Roberson, E.B., Firestone, M.K., 1992. Relationship between desiccation and exopolysaccharide production in a soil *Pseudomonas* sp. Appl. Environ. Microbiol. 58, 1284–1291.

Romaní, A.M., Sabater, S., 1997. Metabolism recovery of a stromatolitic biofilm after drought in a Mediterranean stream. Arch. Hydrobiol. 140, 261–271.

Romaní, A.M., Fischer, H., Mille-Lindblom, C., Tranvik, L.J., 2006a. Interactions of bacteria and fungi on decomposing litter: differential extracellular enzyme activities. Ecology 87, 2559–2569.

Romaní, A.M., Vázquez, E., Butturini, A., 2006b. Microbial availability and size fractionation of dissolved organic carbon after drought in an intermittent stream: biogeochemical link across the stream-riparian interface. Microb. Ecol. 52, 501–512.

Romaní, A.M., Amalfitano, S., Artigas, J., Fazi, S., Sabater, S., Timoner, X., et al., 2013. Microbial biofilm structure and organic matter use in mediterranean streams. Hydrobiologia 719, 43–58.

Rothrock Jr., M.J., Garcia-Pichel, F., 2005. Microbial diversity of benthic mats along a tidal desiccation gradient. Environ. Microbiol. 7, 593–601.

Salvadó, H., Mas, M., Menéndez, S., Gracia, M.P., 2001. Effects of shock loads of salt on protozoan communities of activated sludge. Acta Protozool. 40, 177–185.

Schlief, J., Mutz, M., 2007. Response of aquatic leaf associated microbial communities to elevated leachate DOC: a microcosm study. Int. Rev. Hydrobiol. 92, 146–155.

Schlief, J., Mutz, M., 2009. Effect of sudden flow reduction on the decomposition of Alder leaves (*Alnus glutinosa* [L.] Gaertn.) in a temperate lowland stream: a mesocosm study. Hydrobiologia 624, 205–217.

Schlief, J., Mutz, M., 2011. Leaf decay processes during and after a supra-seasonal hydrological drought in a temperate lowland stream. Int. Rev. Hydrobiol. 96, 633–655.

Selosse, M.-A., Vohník, M., Chauvet, E., 2008. Out of the rivers: are some aquatic hyphomycetes plant endophytes? New Phytol. 178, 3–7.

Shearer, C.A., Descals, E., Kohlmeyer, B., Kohlmeyer, J., Marvanová, L., Padgett, D., et al., 2007. Fungal biodiversity in aquatic habitats. Biodivers. Conserv. 16, 49–67.

Smurov, A.O., Fokin, S.I., 1999. Resistance of *Paramecium* species (Ciliophora, Peniculia) to salinity of environment. Protistology 1, 43–53.

Soule, T., Anderson, I.J., Johnson, S.L., Bates, S.T., Garcia-Pichel, F., 2009. Archaeal populations in biological soil crusts from arid lands in North America. Soil Biol. Biochem. 41, 2069–2074.

Souza, V., Espinosa-Asuar, L., Escalante, A.E., Eguiarte, L.E., Farmer, J., Forney, L., et al., 2006. An endangered oasis of aquatic microbial biodiversity in the Chihuahuan desert. Proc. Natl. Acad. Sci. U. S. A. 103, 6565–6570.

Sridhar, K.R., Bärlocher, F., 1993. Aquatic hyphomycetes on leaf litter in and near a stream in Nova Scotia, Canada. Mycol. Res. 97, 1530–1535.

Sridhar, K.R., Bärlocher, F., 1994. Viability of aquatic hyphomycete conidia in foam. Can. J. Bot. 72, 106–110.

Stanish, L.F., O'Neill, S.P., Gonzalez, A., Legg, T.M., Knelman, J., McKnight, D.M., et al., 2013. Bacteria and diatom co-occurrence patterns in microbial mats from polar desert streams. Environ. Microbiol. 15, 1115–1131.

Sudheep, N., Sridhar, K., 2012. Aquatic hyphomycetes in hyporheic freshwater habitats of southwest India. Limnologica 42, 87–94.

Tank, J.L., Rosi-Marshall, E.J., Griffiths, N.A., Entrekin, S.A., Stephen, M.L., 2010. A review of allochthonous organic matter dynamics and metabolism in streams. J. N. Am. Benthol. Soc. 29, 118–146.

Thion, C., Prosser, J.I., 2014. Differential response of nonadapted ammonia-oxidising archaea and bacteria to drying-rewetting stress. FEMS Microbiol. Ecol. 90, 380–389.

Timoner, X., Acuña, V., von Schiller, D., Sabater, S., 2012. Functional responses of stream biofilms to flow cessation, desiccation and rewetting. Freshw. Biol. 57, 1565–1578.

Timoner, X., Borrego, C.M., Acuña, V., Sabater, S., 2014a. The dynamics of biofilm bacterial communities is driven by flow wax and wane in a temporary stream. Limnol. Oceanogr. 59, 2057–2067.

Timoner, X., Acuña, V., Frampton, L., Pollard, P., Sabater, S., Bunn, S.E., 2014b. Biofilm functional responses to the rehydration of a dry intermittent stream. Hydrobiologia 727, 185–195.

Tzoraki, O., Nikolaidis, N.P., Amaxidis, Y., Skoulikidis, N.T.H., 2007. In-stream biogeochemical processes of a temporary river. Environ. Sci. Technol. 41, 1225–1231.

Uselman, S.M., Snyder, K., Blank, R.R., Jones, T.J., 2011. UVB exposure does not accelerate rates of litter decomposition in a semi-arid riparian ecosystem. Soil Biol. Biochem. 43, 1254–1265.

Verni, F., Rosati, G., 2011. Resting cysts: a survival strategy in Protozoa Ciliophora. Ital. J. Zool. 78, 134–145.

Wang, W., Shor, L.M., LeBoeuf, E.J., Wikswo, J.P., Kosson, D.S., 2005. Mobility of protozoa through narrow channels. Appl. Environ. Microbiol. 71, 4628–4637.

Webster, J.R., Benfield, E.F., 1986. Vascular plant breakdown in freshwater ecosystems. Annu. Rev. Ecol. Syst. 17, 567–594.

Weitere, M., Arndt, H., 2003. Structure of the heterotrophic flagellate community in the water column of the River Rhine (Germany). Eur. J. Protistol. 39, 287–300.

Wey, J.K., Norf, H., Arndt, H., Weitere, M., 2009. Role of dispersal in shaping communities of ciliates and heterotrophic flagellates within riverine biofilms. Limnol. Oceanogr. 54, 1615–1626.

Wey, J.K., Jürgens, K., Weitere, M., 2012. Seasonal and successional influences on bacterial community composition exceed that of protozoan grazing in river biofilms. Appl. Environ. Microbiol. 78, 2013–2024.

Whitman, W.B., Coleman, D.C., Wiebe, W.J., 1998. Prokaryotes: the unseen majority. Proc. Natl. Acad. Sci. U. S. A. 95, 6578–6583.

Wurzbacher, C., Grimmett, I.J., Bärlocher, F., 2015. Metabarcoding-based fungal diversity on coarse and fine particulate organic matter in a first-order stream in Nova Scotia, Canada. F1000Res 4, 1378.

Ylla, I., Sanpera-Calbet, I., Vázquez, E., Romaní, A.M., Muñoz, I., Butturini, A., et al., 2010. Organic matter availability during pre- and post-drought periods in a Mediterranean stream. Hydrobiologia 657, 217–232.

Young, R.G., Matthaei, C.D., Townsend, C.R., 2008. Organic matter breakdown and ecosystem metabolism: functional indicators for assessing river ecosystem health. J. N. Am. Benthol. Soc. 27, 605–625.

Zeglin, L.H., 2015. Stream microbial diversity responds to environmental changes: review and synthesis of existing research. Front. Microbiol. 6, 454.

Zeglin, L.H., Dahm, C.N., Barrett, J.E., Gooseff, M.N., Fitpatrick, S.K., Takacs-Vesbach, C.D., 2011. Bacterial community structure along moisture gradients in the parafluvial sediments of two ephemeral desert streams. Microb. Ecol. 61, 543–556.

Zoppini, A., Marxsen, J., 2011. Importance of extracellular enzymes for biogeochemical processes in temporary river sediments during fluctuating dry-wet conditions. In: Shukla, G., Varma, A. (Eds.), Soil Enzymology, Soil Biology, vol. 22. Springer, Dordrecht, pp. 103–117.

Zoppini, A., Amalfitano, S., Fazi, S., Puddu, A., 2010. Dynamics of a benthic microbial community in a riverine environment subject to hydrological fluctuations (Mulargia River, Italy). Hydrobiologia 657, 37–51.

THE BIOTA OF INTERMITTENT RIVERS AND EPHEMERAL STREAMS: ALGAE AND VASCULAR PLANTS

4.2

Sergi Sabater*,[†], Xisca Timoner[†], Gudrun Bornette[‡], Mélissa De Wilde[§],
Juliet C. Stromberg[¶], John C. Stella[‖]

University of Girona, Girona, Spain ICRA, Girona, Spain[†] Université de Franche-Comté, Besançon Cedex, France[‡]
Institut Méditerranéen de Biodiversité et d'Ecologie (IMBE), Université d'Avignon et des Pays de Vaucluse,
Aix Marseille Université, IUT d'Avignon, Avignon, France[§] Arizona State University, Tempe, AZ, United States[¶] State
University of New York, Syracuse, NY, United States[‖]*

IN A NUTSHELL

– Algae, macrophytes, and riparian plants are the main groups of primary producers in intermittent rivers and ephemeral streams (IRES). Their abundance and community composition depend on the hydrological phases and degree of flow intermittence in IRES

– All of them show particular adaptations (morphological and physiological) that allow them to thrive under a harsh environment

– The abundance of primary producers configures the autotrophic character of IRES, but heterotrophic conditions prevail during the drying phase. Primary producers constitute a food source to consumers in nearby terrestrial ecosystems during the dry period

– Conservation and management of IRES must consider the relevance of primary producers

4.2.1 INTRODUCTION: PRIMARY PRODUCERS IN IRES

Primary producers have an essential role in the functioning of intermittent and ephemeral streams (IRES) as the basis of the trophic network in spite of the harsh environment (Box 4.2.1). Primary producers encompass numerous taxonomic groups and occupy diverse habitats. The streambed supports algae, cyanobacteria, mosses, and macrophytes. Adjacent to the stream, riparian vegetation occupies the bank, obtaining water and nutrients from the stream, shallow water table, and unsaturated subsoil.

Primary producers collectively form the functional foundation of the stream ecosystem because they fix solar energy into organic matter while using inorganic nutrients from upstream and from lateral contributions. Their high diversity contrasts sharply with the numerous environmental challenges associated with living in IRES, including long dry periods and frequent thermal and hydric stress. To overcome these challenges, primary producers display a multitude of adaptations, expressed in particular morphologies, life cycles, and physiological traits, with fascinating variety across the biological kingdoms.

BOX 4.2.1 THE SIGNIFICANCE OF PRIMARY PRODUCERS IN RIVERINE ECOSYSTEMS

Almost all life on Earth relies on primary producers, and they directly or indirectly support the food chain of all consumers and detritivores. Primary producers use inorganic nutrients and atmospheric or aqueous carbon dioxide for photosynthesis, powered by light. In river ecosystems, riparian vegetation dominates the terrestrial-aquatic ecotone, while algae and macrophytes colonize the aquatic environment. The algal, macrophytic, and riparian compartments provide the allochthonous and autochthonous organic matter which is the energy basis for higher trophic levels in streams.

The primary producers inhabiting IRES are submitted to hydric stress, associated with the strong oscillations in the water table and, in many IRES, extended dry periods. Water is essential for photosynthesis because it acts as a reducing agent and is involved in the synthesis of sugars. Water also mediates the osmotic equilibrium of cells and is at the base of numerous other metabolic processes. Because of this, the flow intermittence characterizing IRES requires specific physical and physiological adaptations by the primary producers in IRES and largely determines their community composition as well as their functioning in the ecosystem.

In this chapter, we review the main groups of primary producers and their adaptations to flow intermittence and drying in IRES. Their contributions to stream ecosystem function during different hydrological phases are described, with examples from IRES around the world. We conclude with a discussion of management and conservation of primary producers, emphasizing the importance of protecting their crucial contributions to biodiversity, habitat structure, trophic organization, and ecosystem function in IRES.

4.2.2 MICROBIAL PRIMARY PRODUCERS IN IRES: CYANOBACTERIA AND ALGAE

Cyanobacteria and algae vary in IRES according to their different climatic and geological settings, hydrological patterns, water chemistry, and light irradiance regime (Margalef, 1983; Sabater, 1989; Sheath and Müller, 1997; Chapter 4.1). Accordingly, the assemblage composition of algae in Mediterranean IRES differs from cold or warm arid-zone IRES, but communities share their tight links to flow intermittence. During periods of flow, diversity and even composition may resemble those in perennial systems, at least in the less extreme IRES. However, during the drying phase, a major shift in the composition of the algal community occurs; only those adapted or able to resist the stress associated with drying may persist. Intermittence affects the density and biomass of riverine algae through the direct effect of water stress on algal cell structures, also affecting algal groups and their overall diversity. The dry phase of IRES has lasting effects, in terms of lagged response of both algal diversity and biomass, on communities of subsequent hydrological phases. This legacy may be less important when flows are long enough to allow algal communities to fully recover their original state.

Algae in IRES show various adaptations to drying. Most IRES support desiccation-adapted algal communities such as those that persist dormant on the dry streambeds until flow resumes, enabling quick reactivation. However, resistance structures (e.g., spores, thickened cells) are lacking in systems experiencing unexpected drying. Excessively rapid drying may also prevent the formation of resistance structures (Stanley et al., 2004) so that random processes then become important, with community recovery dependent on dispersal and recolonization from nearby permanent waters (if they exist).

Outcomes of this process of survival-recolonization may depend on the origin of the inocula available for the recolonization (Robson et al., 2008). Remnant pools offer the best refuge, but when they dry, inocula may persist in temporarily wet microenvironments such as below leaf packs. Dry biofilms on cobbles or subsurface sediments also provide algal refugia.

In drying pools, lotic (benthic) algae are gradually replaced by lentic (tychoplanktonic) communities (Fazi et al., 2013). In larger dryland IRES in Australia, waterholes are an especially important refuge where, during the flow contraction period (the "bust" phase), production is limited to a littoral band of benthic algae as well as to phytoplankton production (Bunn et al., 2003), the base of the food webs in these rivers (Sternberg et al., 2008). When the drying is not complete, lotic, lentic, and terrestrial habitats cooccur sequentially in a longitudinal direction, providing beta diversity higher than expected (Tornés and Sabater, 2010).

After pools dry, the exposed substratum becomes the ultimate refuge for algae until flow returns. Refuge prevalence depends on substratum type and its ability to retain moisture but also on the severity of drying. Some substrata have greater ability to retain moisture than others and may be where algae better resist desiccation. Cobbles are usually the most exposed and algae are most affected, with the highest biomass reductions. Sandy sediments have higher capacity to preserve moisture and most likely allow better development and survival of microorganisms (McKew et al., 2010). No refuges persist in IRES where desiccation lasts too long. In these situations, only some subaerial algae (e.g., diatoms *Luticola mutica*, *Hantzschia amphioxys*) or adapted cystic forms (Timoner et al., 2014) remain after drying is complete. Algae are therefore forced to enter dormancy as an alternative to death and remain in this condition until flow returns.

Flow intermittence is a selective force on algal community composition, usually resulting in the decrease of alpha diversity. For example, diatom communities in IRES have fewer species and a lower proportion of specialist taxa than perennial streams (Tornés and Ruhí, 2013). A long drying phase can eliminate most of the sensitive species and reduce diversity because few species are able to resist drying and there are few pioneer species able to recolonize when flow resumes. Pioneer species are generally limited to a few diatom species, whereas those more resistant to desiccation are primarily green algae and cyanobacteria. Diatom genera such as *Achnanthes*, *Fragilaria*, and *Nitzschia* may attach to bare substrata in a few days.

MORPHOLOGICAL AND PHYSIOLOGICAL ADAPTATIONS TO DRYING

Responses of cyanobacteria and algae to flow intermittence include specific growth forms, cellular structures, or even genetic adaptations. Mucilage-forming species (e.g., *Cymbella*, *Gomphonema*) form stalks or tubes that host the cells in a protective filament. Other diatoms not able to resist complete drying can migrate from superficial to deeper sediments (McKew et al., 2010). Some algal groups, such as Rhodophyta and Phaeophyta, have thick cell walls. The encrusting red alga *Hildenbrandia rivularis* is able to attach extremely tightly to the substrata and resist long dry periods, immediately resuming photosynthetic activity after flow returns. These algae cannot be peeled off easily from the substratum but this capability is traded off with a very low growth rate. The chlorophyte *Gongrosira* sp. is another encrusting species that corrodes the substrata onto which it establishes, and this allows it to survive long desiccation but at the cost of losing many viable cells (Ledger et al., 2008). Encrusting forms are not unique to IRES and also occur in deserts where biological soil crusts can become active immediately after short pulses of moisture (Belnap and Lange, 2003).

During flow cessation, unattached algae adapted to drift may be able to reach pool refuges where they can survive and recolonize the system afterward. These algae are exposed to genetic drift and therefore to reduced genetic variation (Margalef, 1983). Some address this risk with polyploidy (several sets of chromosomes), an evolutionary success in intermittent habitats. Polyploidy is common among filamentous green algae such as *Oedogonium*, *Zygnema*, and *Spirogyra*. These taxa are also able to form resistant sexual eggs (zygospores) which lie dormant until favorable conditions return.

The ability of algae and cyanobacteria to photosynthesize is negatively related to moisture content of their habitat (Timoner et al., 2012) and positively related to their ability to recover their optimum hydric condition. Cyanobacteria, especially those forms having sheaths that quickly rehydrate, are extremely quick to react to increased air moisture or resumption of stream flow (Sabater et al., 2000). The extent of chloroplast reduction or the generalized bleaching of algal mats (Fig. 4.2.1A and H) is a direct expression of hydric stress effects on algae. Active chlorophyll (Chl-*a*) may decline by 60%–90% during streambed desiccation (Fig. 4.2.2A; Timoner et al., 2014), degrading because of hydric stress and the effects of high air temperatures and light irradiance (Fig. 4.2.2B). Degradation of chlorophyll (quantified as the Chl-*a* degradation index; Timoner et al., 2014) is very high during desiccation, indicating that algae remained photosynthetically inactive during the drying and nonflow phases. However, some algae (e.g., epilithic types) were able to resume activity immediately after flow resumption (Fig. 4.2.2B).

The decrease in active chlorophyll following hydric stress may accompany the increase in protective carotenoids. These protect the algal cell so that it becomes better adapted to desiccation stress. Two basic types of pigments are produced to protect algal tissues from UV degradation (Table 4.2.1). Scytonemin is a colored pigment in the extracellular polysaccharide sheaths of cyanobacteria (Garcia-Pichel and Castenholz, 1991) in microbial mats and soil crust biofilms exposed to desiccation and high solar radiation. Another group of pigments captures free radicals generated by UV penetration within the cell, and includes echinenone, canthaxanthin, β-carotene, lutein, zeaxanthin, and myxoxanthophyll. These carotenoids have also been observed under low light availability (Timoner et al., 2014), implying their synthesis is to protect biofilms not only from UV radiation but also desiccation and higher temperatures.

ALGAL-DERIVED STREAM METABOLISM IN IRES

During nonflowing periods, metabolic activity declines to a minimum. In contrast, primary production becomes important in IRES when they flow. During these periods, net primary production (NPP) is mainly determined by availability of light and nutrients. In forested IRES that are light-limited, periods of light availability (e.g., before leaf emergence and after leaf fall) coincide with pulses of autotrophic production (Sabater et al., 2011). As many forested IRES are nutrient-limited, any surplus significantly enhances NPP during the short periods of light availability (i.e., windows of opportunity in these systems; Sabater et al., 2011). These production pulses in heterotrophic streams provide high-quality resources to downstream reaches, which might be essential for consumers and spur their production and increase in biomass (Sabater et al., 2011).

In open-canopy systems, light is never limiting, and algae require adaptations to excessive light (Roberts et al., 2004). This is particularly important given the high temperatures in shrinking water channels and pools when fast-growing producers such as the filamentous Zygnematales dominate metabolism. In cold-water systems, the window of opportunity for primary producers is the melting period

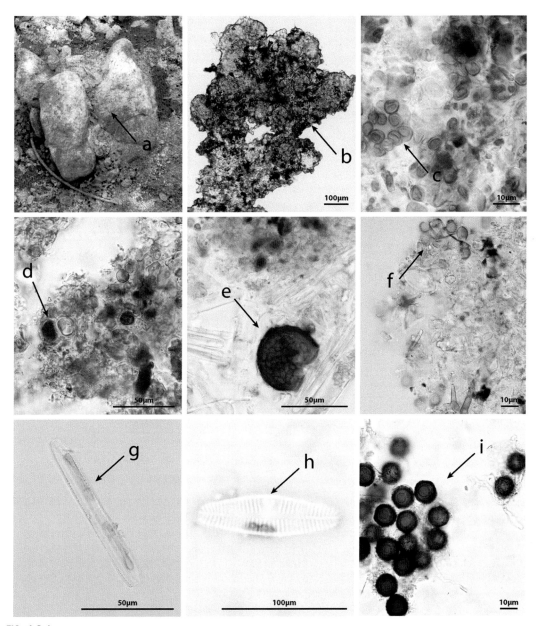

FIG. 4.2.1

Cell resistance structures observed in algal assemblages during the nonflowing phase. In the epilithon: (a) bleached dry biofilm, (b) general view of the crust formation, (c) thickened membrane walls, (d) change in coloration, (e) a resistant spore of *Oedogonium* sp., and (f) spores. In the epipsammon and hyporheos: (g and h) chloroplast reduction and (i) spores. Photos: X. Timoner. Panels b and c as originally published in Timoner et al. (2014) with permission of Springer. Panels e, g, and h as originally published in Sabater, S., Timoner, X., Borrego, C., Acuña, V., 2016. Stream biofilm responses to flow intermittency: from cells to ecosystems. Front. Environ. Sci. 4, 14.

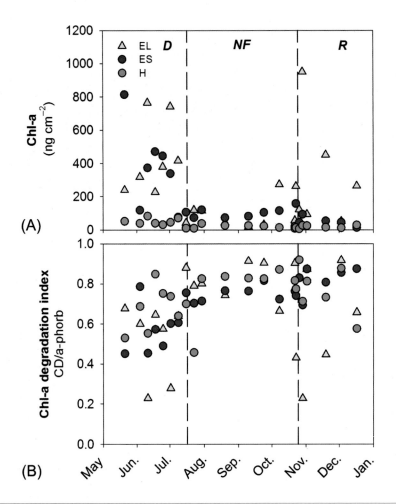

FIG. 4.2.2

Temporal changes in (A) chlorophyll *a* (Chl-*a*) and (B) Chl-*a* degradation (CD/a-phorb) index for each biofilm compartment (*EL*, epilithon; *ES*, epipsammon; *H*, hyporheic zone) in a Mediterranean IRES (Fuirosos, Spain) (Timoner et al., 2014 with permission of Springer). Hydrological phases (*D*, drying; *NF*, nonflowing; *R*, rewetting) are differentiated with vertical dashed lines. Chl-*a* degradation index is the ratio between the Chl-*a* derivatives (CD), which are the Chl-*a* degradation products (chlorophyll-*a* allomer 1, chloro-phyllide-a, phaeophorbide-a and phaeophythin-a1-a2), and the a-phorbins (a-phorb), the sum of Chl-*a* and CD. For this ratio, a value near 1.0 means complete degradation of Chl-*a*.

when the system unfreezes and algae return to activity. In IRES in the McMurdo Dry Valley, Antarctica, this period is only 4–10 weeks each summer. Terrestrial organic inputs to the stream are minimal, and primary production is the driver of stream metabolism, dominated by perennial algal mats. These mats, usually defined as microbial mats, are hot spots of biological activity (photosynthesis and metabolism) during the short flow season (McKnight and Tate, 1997). Flow intermittence reduces grazing pressure, and when flow resumes, dietary-generalist invertebrates are the first to recover (Woodward et al., 2016),

Table 4.2.1 Distribution of protective carotenoids associated with flow intermittence in different algal groups (Rowan, 1989; Jeffrey et al., 1997; Timoner et al., 2014)

	Bacillariophyta (diatoms)	Chlorophyta (green algae)	Chrysophyta (golden-brown algae)	Cryptophyta (cryptomonads)	Cyanophyta (cyanobacteria)	Euglenophyta (euglenids)	Rhodophyta (red algae)
Extracellular							
Scytonemin					x		
Intracellular							
β, β-carotene	x	x	x	x	x	x	x
Cantaxanthin					x		
Echinenone					x		
Lutein		x					x
Myxoxanthophyll					x		
Zeaxanthin	x		x	x	x	x	x

Table as originally published in Sabater, S., Timoner, X., Borrego, C., Acuña, V., 2016. Stream biofilm responses to flow intermittency: from cells to ecosystems. Front. Environ. Sci. 4, 14.

therefore exerting a relatively weak top-down pressure on the algal assemblage. When available, diatoms provide readily digestible autotrophic biomass to grazers such as tadpoles (Peterson and Boulton, 1999). In contrast, long nonflow periods favor cyanobacteria which are much less palatable to grazers than green algae and diatoms (Robson et al., 2008).

Algae play key roles in the biogeochemical and metabolic functions of desert IRES. Algal growth is very high during interflood flow periods, and the production:respiration ratio (P:R) of the stream can exceed 1 (i.e., production is higher than respiration). In Sycamore Creek, Arizona, rapid production coincided with high Chl-*a* content (up to 190 mg m^{-2} in some patches) and P:R was maintained at 1.5 for 60 days postflood (Fisher et al., 1982). Flash floods in these systems scour algae from the stream, reducing the pool of organic matter. However, wherever this organic material is collected and/or becomes buried by stream sediment, subsurface respiration associated with this buried biomass is high. In Sycamore Creek, subsurface bacterial densities were dependent on the import of algal-derived organic matter from the surface (Holmes et al., 1998).

IRES metabolism is tightly associated with the dynamics of the drying and rewetting phases (Sabater et al., 2016). Among effects reported in the literature, the most remarkable is the inconsistent effect of exposure to nonflowing conditions on autotrophic and heterotrophic processes within the stream. Algae may be more affected in their survival and functioning than heterotrophs (Timoner et al., 2012), and this affects long-term responses of stream metabolism. Acuña et al. (2015) showed that the duration of the nonflowing period differentially influences the metabolism of production (autotrophy) and respiration (heterotrophy) of biofilm assemblies. The longer the drying phase, the higher the relative increase of heterotrophy, because autotrophic processes (primary production) are more seriously impaired during dry periods of all durations whereas heterotrophs persist under some degrees of saturation. These differences may explain the complex metabolism patterns observed in IRES experiencing different durations of drying (Sabater et al., 2016); however, these processes are still not well understood.

LATITUDINAL VARIATION IN ALGAL COMMUNITY COMPOSITION

Both algal community composition and the type of prevailing algal growth differ across types of IRES. Some of this variation is evident when comparing temperate (Mediterranean) IRES to hot- and polar-desert IRES, all three encompassing wide variations in hydrology and temperature that challenge adaptation and survival of algal communities.

Mediterranean IRES

Mediterranean IRES have a strongly seasonal hydrological regime. Rainfall declines during summer when many IRES cease flow and even dry completely. Mediterranean calcareous streams are usually highly productive and form persistent algal structures. In some cases, these are thin and enduring structures that cover the streambed, having a rather uniform composition of a few cyanobacteria (*Phormidium* sp., *Schizothrix* spp., *Rivularia* spp.) and algae (*Gongrosira* sp., diatoms) (Sabater, 1989). Other cases have stromatolite-like formations on the streambed, mainly comprising cyanobacteria with some contributions from green algae and diatoms. These usually form in open-canopy systems with intense photosynthetic production and associated calcium carbonate precipitation (Sabater et al., 2000). Stromatolitic crusts may even crack with complete dehydration but can reactivate in a few hours when flow returns (Romaní and Sabater, 1997). Filamentous green algae (Zygnematales), diatoms, and cyanobacteria colonies occupy the upper layers of the crust, forming different patches that change in composition according to the different hydrological phases (Table 4.2.2).

In siliceous forested Mediterranean streams with dim light, red algae such as the encrusting *H. rivularis* or the filamentous *Lemanea* spp. are able to resist drying as well as survive substantial autumnal floods. They may coexist with mats of mosses such as *Fontinalis antipyretica*, which may also be substantial primary producers in these systems. Diatoms dominate communities of low diversity made up of adnate taxa such as *Achnanthidium minutissimum* or *Cocconeis placentula*, often interspersed with small mats of green algae.

Hot-desert IRES

Flows in hot-desert IRES are tightly linked to episodic rainfall episodes, and most of these streams are ephemeral. Flash floods associated with thunderstorms or monsoonal rains are a hydrological characteristic of these IRES (Tornés et al., 2015). Following these events, the streams progressively dwindle until they dry completely, sometimes for many months until the next extreme rainfall event. Especially in small tributaries, flash floods result in an abrupt loss of energy associated with the vast deposition and movement of sediments as well as the rapid recession in stream flow (Peterson and Boulton, 1999), all of which impair algal recolonization. This dynamic is not so immediate in larger channels where the establishment of algal communities may take several weeks after floods (Stanley et al., 2004). Tornés et al. (2015) observed that the algal communities in larger and smaller tributaries in the Gila River basin were the most affected by the floods. In contrast, medium-sized systems showed only minor changes in algal biomass compared with premonsoon periods, mostly because they did not experience streambed disruption. The recovery of the diatom community (algal biomass and diatom richness to

Table 4.2.2 Main groups of algae in IRES occurring in different biomes		
Antarctic Streams	• Microbial mats of filamentous cyanobacteria (*Oscillatoria* spp., *Phormidium* spp., *Nostoc* spp.) • Colonies of Cyanobacteria (*Nostoc commune*) • Green algae mats (*Prasiola*) • Small diatoms (*Humidophila* sp., *Fistulifera* sp.) • Aquatic mosses (*Drepanocladus longifolius*)	Izaguirre and Pizarro (1998) Sheath and Müller (1997) Toro et al. (2007)
Calcareous Mediterranean streams	• Encrusting communities: *Phormidium* sp., *Gongrosira* sp. • Stromatolite-like formations of Cyanobacteria: *Rivularia biasolettiana*, *Schizothrix* spp. • Drifting mats of Zynematales: *Zygnema*, *Spirogyra* and *Mougoetia* filaments. • Rich diatom periphyton in spring and winter • Diatom bloom community (stalked diatom *Cymbella* spp. and *Gomphonema* spp.) in spring	Sabater (1989) and Aboal (1989)
Siliceous Mediterranean streams	• Adnate diatom taxa: *Achnanthidium minutissimum*, *Cocconeis placentula* • Encrusting communities of *Hildenbrandia rivularis* • Filamentous *Lemanea* spp. • Mats of aquatic mosses (*Fontinalis antipyretica*)	Sabater (1989)
Desert arid streams	• *Cladophora glomerata* and its epiphytes • Filamentous cyanobacteria: *Schizothrix* and *Anabaena* • Diatom communities: *A. minutissimum*, *Mayamaea permitis*, *Reimeria sinuata*, *Gomphonema pumilum*, *Nitzschia inconspicua*, *Cocconeis* spp., *Nitzschia dissipata*	Fisher et al. (1982) Tornés et al. (2015)
Information from published data in Antarctic, Mediterranean, and desert IRES. References are provided.		

the predisturbance state) after extreme floods in the Gila basin took only one month after monsoonal floods. This strong resilience of the diatom assemblage after catastrophic events is characteristic of dryland stream ecosystems (Fisher et al., 1982).

In hot-desert IRES, hydrological and associated sedimentological dynamics drive the arrangement of algal communities in patches. Fisher et al. (1982) defined several distinguishable algal patches in Sycamore Creek, a desert stream in Arizona, which were characterized by different proportions of small- to medium-sized adnate diatoms, the green alga *Cladophora glomerata* and its epiphytes, and/ or some filamentous cyanobacteria such as *Schizothrix* spp. and *Anabaena* spp. Diatoms were the first to colonize the stream after floods, while green algae and cyanobacteria persisted during the low-flow period until the stream dried completely. The spatial distribution of the algal patches matched pathways of different water velocity in the channel, with *C. glomerata* occupying places with slower currents and diatoms dominating areas of higher velocity.

Cold-desert IRES

In Arctic and Antarctic areas, snowmelt streams are the most common IRES and host perennial microbial mats mostly consisting of cyanobacteria and diatoms (Esposito et al., 2006). Flow intermittence limits grazing pressure on these mats (Kohler et al., 2015a), differing from the pressures on equivalent algal mats in temperate streams. The dynamics of the mats are constrained by the brief growth period, as well as the hydrological and sedimentological patterns. High flows scour the mats and intraseasonal drying slows their growth (Kohler et al., 2015b). The mats may cover up to 100% of the streambed when low flow velocities and moderate sediment abrasion allow their accumulation over the years; in more dynamic systems, the cover is much lower (Kohler et al., 2015b). Because of the slow growth rate of these mats, multiple summers may be required to recover the initial biomass after disturbance. On experimentally scoured rocks from the ephemeral McMurdo Dry Valley stream, regrowth reached 18%–47% of predisturbance Chl-*a* and 27%–40% of ash-free dry mass by the end of summer (Kohler et al., 2015a).

The arrangement of the microbial mats on streambeds in cold-desert IRES is not uniform. Kohler et al. (2015b) describe conspicuous green and orange-red mats growing in the channel, and black mats growing at the stream margins. The green mats are dominated by the chlorophyte *Prasiola*. The orange-red mats, whose color indicates protective pigments to UV damage, comprise filamentous cyanobacteria and can achieve a thickness of up to 5 mm. The upper layer of the microbial mat may protect other organisms in the mat community from desiccation and UV radiation. The black mats are mostly formed by *Nostoc*.

These microbial mats have a remarkable ability to resist very low temperatures, and their structure enables them to persist in a freeze-dried state during winter (Kohler et al., 2015b), even for many years prior to flow returning. The cyanobacteria component of the mat reactivates within minutes of rehydration due to the ability of their mucilage to retain water. Other algal components also can resist long-term desiccation. Diatoms in Antarctic IRES are well adapted to periodic freezing and desiccation (Verleyen et al., 2003), and many are endemic species (Van de Vijver et al., 2010).

4.2.3 VASCULAR AQUATIC PLANTS

Flow intermittence drives a major shift in vascular macrophyte vegetation in aquatic ecosystems (Sand-Jensen and Frost-Christensen, 1998; De Wilde et al., 2014). The disappearance of surface and interstitial water causes dramatic environmental changes such as sharp increases in water limitation, intensification of solar radiation, and increases in the force of gravity on aquatic plants exposed to air, as well as modifications to carbon and oxygen availability (Baldwin and Mitchell, 2000; Rascio, 2002). The length and the intensity of drying events will consequently govern plant survival and/or the community's resilience after the event (Brock and Casanova, 1997).

Drying also modifies substratum properties by increasing the availability of nutrients such as phosphorus and nitrogen (Song et al., 2007). These, in turn, influence plant growth by increasing fertility and organic matter decomposition but also ammonia availability and resulting anoxia when flow resumes. Sediment characteristics such as particle size and organic matter content influence water retention (Walczak et al., 2002; De Wilde et al., 2017). Sediments rich in organic matter are better than mineral substrata at retaining residual moisture during drying. Silt and clay sediments also have a higher capacity to retain water compared to coarse sediment (Walczak et al., 2002). Consequently, for the same duration of drying, the intensity of water stress differs among sediment types, thus influencing resistance and resilience of vascular plants (De Wilde et al., 2017).

VASCULAR MACROPHYTES IN IRES

Evolutionary shifts between terrestrial and aquatic environments have occurred many times in the history of angiosperms (Cook, 1999; Rascio, 2002; Fig. 4.2.3). Ancestral traits tend to disappear during evolution when their expression is no longer required (i.e., lower selection strength). Consequently, taxonomic groups that evolved early toward aquatic forms should have lost more traits to terrestrial environments compared to those groups that evolved later toward aquatic forms (Peredo et al., 2013). The very few studies comparing the ability of aquatic plants to withstand drying stress suggest that resistance and resilience vary according to phylogenetic position. For example, dicots tend to have higher survival under drying conditions than many monocots (De Wilde et al., 2014), a group that shifted to aquatic habitats earlier in their evolutionary history (Cook, 1999; Chambers et al., 2008). Monocots also require a more gradual drying transition than dicots for producing a terrestrial growth form through phenotypic plasticity (De Wilde et al., 2014). Some studies suggest that aquatic dicots are potentially able to better survive as vegetative forms in ephemeral situations (due to more effective drying-resistance strategies, e.g., Touchette et al., 2007), whereas monocots tend to survive unfavorable periods as seeds or dormant propagules (e.g., tubers or rhizomes, Fox et al., 2014).

VASCULAR PLANT ADAPTATIONS TO DRYING

Functional groups

Among species that are able to withstand drying, several functional categories of response may occur (Brock and Casanova, 1997; Table 4.2.3). Some species are able to fully withstand drying, with the persistence of the aboveground plant parts either with or without modifications of the plant growth form (e.g., leaf plasticity; small size; and tissue modifications such as cuticles, stomata, aerenchyma, and lignin, Kuwabara et al., 2001; Touchette et al., 2007; De Wilde et al., 2014). Some helophyte and amphiphyte species are able to develop such plastic adjustments, even if the latter need substrata with a high capacity for water retention to survive (De Wilde et al., 2017). In other species, the aboveground biomass decays and the plant remains alive in the substratum through its subterranean organs (Engelhardt, 2006; Fox et al., 2014). In these cases, the survival of subterranean plant parts will depend on the moisture content of the sediment (Walczak et al., 2002). A large group of species flowers during the drying period and subsequently dies, leaving a dormant seed bank that germinates when the water returns (Brock et al., 2003).

In ecosystems experiencing seasonal drying, the aquatic plant communities achieve resilience through sexual reproduction (Brock et al., 2003). In situations of infrequent and low-intensity drying (e.g., winter drying, short-term drying, or partial drying), the plant community may resist desiccation through plastic adjustments and dormant vegetative diaspores. When drying is frequent and intense, these strategies may be counter-selected in favor of seed production. In situations characterized by alternating dry and wet periods, a cyclic seasonal succession may occur with submerged plants replaced by floating-leaf species and finally by emergent plants. However, the timing and duration of drying may modify this pattern (Riis and Hawes, 2002; Van der Valk, 2005).

Life History Traits Involved in Tolerance to Drying

Among plants that tolerate drying, several key morpho-anatomical traits are systematically modified (Table 4.2.4). These include a decrease in plant size, water content, and total dry mass. More energy is allocated to roots to enhance water acquisition. Leaf area typically decreases to reduce

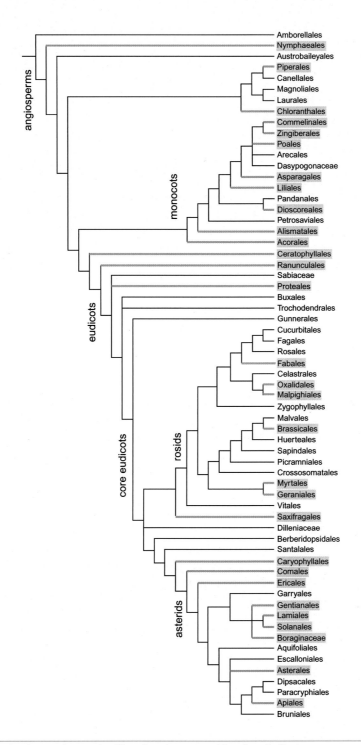

FIG. 4.2.3

Phylogenetic tree of orders and some families of angiosperms. *Blue lines* indicate groups containing aquatic species. According to Cook (1999), Germ and Gaberščik (2003), Chambers et al. (2008), and De Wilde et al. (2014), the more recent the taxonomic group, the higher the species' tolerance to drying.

Modified after The Angiosperm Phylogeny Group, 2009. An update of the Angiosperm Phylogeny Group classification for the orders and families of flowering plants: APG III. Bot. J. Linn. Soc., 161, 105–121., John Wiley and Sons, with permission.

Table 4.2.3 Plant functional groups and their response to drying

Category	Niche	Description	Part of the plant that tolerates drying	Examples	Intensity and duration of drying events
Helophytes	Emergent	Plants tolerate variation in water level without total submersion, belowground organs under water reproduce out of the water	Whole	Cyperaceae, Juncaceae	
Aquatic species without vegetative diaspores	Submerged	Plants need submerged situations and die during drying	Seeds	Some *Potamogeton* species, Najadaceae, *Elodea* sp., Ceratophyllaceae	
Amphiphytes	Indifferently emergent or submerged	Plants tolerate emersion, but need wet substrate; phenotypic plasticity	Whole, tissue renewal	*Veronica* sp., Sparganiaceae, Alismataceae, Hippuridaceae, Apiaceae, Callitrichaceae, Ranunculaceae	
Aquatic species with vegetative diaspores	Submerged	Needs submerged situations, diaspores need wet substrate, aboveground part of the plant decays during drying	Subterranean organs, dormant apices	*Vallisneria* sp., some *Potamogeton* species	
Plastic aquatic species	Preferentially submerged	Grow mostly in submerged situations, but able to tolerate short-term emersion; moderate phenotypic plasticity	Whole, tissue renewal for short-term drying, subterranean organs or seeds for longer periods of drying	Nymphaeaceae, some *Potamogeton* species	

Helophytes are emergent plants that can tolerate partial sediment drying, amphiphytes can tolerate emersion if the sediment remains wet, and aquatic species are plants that are strictly aquatic in their vegetative phase (Brock and Casanova, 1997; Lacoul and Freedman, 2006; Hrivnák et al., 2009).

evapotranspiration, and leaves increase in thickness and decline in water content, enabling them to better withstand gravity. Stomatal density and cuticle thickness increase, improving the regulation of water and gas exchange. There is a greater investment in sexual reproduction, expressed in the high seed density observed in aquatic plants in IRES, while vegetative reproduction declines because of the priority to allocate biomass toward tissue renewal for producing emergent growth forms.

Table 4.2.4 Principal morpho-anatomical adjustments observed in aquatic angiosperms subjected to drying

Aspect of plant strategy	Traits	Plant response to drying	Species (common name)	References
Morphology	Leaf thickness	Increase	*Myosotis scorpioides* (forget-me-not)	Germ and Gaberščik (2003)
			Potamogeton wrightii; P. anguillanus	Iida et al. (2007)
			Ranunculus flabellaris (yellow water buttercup)	Bruni et al. (1996)
	Plant size	Decrease	*Elodea nuttallii* (western waterweed)	Barrat-Segretain and Cellot (2007)
			Cyperus esculentus (yellow nutsedge)	Li et al. (2001)
			Acorus americanus (sweet flag)	Romanello et al. (2008)
			Ranunculus peltatus (pond water-crowfoot)	Volder et al. (1997)
Resource allocation	Total dry mass	Decrease	*Acorus americanus* (sweet flag)	Romanello et al. (2008)
			Typha domingensis (southern cattail)	Deegan et al. (2007)
			Cyperus vaginatus (stiff flat-sedge)	Deegan et al. (2007)
	Root dry mass	Decrease	*Carex lasiocarpa* (woollyfruit sedge); *C. limosa* (mud sedge); *Deyeuxia angustifolia*	Luo et al. (2008)
			Deyeuxia angustifolia	Luo et al. (2008)
			Mentha aquatic (water mint)	Lenssen et al. (2000)
			Lobelia dortmanna (water lobelia)	Pedersen and Sand-Jensen (1992)
			Acorus americanus (sweet flag)	Romanello et al. (2008)
	Rhizome dry mass	Decrease	*Acorus americanus* (sweet flag)	Romanello et al. (2008)
			Mentha aquatica (water mint)	Lenssen et al. (2000)
	Leaf dry mass	Decrease	*Lobelia dortmanna* (water lobelia)	Pedersen and Sand-Jensen (1992)
			Epilobium hirsutum (great willowherb)	Lenssen et al. (2000)
			Mentha aquatica (water mint)	
	Root/shoot ratio	Increase	*Acorus americanus* (sweet flag)	Romanello et al. (2008)
			Berula erecta (lesser water-parsnip)	Sand-Jensen and Frost-Christensen (1999)
	Plant water content	Decrease	*Mentha aquatica* (water mint)	Sand-Jensen and Frost-Christensen (1999)
			Berula erecta (lesser water-parsnip)	
			Acorus americanus (sweet flag)	Romanello et al. (2008)
			Lobelia dortmanna (water lobelia)	Pedersen and Sand-Jensen (1992)
			Littorella uniflora (shoreweed)	Robe and Griffiths (2000)

Table 4.2.4 Principal morpho-anatomical adjustments observed in aquatic angiosperms subjected to drying—cont'd

Aspect of plant strategy	Traits	Plant response to drying	Species (common name)	References
Reproduction	Flower number	Decrease	*Ranunculus peltatus* (pond water-crowfoot)	Volder et al. (1997)
	Seed number	No change	*Ranunculus peltatus* (pond water-crowfoot)	Volder et al. (1997)
	Percentage of flowering stems	Increase	*Cyperus esculentus* (yellow nutsedge) *Schoenoplectus lineolatus*	Li et al. (2001) Ishii and Kadono (2004)
	Fruit number	Decrease	*Ranunculus peltatus* (pond water-crowfoot)	Volder et al. (1997)
Anatomy	Stoma density	Increase	*Potamogeton anguillanus* *P. wrightii*	Iida et al. (2007) Germ and Gaberščik (2003)
			Ranunculus flabellaris (yellow water buttercup)	Bruni et al. (1996)
			Ludwigia arcuata (piedmont primrose-willow)	Kuwabara et al. (2001)
	Aerenchyma density			
	–roots	Decrease	*Paspalum dilatatum* (dallisgrass)	Vasellati et al. (2001)
		No change	*Littorella uniflora* (shoreweed)	Robe and Griffiths (1998)
	–stem	No change	*L. uniflora* (shoreweed)	Robe and Griffiths (1998)
	–leaves	Decrease	*P. dilatatum* (dallisgrass)	Vasellati et al. (2001)
			Lobelia dortmanna (water lobelia)	Pedersen and Sand-Jensen (1992)
			L. uniflora (shoreweed)	Robe and Griffiths (1998)
Biomechanical properties	Stem stiffness	Increase	*Ranunculus repens* (creeping buttercup), *R. acris* (meadow buttercup)	Usherwood et al. (1997)
			Berula erecta (lesser water-parsnip), *Hippuris vulgaris* (common mare's tail), *Juncus articulatus* (jointleaf rush), *Lythrum salicaria* (purple loosestrife), *Mentha aquatica* (water mint), *Myosotis scorpioides* (forget-me-not), *Nuphar lutea* (yellow pond-lily) and *Sparganium emersum* (unbranched bur-reed)	Hamann and Puijalon (2013)

IMPLICATIONS OF DRYING FOR VASCULAR PLANT SPECIES RICHNESS

The composition and richness of macrophyte communities rely on their survival and/or resilience during and after drying events as well as the duration and predictability of these events. IRES with recurrent drying may support higher plant biodiversity than systems with stable water levels (Havens et al., 2005; van Geest et al., 2005). Under these situations, there exists an alternation between aquatic and emergent communities with only limited competition. Under such wet-dry cycles, many species are annual (Arthaud et al., 2012). Strictly aquatic species are eliminated when the wetland dries out (Liu et al., 2006) while the emergent species do not resist the reinundation (Van der Valk, 2005).

Occasional drying events typically have more drastic effects on plant diversity than recurrent drying (Riis and Hawes, 2002). However, the intrinsic ability of species to tolerate drying leads to a variety of responses to different situations (Holmes, 1999), strongly ruled by the dominant texture of the substratum. IRES with coarse-grained sediments suffer from high hydraulic stress, often excluding macrophytes except in protected areas along banks. In contrast, when finer, more organic sediments predominate, hydraulic stress is lower and plants may better survive dewatering.

Finally, eutrophication may decrease vascular plant biomass and diversity in IRES while favoring planktonic and filamentous algae that shade out and outcompete the angiosperms (Fox et al., 2011). When these systems dry, a window of opportunity is opened for invasive species to recruit and establish (Davis et al., 2000).

4.2.4 VASCULAR RIPARIAN PLANTS
RIPARIAN VEGETATION ALONG IRES IN DIFFERENT ECOREGIONS

Riparian plant communities in IRES greatly differ, depending on regional hydrological characteristics (Fig. 4.2.4). In desert and dryland regions, plant productivity is limited by water availability, usually extending throughout the growing season. As streams transition from perennial to intermittent, tall, fast-growing, and large-leaved riparian trees give way to shrubs and trees with shorter stature and smaller leaves, and from herbaceous perennials to annual species (Dodds et al., 2004). Canopies become sparser, with shifts from riparian forest to more open riparian woodlands. The resultant change in productivity and plant growth form cascades to other trophic levels, including leaf litter and soil organic matter (Dodds et al., 2004, Sponseller and Fisher, 2006) and riparian bird life (Brand et al., 2010).

The relationship between resource availability and species diversity is complex. Empirical studies of IRES in American Southwest drylands (Olson et al., 2001) show that plant species richness peaks where flow is intermittent. This is consistent with the intermediate productivity hypothesis: richness in the riparian zone increases as flow changes from perennial to intermittent, and then declines to lowest levels as flow becomes ephemeral (Katz et al., 2012). The high species number along IRES, drolly referred to as the "Goldilocks effect," is because conditions are just right; without permanent water sources to sustain dense cover of large-canopied trees and clonal perennial herbs, open patches of bare mineral soil are seasonally colonized by large numbers of annual plant species (Stromberg et al., 2009b).

Shallow water tables play a strong role in shaping riparian plant communities. Many riparian species are phreatophytes, plants that depend for their water supply on perennial groundwater that lies within their rooting zones. Perennial streams are sustained by inflowing groundwater and have shallow

FIG. 4.2.4

Vascular and riparian plants in IRES. (a) Plants in an ephemeral stream (Forcall, fall 2015, Ebro basin, NE Spain), and (b) an intermittent river (Cosumnes River, Central California, fall 2015, (c) *Apium* plants growing during the flowing period in the Matarranya River, (d) *Ranunculus* plant in a dry streambed, (e) flowering *Tamarix* (salt cedar), and (f) macrophytes and riparian vegetation in a Mediterranean IRES during spring (Matarranya River, Ebro Basin, NE Spain).

Photos: Courtesy S. Sabater (a, c, e, and f), J.C. Stella (b), and G. Bornette (d).

and stable water tables. IRES typically occur in hydrologically losing reaches (Chapter 2.3) and have deeper, fluctuating water tables. Water tables associated with ephemeral streams typically occur well below plant root zones. There are exceptions, however. Some stream reaches with infrequent flow do have shallow water tables and thus sustain the same productive riparian phreatophytes as found on perennial rivers (Stromberg and Merritt, 2016).

In temperate regions, IRES commonly occur in the headwaters of stream networks, whereas the middle and lower reaches are often perennial (McDonough et al., 2011). Headwater streams in temperate regions differ greatly from those in arid and dryland regions because they generally support trees, even the IRES (Dodds et al., 2004). Often, the majority of stream length in a catchment network may be intermittent, even in regions with abundant rainfall (Hansen, 2001). Headwater reaches typically drain small contributing catchment areas, have thin soils (e.g., glacial till) with limited permeability and moisture-holding capacity, and support high forest cover and evapotranspiration rates. All of these features can contribute to a limited capacity to buffer seasonal water deficits and promote streamflow intermittence (McDonough et al., 2011). In addition, karstic (limestone) regions and other areas with fractured bedrock often support streams with spatially intermittent surface flow (Perrin and Tournoud, 2009).

Riparian vegetation studies along IRES in temperate regions are uncommon, and it is likely that the influence of flow intermittence on riparian vegetation structure and composition is less than in arid regions. Headwaters in temperate regions, where most intermittent stream reaches occur, often have continuous forest cover that does not vary significantly in tree composition from adjacent forested areas (Hagan et al., 2006). This contrasts with larger order perennial streams, many of which have distinct riparian tree species and wetland indicators (Williams et al., 1999). Herbaceous plants in headwater riparian zones contribute uniquely to plant diversity and typically form distinct and narrow zones that contain wetland obligate species bordering forest streams. These species may be adapted not only to wet environments but also to particular bedrock, soil biogeochemistry, and the low light levels that occur within closed-canopy forests. Hagan et al. (2006) found that for small first-order headwater streams in Maine, herbaceous plants, but not trees or shrubs, were responsible for plant community differences between the stream and adjacent uplands. Therefore, it is likely that flow intermittence in headwater areas may have an effect on riparian herbaceous communities, but the nature of the relationship is not as well studied as it is in dryland IRES.

RIPARIAN PLANT ADAPTATIONS IN IRES

The seasonal pattern of wetting and drying varies from unimodal in regions such as the winter-rain dominated Mediterranean forests and woodlands to bimodal in regions that experience both summer and winter rains in southwestern North America. In the semiarid region of northeastern Brazil, IRES are characterized by flash floods that rewet small channels, more sustained wet periods that support larger streams, and long dry periods without surface flow (Maltchik and Medeiros, 2006). Surprisingly, research is still needed to tease apart effects of direct rainfall from those of flood pulses on the wet-season productivity-diversity pulse. Most plants along IRES are annuals or short-lived herbaceous perennials and likely respond to the combination of moisture sources to create dense seasonal forelands (Bagstad et al., 2005).

Riparian plants along IRES possess a suite of adaptations to the physical disturbances and transient resource availability typical of fluvial corridors (Naiman et al., 2005). Bendix and Stella (2013) outlined several broad mechanisms of drivers associated with periodic floods and the dynamic hydrology characteristic of near-channel environments. These include (a) flood energy, which contains vegetation distribution due to scour and stem breakage; (b) sedimentation, which can exert influence through creation of new habitat for plant colonization and reduction of competition, burial of existing vegetation, and/or sediment texture effects on water availability; (c) water table depth and dynamics, which regulate soil moisture; (d) soil chemistry influences, including mineral nutrition, salinity, and pollutants; and (e) fluvial controls on propagule dispersal. In desert and dryland regions, plants along IRES may require adaptations to both stressors such as poor soils and elevated soil salinity, and to disturbance agents such as infrequent high-energy floods (Stella and Battles, 2010). Understory plants in temperate IRES headwaters must also tolerate shade as IRES in this setting typically occur within closed canopy forests.

Rooting depths for riparian phreatophytes vary widely among species. Among the trees along IRES that have roots <5 m are members of the Salicaceae such as *Populus* (poplar, cottonwood) and *Salix* (willow) and Oleaceae such as *Fraxinus* (ash). Trees and shrubs that root to depths of 5 m or more can be found within several families including Asteraceae (e.g., *Isocoma* [goldenbush]); Fabaceae (e.g., *Acacia* [acacia], *Alhagi* [camelthorn], *Parkinsonia* [paloverde], *Prosopis* [mesquite]); Proteaceae (e.g., *Banksia* [banksia]); and Tamaricaeae (e.g., *Tamarix* [saltcedar]) (Stromberg, 2013; Hooke and Mant, 2015). Production of an extensive shallow root system represents another adaption of riparian plants to intermittent moisture conditions. Many phreatophytes show a high degree of intraspecific plasticity in root architecture in response to differences in water table conditions (Rood et al., 2011). Some riparian bunch grasses, such as *Sporobolus wrightii* (big sacaton), also can have roots several meters deep. Extreme adaptations to very dry stream beds include succulent roots, such as in *Citrullus colocynthis* (colocynth), that allow for persistence through the dry season (Mahmoud, 2010).

REPRODUCTIVE TRAIT ADAPTATIONS AND TRADE-OFFS

Many reproductive strategies allow plants to persist in environments that have temporally varying resources (Chesson et al., 2004). Plants including shrubs and trees that have a long life span and disperse their seeds yearly can endure poor recruitment years until hydrological conditions improve (Naiman et al., 2005). In water-limited systems, the timing of seed release for some species coincides with wet periods that ensure surface flow to assist dispersal through hydrochory (water transport of propagules) and/or high soil moisture that promotes germination and survival of seedlings (Stella et al., 2006). Other species, including annual plants and short-lived perennials, store diaspores in the environment via a "bank." Soil seed banks, consisting of the viable seeds in the soil and leaf litter, are an essential strategy of many herbaceous riparian plants that grow along IRES (O'Donnell et al., 2014). A study of the intermittent Wannon River, Australia, found that seed banks were similar between channel and bank areas, indicating that the system retained a high degree of longitudinal and lateral connectivity as well as the capacity to store propagules for wet periods (Casanova, 2015). A persistent question relevant to conservation, however, is how long the stored seeds can remain viable (Stella et al., 2006; Boudell and Stromberg, 2008).

Vegetative reproduction constitutes a further regeneration strategy common in riparian zones, particularly within semiarid systems (Stella et al., 2013a). Along IRES, periods of drought stress often result in significant dieback of aboveground biomass (Rood et al., 2003; Stella et al., 2013b). The ability to resprout vigorously when groundwater or stream levels rebound allows perennial plants to persist across these variable periods (Bond and Midgley, 2003, Rodríguez-González et al., 2010). The flashy nature of dryland IRES often causes toppling, shearing, and layering by burial of aboveground plant parts. Uprooted stems and plant fragments transported during high flows can serve as vegetative propagules, leading to clonal reestablishment downstream (Rood et al., 2003). In a South African river where a centennial flood toppled and uprooted trees throughout the riparian zone, half of the woody debris piles within the active channel later resprouted and supported tree regeneration (Pettit et al., 2006).

Across riparian plant species adapted to IRES, there are often trade-offs between traits that allow propagules to disperse widely (e.g., prolific output, light weight, buoyancy) and those that ensure persistence in seed banks and adequate reserves for early growth (e.g., protective seed coats, prodigious endosperm). Therefore, there is no single successful regeneration strategy, but rather alternative ones that exploit particular aspects of the physical regime and fluvial environment (Naiman et al., 2005; Rodríguez-González et al., 2010).

LANDSCAPE BIODIVERSITY PATTERNS

Landscape context and connectivity

Riparian plant community composition of IRES depends upon local and regional factors. Important regional factors include size of the catchment and heterogeneity of the stream network. Some IRES drain small watersheds and thus are regulated largely by the local climate whereas those that drain larger watersheds are also regulated by climatic patterns of the distant mountains that provide water flows (Shaw and Cooper, 2008). IRES at the far downstream end of a dendritic stream network tend to have high species richness, owing to the prevalence of downstream (vs. upstream) dispersal of plant propagules and to the arrival of propagules from diverse upland vegetation types (Kehr et al., 2014). Along "interrupted perennial rivers" (*sensu* Meinzer, 1923), proximity to a perennial reach is important owing to downstream dispersal of propagules. During wet years, flood water can trigger germination of these in-flowing or stored seeds and create ephemeral wetlands in the typically dry stream beds located in the "seed shadow" of perennial reaches (Stromberg et al., 2009a).

Rare and endemic riparian species

Although IRES are not known for being hot spots of rare and endemic plant species, those that do occur in such habitats tend to be herbaceous perennials. Within the Interior Highlands of eastern United States, *Amsonia hubrichtii* (Hubricht's bluestar) and *Vernonia lettermannii* (narrowleaf ironweed) are two of several endemic herbaceous perennials affiliated with rocky open areas along IRES (Zollner et al., 2005). The endangered *Amsonia kearneyana* (Kearney's bluestar) is an herbaceous perennial that is restricted to IRES and adjacent rocky uplands of a small isolated mountain of Arizona. Ponds within IRES of a semiarid portion of Brazil provide habitat for *Gossypium mustelinum* (upland cotton), a wild relative of cotton (Alves et al., 2013). The endangered *Carex helodes* (sedge), an herbaceous perennial, is dependent upon IRES of the South Iberian and North African cork woodlands for persistence (Narbona et al., 2013). The very rare and beautiful *Brodiaea matsonii* (Matson's brodiaea), restricted to rocky crevices of a single IRES in California, was only recently documented as a new species (Preston, 2010).

Riparian community context

Plant community composition of IRES riparian zones is influenced by the surrounding vegetation (Levi and Fehmi, 2014). In addition to riparian obligate species, the floodplains and terraces of IRES also are populated by facultative riparian species, so named because they occur more broadly in upland habitats (Santos, 2010). Such taxa often have xeric adaptations such as sclerophylly and are particularly common in the outer and drier portions of the riparian zone that are least affected by scouring floods and inundation. These species contribute to the overall diversity of the riparian zone, and, conversely, the riparian habitat may function as a regional diversity reservoir for drought-sensitive upland species.

Herbaceous perennials and shrubs can become locally dependent on IRES for regional persistence during extreme drought (Miriti et al., 2007).

Natural and anthropogenic pressures on IRES in agricultural and urban settings commonly influence riparian composition and vegetation dynamics. Many large and small mammals rely on riparian ecosystems for food and shelter, and their actions (e.g., herbivory) can enhance physical heterogeneity and floristic diversity (Naiman and Rogers, 1997; Chapter 4.6). However, intense grazing pressure from ungulates on poorly regulated ranch lands can desertify the riparian ecosystem and cause shifts within plant communities to assemblages more characteristic of upland settings (Allsopp et al., 2007). In urban IRES settings, multiple stressors can give rise to novel plant communities, consisting of mixtures of regional species and cosmopolitan cultivars. Despite strong land use pressures from urbanization and degradation of the overall stream channel environment, urban riparian systems can continue to support ecosystem functions such as shade, allochthonous inputs, and wildlife habitat. Exploring these different expressions of resilience further may enhance our understanding of the responses of IRES to global change.

4.2.5 CONSERVATION AND MANAGEMENT ISSUES OF PRIMARY PRODUCERS IN IRES

As humans continue to create infrastructure to supply cities and farms with water, hydropower, and flood control, natural flows in rivers and streams increasingly are altered. In some cases, IRES can become perennial owing to discharge of urban effluent or agricultural runoff (Chapter 5.1). In many other cases, perennially flowing streams become intermittent. Perennial streams historically have been a locus for urban expansion, and in arid regions, water diversion has considerably reduced stream baseflows while upstream dams have altered flow timing and variability. This raises the need for conservation of these systems that are indeed under threat.

IRES are resilient systems, as the responses of primary producers to floods and drying demonstrate. Algae return to activity quickly after streamflow returns, even after very long dry periods. Macrophytes and riparian plants shift in community composition to sustain analogous functions to those in corresponding plant communities in temperate systems. However, we still know little about the adaptations and particular functions of primary producers in many IRES, and this is particularly relevant since they account for most of the basal source of the metabolism in these systems. Primary producers have crucial implications for energy flow to IRES food webs (Chapter 4.7) yet we lack sufficient knowledge of the particular adaptations of primary producers in IRES, particularly those in arid systems, where their occurrence is limited in space and time. Their role might be extremely important for the biological pulses that characterize those systems, and therefore is a focus for effective management of IRES.

Managing invasive species and conserving sensitive and endemic species should also be priorities in IRES because these species are ultimately linked to habitat quality. For example, some species of algae and vascular plants are known to be endemic to IRES in some regions (e.g., Antarctic region, Mediterranean areas) and suspected in others, but their contributions have not yet been quantified. Many of the primary producers colonizing IRES have unique adaptations and life history traits. Therefore, protecting and managing biodiversity of primary producers in IRES is a particularly urgent priority.

In many areas, IRES are infrequently monitored, particularly during the long nonflowing periods. While some species of algae, macrophytes, and riparian plants may be unambiguous sentinels for their responses to the effects of climate change and global change to IRES, our understanding of their autecology is still poor. Expanding this understanding can increase only with careful and specific monitoring. Present monitoring networks are not sufficient and often neglect headwater systems, particularly IRES. The survival and composition of riparian vegetation and the community composition and mass development of algae and macrophytes are directly affected by poor management of headwaters, including both the land uses affecting the basin as well as the direct in-stream interventions such as dams and weirs.

Finally, the dynamic balance between the different primary producers in supporting energy fluxes in IRES is also yet to be understood. Under which physical and hydrological conditions are riparian plants favored in contrast to macrophytes or algae? What are their respective contributions to the food webs of IRES in different climatic regimes? Answering these and other questions requires addressing IRES not only as the sums of the different parts but as single entities and certainly requires using complementary skills and integrating diverse approaches. The understanding gleaned from these multiple approaches will better inform management and conservation of primary producers in IRES.

REFERENCES

Aboal, M., 1989. Epilithic algal communities from River Segura basin, southeastern Spain. Arch. Hydrobiol. 116, 113–124.

Acuña, V., Casellas, M., Corcoll, N., Timoner, X., Sabater, S., 2015. Increasing duration of flow intermittency in temporary waterways promotes heterotrophy. Freshw. Biol. 60, 1810–1823.

Allsopp, N., Gaika, L., Knight, R., Monakisi, C., Hoffman, M.T., 2007. The impact of heavy grazing on an ephemeral river system in the succulent karoo, South Africa. J. Arid Environ. 71, 82–96.

Alves, M.F., Barroso, P.A.V., Ciampi, A.Y., Hoffmann, L.V., Azevedo, V.C.R., Cavalcante, U., 2013. Diversity and genetic structure among subpopulations of *Gossypium mustelinum* (Malvaceae). Genet. Mol. Res. 12, 597–609.

Arthaud, F., Vallod, D., Robin, J., Bornette, G., 2012. Eutrophication and drought disturbance shape functional diversity and life-history traits of aquatic plants in shallow lakes. Aquat. Sci. 74, 471–481.

Bagstad, K.J., Stromberg, J.C., Lite, S.J., 2005. Response of herbaceous riparian plants to rain and flooding on the San Pedro River, Arizona, USA. Wetlands 25, 210–223.

Baldwin, D.S., Mitchell, A.M., 2000. The effects of drying and re-flooding on the sediment and soil nutrient dynamics of lowland river-floodplain systems: a synthesis. Regul. Rivers Res. Manag. 16, 457–467.

Barrat-Segretain, M.-H., Cellot, B., 2007. Response of invasive macrophyte species to drawdown: the case of *Elodea* sp. Aquat. Bot. 87, 255–261.

Belnap, J., Lange, O.L., 2003. Structure and functioning of biological soil crusts: a synthesis. In: Belnap, J., Lange, O.L. (Eds.), Structure and Functioning of Biological Soil Crusts. Springer-Verlag, Berlin, pp. 471–479.

Bendix, J., Stella, J.C., 2013. Riparian vegetation and the fluvial environment: a biogeographic perspective. In: Shroder, J., Butler, D., Hupp, C. (Eds.), Treatise on Geomorphology, vol.12. Academic Press, San Diego, CA, pp. 53–74.

Bond, W.J., Midgley, J.J., 2003. The evolutionary ecology of sprouting in woody plants. Int. J. Plant Sci. 164, S103–S114.

Boudell, J.A., Stromberg, J.C., 2008. Propagule banks: potential contribution to restoration of an impounded and dewatered riparian ecosystem. Wetlands 28, 656–665.

Brand, L.A., Stromberg, J.C., Noon, B.R., 2010. Avian density and nest survival on the San Pedro River: importance of vegetation type and hydrologic regime. J. Wildl. Manag. 74, 739–754.

Brock, M.A., Casanova, M.T., 1997. Plant life at the edge of wetlands: ecological responses to wetting and drying patterns. In: Klomp, N., Lunt, I. (Eds.), Frontiers in Ecology: Building the Links. Elsevier, Oxford, pp. 181–192.

Brock, M.A., Nielsen, D.L., Shiel, R.J., Green, J.D., Langley, J.D., 2003. Drought and aquatic community resilience: the role of eggs and seeds in sediments of temporary wetlands. Freshw. Biol. 48, 1207–1218.

Bruni, N.C., Dengler, N.G., Young, J.P., 1996. Leaf developmental plasticity of *Ranunculus flabellaris* in response to terrestrial and submerged environments. Can. J. Bot. 74, 823–837.

Bunn, S.E., Davies, P.M., Winning, M., 2003. Sources of organic carbon supporting the food web of an arid zone floodplain river. Freshw. Biol. 48, 619–635.

Casanova, M.T., 2015. The seed bank as a mechanism for resilience and connectivity in a seasonal unregulated river. Aquat. Bot. 124, 63–69.

Chambers, P.A., Lacoul, P., Murphy, K.J., Thomaz, S.M., 2008. Global diversity of aquatic macrophytes in freshwater. Hydrobiologia 595, 9–26.

Chesson, P., Gebauer, R.L.E., Schwinning, S., Huntly, N., Wiegand, K., Ernest, M.S.K., 2004. Resource pulses, species interactions, and diversity maintenance in arid and semi-arid environments. Oecologia 141, 236–253.

Cook, C.D.K., 1999. The number and kinds of embryo-bearing plants which have become aquatic. Perspect. Plant Ecol. Evol. Syst. 2, 79–102.

Davis, M.A., Grime, J.P., Thompson, K., 2000. Fluctuating resources in plant communities: a general theory of invasibility. J. Ecol. 88, 528–534.

De Wilde, M., Sebei, N., Puijalon, S., Bornette, G., 2014. Responses of macrophytes to dewatering: effects of phylogeny and phenotypic plasticity on species performance. Evol. Ecol. 28, 1155–1167.

De Wilde, M., Puijalon, S., Bornette, G., 2017. Sediment type rules the response of aquatic plant communities to dewatering in wetlands. J. Veg. Sci. 28, 172–183.

Deegan, B.M., White, S.D., Ganf, G.G., 2007. The influence of water level fluctuations on the growth of four emergent macrophyte species. Aquat. Bot. 86, 309–315.

Dodds, W.K., Gido, K., Whiles, M.R., Fritz, K.M., Matthews, W.J., 2004. Life on the edge: the ecology of Great Plains prairie streams. Bioscience 54, 205–216.

Engelhardt, K.A.M., 2006. Relating effect and response traits in submersed aquatic macrophytes. Ecol. Appl. 16, 1808–1820.

Esposito, R.M.M., Horn, S.L., McKnight, D.M., Cox, M.J., Grant, M.C., Spaulding, S.A., et al., 2006. Antarctic climate cooling and response of diatoms in glacial meltwater streams. Geophys. Res. Lett. 33, L07406.

Fazi, S., Vazquez, E., Casamayor, E.O., Amalfitano, S., Butturini, A., 2013. Stream hydrological fragmentation drives bacterioplankton community composition. PLoS One 8, e64109.

Fisher, S.G., Gray, L.J., Grimm, N.B., Busch, D.E., 1982. Temporal succession in a desert stream ecosystem following flash flooding. Ecol. Monogr. 52, 93–110.

Fox, A.D., Cao, L., Zhang, Y., Barter, M., Zhao, M.J., Meng, F.J., 2011. Declines in the tuber-feeding waterbird guild at Shengjin Lake National Nature Reserve, China—a barometer of submerged macrophyte collapse. Aquat. Conserv. Mar. Freshwat. Ecosyst. 21, 82–91.

Fox, A.D., Meng, F., Liu, J., Yang, W., Shan, K., Cao, L., 2014. Effects of the length of inundation periods on investment in tuber biomass and sexual reproduction by *Vallisneria spinulosa* SZ Yan Ramets. Knowl. Manag. Aquat. Ecosyst. 414, Article 3: 1–10.

Garcia-Pichel, F., Castenholz, R.W., 1991. Characterization and biological implications of scytonemin, a cyanobacterial sheath pigment. J. Phycol. 27, 395–409.

Germ, M., Gaberščik, A., 2003. Comparison of aerial and submerged leaves in two amphibious species, *Myosotis scorpioides* and *Ranunculus trichophyllus*. Photosynthetica 41, 91–96.

Hagan, J.M., Pealer, S., Whitman, A.A., 2006. Do small headwater streams have a riparian zone defined by plant communities? Can. J. For. Res. 36, 2131–2140.

Hamann, E., Puijalon, S., 2013. Biomechanical responses of aquatic plants to aerial conditions. Ann. Bot. 112, 1869–1878.

Hansen, W.F., 2001. Identifying stream types and management implications. For. Ecol. Manag. 143, 39–46.

Havens, K.E., Fox, D., Gornak, S., Hanlon, C., 2005. Aquatic vegetation and largemouth bass population responses to water-level variations in Lake Okeechobee, Florida (USA). Hydrobiologia 539, 225–237.

Holmes, N.T.H., 1999. Recovery of headwater stream flora following the 1989-1992 groundwater drought. Hydrol. Process. 13, 341–354.

Holmes, R.M., Fisher, S.G., Grimm, N.B., Harper, B.J., 1998. The impact of flash floods on microbial distribution and biogeochemistry in the parafluvial zone of a desert stream. Freshw. Biol. 40, 641–654.

Hooke, J., Mant, J., 2015. Morphological and vegetation variations in response to flow events in rambla channels of SE Spain. In: Dykes, A.P., Mulligan, M., Wainwright, J. (Eds.), Monitoring and Modelling Dynamic Environments. Wiley, Chichester, pp. 61–98.

Hrivnák, R., Oťaheľová, H., Gömöry, D., 2009. Seasonal dynamics of macrophyte abundance in two regulated streams. Cent. Eur. J. Biol. 4, 241–249.

Iida, S., Yamada, A., Amano, M., Ishii, J., Kadono, Y., Kosuge, K., 2007. Inherited maternal effects on the drought tolerance of a natural hybrid aquatic plant, *Potamogeton anguillanus*. J. Plant Res. 120, 473–481.

Ishii, J., Kadono, Y., 2004. Sexual reproduction under fluctuating water levels in an amphibious plant *Schoenoplectus lineolatus* (Cyperaceae): a waiting strategy? Limnology 5, 1–6.

Izaguirre, I., Pizarro, H., 1998. Epilithic algae in a glacial stream at Hope Bay (Antarctica). Polar Biol. 19, 24–31.

Jeffrey, S.W., Mantoura, R.F.C., Wright, S.W., 1997. Phytoplankton Pigments in Oceanography: Guidelines to Modern Methods. UNESCO, Paris.

Katz, G.L., Denslow, M.W., Stromberg, J.C., 2012. The Goldilocks effect: intermittent stream reaches sustain more plant species than those with perennial or ephemeral flow. Freshw. Biol. 57, 467–480.

Kehr, J.M., Merritt, D.M., Stromberg, J.C., 2014. Linkages between primary seed dispersal, hydrochory, and flood timing in a dryland river. J. Veg. Sci. 25, 287–300.

Kohler, T.J., Chatfield, E., Gooseff, M.N., Barrett, J.E., McKnight, D.M., 2015a. Recovery of Antarctic stream epilithon from simulated scouring events. Antarct. Sci. 27, 341–354.

Kohler, T.J., Stanish, L.F., Crisp, S.W., Koch, J.C., Liptzin, D., Baeseman, J.L., et al., 2015b. Life in the main channel: long-term hydrologic control of microbial mat abundance in McMurdo Dry Valley Streams, Antarctica. Ecosystems 18, 310–327.

Kuwabara, A., Tsukaya, H., Nagata, T., 2001. Identification of factors that cause heterophylly in *Ludwigia arcuata* Walt. (Onagraceae). Plant Biol. 3, 98–105.

Lacoul, P., Freedman, B., 2006. Environmental influences on aquatic plants in freshwater ecosystems. Environ. Rev. 14, 89–136.

Ledger, M.E., Harris, R.M.L., Armitage, P.D., Milner, A.M., 2008. Disturbance frequency influences patch dynamics in stream benthic algal communities. Oecologia 155, 809–819.

Lenssen, J.P.M., Menting, F.B.J., Van der Putten, W.H., Blom, C.W.P.M., 2000. Vegetative reproduction by species with different adaptations to shallow-flooded habitats. New Phytol. 145, 61–70.

Levi, E.M., Fehmi, J.S., 2014. Landscape variables as predictors of characteristics of native-grass communities in xeroriparian areas of the Sonoran Desert. Southwest. Nat. 59, 103–109.

Li, B., Shibuya, T., Yogo, Y., Hara, T., Matsuo, K., 2001. Effects of light quantity and quality on growth and reproduction of a clonal sedge, *Cyperus esculentus*. Plant Species Biol. 16, 69–81.

Liu, G.-H., Li, W., Zhou, J., Liu, W.-Z., Yang, D., Davy, A.J., 2006. How does the propagule bank contribute to cyclic vegetation change in a lakeshore marsh with seasonal drawdown? Aquat. Bot. 84, 137–143.

Luo, W., Song, F., Xie, Y., 2008. Trade-off between tolerance to drought and tolerance to flooding in three wetland plants. Wetlands 28, 866–873.

Mahmoud, T., 2010. Desert Plants of Egypt's Wadi El Gemal National Park. The American University in Cairo Press, Cairo.

Maltchik, L., Medeiros, E.S.F., 2006. Conservation importance of semi-arid streams in north-eastern Brazil: implications of hydrological disturbance and species diversity. Aquat. Conserv. Mar. Freshwat. Ecosyst. 16, 665–677.

Margalef, R., 1983. Limnología. Omega, Barcelona.

McDonough, O.T., Hosen, J.D., Palmer, M.A., 2011. Temporary streams: the hydrology, geography, and ecology of non-perennially flowing waters. In: Elliott, H.S., Martin, L.E. (Eds.), River Ecosystems: Dynamics, Management and Conservation. Nova Science, New York, pp. 259–290.

McKew, B., Taylor, J., McGenity, T., Underwood, G., 2010. Resistance and resilience of benthic biofilm communities from a temperate saltmarsh to desiccation and rewetting. ISME J. 5, 30–41.

McKnight, D.M., Tate, C.M., 1997. Canada stream: a glacial meltwater stream in Taylor Valley, South Victoria Land, Antarctica. J. N. Am. Benthol. Soc. 16, 14–17.

Meinzer, O.E., 1923. Outline of ground-water hydrology, with definitions. U. S. Geol. Surv. Water Supply Pap. 494, 48–59.

Miriti, M.N., Rodríguez-Buriticá, S., Wright, S.J., Howe, H.F., 2007. Episodic death across species of desert shrubs. Ecology 88, 32–36.

Naiman, R.J., Rogers, K.H., 1997. Large animals and system-level characteristics in river corridors. Bioscience 47, 521–529.

Naiman, R.J., Decamps, H., McClain, M.E., 2005. Riparia: Ecology, Conservation, and Management of Streamside Communities. Elsevier Academic Press, London.

Narbona, E., Delgado, A., Encina, F., Miguez, M., Buide, M., 2013. Seed germination and seedling establishment of the rare *Carex helodes* Link depend on the proximity to water. Aquat. Bot. 110, 55–60.

O'Donnell, J., Fryirs, K., Leishman, M.R., 2014. Digging deep for diversity: riparian seed bank abundance and species richness in relation to burial depth. Freshw. Biol. 59, 100–113.

Olson, D.M., Dinerstein, E., Wikramanayake, E.D., Burgess, N.D., Powell, G.V.N., Underwood, E.C., et al., 2001. Terrestrial ecoregions of the world: a new map of life on Earth. Bioscience 51, 933–938.

Pedersen, O., Sand-Jensen, K., 1992. Adaptations of submerged *Lobelia dortmanna* to aerial life form: morphology, carbon sources and oxygen dynamics. Oikos 65, 89–96.

Peredo, E.L., King, U.M., Les, D.H., 2013. The plastid genome of *Najas flexilis*: adaptation to submersed environments is accompanied by the complete loss of the NDH complex in an aquatic angiosperm. PLoS One 8, e68591.

Perrin, J.-L., Tournoud, M.-G., 2009. Hydrological processes controlling flow generation in a small Mediterranean catchment under karstic influence. Hydrol. Sci. J. 54, 1125–1140.

Peterson, C.G., Boulton, A.J., 1999. Stream permanence influences microalgal food availability to grazing tadpoles in arid-zone springs. Oecologia 118, 340–352.

Pettit, N.E., Latterell, J.J., Naiman, R.J., 2006. Formation, distribution and ecological consequences of flood-related wood debris piles in a bedrock confined river in semi-arid South Africa. River Res. Appl. 22, 1097–1110.

Preston, R.E., 2010. *Brodiaea matsonii* (Asparagaceae: Brodiaeoideae), a new species from Shasta County, California. Madrono 57, 261–267.

Rascio, N., 2002. The underwater life of secondarily aquatic plants: some problems and solutions. Crit. Rev. Plant Sci. 21, 401–427.

Riis, T., Hawes, I., 2002. Relationships between water level fluctuations and vegetation diversity in shallow water of New Zealand lakes. Aquat. Bot. 74, 133–148.

Robe, W.E., Griffiths, H., 1998. Adaptations for an amphibious life: changes in leaf morphology, growth rate, carbon and nitrogen investment, and reproduction during adjustment to emersion by the freshwater macrophyte *Littorella uniflora*. New Phytol. 140, 9–23.

Robe, W.E., Griffiths, H., 2000. Physiological and photosynthetic plasticity in the amphibious, freshwater plant, *Littorella uniflora*, during the transition from aquatic to dry terrestrial environments. Plant Cell Environ. 23, 1041–1054.

Roberts, S., Sabater, S., Beardall, J., 2004. Benthic microalgal colonization in streams of differing riparian cover and light availability. J. Phycol. 40, 1004–1012.

Robson, B.J., Matthews, T.G., Lind, P.R., Thomas, N.A., 2008. Pathways for algal recolonization in seasonally-flowing streams. Freshw. Biol. 53, 2385–2401.

Rodríguez-González, P.M., Stella, J.C., Campelo, F., Ferreira, M.T., Albuquerque, A., 2010. Subsidy or stress? Tree structure and growth in wetland forests along a hydrological gradient in Southern Europe. For. Ecol. Manag. 259, 2015–2025.

Romanello, G.A., Chuchra-Zbytniuk, K.L., Vandermer, J.L., Touchette, B.W., 2008. Morphological adjustments promote drought avoidance in the wetland plant *Acorus americanus*. Aquat. Bot. 89, 390–396.

Romaní, A., Sabater, S., 1997. Metabolism recovery of a stromatolitic biofilm after drought in a Mediterranean stream. Arch. Hydrobiol. 140, 261–271.

Rood, S.B., Kalischuk, A.R., Polzin, M.L., Braatne, J.H., 2003. Branch propagation, not cladoptosis, permits dispersive, clonal reproduction of riparian cottonwoods. For. Ecol. Manag. 186, 227–242.

Rood, S.B., Bigelow, S.G., Hall, A.A., 2011. Root architecture of riparian trees: river cut-banks provide natural hydraulic excavation, revealing that cottonwoods are facultative phreatophytes. Trees 25, 907–917.

Rowan, K.S., 1989. Photosynthetic Pigments of Algae. Cambridge University Press, Cambridge.

Sabater, S., 1989. Encrusting algal assemblages in a Mediterranean river basin. Arch. Hydrobiol. 114, 555–573.

Sabater, S., Guasch, H., Romaní, A., Muñoz, I., 2000. Stromatolitic communities in Mediterranean streams: adaptations to a changing environment. Biodivers. Conserv. 9, 379–392.

Sabater, S., Artigas, J., Gaudes, A., Muñoz, I., Urrea, G., Romaní, A.M., 2011. Moderate long-term nutrient input enhances autotrophy in a forested Mediterranean stream. Freshw. Biol. 56, 1266–1280.

Sabater, S., Timoner, X., Borrego, C., Acuña, V., 2016. Stream biofilm responses to flow intermittency: from cells to ecosystems. Front. Environ. Sci. 4, 14.

Sand-Jensen, K., Frost-Christensen, H., 1998. Photosynthesis of amphibious and obligately submerged plants in CO_2-rich lowland streams. Oecologia 117, 31–39.

Sand-Jensen, K., Frost-Christensen, H., 1999. Plant growth and photosynthesis in the transition zone between land and stream. Aquat. Bot. 63, 23–35.

Santos, M.J., 2010. Encroachment of upland Mediterranean plant species in riparian ecosystems of southern Portugal. Biodivers. Conserv. 19, 2667–2684.

Shaw, J., Cooper, D., 2008. Linkages among watersheds, stream reaches, and riparian vegetation in dryland ephemeral stream networks. J. Hydrol. 350, 68–82.

Sheath, R.G., Müller, K.M., 1997. Distribution of stream macroalgae in four high arctic drainage basins. Arctic 50, 355–364.

Song, K.-Y., Zoh, K.-D., Kan, H., 2007. Release of phosphate in a wetland by changes in hydrological regime. Sci. Total Environ. 380, 13–18.

Sponseller, R.A., Fisher, S.G., 2006. Drainage size, stream intermittency, and ecosystem function in a Sonoran Desert landscape. Ecosystems 9, 344–356.

Stanley, E.H., Fisher, S.G., Jones, J.B., 2004. Effects of water loss on primary production: a landscape-scale model. Aquat. Sci. 66, 130–138.

Stella, J.C., Battles, J.J., 2010. How do riparian woody seedlings survive seasonal drought? Oecologia 164, 579–590.

Stella, J.C., Battles, J.J., Orr, B.K., McBride, J.R., 2006. Synchrony of seed dispersal, hydrology and local climate in a semi-arid river reach in California. Ecosystems 9, 1200–1214.

Stella, J.C., Rodríguez-González, P.M., Dufour, S., Bendix, J., 2013a. Riparian vegetation research in Mediterranean-climate regions: common patterns, ecological processes, and considerations for management. Hydrobiologia 719, 291–315.

Stella, J.C., Riddle, J., Piégay, H., Gagnage, M., Trémélo, M.L., 2013b. Climate and local geomorphic interactions drive patterns of riparian forest decline along a Mediterranean Basin river. Geomorphology 202, 101–114.

Sternberg, D., Balcombe, S., Marshall, J., Lobegeiger, J., 2008. Food resource variability in an Australian dryland river: evidence from the diet of two generalist native fish species. Mar. Freshw. Res. 59, 137–144.

Stromberg, J.C., 2013. Root patterns and hydrogeomorphic niches of riparian plants in the American Southwest. J. Arid Environ. 94, 1–9.

Stromberg, J.C., Merritt, D., 2016. Riparian plant guilds of ephemeral, intermittent and perennial rivers. Freshw. Biol. 61, 1259–1275.

Stromberg, J.C., Hazelton, A.F., White, M.S., White, J.M., Fischer, R.A., 2009a. Ephemeral wetlands along a spatially intermittent river: temporal patterns of vegetation development. Wetlands 29, 330–342.

Stromberg, J.C., Hazelton, A.F., White, M.S., 2009b. Plant species richness in ephemeral and perennial reaches of a dryland river. Biodiversity and Conservation 18, 663–677.

Timoner, X., Acuña, V., Von Schiller, D., Sabater, S., 2012. Functional responses of stream biofilms to flow cessation, desiccation and rewetting. Freshw. Biol. 57, 1565–1578.

Timoner, X., Buchaca, T., Acuña, V., Sabater, S., 2014. Photosynthetic pigment changes and adaptations in biofilms in response to flow intermittency. Aquat. Sci. 76, 565–578.

Tornés, E., Ruhí, A., 2013. Flow intermittency decreases nestedness and specialisation of diatom communities in Mediterranean rivers. Freshw. Biol. 58, 2555–2566.

Tornés, E., Sabater, S., 2010. Variable discharge alters habitat suitability for benthic algae and cyanobacteria in a forested Mediterranean stream. Mar. Freshw. Res. 61, 441–450.

Tornés, E., Acuña, V., Dahm, C.N., Sabater, S., 2015. Flood disturbance effects on benthic diatom assemblage structure in a semiarid river network. J. Phycol. 51, 133–143.

Toro, M., Camacho, A., Rochera, C., Rico, E., Bañón, M., Fernández-Valiente, E., et al., 2007. Limnological characteristics of the freshwater ecosystems of Byers Peninsula, Livingston Island, in maritime Antarctica. Polar Biol. 30, 635–649.

Touchette, B.W., Lannacone, L.R., Turner, G.E., Frank, A.R., 2007. Drought tolerance versus drought avoidance: a comparison of plant-water relations in herbaceous wetland plants subjected to water withdrawal and repletion. Wetlands 27, 656–667.

Usherwood, J.R., Ennos, A.R., Ball, D.J., 1997. Mechanical and anatomical adaptations in terrestrial and aquatic buttercups to their respective environments. J. Exp. Bot. 48, 1469–1475.

Van De Vijver, B., Mataloni, G., Stanish, L., Spaulding, S.A., 2010. New and interesting species of the genus *Muelleria* (Bacillariophyta) from the Antarctic region and South Africa. Phycologia 49, 22–41.

Van der Valk, A.G., 2005. Water level fluctuations in North American prairie wetlands. Hydrobiologia 539, 171–188.

Van Geest, G.V., Wolters, H., Roozen, F.C.J.M., Coops, H., Roijackers, R.M.M., Buijse, A.D., et al., 2005. Water-level fluctuations affect macrophyte richness in floodplain lakes. Hydrobiologia 539, 239–248.

Vasellati, V., Oesterheld, M., Medan, D., Loreti, J., 2001. Effects of flooding and drought on the anatomy of *Paspalum dilatatum*. Ann. Bot. 8, 355–360.

Verleyen, E., Hodgson, D.A., Vyverman, W., Roberts, D., McMinn, A., Vanhoutte, K., et al., 2003. Modelling diatom responses to climate induced fluctuations in the moisture balance in continental Antarctic lakes. J. Paleolimnol. 30, 195–215.

Volder, A., Bonis, A., Grillas, P., 1997. Effects of drought and flooding on the reproduction of an amphibious plant, *Ranunculus peltatus*. Aquat. Bot. 58, 113–120.

Walczak, R., Rovdan, E., Witkowska-Walczak, B., 2002. Water retention characteristics of peat and sand mixtures. Int. Agrophys. 16, 161–165.

Williams, C.E., Moriarity, W.J., Walters, G.L., Hill, L., 1999. Influence of inundation potential and forest overstory on the ground-layer vegetation of Allegheny Plateau riparian forests. Am. Midl. Nat. 141, 323–338.

Woodward, G., Bonada, N., Brown, L.E., Death, R.G., Durance, I., Gray, C., et al., 2016. The effects of climatic fluctuations and extreme events on running water ecosystems. Philos. Trans. R. Soc. B 371, 20150274.

Zollner, D., MacRoberts, M.H., MacRoberts, B.R., Ladd, D., 2005. Endemic vascular plants of the Interior Highlands, USA, SIDA. Contrib. Bot. 21, 1781–1791.

FURTHER READING

Morison, M.O., Sheath, R.G., 1985. Responses to desiccation stress by *Klebsormidium rivulare* (Ulotricales, Chlorophyta) from a Rhode Island stream. Phycologia 24, 129–145.

Stanley, E.H., Fisher, S.G., Grimm, N.B., 1997. Ecosystem expansion and contraction in streams. Bioscience 47, 427–435.

The Angiosperm Phylogeny Group, 2009. The Angiosperm Phylogeny Group, 2009. An update of the Angiosperm Phylogeny Group classification for the orders and families of flowering plants: APG III. Bot. J. Linn. Soc. 161, 105–121.

Zohary, M., 1961. On hydro-ecological relations of the Near East desert vegetation. Arid Zone Res. 16, 199–212.

THE BIOTA OF INTERMITTENT RIVERS AND EPHEMERAL STREAMS: AQUATIC INVERTEBRATES

Rachel Stubbington[*], Michael T. Bogan[†], Núria Bonada[‡], Andrew J. Boulton[§], Thibault Datry[¶], Catherine Leigh[¶,,††], Ross Vander Vorste[‡‡]**

Nottingham Trent University, Nottingham, United Kingdom[*] University of Arizona, Tucson, AZ, United States[†]
Universitat de Barcelona (UB), Barcelona, Spain[‡] University of New England, Armidale, NSW, Australia[§]
Irstea, UR MALY, centre de Lyon-Villeurbanne, Villeurbanne, France[¶] CESAB-FRB, Immeuble Henri Poincaré,
Aix-en-Provence, France[**] Griffith University, Nathan, QLD, Australia[††] Virginia Water Resources Research Center,
Blacksburg, VA, United States[‡‡]

IN A NUTSHELL

- Intermittent rivers and ephemeral streams (IRES) are temporally dynamic ecosystems that can support a diverse and distinctive aquatic invertebrate fauna
- Resistance and resilience mechanisms allow species and communities to persist in IRES during dry phases and to recolonize quickly once flow returns
- Human influences including climate change and water abstraction alter natural patterns of flow intermittence, with increasing intermittence typically reducing aquatic invertebrate diversity
- Effective management, restoration, and legislation are needed to safeguard the diversity of IRES invertebrate communities

4.3.1 INTRODUCTION

Aquatic invertebrates are a diverse group of organisms that inhabit IRES throughout the world, in regions spanning alpine, arid, Mediterranean, polar, temperate, and tropical climates. IRES invertebrates include benthic, planktonic, and stygobitic taxa (those associated with surface sediments, open water, and groundwater, respectively) and range from widespread taxa to endemic IRES specialists. Classified by size as either macroinvertebrates (those visible without magnification) or smaller meiofauna, macroinvertebrates are more widely studied (Robertson et al., 2000), although meiofauna can also be diverse and abundant in IRES communities (Meyer et al., 2007).

The roles that IRES invertebrates play in ecosystem functioning are strongly controlled by flow intermittence. For example, aquatic invertebrates are important leaf litter processors, but breakdown rates can be reduced in IRES compared to perennial rivers due to fewer "shredders" that feed on such coarse particulate organic material (Richardson, 1990), reducing the release of finer particulate organic material consumed by other feeding groups. Aquatic invertebrates provide prey for IRES predators such as fish (Courtwright and May, 2013; Chapter 4.5) and are also consumed by terrestrial predators including insects, reptiles, birds, and mammals (Leigh et al., 2013), particularly when stranded in isolated pools or on drying channels. In addition, adult insects emerging from IRES transport aquatic material to terrestrial habitats, providing an important energy source for riparian predators (Lynch et al., 2002; Progar and Moldenke, 2002; Chapters 4.4, 4.6, and 4.7).

Research into IRES invertebrates grows annually (Leigh et al., 2016a), and their diversity and ubiquity make them ideal organisms for addressing many research and management challenges in IRES (Datry et al., 2014a). This chapter reviews aquatic invertebrates in IRES, describing the ecosystems they inhabit, examining how their taxonomic and functional diversity change in response to varying environmental conditions, and considering the adaptations that promote survival in these dynamic ecosystems. We then highlight human activities that threaten aquatic invertebrate communities, explore ways to protect them, and suggest priorities for future research.

4.3.2 IRES AS HABITATS FOR AQUATIC INVERTEBRATES

IRES are complex, dynamic, and diverse ecosystems (Chapters 2.1 and 2.3) in which the shifting habitat mosaics (*sensu* Datry et al., 2016a) associated with intermittence drive the composition, abundance, and diversity of aquatic invertebrate communities (Datry et al., 2014b; Leigh and Datry, 2016; Sections 4.3.3 and 4.3.4). Aquatic invertebrates inhabiting IRES experience highly variable environmental conditions, and critical thresholds at which habitats change profoundly are crossed as IRES transition between flowing, pool, and dry phases (Boulton, 2003). During drying, lateral connections between the main channel and riparian habitats are severed first, then flow ceases and lentic (standing water) pools form. Longitudinal (upstream-downstream) connectivity can be lost, disconnecting pools that may then contract over time. Pools typically have high and variable water temperatures, increasing nutrient concentrations and declining oxygen availability (Lake, 2011, Chapters 3.1 and 3.2). Contracting pools are also characterized by intensifying biotic interactions including predation and competition for limited resources (Boulton and Lake, 1992; Jackson et al., 1999). Ultimately, contraction of disconnected pools may lead to complete surface water loss.

Where flowing surface water is lost, invertebrates may persist in refuges: habitats in which their survival is enhanced (Section 4.3.5). Perennial flowing-water refuges include reaches upstream and downstream of an intermittent reach as well as nearby rivers (Fig. 4.3.1; Boulton, 1989; Chester and Robson, 2011). Persistent pools and other perennial lentic waters are also important refuges (Paltridge et al., 1997), and moist microhabitats including leaf litter, algal mats, woody debris, and damp sediment beneath large stones provide localized refuges in surface channels (Fig. 4.3.1; Boulton, 1989; Robson et al., 2011). Below the drying benthic zone, the saturated subsurface sediments of the hyporheic zone may provide an extensive refuge, particularly when cool, well-oxygenated water fills spacious interstices (Stubbington, 2012). These refuges allow aquatic invertebrates to recolonize intermittent reaches after flow resumes (Datry et al., 2016b; Section 4.3.5).

4.3.3 TAXONOMIC DIVERSITY OF IRES INVERTEBRATE COMMUNITIES

Most studies, typically conducted at the reach scale, have indicated that alpha diversity (defined in Box 4.3.1) declines with increasing flow intermittence, and that communities at sites with greater

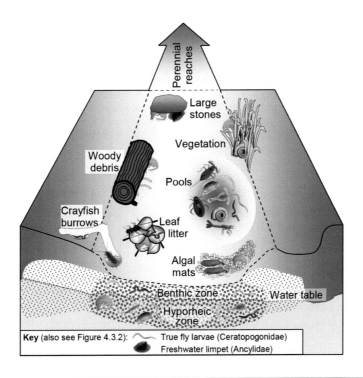

FIG. 4.3.1

In-stream refuges in a conceptualized IRES. Placement of example taxa in refuges is informed by Boulton (1989) and Chester and Robson (2011); other taxa also occur.

Adapted from Stubbington, R., 2012. The hyporheic zone as an invertebrate refuge: a review of variability in space, time, taxa and behaviour. Marine Freshw. Res. 63, 293-311, with permission from CSIRO Publishing. Invertebrate drawings: Courtesy: Pau Fortuño Estrada. Not to scale.

intermittence are highly nested (Box 4.3.1), comprising subsets of taxa found at less intermittent or perennial sites (Box 4.3.2, Fig. 4.3.2A; Datry et al., 2014b). A comprehensive metaanalysis of perennial river-IRES comparisons revealed that alpha diversity was significantly lower in IRES than in perennial rivers, regardless of climate, season, sampling methodology, or the habitats or river sections sampled (Soria et al., 2017). However, there are exceptions to this pattern: Bonada et al. (2007a) recorded comparable richness at perennial and intermittent sites in a Mediterranean river network, and Dieterich and Anderson (2000) attributed higher richness at forested IRES compared to perennial headwater streams to greater within-reach environmental heterogeneity.

TEMPORAL VARIABILITY IN TAXONOMIC DIVERSITY

The typical decline in richness with increasing intermittence (as characterized by comparing communities present during flowing phases; Datry et al., 2014b) reflects the loss of taxa at critical thresholds as drying proceeds (Fig. 4.3.4; Boulton, 2003; Section 4.3.2). As water levels decline, taxa associated with marginal vegetation are lost first, such as Hydroptilidae caddisfly larvae that feed by piercing aquatic plant cells, and Sisyridae spongilla-fly larvae found on sponge-encrusted riparian tree roots (Boulton and Lake, 2008; Bogan et al., 2015). As flow velocities fall, rheophilic (flow-loving) taxa with high

BOX 4.3.1 DEFINING DIVERSITY: TAXONOMIC, PHYLOGENETIC, AND FUNCTIONAL APPROACHES

Taxonomic diversity refers to the richness (number of taxa) and evenness (relative abundance of taxa) of a community and comprises alpha, beta, and gamma components. Alpha diversity is the local diversity within a community present in particular habitat. Beta diversity refers to differences in taxonomic composition among habitats and over time. This measure includes nestedness (which indicates that at sites with fewer taxa, those present are subsets of taxa at richer sites; Box 4.3.2; Fig. 4.3.2A) and turnover (replacement of some taxa by others, for example, due to environmental changes; Fig. 4.3.2B). Gamma diversity is determined by alpha and beta components and indicates the combined diversity across all habitats in a landscape.

A limitation of taxonomic diversity measures is that assemblages with comparable richness and evenness may comprise species of variable relatedness. To remedy this, taxonomic diversity can be used alongside phylogenetic diversity (which considers evolutionary relationships among taxa; Chapter 4.10) to determine the taxonomic distinctness of a community (Warwick and Clarke, 1995).

Functional diversity links taxonomic diversity to ecosystem functions provided by a community (Chapter 4.10) and complements measures of taxonomic and phylogenetic diversity by enabling comparison of taxonomically distinct but functionally similar communities (Boersma et al., 2016). The functional diversity of a community refers to the range of functional traits possessed by its taxa, where functional traits are characteristics that influence an organism's response to the environment and/or its effects on an ecosystem (Díaz and Cabido, 2001; Cadotte et al., 2011; Chapter 4.9). Traits include body size, feeding mode, dispersal mechanism, and desiccation-tolerant life stages, and each includes multiple states (e.g., "low" or "high" dispersal, Poff et al., 2006). Invertebrate trait databases for Europe (Tachet et al., 2002) and North America (Poff et al., 2006) have facilitated exploration of the functional diversity of IRES aquatic invertebrate communities (Vander Vorste et al., 2016a; Leigh et al., 2016b), including regional comparisons (Statzner and Bêche, 2010), although a lack of trait-related information can necessitate exclusion of meiofauna from analyses (Boersma et al., 2014).

BOX 4.3.2 NESTEDNESS AND TURNOVER INFLUENCE AQUATIC INVERTEBRATE DIVERSITY IN IRES

The typical decline in alpha diversity with increasing flow intermittence, as characterized by comparison of flowing-phase communities, reflects community nestedness (Box 4.3.1; Fig. 4.3.2A), with taxa at more intermittent sites (group C) being subsets of those at less intermittent (group B) and perennial sites (group A; Datry et al., 2014a, but see Bogan et al., 2013). Although variation occurs within each taxon and among systems, perennial sites include a wide range of taxa, such as many mayfly and stonefly nymphs, some caddisfly larvae, riffle beetles, amphipod and isopod crustaceans, true fly larvae, oligochaete worms, and copepod and ostracod microcrustaceans. As flow intermittence increases, rheophilic (flow-loving) and desiccation-sensitive taxa progressively disappear (from group A to B to C, Fig. 4.3.2A; Datry et al., 2014a; Section 4.3.3).

Despite lower flowing-phase alpha diversity, temporal variation in IRES habitats provides opportunities for taxon turnover (Box 4.3.1), with different taxa thriving at different times (Fig. 4.3.2B, Fig. 4.3.3; Bêche et al., 2006; Bonada et al., 2007a). During flowing phases, resilient lotic taxa (group B, Fig. 4.3.2A and B) dominate and may be joined by perennial stream (group A) taxa. When flow ceases, these taxa are replaced by macroinvertebrates characteristic of lentic pools, including damselfly nymphs, diving beetles, true bugs, and snails (Fig. 4.3.3; Bonada et al., 2006, 2008), and a meiofauna assemblage including rotifers, copepods, nematodes, and ostracods (Group D, Fig. 4.3.2B; Smith et al., 2001; Storey and Quinn, 2008). Other taxa, including many true fly larvae and Oligochaeta and certain caddisfly larvae, are generalists that can persist in lotic and lentic conditions (group E, Fig. 4.3.2B). This taxonomic "time-sharing" (Bogan and Lytle, 2007) increases the overall, year-round diversity (i.e., temporal beta diversity) (Bogan et al., 2015) of IRES. Beta diversity increases further when aquatic and terrestrial contributions are combined to estimate total IRES invertebrate richness (Corti and Datry, 2016; Chapter 4.4).

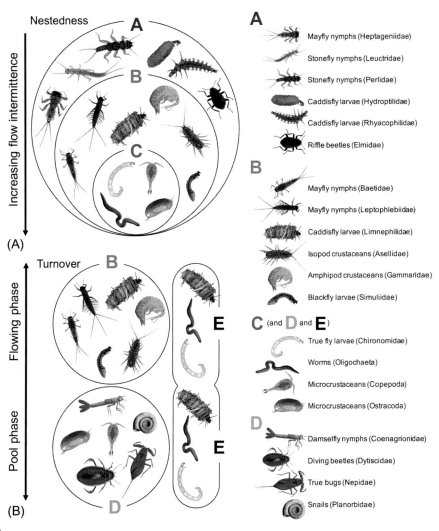

FIG. 4.3.2

Flowing-phase invertebrate community richness typically declines with increasing flow intermittence, with communities at more intermittent sites being nested subsets of those present at sites with greater flow permanence (A). Invertebrate communities exhibit turnover as in-stream conditions transition between flowing and pool phases (B).

Invertebrate drawings: Courtesy: Pau Fortuño Estrada. Not to scale.

oxygen requirements lose their preferred habitat and most are eliminated after flow ceases, including many species of EPT (Ephemeroptera, Plecoptera, and Trichoptera, i.e., mayfly nymphs, stonefly nymphs, and caddisfly larvae; Fig. 4.3.4; Boulton, 2003; Stubbington et al., 2009a). Filter feeders such as Simuliidae blackfly larvae require flow to feed (by filtering fine particles from moving water) and are often also eliminated by flow cessation. In contracting pools, taxa risk elimination if water quality becomes intolerable (Chapter 3.1) or if finite prey resources are consumed (Fig. 4.3.4; Boersma

FIG. 4.3.3

Aquatic invertebrates in each IRES phase: flowing phases feature flow-loving caddisfly larvae (a), stonefly (b), and mayfly nymphs (c); the pool phase is characterized by damselfly nymphs (d), true bugs (e), and beetles (f); in dry phases, mussels, cased caddisfly larvae (g) and snails (h) may persist if sediments remain moist; when flow resumes, blackfly larvae (i), dobsonfly larvae (j) and stonefly nymphs (k) are early colonizers.

Photos: Courtesy M. Bogan.

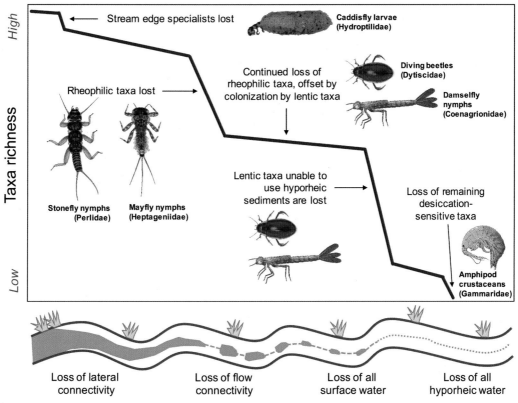

FIG. 4.3.4

The decline in richness with increasing intermittence, indicating critical thresholds in relation to habitat types and availability, as illustrated in the upper panel (adapted from Boulton, 2003).

Invertebrate drawings: Courtesy Pau Fortuño Estrada. Not to scale.

et al., 2014). Taxon-specific oxygen requirements and temperature tolerances are commonly exceeded as pools contract. For example, temperatures in disconnected pools in the temperate Selwyn River, New Zealand, reached 36.6°C (Drummond et al., 2015), higher than the 24-hour tolerance threshold of many aquatic invertebrates (Quinn et al., 1994). If pools dry, complete surface water loss eliminates desiccation-sensitive taxa. This causes steep reductions in taxa richness (Fig. 4.3.4; Boulton and Lake, 2008; Lake, 2011), despite some taxa persisting in refuges (Fig. 4.3.1, Sections 4.3.2 and 4.3.5).

Despite the temporal changes described earlier, and despite interannual variability in rainfall and temperature, total richness can remain stable among years (Bêche et al., 2009; Storey, 2015). In contrast, environmental variability drives differences in community composition among years, in particular between "wet" and "dry" years (i.e., years with different rainfall inputs, resulting in system-specific in-stream conditions ranging from reduced water depth to complete drying; Resh et al., 2013). For example, Bêche and Resh (2007) found that in Mediterranean-climate Californian streams, caddisfly larvae, beetles, and true bugs characterized dry years, whereas several stonefly taxa were associated with wet years. Similarly, in temperate New Zealand IRES, EPT richness was lower while richness of Cladocera and other meiofauna was higher in dry years, due to slower flows (Storey, 2015). Such interannual variability may have lagged effects, with altered recruitment in one year affecting population densities in subsequent years (Boulton and Lake, 1992).

SPATIAL VARIABILITY IN TAXONOMIC DIVERSITY

Environmental characteristics vary longitudinally in river ecosystems. In response to this environmental variation, alpha diversity of perennial-river aquatic invertebrate communities is predicted to peak in the mid-reaches (Fig. 4.3.5A; Vannote et al., 1980; Chapter 4.9), whereas beta diversity (defined in Box 4.3.1) is typically higher in headwaters due to isolation and habitat diversity (Sánchez-Fernández et al., 2008; Finn et al., 2011; Sarremejane et al., 2017). Whether similar patterns characterize aquatic invertebrate community diversity along IRES is unknown (Datry et al., 2014a). However, patterns may vary depending on the spatial arrangement of intermittent and perennial sections (Fig. 4.3.5), specifically whether intermittent sections occur in headwaters, mid-reaches, or lower reaches; occur as patches; or extend throughout the entire network (Fig. 4.3.5B–E; Datry et al., 2016c; Chapter 2.3).

In IRES with perennial headwaters, richness may decrease with distance downstream if intermittence increases (Fig. 4.3.5B), due to a reduced quantity and quality of food resources (Datry, 2012) as well as increasing distance from upstream recolonists. In contrast, richness may increase from upstream to downstream in catchments with intermittent headwaters, because downstream perennial reaches are refuges that provide recolonists once flow resumes (Fig. 4.3.5C; Sarremejane et al., 2017). Where drying is patchy, richness may remain stable along a river, because dry patches can be recolonized by processes including drift, benthic migrations, and multidirectional dispersal by flying adult insects (Fig. 4.3.5D); the extent of dispersal by flying insects may also influence diversity in systems with intermittent headwaters. Where complete networks dry, richness may decrease with distance upstream if headwaters are isolated and adult insect flight is the dominant recolonization mechanism (Fig. 4.3.5E). Research is needed to confirm these suggested patterns, which are likely to vary depending on local recolonist sources.

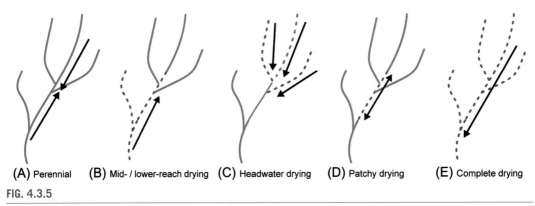

(A) Perennial   (B) Mid- / lower-reach drying   (C) Headwater drying   (D) Patchy drying   (E) Complete drying

FIG. 4.3.5

Predicted changes in the taxonomic diversity of aquatic invertebrates in a network with different arrangements of perennial (*blue*) and dry (*orange*) reaches. Arrow directions indicate expected increases in alpha diversity according to Vannote et al. (1980) (A) and considering the positions of sources of recolonists following flow resumption (B–E).

As noted by Dieterich and Anderson (2000) for alpha diversity, invertebrate community richness can be higher at intermittent compared to perennial sites due to habitat heterogeneity. In addition, spatial variation in environmental conditions among intermittent sites can mean that spatial beta diversity increases with intermittence (Schriever and Lytle, 2016). Similarly, in subtropical Australian IRES networks, higher gamma diversity (defined in Box 4.3.1) at "strongly intermittent" compared to "mildly intermittent" and "near-perennial" sites has been attributed to greater habitat diversity (Box 4.3.2; Rolls et al., 2016). The few among-region comparisons of IRES richness come from Mediterranean-climate regions. For example, Bonada et al. (2008) found that richness differed among Californian, South African, southwestern Australian, and Mediterranean Basin regions: northern hemisphere regions had higher gamma diversity, probably due to past geological events such as glaciations promoting evolution of new species.

PHYLOGENETIC DIVERSITY

Although Williams (1996) highlighted the wide phylogenetic separation of taxa present in temporary freshwaters, little research has examined the phylogenetic diversity or taxonomic distinctness (defined in Box 4.3.1) of IRES communities. Some evidence from terrestrial ecosystems suggests that phylogenetic diversity may increase along disturbance gradients (Giehl and Jarenkow, 2015), whereas the phylogenetic diversity of marine macroinvertebrate communities may decline as disturbance increases due to species-poor higher taxa being lost before those with many representatives (Clarke and Warwick, 2001). Taxonomic distinctness can also characterize disturbance impacts on community composition (Marchant, 2007) and may therefore differ between IRES and perennial-river invertebrate communities. Patterns require confirmation for lotic systems including IRES (Campbell et al., 2008).

4.3.4 FUNCTIONAL DIVERSITY OF IRES INVERTEBRATE COMMUNITIES

The functional diversity (defined in Box 4.3.1) of IRES invertebrate communities may differ from that in perennial rivers due to greater flow variability and associated changes in resource and habitat availability (Bêche et al., 2006; Bogan and Lytle, 2007). Taxon loss from IRES communities as drying progresses can reduce functional diversity during certain phases (e.g., after flow ceases and following surface water loss) if the lost taxa have trait states such as rheophily and filter feeding that are not possessed by taxa that remain. The functional diversity of IRES communities may also be lower than that of perennial river communities due to a reduction in large-bodied and long-lived taxa, K-selected taxa (which invest heavily in a few offspring), predators, and primary consumers (Ledger et al., 2013; Schriever et al., 2015). In contrast to these observations, Leigh et al. (2016b) synthesized studies conducted at small spatial scales (i.e., habitats, reaches, or streams) from temperate, Mediterranean, and arid regions and found no differences in functional diversity, richness, or redundancy (defined later) between intermittent and perennial systems among regions. Functional diversity may even be higher in IRES than in perennial streams when multiple seasons are considered, due to the contrasting functional traits of taxa associated with different hydrological phases (Fig. 4.3.3; Bonada et al., 2007b).

Differences in functional diversity between invertebrate communities in IRES and perennial rivers may be less pronounced than differences in taxonomic diversity, due to functional redundancy among taxa. In IRES, functional redundancy occurs when different taxa perform similar functions but have contrasting sensitivity to intermittence (Rosenfeld, 2002; Boersma et al., 2014). This redundancy seems most apparent in environmentally harsh systems, which may reflect the prevalence of taxa with traits that promote resistance and resilience to disturbance (Vander Vorste et al., 2016a). For example, Boersma et al. (2014) used experimental pools with different water depths to replicate different phases of the drying process and recorded considerable overlap in the traits of dryland IRES invertebrate assemblages, including body size, feeding mode, and respiration mode. Similarly, Vander Vorste et al. (2016b) reported stable functional diversity despite decreased taxa richness following drying in harsh, braided rivers.

The changes in invertebrate community composition that accompany transitions between flowing, pool, and dry conditions in IRES (Section 4.3.2) can alter functional diversity. For example, Bêche et al. (2006) found that 39 of 73 macroinvertebrate trait states varied between flowing phases and pool/dry phases in the intermittent headwaters of a Californian stream, whereas no traits differed between seasons in a perennial river. Traits that varied in the intermittent stream were associated with survival during dry phases, such as dormancy and desiccation-tolerant life stages (Section 4.3.5), and were more common in the dry season (Bêche et al., 2006).

Feeding groups can be an informative aspect of functional diversity in IRES, with dominant groups changing as drying proceeds (Bogan and Lytle, 2007). The proportion of predators typically increases as flow ceases and prey become concentrated in contracting pools, with beetles and true bugs sometimes accounting for >75% of invertebrates in a "predator soup" (Bogan and Lytle, 2007; Boulton and Lake, 2008). These increases are accompanied by corresponding declines in feeding modes associated with flowing phases, including filter feeding and shredding (Bogan and Lytle, 2007). In contrast, predators are scarce when flow first resumes, while collector-gatherer and scraper/grazer taxa may thrive, feeding on the biofilm-coated detritus that has accumulated on the bed during pool and dry phases (Boulton and Lake, 1992; Bogan et al., 2013). The early stages soon after flow resumption also allow "pioneer" taxa to complete their life cycles relatively free of competition with later colonizers (Langton and Casas, 1999; Williams, 2006). In particular, many dryland streams have distinct pioneer assemblages dominated by winter stonefly, nonbiting midge, blackfly, and dobsonfly juveniles after flow resumes (Fig. 4.3.3; Bogan et al., 2013).

4.3.5 INVERTEBRATE ADAPTATIONS TO FLOW INTERMITTENCE

Adaptations are characteristics that promote an organism's survival in the prevailing environmental conditions, and include behaviors, physiological and morphological characteristics, and life history strategies (Lytle and Poff, 2004; Chapter 4.10). Each taxon that persists in IRES must possess adaptations that interact to promote resistance (the ability to tolerate conditions *in situ*) and/or resilience (the capacity to recolonize after flow resumes) to flow intermittence (Nimmo et al., 2015; Chapters 4.8 and 4.10). Adaptations typically result in high aquatic invertebrate community resistance in IRES that retain pools (Boersma et al., 2014; Verdonschot et al., 2015), whereas resilience traits are the main mechanisms promoting persistence in IRES that dry (Chester and Robson, 2011; Datry et al., 2014b).

Many aquatic invertebrates in IRES are generalists which have broad environmental preferences and which also inhabit perennial rivers. Some major orders present in IRES include many generalists and fewer specialists. For example, aquatic juveniles of most EPT species are associated with perennial flow (Fig. 4.3.2; Bonada et al., 2007a, 2008), but a few are IRES specialists, including the caddisfly

Mesophylax aspersus and the stonefly *Mesocapnia arizonensis* (Salavert et al., 2008; Bogan et al., 2013). Adaptation to flow intermittence therefore varies at multiple taxonomic levels and may also differ among populations within a species (Lytle et al., 2008a; Chapter 4.10).

REFUGE USE PROMOTES PERSISTENCE IN IRES

Reaches with perennial flow are refuges that typically support many taxa associated with flowing phases in IRES (Boulton, 1989; Chester and Robson, 2011). Flowing upstream refuges allow energy-efficient recolonization of downstream intermittent reaches by taxa that actively enter the drift to disperse to new areas (Fig. 4.3.5B; Brittain and Eikeland, 1988; Paltridge et al., 1997), and downstream drifters may initially dominate assemblages in newly flowing reaches (Townsend and Hildrew, 1976; but see Vander Vorste et al., 2016a). Although many invertebrates may persist in upstream perennial reaches by chance and then drift passively to intermittent downstream reaches, active movement into upstream refuges requires an initial, energetically expensive migration chasing a receding water line (Table 4.3.1; Lytle et al., 2008b; Chapter 4.8). Equally, Boersma and Lytle (2014) followed a giant belostomatid water bug crawling downstream in a dry channel, their observations indicating that drying can initiate refuge-seeking migrations over land. Perennial refuges downstream of intermittent reaches allow taxa with strong swimming ability and a tendency to move into the current (e.g., gammarid amphipods) to recolonize by active upstream movement after flow returns (Fig. 4.3.5C, Table 4.3.1; Stanley et al., 1994). That both downstream and upstream migrations contribute to community recovery is evidenced by comparable resilience in IRES with and without perennial headwaters (Fig. 4.3.5B and C) within months of flow resuming (Datry et al., 2014b), and by genetically comparable populations in adjacent perennial and intermittent reaches (Zickovich and Bohonak, 2007).

Contracting pools are effective refuges for IRES invertebrates that can thrive in these transitional habitats despite declining water quality and lack of flow (Fig. 4.3.1). For example, Dytiscidae diving beetles and Belostomatidae water bugs are air breathers (Table 4.3.1) that surface to replenish oxygen stores (Lake, 2011; Bogan et al., 2015). Some copepods and ostracods can tolerate low oxygen concentrations and can therefore achieve high densities in pools (Storey and Quinn, 2008). Higher taxa typically associated with well-oxygenated water may also have some species that persist in pools, such as *Gumaga* caddisfly larvae that become sedentary to reduce their oxygen requirements (Fig. 4.3.6a; Jackson et al., 1999). If the refuge capacity of diminishing pools declines due to poor water quality or the consumption of finite prey resources, aquatic insects with flying adult life stages (notably true flies and beetles; Bogan and Boersma, 2012) or that crawl in terrestrial habitats (e.g., giant water bugs; Boersma and Lytle, 2014) may emigrate in search of new refuges.

Where perennial surface-water refuges are unavailable, desiccation-sensitive taxa may respond to surface water loss by moving into the wet subsurface habitat of the hyporheic zone (Fig. 4.3.1; Stubbington, 2012). Adaptations to use the hyporheic refuge include active migration as the water table falls and an appropriate morphology: elongate true fly larvae, robust amphipods, and small microcrustaceans are all common subsurface refugees (Table 4.3.1). In addition, smaller individuals of any taxon may be particularly well adapted to hyporheic refuge use (Table 4.3.1). For example, Vander Vorste et al. (2016a) found that invertebrates recolonizing the channel from subsurface sediments were smaller than those present before drying. To persist in subsurface sediments, invertebrates must also be physiologically tolerant of interstitial water quality, including dissolved oxygen concentrations that typically decline with increasing depth below the streambed.

Table 4.3.1 Aquatic invertebrate trait states that promote resistance and/or resilience to intermittence

Resistance traits	Trait state	Example taxa	Justification
Body armoring[a]	Strong	Cased caddisfly larvae, e.g., Limnephilidae	Limits water loss from body during dry phases
Desiccation-tolerant life stages[a,b]	Present	Stonefly nymphs, e.g., Capniidae and eggs, e.g., Nemouridae	Promotes survival during dry phases
Maximum size[b]	<5 mm	True fly larvae, e.g., Chironomidae	Promotes movement into the hyporheic zone during drying
Respiration[a,b]	Air breathers (spiracle/plastron)	Adult true bugs, e.g., Belostomatidae	Enhances survival during dry phases and in low-oxygen pools
Rheophily	Low	Adult beetles, e.g., Dytiscidae	Enhances survival during pool phases

Resilience traits	Trait state	Example taxa	Justification
Adult flying ability	Strong	Caddisfly adults, e.g., Leptoceridae	Facilitates recolonization over larger areas
Adult life span[a]	Long (>1 month)	Diving beetles, e.g., Dytiscidae	Adult females more likely to survive long enough to lay eggs when flow resumes
Female dispersal[a]	High	Dragonfly adults, e.g., Cordulegastridae	Facilitates recolonization from distant perennial refuges
Life cycle duration[b]	Short (≤1 year)	True fly larvae, e.g., Chironomidae	Allows life cycle completion and faster recolonization
Occurrence in drift[a]	Common (typically observed)	True fly larvae, e.g., Chironomidae	Facilitates recolonization from upstream perennial reaches
Maximum crawling rate[a]	High	Crustaceans, e.g., Gammaridae	Promotes movement into the hyporheic zone during drying
		Adult true bugs, e.g., Belostomatidae	Facilitates entrance into upstream and downstream refuges
Reproduction[b]	Giving birth to live young	Crustaceans, e.g., Gammaridae	Allows faster recolonization and population reestablishment
	Terrestrial egg clutches	Beetles, e.g., Hydraenidae	Allows avoidance of dry phases by aquatic life stages
	Asexual reproduction	Worms, e.g., Naididae	Allows faster recolonization and population reestablishment
Relationship to substrate[b]	Inhabiting subsurface interstices	Groundwater crustaceans, e.g., Niphargidae	Survival in saturated interstices after surface water loss
Swimming ability[a]	Strong	Crustaceans, e.g., Gammaridae	Facilitates recolonization from upstream and downstream reaches

Trait states adapted from footnotes "a" and "b" with trait categories translated in Charvet et al. (2000).
Traits and their justification were informed by Bonada et al. (2007a), Robertson and Wood (2010), Robson et al. (2011), and Datry et al. (2014a)
[a]Poff et al. (2006).
[b]Tachet et al. (2002).

FIG. 4.3.6

Invertebrates with adaptations promoting persistence in IRES: (a) sedentary *Gumaga* caddisfly larvae congregate in pools; (b) the stonefly *Mesocapnia arizonensis* enters dormancy as nymphs; (c) dormant *Hydrobaenus* chironomid larvae inhabit protective cases; (d) Limnephilidae caddisfly larvae enter dormancy in humidity-trapping cases.

Photos: Courtesy M. Bogan (a, b, and d) and R. Vander Vorste (c).

If benthic and hyporheic sediments dry, only taxa with desiccation-resistant life stages can persist *in situ* (Fig. 4.3.1; Datry, 2012). Few taxa are physiologically adapted to tolerate complete drying (Stanley et al., 1994; Alpert, 2006); exceptions include some microcrustaceans and true flies (Strachan et al., 2015). Notably, larvae of the African "sleeping chironomid" *Polypedilum vanderplanki* can survive drying to a moisture content of <8% in a desiccation-tolerant state in which metabolism virtually ceases (Hinton, 1960). Such "anhydrobiosis" is far more common in meiofauna than macroinvertebrates. For example, many tardigrades and nematodes can become metabolically inactive at any life stage (Watanabe, 2006).

Where interstices in a "dry" bed remain humid, taxa with partial desiccation tolerance can persist in the invertebrate "seedbank" (Table 4.3.1; *sensu* Stubbington and Datry, 2013). For example, Datry et al. (2012) recorded 46 aquatic taxa in a temperate IRES seedbank, including beetles, caddisfly and true fly larvae, and many meiofauna. Some of these taxa can survive long periods in moist sediments, notably meiofauna including rotifers and nematodes which remain viable for decades as dormant eggs or developed individuals (Pennak, 1989). Macroinvertebrate traits enhancing physiological desiccation tolerance include body armoring; for example, case-building Limnephilidae caddisfly larvae can persist in interstices for weeks within humidity-trapping cases (Fig. 4.3.6d, Table 4.3.1; Ruiz-García and Ferreras-Romero, 2007). However, the greater physiological demands of surviving in drier sediments (Wickson et al., 2012) mean that seedbanks may become graveyards as aridity increases (Stubbington and Datry, 2013). Only a few specialists such as dormant *M. arizonensis* stonefly nymphs (Bogan et al., 2015) (Fig. 4.3.6b, Table 4.3.1) and *Hydrobaenus* nonbiting midge larvae (Fig. 4.3.6c) persist in dryland IRES seedbanks.

LIFE CYCLE ADAPTATIONS AND REFUGE USE INTERACT TO PROMOTE SURVIVAL

IRES insects may have life cycle adaptations that involve synchronizing aquatic life stages with flowing phases, and terrestrial adults and dormant eggs with dry phases, to promote persistence in systems with temporally predictable dry phases. For example, desiccation-resistant eggs of the IRES-associated caddisfly *Stenophylax permistus* hatch following flow resumption (Ruiz-García and Ferreras-Romero, 2007). Larvae then develop and pupate during the flowing phase, and adults emerge before the stream dries and spend the dry phase in terrestrial environments (Crichton, 1971). Riparian and terrestrial habitats can therefore provide refuges that promote insect persistence in IRES (Robson et al., 2011; Chapter 4.8, Section 4.3.7). Life cycle synchronization can also involve varying life stage duration in wet and dry years. For example, semivoltine *Neohermes filicornis* fishfly larvae emerge from IRES earlier in dry years (Cover et al., 2015). Such semivoltine insects (having a life cycle exceeding one year) must persist during dry phases as juveniles, typically in a dormant state. *Guadalgenus franzi* stonefly nymphs and *N. filicornis* larvae are among those thought to burrow into hyporheic sediments and become dormant as juveniles during dry phases in Mediterranean-climate IRES (Agüero-Pelegrín and Ferreras-Romero, 2002; Cover et al., 2015).

Following flow resumption, nearby perennial waters are potential sources of IRES recolonists, notably for insects that complete pupation and emergence during flowing phases then fly overland as aerial adults (Bogan and Boersma, 2012). As flow duration lengthens, the terrestrial adults of an increasing number of taxa with desiccation-sensitive aquatic juveniles (e.g., mayflies and dragonflies) have time to recolonize IRES by flight from perennial waters, until community composition comes to resemble its predrying state (Bogan et al., 2015). High female dispersal capacity is therefore a trait associated with resilient IRES insects (Table 4.3.1), as reflected by extensive gene flow between populations and a lack of genetic structure in taxa such as the Australian caddisfly *Lectrides varians* (Wickson et al., 2014; Chester et al., 2015). Genetic analyses also provide evidence of dispersal by flight between isolated IRES, for example, in mayflies and caddisflies (Chester et al., 2015).

Aerial adult insects capable of overland flight recolonize IRES by laying eggs on either water or dry sediments. Eggs deposited on water may develop immediately or may not hatch until triggered by cues such as low temperatures (Hynes, 1976). Such delayed hatching increases the likelihood of aquatic life stages completing development during an uninterrupted flowing phase. Additional "bet-hedging" strategies (Lytle and Poff, 2004), including asynchronous egg hatching (Sandberg and Stewart, 2004), can also increase the chance of at least a partial cohort developing while immersed, despite "false starts" (short periods of immersion; Strachan et al., 2016). Following egg hatching, rapid development facilitates completion of aquatic life stages. For example, chironomid adults may emerge 19 days after larval development begins (Table 4.3.1; Jackson and Sweeney, 1995). For taxa completing their entire life cycle in water, ovoviviparity (giving birth to live young) and asexual reproduction are reproductive traits that facilitate rapid recolonization and population reestablishment following flow resumption in intermittent reaches (Table 4.3.1).

ADAPTATIONS TO IRES ARE TRADE-OFFS THAT ALSO INFLUENCE OTHER ASPECTS OF SURVIVAL

Adaptations that promote persistence during pool and dry phases may increase exposure to other threats; for example, dormant individuals are vulnerable to predation (Robson et al., 2011). Equally, refuge inhabitation may reduce fitness, for example, where drying refuges such as the hyporheic zone have limited food resources (Vander Vorste et al., 2016c). In addition, desiccation tolerance entails metabolic costs that may be offset by slower growth (Alpert, 2006) and by reduced fitness of dormant individuals (Wickson et al., 2014). Invertebrate traits may therefore reflect trade-offs between dry-phase persistence and exposure to other threats, with adaptations differing within and among IRES and climate zones depending on environmental conditions (Welborn et al., 1996), and with genetically distinct populations within a species exhibiting contrasting behavioral responses to different hydrological regimes (Phillipsen and Lytle, 2013). For example, where complete drying occurs, desiccation tolerance may be more common in temperate than dryland fauna, because sediment moisture content remains relatively high and a physiology closer to the "norm" is sufficient for survival (Lytle and Poff, 2004; Stubbington and Datry, 2013). With decreasing humidity, taxon-specific desiccation-tolerance thresholds are exceeded and most dryland IRES taxa instead use resilience traits that facilitate movement into (Stanley et al., 1994; Lytle et al., 2008b) and subsequent recolonization from perennial refuges (Boulton et al., 1992).

The trait combinations that promote persistence in IRES may also enhance survival of other flow extremes. For example, *Gammarus* amphipods may enter the hyporheic zone in response to both increasing and decreasing discharge, and so may be early recolonists after both flooding and drying events (Stubbington, 2012). Equally, physiological adaptations may promote cotolerance of multiple stressors; for example, a strong positive correlation between salinity tolerance and desiccation tolerance has been recorded in beetles from saline IRES (Arribas et al., 2014). Therefore, the trait combinations of IRES invertebrates may not be unique to intermittent systems but reflect the filtering of traits by harsh environmental conditions that can also occur in perennial rivers (Vander Vorste et al., 2016b).

4.3.6 THREATS TO IRES INVERTEBRATE COMMUNITIES

Freshwaters are among the world's most threatened ecosystems (Strayer and Dudgeon, 2010), and IRES are at particular risk due to their poor public perception (Boulton, 2014; Datry et al., 2014a; Chapters 5.2 and 5.3). Aquatic invertebrates face multiple threats relating to human alteration of the hydrological, physical, chemical, and biological characteristics of IRES (Chapter 5.1). In-stream communities are particularly vulnerable during dry phases, when channels may be viewed as biologically inactive and exploited, for example, for recreation and agriculture (Fig. 4.3.7; Hans et al., 1999; Steward et al., 2012).

Increases in the duration, frequency, and intensity of flow cessation and drying events resulting from climate change and water abstraction severely threaten IRES and their aquatic invertebrates (Chapter 5.1), typically causing alpha and beta diversity to decline as a smaller range of generalist taxa become dominant (Datry et al., 2014b; Section 4.3.3). Equally, artificial perennialization of naturally intermittent systems is altering invertebrate communities such as in dryland and Mediterranean regions where sewage, urban, agricultural, and industrial effluents are released into IRES channels (Múrria et al., 2008; Steward et al., 2012; Chapter 5.1). Artificially perennial flow may be viewed as "improving" ecological quality, but may lead to local extinction of specialist IRES invertebrates which survive because intermittence eliminates their competitors or predators, or because IRES provide conditions that

FIG. 4.3.7

Río Seco (Málaga, Spain) is an intermittent river channel used for recreation during dry phases; a car can be seen driving along the channel.

Photo: Courtesy N. Bonada.

match their functional traits (Section 4.3.4). Loss of intermittence may also threaten taxa associated with isolated pools and dry phases (Box 4.3.2; Section 4.3.3). Other patterns of altered IRES hydrology may be more complex. For example, seasonal and between-year flow variability may be subdued by flow regulation associated with agricultural land use (Belmar et al., 2013); this hydrological homogenization causes community diversity to decline in groups such as aquatic beetles (Bruno et al., 2014).

The consequences of altered flow permanence on aquatic invertebrate biodiversity are poorly understood, but shifts between perennial and intermittent flow may be tipping points that drive communities to alternate states. Artificial intermittence is likely to eliminate taxa from entire networks (Bogan and Lytle, 2011; Datry et al., 2016c), and where lost taxa played key ecological roles, ecosystem processes can be affected. For example, *Gammarus* amphipods release energy for other feeding groups by shredding leaf litter (Section 4.3.1), but are desiccation sensitive (Stubbington et al., 2009b; Vander Vorste et al., 2016c) and their populations decrease as intermittence increases (Datry et al., 2011). Consequently, flow cessation and drying may reduce leaf litter decomposition (Corti et al., 2011), with ecosystem function sometimes remaining altered months after flow resumes (Maamri et al., 1997; Datry et al., 2011). Similarly, net-spinning caddisfly larvae are "ecosystem engineers" that stabilize substrates (Johnson et al., 2009) but may be lost if intermittence increases, which could exacerbate sediment mobilization during floods. Finally, top invertebrate predators can be eliminated by shifts to intermittent flow regimes, triggering cascading effects that extend through food webs and alter community structure (Bogan and Lytle, 2011).

Aquatic invertebrates are also threatened by pollutants, including sewage, pesticides, and fine sediment. The negative effects of such pollutants on invertebrate communities may be exacerbated in IRES by natural transitions between flowing, pool, and dry phases. As flow recedes and pools contract, the capacity to dilute pollutants may decline and salinization (elevation of dissolved inorganic salt concentrations) may increase (Kefford et al., 2016; Chapters 3.1 and 3.2) compared to perennial streams. In contrast, the natural salinity of some Mediterranean IRES has been diluted by freshwater inputs from irrigated agricultural land, threatening saline-tolerant IRES invertebrates including true fly larvae and beetles (Velasco et al., 2006). Pollutants that accumulate on the bed during dry phases (Hans et al., 1999) are mobilized by flow resumption, with pulses of contaminated water reducing downstream water quality (Datry et al., 2014a; Chapter 3.2); impacts on IRES invertebrates are typically taxon- and contaminant-specific. The negative effects of fine sediment pollution, which include clogging respiratory structures such as gills, may be greater in IRES than in perennial rivers: first, more fine particles are deposited on the bed as flow velocities decline and as flow ceases in IRES; second, blocking of interstitial pathways may prevent invertebrate migrations into the hyporheic zone refuge (Vadher et al., 2015; Section 4.3.5).

Dry riverbeds are sometimes mined for sand and gravel (Chapter 5.1), and this sediment extraction can alter the channel shape and physical characteristics (Fig. 4.3.8), reducing in-stream habitat diversity. Dry-phase assemblages are impacted through removal of desiccation-resistant life stages present in the invertebrate seedbank, which may impair community recovery after flow resumes (Section 4.3.5). In one Bolivian IRES, intensive sediment mining reduced flowing-phase aquatic invertebrate density and richness, and during dry phases, sediments from mining-impacted reaches contained 0%–50% of the taxa found in nonimpacted reaches (T. Datry, personal communication, Fig. 4.3.8).

Outside the channel, clearance of IRES floodplain and riparian zone vegetation for agriculture may occur alongside fertilizer use and water abstraction (Fig. 4.3.9A), with multiple interacting impacts on

FIG. 4.3.8

Rio Biloma (Cochabamba, Bolivia): an intermittent river channel after excavation for sediment mining.

Photo: Courtesy T. Datry.

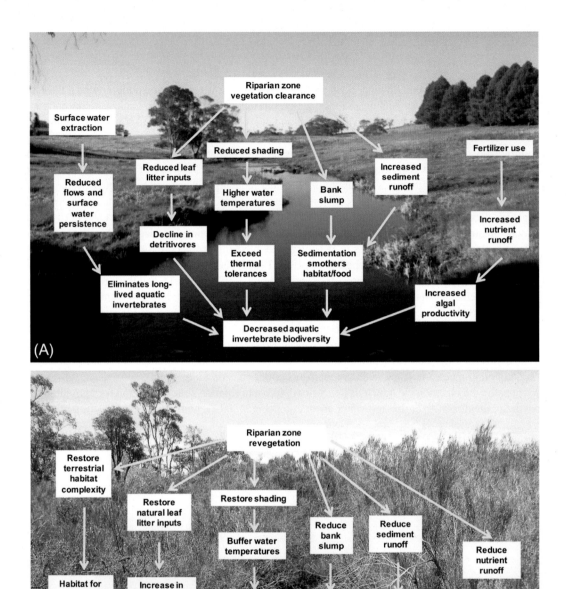

FIG. 4.3.9

Multiple interacting impacts on invertebrate biodiversity resulting from clearance of riparian zone vegetation, fertilizer use and surface water abstraction for agriculture in IRES catchments (A) and the potential reversal of many of these effects by riparian zone restoration (B).

Photos: Courtesy A.J. Boulton.

aquatic invertebrate communities. Reduced vegetation cover can lead to bank erosion and sediment inputs, with fine particles smothering benthic sediments and restricting access to subsurface sediments. Trees, shrubs, and associated habitats, used by terrestrial adult life stages of IRES insects, are removed. Loss of channel shading promotes algal growth in nutrient-rich waters, and water temperatures may exceed the thermal tolerance of some invertebrates, especially when pools contract (Lake, 2011). During dry phases, sediment temperatures increase, potentially exceeding the thermal tolerances of seedbank inhabitants and increasing mortality of desiccation-tolerant life stages (Storey and Quinn, 2013). Finally, loss of riparian vegetation alters the type and amount of organic matter entering IRES channels, shifting aquatic food web structure from dominance by leaf litter shredders to algal herbivores. Collectively, these impacts may reduce aquatic invertebrate diversity (Fig. 4.3.9A).

IRES and their aquatic invertebrate communities are typically exposed to multiple interacting threats because flow regime alteration, pollutant inputs, and land use changes are all concentrated into areas of high human activity (Tockner et al., 2010). For example, groundwater abstraction to irrigate crops is a major cause of artificial intermittence (Vörösmarty and Sahagian, 2000; Chapter 5.1) and is often associated with high pesticide loads, with likely interacting impacts on invertebrate diversity. In addition, altered hydrology may enhance the invasion success of nonnative species. For example, artificial perennialization may facilitate the spread of desiccation-sensitive invasive species such as *Dikerogammarus* "killer shrimps" (Poznańska et al., 2013) and the mudsnail *Potamopyrgus antipodarum* (Múrria et al., 2008). Equally, reduced flow permanence can favor desiccation-tolerant invasive species over desiccation-sensitive natives, as observed for *Orconectes* crayfish (Larson et al., 2009). Novel experiments (e.g., Elbrecht et al., 2016) are needed to examine the cumulative effects of multiple interacting stressors on IRES invertebrates.

4.3.7 MANAGING IRES TO PROMOTE AQUATIC INVERTEBRATE BIODIVERSITY

As described in Section 4.3.6, the main threats to aquatic invertebrate biodiversity in IRES are altered hydrology including changes to flow permanence, deterioration of water quality due to pollution, changes to channel morphology due to sediment extraction, loss or alteration of riparian vegetation, and arrival of invasive taxa (Chapter 5.1). That these threats and their causes typically coincide (Tockner et al., 2010) is highly relevant to ecosystem management because a single, carefully selected management strategy may simultaneously address multiple threats, with widespread benefits for biodiversity.

One management strategy to address multiple threats to IRES invertebrates is to restore native riparian vegetation (Sweeney and Newbold, 2014; Chapter 5.4). Such vegetation stabilizes banks and intercepts runoff containing pesticides, nutrients, fine sediment, and pollutants (Fig. 4.3.9B). Denser shading and decreased nutrient inputs reduce algal productivity, and increased leaf litter inputs may allow shredder taxa to recover their dominance in IRES food webs (Kupilas et al., 2016). Riparian vegetation also provides habitat for the terrestrial adult life stages of insects with aquatic juveniles, and its restoration may promote recolonization of IRES by flying adults following flow resumption (Chester et al., 2015). Provided that recolonization is possible, this management strategy may substantially restore aquatic invertebrate biodiversity (Chapter 4.8), favoring life stages in inundated aquatic habitats as well as aerial adults that use the terrestrial habitat provided by riparian vegetation (Fig. 4.3.9B).

There are, however, few documented cases of IRES restoration (Chapter 5.4) and, seemingly, none published that have specifically targeted invertebrate biodiversity. In contrast to the widespread recognition of the conservation value of temporary pond biota (Van den Broek et al., 2015), equivalent protection of IRES faunas is minimal. Very few aquatic invertebrate species that are solely or largely restricted to IRES are listed for protection, contrasting with the far greater number of species in perennial rivers. One rare example is the snail *Glacidorbis occidentalis*, which is largely restricted to IRES in native eucalypt (jarrah) forest in Western Australia (Bunn et al., 1989), and is listed as Vulnerable on the IUCN Red List of Threatened Species (IUCN, 2015). At a national level, the mayfly *Paraleptophlebia werneri*, which typically inhabits the "winterbourne" headwaters of chalk streams in southern England, is listed as Nationally Scarce (Macadam, 2016). The lack of protection for other rare and potentially threatened aquatic IRES invertebrates probably reflects our limited knowledge of their distribution and ecology, rather than an absence of species needing protection. Regional surveys of temporary ponds (e.g., Karaouzas et al., 2015) have indicated that many rare taxa are restricted to only a few ponds, and protecting habitat diversity and connectivity is crucial for aquatic invertebrate conservation in pond networks (Florencio et al., 2014). Regional conservation strategies are therefore vital to protect aquatic invertebrate diversity in temporary pond networks and the same may be true in IRES.

Crucially, although perennial rivers often support higher aquatic invertebrate diversity during flowing phases (Section 4.3.3), several studies have demonstrated that intermittent reaches support unique, endemic, or locally rare species, contributing to regional-scale (gamma) diversity and warranting inclusion in conservation planning (Sánchez-Fernández et al., 2008; Bogan et al., 2013). Protection measures such as conserving natural flow regimes, riparian vegetation (Fig. 4.3.9B), and perennial refuges have been advocated for maintaining aquatic invertebrate diversity (Chester and Robson, 2011; Cañedo-Argüelles et al., 2015). However, effective management is hampered by limited understanding of the ecology of IRES invertebrates, especially rare species. In addition, management and conservation strategies must be implemented at suitably broad spatial scales that both acknowledge the importance of dispersal dynamics and preserve the natural landscape fragmentation that hinders dispersal of the competitors and predators that exclude certain IRES invertebrates from perennial rivers.

Phylogenetic as well as taxonomic diversity should be considered when identifying nature conservation priorities, as priorities identified using these two approaches can differ (Vane-Wright et al., 1991). In addition, conservation efforts should recognize and protect genetic variability within and among IRES invertebrate taxa. Species with limited dispersal abilities (e.g., the giant water bug *Abedus herberti* and *Geocharax* crayfish) may have genetically distinct populations even at small spatial scales (Phillipsen and Lytle, 2013; Chester et al., 2015), and species with strong dispersal (e.g., the caddisfly *Lectrides varians*) may also be paraphyletic (i.e., comprising multiple distinct genetic lineages; Wickson et al., 2014). Such genetic distinctions among populations within a putative species sometimes warrant taxonomic revisions, and reclassification may increase the rarity of the resultant new species (Wickson et al., 2014).

4.3.8 CONCLUSIONS AND FUTURE RESEARCH PRIORITIES

With climate change predictions of intensifying droughts interacting with increasing water resource pressures and land-use alteration in many regions, flow regimes are changing and intermittence may be increasing (Chapter 1). Consequently, IRES invertebrates face increasing and unpredictable challenges. Adaptation to flow intermittence typically necessitates adaptation to extremity and unpredictability, and IRES taxa may therefore survive some environmental change. However, biotic responses will vary among and within taxa, and populations are facing pressures to adapt at rates that may prove unachievable. In particular, where the spatial extent of drying increases, genetic bottlenecks stemming from increasing isolation may reduce the ability of populations to adapt, particularly for taxa with limited dispersal (Zickovich and Bohonak, 2007). Restoration initiatives that address multiple interacting stressors are therefore needed to safeguard IRES invertebrate communities. Regulatory agencies need to incorporate IRES and their aquatic invertebrates into long-term, routine biomonitoring programs to ensure changes in community health are documented (Cid et al., 2016). Indices developed for perennial streams are potentially appropriate for such biomonitoring (Mazor et al., 2014), particularly in systems with longer flowing phases (Prat et al., 2014). Finally, the conservation status of specialist IRES invertebrates requires examination at local to international scales, and at genetic to metapopulation levels, to identify populations and species requiring targeted conservation measures.

With IRES becoming an increasingly dominant feature of many river networks, aquatic invertebrates are effective study organisms for addressing wider IRES research challenges, due to their abundance, diversity, and ubiquity. Research priorities that invertebrates could be used to address include exploration of the pathways linking IRES to their catchments and connecting neighboring streams, the long-term (e.g., decadal) temporal dynamism of IRES habitats and communities, and community responses to multiple interacting human activities (Leigh et al., 2016a). Further research is also needed to characterize the contributions that aquatic invertebrates make to IRES biodiversity through their genetic, taxonomic, and phylogenetic diversity, and to highlight the importance of IRES invertebrates in relation to the ecosystem functions they provide.

REFERENCES

Agüero-Pelegrín, M., Ferreras-Romero, M., 2002. The life cycle of *Guadalgenus franzi* (Aubert, 1963) (Plecoptera: Perlodidae) in the Sierra Morena mountains (southern Spain): semivoltinism in seasonal streams of the Mediterranean basin. Aquat. Insects 24, 237–245.

Alpert, P., 2006. Constraints of tolerance: why are desiccation-tolerant organisms so small or rare? J. Exp. Biol. 209, 1575–1584.

Arribas, P., Andújar, C., Abellán, P., Velasco, J., Millán, A., Ribera, I., 2014. Tempo and mode of the multiple origins of salinity tolerance in a water beetle lineage. Mol. Ecol. 23, 360–373.

Bêche, L.A., Resh, V.H., 2007. Short-term climatic trends affect the temporal variability of macroinvertebrates in California 'Mediterranean' streams. Freshw. Biol. 52, 2317–2339.

Bêche, L.A., McElravy, E.P., Resh, V.H., 2006. Long-term seasonal variation in the biological traits of benthic-macroinvertebrates in two Mediterranean-climate streams in California, U.S.A. Freshw. Biol. 51, 56–75.

Bêche, L.A., Connors, P.G., Resh, V.H., Merelender, A.M., 2009. Resilience of fishes and invertebrates to prolonged drought in two California streams. Ecography 32, 778–788.

Belmar, O., Bruno, D., Martínez-Capel, F., Barquín, J., Velasco, J., 2013. Effects of flow regime alteration on fluvial habitats and riparian quality in a semiarid Mediterranean basin. Ecol. Indic. 30, 52–64.

Boersma, K.S., Lytle, D.A., 2014. Overland dispersal and drought-escape behavior in a flightless aquatic insect, *Abedus herberti* (Hemiptera: Belostomatidae). Southwest. Nat. 59, 301–302.

Boersma, K.S., Bogan, M.T., Henrichs, B.A., Lytle, D.A., 2014. Invertebrate assemblages of pools in arid-land streams have high functional redundancy and are resistant to severe drying. Freshw. Biol. 59, 491–501.

Boersma, K.S., Dee, L.E., Miller, S.J., Bogan, M.T., Lytle, D.A., Gitelman, A.I., 2016. Linking multidimensional functional diversity to quantitative methods: a graphical hypothesis-evaluation framework. Ecology 97, 583–593.

Bogan, M.T., Boersma, K.S., 2012. Aerial dispersal of aquatic invertebrates along and away from arid-land streams. Freshw. Sci. 31, 1131–1144.

Bogan, M.T., Lytle, D.A., 2007. Seasonal flow variation allows 'time-sharing' by disparate aquatic insect communities in montane desert streams. Freshw. Biol. 52, 290–304.

Bogan, M.T., Lytle, D.A., 2011. Severe drought drives novel community trajectories in desert stream pools. Freshw. Biol. 56, 2070–2081.

Bogan, M.T., Boersma, K.S., Lytle, D.A., 2013. Flow intermittency alters longitudinal patterns of invertebrate diversity and assemblage composition in an arid-land stream network. Freshw. Biol. 58, 1016–1028.

Bogan, M.T., Boersma, K.S., Lytle, D.A., 2015. Resistance and resilience of invertebrate communities to seasonal and supraseasonal drought in arid-land headwater streams. Freshw. Biol. 60, 2547–2558.

Bonada, N., Rieradevall, M., Prat, N., Resh, V.H., 2006. Benthic macroinvertebrate assemblages and macrohabitat connectivity in Mediterranean-climate streams of northern California. J. N. Am. Benthol. Soc. 25, 32–43.

Bonada, N., Dolédec, S., Statzner, B., 2007a. Taxonomic and biological trait differences of stream macroinvertebrate communities between mediterranean and temperate regions: implications for future climatic scenarios. Glob. Change Biol. 13, 1659–1671.

Bonada, N., Rieradevall, M., Prat, N., 2007b. Macroinvertebrate community structure and biological traits related to flow permanence in a Mediterranean river network. Hydrobiologia 589, 91–106.

Bonada, N., Rieradevall, M., Dallas, H., Davis, J., Day, J., Figueroa, R., et al., 2008. Multi-scale assessment of macroinvertebrate richness and composition in Mediterranean-climate rivers. Freshw. Biol. 53, 772–788.

Boulton, A.J., 1989. Over-summering refuges of aquatic macroinvertebrates in two intermittent streams in central Victoria. Trans. R. Soc. S. Aust. 113, 23–34.

Boulton, A.J., 2003. Parallels and contrasts in the effects of drought on stream macroinvertebrate assemblages. Freshw. Biol. 48, 1173–1185.

Boulton, A.J., 2014. Conservation of ephemeral streams and their ecosystem services: what are we missing? Aquat. Conserv. Mar. Freshwat. Ecosyst. 24, 733–738.

Boulton, A.J., Lake, P.S., 1992. The ecology of two intermittent streams in Victoria, Australia. III. Temporal changes in faunal composition. Freshw. Biol. 27, 123–138.

Boulton, A.J., Lake, P.S., 2008. Effects of drought on stream insects and its ecological consequences. In: Lancaster, J., Briers, R.A. (Eds.), Aquatic Insects: Challenges to Populations. CAB International, Wallingford, UK, pp. 81–102.

Boulton, A.J., Stanley, E.H., Fisher, S.G., Lake, P.S., 1992. Over-summering strategies of macroinvertebrates in intermittent streams in Australia and Arizona. In: Robarts, R.D., Bothwell, M.L. (Eds.), Aquatic Ecosystems in Semi-arid Regions: Implications for Resource Management. NHRI Symposium Series, 7. Environment Canada, Saskatoon, pp. 227–237.

Brittain, J.E., Eikeland, T.J., 1988. Invertebrate drift—a review. Hydrobiologia 166, 77–93.

Bruno, D., Belmar, O., Sánchez-Fernández, D., Guareschi, S., Millán, A., Velasco, J., 2014. Responses of Mediterranean aquatic and riparian communities to human pressures at different spatial scales. Ecol. Indic. 45, 456–464.

Bunn, S.E., Davies, P.M., Edward, D.H., 1989. The association of *Glacidorbis occidentalis* Bunn and Stoddard 1983 (Gastropoda: Glacidorbidae) with intermittently-flowing, forest streams in south-western Australia. J. Malacological Soc. Aust. 10, 25–34.

Cadotte, M.W., Carscadden, K., Mirotchnick, N., 2011. Beyond species: functional diversity and the maintenance of ecological processes and services. J. Appl. Ecol. 48, 1079–1087.

Campbell, W.B., Arce-Perez, R., Gomez-Anaya, J.A., 2008. Taxonomic distinctness and aquatic Coleoptera: comparing a perennial and intermittent stream with differing geomorphologies in Hidalgo, Mexico. Aquat. Ecol. 42, 103–113.

Cañedo-Argüelles, M., Boersma, K.S., Bogan, M.T., Olden, J.D., Phillipsen, I., Schriever, T.A., et al., 2015. Dispersal strength determines meta-community structure in a dendritic riverine network. J. Biogeogr. 42, 778–790.

Charvet, S., Statzner, B., Usseglio-Polatera, P., Dumont, B., 2000. Traits of benthic macroinvertebrates in semi-natural French streams: an initial application to biomonitoring in Europe. Freshw. Biol. 43, 277–296.

Chester, E.T., Robson, B.J., 2011. Drought refuges, spatial scale and recolonisation by invertebrates in non-perennial streams. Freshw. Biol. 56, 2094–2104.

Chester, E.T., Miller, A.D., Valenzuela, I., Wickson, S.J., Robson, B.J., 2015. Drought survival strategies, dispersal potential and persistence of invertebrate species in an intermittent stream landscape. Freshw. Biol. 60, 2066–2083.

Cid, N., Verkaik, I., García-Roger, E., Rieradevall, M., Bonada, N., Sánchez-Montoya, M., et al., 2016. A biological tool to assess flow connectivity in reference temporary streams from the Mediterranean Basin. Sci. Total Environ. 540, 178–190.

Clarke, K.R., Warwick, R.M., 2001. A further biodiversity index applicable to species lists: variation in taxonomic distinctness. Mar. Ecol. Prog. Ser. 216, 265–278.

Corti, R., Datry, T., 2016. Terrestrial and aquatic invertebrates in the riverbed of an intermittent river: parallels and contrasts in community organisation. Freshw. Biol. 61, 1308–1320.

Corti, R., Datry, T., Drummond, L., Larned, S.T., 2011. Natural variation in immersion and emersion affects breakdown and invertebrate colonization of leaf litter in a temporary river. Aquat. Sci. 73, 537–550.

Courtwright, J., May, C.L., 2013. Importance of terrestrial subsidies for native brook trout in Appalachian intermittent streams. Freshw. Biol. 58, 2423–2438.

Cover, M.R., Seo, J.H., Resh, V.H., 2015. Life history, burrowing behavior, and distribution of *Neohermes filicornis* (Megaloptera: Corydalidae), a long-lived aquatic insect in intermittent streams. West. North Am. Nat. 75, 474–490.

Crichton, M., 1971. A study of caddis flies (Trichoptera) of the family Limnephilidae, based on the Rothamsted Insect Survey, 1964-68. J. Zool. 163, 533–563.

Datry, T., 2012. Benthic and hyporheic invertebrate assemblages along a flow intermittence gradient: effects of duration of dry events. Freshw. Biol. 57, 563–574.

Datry, T., Corti, R., Claret, C., Philippe, M., 2011. Flow intermittence controls leaf litter breakdown in a French temporary alluvial river: the "drying memory". Aquat. Sci. 73, 471–483.

Datry, T., Corti, R., Philippe, M., 2012. Spatial and temporal aquatic–terrestrial transitions in the temporary Albarine River, France: responses of invertebrates to experimental rewetting. Freshw. Biol. 57, 716–727.

Datry, T., Larned, S.T., Fritz, K.M., Bogan, M.T., Wood, P.J., Meyer, E.I., et al., 2014a. Broad-scale patterns of invertebrate richness and community composition in temporary rivers: effects of flow intermittence. Ecography 37, 94–104.

Datry, T., Larned, S.T., Tockner, K., 2014b. Intermittent rivers: a challenge for freshwater ecology. Bioscience 64, 229–235.

Datry, T., Bonada, N., Heino, J., 2016a. Towards understanding the organisation of metacommunities in highly dynamic ecological systems. Oikos 125, 149–159.

Datry, T., Moya, N., Zubieta, J., Oberdorff, T., 2016b. Determinants of local and regional communities in intermittent and perennial headwaters of the Bolivian Amazon. Freshw. Biol. 61, 1335–1349.

Datry, T., Pella, H., Leigh, C., Bonada, N., Hugueny, B., 2016c. A landscape approach to advance intermittent river ecology. Freshw. Biol. 61, 1200–1213.

Díaz, S., Cabido, M., 2001. Vive la différence: plant functional diversity matters to ecosystem processes. Trends Ecol. Evol. 16, 646–655.

Dieterich, M., Anderson, N.H., 2000. The invertebrate fauna of summer-dry streams in western Oregon. Arch. Hydrobiol. 147, 273–295.

Drummond, L.R., McIntosh, A.R., Larned, S.T., 2015. Invertebrate community dynamics and insect emergence in response to pool drying in a temporary river. Freshw. Biol. 60, 1596–1612.

Elbrecht, V., Beermann, A.J., Goessler, G., Neumann, J., Tollrian, R., Wagner, R., 2016. Multiple-stressor effects on stream invertebrates: a mesocosm experiment manipulating nutrients, fine sediment and flow velocity. Freshw. Biol. 61, 362–375.

Finn, D.S., Bonada, N., Múrria, C., Hughes, J.M., 2011. Small but mighty: headwaters are vital to stream network biodiversity at two levels of organization. J. N. Am. Benthol. Soc. 30, 963–980.

Florencio, M., Díaz-Paniagua, C., Gómez-Rodríguez, C., Serrano, L., 2014. Biodiversity patterns in a macroinvertebrate community of a temporary pond network. Insect Conserv. Divers. 7, 4–21.

Giehl, E.L., Jarenkow, J.A., 2015. Disturbance and stress gradients result in distinct taxonomic, functional and phylogenetic diversity patterns in a subtropical riparian tree community. J. Veg. Sci. 26, 889–901.

Hans, R.K., Farooq, M., Babu, G.S., Srivastava, S.P., Joshi, P.C., Viswanathan, P.N., 1999. Agricultural produce in the dry bed of the River Ganga in Kanpur, India—a new source of pesticide contamination in human diets. Food Chem. Toxicol. 37, 847–852.

Hinton, H.E., 1960. Cryptobiosis in the larvae of *Polypedilum vanderplanki* Hint. (Chironomidae). J. Insect Physiol. 5, 286–300.

Hynes, H.B.N., 1976. Biology of Plecoptera. Annu. Rev. Entomol. 21, 135–153.

IUCN (International Union for Conservation of Nature), 2015. The IUCN Red List of Threatened Species: *Glacidorbis occidentalis*. www.iucnredlist.org/details/9201/0 (Accessed 10.09.15).

Jackson, J.K., Sweeney, B.W., 1995. Egg and larval development times for 35 species of tropical stream insects from Costa Rica. J. N. Am. Benthol. Soc. 14, 115–130.

Jackson, J., McElravy, E., Resh, V., 1999. Long-term movements of self-marked caddisfly larvae (Trichoptera: Sericostomatidae) in a California coastal mountain stream. Freshw. Biol. 42, 525–536.

Johnson, M.F., Reid, I., Rice, S.P., Wood, P.J., 2009. Stabilization of fine gravels by net-spinning caddisfly larvae. Earth Surf. Process. Landf. 34, 413–423.

Karaouzas, I., Dimitriou, E., Lampou, A., Colombari, E., 2015. Seasonal and spatial patterns of macroinvertebrate assemblages and environmental conditions in Mediterranean temporary ponds in Greece. Limnology 16, 41–53.

Kefford, B.J., Buchwalter, D., Cañedo-Argüelles, M., Davis, J., Duncan, R.P., Hoffmann, A., 2016. Salinized rivers: degraded systems or new habitats for salt-tolerant faunas? Biol. Lett. 12, 20151072.

Kupilas, B., Friberg, N., McKie, B.G., Jochmann, M.A., Lorenz, A.W., Hering, D., 2016. River restoration and the trophic structure of benthic invertebrate communities across 16 European restoration projects. Hydrobiologia 769, 105–120.

Lake, P.S., 2011. Drought and Aquatic Ecosystems: Effects and Responses. Wiley-Blackwell, West Sussex.

Langton, P.H., Casas, J., 1999. Changes in chironomid assemblage composition in two Mediterranean mountain streams over a period of extreme hydrological conditions. Hydrobiologia 390, 37–49.

Larson, E.R., Magoulick, D.D., Turner, C., Laycock, K.H., 2009. Disturbance and species displacement: different tolerances to stream drying and desiccation in a native and an invasive crayfish. Freshw. Biol. 54, 1899–1908.

Ledger, M.E., Brown, L.E., Edwards, F.K., Milner, A.M., Woodward, G., 2013. Drought alters the structure and functioning of complex food webs. Nat. Clim. Change 3, 223–227.

Leigh, C., Datry, T., 2016. Drying as a primary determinant of biodiversity in river systems: a broad-scale analysis. Ecography 40, 487–499.

Leigh, C., Reis, T.M., Sheldon, F., 2013. High potential subsidy of dry-season aquatic fauna to consumers in riparian zones of wet–dry tropical rivers. Inland Waters 3, 411–420.

Leigh, C., Bonada, N., Boulton, A.J., Hugueny, B., Larned, S.T., Vander Vorste, R., et al., 2016a. Invertebrate assemblage responses and the dual roles of resistance and resilience to drying in intermittent rivers. Aquat. Sci. 78, 291–301.

Leigh, C., Boulton, A.J., Courtwright, J.L., Fritz, K., May, C.L., Walker, R.H., et al., 2016b. Ecological research and management of intermittent rivers: an historical review and future directions. Freshw. Biol. 61, 1181–1199.

Lynch, R.J., Bunn, S.E., Catterall, C.P., 2002. Adult aquatic insects: potential contributors to riparian food webs in Australia's wet-dry tropics. Austral Ecol. 27, 515–526.

Lytle, D.A., Poff, N.L., 2004. Adaptation to natural flow regimes. Trends Ecol. Evol. 19, 94–100.

Lytle, D.A., Bogan, M.T., Finn, D.S., 2008a. Evolution of aquatic insect behaviours across a gradient of disturbance predictability. Proc. R. Soc. B 275, 453–462.

Lytle, D.A., Olden, J.D., McMullen, L.E., 2008b. Drought-escape behaviors of aquatic insects may be adaptations to highly variable flow regimes characteristic of desert rivers. Southwest. Nat. 53, 399–402.

Maamri, A., Chergui, H., Pattee, E., 1997. Leaf litter processing in a temporary northeastern Moroccan river. Arch. Hydrobiol. 140, 513–531.

Macadam, C.R., 2016. A review of the status of mayflies (Ephemeroptera) of Great Britain—Species Status No. 28: Natural England Commissioned Reports, Number 193. Natural England, http://publications.naturalengland.org.uk/publication/4635857668538368 (Accessed 31.10.16).

Marchant, R., 2007. The use of taxonomic distinctness to assess environmental disturbance of insect communities from running water. Freshw. Biol. 52, 1634–1645.

Mazor, R.D., Stein, E.D., Ode, P.R., Schiff, K., 2014. Integrating intermittent streams into watershed assessments: applicability of an index of biotic integrity. Freshw. Sci. 33, 459–474.

Meyer, J.L., Strayer, D.L., Wallace, J.B., Eggert, S.L., Helfman, G.S., Leonard, N.E., 2007. The contribution of headwater streams to biodiversity in river networks. J. Am. Water Resour. Assoc. 43, 86–103.

Múrria, C., Bonada, N., Prat, N., 2008. Effects of the invasive species *Potamopyrgus antipodarum* (Hydrobiidae, Mollusca) on community structure in a small Mediterranean stream. Fundam. Appl. Limnol. 171, 131–143.

Nimmo, D., Mac Nally, R., Cunningham, S., Haslem, A., Bennett, A., 2015. Vive la résistance: reviving resistance for 21st century conservation. Trends Ecol. Evol. 30, 516–523.

Paltridge, R.M., Dostine, P.L., Humphrey, C.L., Boulton, A.J., 1997. Macroinvertebrate recolonization after re-wetting of a tropical seasonally-flowing stream (Magela Creek, Northern Territory, Australia). Mar. Freshw. Res. 48, 633–645.

Pennak, R.W., 1989. Freshwater Invertebrates of the United States, third ed. John Wiley and Sons, New York.

Phillipsen, I.C., Lytle, D.A., 2013. Aquatic insects in a sea of desert: population genetic structure is shaped by limited dispersal in a naturally fragmented landscape. Ecography 36, 731–743.

Poff, N.L., Olden, J.D., Vieira, N.K.M., Finn, D.S., Simmons, M.P., Kondratieff, B.C., 2006. Functional trait niches of North American lotic insects: traits-based ecological applications in light of phylogenetic relationships. J. N. Am. Benthol. Soc. 25, 730–755.

Poznańska, M., Kakareko, T., Krzyżyński, M., Kobak, J., 2013. Effect of substratum drying on the survival and migrations of Ponto-Caspian and native gammarids (Crustacea: Amphipoda). Hydrobiologia 700, 47–59.

Prat, N., Gallart, F., Von Schiller, D., Polesello, S., García-Roger, E.M., Latron, J., et al., 2014. The MIRAGE toolbox: an integrated assessment tool for temporary streams. River Res. Appl. 30, 1318–1334.

Progar, R.A., Moldenke, A.R., 2002. Insect production from temporary and perennially flowing headwater streams in western Oregon. J. Freshw. Ecol. 17, 391–407.

Quinn, J.M., Steele, G.L., Hickey, C.W., Vickers, M.L., 1994. Upper thermal tolerances of twelve New Zealand stream invertebrate species. N. Z. J. Mar. Freshw. Res. 28, 391–397.

Resh, V.H., Bêche, L.A., Lawrence, J.E., Mazor, R.D., McElravy, E.P., O'Dowd, A.P., et al., 2013. Long-term population and community patterns of benthic macroinvertebrates and fishes in Northern California Mediterranean-climate streams. Hydrobiologia 719, 93–118.

Richardson, W.B., 1990. A comparison of detritus processing between permanent and intermittent headwater streams. J. Freshw. Ecol. 5, 341–357.

Robertson, A., Rundle, S., Schmid-Araya, J., 2000. Putting the meio- into stream ecology: current findings and future directions for lotic meiofaunal research. Freshw. Biol. 44, 177–183.

Robertson, A.L., Wood, P.J., 2010. Ecology of the hyporheic zone: origins, current knowledge and future directions. Fundam. Appl. Limnol. 176, 279–289.

Robson, B., Chester, E., Austin, C., 2011. Why life history information matters: drought refuges and macroinvertebrate persistence in non-perennial streams subject to a drier climate. Mar. Freshw. Res. 62, 801–810.

Rolls, R.J., Heino, J., Chessman, B.C., 2016. Unravelling the joint effects of flow regime, climatic variability and dispersal mode on beta diversity of riverine communities. Freshw. Biol. 61, 1350–1364.

Rosenfeld, J.S., 2002. Functional redundancy in ecology and conservation. Oikos 98, 156–162.

Ruiz-García, A., Ferreras-Romero, M., 2007. The larva and life history of *Stenophylax crossotus* McLachlan, 1884 (Trichoptera: Limnephilidae) in an intermittent stream from the southwest of the Iberian Peninsula. Aquat. Insects 29, 9–16.

Salavert, V., Zamora-Muñoz, C., Ruiz-Rodríguez, M., Fernández-Cortés, A., Soler, J.J., 2008. Climatic conditions, diapause and migration in a troglophile caddisfly. Freshw. Biol. 53, 1606–1617.

Sánchez-Fernández, D., Bilton, D.T., Abellán, P., Ribera, I., Velasco, J., Millán, A., 2008. Are the endemic water beetles of the Iberian Peninsula and the Balearic Islands effectively protected? Biol. Conserv. 141, 1612–1627.

Sandberg, J.B., Stewart, K.W., 2004. Capacity for extended egg diapause in six *Isogenoides* Klapálek species (Plecoptera: Perlodidae). Trans. Am. Entomol. Soc. 130, 411–423.

Sarremejane, R., Mykrä, H., Bonada, N., Aroviita, J., Muotka, T., 2017. Habitat connectivity and dispersal ability drive the assembly mechanisms of macroinvertebrate communities in river networks. Freshw. Biol. http://dx.doi.org/10.1111/fwb.12926.

Schriever, T.A., Lytle, D.A., 2016. Convergent diversity and trait composition in temporary streams and ponds. Ecosphere 7, e01350.

Schriever, T.A., Bogan, M.T., Boersma, K.S., Cañedo-Argüelles, M., Jaeger, K.L., Olden, J.D., et al., 2015. Hydrology shapes taxonomic and functional structure of desert stream invertebrate communities. Freshw. Sci. 34, 399–409.

Smith, F., Brown, A., Pope, M., 2001. Meiofauna in intermittent streams differ among watersheds subjected to five methods of timber harvest. Hydrobiologia 464, 1–8.

Soria, M., Leigh, C., Datry, T., Bini, L.M., Bonada, N., 2017. Biodiversity in perennial and intermittent rivers: a meta-analysis. Oikos. http://dx.doi.org/10.1111/oik.04118.

Stanley, E.H., Buschman, D.L., Boulton, A.J., Grimm, N.B., Fisher, S.G., 1994. Invertebrate resistance and resilience to intermittency in a desert stream. Am. Midl. Nat. 131, 288–300.

Statzner, B., Bêche, L.A., 2010. Can biological invertebrate traits resolve effects of multiple stressors on running water ecosystems? Freshw. Biol. 55, 80–119.

Steward, A.L., von Schiller, D., Tockner, K., Marshall, J.C., Bunn, S.E., 2012. When the river runs dry: human and ecological values of dry riverbeds. Front. Ecol. Environ. 10, 202–209.

Storey, R., 2015. Macroinvertebrate community responses to duration, intensity and timing of annual dry events in intermittent forested and pasture streams. Aquat. Sci. 78, 395–414.

Storey, R.G., Quinn, J.M., 2008. Composition and temporal changes in macroinvertebrate communities of intermittent streams in Hawke's Bay, New Zealand. N. Z. J. Mar. Freshw. Res. 42, 109–125.

Storey, R.G., Quinn, J.M., 2013. Survival of aquatic invertebrates in dry bed sediments of intermittent streams: temperature tolerances and implications for riparian management. Freshwater Sci. 32, 250–266.

Strachan, S.R., Chester, E.T., Robson, B.J., 2015. Freshwater invertebrate life history strategies for surviving desiccation. Springer Sci. Rev. 3, 57–75.

Strachan, S.R., Chester, E.T., Robson, B.J., 2016. Habitat alters the effect of false starts on seasonal-wetland invertebrates. Freshw. Biol. 61, 680–692.

Strayer, D.L., Dudgeon, D., 2010. Freshwater biodiversity conservation: recent progress and future challenges. J. N. Am. Benthol. Soc. 29, 344–358.

Stubbington, R., 2012. The hyporheic zone as an invertebrate refuge: a review of variability in space, time, taxa and behaviour. Mar. Freshw. Res. 63, 293–311.

Stubbington, R., Datry, T., 2013. The macroinvertebrate seedbank promotes community persistence in temporary rivers across climate zones. Freshw. Biol. 58, 1202–1220.

Stubbington, R., Greenwood, A.M., Wood, P.J., Armitage, P.D., Gunn, J., Robertson, A.L., 2009a. The response of perennial and temporary headwater stream invertebrate communities to hydrological extremes. Hydrobiologia 630, 299–312.

Stubbington, R., Wood, P.J., Boulton, A.J., 2009b. Low flow controls on benthic and hyporheic macroinvertebrate assemblages during supra-seasonal drought. Hydrol. Process. 23, 2252–2263.

Sweeney, B.W., Newbold, J.D., 2014. Streamside forest buffer width needed to protect stream water quality, habitat, and organisms: a literature review. J. Am. Water Resour. Assoc. 50, 560–584.

Tachet, H., Richoux, P., Bournaud, M., Usseglio-Polatera, P., 2002. Invertébrés d'eau Douce: Systématique, Biologie, Écologie. CNRS Éditions, Paris.

Tockner, K., Pusch, M., Borchardt, D., Lorang, M.S., 2010. Multiple stressors in coupled river–floodplain ecosystems. Freshw. Biol. 55, 131–151.

Townsend, C.R., Hildrew, A.G., 1976. Field experiments on the drifting, colonization and continuous redistribution of stream benthos. J. Anim. Ecol. 45, 759–772.

Vadher, A.N., Stubbington, R., Wood, P.J., 2015. Fine sediment reduces vertical migrations of *Gammarus pulex* (Crustacea: Amphipoda) in response to surface water loss. Hydrobiologia 753, 61–71.

Van den Broek, M., Waterkeyn, A., Rhazi, L., Grillas, P., Brendonck, L., 2015. Assessing the ecological integrity of endorheic wetlands, with focus on Mediterranean temporary ponds. Ecol. Indic. 54, 1–11.

Vander Vorste, R., Corti, R., Sagouis, A., Datry, T., 2016a. Invertebrate communities in gravel-bed, braided rivers are highly resilient to flow intermittence. Freshwater Sci. 35, 164–177.

Vander Vorste, R., Malard, F., Datry, T., 2016b. Is drift the primary process promoting the resilience of river invertebrate communities? A manipulative field experiment in an alluvial, intermittent river. Freshw. Biol. 61, 1276–1292.

Vander Vorste, R., Mermillod-Blondin, F., Hervant, F., Mons, R., Forcellini, M., Datry, T., 2016c. Increased depth to the water table during river drying decreases the resilience of *Gammarus pulex* and alters ecosystem function. Ecohydrology 9, 1177–1186.

Vane-Wright, R.I., Humphries, C.J., Williams, P.H., 1991. What to protect?—systematics and the agony of choice. Biol. Conserv. 55, 235–254.

Vannote, R.L., Minshall, G.W., Cummins, K.W., Sedell, J.R., Cushing, C.E., 1980. The river continuum concept. Can. J. Fish. Aquat. Sci. 37, 130–137.

Velasco, J., Millán, A., Hernández, J., Gutiérrez, C., Abellán, P., Sánchez, D., et al., 2006. Response of biotic communities to salinity changes in a Mediterranean hypersaline stream. Saline Syst. 2, 12.

Verdonschot, R., van Oosten-Siedlecka, A.M., ter Braak, C.J.F., Verdonschot, P.F.M., 2015. Macroinvertebrate survival during cessation of flow and streambed drying in a lowland stream. Freshw. Biol. 60, 282–296.

Vörösmarty, C.J., Sahagian, D., 2000. Anthropogenic disturbance of the terrestrial water cycle. Bioscience 50, 753–765.

Warwick, R.M., Clarke, K.R., 1995. New 'biodiversity' measures reveal a decrease in taxonomic distinctness with increasing stress. Mar. Ecol. Prog. Ser. 129, 301–305.

Watanabe, M., 2006. Anhydrobiosis in invertebrates. Appl. Entomol. Zool. 41, 15–31.

Welborn, G.A., Skelly, D.K., Werner, E.E., 1996. Mechanisms creating community structure across a freshwater habitat gradient. Annu. Rev. Ecol. Syst. 27, 337–363.

Wickson, S., Chester, E.T., Robson, B.J., 2012. Aestivation provides flexible mechanisms for survival of stream drying in a larval trichopteran (Leptoceridae). Mar. Freshw. Res. 63, 821–826.

Wickson, S.J., Chester, E.T., Valenzuela, I., Halliday, B., Lester, R.E., Matthews, T.G., 2014. Population genetic structure of the Australian caddisfly *Lectrides varians* Mosely (Trichoptera: Leptoceridae) and the identification of cryptic species in south-eastern Australia. J. Insect Conserv. 18, 1037–1046.

Williams, D.D., 1996. Environmental constraints in temporary fresh waters and their consequences for the insect fauna. J. N. Am. Benthol. Soc. 15, 634–650.

Williams, D.D., 2006. The Ecology of Temporary Waters. Oxford University Press, Oxford.

Zickovich, J.M., Bohonak, A.J., 2007. Dispersal ability and genetic structure in aquatic invertebrates: a comparative study in southern California streams and reservoirs. Freshw. Biol. 52, 1982–1996.

THE BIOTA OF INTERMITTENT RIVERS AND EPHEMERAL STREAMS: TERRESTRIAL AND SEMIAQUATIC INVERTEBRATES

Alisha L. Steward[*,†], **Simone D. Langhans**[‡], **Roland Corti**[‡,§], **Thibault Datry**[§]

Queensland Government, Brisbane, QLD, Australia[*] *Griffith University, Nathan, QLD, Australia*[†] *Leibniz-Institute of Freshwater Ecology and Inland Fisheries (IGB), Berlin, Germany*[‡] *Irstea, UR MALY, centre de Lyon-Villeurbanne, Villeurbanne, France*[§]

IN A NUTSHELL

- Intermittent rivers and ephemeral streams (IRES) can support a diverse and often abundant terrestrial and semiaquatic invertebrate (TSAI) fauna during both wet and dry hydrological phases
- TSAIs can be found at the shoreline, on the surface of exposed gravel bars, within unsaturated gravels, in dry riverbeds, in riparian zones, and on floodplains
- TSAIs play a key role in processing organic matter and nutrients. Drying events provide a pulsed subsidy of aquatic material to be consumed by terrestrial predators and scavengers, while some TSAIs that are washed downstream during rewetting events represent a valuable food resource for aquatic consumers in the receiving waters. These reciprocal subsidies represent a unique linkage between the aquatic and terrestrial phases of IRES
- The TSAI fauna is typically ignored in monitoring and management programs of IRES. Explicit inclusion of IRES in management and legislation will help safeguard the diversity of TSAIs

4.4.1 INTRODUCTION

Flow intermittence involves alternating "terrestrial" and "aquatic" phases in river channels (Chapters 2.3 and 4.9; Stanley et al., 1997; Datry et al., 2016), promoting a temporal mosaic of habitat types for invertebrates (Tockner et al., 2006). Much less is known about the species composition and ecological roles of terrestrial and semiaquatic invertebrates (TSAIs) of intermittent rivers and ephemeral streams (IRES) than their aquatic counterparts (Steward et al., 2011).

TSAIs of IRES are represented by a wide range of taxonomic groups (Box 4.4.1). They are typically ground-dwelling, and many are predators or scavengers. Wolf spiders (Lycosidae), ground beetles (Carabidae), rove beetles (Staphylinidae), and ants (Formicidae) are commonly found along the shoreline of rivers and streams during wet phases, and on the riverbed surface and within the channel

Intermittent Rivers and Ephemeral Streams. http://dx.doi.org/10.1016/B978-0-12-803835-2.00008-5

245

BOX 4.4.1 TERRESTRIAL AND SEMIAQUATIC INVERTEBRATES OF IRES

The terrestrial and semiaquatic invertebrates (TSAIs) of IRES are represented by a wide range of taxonomic groups (Fig. 4.4.1). Species richness and abundance are typically dominated by insects, but IRES are also home to arachnids, centipedes, crustaceans, millipedes, snails, springtails, and worms. Predatory beetles, such as ground beetles (Carabidae, Fig. 4.4.1a) and rove beetles (Staphylinidae, Fig. 4.4.1b), are common inhabitants of IRES, and include some specialized species that are not found anywhere else. Ants (Formicidae, Fig. 4.4.1c) can be found in high numbers and prey on aquatic biota that have been swept to the shoreline or stranded on the riverbed. Ants are also common after IRES have dried up. Wolf spiders (Lycosidae, Fig. 4.4.1d) do not make webs; instead, they ambush their prey. These spiders can obtain much of their food from the water's edge, including emerging aquatic insects. Springtails (Collembola, Fig. 4.4.1e) are tiny invertebrates that prefer moist environments and can be highly abundant in IRES. Semiaquatic invertebrates, such as velvet shore bugs (Ochteridae, Fig. 4.4.1f), inhabit the water's edge, and are not typically found once IRES have dried up. Pygmy mole crickets (Tridactylidae, Fig. 4.4.1g) also live at the shoreline in moist sand and feed on detritus and plant material, even ingesting sand grains coated with algae. Pill bugs (Isopoda, Fig. 4.4.1h) are terrestrial crustaceans sometimes found in moist riverbeds; they can roll themselves into a ball for defense or to prevent water loss.

substrate during dry phases (Steward et al., 2011; Corti and Datry, 2012, 2016; Steward, 2012; Dell et al., 2014; Langhans and Tockner, 2014a,b) (Figs. 4.4.1 and 4.4.2). Other terrestrial invertebrates, such as springtails (Collembola), pygmy mole crickets (Tridactylidae), and pill bugs (Isopoda), can also be common inhabitants of IRES (Blackith and Guillet, 1995) (Fig. 4.4.1). Semiaquatic invertebrates, such as velvet shore bugs (Ochteridae, Fig. 4.4.1), can also inhabit the shorelines of IRES; however, they require water for part of their life cycle or depend heavily on aquatic food, and may therefore need to utilize specialized life history strategies such as desiccation resistance to persist until the channel is rewetted (Robson et al., 2011). Aquatic invertebrates (Chapter 4.3) also can be found in a resting state in the dry beds of IRES.

TSAIs play a major role in processing organic matter and nutrients, influenced by riverbed characteristics such as moisture content, litter quality, and time since the bed was last inundated. Ants, for example, can move large volumes of soil within the bed and can store organic material within their nest chambers, providing areas higher in carbon, nitrogen, and phosphorus than in surrounding areas (Hölldobler and Wilson, 1990; Wishart, 2000). Drying events provide a pulsed subsidy of dead and dying aquatic material to be consumed by a terrestrial "clean-up crew" of scavengers and predators, while TSAIs washed downstream during rewetting events represent a valuable food resource for aquatic consumers in the receiving waters. Wishart (2000) found that a large proportion of the terrestrial invertebrate biomass sampled in the dry beds of South African IRES would be unable to escape the rapid onset of flow due to the lack of flight, contributing to the particulate organic matter input of these rivers. These reciprocal subsidies represent an important temporal linkage between the aquatic and terrestrial phases of IRES that is not strongly developed in perennial rivers.

TSAIs play important roles in other ecosystem functions. TSAI predators and scavengers consume aquatic invertebrates, with emerging adult insects providing seasonal aquatic subsidies to riparian zones (Paetzold et al., 2005, 2006) (Fig. 4.4.2). Conversely, TSAIs swept downstream in advancing wetted fronts can become food for aquatic consumers. TSAIs may also be vectors of parasites, such as nematomorph worms that reside in terrestrial invertebrates but require water to complete their life cycle. A study of a Japanese headwater stream found that camel crickets and grasshoppers (Orthoptera) were 20 times more likely to enter a stream if infected by a nematomorph parasite (*Gordionus* spp.), and that the orthopterans provided 60% of the annual energy intake of the resident trout population (Sato et al.,

FIG. 4.4.1

Microscopy photographs of TSAIs collected from IRES: (a) head of a ground beetle (Carabidae, total body length ~12 mm), (b) rove beetle (Staphylinidae, length ~5 mm), (c) ants (Dolichoderinae: Formicidae, length ~4 mm), (d) wolf spider (Lycosidae, length ~5 mm), (e) springtail (Collembola, length <2 mm), (f) velvet shore bug nymph (Ochteridae, length <2 mm), (g) pygmy mole cricket (Tridactylidae, length ~10 mm), and (h) pill bug (Isopoda, length ~5 mm).

Photos: Courtesy A. Steward.

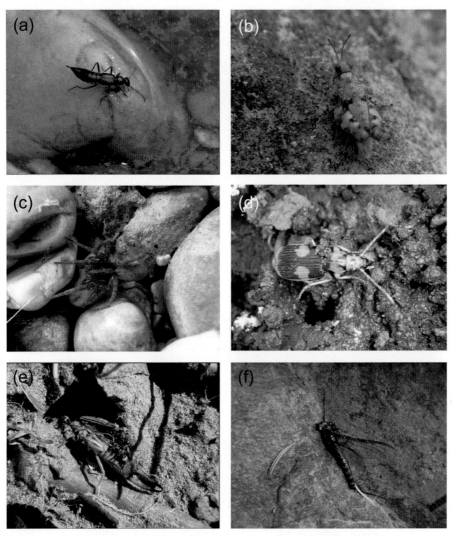

FIG. 4.4.2

Photographs of TSAIs *in situ*: (a) rove beetle (Staphylinidae) at the shoreline of the Tagliamento River, Italy, (b) grasshopper (Orthoptera) at the shoreline of the Einasleigh River, Australia, (c) wolf spider (Lycosidae) at the shoreline of the Tagliamento River, Italy, (d) ground beetle (Carabidae) on moist sediment of an unnamed riverbed, Australia, (e) ants carrying away an earwig (Dermaptera) from a dry channel in the Tagliamento River, Italy, and (f) an ant (Formicidae) dragging an emerged mayfly (Ephemeroptera) from the shoreline of Wild Cattle Creek, Australia.

Photos: Courtesy A. Steward.

2011). Other ecosystem functions provided by TSAIs include increasing the complexity of riverbeds through tunneling and burrowing, which can increase the water infiltration capacity of the substrate.

In this chapter, we review the TSAI communities of IRES, examining their habitat requirements, taxonomic diversity, adaptations to wet/dry phases, and their functional and ecological roles. We identify threats to TSAI populations in IRES and explore opportunities to protect this fauna, suggesting ways in which they can be incorporated into management and monitoring programs of IRES.

4.4.2 HABITAT REQUIREMENTS OF IRES INVERTEBRATES

IRES are dynamic ecosystems, with aquatic and terrestrial habitats expanding, contracting, and fragmenting through time (Stanley et al., 1997; Datry et al., 2016; Chapters 2.3 and 4.9). They provide a variety of habitats for terrestrial and semiaquatic invertebrates during both wet and dry phases. When IRES are flowing, TSAIs can be found at the shoreline, on the surface of exposed gravel bars, within the unsaturated gravel, in riparian zones, and on floodplains (Figs. 4.4.2–4.4.4; Langhans and Tockner, 2014a,b).

As a river dries, riffles cease to exist, followed by the disappearance of pools, surface water, and finally, drought refuges for aquatic biota (Boulton, 2003). Drying of IRES removes habitat for aquatic invertebrates (Stanley et al., 1997) but represents an expansion of habitat for terrestrial invertebrates to colonize as more and more riverbed area is exposed (Fig. 4.4.3). Different spatial and temporal drying patterns may favor different functional groups of TSAIs. The progression from wet to dry is usually gradual, although in the wide channels of some desert rivers, entire riffles, runs, and pools can rapidly dry in a day (Stanley et al., 1997). Similarly, 100–300 m sections of gravel-bed rivers in New Zealand have been observed to dry in a day (Davey et al., 2006).

TSAIs can inhabit coarse and fine riverbed substrates, and substrates that are a combination of the two (Fig. 4.4.4). Coarse substrates such as boulders, cobbles, gravels, woody debris, and leaf litter have larger interstitial spaces available for invertebrates to inhabit than finer substrates such as sand and silt. Moring and Stewart (1994) found that the abundance of prey for streamside and riparian wolf spiders (Lycosidae) was higher in coarser (rock-cobble) than finer (sand-cobble) substrates and, as a result, the coarser substrates supported a higher abundance and higher species richness of wolf spiders. Fine substrates, however, can be preferred by taxa capable of digging. For example, beetles from the anthicid genus *Mecynotarsus* (Fig. 4.4.5) can be abundant in sandy riverbeds (Steward, 2012), using their pronotal horn to dig through sand (Hashimoto and Hayashi, 2012). These beetles may dig to avoid predators, extreme temperatures or humidity, or to forage (Hashimoto and Hayashi, 2012).

TSAIs often need to withstand extreme environmental conditions because dry riverbeds are often harsh exposed places devoid of vegetation that experience higher fluctuations in air temperature and solar radiation, and lower humidity than that provided by shaded riparian zones (Steward, 2012). For most eukaryotic organisms, 60°C is their upper thermal tolerance (Tansey and Brock, 1972). A study of Australian dry riverbeds found that ground-level air temperatures exceeded 60°C more frequently and for a longer duration than in adjacent riparian zones, and that dry riverbeds were sometimes up to 20°C hotter than riparian zones (Fig. 4.4.6). Similarly, smaller diel temperature ranges have been recorded from riparian zones than from exposed riverbed gravels in Italy and Switzerland (Tonolla et al., 2010). Ground temperatures of over 60°C have also been recorded in the dry riverbed of the Kuiseb River, Israel (Holm and Edney, 1973).

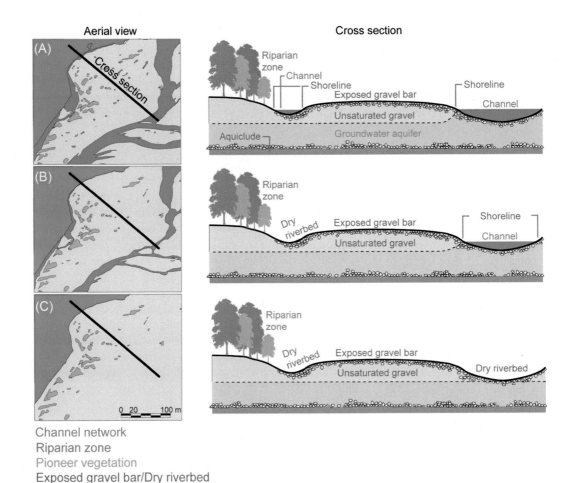

Channel network
Riparian zone
Pioneer vegetation
Exposed gravel bar/Dry riverbed

FIG. 4.4.3

Conceptual model showing habitats available to TSAIs in IRES during (A) high flow, (B) low flow, and (C) dry conditions.

Figure modified from Langhans, S.D., Tockner, K., 2014b. Edge effects are important in supporting beetle biodiversity in a gravel-bed river floodplain. PLoS One, 9, e114415.

Dry riverbeds are physically harsher places for biota than riparian zones and may influence the spatial and temporal activity of TSAI taxa. Invertebrates may have behavioral adaptations that involve the use of daytime refuges; for example, some desert beetles can find tolerable temperatures by retreating underground or moving to the base of vegetation during the hottest times of the day (Holm and Edney, 1973). Langhans and Tockner (2014a) found terrestrial arthropods inhabiting the unsaturated sediments of a gravel-bed river down to 1.1 m all year round and suggest that temperatures within the sediments are more stable throughout the year than those at the surface. Sediments, therefore, may act as a temperature refuge for TSAIs. It is also likely that humidity is more favorable within the sediments than above them.

FIG. 4.4.4

Examples of TSAI habitats in IRES: (a) shoreline habitat along Reynolds Creek, Australia, (b) algal mat on the dry riverbed of Oaky Creek, Australia, (c) cracking clay riverbed, Mungallala Creek, Australia, (d) leaf litter on the sandy riverbed of Rosser Creek, Australia, (e) woody debris in the Tagliamento River, Italy, (f) pebble, gravel and sand in the Walsh River, Australia, (g) cobble, pebble and gravel in the Tagliamento River, Italy, and (h) bedrock in the Walsh River, Australia.

Photos: Courtesy A. Steward (a, b, and d–h); J. Marshall (c).

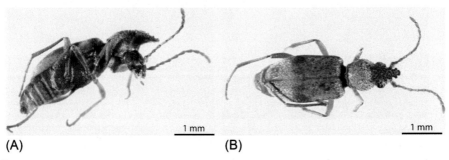

(A) **(B)**

FIG. 4.4.5

Anthicid beetles, such as this *Mecynotarsus* sp., can be common and highly abundant in sandy riverbeds. Note the large, shovel-like pronotal horn used for digging through sand: (A) lateral view, (B) dorsal view.

Photos: Courtesy A. Steward.

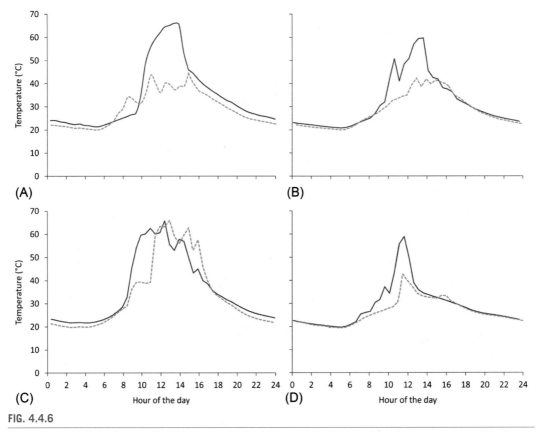

(A) **(B)**

(C) **(D)**

FIG. 4.4.6

Air temperature data at ground level from dry-riverbed (brown solid line) and riparian (green dashed line) habitats in IRES of Queensland, Australia: (A) Reynolds Creek, (B) Wild Cattle Creek, (C) Oaky Creek, and (D) Purga Creek. Data were logged at 30-minute intervals for approximately 24 h in Dec. 2009 (austral summer).

Sites and data from Steward, A.L., 2012. When the River Runs Dry: The Ecology of Dry River Beds. PhD Thesis, Griffith University, Brisbane.

Large mats of algae can cover the dried beds of IRES (Strandine, 1941; Fig. 4.4.4). The micro-habitat underneath such algal mats can remain cool and moist, and may act as a temperature and/or humidity refuge for TSAIs, prolonging the survival of aquatic and semiaquatic invertebrates lacking physiological or morphological adaptations to resist desiccation (Strandine, 1941). Although Strandine (1941) found that algal mats were inhabited by an average of 6600 living aquatic snails per m^2 of algae, the extent to which algal mats are used by TSAIs remains unknown.

4.4.3 TAXONOMIC DIVERSITY OF TSAI COMMUNITIES

The taxonomic composition of TSAI communities depends on flow regime (Chapter 2.2) and hydro-logical connectivity (Chapter 2.3) of IRES, with a different suite of fauna being present during wet, drying, and dry phases. Succession occurs as the riverbed dries and rewets, with semiaquatic inverte-brate taxa, such as velvet water bugs (Hebridae), typically present until the bed undergoes complete drying (Steward, 2012). As the dry spell continues, the community composition becomes more "ter-restrial" with fewer or no semiaquatic taxa present in an active stage, although aquatic and semiaquatic taxa may persist as desiccation-resistant life stages (Chapters 4.3 and 4.8). The invertebrates of drying riverbeds do not appear to be a subset of those in the riparian zone, and some riverbed communities remain unique from adjacent riparian forest communities (Lalley et al., 2006; Steward et al., 2011; Steward, 2012).

A study by Steward (2012) on a drying ephemeral stream in subtropical Australia found that the first TSAI taxa to be collected from the shoreline of a drying pool included anthicid beetles (Anthicidae), ants (Formicidae), velvet water bugs (Hebridae), true flies (Diptera), mites (Acarina), springtails (Collembola), and wolf spiders (Lycosidae). Once the isolated pool in the study dried up, additional taxa were recorded, including Amphipoda, beetle taxa such as minute fungus beetles (Corylophidae), featherwing beetles (Ptiliidae) and rove beetles (Staphylinidae), some extra mite taxa (Acarina), and further hemipteran bug taxa including toad bugs (Gelastocoridae) and aphids (Aphididae). Additional taxa were found in the dry riverbed after it had been dry for 4 months, including jumping spiders (Salticidae), a beetle from the genus *Mecynotarsus* (Anthicidae), latridiid beetles (Latridiidae), thrips (Thysanoptera), and yet more ant taxa. Semiaquatic velvet water bugs (Hebridae) and velvet shore bugs (Ochteridae) were not found after the pool dried until it rewetted the following year. The resumption of flow coincided with the loss of some ant, beetle, bug, mite, and springtail taxa (Steward, 2012).

TSAI taxa recorded from IRES in Australia, Canada, France, Greece, Italy, New Zealand, Portugal, Spain, and South Africa are listed in Table 4.4.1. A wide range of invertebrate taxa is recognized from several phyla: Annelida, Arthropoda, Mollusca, and Nematoda. The list, however, is dominated by insects (Class Insecta), from the Orders Archaeognatha (bristletails), Blattodea (cockroaches), Coleoptera (beetles), Diptera (true flies), Hemiptera (true bugs), Hymenoptera (wasps and ants), Mantodea (mantids), Neuroptera (lacewings), Orthoptera (grasshoppers, crickets), Psocoptera (book lice), Thysanoptera (thrips), and Zygentoma (silverfish). Other arthropods listed include Arachnida (spiders, mites, harvestmen, pseudoscorpions), Chilopoda (centipedes), Collembola (springtails), Diplopoda (millipedes), Diplura, and Protura.

Globally, TSAI assemblages in IRES share similar taxonomic patterns, such as the dominance of predators and scavengers at the shoreline, and the composition of the "clean-up crew" consisting of opportunistic predators and scavenging taxa during the drying phase (e.g., Williams and Hynes,

Table 4.4.1 List of TSAI taxonomic groups found in IRES

Phylum/division	Class/subclass	Superorder/order	Superfamily/family	Subfamily
Annelida	Oligochaeta[a,b,n,p]			
Arthropoda	Arachnida[b,f,h]	Acariformes[e,i,k,l,o,p], Oribatida[p]		
		Araneae[b,f,k,l,o,p]	Agelenidae[n], Clubionidae[n], Corinnidae[n], Linyphiidae[j,n], Liocranidae[n], Lycosidae[c,i,j,k,n,p], Micropholcommatidae[p], Miturgidae[p], Oecobiidae[p], Salticidae[k,p], Theridiidae[n]	
		Opiliones[n]		
		Pseudoscorpionida[e,k,p]		
		Scorpiones[k]		
		Solifugae	Solpugidae[e]	
	Chilopoda[h,p]	Geophilomorpha, Scolopendromorpha	Geophilidae[l], Scolopendridae[l]	
	Collembola[h,i,k,n,o,p]	Entomobryomorpha	Entomobryidae[l,p], Entomobryoidea[p], Isotomidae[p]	
		Poduromorpha	Paronellidae[l], Poduroidea[p]	
		Symphypleona[p]	Sminthuridae[p]	
	Diplopoda[a,k,p]	Diplura[k,p]		
	Entognatha	Protura[h]		
	Insecta	Archaeognatha[e,i,o]		
		Blattodea[e,k,p]	Blattidae[l]	
		Coleoptera[e,i,o,p]	Aderidae[p], Anobiidae[l], Anthicidae[g,i,j,k,l,m,n,p,q], Anthribidae[p], Bostrichidae[k,p], Bothrideridae[l,p], Brentidae[p], Byrrhidae[m], Cantharidae[m], Carabidae[d,e,g,i,k,l,m,n,p,q], Cerambycidae[p], Chrysomelidae[l,m,n,p,q], Clambidae[n,q], Coccinellidae[m,p], Corylophidae[k,n,p], Cryptophagidae[m,n], Curculionidae[m,n,p,q], Dasytidae[n], Dryopidae[m,n], Dytiscidae[m,q], Elateridae[j,k,l,m,n,p,q], Elmidae[m,n], Georissidae[m], Gyrinidae[m], Haliplidae[m,n], Heteroceridae[a,b,p], Histeridae[m,p], Hydraenidae[c,l,m], Hydrophilidae[a,l,m,n,p], Laemophloeidae[m,p], Latrididae[k,m,n,p], Leiodidae[m,n], Limnichidae[m], Lucanidae[m,n], Melyridae[p], Monotomidae[m], Mycetophagidae[p], Nitidulidae[k,m,p], Oedermeridae[m], Passalidae[l], Pselaphidae[l,n,p], Psephenidae[l], Ptiliidae[k,m,p,q], Scarabaeidae[e,m,n,q], Scarabaeoidea[b,k,p], Scydmaenidae[m,p], Silphidae[n,q], Silvanidae[l,p], Spercheidae[p], Staphylinidae[a,b,h,j,k,l,m,n,p,q], Tenebrionidae[p], Throscidae[m,p]	

Continued

	Taxon	Families	Subfamilies
	Dermaptera[e,k,p]		
	Diptera[i,k,p]	Anthomyiidae[a], Cecidomyiidae[p], Ceratopogonidae[b,l], Chironomidae[h,p], Diopsidae[d], Dolichopodidae[a], Heleomyzidae[l], Mycetophilidae[l,p], Phoridae[p], Psychodidae[l], Sciaridae[l], Sepsidae[b], Sphaeroceridae[a,b], Tipulidae[a,l,p]	
	Hemiptera[e,i,k,n]	Aleyrodidae[p], Anthocoridae[p], Aphididae[h,k,p], Aphidoidea[n], Cicadellidae[n,p], Coccoidea[k,p], Cydnidae[p], Dipsocoridae[p], Fulgoroidea[p], Gelastocoridae[p], Hebridae[k,l,p], Hydrometridae[p], Lygaeidae[l,p], Miridae[p], Ochteridae[k], Peloridiidae[l], Psylloidea[l,p], Reduviidae[l,p], Saldidae[l]	
	Hymenoptera[e,k,l]	Formicidae[b,c,e,f,h,i,j,k,n,o,p], Mutillidae[k,p], Scelionidae[k], Sphecidae[p]	Cerapachyinae[p], Dolichoderinae[k,p], Dorylinae[k], Ectatomminae[k,p], Formicinae[k,l,n,p], Myrmicinae[k,n,p], Ponerinae[n,p], Pseudomyrmecinae[l,p]
	Lepidoptera[e,k,l,p]	Noctuidae[j]	
	Mantodea[e]	Mantidae[l,p]	
	Mecoptera	Choristidae[l]	
	Neuroptera	Chrysopidae[p]	
	Orthoptera[e,i,k]	Acrididae[l], Gryllidae[j,k,l,n,p], Gryllotalpidae[p], Rhaphidophoridae[j], Tetrigidae[d,j,p], Tettigoniidae[d,j,p], Tridactylidae[d,j,p]	
	Psocoptera[k,o,p]	Epipsocidae[l]	
	Thysanoptera[k,p]	Phlaeothripidae[l]	
	Trichoptera[p]		
	Zygentoma[e]		
	Amphipoda[c,e,k,p]		
Malacostraca	Isopoda[e,k,n,p]	Oniscidae[l]	

Table 4.4.1 List of TSAI taxonomic groups found in IRES—cont'd

Phylum/division	Class/subclass	Superorder/order	Superfamily/family	Subfamily
Mollusca Nematoda[p]	Gastropoda[e,k,n]		Arionoidea[i], Cochlicopidae[b], Limacidae[b], Planorbidae[l]	

[a]Moon (1956).
[b]Williams and Hynes (1977).
[c]Boulton and Suter (1986).
[d]Blackith and Guillet (1995).
[e]Wishart (2000).
[f]Lalley et al. (2006).
[g]Langhans (2006).
[h]Larned et al. (2007).
[i]Steward et al. (2011).
[j]McCluney and Sabo (2012).
[k]Steward (2012).
[l]Dell et al. (2014).
[m]Langhans and Tockner (2014b).
[n]Corti and Datry (2016).
[o]Sánchez-Montoya et al. (2016).
[p]Steward (unpublished data).
[q]von Schiller (unpublished data).

1977; Wishart, 2000; Lalley et al., 2006; Langhans, 2006; Larned et al., 2007; Steward, 2012; Dell et al., 2014; Langhans and Tockner, 2014b; Corti and Datry, 2016; Sánchez-Montoya et al., 2016). Additionally, many TSAI taxa are ground dwelling. Beetles, ants, and spiders, for example, are common to IRES in most of the countries that have been sampled (Table 4.4.1), with many shared taxonomic groups even at the family level. Ground beetles (Carabidae) were noted from Australia, France, Greece, Italy, Portugal, and South Africa; rove beetles (Staphylinidae) from Australia, Canada, France, New Zealand, Portugal, and Italy; and wolf spiders (Lycosidae) from Australia, Canada, France, and South Africa. These data represent a wide range of climatic zones—wet-dry tropical, subtropical, semiarid, arid, humid continental, Mediterranean, and temperate—and suggest that a similar suite of taxonomic groups (at the family level, at least) is found in IRES habitats worldwide, regardless of the local climate experienced.

The taxonomic diversity and abundance of invertebrates in dry riverbeds of IRES can be high, and communities are typically dominated by ants, beetles, and spiders (Wishart, 2000; Steward et al., 2011). Similar patterns in invertebrate composition have been found in dry riverbeds around the world. High abundances of ants, beetles, and spiders were recorded in Australia (Steward, 2012), Italy (Steward, 2012), Namibia (Lalley et al., 2006), and South Africa (Wishart, 2000), and high abundances of ants and springtails in New Zealand (Larned et al., 2007). Wishart (2000) found that the invertebrate composition in dry riverbeds was diverse, with 19 invertebrate orders identified from just three sites. Lalley et al. (2006) sampled invertebrates in a Namibian desert and reported that the highest number of species occurred in a temporary river channel with shrub patches, compared to six other habitat types they sampled. The dry river channel also contained the highest number of unique species. Sampling from 22 sites, Steward et al. (2011) collected 66 morphospecies unique to dry riverbeds that were not found in adjacent riparian zones. These morphospecies were from the following groups: Acarina, Coleoptera, Collembola, Dermaptera, Diptera, Formicidae, Hemiptera, Hymenoptera (non-Formicidae taxa), Isoptera, Lepidoptera, and Orthoptera.

4.4.4 FUNCTIONAL DIVERSITY OF TSAI COMMUNITIES

TSAIs are involved in diverse ecosystem functions in IRES thanks to a high functional diversity derived intrinsically from the dual aquatic-terrestrial nature of IRES. IRES support a wide range of feeding groups, including predators, scavengers, herbivores, detritivores, and parasites (Williams, 1993). Predators and scavengers are often the dominant functional classes of ground-dwelling invertebrates in IRES, especially in rivers with extensive areas of exposed gravel bars such as braided rivers (Fig. 4.4.3; Corti and Datry, unpublished data). Ground-dwelling predators and scavengers are the dominant feeding groups along river shores due to the presence of aquatic subsidies with high energy value. Terrestrial riparian predators inhabiting river banks such as wolf spiders (Lycosidae) and ground beetles (Carabidae) (Fig. 4.4.2) typically prey on emerging aquatic insects, which can make up 85% of their diets (Boumezzough and Musso, 1983; Paetzold et al., 2005; Greenwood and McIntosh, 2010). Aquatic insects may be consumed due to failed emergence; consumption may also include pupal skins. Aquatic insects probably represent a crucial link for dry-riverbed food webs, but effects resulting from the loss of these invertebrates as IRES dry up have not yet been explored.

Herbivores can be abundant in IRES. Many herbivorous species such as pygmy grasshoppers (e.g., *Paratettix aztecus* and *P. mexicanus*; Tetrigidae) inhabit the shoreline of rivers and are likely to colonize the riverbed as it progressively dries. Algal-mat development (Fig. 4.4.4) and vegetation growth may constitute valuable energy sources for these herbivores. Similarly, detritivorous species are likely to be attracted by stranded organic matter such as dead and dying fish and aquatic invertebrates, with substantial consequences for nutrient cycling (Chapter 3.2) and trophic food webs (Chapter 4.7). An experimental study by Steward (2012) using dead fish placed on dry riverbeds in Italy and Australia found that the fish were actively consumed or foraged by ants (Formicidae), beetles (Coleoptera), European wasps (*Vespula germanica*; Vespidae), fly larvae (Diptera), and slugs (Gastropoda). Other taxa associated with the fish but that did not actively consume them included ground beetles (Carabidae), Scarabaeoidea beetles, rove beetles (Staphylinidae), and adult flies (Diptera). Some taxa found on carcasses are, in fact, predators or parasites of the scavenging taxa (Catts and Goff, 1992).

TSAIs can influence the diet of the aquatic fauna after rewetting. The large number of terrestrial invertebrates washed downstream in the advancing wetted fronts probably also influences the diet of downstream fish by providing important energy subsidies and by cascade chains, which could affect the river food web dynamics (Nakano et al., 1999; Chapter 4.7). For example, during flow pulses and submersion of riverine habitats, fish move toward the inundated floodplain and seasonal tributaries, benefiting from the higher densities of terrestrial invertebrate prey (Limm and Marchetti, 2009; Eberle and Stanford, 2010). In IRES, fish survive drying periods by migrating toward perennial refugia, from which they can recolonize surrounding habitats upon rewetting (Davey and Kelly, 2007). Importantly, the amount of terrestrial invertebrates and particulate organic matter swept by advancing wetted fronts may constitute valuable food resources for aquatic food webs at downstream confluences and reservoirs, as well as a feeding bonanza for birds, reptiles, and small mammals (Chapter 4.6), potentially providing a substantial transfer of energy into the adjacent riparian and terrestrial zones. Thus, the exchanges of material and organisms across aquatic and terrestrial ecosystems may occur far away from the zones where they originated.

TSAIs also play significant functional roles in IRES by processing organic matter and altering the structural complexity of the substrate. TSAIs such as earthworms (Oligochaeta) can recycle nutrients within the riverbed during the dry phase. Earthworms also enhance water infiltration through tunneling and burrowing (Pimentel et al., 1995). As mentioned earlier, *Mecynotarsus* (Anthicidae) beetles use their pronotal horn to dig through sand (Fig. 4.4.5; Hashimoto and Hayashi, 2012), and it is expected that these beetles and other TSAI taxa that tunnel, burrow, or dig into riverbeds have similar effects to those of earthworms.

4.4.5 ADAPTATIONS OF TSAI TO FLOW INTERMITTENCE

While drying is a primary driver of aquatic invertebrate community composition (Datry et al., 2014a), inundation is arguably a strong driver of TSAI composition in IRES, as reported in riparian zones (Lambeets et al., 2008), alluvial forests (Bonn and Schröder, 2001), floodplains (Junk et al., 1989; Ballinger et al., 2005; Datry et al., 2014a), and wetlands (Plum, 2005). Flow velocities during the first flush of river rewetting can be as high as $3\,\mathrm{ms}^{-1}$, with most values in the range of $1–2\,\mathrm{ms}^{-1}$ (Jahns, 1949; Sharma et al., 1984; Reid et al., 1998; Jacobson et al., 2000; Doering et al., 2007). During these sudden terrestrial-aquatic transitions, terrestrial invertebrates can be entrained by swiftly advancing wetted fronts (Jacobson et al., 2000; Larned et al., 2010) that move too fast to allow terrestrial invertebrates to escape to adjacent riparian areas (Boumezzough and Musso, 1983). These events are catastrophic for TSAIs if they are not followed rapidly by drying because some TSAI taxa have a relatively low resistance to submersion (Chapter 4.8), and nonaerial TSAIs often have poor dispersal capacities in aquatic environments.

However, rewetting is not necessarily catastrophic for all TSAIs, and semiaquatic taxa probably survive rewetting better than fully terrestrial taxa. Resistance and resilience traits to flooding have been identified for many terrestrial taxa from riparian habitats such as ground beetles (Carabidae), spiders (Araneae), and true bugs (Hemiptera), and these traits appear to also apply in IRES. Resistance traits include mechanisms such as flotation, swimming, flight, climbing onto floating organic matter, and respiration through air bubbles (Andersen, 1968; Boumezzough and Musso, 1983; Lytle and White, 2007; Lambeets et al., 2008). Some ground beetles (Carabidae) can survive more than 20 hours completely submerged (Boumezzough and Musso, 1983; Lambeets et al., 2008; Kolesnikov et al., 2012). In the Albarine River in France, at least one-third of the 71 terrestrial taxa washed downstream in an advancing wetted front survived submersion and could potentially colonize downstream riparian habitats (Corti and Datry, 2012).

Resistance can also involve behavioral mechanisms. For example, in desert streams, some hemipterans use rainfall cues to anticipate and escape flash floods (Lytle, 1999; Lytle and Smith, 2004). In addition, many terrestrial taxa are strong fliers (Bonn, 2000); most ground beetles (Carabidae) are able to fly (Boiteau et al., 2000), some for distances exceeding 1 km (Meijer, 1974; Kotze, 2008). Therefore, reestablishment of terrestrial communities upon drying may be rapid (Hering et al., 2004). In the Selwyn River (New Zealand) for instance, the density of ground-dwelling invertebrates showed a steady fivefold increase during the 90 days following complete riverbed drying, whereas taxonomic richness rapidly increased 1.5-fold within the first 7 days of drying (Corti et al., unpublished data).

River drying and decreasing soil moisture also have predictable consequences for TSAI physiology (Hadley, 1994) and behavior (Davis and DeNardo, 2009), and may particularly influence communities in dryland regions (McCluney and Sabo, 2012). For instance, drying reduced the diversity and altered the composition of riparian terrestrial communities in a desert stream, primarily as a result of decreasing water availability (McCluney and Sabo, 2012). However, drying also affects TSAIs by reducing aquatic prey availability. For example, as aquatic prey became rarer during flow cessation in IRES in southern New Zealand, there was increased competition and presumably cannibalism, decreases in daily prey consumption, as well as a subsequent change in size class and spatial structures of pisaurid spider (*Dolomedes aquaticus*) populations (Greenwood and McIntosh, 2010). During dry periods, TSAIs may switch to a diet composed mainly of terrestrial prey (Sabo and Power, 2002; Briers et al., 2005). River regulation of the Rio Grande, New Mexico, led to a twofold decrease in the densities of aquatic invertebrates, and this subsequently induced terrestrial communities to be less diverse and abundant along riparian zones of regulated compared with unregulated sections (Kennedy and Turner, 2011). It is likely that direct and indirect effects of drying interact synergistically. For instance, drying directly reduces the number of terrestrial riparian prey and predators and indirectly reduces the number of terrestrial predators through the reduction of aquatic prey (Hagen and Sabo, 2012; Allen et al., 2013). Hence, the effects of drying may be particularly significant for predaceous species.

The effects of river drying on TSAIs are likely to depend on climate. In the temperate Albarine River in France, Corti and Datry (2014) found only weak effects of river drying on riparian arthropods at the community level, and taxonomic richness was even higher at intermittent sites compared to perennial sites. Sufficient inputs of rainwater combined with hot summer temperatures in the Albarine River may stimulate a pulse in riparian primary production, which could in turn support riparian primary consumers (Marczak et al., 2007; Klemmer and Richardson, 2013). Abundant primary consumers could allow some predators to be less dependent upon aquatic prey or to shift their diet from aquatic to terrestrial prey as surface water disappears (Paetzold et al., 2005). In addition, rain, dew, and vegetation cover can provide sufficient moisture and drinking water for riparian arthropods during the dry phase. Another possible reason for the weak effect of river drying in the Albarine River could be associated with life history characteristics. Activity patterns of TSAIs may align with unfavorable seasonal conditions, where species may have completed their seasonal life cycle in preparation for subsequent diapause before the dry phase (Lovei and Sunderland, 1996). Life history characteristics may thus favor the resistance of TSAIs to the decline in aquatic resources and changes in environmental conditions when the river dried.

TSAIs may not have strong relationships with the spatial or temporal variability of drying events, unlike aquatic invertebrates (Chapter 4.3). This difference is due to the resistance and resilience traits of TSAIs, which enable the invertebrates to seek refuge in, and recolonize from, relatively stable upland habitats. Nonetheless, flow cessation and rewetting events affect TSAI assemblages in predictable ways. For example, long flow durations may eliminate TSAIs that lack inundation resistance, whereas rewetting and drying events interrupt and reset terrestrial taxa succession.

4.4.6 THREATS TO THE TSAI COMMUNITIES OF IRES

Human activities that change the environmental conditions of IRES are likely to influence TSAI communities (Chapter 5.1). Grazing, weed invasion, siltation, and hydrological alterations can impact rivers and streams, the shoreline, gravel bars, and riparian zones during the wet phase (Balneaves and Hughey, 1990; Wood and Armitage, 1997; Nilsson et al., 2005; Bates et al., 2007; Sadler and Bates, 2008), and are likely stressors on dry riverbeds during the dry phase (Fig. 4.4.7). Trampling by livestock during the dry phase may alter and compact the riverbed sediments, siltation may reduce interstitial spaces through in-filling, and weed invasion can alter the vegetation and litter cover of the riverbed, possibly affecting the quality of dry riverbeds as habitats. IRES can also be degraded when their channels are used as roads (Fig. 4.3.7 in Chapter 4.3) or for recreational purposes such as walking tracks.

Reduced flood frequency has negatively impacted the resting stages of aquatic biota of IRES (Jenkins and Boulton, 2007) and may have negative effects on the habitat and diversity of TSAIs through increased vegetation cover. Conversely, permanent wetting ("perennialization," Chapter 5.1) after the construction of instream barriers (dams or weirs), interbasin transfers, wastewater discharge, and runoff will be detrimental to TSAIs, particularly the terrestrial invertebrates of dry riverbeds because dry-riverbed habitat is eliminated altogether. Similarly, the increased frequency and duration of dry periods may impact dry-riverbed invertebrates by reducing the opportunities for predatory and scavenging taxa to consume stranded aquatic animals and plants, which may be important for their survival or recruitment. Shifts from perennial to intermittent flow as a result of climate change and water extraction (Chapter 5.1) may favor the TSAIs of dry riverbeds as the extent of aquatic habitat declines and terrestrial habitat expands.

Results of climate change modeling scenarios predict global surface temperatures to increase by 1–4°C during the 21st century (Meehl et al., 2007). These changes may ultimately impact TSAI assemblages because temperatures in dry riverbeds can exceed the thermal tolerances of many species; therefore, future temperature increases may extend the duration of periods when dry riverbeds are

FIG. 4.4.7

Threats to TSAIs of IRES: (a) cattle grazing along a riverbed, (b) cattle hoof prints in a dry riverbed, (c) weed infestation (*Lantana camara*) (road bridge in background), and (d) a road crossing as a barrier between dry and wet habitats.

Photos: Courtesy A. Steward (a, c, and d); J. Marshall (b).

inhospitable to most invertebrates. The combined effects of climate change and water management practices may increase or decrease the duration of the wet and dry phases in rivers (Jackson et al., 2001; Chiew and McMahon, 2002; Lehner et al., 2006), with unknown impacts on TSAI taxa. One reason that IRES are at risk of degradation is that the diversity of habitats and their associated fauna are not recognized in the majority of river management policies (Chapter 5.3). Understanding of the responses of TSAIs to prolonged dry and wet conditions and higher temperatures is currently limited, and many knowledge gaps exist (Section 4.4.8).

4.4.7 MANAGING IRES TO PRESERVE TSAI DIVERSITY AND THEIR ECOLOGICAL FUNCTIONS

Although IRES currently comprise more than half of the global river network (Datry et al., 2014b) and provide unique habitats that support TSAI species of high conservation value (Sadler et al., 2004; Bates et al., 2009; Langhans and Tockner, 2014b), recognition of their conservation significance is still limited (Acuña et al., 2014). Consequently, examples of catchment-scale conservation and adequate management of IRES are rare (Leigh et al., 2016; Chapters 5.4 and 5.5). This is because neither the diversity of habitats nor the associated fauna (aquatic, semiaquatic, and terrestrial) of IRES are currently recognized in the majority of protective policies and legislations (Acuña et al., 2014; Chapter 5.3). The European Water Framework Directive (WFD; European Commission, 2000), for instance, does not discriminate between temporary and perennial rivers, thereby ignoring the existence of IRES altogether. As a first step toward acknowledging the value of IRES and their characteristics, the set of river types in the WFD needs to be extended to explicitly include IRES, and adaptive assessment methods need to be developed that consider the variability in hydrological conditions (Nikolaidis et al., 2013; Prat et al., 2014).

IRES have mostly been ignored in ecological river assessment and monitoring programs. Assessing the ecological condition of these systems remains difficult (Datry et al., 2011) as is determining their natural or "reference" conditions (Prat et al., 2014). This, together with the current lack of knowledge of how to differentiate between the effect of natural and anthropogenic causes of intermittent flow on biodiversity, may be reasons why IRES have so far mostly been ignored (Boulton, 2014; Mazor et al., 2014; Chapter 5.4). Many studies stress the need for new or modified assessment methods that can deal with these challenges (Boulton et al., 2000; Sheldon, 2005; Dallas, 2013). Leigh et al. (2016) list TSAIs as well as microbial biomarkers (Wilkes et al., 2013) and hyporheic invertebrates (Leigh et al., 2013) as promising bioindicators of ecological condition for IRES.

Since TSAIs persist long after hyporheic refuges cease to exist, they may be the most promising avenue for the future biological assessment of IRES, together with assessing the tolerance and preferences of aquatic invertebrates (Chessman and Royal, 2004). Criteria for selecting appropriate biological indicators (Andersen, 1999) relate to an indicator's (1) distribution, abundance, and richness; (2) functional importance in ecosystems; (3) sensitivity to environmental change; (4) ability to be sampled, sorted, and identified; and (5) responses to change. TSAIs meet almost all of these criteria as they are found in IRES all over the world, can be present in high abundances, are taxonomically diverse, and can be easily and cheaply sampled (Box 4.2.2; Fig. 4.4.8). Andersen (1999: 61) argues that "a primary challenge with indicator taxa consequently lies in distinguishing anthropogenic perturbation (i.e., an ecologically meaningful "signal") from natural variability (background "noise")." It is therefore pertinent to investigate whether TSAIs respond to anthropogenic disturbance in order to meet the fifth criterion discussed previously. Recent results show that this is the case, with TSAIs negatively responding to the impact of livestock and feral mammals (Steward et al., 2016).

One step toward using TSAIs as biological indicators is to gather spatial data on their distribution across different biomes. These data can be used to analyze the links between the occurrence of TSAI and the health status of IRES. The flow regime, water quality, and catchments of many IRES have been subject to profound, long-term human modification (Chapter 5.1), and the systems that we assess now and into the future will probably have the characteristics of "novel" ecosystems (Hobbs et al., 2006) and differ substantially from their unmodified condition. As restoring such systems back to their "natural conditions" is usually impossible, Leigh et al. (2016) suggest establishing realistic policy goals that acknowledge this fact. Nonetheless, managing IRES adequately will benefit TSAIs. Efforts should focus on improving or maintaining local and regional habitat heterogeneity through the promotion of natural flow and sediment regimes (Bonn and Kleinwächter, 1999; Eyre et al., 2001, 2002; Manderbach and Framenau, 2001; Adis and Junk, 2002; Sadler et al., 2004). A catchment-level approach seems to be especially important; meta-populations of IRES specialists will only be sustainable if unconstrained floodplain segments, multiple gravel bars, and associated ecotones are available (Sadler et al., 2004).

BOX 4.4.2 SAMPLING APPROACHES FOR TSAIs

TSAIs can be sampled actively by hand or passively using pitfall traps (e.g., Steward et al., 2011; Steward, 2012; Corti et al., 2013). Each method has its advantages and disadvantages (Table 4.4.2). Both methods are relatively inexpensive and easy to use, and can be performed along the water's edge, in gravel bars, in riparian zones, or within dry riverbeds.

Hand collection is performed in a defined area for a defined period of time and is done using forceps, and an aspirator (Fig. 4.4.8a) to collect the invertebrates. Head torches can be used at night. The area can be defined by pushing a quadrat into the substrate. Invertebrates can be found by turning over the substrate and debris, such as rocks and leaves, down to a sediment depth of approximately 10cm.

Pitfall traps are plastic containers dug into the substrate until they are flush with the surrounding surface (Fig. 4.4.8b). Plastic plates mounted on sticks can be assembled over the openings (Fig. 4.4.8b) to keep rain and debris out. Ethylene glycol or ethanol with added glycerol can be added to the containers as a preservative and a killing agent. Saline solution can also be used. Hard substrates, such as bedrock or concrete, may require a modified version of the method, such as providing a ramp up to the traps. Replicate traps need to be used at each site.

Samples from both methods are taken back to the laboratory for analysis. A microscope is used to identify the specimens.

Table 4.4.2 Sampling approaches for TSAIs of IRES and their advantages and disadvantages

	Hand collection	Pitfall trap collection
Equipment	Quadrat frame, forceps, aspirator (Fig. 4.4.8a), collection vials, preservative, labels	Small shovel, plastic containers, preservative, plastic plates and skewers for rain covers, labels (Fig. 4.4.8b)
Cost	Equipment is inexpensive	Equipment is inexpensive
		Requires a return visit to the site, doubling travel costs
Effort and efficiency	Invertebrates are collected from a defined area, for a defined period of time	Time is required to set up and retrieve each trap. Traps are typically set for 24h or longer
		Can be time consuming in hard substrates where trap placement can take longer
		Some traps may not "work" and may need to be excluded from analysis due to interference by animals, etc.
Interference		Animals, vandals, and storms can interfere with and damage the traps
Bias	Only collects invertebrates that can be easily seen and easily collected—may not collect small invertebrates such as springtails (Collembola) and mites (Acarina). May miss invertebrates living in cracks, etc.	Only collects active invertebrates
		Some killing agents and/or preservatives can attract or repel invertebrates
		Traps could interfere with each other
Weather	Difficult to hand-collect in bad weather due to reduced visibility and activity of invertebrates, e.g., rain, extreme heat	Traps can collect animals during bad weather; however, the resulting sample may differ to samples conducted during more favorable times

Continued

BOX 4.4.2 SAMPLING APPROACHES FOR TSAIs *(Cont'd)*

Table 4.4.2 Sampling approaches for TSAIs of IRES and their advantages and disadvantages—cont'd

	Hand collection	Pitfall trap collection
Timing	Typically restricted to day time. Difficult to do at night—would require a strong light source/head torches in order to see, which could attract or repel invertebrates	Collects invertebrates during the day and night
	Morning collections may collect a different assemblage than midday or afternoon collections	
Dangers and safety	Risk of being bitten by invertebrates, e.g., spiders	Injuries caused from digging holes for traps
	Can be physically demanding, particularly in hot, humid climates. Can cause sore knees—need to wear knee protection. Can also cause sore backs from bending over	

Modified from Steward, A.L., 2012. When the River Runs Dry: The Ecology of Dry River Beds. PhD Thesis, Griffith University, Brisbane.

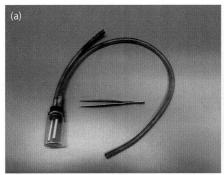 

FIG. 4.4.8

Equipment used to collect TSAIs: (a) forceps and an aspirator, consisting of a glass vial, rubber bung, aluminum intake, and suction tubing, and (b) a pitfall trap set into a cobble-dominated riverbed with a plastic cover above it.

Photos: Courtesy A. Steward

Successful management of IRES should preserve the ecological functions that TSAIs provide. Aside from TSAIs contributing to the biological diversity of IRES, TSAIs have been shown in this chapter to play other important roles such as providing food for terrestrial mammals, birds, reptiles, and amphibians during the dry phase, and food for aquatic biota once the riverbed has rewetted. In this way, TSAIs underpin crucial trophic linkages between the wet and dry phases of IRES. TSAIs can consume stranded and emerging aquatic biota, transferring energy from aquatic systems to the adjacent riparian zones. TSAIs also recycle nutrients within the riverbed substrate and enhance water infiltration through tunneling, burrowing, and digging. Thus, anthropogenic impacts on IRES that reduce the abundance and diversity of TSAIs will alter the linkages between aquatic and terrestrial food webs and limit the availability of TSAIs as food for terrestrial animals. The consequence of alterations to TSAI communities and the ecological functions that they perform has yet to be investigated.

4.4.8 KNOWLEDGE GAPS AND RESEARCH NEEDS

To advance our understanding of terrestrial community ecology in IRES, several research gaps need to be addressed (Table 4.4.3). TSAIs have only been collected from a handful of sites and countries around the world, despite IRES being present on every continent (Chapter 1). The unique diversity of

Table 4.4.3 Summary of knowledge gaps and related research questions for TSAIs of IRES

Knowledge gap	Research questions
Unique diversity	How does the diversity of TSAIs differ between countries and climate types that have previously been overlooked?
	How does TSAI community composition differ through time?
	Are there seasonal or annual differences in taxonomic richness or abundance?
	How do TSAI assemblages differ between habitat types?
	Can TSAIs inhabit impervious substrates such as bedrock systems or artificial substrates such as concrete channels?
Corridors for TSAIs	How important is the channel as a dispersal corridor for TSAIs?
	Is rafting during flood events an important dispersal mode for maintaining the viability of populations of TSAIs?
Refuge for TSAIs	To what depth do TSAIs inhabit riverine sediments?
	Do IRES sediments act as a refuge for TSAIs during floods?
	Are algal mats on dry riverbeds used by TSAIs as a temperature and/or humidity refuge?
Effects of drying and flooding on TSAIs	How long can quiescent stages of semiaquatic invertebrate biota remain viable in dry riverbeds?
	How are aquatic–terrestrial interactions affected by drying events in IRES?
	During drying, is there a pulse of terrestrial invertebrates moving in from adjoining habitats to take advantage of the increased food availability?
	Are there critical thresholds in the duration, spatial extent, and severity of drying in IRES that may lead to fundamental shifts in the community structure of TSAIs?
	How long can terrestrial invertebrates tolerate inundation?
TSAIs in a landscape setting	How does the diversity of TSAIs differ throughout a river network (headwaters to lowlands)? Is this variation consistent over time?
Impacts on TSAIs and IRES	How do TSAIs respond to natural and human disturbances?
	How will climate change affect the distribution of TSAIs?
	Do changing flow regimes increase the susceptibility of IRES to invasions from exotic TSAI species?
	If TSAI assemblages are altered, how does that affect the ecological functioning of IRES?
TSAIs as biological indicators	Can TSAIs be used to monitor and assess the health of IRES?
	What biological indices should be calculated for monitoring and assessment purposes?

TSAIs should be further investigated by sampling in other countries and climate types, and also from habitats that have previously been overlooked. For instance, we do not know to what depth TSAIs inhabit riverine sediments. We also do not know how TSAI assemblages differ between habitat types, or whether TSAIs can inhabit impervious substrates (such as bedrock-controlled IRES) or artificial substrates such as concrete channels. TSAI community composition probably varies through time. Are there seasonal or annual differences in taxonomic richness or abundance? How will climate change affect these changes and the resulting distribution of TSAIs?

IRES sediments may act as a refuge for TSAIs during floods. Once the floodwaters subside, does recolonization of the riverbed occur from within the riverine sediments, from the riparian zone, or from somewhere else? We need to explore the traits that TSAIs utilize to survive both inundation and drying. How long can terrestrial invertebrates tolerate inundation, and how long can quiescent stages of semiaquatic invertebrate biota remain viable in dry riverbeds? How will changes in flow regime and hydrological connectivity in IRES influence these taxa? The importance of the channel as a dispersal corridor for TSAIs also needs to be determined. Is rafting during flood events an important dispersal mode for maintaining the viability of populations of TSAIs?

How are aquatic-terrestrial interactions affected by drying events in IRES? This could be investigated by combining stable isotope analysis with manipulative experiments. We may expect to see a pulse of terrestrial invertebrates moving in from adjoining habitats to take advantage of the increased food availability. We also need to understand how TSAI communities differ at the catchment scale. How does the diversity of TSAIs differ throughout a river network from headwaters to lowlands? Are these differences consistent among IRES with similar spatial patterns of drying?

Finally, the use of TSAIs as biological indicators of IRES health has been suggested (Steward et al., 2011, 2012). We first need to understand how TSAIs respond to natural and human disturbances. If TSAI assemblages respond differentially, how do these different responses affect the ecological functioning of IRES? Maybe manipulative experiments that change the taxonomic composition or abundance of TSAIs in an IRES could illustrate roles played by potential bioindicators?

4.4.9 CONCLUSIONS

TSAIs are important ecological, functional, and taxonomic components of IRES that have previously been overlooked. Recent studies have revealed a unique diversity of TSAIs that inhabit IRES during flow, no-flow, and dry events. However, TSAI faunal communities are at risk of degradation due to limited policies and legislation to protect them and their IRES habitats. TSAIs are currently being considered as biological indicators to monitor and assess the health of IRES. Future strategies in IRES research and management need to consider the TSAI fauna in addition to the routinely sampled aquatic fauna. Research questions addressing the response of TSAIs to anthropogenic impacts should be given high priority and the results used to inform IRES policy and management.

REFERENCES

Acuña, V., Datry, T., Marshall, J., Barceló, D., Dahm, C.N., Ginebreda, A., et al., 2014. Why should we care about temporary waterways? Science 343, 1080–1081.

Adis, J., Junk, W.J., 2002. Terrestrial invertebrates inhabiting lowland river floodplains of Central Amazonia and Central Europe: a review. Freshw. Biol. 47, 711–731.

Allen, D.C., McCluney, K.E., Elser, S.R., Sabo, J.L., 2013. Water as a trophic currency in dryland food webs. Front. Ecol. Environ. 12, 156–160.

Andersen, J., 1968. The effect of inundation and choice of hibernation sites of Coleoptera living on river banks. Norsk Entomologisk Tidsskrift 14, 115–133.

Andersen, A.N., 1999. My bioindicator or yours? Making the selection. J. Insect Conserv. 3, 61–64.

Ballinger, A., Nally, R.M., Lake, P.S., 2005. Immediate and longer-term effects of managed flooding on floodplain invertebrate assemblages in south-eastern Australia: generation and maintenance of a mosaic landscape. Freshw. Biol. 50, 1190–1205.

Balneaves, J.M., Hughey, K., 1990. The need for control of exotic weeds in braided river beds for conservation of wildlife. In: Proceedings of the 9th Australian Weeds Conference, Adelaide, South Australia, pp. 103–108.

Bates, A.J., Sadler, J.P., Fowles, A.P., 2007. Livestock trampling reduces the conservation value of beetle communities on high quality exposed riverine sediments. Biodivers. Conserv. 16, 1491–1509.

Bates, A.J., Sadler, J.P., Henshall, S., Hannah, D.M., 2009. Ecology and conservation of arthropods of exposed riverine sediments (ERS). Terrestrial Arthropod Rev. 2, 77–98.

Blackith, R., Guillet, A., 1995. The insect fauna of an erratically dry African river bed. Entomol. Monthly Mag. 131, 65–67.

Boiteau, G., Bousquet, Y., Osborn, W., 2000. Vertical and temporal distribution of Carabidae and Elateridae in flight above an agricultural landscape. Environ. Entomol. 29, 1157–1163.

Bonn, A., 2000. Flight activity of carabid beetles on a river margin in relation to fluctuating water levels. In: Brandmayr, P., Lövey, G.L., Zetto Brandmayr, T., Casale, A., Vigna Taglianti, A. (Eds.), Natural History and Applied Ecology of Carabid Beetles. Pensoft, Sofia, pp. 147–160.

Bonn, A., Kleinwächter, M., 1999. Microhabitat distribution of spider and ground beetle assemblages (Araneae, Carabidae) on frequently inundated river banks of the River Elbe. Zeitschrift für Ökologie und Naturschutz 8, 109–123.

Bonn, A., Schröder, B., 2001. Habitat models and their transfer for single and multi species groups: a case study of carabids in an alluvial forest. Ecography 24, 483–496.

Boulton, A.J., 2003. Parallels and contrasts in the effects of drought on stream macroinvertebrate assemblages. Freshw. Biol. 48, 1173–1185.

Boulton, A.J., 2014. Conservation of ephemeral streams and their ecosystem services: what are we missing? Aquat. Conserv. Mar. Freshwat. Ecosyst. 24, 733–738.

Boulton, A.J., Suter, P.J., 1986. Ecology of temporary streams–an Australian perspective. In: De Deckker, P., Williams, W.D. (Eds.), Limnology in Australia. CSIRO/Dr. W. Junk, Melbourne/Dordrecht, pp. 313–327.

Boulton, A.J., Sheldon, F., Thoms, M.C., Stanley, E.H., 2000. Problems and constraints in managing rivers with variable flow regimes. In: Boon, P.J., Davies, B.R., Petts, G.E. (Eds.), Global Perspectives on River Conservation: Science, Policy and Practice. John Wiley and Sons, London, pp. 411–426.

Boumezzough, A., Musso, J.-J., 1983. Etude des communautés animales ripicoles du bassin de la rivière Aille (Var-France). I. Aspects biologiques et éco-éthologiques. Ecologia Mediterranea 9, 31–56.

Briers, R.A., Cariss, H.M., Geoghegan, R., Gee, J.H.R., 2005. The lateral extent of the subsidy from an upland stream to riparian lycosid spiders. Ecography 28, 165–170.

Catts, E.P., Goff, M.L., 1992. Forensic entomology in criminal investigations. Annu. Rev. Entomol. 37, 253–272.

Chessman, B.C., Royal, M.J., 2004. Bioassessment without reference sites: use of environmental filters to predict natural assemblages of river macroinvertebrates. J. N. Am. Benthol. Soc. 23, 599–615.

Chiew, F.H.S., McMahon, T.A., 2002. Modelling the impacts of climate change on Australian streamflow. Hydrol. Process. 16, 1235–1245.

Corti, R., Datry, T., 2012. Invertebrates and sestonic matter in an advancing wetted front travelling down a dry river bed (Albarine, France). Freshwater Sci. 31, 1187–1201.

Corti, R., Datry, T., 2014. Drying of a temperate, intermittent river has little effect on adjacent riparian arthropod communities. Freshw. Biol. 59, 666–678.

Corti, R., Datry, T., 2016. Terrestrial and aquatic invertebrates in the riverbed of an intermittent river: parallels and contrasts in community organisation. Freshw. Biol. 61, 1308–1320.

Corti, R., Larned, S.T., Datry, T., 2013. A comparison of pitfall-trap and quadrat methods for sampling ground-dwelling invertebrates in dry riverbeds. Hydrobiologia 717, 13–26.

Dallas, H., 2013. Ecological status assessment in Mediterranean rivers: complexities and challenges in developing tools for assessing ecological status and defining reference conditions. Hydrobiologia 719, 483–507.

Datry, T., Arscott, D.B., Sabater, S., 2011. Recent perspectives on temporary river ecology. Aquat. Sci. 73, 453–457.

Datry, T., Corti, R., Belletti, B., Piégay, H., 2014a. Ground-dwelling arthropod communities across braided river landscape mosaics: a Mediterranean perspective. Freshw. Biol. 59, 1308–1322.

Datry, T., Larned, S.T., Tockner, K., 2014b. Intermittent rivers: a challenge for freshwater ecology. Bioscience 64, 229–235.

Datry, T., Pella, H., Leigh, C., Bonada, N., Hugueny, B., 2016. A landscape approach to advance intermittent river ecology. Freshw. Biol. 61, 1200–1213.

Davey, A.J.H., Kelly, D.J., 2007. Fish community responses to drying disturbances in an intermittent stream: a landscape perspective. Freshw. Biol. 52, 1719–1733.

Davey, A.J.H., Kelly, D.J., Biggs, B.J.F., 2006. Refuge-use strategies of stream fishes in response to extreme low flows. J. Fish Biol. 69, 1047–1059.

Davis, J.R., DeNardo, D.F., 2009. Water supplementation affects the behavioral and physiological ecology of Gila monsters (*Heloderma suspectum*) in the Sonoran Desert. Physiol. Biochem. Zool. 82, 739–748.

Dell, A.I., Alford, R.A., Pearson, R.G., 2014. Intermittent pool beds are permanent cyclic habitats with distinct wet, moist and dry phases. PLoS One 9, e108203.

Doering, M., Uehlinger, U., Rotach, A., Schlaepfer, D.R., Tockner, K., 2007. Ecosystem expansion and contraction dynamics along a large Alpine alluvial corridor (Tagliamento River, Northeast Italy). Earth Surf. Process. Landf. 32, 1693–1704.

Eberle, L.C., Stanford, J.A., 2010. Importance and seasonal availability of terrestrial invertebrates as prey for juvenile salmonids in floodplain spring brooks of the Kol River (Kamchatka, Russian Federation). River Res. Appl. 26, 682–694.

European Commission, 2000. Directive 2000/60/EC of the European Parliament and of the Council of 23 October 2000 establishing a framework for community action in the field of water policy. Off. J. Eur. Communities 43 (L327), 1–75.

Eyre, M.D., Luff, M.L., Phillips, D.A., 2001. The ground beetles (Coleoptera: Carabidae) of exposed riverine sediments in Scotland and northern England. Biodivers. Conserv. 10, 403–426.

Eyre, M.D., Woodward, J.C., Luff, M.L., 2002. The spider assemblages (Araneae) of exposed riverine sediments in Scotland and northern England. Bull. Br. Arachnol. Soc. 12, 287–294.

Greenwood, M.J., McIntosh, A.R., 2010. Low river flow alters the biomass and population structure of a riparian predatory invertebrate. Freshw. Biol. 55, 2062–2076.

Hadley, N.F., 1994. Water Relations of Terrestrial Arthropods. Academic Press, San Diego, CA.

Hagen, E.M., Sabo, J.L., 2012. Influence of river drying and insect availability on bat activity along the San Pedro River, Arizona (USA). J. Arid Environ. 84, 1–8.

Hashimoto, K., Hayashi, F., 2012. Structure and function of the large pronotal horn of the sand-living anthicid beetle *Mecynotarsus tenuipes*. Entomol. Sci. 15, 274–279.

Hering, D., Gerhard, M., Manderbach, R., Reich, M., 2004. Impact of a 100-year flood on vegetation, benthic invertebrates, riparian fauna and large woody debris standing stock in an alpine floodplain. River Res. Appl. 20, 445–457.

Hobbs, R.J., Arico, S., Aronson, J., Baron, J.S., Bridgewater, P., Cramer, V.A., et al., 2006. Novel ecosystems: theoretical and management aspects of the new ecological world order. Glob. Ecol. Biogeogr. 15, 1–7.

Hölldobler, B., Wilson, E.O., 1990. The Ants. Harvard University Press, Harvard, MA.

Holm, E., Edney, E.B., 1973. Daily activity of Namib Desert arthropods in relation to climate. Ecology 54, 45–56.

Jackson, R.B., Carpenter, S.R., Dahm, C.N., McKnight, D.M., Naiman, R.J., Postel, S.L., et al., 2001. Water in a changing world. Ecol. Appl. 11, 1027–1045.

Jacobson, P.J., Jacobson, K.M., Angermeier, P.L., Cherry, D.S., 2000. Variation in material transport and water chemistry along a large ephemeral river in the Namib Desert. Freshw. Biol. 44, 481–491.

Jahns, R.H., 1949. Desert floods. Eng. Sci. 12, 10–14.

Jenkins, K.M., Boulton, A.J., 2007. Detecting impacts and setting restoration targets in arid-zone rivers: aquatic micro-invertebrate responses to reduced floodplain inundation. J. Appl. Ecol. 44, 823–832.

Junk, W.J., Bayley, P.B., Sparks, R.E., 1989. The flood pulse concept in river-floodplain systems. Can. Spec. Publ. Fish. Aquat. Sci. 106, 110–127.

Kennedy, T.L., Turner, T.F., 2011. River channelization reduces nutrient flow and macroinvertebrate diversity at the aquatic terrestrial transition zone. Ecosphere 2, 1–13.

Klemmer, A.J., Richardson, J.S., 2013. Quantitative gradient of subsidies reveals a threshold in community-level trophic cascades. Ecology 94, 1920–1926.

Kolesnikov, F.N., Karamyan, A.N., Hoback, W.W., 2012. Survival of ground beetles (Coleoptera: Carabidae) submerged during floods: field and laboratory studies. Eur. J. Entomol. 109, 71–76.

Kotze, D.J., 2008. The occurrence and distribution of carabid beetles (Carabidae) on islands in the Baltic Sea: a review. J. Insect Conserv. 12, 265–276.

Lalley, J.S., Viles, H.A., Henschel, J.R., Lalley, V., 2006. Lichen-dominated soil crusts as arthropod habitat in warm deserts. J. Arid Environ. 67, 579–593.

Lambeets, K., Vandegehuchte, M.L., Maelfait, J.-P., Bonte, D., 2008. Understanding the impact of flooding on trait-displacements and shifts in assemblage structure of predatory arthropods on river banks. J. Anim. Ecol. 77, 1162–1174.

Langhans, S.D., 2006. Riverine Floodplain Heterogeneity as a Controller of Organic Matter Dynamics and Terrestrial Invertebrate Distribution. Swiss Federal Institute of Technology, Zurich. PhD Thesis.

Langhans, S.D., Tockner, K., 2014a. Is the unsaturated sediment a neglected habitat for riparian arthropods? Evidence from a large gravel-bed river. Glob. Ecol. Conserv. 2, 129–137.

Langhans, S.D., Tockner, K., 2014b. Edge effects are important in supporting beetle biodiversity in a gravel-bed river floodplain. PLoS One 9, e114415.

Larned, S.T., Datry, T., Robinson, C.T., 2007. Invertebrate and microbial responses to inundation in an ephemeral river reach in New Zealand: effects of preceding dry periods. Aquat. Sci. 69, 554–567.

Larned, S.T., Datry, T., Arscott, D.B., Tockner, K., 2010. Emerging concepts in temporary-river ecology. Freshw. Biol. 55, 717–738.

Lehner, B., Doell, P., Alcamo, J., Henrichs, T., Kaspar, F., 2006. Estimating the impact of global change on flood and drought risks in Europe: a continental, integrated analysis. Clim. Chang. 75, 273–299.

Leigh, C., Stubbington, R., Sheldon, F., Boulton, A.J., 2013. Hyporheic invertebrates as bioindicators of ecological health in temporary rivers: a meta-analysis. Ecol. Indic. 32, 62–73.

Leigh, C., Boulton, A.J., Courtwright, J.L., Fritz, K., May, C.L., Walker, R.H., et al., 2016. Ecological research and management of intermittent rivers: an historical review and future directions. Freshw. Biol. 61, 1181–1199.

Limm, M.P., Marchetti, M.P., 2009. Juvenile Chinook salmon (*Oncorhynchus tshawytscha*) growth in off-channel and main-channel habitats on the Sacramento River, CA using otolith increment widths. Environ. Biol. Fish 85, 141–151.

Lovei, G.L., Sunderland, K.D., 1996. Ecology and behavior of ground beetles (Coleoptera: Carabidae). Annu. Rev. Entomol. 41, 231–256.

Lytle, D.A., 1999. Use of rainfall cues by *Abedus herberti* (Hemiptera: Belostomatidae): a mechanism for avoiding flash floods. J. Insect Behav. 12, 1–12.

Lytle, D.A., Smith, R.L., 2004. Exaptation and flash flood escape in the giant water bugs. J. Insect Behav. 17, 169–178.

Lytle, D.A., White, N.J., 2007. Rainfall cues and flash-flood escape in desert stream insects. J. Insect Behav. 20, 413–423.

Manderbach, R., Framenau, V.W., 2001. Spider (Arachnida: Araneae) communities of riparian gravel banks in the northern parts of the European Alps. Bull. Br. Arachnol. Soc. 12, 1–9.

Marczak, L.B., Hoover, T.M., Richardson, J.S., 2007. Trophic interception: how a boundary-foraging organism influences cross-ecosystem fluxes. Oikos 116, 1651–1662.

Mazor, R.D., Stein, E.D., Ode, P.R., Schiff, K., 2014. Integrating intermittent streams into watershed assessments: applicability of an index of biotic integrity. Freshwater Sci. 33, 459–474.

McCluney, K.E., Sabo, J.L., 2012. River drying lowers the diversity and alters the composition of an assemblage of desert riparian arthropods. Freshw. Biol. 57, 91–103.

Meehl, G.A., Stocker, T.F., Collins, W.D., Friedlingstein, P., Gaye, A.T., Gregory, J.M., et al., 2007. Global climate projections. In: Solomon, S., Qin, D., Manning, M., Chen, Z., Marquis, M., Averyt, K.B., Tignor, M., Miller, H.L. (Eds.), Climate Change 2007: The Physical Science Basis: Contribution of Working Group I to the Fourth Assessment Report of the Intergovernmental Panel on Climate Change (IPCC). Cambridge, Cambridge University Press, pp. 747–845.

Meijer, J., 1974. A comparative study of the immigration of carabids (Coleoptera, Carabidae) into a new polder. Oecologia 16, 185–208.

Moon, H.P., 1956. Observations on a small portion of a drying chalk stream. Proc. Zool. Soc. London 126, 327–334.

Moring, J.B., Stewart, K.W., 1994. Habitat partitioning by the wolf spider (Araneae, Lycosidae) guild in streamside and riparian vegetation zones of the Conejos River, Colorado. J. Arachnol. 22, 205–217.

Nakano, S., Miyasaka, H., Kuhara, N., 1999. Terrestrial-aquatic linkages: riparian arthropod inputs alter trophic cascades in a stream food web. Ecology 80, 2435–2441.

Nikolaidis, N.P., Demetropoulou, L., Froebrich, J., Jacobs, C., Gallart, F., Prat, N., et al., 2013. Towards sustainable management of Mediterranean river basins: policy recommendations on management aspects of temporary streams. Water Policy 15, 830–849.

Nilsson, C., Reidy, C.A., Dynesius, M., Revenga, C., 2005. Fragmentation and flow regulation of the world's large river systems. Science 308, 405–408.

Paetzold, A., Schubert, C., Tockner, K., 2005. Aquatic terrestrial linkages along a braided-river: riparian arthropods feeding on aquatic insects. Ecosystems 8, 748–759.

Paetzold, A., Bernet, J.F., Tockner, K., 2006. Consumer-specific responses to riverine subsidy pulses in a riparian arthropod assemblage. Freshw. Biol. 51, 1103–1115.

Pimentel, D., Harvey, C., Resosudarmo, P., Sinclair, K., Kurz, D., McNair, M., et al., 1995. Environmental and economic costs of soil erosion and conservation benefits. Science 267, 1117–1122.

Plum, N., 2005. Terrestrial invertebrates in flooded grassland: a literature review. Wetlands 25, 721–737.

Prat, N., Gallart, F., Von Schiller, D., Polesello, S., García-Roger, E.M., Latron, J., et al., 2014. The mirage toolbox: an integrated assessment tool for temporary streams. River Res. Appl. 30, 1318–1334.

Reid, I., Laronne, J.B., Powell, D.M., 1998. Flash-flood and bedload dynamics of desert gravel-bed streams. Hydrol. Process. 12, 543–557.

Robson, B.J., Chester, E.T., Austin, C.M., 2011. Why life history information matters: drought refuges and macroinvertebrate persistence in non-perennial streams subject to a drier climate. Mar. Freshw. Res. 62, 801–810.

Sabo, J.L., Power, M.E., 2002. Numerical response of lizards to aquatic insects and short-term consequences for terrestrial prey. Ecology 83, 3023–3036.

Sadler, J.P., Bates, A.J., 2008. The ecohydrology of invertebrates associated with exposed riverine sediments. In: Wood, P.J., Hannah, D.M., Sadler, J.P. (Eds.), Hydroecology and Ecohydrology. John Wiley and Sons, Chichester, pp. 37–56.

Sadler, J.P., Bell, D., Fowles, A., 2004. The hydroecological controls and conservation value of beetles on exposed riverine sediments in England and Wales. Biol. Conserv. 118, 41–56.

Sánchez-Montoya, M.D.M., Schiller, D., Ruhí, A., Pechar, G.S., Proia, L., Miñano, J., et al., 2016. Responses of ground-dwelling arthropods to surface flow drying in channels and adjacent habitats along Mediterranean streams. Ecohydrology 9, 1376–1387.

Sato, T., Watanabe, K., Kanaiwa, M., Niizuma, Y., Harada, Y., Lafferty, K.D., 2011. Nematomorph parasites drive energy flow through a riparian ecosystem. Ecology 92, 201–207.

Sharma, K.D., Vangani, N.S., Choudhari, J.S., 1984. Sediment transport characteristics of the desert streams in India. J. Hydrol. 67, 261–272.

Sheldon, F., 2005. Incorporating natural variability into the assessment of ecological health in Australian dryland rivers. Hydrobiologia 552, 45–56.

Stanley, E.H., Fisher, S.G., Grimm, N.B., 1997. Ecosystem expansion and contraction in streams. Bioscience 47, 427–435.

Steward, A.L., 2012. When the River Runs Dry: The Ecology of Dry River Beds. Griffith University, Brisbane. PhD Thesis.

Steward, A.L., Marshall, J.C., Sheldon, F., Harch, B., Choy, S., Bunn, S.E., et al., 2011. Terrestrial invertebrates of dry river beds are not simply subsets of riparian assemblages. Aquat. Sci. 73, 551–566.

Steward, A.L., Von Schiller, D., Tockner, K., Marshall, J.C., Bunn, S.E., 2012. When the river runs dry: human and ecological values of dry riverbeds. Front. Ecol. Environ. 10, 202–209.

Steward, A.L., Marshall, J.C., Negus, P., Clifford, S.E., Dent, C., 2016. How do we assess the health of rivers when they are dry? In: Presentation for the Society for Freshwater Science (SFS), Sacramento, USA. 21–26th May, 2016. http://sfsannualmeeting.org/archive/2016/Schedule/grid_Topics.cfm?dtid=658,659,660,661,662,663&pdtid=657&rid=295 (accessed 27.06.16).

Strandine, E.J., 1941. Effects of soil moisture and algae on the survival of a pond snail during periods of relative dryness. Nautilus 54, 128–130.

Tansey, M.R., Brock, T.D., 1972. The upper temperature limit for eukaryotic organisms. Proc. Natl. Acad. Sci. U. S. A. 69, 2426–2428.

Tockner, K., Paetzold, A., Karaus, U., Claret, C., Zettel, J., 2006. Ecology of braided rivers. In: Sambroock Smith, G.H., Best, J.L., Bristow, C.S., Petts, G. (Eds.), Braided Rivers–IAS Special Publication. Blackwell, Oxford, pp. 332–352.

Tonolla, D., Acuña, V., Uehlinger, U., Frank, T., Tockner, K., 2010. Thermal heterogeneity in river floodplains. Ecosystems 13, 727–740.

Wilkes, G., Brassard, J., Edge, T.A., Gannon, V., Jokinen, C., Jones, T.H., et al., 2013. Coherence among different microbial source tracking markers in a small agricultural stream with or without livestock exclusion practices. Appl. Environ. Microbiol. 79, 6207–6219.

Williams, K.S., 1993. Use of terrestrial arthropods to evaluate restored riparian woodlands. Restor. Ecol. 1, 107–116.

Williams, D.D., Hynes, H.B., 1977. The ecology of temporary streams II. General remarks on temporary streams. Internationale Revue der Gesamten Hydrobiologie und Hydrographie 62, 53–61.

Wishart, M.J., 2000. The terrestrial invertebrate fauna of a temporary stream in southern Africa. Afr. Zool. 35, 193–200.

Wood, P.J., Armitage, P.D., 1997. Biological effects of fine sediment in the lotic environment. Environ. Manag. 21, 203–217.

THE BIOTA OF INTERMITTENT RIVERS AND EPHEMERAL STREAMS: FISHES

4.5

Adam Kerezsy*,†, Keith Gido‡, Maria F. Magalhães§, Paul H. Skelton¶

Dr Fish Contracting, Lake Cargelligo, NSW, Australia* Griffith University, Nathan, QLD, Australia† Kansas State University, Manhattan, KS, United States‡ University of Lisbon, Lisbon, Portugal§ SAIAB, Grahamstown, South Africa¶

IN A NUTSHELL

– Most fish species in intermittent rivers and ephemeral streams (IRES) survive by using a combination of adaptable colonization and recruitment strategies.
– A few species possess specialized adaptations such as air-breathing that allow them to persist in extreme IRES.
– Fish in IRES are threatened by increased habitat fragmentation (due to factors such as river regulation, water abstraction, and climate change) and the presence of alien species.
– Preservation of IRES with natural flow patterns should be a global management priority to conserve specialized biota such as fish.

4.5.1 INTRODUCTION

There is perhaps no organism more associated with water than a fish, but not all fishes have the same association with water. Although the vast majority of fishes—both marine and freshwater—relies on constant immersion in permanent water, a very small subset has overcome the obvious ecological challenges presented by intermittent rivers and ephemeral streams (IRES) that cease to flow and, in many cases, dry to pools or completely. This chapter is an overview of fishes that reside in waters with temporary, irregular, or ephemeral flow regimes (Chapter 2.2). It considers IRES on four continents and describes aspects of fish presence and diversity within them. A very basic question underlies this chapter: why and how would organisms that breed and feed in water and rely on water for all aspects of their life history live in areas where water presence (Fig. 4.5.1) is seldom guaranteed?

As in many plant and animal groups, there are exceptions to the general fish-in-water rule, and several fish species possess adaptations that enable them to survive out of water for short periods. Climbing perch *Anabas testudineus*, so named for its ability to "crawl" short distances overland, and other members of the family Anabantidae extract oxygen from the air, as do various members of the speciose Gobiidae family and certain catfishes such as the Clariidae. Mudskippers (subfamily Oxudercinae) extract atmospheric oxygen through their skin whereas snakeheads (Channidae) can survive without water by estivating in burrows (Lévêque, 1997; Allen et al., 2002).

As interesting as these fish are, this chapter takes a broader view because most riverine fishes lack specialized breathing organs, much less the ability to estivate. Although understandable in the context of

FIG. 4.5.1

Lake Namabooka, in the Georgina catchment of the Lake Eyre Basin in central Australia, only fills following unpredictable rainfall once every 5–10 years and exemplifies the central theme of this chapter: how and why would fish live in such an area?

Photos: Courtesy A. Kerezsy.

permanent rivers with constant or near-constant flow, it is salient to note that most fishes that occur in IRES also lack specific physiological adaptations for flow intermittence or drying. Instead, they are riverine fish that respire using their gills and that use their fins and tail for locomotion. They are adapted to live in water, yet they persist in rivers that can be wet one season and dry the next, wet one year and dry the next, or—in extreme circumstances—dry for long and unpredictable periods interrupted by short-lived floods. When the water comes, the fishes appear to come with it or, at the very least, follow soon afterward.

The presence and distribution of fish in IRES is discussed, as well as the relevant abiotic characteristics of the systems on four continents (Africa, Australia, North America, and the Mediterranean area of Europe) to identify global themes and similarities. We then review fish adaptability—particularly with regard to colonization/dispersal ability, recruitment, and specialized adaptations to drying—and the advantages of living in ephemeral waters. We conclude by considering present and future threats to the fishes in IRES and propose a priority list of species and systems that require conservation. Where relevant, we suggest recommendations for management. Despite the potential complexity of such a huge and varied topic, the primary goal of our chapter is to address four simple questions: what kinds of fishes live in IRES, why do they live there, how do they live there, and what threatens their existence?

4.5.2 THE FISH FAUNA OF IRES
AFRICAN IRES AND THEIR FISHES

The majority of the African continent is generally dry (annual rainfall <500 mm), and as a consequence, xeric and seasonal savannah aquatic ecosystems dominate the continent and surround the moist forests of Central and West African lowlands and coasts (Thieme et al., 2005). These ecosystems are characterized by IRES that flow briefly but are otherwise often dry (Fig. 4.5.2).

Five families of African freshwater fishes (out of a total of approximately 30) are regularly associated with IRES (Lévêque, 1997): the lungfishes (Protopteridae), the annual killifishes (Nothobranchiidae), the air-breathing catfishes (Clariidae, Fig. 4.5.3a), the Cichlidae (Fig. 4.5.3b), and the Cyprinidae (Fig. 4.5.3c). Fishes in these families are all widespread (Lévêque, 1997; Thieme et al., 2005; Snoeks et al., 2011).

FIG. 4.5.2

South African examples of IRES: (a) the Buffels River in the Karoo environs and (b) a dry savannah tributary of the Limpopo River system.

Photos: Courtesy P. Skelton.

FIG. 4.5.3

Examples of African fish species from IRES include: (a) the air-breathing African sharp-tooth catfish *Clarias gariepinus*, (b) the Mozambique tilapia *Oreochromis mossambicus*, and (c) the chubbyhead minnow *Enteromius anoplus*.

Photos: Courtesy P. Skelton.

AUSTRALIAN IRES AND THEIR FISHES

Australia is the driest inhabited continent and has a climate that renders most of its rivers intermittent to some degree. This comparative lack of water means the continent contains a small total number of freshwater fish species (approximately 300: Allen et al., 2002). Given Australia's aridity and the unpredictability of flows, it is not surprising that an even smaller subset of species (fewer than 30) occurs in the continent's center where annual rainfall averages far less than 500 mm (Wager and Unmack, 2000). Indeed, catchments in Australia's arid interior represent the realistic limit of habitable waterways for fish, with many rivers entirely ephemeral and reliant solely on unpredictable flooding and rainfall. A good example is the Todd River, which flows—very rarely—through the central Australian town of Alice Springs but most of the time is a dry sandy channel. The characterization of Australia as a desert surrounded by an ocean with a few small rivers on the periphery is very appropriate; a large section of the western interior contains no large rivers at all and the two largest rivers to the east of this area—the Diamantina and the Cooper—have among the most unpredictable flow regimes on Earth (Puckridge et al., 1998).

As in Africa, the fish fauna of Australia's most extreme IRES is limited to a small number of families. Catfishes (Plotosidae, Fig. 4.5.4a), perches (Terapontidae and Percichthyidae, Fig. 4.5.4b), and the bony herring *Nematolosa erebi* are the larger species, although few exceed 40 cm in total length.

FIG. 4.5.4

Fish species found in Australian IRES include: (a) silver tandan *Porochilus argenteus*, (b) yellowbelly *Macquaria* sp., (c) desert rainbowfish *Melanotaenia splendida tatei*, and (d) carp gudgeon *Hypseleotris* sp.

Photos: Courtesy A. Kerezsy (a, c and d) and M. Brigden (b).

The smaller species include glassfishes (Ambassidae), gobies (Gobiidae), gudgeons (Eleotridae, Fig. 4.5.4d), and rainbowfishes (Melanotaenidae, Fig. 4.5.4c).

NORTH AMERICAN IRES AND THEIR FISHES

In North America, the greatest concentration of IRES with large basin areas and fish populations occurs in western deserts or central grassland biomes, where variable climate and hydrology (both seasonal and stochastic) results in stream drying for much of the year. Opportunistic fish species are common in such habitats (Poff and Allan, 1995; McManamay et al., 2014), and examples include central stoneroller *Campostoma anomalum*, creek chub *Semotilus atromaculatus*, Sonora sucker *Catostomus insignis*, and logperch *Percina* spp. (Fig. 4.5.5a–d, respectively). In contrast, true desert fishes such as pupfish in the genus *Cyprinodon* are restricted to drying habitats or isolated pools (Minckley and Deacon, 1991). IRES can also be important for economically important species such as coho salmon *Oncorhynchus kisutch* which move from permanent rivers into these habitats for spawning and juvenile rearing (Wigington et al., 2006).

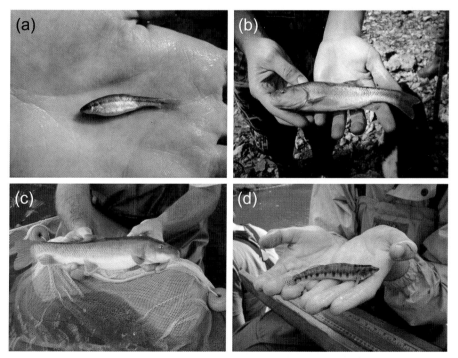

FIG. 4.5.5

North American fish species found in IRES include: (a) central stoneroller *Campostoma anomalum*, (b) creek chub *Semotilus atromaculatus*, (c) Sonora sucker *Catostomus insignis*, and (d) logperch *Percina caprodes*.

Photos: Courtesy K. Gido.

IRES OF MEDITERRANEAN EUROPE AND THEIR FISHES

IRES dominate surface runoff in semiarid areas of Mediterranean Europe (Cushing et al., 1995; Estrela et al., 1996). These Mediterranean rivers are characterized by sequential, seasonally predictable flooding (late autumn-winter) and drying (summer-early autumn) phases that vary markedly in intensity from year to year (Gasith and Resh, 1999). This hydrological variability means that fish in Mediterranean rivers persist in some of the most hydrologically unstable and diversely stressed river systems worldwide (Bonada and Resh, 2013).

Mediterranean rivers contribute significantly to the diversity of European freshwater fish because most endemic fish in Europe occur in areas with Mediterranean climates (Griffiths, 2006; Reyjol et al., 2007; Freyhof and Brooks, 2011). Levels of endemism are especially high in southern peninsulas, presumably because these regions have acted as refugia during glacial periods (review in Blondel et al., 2010). Moreover, these basins show considerable faunal similarities with North African and Middle Eastern basins (Reyjol et al., 2007). Although most endemic fishes come from the Cyprinidae (Fig. 4.5.6),

FIG. 4.5.6

Fish in IRES in Mediterranean Europe include endemic cyprinids such as (a) *Luciobarbus microcephalus* and (b) the hybrid complex *Squalius alburnoides*.

Photos: Courtesy F. Ribeiro (a) and S. Carona (b).

endemic species also occur in other families such as the Cobitidae, Cyprinodontidae, Gobiidae, and Petromyzontidae (but see review in Tierno de Figueroa et al., 2013). Overall, peri-Mediterranean areas are currently considered biodiversity hot spots for freshwater fish in Europe (Reyjol et al., 2007).

4.5.3 WHY DO FISH LIVE IN IRES?

The fundamental reasons that fish live in IRES are access to resources and the ability to maintain their populations across large areas. Given that a subset of riverine species colonizes IRES, it follows that these species have evolved to capitalize on the specific or general resources provided by each basin or catchment.

Despite their variable hydrology (Chapters 2.2 and 2.3), IRES contribute substantially to the overall functioning of river ecosystems, and the sometimes erratic wetting and drying cycles create massive increases in resources in recently inundated areas (Arthington et al., 2005; Kingsford et al., 2006; Balcombe et al., 2007). Review papers on the significance of headwater streams, many of which are intermittent, emphasize their importance to aquatic animals by providing an interface for trophic subsidies (Chapter 4.7) and a viable habitat for migratory species (Freeman et al., 2007; Meyer et al., 2007; Chapter 4.8). These systems often have high levels of resource availability through terrestrial subsidies (Courtwright and May, 2013) or through reduced competition for existing resources.

Although many IRES are highly stochastic environments, the dynamics of their fish communities can often be predicted by where the IRES occurs in the landscape and the climatic and topographic factors that determine the amount, location, and timing of surface flow (e.g., Magalhães et al., 2002b; Filipe et al., 2010; Oliveira et al., 2012). Within basins, species richness, abundance, and biomass generally increase downstream, with summer shortage of water often limiting fish occurrence in headwaters (Pires et al., 1999; Corbacho and Sanchez 2001; Magalhães et al., 2002b; Filipe et al., 2010). Nevertheless, the mechanisms driving increasing diversity have been found to vary. For example, Ostrand and Wilde (2002) reported species turnover from highly tolerant, cyprinodontid-dominated communities upstream to less-tolerant cyprinid-dominated communities downstream, whereas Whitney et al. (2015) found community structure was nested and the few species that occurred in upstream habitats were a subset of downstream communities. Regardless, patterns of higher fish diversity downstream are generally driven by greater stability and proximity to refugia habitats in larger rivers (Schlosser, 1987; Roberts and Hitt, 2010) and illustrate the importance of maintaining hydrological connectivity among habitats in these dynamic systems.

Examples from the four continents considered in this review illustrate that many species of fish in IRES are able to take advantage of either seasonal population booms of prey species or the exclusivity of isolated (and drying) wetlands. For example, the Afro-Asian air-breathing catfish family Clariidae includes widespread large-bodied African species such as *Clarias gariepinus* and *Clarias anguillaris*, both of which are predominantly associated with extreme environments (Bruton, 1979a,b). Indeed, as waterholes recede and dry in African systems, large numbers of Clariidae catfish are frequently the only and last aquatic survivors (Fig. 4.5.7).

The term "boom and bust" is frequently used to describe the ecology of both aquatic and terrestrial systems in Australia's dry inland (Kingsford et al., 2006; Wardle et al., 2013) and is also applicable to the other areas considered in this overview. The primary driver is rainfall leading to population "booms" followed by drought resulting in "busts." Studies in central Australia report the occurrence of

FIG. 4.5.7

Air-breathing catfish *Clarias gariepinus* in a drying river pool, South Africa.

Photo: Courtesy P. Skelton.

several fish species in rivers that flow only briefly and subsequently become dry for prolonged periods (Glover and Sim, 1978; Bailey and Long, 2001; Kerezsy et al., 2013). Similarly, Lake Eyre, the large terminal lake in South Australia into which the Cooper and Diamantina rivers drain intermittently, is a temporary haven for massive numbers of desert species such as the sardine-like Lake Eyre hardyhead *Craterocephalus eyresii* and the bony herring. However, these populations die in their millions, often soon after the flows recede and the lake begins evaporating (Ruello, 1976). Temporary "booms" of algae, zooplankton, and invertebrates (Chapters 4.2 and 4.3) in these mostly dry areas explain the sudden influx of massive numbers of fish.

Although it is tempting to consider that IRES may be inferior aquatic systems due to their fluctuating hydrology, this anthropocentric notion is probably misguided. Instead, it appears that IRES provide ecological advantages to resident fish species, as indicated by a study of juvenile Arkansas darter *Etheostoma cragini* in Colorado. Labbe and Fausch (2000) found that individuals that had migrated into ephemeral reaches of Big Sandy Creek were in better condition than individuals that remained in perennial reaches. Another characteristic of IRES that influences the structure of fish communities in some areas is the tight linkage with terrestrial subsidies. For example, Courtwright and May (2013) showed that the diets of brook trout *Salvelinus fontinalis* in intermittent headwater streams primarily consisted of terrestrially derived insects. Even after the supply of terrestrial insects was experimentally removed from these streams, fish did not switch to consuming aquatic insects.

Population and species maintenance is also likely facilitated by a dispersed metapopulation (Chapter 4.9) spread throughout a network of IRES. In Mediterranean IRES, for example, after flows decline in late spring, fish such as barbels either disperse along river segments in search of suitable dry-season refuges (*Luciobarbus bocagei*—Alexandre et al., 2016) or stay in deep pools most likely to persist through summer (*Barbus haasi*—Aparicio and Sostoa, 1999). Following flow resumption, chub *Squalius torgalensis* leave dry-season refuges and move into previously dry areas, in the process promoting the mixing of populations that were previously separated (Pires et al., 2014). This behavior is likely to facilitate demographic rescue in the most isolated habitats as well as contributing to local population recovery and overall persistence.

4.5.4 HOW DO FISH SURVIVE IN IRES?

Fish communities in IRES fall into two main categories: a large group of riverine species that maintain populations by migratory behavior, and a small group with specialized adaptations enabling them to persist in water-limited habitats. Fish adaptations to IRES are diverse and span a range of life history attributes as well as tolerance to extreme or unusual water quality parameters and climatic conditions. In extreme cases (Box 4.5.1), fishes have specific ways of surviving in areas where complete desiccation occurs. Nevertheless, most fishes in IRES are riverine species that have been able to adapt to and benefit from the conditions imposed by ephemerality.

Fish dispersal in seasonal rivers is a widespread phenomenon with a clear adaptive element that allows dispersal and occupation of available habitats when aquatic connections exist (Cambray, 1990; Gratwicke and Marshall, 2005). In African IRES, fish migrations include both upstream and downstream longitudinal movements and lateral migrations from the main channel into flooded margins (Table 4.5.1). The lateral migrations allow for rapid feeding and breeding by smaller species in order for the majority of the fish populations to return to the main channel with the receding floodwaters. Many fish are trapped in residual pans and floodplain pools, and generally eaten to extinction by birds, mammals, and humans (van der Waal, 1998; Barson and Nhiwatiwa, 2010; Tweddle et al., 2015). A similar fate awaits the residual populations in the pools of drying rivers (Minshull, 2008). Species found surviving in sandy bed pools of the Motloutse tributary of the Limpopo in South Africa include the air-breathing sharp-tooth catfish *C. gariepinus*, various cyprinids, and the cichlid *Oreochromis mossambicus* (van der Waal, 1997). The same group of species has been found to recolonize floodplain pans on the Save River in the Zimbabwean lowveld (Barson and Nhiwatiwa, 2010), and a similar community survived in residual pools of the Kunyeri River, the feeder river to Lake Ngami, which dried in 1982 (Paul Skelton, personal observations).

In Australian IRES, fishes similarly migrate throughout ephemeral systems when flows occur, with the majority also colonizing floodplains and other off-channel areas (Balcombe et al, 2007). In the east Simpson Desert in arid Australia, Kerezsy et al. (2013) identified two main groups of colonists (Table 4.5.1) that migrated through floodwaters to temporary habitats in the highly ephemeral Mulligan River. Extreme colonizing species migrated at least 300 km upstream from the closest permanent water and mostly comprised small-bodied species such as desert rainbowfish, glassfish, the plotosid catfish silver tandan, the terapontid spangled perch *Leiopotherapon unicolor*, and bony herring. In contrast, conservative colonizing species, including the percichthyid yellowbelly and the terapontids Barcoo grunter *Scortum barcoo* and Welch's grunter *Bidyanus welchi*, generally stayed within 150 km of the closest permanent water and remained in deeper sections of inundated channel. Conservative colonists were generally larger-bodied species with longer life spans.

In North American prairie streams, high flows and floods similarly facilitate connectivity of habitats and allow fishes from downstream perennial reaches to colonize upstream intermittent habitats (Franssen et al., 2006; Table 4.5.1). However, floods in these systems can be infrequent (2–5 years; Franssen et al., 2006). Whitney et al. (2015) reported complete extirpation of fishes in a small desert stream following wildfire and noted that recolonization was delayed because of the presence of an intermittent reach between the perennial habitat in this tributary stream and the potential refugia population in the main streams of the river network. A recent survey of this system shows the fish communities have yet to return to this area a full two years after the wildfire (K. Gido, unpublished data). Thus, connectivity of fish habitat in IRES is linked to hydrology and recovery from disturbance can take weeks, months, or

BOX 4.5.1 AFRICAN EXTREMOPHILES

The lungfishes (Protopteridae; Fig. 4.5.8) represent the ultimate "extremophile" survivors of temporary waters (Greenwood, 1986; Polačik and Podrabsky, 2015) and are commonly found in temporary floodplain habitats of savannah rivers (Jubb, 1967). Air-breathing and the ability to estivate in cocoons in the completely dried substrate allow lungfish to survive over the dry season until the habitat is again filled with water. The estivation process begins as fish stress levels rise, triggered when the water level recedes, water quality drops, and temperature rises. The fish burrow into the substrate and, when the water level drops below the substrate surface, the fish exudes mucus that dries to form a cocoon with a breathing pore at the upper end (Johnels and Svensson, 1954; Fishman et al., 1986).

In contrast, embryonic diapause within the egg is the main survival mechanism of the annual killifish *Nothobranchius* (Watters, 2009; Polačik and Podrabsky, 2015; Fig. 4.5.8). The essential habitat for *Nothobranchius* killifish is seasonal water bodies that dry in the dry season and refill in the rainy season, and with a substrate with thick black or gray vertisol clay (Watters, 2009). The life history characteristics that allow *Nothobranchius* fishes to succeed in such environments include small adult size, rapid growth and maturity, and daily one-on-one spawning, with eggs fertilized and deposited into the substrate regularly throughout the rainy season when there is water in the habitat (Watters, 2009). The eggs have a tough chorion able to withstand the rigors of drying pools where disturbance of the substrate is likely from terrestrial animals seeking the last available surface water. *Nothobranchius* ova undergo several stages of development between deposition and hatching, interspersed by phases of diapause induced by environmental stress such as anaerobic conditions (Watters, 2009). Development proceeds through the stages to a point of near hatching where a final diapause occurs. Hatching is induced by substantial flooding of the habitat after the first rains occur.

FIG. 4.5.8

African extremophiles include: (a) an African lungfish *Protopterus annectens* and (b) the southern rainbow killifish *Nothobranchius pienaari*.

Photos: Courtesy R. Bills (a) and B. Watters (b).

even years. Genetic evidence can provide a longer term evaluation of the importance of connectivity in IRES (Chapter 4.9). For example, genetic analysis of Rio Grande sucker *Pantosteus plebeius* in desert stream networks shows that low-elevation reaches separated by dry reaches are greater barriers to gene flow than higher-elevation perennial reaches (Turner et al., 2015).

In Mediterranean rivers, large numbers of chub *S. torgalensis* move away from dry season pools following flow resumption. Movements continue throughout the period of hydrological connection, with some individuals migrating long distances upstream. This movement pattern is likely to facilitate the mixing of individuals that were spatially separated, as well as rapid colonization of dewatered areas at variable distances from the dry season refugia and the flux of individuals and genes throughout river segments (Pires et al., 2014).

Table 4.5.1 Examples of fish adaptations in IRES in Australia, North America, Mediterranean Europe, and Africa

Adaptation	Examples
Specific adaptations to drying habitats	
Australia	Gobies (*Chlamydogobius* spp.) occur throughout Australia's arid areas and extract oxygen from the atmosphere using a pharyngeal organ (Thompson and Withers, 2002)
North America	Physiology of pupfish (*Cyprinodon* spp.) allows them to withstand extreme temperatures and salinities (Minckley and Deacon, 1991). Species such as bowfin *Amia calva* can gulp air during periods of hypoxia (Johansen et al., 1970)
Europe	Fishes in Mediterranean rivers lack specific adaptations to survive in drying habitats but are tolerant of environmental variability (Ferreira et al., 2007)
Africa	Lungfish breathe air using a lung and estivate in cocoons during the dry period (Greenwood, 1986). Air-breathing catfish (Clariidae) have an epibranchial arborescent organ to extract oxygen from the atmosphere and survive longer than other species in drying habitats (Bruton, 1979a)
Recruitment strategies	
Australia	In central Australia, seven species have been demonstrated to spawn continuously to take advantage of irregular flows: bony herring *Nematolosa erebi*, banded grunter *Amniataba percoides*, carp gudgeon *Hypseleotris* spp., yellowbelly *Macquaria* sp., rainbowfish *Melanotaenia splendida tatei*, glassfish *Ambassis* sp., and spangled perch (Kerezsy et al., 2011). Ephemeral areas such as floodplains provide nursery habitat for juvenile fish across the continent
North America	Pelagic spawning fishes in the genera *Macrhybopsis*, *Notropis*, *Platygobio*, and *Hybognathus* broadcast eggs during floods to decrease likelihood of desiccation and potentially reduce predation of eggs and larvae (Perkin et al., 2014)
Europe	Fishes in Mediterranean rivers generally show early maturity and high fecundity, and some are multiple spawners (Ferreira et al., 2007)
	Species spawning early in the season are less prone to recruitment failures associated with drought but may be negatively affected by floods (Magalhães et al., 2007)
Africa	*Nothobranchius* killifish grow and mature rapidly, spawn continuously and lay tough, drought-resistant eggs in the substrate. Embryonic development is phased and suspended for variable periods (diapause) in tune with drought and rainfall periodicity (Nagy, 2014). Sharptooth catfish spawn on flooded grassy shores of lakes and rivers with the onset of rains and local floods (Bruton, 1979b). Floodplains of savannah rivers are important breeding and nursery areas for many species during flooding (Lowe-McConnell, 1985)

Continued

Table 4.5.1 Examples of fish adaptations in IRES in Australia, North America, Mediterranean Europe, and Africa—cont'd

Adaptation	Examples
Migration and dispersal capabilities	
Australia	In Australia's desert rivers, at least five species (bony herring, spangled perch, rainbowfish, glassfish, and silver tandan) undertake extreme migration to ephemeral areas that occasionally become inundated, and another subset of larger-bodied species exhibits more conservative migration behavior, moving to deeper areas (Kerezsy et al., 2013)
North America	North American fishes generally recolonize intermittent reaches from perennial habitats (Whitney et al., 2015). Other species, such as longnose gar *Lepisosteus ossculus* and suckers (Catostomidae) can migrate large distances into intermittent habitats to spawn. Arctic grayling *Thymallus arcticus* move from lakes into IRES in arctic regions (Betts and Kane, 2015)
Europe	Fishes in Mediterranean rivers recolonize dewatered reaches from persistent dry-season pools and aquifer-fed runs (Pires et al., 2014)
	Spawning movements and larvae drift may also play a role in the recolonization process (Skoulikidis et al., 2011)
Africa	Longitudinal and lateral migration of fishes is an essential feature of IRES in Africa. Diurnal migration of juvenile and adult cyprinids (*Labeo umbratus* and *Pseudobarbus asper*) in flooding Groot-Gamtoos is seen as dispersal after drought (Cambray, 1990). Migration in and out of terminal headwater habitats by minnows and other fishes is a regular feature of these temporary wetlands (Gratwicke and Marshall, 2005)

The role of flows and flooding as drivers of biological processes such as fish recruitment was encapsulated in the original Flood Pulse Concept by Junk et al. (1989), and holds true for many IRES, particularly those where flows occur regularly (e.g., at a particular time of year) despite having variable volumes (Table 4.5.1). The annual flood cycle of tropical savannah rivers in Africa is described by Lowe-McConnell (1985, 1987), Welcomme (1979, 1986) and Lévêque (1997). In these rivers, sustained rainfall over several weeks or months results in a well-defined single or multiple flood peak that overflows into shallow floodplains; fishes respond by migrating upstream to breed (van der Waal, 1997; Gratwicke and Marshall, 2005). Similarly predictable responses occur in Mediterranean rivers (e.g., assemblages in some intermittent reaches are composed chiefly of young-of-year fish soon after flow resumes; Skoulikidis et al., 2011) and in North American IRES.

At local scales, habitat size is an important driver of fish community structure in IRES. In many IRES, juvenile fishes occupy shallow habitats and likely benefit from warmer water temperatures and increased resources (Labbe and Fausch, 2000; Martin et al., 2013). As waters recede, simplification of the food web occurs, with smaller habitat areas associated with simpler food webs and shorter food chains (McHugh et al., 2015; Chapter 4.7). These patterns are similar in IRES worldwide: predictable flooding occurs, fish breed, and small and/or shallow areas function as nurseries.

The Flood Pulse Concept has been modified for application to less-predictable systems such as boom-and-bust IRES in Australia. Rather than being driven by an annual, predictable flood, many Australian rivers have far more variable hydrological conditions. Working in central Australia, Puckridge et al. (1998) shifted the conceptual emphasis from floods to flow regimes and demonstrated that flows of any magnitude and/or duration could elicit an ecological response in unpredictable rivers. Concentrating on fish recruitment in more temperate systems in the southeast, Humphries et al. (1999) developed the low-flow recruitment hypothesis, concluding that for many Australian fishes—and specifically smaller species—over-bank flooding was not a requirement for successful recruitment. Subsequent studies continue to redefine the role of flow in recruitment success for fish that are adapted to living in variable Australian rivers (King et al., 2003; Cockayne et al., 2015). The general conclusion remains that Australian fishes, persisting as they do in unpredictable rivers, appear to have variable and sometimes elastic reproductive strategies in order to capitalize on the fickle conditions. For example, most fish species living in arid regions have been demonstrated to breed continually—a useful trait when an annual flow is not guaranteed (Kerezsy et al., 2011; Table 4.5.1).

Although movement and recruitment allow the majority of fishes in IRES to successfully utilize such areas to complete their life cycles, certain species have evolved specialized adaptations or increased tolerances to survive when areas dry to such a degree that oxygen and water itself become limiting factors (Table 4.5.1). In extreme environments such as desert rivers in Africa, fish diversity declines toward peripheral habitats prone to severe and prolonged desiccation. The diverse assemblage of African cichlids includes several eurytopic genera such as *Oreochromis*, *Sarotherodon*, and *Tilapia* that frequent seasonally fluctuating habitats of savannah ecosystems (Trewavas, 1983). *Oreochromis* species, in particular, are physiologically adapted to extremely harsh conditions, including hypersaline lakes and warm springs (Trewavas, 1983; Hecht and Zway, 1984; Nyingi and Agnèse, 2011).

The Anabantidae, commonly referred to as "labyrinth" fishes due to their superbranchial air-breathing organ, is another African family that survives in temporary water bodies. However, the air-breathing and terrestrial habits of African anabantid species such as the many-spined climbing perch *Ctenopoma multispine* are less associated with the desiccation of the environment and more with tolerating deoxygenation of swampy environments and undertaking overland breeding migrations during wet periods (Benl and Foersch, 1978; Bruton and Kok, 1980; Sayer and Davenport, 1991).

Air-breathing is a vital adaptation for survival in extreme drying habitats. The last species present in drying pools of African rivers and floodplains is the sharp-tooth catfish (e.g., Bell-Cross and Minshull, 1988; Barson and Nhiwatiwa, 2010), and provided the substrate is not completely desiccated, this species' capabilities, including the ability to leave a drying water body and move overland, facilitate its survival (Bruton, 1979a,b; van der Waal, 1998). In contrast, complete desiccation of the environment can be survived by two groups of African fish: the lungfishes and the annual killifishes (Box 4.5.1). Although the species encountered in the harsh peripheral pools are characteristically ecological "pioneers" and "residuals" such as *Clarias* catfish (van der Waal, 1998), only "extremophile" species with special adaptations to surviving desiccation (Greenwood, 1986; Watters, 2009) exist in remote disconnected rain-filled pans.

Extreme IRES in Australia similarly provide habitat for fish species with specific adaptations that allow them to survive. Small gobies of the genus *Chlamydogobius* are distributed throughout Australia's arid interior, primarily in shallow artesian springs and residual waterholes that persist following flooding. In these hot, extremely shallow (frequently <3 cm deep, A. Kerezsy, personal observation) and oxygen-poor habitats, the gobies extract atmospheric oxygen using a pharyngeal organ (Thompson and Withers, 2002). Another specialized Australian fish, the salamanderfish *Lepidogalaxias salamandroides*, is found only in acidic peat flats in Western Australia and survives without water for several months by estivating in burrows (Allen et al., 2002). Although most of Australia's IRES fish are pioneers rather than extremophiles, their survival capabilities in some of the world's most unpredictable rivers are indicative of their adaptations, exemplified by "Australia's toughest freshwater fish" (Box 4.5.2).

Unlike Australia and Africa, there are no North American fishes capable of estivation. Instead, fishes occupying IRES in this region are often capable of breathing air (gar), withstanding extreme physiological

BOX 4.5.2 AUSTRALIA'S TOUGHEST FRESHWATER FISH

Temperatures in inland Australian waterways can reach close to 40°C in summer yet cool to almost freezing at night in winter. Salinity levels can exceed seawater as some waterholes dry. Australia's most widespread freshwater fish, the spangled perch *Leiopotherapon unicolor* (Fig. 4.5.9), epitomizes the attributes that an animal requires to succeed in such environments. Spangled perch, as the name suggests, is a perch-shaped fish of the terapontid family and occurs across northern and central Australia. Growing to a maximum length of 30cm, spangled perch is a generalist carnivore, a fecund spawner and has been reported migrating in flooded wheel ruts after rain (Lintermans, 2007). The species has been seen in rapidly drying desert waterholes scarcely deeper than the fish's body depth, as well as in the many bores and other artificial water points scattered throughout the inland. Its ability to disperse, successfully colonize recently wet areas and ensure a steady supply of juveniles are the hallmarks of success of "Australia's toughest fish" and, by extrapolation, the attributes most likely to ensure survival of freshwater fishes in Australian IRES.

FIG. 4.5.9

Spangled perch, Australia's most widespread freshwater fish, is common in IRES across the north and center of the continent and possesses a suite of traits that characterizes fishes from these tropical and arid areas: wide dietary preferences, tolerance of fluctuating water quality parameters, opportunistic recruitment, and extreme dispersal ability.

Photo: Courtesy A. Kerezsy.

conditions (some minnows and pupfish), rapid population growth (mosquito fish), or a combination of these traits. Perhaps the ultimate survivor of naturally extreme environments is the desert pupfish *Cyprinodon macularius*, which can withstand temperatures up to 45°C and salinity to 70ppt (Lowe and Heath 1969).

4.5.5 THREATS TO FISHES IN INTERMITTENT RIVERS

Extended drying periods and below-average rainfall are normal climatic perturbations in arid areas. Although the consequences are often catastrophic for a particular cohort of fishes in a catchment, recovery occurs when a series of wetter-than-average months or years replenish formerly dry reaches and rejuvenate catchments. Examples discussed previously, such as the mass die-off of fish in outback Australia or the failure of fish to recolonize IRES in North America following wildfire, are not permanent changes but merely examples of the immense variability within such systems. Given time and flows, fish are likely to recolonize both areas when conditions are suitable. However, natural perturbations such as these can be amplified by anthropogenic changes such as river regulation (and resulting fragmentation of habitats), the imposition of alien species, and the looming problems associated with climate change (Table 4.5.2).

Table 4.5.2 Examples of threats to fish in IRES in Australia, North America, Mediterranean Europe, and Africa

Threat	Examples
Fragmentation	
Australia	Rivers throughout Australia's Murray-Darling Basin have been regulated since colonization, resulting in declines of many native species
North America	Fragmentation in IRES caused by poorly designed road crossings as well as heavily engineered dams. Many fish species have been extirpated upstream of impoundments
Europe	Habitat loss and fragmentation associated with damming contribute significantly to the decline of native fishes
Africa	Instream dams and weirs in dry areas are a major development in Africa and probably cause the decline of fish populations in these areas by interrupting fish migrations
Alien species	
Australia	Carp, redfin perch, and goldfish threaten the Murray-Darling Basin; *Tilapia* spp. threaten native assemblages in the tropics; and gambusia are problematic across the continent
North America	IRES in the western United States are most prone to invasive fish species, including many sport fishes native to eastern United States such as bass (*Micropterus* spp.), sunfish (*Lepomis* spp.), and catfish (*Ictalurus* spp. and *Amieurus* spp.). Small cyprinids and gambusia are also common invaders of IRES. Tilapia (*Oreochromis* spp.) escaping from aquaculture facilities have exploited IRES in southern latitudes
Europe	Mediterranean rivers are a hot spot for fish invasions, including highly successful invaders such as common carp, pumpkinseed *Lepomis gibbosus*, and gambusia
Africa	In southern Africa, widespread translocation of sharp-tooth catfish via interbasin transfers and humans threaten many fish species in IRES. Carp and Mozambique tilapia (*Oreochromis mossambicus*) are invasive in dry areas of South Africa and Namibia
Groundwater extraction	
Australia	Long-term extraction of groundwater from Australia's Great Artesian Basin has likely impacted springs and IRES waterholes reliant on this groundwater in the arid zone
North America	Massive declines in surface water in IRES can be attributed to groundwater extraction, particularly in the Great Plains region of the United States
Europe	Water extraction threatens native fish in most Mediterranean rivers
Africa	The cave catfish in Namibia (*Clarias cavernicola*), restricted to a single cave lake, is threatened by a declining water table due to groundwater extraction
Climate change	
Australia	More intense droughts are likely to reduce the number of IRES that receive periodic flows, resulting in reduced habitat for fish species
North America	Same as for Australia
Europe	Mediterranean rivers already experience substantial reductions in water availability, which will worsen with longer and more severe droughts expected under altered future climates
Africa	Climate change could threaten source tributaries in montane regions where fish species such as the Maloti minnow *Pseudobarbus quathlambae* is relict and fragmented into small isolated populations. Increased drought frequency in mesic areas in southern Africa such as the Karoo could threaten IRES

DROUGHT AND CLIMATE CHANGE

Within the context of the four continents considered in this overview, summer drying and drought is particularly significant for fish in Mediterranean rivers and North American IRES because, unlike some species from Africa and Australia, the persistence of fish in Mediterranean and North America IRES is strictly dependent upon perennial surface waters which act as refuges and sources of colonists after flow resumes (Robson et al., 2013; Chapter 4.8). Although large pools with well-developed canopies generally hold the richest assemblages and the highest overall abundances of native species (Magalhães et al., 2002a; Pires et al., 2010), the presence of networks of heterogeneous refuges may be critical in promoting adequate conditions for the persistence of species and life stages with different habitat requirements.

Fish assemblages in IRES on all four continents show considerable stability in the face of present-day droughts, probably due to the way historical hydrological filters have winnowed the pools of native species to those best adapted to cope with the prevailing patterns of variability (see also Hershkovitz and Gasith, 2013). Although species abundances can decline significantly in dry years (Bernardo et al., 2003; Skoulikidis et al., 2011), small and transient changes in species richness, composition, and rank abundance are normal (Magalhães et al., 2007). Indeed, except in some upstream reaches, local assemblages appear to be resilient to droughts (Chapter 4.8) and are capable of recovering to their former structure.

More frequent and severe droughts associated with climate change are likely to exacerbate intermittence and loss of hydrological connectivity, especially in small- and middle-sized streams (Giorgi and Lionello, 2008). This is likely to have severe impacts on native fishes, with species facing a trade-off between adaptation to new conditions or dispersal to new habitats (Filipe et al., 2013). Prolonged and intensified drought in Mediterranean regions of North America may result in dramatic changes or gradual shifts in freshwater populations and communities (Resh et al., 2013), and climate warming will increase periods of intermittence in arctic rivers, further fragmenting populations of Arctic grayling *Thymallus arcticus* (Betts and Kane, 2015). At broader scales, it seems likely that there will be considerable shifts in species richness and composition, with declines or local extinctions of the most sensitive species and their replacement by more resistant species (Magalhães et al., 2007; Filipe et al., 2013; Table 4.5.2).

WATER EXTRACTION, RIVER REGULATION, AND FRAGMENTATION OF HABITAT

Arguably the greatest threat to IRES biota is the fragmentation of their habitats. Fish ecology in IRES is driven by source-sink dynamics, and therefore long-term survival is underpinned by colonization opportunities. Barriers that inhibit these movements are particularly detrimental to the natural functioning of species and ecosystems. Perkin et al. (2014) proposed an ecological "ratchet mechanism" to describe how dams inhibit the ability of fish populations to recolonize areas following drying, often leading to basin-wide extirpations (Fig. 4.5.10). Increasing human populations are causing increases in disturbance frequencies and more severe drying, which is predicted to increase extirpations in these highly fragmented systems.

Historically, management of Australia's rivers was primarily concerned with mitigating their intermittence in the century immediately following settlement of the continent by Anglo-Europeans, and provides a useful case study of the negative effects of human-induced fragmentation, particularly on fish populations. The development of river regulation and irrigation infrastructure is best exemplified by the highly modified Murray-Darling Basin, where the rivers have been "controlled" by large headwater reservoirs as well as a series of smaller weirs and locks further downstream (Lintermans, 2007). These structures

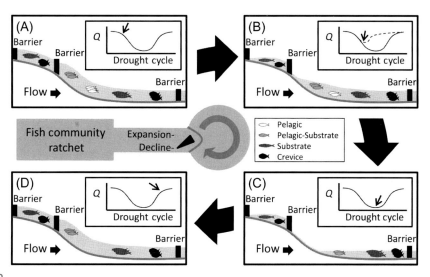

FIG. 4.5.10

Conceptual diagram of the ecological ratchet mechanism caused by interactions between habitat fragmentation and drought. Ratchets can operate as mechanisms for population expansion or decline (shown here) and involve forward movement without reciprocated reverse movement through four states in a fragmented riverscape: (A) before drought, (B) onset of drought, (C) height of drought, and (D) following drought. Fish abundance is equal to the number of symbols. If a drought subsides (dashed line of insert in B), then a "pseudo-ratchet cycle" ensues.

Reproduced with permission from Perkin et al. (2014).

achieved their intended purpose to allow towns and agricultural districts to develop, but it transpired—much later—that they had unfavorable outcomes for native biota. Native fishes in these IRES evolved with adaptations to boom-and-bust cycles permitting temporary migration and access to floodplains and anabranches. However, river regulation created a series of permanent reaches separated by barriers (weirs) that effectively stifled such behavior. As a result, native fishes declined, their migratory behavior was compromised, and the rivers deteriorated in health (Reynolds, 1983; Harris and Gehrke, 1997).

Many of the largest species, such as the Murray cod *Maccullochella peelii*, its close relative the trout cod *Maccullochella macquariensis*, the Macquarie perch *Macquaria australasica*, the silver perch *Bidyanus bidyanus*, and the freshwater catfish *Tandanus tandanus*, are now listed as threatened species, and river rehabilitation and remediation are now recognized as crucial to the recovery of the Murray-Darling Basin. Removal of weirs, the installation of fishways and fish ladders, and the provision of environmental watering by state and federal management agencies are some of the measures that seek to redress the river regulation of yesteryear and aim to restore the ecological balance in these systems.

ALIEN SPECIES

Alien fish species are present in all four of the continental areas considered in this overview. The most widespread alien fishes are small-bodied live-bearers such as gambusia *Gambusia holbrooki*, large-bodied cyprinids such as carp *Cyprinus carpio*, and the slightly smaller cichlids (predominantly *O. mossambicus*; Table 5.4.2). Like their terrestrial counterparts, nonnative fishes that succeed in foreign environments possess broad dietary and habitat requirements. Unsurprisingly, those that inhabit IRES share traits common to native species inhabiting the same environments: the ability to colonize during flows, the ability to recruit successfully, and the ability to survive during periods of suboptimal water quality. In Australia's Murray-Darling Basin, for example, carp breeding "hot spots" have been demonstrated to occur in shallow off-river habitats. An unlucky consequence of environmental flows intended to ameliorate the effects of river regulation is that their provision seems to benefit this alien species (Rayner et al., 2009; Conallin et al., 2012).

Worldwide, alien fishes generally originated from importation of foreign species to bolster local food, provide sport, or as ornamental fishes (Table 4.5.2). An exception is the widespread presence of gambusia which was originally transferred around the world in the early 20th century to combat malaria. However, the unfortunate results are no different and this species has been implicated in the decline of many native fishes (Howe et al., 1997; Fairfax et al., 2007). The translocation of native species to IRES outside their natural range is an equally problematic phenomenon, with salient examples including bass (*Micropterus* spp.) in the United States and sharp-tooth catfish throughout southern Africa (Table 4.5.2).

4.5.6 CONSERVATION PRIORITIES FOR FISH IN IRES

Conservation of fish in IRES relies upon habitat conservation to maintain periodic flow connectivity and enable ephemeral areas to experience regular and/or irregular wetting episodes. As such, and as a first priority, areas that currently experience natural or near-natural flow regimes should be preserved to maintain their intact ecological functions (Table 4.5.3). Progress toward this goal is piecemeal at best in the four areas considered in this overview (Table 4.5.3). This is concerning, particularly as flow

Table 4.5.3 Examples of conservation priorities in IRES in Australia, North America, Mediterranean Europe, and Africa

Priorities	Examples
Preservation of IRES with natural flows	
Australia	The arid-zone Lake Eyre Basin in central Australia (all IRES) and most of the rivers in Australia's tropical north remain unregulated but face calls for regulation to address water shortages for uses such as agriculture and mining
North America	Very little has been accomplished on large IRES. Some watersheds in smaller basins have been purchased by conservation or government agencies
Europe	There are needs for preservation of IRES inhabited by endangered and range-restricted species, for improved connectivity between protected areas and for increased protection against threats
Africa	The preservation of IRES in desert areas such as the Sahara is vital for the sustenance of all life, including humans

Table 4.5.3 Examples of conservation priorities in IRES in Australia, North America, Mediterranean Europe, and Africa—cont'd

Priorities	Examples
Repair/remediation of regulated rivers to reinstate natural flows and improve habitat	
Australia	Remediation activities including environmental flows and modification of structures to allow fish passage should continue in regulated rivers including IRES in the Murray-Darling Basin, southeastern Australia
North America	Several fishways on IRES have been built and there is a program to retrofit road crossings to allow fish passage. Some dam removal projects have recently occurred
Europe	Fish passage and environmental flows that mimic natural flow variation are increasingly recognized as critical to maintain biodiversity and river health
Africa	Installation of fishways in IRES is advocated but not widely practiced in Africa
Priority native species	
Australia	Recovery actions should focus on the requirements of listed endangered species (e.g., silver perch and freshwater catfish in the southeast, Balston's pygmy perch *Nannatherina balstoni* in the southwest)
North America	Many species are candidates, including large-river pelagic spawning fish (*Hybognathus* spp., *Macrhybopsis* spp.), pupfishes (*Cyprinodon* spp.), and live-bearers (*Poecilia* spp.) occupying intermittent spring habitats
Europe	Although endemic and threatened species with restricted distributions are strong candidates, conservation of biodiversity is increasingly recognized to depend on assemblage rather than species-specific conservation plans
Africa	Annual killifish are often localized and face threats from habitat destruction and agriculture, including overgrazing of riverine floodplains and marginal swamps. Lungfish are systematically exploited by subsistence fishers; increased human population pressures could threaten their existence
Priority alien species	
Australia	Control actions should focus on carp and redfin perch in the south, tilapia in the north, and gambusia across the continent
North America	Control of alien species should focus on isolated habitats where control is practical and on all forms of nonnatives. Predatory fish (e.g., largemouth bass and tilapia) probably have the most detrimental effects
Europe	Control of invasive species such as common carp and topmouth gudgeon *Pseudorasbora parva* in source habitats deserve special attention, but effective legislation and public awareness are also critical to reduce introductions
Africa	Genetic contamination of indigenous *Oreochromis* species is a major threat to isolated and temporary populations. Deliberate introduction of aliens into refuge sites is a potential threat in many IRES; air-breathing catfish (*Clarias*) are especially problematic in this respect

regulation continues to compromise IRES in areas such as Africa (Table 4.5.2) and climate change is likely to exacerbate anthropogenic water needs worldwide (Chapters 1 and 5.1).

In areas where flow and/or connectivity has been altered, recovery efforts for IRES and their fish species (especially those that are listed as endangered) should focus on ameliorating the effects of fragmentation, alien species, and other threats so that populations can recover and persist over time (Table 4.5.3). Recovery strategies may seek to conserve native species; eradicate alien species; and/or restore IRES, their catchments, and their flow regimes at multiple spatial scales. A more formal investigation of global conservation priorities with a focus on fish in IRES is warranted given the pervasive themes of native fish decline, alien fish invasion, and the continued fragmentation of IRES that are so obvious in this overview.

4.5.7 CONCLUSIONS

This chapter has sought to profile fish species and communities in IRES on four continents and should be considered an overview rather than a comprehensive summary. Several key themes emerge about the characteristics of IRES fishes, the roles they play in these rivers, the threats they face, and some potential strategies for their conservation.

Fish presence and persistence in IRES is underpinned by two key factors: the adaptations of the fish species that enable them to survive in such systems, and the integrity of the catchments themselves. The breadth of adaptations, and especially the convergence of similar traits across different families and on different continents, demonstrates evolution by natural selection in progress. They also remind us that fish, already highly diverse, are capable of adapting to living in virtually *any* area where water is either permanently or occasionally present.

IRES are, by their very nature, fickle and unpredictable environments. In years when floods do not occur or when rainfall is below average, fish populations will decline and may become locally extirpated. If these conditions persist, recovery might not occur for several years. However, as long as IRES or at least some sections of them remain in reasonable ecological health and are physically capable of reconnecting with parent rivers or more permanent reaches, one aspect of their ecology that is predictable is that when water returns, fish will reappear. Different flow regimes (Chapter 2.2) may elicit different biological responses between seasons but over long time frames, fish persistence in such systems is driven by the regular or irregular patterns of wetting and drying. No matter how different each flood or flow seems, there will be ecological similarities.

One similarity is how IRES fishes contribute significantly to food webs and secondary production. For example, the contribution that fish in IRES can make to arid-zone food webs is superbly demonstrated throughout the Australian inland where a consistent pattern of ecological succession plays out after high flows. Erratic floodwaters provide migration pathways that are quickly utilized by colonizing fish species. Migratory fish-eating birds, notably large-bodied species such as pelicans and cormorants (Fig. 4.5.11), arrive in huge numbers shortly afterward. Then, as the waterholes dry, terrestrial predators and scavengers including rodents, reptiles, and both native and introduced large mammals consume the stranded survivors (R. Kingsford and A. Kerezsy, personal observations; Chapter 4.6).

Intensive human settlement and the subsequent alterations to river catchments have had the most severe impacts on IRES (Chapter 5.1), because even the most hardy and adapted fish species

FIG. 4.5.11

Fish-eating birds such as these Australian pelicans are big beneficiaries when IRES experience flows and floods. Soon after flow resumes, pelicans arrive and feed on the newly arrived cohorts of bony herring and other fish species.

Photo: Courtesy A. Kerezsy.

cannot survive the effects of rivers that cease to flow entirely, dam walls that are too high to scale, or the diseases, predation, and displacement that can occur when alien species are liberated in a reach or catchment. Ameliorating such disruptions is likely to remain a key challenge for river managers worldwide, especially as our global population and commensurate demand for water grow. Although IRES are often overlooked by regulatory agencies as essential habitats for fishes, this chapter illustrates the wide diversity of these habitats and their importance in maintaining healthy river ecosystems.

Our recommendations provide a comparatively simple approach to conserving IRES and their fish communities, even though the implementation of strategies and actions is likely to be extremely complicated. In the first instance, IRES that retain their natural flow regime and that are unaffected by river regulation and water extraction should be identified and allowed to remain in their natural state. Such areas have obvious conservation and research potential, and also provide reference areas for remediation of other rivers, reaches, and areas (whether intermittent or permanent).

In areas where human-induced perturbation has occurred, and especially in areas that are known to provide habitat for rare, endangered or unusual species, research and management aimed at restoring or remediating IRES should be undertaken if possible. This multifaceted and onerous task involves research on specific species, their ecological requirements, and specific threats and stressors. Often, on-ground works are needed, integrating disciplines ranging from biology to engineering. Examples include the installation of fish ladders on weirs to facilitate movement, the resnagging of river channels, the reinstatement of the natural flow regime by managing flows, and the removal or control of alien species using techniques such as physical removal and chemical control. Although complex and expensive, this work is necessary to preserve the ecological integrity of aquatic ecosystems where water itself is not always present. In this regard, the fact that humans have a natural tendency to link fish with water is a definite advantage: if we can explain that IRES are facing an uncertain future, the most graphic evidence for many people will be the decline or absence of native fishes.

REFERENCES

Alexandre, C.M., Almeida, P.R., Neves, T., Mateus, C.S., Costa, J.L., Quintela, B.R., 2016. Effects of flow regulation on the movement patterns and habitat use of a potamodromous cyprinid species: fish movement response to flow regulation. Ecohydrology 9, 326–340.

Allen, G.R., Midgley, S.H., Allen, M., 2002. Field Guide to the Freshwater Fishes of Australia. Western Australian Museum, Perth.

Aparicio, E., Sostoa, A., 1999. Pattern of movements of adult *Barbus haasi* in a small Mediterranean stream. J. Fish Biol. 55, 1086–1095.

Arthington, A.H., Balcombe, S.R., Wilson, G.A., Thoms, M.C., Marshall, J., 2005. Spatial and temporal variation in fish assemblage structure in isolated waterholes during the 2001 dry season of an arid-zone river, Cooper Creek, Australia. Mar. Freshw. Res. 56, 25–35.

Bailey, V., Long, P., 2001. Wetland, Fish and Habitat Survey in the Lake Eyre Basin, Queensland: Final Report. Queensland Department of Natural Resources and Mines, Brisbane.

Balcombe, S.R., Bunn, S.E., Arthington, A.H., Fawcett, J.H., McKenzie-Smith, F.J., Wright, A., 2007. Fish larvae, growth and biomass relationships in an Australian arid zone river: links between floodplains and waterholes. Freshw. Biol. 52, 2385–2398.

Barson, M., Nhiwatiwa, T., 2010. Influence of drought and flooding on the colonization of floodplain pans by riverine fishes in the Zimbabwean lowveld. Afr. J. Aquat. Sci. 35, 205–208.

Bell-Cross, G., Minshull, J., 1988. The Fishes of Zimbabwe. National Museums and Monuments of Zimbabwe, Harare.

Benl, G., Foersch, W., 1978. Beitrag zur Kenntnis des *Ctenopoma multispinis* W. Peters, 1844. Spixiana 1, 287–299.

Bernardo, J.M., Ilhéu, M., Matono, P., Costa, A.M., 2003. Interannual variation of fish assemblage structure in a Mediterranean river: implications of streamflow on the dominance of native or exotic species. River Res. Appl. 19, 521–532.

Betts, E.D., Kane, D.L., 2015. Linking North Slope of Alaska climate, hydrology, and fish migration. Hydrol. Res. 46, 578–590.

Blondel, J., Aronson, J., Bodiou, J., Boeuf, G., 2010. The Mediterranean Region: Biological Diversity in Space and Time, second ed. Oxford University Press, Oxford.

Bonada, N., Resh, V.H., 2013. Mediterranean-climate streams and rivers: geographically separated but ecologically comparable freshwater systems. Hydrobiologia 719, 1–29.

Bruton, M.N., 1979a. The survival of habitat desiccation by air-breathing clariid catfishes. Environ. Biol. Fish 4, 273–280.

Bruton, M.N., 1979b. The breeding biology and early development of *Clarias gariepinus* (Pisces: Clariidae) in Lake Sibaya, South Africa, with a review of breeding in the subgenus *Clarias* (*Clarias*). Trans. Zool. Soc. London 35, 1–45.

Bruton, M.N., Kok, H.M., 1980. The freshwater fishes of Maputaland. In: Bruton, M.N., Cooper, K.H. (Eds.), Studies on the Ecology of Maputaland. Rhodes University, Grahamstown and the Natal Branch of the Wildlife Society of Southern Africa, Durban, pp. 210–244.

Cambray, J.A., 1990. Adaptive significance of a longitudinal migration by juvenile freshwater fish in the Gamtoos River System, South Africa. S. Afr. J. Wildl. Res. 20, 148–156.

Cockayne, B.J., Sternberg, D., Schmarr, D.W., Duguid, A.W., Mathwin, R., 2015. Lake Eyre golden perch (*Macquaria* sp.) spawning and recruitment is enhanced by flow events in the hydrologically variable rivers of Lake Eyre Basin, Australia. Mar. Freshw. Res. 66, 822–830.

Conallin, A.J., Smith, B.B., Thwaites, L.A., Walker, K.F., Gillanders, B.M., 2012. Environmental water allocations in regulated lowland rivers may encourage offstream movements and spawning by common carp, *Cyprinus carpio*: implications for wetland rehabilitation. Mar. Freshw. Res. 63, 865–877.

Corbacho, C., Sanchez, J.M., 2001. Patterns of species richness and introduced species in native freshwater fish faunas of a Mediterranean-type basin: the Guadiana River (Southwest Iberian Peninsula). Regul. Rivers Res. Manag. 17, 699–707.

Courtwright, J., May, C.L., 2013. Importance of terrestrial subsidies for native brook trout in Appalachian intermittent streams. Freshw. Biol. 58, 2423–2438.

Cushing, C.E., Cummins, K.W., Minshall, G.W. (Eds.), 1995. Ecosystems of the World 22, River and Stream Ecosystems. Elsevier, Amsterdam.

Estrela, T., Marcuello, C., Iglesias, A., 1996. Water Resources Problem in Southern Europe—An Overview Report. European Environment Agency, Copenhagen.

Fairfax, R., Fensham, R., Wager, R., Brooks, S., Webb, A., Unmack, P., 2007. Recovery of the red-finned blue-eye: an endangered fish from springs of the Great Artesian Basin. Wildl. Res. 34, 156–166.

Ferreira, M.T., Oliveira, J., Caiola, N., Sostoa, A., Casals, F., Cortes, R., et al., 2007. Ecological traits of fish assemblages from Mediterranean Europe and their responses to human disturbance. Fish. Manag. Ecol. 14, 473–481.

Filipe, A.F., Magalhães, M.F., Collares-Pereira, M.J., 2010. Native and introduced fish species richness in Mediterranean streams: the role of multiple landscape influences. Divers. Distrib. 16, 773–785.

Filipe, A.F., Lawrence, J.E., Bonada, N., 2013. Vulnerability of stream biota to climate change in mediterranean climate regions: a synthesis of ecological responses and conservation challenges. Hydrobiologia 719, 331–351.

Fishman, A.P., Pack, A.I., Delaney, R.G., 1986. Estivation in *Protopterus*. J. Morphol. 190 (Suppl. 1), 237–248.

Franssen, N.R., Gido, K.B., Strakosh, T.R., Bertrand, K.N., Franssen, C.M., Paukert, C.P., et al., 2006. Effects of floods on fish assemblages in an intermittent prairie stream. Freshw. Biol. 51, 2072–2086.

Freeman, M.C., Pringle, C.M., Rhett Jackson, C., 2007. Hydrologic connectivity and the contribution of stream headwaters to ecological integrity at regional scales. J. Am. Water Resour. Assoc. 43, 5–14.

Freyhof, J., Brooks, E., 2011. European Red List of Freshwater Fishes. Publications of the Office of the European Union, Luxembourg.

Gasith, A., Resh, V.H., 1999. Streams in mediterranean climate regions: abiotic influences and biotic responses to predictable seasonal events. Annu. Rev. Ecol. Syst. 30, 51–81.

Giorgi, F., Lionello, P., 2008. Climate change projections for the Mediterranean region. Glob. Planet. Chang. 63, 90–104.

Glover, C.J.M., Sim, T.C., 1978. A survey of central Australian ichthyology. Aust. J. Zool. 15, 61–64.

Gratwicke, B., Marshall, B.E., 2005. Fish migrations in two seasonal streams in Zimbabwe. Afr. J. Aquat. Sci. 30, 107–118.

Greenwood, P.H., 1986. The natural history of African lungfishes. J. Morphol. 190 (Suppl. 1), 163–179.

Griffiths, D., 2006. Pattern and process in the ecological biogeography of European freshwater fish. J. Anim. Ecol. 75, 734–751.

Harris, J.H., Gehrke, P.C., 1997. Fish and Rivers in Stress—the NSW Rivers Survey, NSW. Fisheries Office of Conservation and the Cooperative Research Centre for Freshwater Ecology, Cronulla and Canberra.

Hecht, T., Zway, P., 1984. On the stunted Mocambique tilapia *Oreochromis mossambicus* (Peters, 1852) (Pisces: Cichlidae) of the Matiovila Hot Springs, Kruger National Park. Koedoe 27, 25–38.

Hershkovitz, Y., Gasith, A., 2013. Resistance, resilience, and community dynamics in mediterranean-climate streams. Hydrobiologia 719, 59–75.

Howe, E., Howe, C., Lim, R., Burchett, M., 1997. Impact of the introduced poeciliid *Gambusia holbrooki* (Girard, 1859) on the growth and reproduction of *Pseudomugil signifier* (Kner, 1865) in Australia. Aust. J. Mar. Freshwat. Res. 48, 425–434.

Humphries, P., King, A.J., Koehn, J.D., 1999. Fish, flows and flood plains: links between freshwater fishes and their environment in the Murray-Darling River system, Australia. Environ. Biol. Fish 56, 129–151.

Johansen, K., Hanson, D., Lenfant, C., 1970. Respiration in the primitive air-breather *Amia calva*. Respir. Physiol. 9, 162–174.

Johnels, A.G., Svensson, G.S.O., 1954. On the biology of *Protopterus annectens* (Owen). Arkiv Zoologi Stockholm 7, 131–164.

Jubb, R.A., 1967. The Freshwater Fishes of Southern Africa. AA Balkema, Cape Town.

Junk, W.J., Bayley, P.B., Sparks, R.E., 1989. The flood pulse concept in river-floodplain systems. Can. Spec. Publ. Fish Aquat. Sci. 106, 110–127.

Kerezsy, A., Balcombe, S., Arthington, A., Bunn, S., 2011. Continuous recruitment underpins fish persistence in the arid rivers of far western Queensland, Australia. Mar. Freshw. Res. 62, 1178–1190.

Kerezsy, A., Balcombe, S., Tischler, M., Arthington, A., 2013. Fish movement strategies in an ephemeral river in the Simpson Desert, Australia. Austral Ecol. 38, 798–808.

King, A.J., Humphries, P., Lake, P.S., 2003. Fish recruitment on floodplains: the roles of patterns of flooding and life history characteristics. Can. J. Fish. Aquat. Sci. 60, 773–786.

Kingsford, R.T., Georges, A., Unmack, P.J., 2006. Vertebrates of desert rivers: meeting the challenges of temporal and spatial unpredictability. In: Kingsford, R.T. (Ed.), Ecology of Desert Rivers. Cambridge University Press, Cambridge, pp. 154–200.

Labbe, R., Fausch, K.D., 2000. Dynamics of intermittent stream habitat regulate persistence of a threatened fish at multiple scales. Ecol. Appl. 10, 1774–1791.

Lévêque, C., 1997. Biodiversity Dynamics and Conservation. The Freshwater Fish of Tropical Africa. Cambridge University Press, Cambridge.

Lintermans, M., 2007. Fishes of the Murray-Darling Basin: An Introductory Guide. Murray-Darling Basin Authority, Canberra. MDBC Publication No. 10/07.

Lowe, C.H., Heath, W.G., 1969. Behavioral and physiological responses to temperature in the desert pupfish *Cyprinodon macularius*. Physiol. Zool. 42, 53–59.

Lowe-McConnell, R.H., 1985. The biology of the river systems with particular reference to the fishes. In: Grove, A.T. (Ed.), The Niger and its Neighbours. Environmental History and Hydrobiology, Human Use and Health Hazards of the Major West African Rivers. AA Balkems, Rotterdam, pp. 101–142.

Lowe-McConnell, R.H., 1987. Ecological Studies in Tropical Fish Communities. Cambridge University Press, Cambridge.

Magalhães, M.F., Batalha, D.C., Collares-Pereira, M.J., 2002a. Gradients in stream fish assemblages across a Mediterranean landscape: contributions of environmental factors and spatial structure. Freshw. Biol. 47, 1015–1031.

Magalhães, M.F., Beja, P., Canas, C., Collares-Pereira, M.J., 2002b. Functional heterogeneity of dry-season fish refugia across a Mediterranean catchment: the role of habitat and predation. Freshw. Biol. 47, 1919–1934.

Magalhães, M.F., Beja, P., Schlosser, I.J., Collares-Pereira, M.J., 2007. Effects of multi-year droughts on the fish assemblages of seasonally drying Mediterranean streams. Freshw. Biol. 52, 1494–1510.

Martin, E.C., Whitney, J.E., Gido, K.B., Hase, K.J., 2013. Habitat associations of stream fishes in protected tallgrass prairie streams. Am. Midl. Nat. 170, 39–51.

McHugh, R.A., Thompson, R.M., Greig, H.S., Warburton, H.J., McIntosh, A.R., 2015. Habitat size influences food web structure in drying streams. Ecography 38, 700–712.

McManamay, R.A., Bevelhimer, M.S., Frimpong, E.A., 2014. Associations among hydrologic classifications and fish traits to support environmental flow standards. Ecohydrology 8, 460–479.

Meyer, J.L., Strayer, D.L., Wallace, J.B., Eggert, S.L., Helfman, G.S., Leonard, N.E., 2007. The contribution of headwater streams to biodiversity in river networks. J. Am. Water Resour. Assoc. 43, 86–103.

Minckley, W.L., Deacon, J.E. (Eds.), 1991. Battle Against Extinction: Native Fish Management in the American West. University of Arizona Press, Tucson.

Minshull, J.L., 2008. Dry season fish survival in isolated pools and within sand-beds in the Mzingwane River, Zimbabwe. Afr. J. Aquat. Sci. 33, 95–98.

Nagy, B., 2014. Life history and reproduction of *Nothobranchius* fishes. J. Am. Killifish Assoc. 47, 182–192.

Nyingi, D.W., Agnèse, J.-F., 2011. Tilapia in eastern Africa—a friend and foe. In: Darwall, W.R.T., Smith, K.G., Allen, D.J., Holland, R.A., Harrison, I.J., Brooks, E.G.E. (Eds.), The Diversity of Life in African Freshwaters: Under Water, Under Threat. An Analysis of the Status and Distribution of Freshwater Species Throughout Mainland Africa. IUCN, Cambridge, United Kingdom and Gland, Switzerland, pp. 76–78.

Oliveira, J.M., Segurado, P., Santos, J.M., Teixeira, A., Ferreira, M.T., Cortes, R.V., 2012. Modelling stream-fish functional traits in reference conditions: regional and local environmental correlates. PLoS One 7, e45787.

Ostrand, K.G., Wilde, G.R., 2002. Seasonal and spatial variation in a prairie stream-fish assemblage. Ecol. Freshw. Fish 11, 137–149.

Perkin, J.S., Gido, K.B., Costigan, K.H., Daniels, M.D., Johnson, E.R., 2014. Fragmentation and drying ratchet down Great Plains stream fish diversity. Aquat. Conserv. Mar. Freshwat. Ecosyst. 25, 639–655.

Pires, A.M., Cowx, I.G., Coelho, M.M., 1999. Seasonal changes in fish community structure of intermittent streams in the middle reaches of the Guadiana basin, Portugal. J. Fish Biol. 54, 235–249.

Pires, D.F., Pires, A.M., Collares-Pereira, M.J., Magalhães, M.F., 2010. Variation in fish assemblages across dry-season pools in a Mediterranean stream: effects of pool morphology, physicochemical factors and spatial context. Ecol. Freshw. Fish 19, 74–86.

Pires, D.F., Beja, P., Magalhães, M.F., 2014. Out of pools: movement patterns of mediterranean stream fish in relation to dry season refugia. River Res. Appl. 30, 1269–1280.

Poff, N.L., Allan, J.D., 1995. Functional organization of stream fish assemblages in relation to hydrological variability. Ecology 76, 606–627.

Polačik, M., Podrabsky, J.E., 2015. Temporary environments. In: Riesch, R., Tobler, M., Plath, M. (Eds.), Extremophile Fishes. Springer, Switzerland, pp. 217–242.

Puckridge, J.T., Sheldon, F., Walker, K.F., Boulton, A.J., 1998. Flow variability and the ecology of large rivers. Aust. J. Mar. Freshwat. Res. 49, 55–72.

Rayner, T.S., Jenkins, K.M., Kingsford, R.T., 2009. Small environmental flows, drought and the role of refugia for freshwater fish in the Macquarie Marshes, arid Australia. Ecohydrology 2, 440–453.

Resh, V.H., Bêche, L.A., Lawrence, J.E., Mazor, R.D., McElravy, E.P., O'Dowd, A.P., et al., 2013. Long-term population and community patterns of benthic macroinvertebrates and fishes in Northern California Mediterranean-climate streams. Hydrobiologia 719, 93–118.

Reyjol, Y., Hugueny, B., Pont, D., Giorgio Bianco, P., Beier, U., Caiola, N., et al., 2007. Patterns in species richness and endemism of European freshwater fish. Glob. Ecol. Biogeogr. 16, 65–75.

Reynolds, L.F., 1983. Migration patterns of five fish species in the Murray-Darling river system. Aust. J. Mar. Freshwat. Res. 34, 857–871.

Roberts, J.H., Hitt, N.P., 2010. Longitudinal structure in temperate stream fish communities: evaluating conceptual models with temporal data. In: Gido, K.B., Jackson, D.A. (Eds.), Community Ecology of Stream Fishes: Concepts, Approaches, and Techniques. American Fisheries Society, Maryland, pp. 281–299.

Robson, B.J., Chester, E.T., Mitchell, B.D., Matthews, G., 2013. Disturbance and the role of refuges in Mediterranean climate streams. Hydrobiologia 719, 77–91.

Ruello, N.V., 1976. Observations on some massive fish kills in Lake Eyre. Aust. J. Mar. Freshwat. Res. 27, 667–672.

Sayer, M.D.J., Davenport, J., 1991. Amphibious fish: why do they leave water? Rev. Fish Fisheries 1, 159–181.

Schlosser, I.J., 1987. A conceptual framework for fish communities in small warmwater streams. In: Matthews, W.J., Heins, D.C. (Eds.), Community and Evolutionary Ecology of North American Stream Fishes. University of Oklahoma Press, Norman, pp. 17–24.

Skoulikidis, N.T., Vardakas, L., Karaouzas, I., Economou, A.N., Dimitriou, E., Zogaris, S., 2011. Assessing water stress in Mediterranean lotic systems: insights from an artificially intermittent river in Greece. Aquat. Sci. 73, 581–597.

Snoeks, J., Harrison, I.J., Stiassny, M.L.J., 2011. The status and distribution of freshwater fishes. In: Darwall, W.R.T., Smith, K.G., Allen, D.J., Holland, R.A., Harrison, I.J., Brooks, E.G.E. (Eds.), The Diversity of Life in African Freshwaters: Under Water, Under Threat. An Analysis of the Status and Distribution of Freshwater Species Throughout Mainland Africa. IUCN, Cambridge, United Kingdom and Gland, Switzerland, pp. 42–73.

Thieme, M.L., Abell, R., Stiassny, M.L.J., Skelton, P., Lehner, B., Teugels, et al., 2005. Freshwater Ecoregions of Africa and Madagascar. Island Press, Washington, DC.

Thompson, G., Withers, C., 2002. Aerial and aquatic respiration of the Australian desert goby, *Chlamydogobius eremius*. Comp. Biochem. Physiol. A 131, 871–879.

Tierno de Figueroa, J.M., Lopez-Rodrıguez, M.J., Fenoglio, S., Sanchez-Castillo, P., Fochetti, R., 2013. Freshwater biodiversity in the rivers of the Mediterranean Basin. Hydrobiologia 719, 137–186.

Trewavas, E., 1983. Tilapiine Fishes of the Genera *Sarotherodon*, *Oreochromis* and *Danakilia*. British Museum (Natural History), London.

Turner, T.F., Osborne, M.J., McPhee, M.V., Kruse, C.G., 2015. High and dry: intermittent watersheds provide a test case for genetic response of desert fishes to climate change. Conserv. Genet. 16, 399–410.

Tweddle, D., van der Waal, B.C.W., Peel, R.A., 2015. Distribution and migration of the Caprivi killifish *Nothobranchius capriviensis,* Watters, Wildekamp Shidlovskiy 2015, an assessment of its conservation status, and a note on other killifish in the same area. J. Am. Killifish Assoc. 47, 134–151.

van der Waal, B.C.W., 1997. Some observations on the fish life in a seasonal sand river. South. Afr. J. Aquat. Sci. 23, 95–102.

van der Waal, B.C.W., 1998. Survival strategies of sharptooth catfish *Clarias gariepinus* in desiccating pans in the northern Kruger National Park. Koedoe 41, 131–138.

Wager, R., Unmack, P.J., 2000. Fishes of the Lake Eyre catchment in central Australia. Queensland Department of Primary Industries, Brisbane.

Wardle, G., Pavey, C., Dickman, C., 2013. Greening of arid Australia: New insights from extreme years. Austral Ecol. 38, 731–740.

Watters, B., 2009. The ecology and distribution of *Nothobranchius* fishes. J. Am. Killifish Assoc. 42, 37–76.

Welcomme, R.L., 1979. Fisheries Ecology of Floodplain Rivers. Longman, London.

Welcomme, R.L., 1986. Fish of the Niger system. In: Davies, B.R., Walker, K.F. (Eds.), The Ecology of River Systems. Dr W Junk, Dordrecht, pp. 25–48.

Whitney, J.E., Gido, K.B., Martin, E.C., Hase, K.J., 2015. The first to go and the last to leave: colonization and extinction dynamics of common and rare fishes in intermittent prairie streams. Freshw. Biol. 61, 1321–1334.

Wigington, P., Ebersole, J., Colvin, M., Leibowitz, S., Miller, B., Hansen, B., et al., 2006. Coho salmon dependence on intermittent streams. Front. Ecol. Environ. 4, 513–518.

THE BIOTA OF INTERMITTENT AND EPHEMERAL RIVERS: AMPHIBIANS, REPTILES, BIRDS, AND MAMMALS

4.6

María M. Sánchez-Montoya*,†, Marcos Moleón‡, José A. Sánchez-Zapata§, Daniel Escoriza¶

Leibniz-Institute of Freshwater Ecology and Inland Fisheries (IGB), Berlin, Germany University of Murcia, Murcia, Spain† University of Granada, Granada, Spain‡ Miguel Hernández University, Elche, Spain§ University of Girona, Girona, Spain¶*

IN A NUTSHELL

- Intermittent rivers and ephemeral streams (IRES) support a high diversity of amphibian, reptile, bird, and mammal species (hereafter, wildlife)
- Wildlife uses IRES as water and food resources, breeding and nesting sites, movement corridors, migration stopovers, and resting and shelter areas
- In IRES, wildlife plays pivotal ecological roles as consumers and/or prey, seed dispersal agents, landscape engineers, and animal-mediated nutrient cyclers
- Flow intermittence affects wildlife by altering habitat availability, stream connectivity, food resources, and species interactions
- Wildlife shows morphological, physiological, and behavioral adaptations to cope with flow intermittence
- Conservation of wildlife in IRES requires better knowledge of species' flow requirements, and preservation or restoration of natural flow regimes and ecological refuges

4.6.1 INTRODUCTION

Globally, intermittent rivers and ephemeral streams (IRES) support a high diversity of amphibians, reptiles, birds, and mammals (hereafter, wildlife) (e.g., Levick et al., 2008; Box 4.6.1). However, very little attention has been paid to IRES, possibly because this topic "falls into a gap" between the fields of terrestrial and aquatic ecology. As a result, only scattered studies on the ecology and distribution of these vertebrate groups in IRES exist.

In arid and semiarid areas where water is a limiting resource, IRES support more animals and different species compared with the surrounding landscape, as they offer higher moister content and more abundant vegetation (Soykan et al., 2012). For instance, in the arid southwest region of North America, about 80% of all animals use riparian resources and habitats in some life stage, and more than 50% of breeding bird species nest chiefly in riparian habitats (Krueper, 1993). IRES also harbor rich and distinctive biological communities in temperate and humid areas, partly because of the predictability

BOX 4.6.1 DIVERSITY AND DISTRIBUTION OF WILDLIFE IN IRES

Amphibians are characterized by a biphasic cycle, typically with aquatic larvae and terrestrial adults (e.g., Wells, 2010); therefore, they are common in habitats subjected to wet and dry states. Approximately 7511 amphibian species (6623 Anura; 682 Caudata; 206 Gymnophiona) are known (Vences and Köhler, 2008). Of this total, 71% are aquatic and live in water in at least one life history stage, 3% of these species are water dependent, and around 25% are not water dependent (Vences and Köhler, 2008). Anura (frogs and toads) are very rich in species in wet tropical regions like South America, Madagascar, and Southeast Asia. In contrast, in arid environments, anuran diversity is much lower, and the few species present are restricted to canyons, riverbeds, or oases (Bons and Geniez, 1996). Caudata (salamanders) are also present in IRES, mainly in the northern hemisphere (Wells, 2010).

Reptiles are well suited to occupy IRES, given their drought tolerance and low metabolism. They occur on all continents except Antarctica, with approximately 10,272 species described (341 Testudinata; 1 Rhynchocephalia; 9905 Squamata; 25 Crocodylia) (Vitt and Caldwell, 2014). Testudinata (turtles, terrapins, and tortoises) are diverse in tropical, subtropical, and warm-temperate streams. Most of these species are semiaquatic, with part of their life cycle on land. Many Squamata (snakes and lizards) occur in IRES. Streams provide higher moisture conditions for lizards in their distributional margins (e.g., mesic lacertids in Mediterranean peninsulas; Salvador and Ramos, 1998). In tropical and subtropical regions, agamids, iguanids, gerrhosaurids, skinks, and monitors are true semiaquatic lizards. Snakes are also widely diversified in streams, including giant boids, colubrids, elapids, and viperids. Crocodylia (alligators, caimans, crocodiles, and gharials), considered the largest freshwater dwellers, live in tropical and subtropical areas, often in aquatic habitats in IRES (Martin, 2008).

Many bird species are able to cope with the pulsed resources that characterize IRES, thanks to the group's movement abilities and generalist feeding behavior (Ostfeld and Keesing, 2000). Of the 10,000 extant bird species (Clements, 2011), 5% require freshwater habitats for part of their life cycle (Dehorter and Guillemain, 2008). The diversity of these water-dependent species is low in biogeographic regions with severe climatic conditions where IRES are abundant, such as Australia (Dehorter and Guillemain, 2008). In addition to water-dependent birds inhabiting IRES across different regions, a high diversity of terrestrial species uses the associated riparian habitats, with reproduction adjusted to water availability, especially in arid and semiarid regions (Shine and Brown, 2008).

Many mammals depend on IRES to meet their basic needs for standing water and other resources. The 5487 mammal species (5 Monotremata; 331 Metatheria; 5209 Eutheria) are distributed throughout all continents (Veron et al., 2008). The only freshwater monotreme species (platypus: *Ornithorhynchus anatinus*) persists in IRES in eastern Australia and Tasmania (Souter and Williams, 2001). Chiropterans (bats) are commonly found along IRES (e.g., Hagen and Sabo, 2012). Lagomorpha such as the riverine rabbit (*Bunolagus monticularis*) in South Africa and the European rabbit (*Oryctolagus cuniculus*) live in close association with IRES (e.g., Mills and Hes, 1997; Hughes et al., 2008). Rodentia (rodents) includes many semiaquatic species commonly found in IRES, such as beavers (*Castor*; Gibson and Olden, 2014). Carnivora includes species which exploit IRES, like otters (e.g., *Lutra lutra*; Ruiz-Olmo et al., 2007). The presence of many ungulate species is notorious in IRES, and some of the largest mammals, such as elephants (*Elephas maximus* and *Loxodonta africana*) and hippopotamus (*Hippopotamus amphibius*) depend on IRES, especially during dry seasons (Kok and Nel, 1996; Lakshminarayanan et al., 2016).

of dry periods (Meyer et al., 2007). In those areas, headwater IRES are key systems for the diversity of downstream and riparian ecosystems, and for the biological integrity of the entire river network.

4.6.2 IMPORTANCE OF IRES FOR AMPHIBIANS, REPTILES, BIRDS, AND MAMMALS

IRES provide multiple ecological functions for wildlife along channels and riparian habitats in both wet and dry phases (Fig. 4.6.1). Animals use IRES as water and food resources, breeding and nesting sites, movement corridors, migration stopovers, and resting and shelter areas.

Riparian zone	Flowing channel	Dry channel
1. Food resource	5. Food resource	9. Corridor terrestrial spp.
2. Breeding terrestrial spp.	6. Water resource	10. Shelter
3. Corridor terrestrial spp.	7. Breeding aquatic spp.	11. Food resource
4. Shelter	8. Corridor aquatic spp.	

FIG. 4.6.1

IRES provide multiple ecological functions for wildlife along channels and riparian habitats during both wet and dry phases. Even when dry, river beds may provide important ecological functions to semiaquatic and terrestrial vertebrates, while functions along riparian zones can be maintained. This drawing illustrates common terrestrial species found in Mediterranean IRES.

WATER AND FOOD RESOURCES

In flowing channels and isolated pools of IRES, amphibian larvae consume aquatic vegetation, detritus, and benthic algae and may compete with other primary consumers, such as gastropods, insects, and crustaceans (Brönmark et al., 1991; Chapter 4.3). Small aquatic invertebrates are also important prey for salamander larvae and some tadpoles (Parker, 1994). Moreover, reptiles prey on stranded fish in permanent pools along IRES (e.g., western terrestrial garter snakes (*Thamnophis elegans*) in

California; Zale et al., 1989). Other reptile species use riparian habitats of IRES when prey availability is especially high; for example, the Milos viper (*Macrovipera schweizeri*) during bird migrations (Nilson et al., 1999). Some reptiles occupy IRES opportunistically, displaying smaller sizes. One example is the reduction of 50% in size in the Australian freshwater crocodile, *Crocodylus johnsoni* (Webb and Manolis, 1998).

Water birds (i.e., Anatidae, Ardeidae, Ciconiidae, Charadriidae, Scolopacidae) depend on aboveground water to feed on a variety of plants and animals. For instance, herons and kingfishers feed on fish and aquatic invertebrates in pools of IRES (e.g., Tramer and Tramer, 1977). Many water birds track temporal water availability over vast areas in desert and semiarid ecosystems. Such nomadic species include the Australian banded stilt (*Cladorhynchus leucocephalus*) and the grey teal (*Anas gracilis*), which have been particularly well documented in arid ecosystems, where most water birds respond to the pulsed availability of surface water in Australia (Kingsford et al., 2010; Pedler et al., 2014) and elsewhere (Petrie and Rogers, 2004). Riparian bird species may also benefit from insect subsidies from temporary aquatic habitats (Baxter et al., 2005).

For some mammals, IRES are an essential source of standing water. Piscivorous mammals, such the Eurasian otter (*Lutra lutra*) in Mediterranean regions and the clawless otter (*Aonyx capensis*) in Africa can exploit IRES for food all year round provided that some pools remain during the driest season and support if not fish, then crabs, crayfish, frogs, or other aquatic life (Ruiz-Olmo and Delibes, 1998; Skinner and Chimimba, 2005). Some insectivorous mammals, including bats, forage in IRES, consuming both aerial and ground-dwelling insects (Seidman and Zabel, 2001). In the semiarid savannahs of Africa, most grazing ungulates are highly water dependent and their distribution is almost completely confined to areas within 15 km of water (Western, 1975). In desert regions of Namibia, ephemeral rivers and their floodplains provide the habitat on which elephants and other large mammals, such as black rhinos (*Diceros bicornis*) and giraffes (*Giraffa camelopardalis*), rely for long-term survival, and which provide water and food even during the dry season (Viljoen, 1989; Skinner and Chimimba, 2005). Elephants are able to cover requirements for extra sodium in their diet by drinking saline water (Weir, 1972), present in many IRES. At night, hippos (*Hippopotamus amphibius*) forage on the short green grass that grows close to water, normally less than 3 km but up to 10 km depending on the season (Owen-Smith, 1992). Similarly, the Asian elephant is closely associated with IRES and their riparian habitats during the dry season (Lakshminarayanan et al., 2016).

IRES may also be viewed as food hot spots for mammalian carnivores, either in the form of live prey or as carrion because many sick, malnourished, and senescent mammals often die near surface water (normally during the late dry season; Pereira et al., 2014), and their carcasses are readily exploited by a multitude of scavenging carnivores (Moleón et al., 2014). In addition, many mammals find arthropods and fruits associated with the riparian zones of IRES (e.g., Rosalino and Santos-Reis, 2008).

BREEDING AND NESTING SITES

Many amphibians use IRES for reproduction and may complete their development if water remains for at least two months (Zeiner et al., 1988–1990). Many salamanders breed in streams, such as Mediterranean fire salamanders (*Salamandra* spp.) or rough-skinned newts (*Taricha* spp.) in California. Coastal giant salamanders (*Dicamptodon tenebrosus*) deposit their eggs in pools in intermittent headwater reaches where resource competition and predation pressures from fish are reduced by drying (Wilkins and Peterson, 2000). Turtles may breed on the sandy banks or in the meadows close to streams, usually within 2–300 m (Bodie, 2001). The endemic Australian freshwater crocodile (*C. johnsoni*) nests in sandy burrows during the dry season close to isolated mainstream pools (Webb et al., 1983).

Most large colonial water birds (i.e., herons, flamingos, pelicans), that are associated with fluctuating water availability, concentrate in floodplains that provide islands for breeding rather than in narrower aquatic systems, such as river beds, that usually attract smaller and less colonial species like plovers (e.g., *Charadrius* spp.) and ducks (e.g., *Tadorna* spp.) (del Hoyo et al., 1992). Riverbank cliffs in IRES are important for hole-nesting species such as bee-eaters (Meropidae) that dig burrows in earth banks in arid and semiarid ecosystems of Africa, Australia, and Southern Europe. Other hole-nesting birds (sparrows, swifts, owls, jackdaws, and even woodpeckers) benefit from burrow availability in riverbank cliffs which provide them with nest sites, even in the absence of tree cavities, the major nesting habitat for many bird species elsewhere (e.g., Newton, 1994). In addition, riparian vegetation makes a disproportionately positive contribution, in relation to the surrounding area, to breeding populations of birds in different regions (Carlisle et al., 2009). In general, a positive relationship exists between streamside zone width and breeding bird abundance and diversity (Dickson et al., 1995).

Many mammals, from rodents to baboons and elephants, are largely associated with IRES in arid environments and reproduce there. For instance, mating and calving of hippos take place in water (Skinner and Chimimba, 2005).

AQUATIC AND TERRESTRIAL MOVEMENT CORRIDORS

Flowing channels in IRES, reconnected by floods, provide important movement corridors for aquatic species. For example, juvenile crocodiles, turtles, and snakes seasonally migrate along channels to occupy highly productive stream pools (Webb and Manolis, 1998). Hippo movements are restricted to the vicinities of IRES and other bodies of water (Skinner and Chimimba, 2005).

Dry channels can also act as preferential pathways for wildlife and domestic terrestrial fauna during the dry phase (Steward et al., 2012). For example, Eurasian and Nearctic turtles perform seasonal movements of 2–4 km following the appearance of ephemeral pools (Ernst and Barbour, 1989). Nile crocodiles (*Crocodylus niloticus*) (Martin, 2008) and elephants (Kok and Nel, 1996; Jacobson, 1997) use dry channels as routes to access drinkable water in disconnected pools. Recently, dry riverbeds in Mediterranean IRES have been reported as important movement corridors for a wide range of terrestrial vertebrates including lizards, birds, rodents, lagomorphs, carnivores, and ungulates, thus enhancing natural connectivity and supporting biodiversity along rivers and within the entire catchment (Sánchez-Montoya et al., 2016).

In addition, riparian zones, including those of IRES, are preferential pathways for the movement of many terrestrial and semiaquatic fauna (e.g., Hilty et al., 2006). For instance, postmetamorphic amphibians use riparian habitats to migrate up to 5 km (Zeiner et al., 1988–1990). Similarly, many carnivores use riparian habitats in Mediterranean IRES as corridors (Santos et al., 2011).

MIGRATION STOPOVERS AND RESTING AND SHELTER AREAS

Many bird species use IRES as stopovers during their migrations. The use of IRES and oases allows birds to refuel during migration while feeding on fruits and insects (Bairlein, 1988). Although most usually stay for only one or a few days at each site (Lavee et al., 1991), the importance of these stopovers seems essential for their survival as, for example, most birds (60%) would not be able to cross deserts without feeding and drinking (Biebach et al., 1986). Thus, riparian patches in southeastern Arizona are important to route migrants as stopover sites regardless of their size and degree of isolation or connectivity (Skagen et al., 1998). IRES could be also crucial for maintaining the last long-distance migrations of terrestrial mammals on our planet (Thuiller et al., 2006; Harris et al., 2009) as migratory mammals use IRES on their migration routes for water and food supplies (Skinner and Chimimba, 2005).

IRES are also resting sites for many species. Hippos spend most of their daily activity, from predawn hours to dusk, resting in water where most of their social interactions take place (Owen-Smith, 1992). During the driest season, daytime hippo activity is restricted to permanent pools where they can submerge completely. In wetter areas, IRES can serve as resting sites during the wet season if major rivers are flooded (Skinner and Chimimba, 2005).

Shelters in IRES not only provide refuge from predators but also critical protection from extreme heat and aridity (Levick et al., 2008). In some IRES, the riverbank provides small caves and cracks which provide critical refuges for animals such as the desert tortoise (*Gopherus agassizii*) (Martin and Van Devender, 2002). Some crocodiles take refuge in caves along river banks or burrow in the mud of riverbeds (Tucker et al., 1997; Shine et al., 2001). Certain mammals such as megaherbivores seek shade from riparian vegetation along IRES (Owen-Smith, 1992). Riverine forests are shelter habitats for amphibians, including those with no aquatic reproduction (e.g., robber frogs *Eleutherodactylus* spp.) (Rojas-Ahumada and Menin, 2010).

4.6.3 ECOLOGICAL ROLES OF WILDLIFE IN IRES

These four vertebrate groups play important ecological roles in IRES ecosystems. As consumers and prey, seed dispersal agents, landscape engineers, and nutrient suppliers, they provide direct and indirect biotic control of the structure and function of IRES (Fig. 4.6.2). In some cases, one single species can exert major control over an ecosystem by performing multiple different functions (Box 4.6.2).

CONSUMERS, PREY, AND SEED DISPERSAL AGENTS

Wildlife in IRES, acting as consumers, prey or both, may directly control the abundance and dynamics of species at lower and higher trophic levels. Juvenile crocodiles, turtles, semiaquatic lizards, and snakes occupy highly productive pools and become the apex predators as they consume all types of animal prey (Vitt and Caldwell, 2014). For example, the semiaquatic Sonoran mud turtle (*Kinosternon sonoriense*), which inhabits IRES in western US deserts, is one of the few predators to access newly created pools (Stone, 2001). Monitor lizards (*Varanus* spp.) play a key role in these ecosystems as they are large predators that destroy large numbers of nests, which affects the reproductive success of turtles and crocodiles (Limpus, 1971). During the dry season, crocodiles migrate to permanent pools, where they capture terrestrial mammals, fish, and turtles (Webb and Manolis, 1998; Whiting and Whiting, 2011). In addition, predatory birds such as herons and kingfishers feed on fish and invertebrates in the pools of IRES and strongly influence these communities (e.g., Tramer and Tramer, 1977). Some vertebrates are also major prey for aquatic and terrestrial predators. For example, amphibian eggs, larvae, and metamorphlings are important resources for many aquatic and terrestrial predators (Fig. 4.6.2a), such as dragonflies and diving beetles, crustaceans, fish, reptiles, and birds.

FIG. 4.6.2

Wildlife plays key ecological functions in IRES. They exert biotic control of: (i) biodiversity as consumers and/ or prey: (a) Algerian ribbed newt (*Pleurodeles nebulosus*) larvae feeding on Mediterranean painted frog (*Discoglossus pictus*) in Cap Bon peninsula, Tunisia, (ii) habitat structure as ecosystems engineers: (b.1, b.2) white-fronted bee-eater (*Merops bullockoides*) holes dug in the bank of a river in Hluhluwe-Imfolozi Park, South Africa; (c) wild burro (*Equus asinus*) digging a hole in the dry channel of Black Canyon in Arizona, USA, and (iii) nutrient cycling: (d) latrine of European rabbits (*Oryctolagus cuniculus*) in dry riverbeds in Rambla de la Parra, Spain; (e) effect of fecal nutrients from a domestic animal on primary producers in Chicamo stream, Spain.

Photos: Courtesy D. Escoriza (a), D. Carmona-López (b.1), J. Bautista (b.2), E. Lundgren (c), M.M. Sánchez-Montoya (d), and R. Gómez (e).

Some mammals and birds are crucial for many plant species by acting as seed dispersal agents (Herrera and Pellmyr, 2002). In this way, they may exert indirect biotic control of plant communities in IRES.

LANDSCAPE ENGINEERS

Wildlife alters the structure (e.g., geomorphology and vegetation features) and function (e.g., connectivity) in IRES by acting as landscape engineers (Naiman and Rogers, 1997). For instance, the colonial European bee-eater (*Merops apiaster*), a common species in Mediterranean IRES, and the white-fronted

BOX 4.6.2 KEY ECOLOGICAL ROLES OF WILDLIFE IN IRES: THE CASE OF THE AFRICAN ELEPHANT (*L. AFRICANA*) IN THE UGAB RIVER (NAMIBIA)

During the dry season, African elephants (*L. africana*) are dependent upon IRES such as the Ugab River (Namibia) (Fig. 4.6.3a and b), where they play different key ecological roles. One role is as seed dispersal agents of *Acacia albida*. Seeds contained in elephant dung on dry riverbeds (Fig. 4.6.3c) are dispersed by water in the wet phase when the chances of germination increase (Nilsson and Dynesius, 1994). Elephants are also one of the main landscape engineers in IRES. They dig for water in dry riverbeds, creating waterholes (Fig. 4.6.3d) that also provide water and refuge for other animals, and create trail networks during their movement (Fig. 4.6.3e) that are also used by other species such as leopards (*Panthera pardus*) (Fig. 4.6.3f). In addition, they substantially affect vegetation communities (Fig. 4.6.3c), and influence the size structure of trees in riparian areas of IRES (Jacobson, 1997). Finally, they provide animal-mediated nutrient cycling, supplying nutrients to the ecosystem via excretory processes (Fig. 4.6.3c). Elephant dung can be a critical source of nutrients for IRES soil and are the key food for many invertebrates such as dung beetles (Haynes, 1991) and even vertebrates like hippos (Skinner and Chimimba, 2005).

FIG. 4.6.3

A single wildlife species can play multiple roles in an IRES such as the Ugab River in the northwestern Namibia (a). African elephants (*Loxodonta africana*) moving along the dry channel of the river (b) can leave droppings and damage acacia trees by foraging (c). Elephants also dig waterholes (d) and create trails in the riparian zone (e) that may be used by other animals, as shown by these leopard footprints on elephant footprints (f).

Photos: Courtesy M. Moleón (a and d-f) and M.M. Sánchez-Montoya (b and c).

bee-eater (*Merops bullockoides*) in South Africa (Fig. 4.6.2b), can dig burrows in the banks of IRES that provide nesting habitat and food to other species such as other birds, reptiles, small mammals, and invertebrates, particularly beetles and flies (Jones et al., 1994; Casas-Crivillé and Valera, 2005). Beavers are also present in dryland streams of North America where they create dams and ponds, altering IRES flow regimes and sometimes converting them to perennial flow. Beaver dams increase stream flow during dry seasons but also reduce stream velocity and erosive power during peak flow (Gibson and Olden, 2014).

Some mammal species in Africa such as baboons (*Papio* spp.), gemsboks (*Oryx gazelle*), and elephants (*Loxodonta africana*) dig wells in dry riverbeds, providing habitat for aquatic invertebrates (Walker, 1996) and water not only for themselves, but also for many other animals, including humans (Haynes, 1991; Naiman and Rogers, 1997; Ramey et al., 2013). This digging behavior in dry channels has been also reported for the Asiatic wild ass (*Equus hemionus*) in Mongolia (Kaczensky et al., 2006) and the wild introduced burro (*Equus asinus*) in Arizona (Fig. 4.6.2c; E. Lugendre, personal communication), this latter case providing water for many species such as mule deer, coyotes, javelinas, and cattle. Finally, hippos have profound effects on the local hydraulic conditions of African IRES (e.g., flow diversion; McCarthy et al., 1998; Ellery et al., 2003), by creating trails that make channels deeper and reducing roughness by removing aquatic vegetation.

NUTRIENT CYCLING

Wildlife can greatly influence nutrient cycling in IRES. They can supply nutrients via excretion, recycle nutrients within a habitat, and/or translocate them across habitats or ecosystems (Vanni, 2002). The feces of many species supply nutrients to primary producers via decomposition and remineralization by microbes (e.g., Hansson et al., 1987). For instance, the latrines of European rabbits (Fig. 4.6.2d), commonly observed in dry riverbeds and riparian zones of Mediterranean semiarid IRES (Sánchez et al., 2004), may constitute fertile islands in soils with low concentrations of essential plant macronutrients (Puigdefábregas et al., 1996) and can therefore facilitate plant growth (Delibes-Mateos et al., 2008). Feces of livestock and domestic animals can be also an important source of nutrients for instream primary producers in IRES in the wet phase (Fig. 4.6.2e), which usually require nutrients in a dissolved form (Vanni, 2002; Chapter 4.2). Finally, nutrients sequestered in animal bodies may be released and become available when animals die and decompose. Vertebrate carcasses represent a huge amount of biomass that may have multiple ecological effects in IRES, ranging from scavenger aggregation to soil fertilization (Barton et al., 2013).

4.6.4 FLOW INTERMITTENCE EFFECTS ON WILDLIFE

Flow intermittence affects the presence, abundance, and/or activity of IRES wildlife through direct effects on habitat availability and stream connectivity and through indirect effects on food resource and species interactions.

AMPHIBIANS

Streamflow intermittence favors some amphibians as it reduces the presence of aquatic predators (Holomuzki, 1995). For instance, the terrestrial black salamander (*Aneides flavipunctatus*) preferentially selects intermittent reaches for laying its eggs on stream banks, where flow intermittence has also had a direct effect by increasing habitat availability (Welsh et al., 2005). However, unusual droughts drastically reduce their reproductive output. Adults can become hypogean during these harsh periods and feed on subterranean invertebrates (Price et al., 2012). These harsh conditions can leave their mark on the genetic structure of populations, including some amphibians with good dispersal capacity (Murphy et al., 2010). Different responses have been detected according to species resilience. For instance, the American bullfrog (*Lithobates catesbeianus*) became extinct under drought conditions, while the northern cricket frog (*Acris crepitans*) recovered rapidly (Blair, 1957).

REPTILES

During drying, freshwater turtle species either estivate in the mud of IRES stream beds and in terrestrial refugia such as leaf litter and soil (e.g., Ligon and Peterson, 2002; Buhlmann et al., 2009), or move to other aquatic habitats, mainly isolated pools (e.g., Gibbons et al., 1983; Rees et al., 2009). In the former case, the turtles that depart sooner due to earlier drying can have significantly smaller carapace lengths and body mass, which suggests a morphological population response to drying (Bondi and Mark, 2013). Also, they can alter their reproduction pattern in response to drought conditions, for instance, by reducing their reproductive output in dry years (Gibbons et al., 1983). Water snakes usually respond to drought periods by migrating to permanent pools, with estivation documented only for the black swamp snake (*Seminatrix pygaea*) from southeastern United States (Winne et al., 2006). In the Mediterranean region, the dice snake (*Natrix tessellata*) adapted its diet to include more fish than amphibians, as a response to stream bed desiccation (Luiselli et al., 2007). Ten species of crocodiles can estivate as their metabolic rates decline (Christian et al., 1996).

BIRDS

River drying can be an important limiting factor to bird populations in arid and semiarid systems (Carothers et al., 1974). Declines in bird populations linked to loss of riparian habitats owing to river drying have been long described as a major effect (e.g., Colorado River; Rosenberg et al., 1991). For example, along a hydrological gradient in the San Pedro River in Arizona, Brand et al. (2010) found that the hydrological regime interacted with vegetation type which, in turn, influenced both the breeding bird density and nest survival of bird species.

MAMMALS

Mammals are influenced by loss of standing water and food availability during the dry phase of IRES. For instance, river drying can lead to reduced bat activity via reductions in insect prey availability which is tied to the timing of peak emergence and duration of emergence, and also to the extent of riparian vegetation (Hagen and Sabo, 2011, 2012, 2014). Seasonal river drying initially causes imperfect tracking by consumers of localized concentrations of resources, but later results in the disappearance of insects and bats after reaches have completely dried (Hagen and Sabo, 2011). Timing of breeding of the Eurasian otter occurs earlier in Mediterranean streams than in temperate environments as a response to summer drought (Ruiz-Olmo et al., 2002).

4.6.5 WILDLIFE ADAPTATIONS TO COPE WITH FLOW INTERMITTENCE

Wildlife possesses many morphological, physiological, and behavioral adaptations to cope with flow intermittence or to live in and around IRES, particularly in arid and semiarid areas (Fig. 4.6.4).

MORPHOLOGICAL ADAPTATIONS

Amphibian larvae in some cases undergo morphological changes during river drying, such as broadening their heads and developing jaw muscles, which enable them to prey on larger items (Pfennig, 1999). This provides access to prey of high nutritional value (such as anostracans or other tadpoles) and exploitation of a trophic resource not used by tadpoles with generalist phenotypes (Pfennig, 1999). One of the most outstanding examples of morphological adaptation to arid conditions is represented by camels, which inhabit most deserts and steppes in Asia, Africa, and South America (Wilson and

FIG. 4.6.4

Wildlife presents morphological, physiological, and behavioral adaptations for coping with flow intermittence in IRES. Reproduction in salamanders such as Corsican fire salamander (a, *Salamandra corsica*; central Corsica) is synchronous with the first rains. European pond turtles (b, *Emys orbicularis;* Khroumire, Tunisia) congregate in the remaining pools or estivate under the mud. Double-banded sandgrouse (c, *Pterocles bicinctus*) live in arid ecosystems from Eurasia and Africa and move large distances regularly to drink at waterholes. An African elephant (*L. africana*) sprays sand with its trunk (d) to protect its skin from the sun in IRES in Namibia.

Photos: Courtesy D. Escoriza (a and b), J.A. Sánchez-Zapata (c), and S. Eguía (d).

Mittermeier, 2009). The dromedary (*Camelus dromedaries*) and bactrian camels (*C. bactrianus*) have one or two humps, respectively, that store fat and provide energy and metabolic water, offering resistance to long periods without drinking.

PHYSIOLOGICAL ADAPTATION

Many examples of this type of adaptation are present in wildlife inhabiting IRES. In amphibians, some frogs have developed mechanisms to reduce desiccation-related larval mortality. Larvae tolerate a wide range of fluctuations in water temperature and chemical composition, and their development accelerates when ponds start to desiccate (Brady and Griffiths, 2000). Fast larval development favors long-term stable populations. For instance, the painted frogs (*Discoglossus* spp.) that breed in highly seasonal streams in the western Mediterranean region (Escoriza and Boix, 2014) are among the species with the shortest larval period in Europe (Díaz-Paniagua et al., 2005).

Protective cocoons that increase resistance to drying are developed in the adults of at least 12 frog species around the world—for example, in Australia (sandhill frog: *Arenophryne rotunda*; northern burrowing frog: *Neobatrachus aquilonius*), North America (lowland burrowing treefrog: *Smilisca fodiens*), and South America (escuerzo: *Lepidobatrachus llanensis*). For some frogs (e.g., *Ceratophrys* spp., *Notaden* spp., *Scaphiopus* spp.) and salamanders (tiger salamander: *Ambystoma trigrinum*), their metabolic rates decline and plasma osmolarity increases in order to reduce water loss (Cartledge et al., 2006; Groom et al., 2013). Although most amphibians are ureotelic, the waxy monkey frog (*Phyllomedusa sauvagei*) from South America and the grey foam-nest tree frog (*Chiromantis xerampelina*) from South Africa have developed uricotely to reduce fluid loss (Balinsky et al., 1976; Campbell et al., 1984).

Birds have developed many physiological adaptations to desert conditions (reviewed by Williams and Tieleman, 2005), many of which are likely to be shared by birds that inhabit IRES in arid ecosystems. Adaptations include a lower field metabolic rate and evaporative water loss in desert versus mesic ecosystems in the larks (Alaudidae) (Williams and Tieleman, 2005). Hyperthermia (body temperature above normal) in small birds (Tieleman and Williams, 1999), and even the suppression of the classical adrenocortical response to stress (Wingfield et al., 1992), have also been proposed as physiological mechanisms under low water availability and extreme heat conditions. Some mammal species such as camels can digest low-quality forage, and dromedaries can also reduce their metabolic rate (Wilson and Mittermeier, 2009).

BEHAVIORAL ADAPTATIONS

Behavioral adaptations are common in all four vertebrate groups. In amphibians, some frogs such as *S. fodiens* have compact bodies and densely ossified skulls to facilitate burrowing as a mechanism to cope with river drying (Wells, 2010). In some amphibian species, reproduction is typically synchronous with the first rains (Fig. 4.6.4a), which usually occur in successive pulses, so that several larval cohorts coexist in the same pool. This strategy is characteristic of the frogs and salamanders exposed to erratic hydrological regimes under steppic and Mediterranean climates (Gómez-Rodríguez et al., 2009; Escoriza and Ben Hassine, 2015). In the eastern Mediterranean, the first cohort of the eastern fire salamanders (*Salamandra infraimmaculata*) exerts intense cannibalistic pressure on later ones, increasing its chances of survival during prolonged summer droughts (Degani, 2016).

Aquatic reptiles also display behavioral adaptations to deal with river drying. For example, estivation is the commonest adaptation of freshwater turtles under semiarid and arid climates (Fig. 4.6.4b), while in wetter environments, species perform short migratory movements (Gibbons et al., 1983). Other adaptations to IRES include hatching triggered by rainfall in the pig-nosed turtle (*Carettochelys insculpta*), a soft-shell turtle which synchronizes the emergence of hatchlings with increased river flow (Webb et al., 1986). Female yacarés (*Caiman yacare*) protect juveniles during seasonal migrations to reach pools that still contain water, even protecting offspring that are not their own (Cintra, 1989).

Many bird species have evolved behavioral mechanisms to allow them to exploit favorable microclimates, even in the most arid places. Nomadism is a major adaptation of birds to track unpredictable resources on large spatial scales in terrestrial and aquatic ecosystems worldwide, including IRES subject to large spatiotemporal variations in resource availability (Allen and Saunders, 2005). Other species such as the double-banded sandgrouse (*Pterocles bicinctus*), which lives in arid ecosystems from Eurasia and Africa, move large distances regularly to drink at waterholes (Fig. 4.6.4c). Behavioral responses to temperature are present in some species (Davies, 1982). Some birds display nocturnal activity and diurnal immobility at low metabolic rates. Diurnal birds confine their activity to cool periods and take long flights to drinking places early in the morning or late in the evening. This behavior has been reported in many Australian and African species, such as Bourke's parrot (*Neopsephotus bourkii*) in Australian arid areas characterized by the presence of IRES (Pavey and Nano, 2009).

Many mammal species in arid lands have developed mechanisms to cope with water scarcity and high temperatures. Elephants are able to move long distances (home ranges of up to $8700\,km^2$ and mean daily distances of 27.5 km) thanks to their exceptional capacity to memorize resource distribution, and can go for up to 4 days without drinking, allowing them to survive in very arid regions and to forage up to 70 km from water (Lindeque and Lindeque, 1991). Many ungulates in these environments can also move long distances (e.g., round-trip distances of up to 550 km for pronghorns in North America and 500 km for Burchell's zebra (*Equus quagga*) in South Africa; Berger, 2004; Naidoo et al., 2016). Nocturnal activity is also a common feature of desert-dwelling mammals (e.g., Skinner and Chimimba, 2005). Many mammals burrow underground at the hottest part of the day to avoid heat and to increase water retention. Some megaherbivores and ungulates wallow in mud or dust themselves (Fig. 4.6.4d) to avoid excessive water loss (Owen-Smith, 1992; Skinner and Chimimba, 2005).

4.6.6 VULNERABILITY, CONSERVATION, AND MANAGEMENT OF WILDLIFE IN IRES

Direct anthropogenic activities combined with the spread of invasive species, emerging diseases, and climate change effects (Chapter 5.1) threaten the local habitats and biological diversity of wildlife in IRES.

FLOW ALTERATION AND HABITAT DEGRADATION

IRES can be particularly susceptible to water diversion (directly or via groundwater withdrawal) and flow regulation in arid and semiarid areas. For amphibians, dams act as propagule reservoirs for large predatory fish and crustaceans (Gehrke and Harris, 2001) and cause many amphibian populations to decline. Birds can also be severely affected. For example, in the Macquarie River in eastern Australia, the reduction of marshes to 40%-50% of their original size by flow diversions and weirs has led to a sharp drop in the number of colonially nesting water birds (Kingsford and Thomas, 1995).

Degradation and fragmentation of riparian habitats in IRES jeopardize the survival of wildlife species (e.g., Levick et al., 2008). For instance, timber harvesting in ephemeral headwater streams in the United States is linked to the reduction in abundance of some species of terrestrial and stream-breeding salamanders, such as slimy salamander (*Plethodon glutinosus*) (Maigret et al., 2014).

WATER POLLUTION

In different regions of the world, flows of IRES have been turned into perennial flows dominated by effluent discharges, particularly in urbanized watersheds (Brooks et al., 2006). Effluents that have high concentrations of nutrients and contaminants can strongly affect wildlife in IRES. For instance, amphibians are severely affected by nitrogen pollution, with lethal and sublethal effects at nitrate concentrations between 2.5 and $100\,mg\,l^{-1}$ (Rouse et al., 1999). Mercury-contaminated industrial effluents affect riparian insectivorous species such as birds and bats (Powell, 1983). A wide range of pharmaceutical compounds detected in the effluent from wastewater treatment plants (Stackelberg et al., 2004; Hernando et al., 2006) has adverse effects on fertility in wild amphibians (Carlsson et al., 2009; Säfholm et al., 2014), although responses of other terrestrial vertebrate groups to pharmaceuticals and their metabolites are still largely unknown (Daughton and Ternes, 1999). Some pesticides used in agriculture also interfere with the endocrine system of frogs and enzymatic activity (acetylcholinesterase), and reduce juvenile recruitment and cause mortality in early development stages (Sparling et al., 2001).

River salinization is a frequent phenomenon, especially in the rivers of dry and Mediterranean climates with prolonged droughts (e.g., Cañedo-Argüelles et al., 2016). Increasing salinity can have adverse effects on fauna depending on its tolerance. Amphibians are poor regulators of salt, and salinity effects on the embryonic and larval stages of different frog species appear beyond about $6000\,\mu S\,cm^{-1}$ at 25°C (e.g., Smith et al., 2007). Freshwater turtles are generally sensitive to salinities above $5000\,mg\,l^{-1}$ and lower salinities can even affect species that do not possess functional salt glands (Hart et al., 1991). Water birds' sensitivity to salinity varies, mainly due to changes in their food, nesting, and cover, when salinization affects macroinvertebrates and macrophytes. Breeding success for some water bird species declines when salinity exceeds $3000\,mg\,l^{-1}$ (Hart et al., 1991).

SPREAD OF INVASIVE SPECIES AND DISEASES

Introducing invasive species in IRES, frequently favored by flow regulation or anomalous droughts (e.g., Ruhí et al., 2015), can disrupt indigenous communities. For instance, dams promote the invasion by exotic amphibian species such as the American bullfrog (*L. catesbeianus*) and clawed frog (*Xenopus laevis*) (Gehrke and Harris, 2001), which act as competitors, predators, and pathogen vectors, and are major causes of local declines of frogs (Solís et al., 2010), salamanders, and endemic fish (Mueller et al., 2006).

Exotic species in riparian zones of IRES can affect different populations. For instance, the establishment of exotic tamarisks (saltcedars: *Tamarix* spp.), a drought-tolerant shrub species, in native riparian ecosystems in arid and semiarid southwestern United States (Stromberg et al., 2007, 2013) negatively affects migratory bird populations by reducing stopover-site quality (e.g., Walker, 2008). However, tamarisks also have a positive effect on an endangered riparian songbird by providing key nesting habitat (Sogge et al., 2008). The spread of Russian olive (*Elaeagnus angustifolia*) along IRES in eastern Colorado, by replacing some species of Salicaceae (Zale et al., 1989), has negatively affected around 31% of native bird species that depend on larger cottonwoods for cavity nesting or insect prey (Shafroth et al., 1995).

Amphibian extinction is linked to the human-mediated translocation of pathogenic fungi, such as the chytrid fungus. The spread of this fungus has caused the extinction of some tropical IRES-dwelling frogs in otherwise pristine habitats, particularly in Australia and Mesoamerica (Hero and Morrison, 2004; Cheng et al., 2011).

CLIMATE CHANGE

Assessments of global patterns of climate change forecast a general increase in temperature and in intensity and frequency of floods and droughts, including in IRES (Chapter 2.2). Species that inhabit arid conditions may be better adapted to such changes (e.g., Dean, 2004) than those in wetter areas. Even so, climate change in recent decades has affected all levels of ecological organization, including population and life history, phenology and geographical range, and species composition of communities (Berteaux et al., 2006).

Global warming is linked to the extirpation or depletion of some amphibian populations (Araújo et al., 2006), and particularly affects "niche specialists," such as golden toad (*Incilius periglenes*) in Costa Rica (Crump et al., 1992). Global warming causes changes in reproduction timing, such as early breeding in the western toad (*Anaxyrus boreas*), although no widespread pattern has been found (Blaustein et al., 2001). Some species are predicted to decline as a result of smaller and less predictable rainfall which will affect their breeding cycle, such as the quacking frog (*Crinia georgiana*) in Australia (Edwards et al., 2007). For reptiles in IRES, climate warming can impact crocodiles and some turtles which present temperature-dependent sex determination, because the sex ratio of offspring could be altered by only a 1°C shift in incubation temperature (e.g., Janzen, 1994).

The response of many bird species to recent climate change indicates that many have the phenotypic plasticity to cope with such change (e.g., Crick, 2004). However, others could find it difficult to adapt due to, for instance, decoupling in migratory timing and food resources (Both and Visser, 2001). The earlier arrival of migrant birds in response to lower rainfall has been reported in Australia (Chambers, 2008) and South Africa (Simmons et al., 2004), among other geographical areas. For passerine migratory species which usually migrate to Africa, microevolutionary shifts to more sedentary activity have been detected in response to climatic warming (Coppack et al., 2003), and it is speculated that nomadic birds will move more frequently or further (Simmons et al., 2004).

For mammals, reduced availability of standing water in the warmest months has been reported to affect those species that require water for successful reproduction, such as bats in North America (Adams and Hayes, 2008).

RECOMMENDATIONS FOR CONSERVATION AND MANAGEMENT IN THE CONTEXT OF GLOBAL CHANGE

The conservation of wildlife in IRES needs a better knowledge of the species that inhabit these systems. The large knowledge gaps regarding the ecology and threats to these groups in IRES (e.g., Zale et al., 1989; Hero and Morrison, 2004) should be addressed by baseline inventories to describe the distribution of species compared to those of perennial watercourses (Levick et al., 2008), in combination with studies of species' flow requirements and their responses to changing flow regimes. In particular, climate-induced shifts in stream drying and connectivity and their implications for the persistence of vertebrate fauna considering the dispersal capacity of organisms (Jaeger et al., 2014) emerges as one of the most urgent issues to study.

Preservation or restoration of natural flow regimes seems to be vital to support wildlife populations because they typically require the combined aquatic and terrestrial mosaic of habitats in IRES. Reduction in water extractions for irrigation and other human activities is one of the most relevant management steps in the conservation of vertebrate habitats. In addition, natural flows may promote the survival of native species over invasive species (Rahel and Olden, 2008).

Refugia have been suggested as priority sites for conservation under climate change because of their ability to facilitate survival of biota under adverse conditions (Davis et al., 2013). The management of these refuges, both aquatic (permanent pools, perennial reaches, and springs) and terrestrial (riparian and upland areas and dry river beds) for IRES wildlife species, should be done in a landscape context to ensure connectivity among them (Chapter 2.3) and between refuges and the surrounding landscape (Sheldon et al., 2010; Robson et al., 2013) determined by natural pattern of flooding and drying (Sheldon et al., 2010).

Although conservation programs in IRES are uncommon (Leigh et al., 2016; Chapter 5.4), there are several that contribute to maintain wildlife diversity of IRES. For instance, after almost three decades of actions, the Nature Conservancy in Arizona has achieved the protection of more than one-third of the river corridor in the San Pedro River, one of the large undammed rivers in southwestern United States. Interestingly, a citizen science project is helping to map flows of the San Pedro River to allow water managers to monitor flow patterns and adapt strategies to restore year-round flows.

4.6.7 CONCLUSIONS

The ecological value of IRES for vertebrate wildlife has been largely overlooked. However, as evidenced in this chapter, IRES are priority ecosystems for conserving wildlife diversity, not only along riverine landscapes but also within the entire catchment due to the high mobility of these organisms. In addition, we have highlighted the large control that wildlife exerts on both abiotic and biotic components and processes in IRES. Thus, neglecting this influence may hinder obtaining a complete understanding of the functioning of freshwater ecosystems.

The scarce knowledge about the mechanisms linking IRES and wildlife, sometimes heavily reliant on anecdotal observations and speculation, limits our decision-making in conservation strategies. These strategies are especially urgent given the current scenario of climate change and other anthropogenic impacts. Specifically, further investigation on the behavioral, functional, and demographic responses of different vertebrate groups to low flow and dry periods will allow us to determine their tolerance to flow intermittence. The integration of different disciplines such as hydrology, limnology, and wildlife ecology provides a promising platform to fully approach the ecological study of IRES-wildlife interactions.

ACKNOWLEDGMENTS

M.M.S.M. was supported by a Marie-Curie postdoctoral fellowship (CLITEMP Project: 330466; MC-IEF; FP7-people-2012-IEF), M.M. by the Severo Ochoa Program for Centres of Excellence in R+D+I (SEV-2012-0262), and D.E. by a Seed Grant of the British Herpetological Society and a Mohamed bin Zayed Species Conservation Fund (Project 152511965). We would like to thank Joaquín Pajarón for the preparation of figures, and Helen Warburton for revising the English. The valuable comments by two dedicated reviewers and the editors Núria Bonada and Andrew Boulton are greatly appreciated.

REFERENCES

Adams, R.A., Hayes, M.A., 2008. Water availability and successful lactation by bats as related to climate change in arid regions of western North America. J. Anim. Ecol. 77, 1115–1121.

Allen, C.R., Saunders, D.A., 2005. Variability between scales: predictors of nomadism in birds of an Australian Mediterranean climate ecosystem. Ecosystems 5, 348–359.

Araújo, M.B., Thuiller, W., Pearson, R.G., 2006. Climate warming and the decline of amphibians and reptiles in Europe. J. Biogeogr. 33, 1712–1728.

Bairlein, F., 1988. How do migratory songbirds cross the Sahara? Trends Ecol. Evol. 3, 191–194.

Balinsky, J.B., Chemaly, S.M., Currin, A.E., Lee, A.R., Thompson, R.L., Van der Westhuizen, D.R., 1976. A comparative study of enzymes of urea and uric acid metabolism in different species of Amphibia, and the adaptation to the environment of the tree frog *Chiromantis xerampelina* Peters. Comp. Biochem. Physiol. B Comp. Biochem. 54, 549–555.

Barton, P.S., Cunningham, S.A., Lindenmayer, D.B., Manning, A.D., 2013. The role of carrion in maintaining biodiversity and ecological processes in terrestrial ecosystems. Oecologia 171, 761–772.

Baxter, C.V., Fausch, K.D., Saunders, W.C., 2005. Tangled webs: reciprocal flows of invertebrate prey link streams and riparian zones. Freshw. Biol. 50, 201–220.

Berger, J., 2004. The last mile: how to sustain long-distance migration in mammals. Conserv. Biol. 18, 320–331.

Berteaux, D., Humphries, M.M., Krebs, C.J., Lima, M., McAdam, A.G., Pettorelli, N., et al., 2006. Constraints to projecting the effects of climate change on mammals. Clim. Res. 32, 151–158.

Biebach, H., Friedrich, W., Heine, G., 1986. Interaction of bodymass, fat, foraging and stopover period in trans-Sahara migrating passerine birds. Oecologia 69, 370–379.

Blair, W.F., 1957. Changes in vertebrate populations under conditions of drought. Cold Spring Harb. Symp. Quant. Biol. 22, 73–275.

Blaustein, A.R., Belden, L.K., Olson, D.H., Green, D.M., Root, T.L., Kiesecker, J.M., 2001. Amphibian breeding and climate change. Conserv. Biol. 15, 1804–1809.

Bodie, J.R., 2001. Stream and riparian management for freshwater turtles. J. Environ. Manag. 62, 443–455.

Bondi, C.A., Mark, S.B., 2013. Differences in flow regime influences the seasonalmigrations, body size, and body conditions of Western Pond Turtles (*Actinemys marmorata*) that inhabit perennial and intermittent riverine sites in northern California. Copeia 2013, 142–153.

Bons, J., Geniez, P., 1996. Amphibiens et Reptiles du Maroc. Asociación Herpetológica Española, Barcelona.

Both, C., Visser, M.E., 2001. Adjustment to climate change is constrained by arrival date in a long-distance migrant bird. Nature 411, 296–298.

Brady, L.D., Griffiths, R.A., 2000. Developmental responses to pond desiccation in tadpoles of the British anuran amphibians (*Bufo bufo, B. calamita* and *Rana temporaria*). J. Zool. 252, 61–69.

Brand, L.A., Stromberg, J.C., Noon, B.R., 2010. Avian density and nest survival on the San Pedro River: importance of vegetation type and hydrologic regime. J. Wildl. Manag. 74, 739–754.

Brönmark, C., Rundle, S.D., Erlandsson, A., 1991. Interactions between freshwater snails and tadpoles: competition and facilitation. Oecologia 87, 8–18.

Brooks, B.W., Riley, T.M., Taylor, R.D., 2006. Water quality of effluent-dominated ecosystems: ecotoxicological, hydrological, and management considerations. Hydrobiologia 556, 365–379.

Buhlmann, K.A., Congdon, J.D., Gibbons, J.W., Greene, J.D., 2009. Ecology of chicken turtles (*Deirochelys reticularia*) in a seasonal wetland ecosystem: exploiting resource and refuge environments. Herpetologica 65, 39–53.

Campbell, J.W., Vorhaben, J.E., Smith, D.D., 1984. Hepatic ammonia metabolism in a uricotelic treefrog *Phyllomedusa sauvagei*. Am. J. Phys. Regul. Integr. Comp. Phys. 246, 805–810.

Cañedo-Argüelles, M., Hawkins, C.P., Kefford, B.J., Schäfer, R.B., Dyack, B.J., Brucet, S., et al., 2016. Saving freshwater from salts. Science 351, 914–916.

Carlisle, J.D., Skagen, S.K., Kus, B.E., Riper III, C.V., Paxtons, K.L., Kelly, J.F., 2009. Landbird migration in the American West: recent progress and future research directions. Condor 111, 211–225.

Carlsson, G., Örn, S., Larsson, D.G., 2009. Effluent from bulk drug production is toxic to aquatic vertebrates. Environ. Toxicol. Chem. 28, 2656–2662.

Carothers, S.W., Johnson, R.R., Aitchison, S.W., 1974. Population structure and social organization of southwestern riparian birds. Am. Zool. 14, 97–108.

Cartledge, V.A., Withers, P.C., McMaster, K.A., Thompson, G.G., Bradshaw, S.D., 2006. Water balance of field-excavated aestivating Australian desert frogs, the cocoon-forming *Neobatrachus aquilonius* and the non-cocooning *Notaden nichollsi* (Amphibia: Myobatrachidae). J. Exp. Biol. 209, 3309–3321.

Casas-Crivillé, A., Valera, F., 2005. The European bee-eater (*Merops apiaster*) as an ecosystem engineer in arid environments. J. Arid Environ. 60, 227–238.

Chambers, L.E., 2008. Trends in timing of migration of south-western Australian birds and their relationship to climate. J. EMU Austral Ornithol. 108, 1–14.

Cheng, T.L., Rovito, S.M., Wake, D.B., Vredenburg, V.T., 2011. Coincident mass extirpation of neotropical amphibians with the emergence of the infectious fungal pathogen *Batrachochytrium dendrobatidis*. Proc. Natl. Acad. Sci. U. S. A. 108, 9502–9507.

Christian, K., Green, B., Kennett, R., 1996. Some physiological consequences of estivation by freshwater crocodiles, *Crocodylus johnstoni*. J. Herpetol. 30, 1–9.

Cintra, R., 1989. Maternal care and daily pattern of behavior in a family of caimans, *Caiman yacare* in the Brazilian Pantanal. J. Herpetol. 23, 320–322.

Clements, J.F., 2011. The Clements Checklist of Birds of the World, sixth ed. Cornell University Press, Ithaca, NY.

Coppack, T., Pulido, F., Czisch, M., Auer, D.P., Berthold, P., 2003. Photoperiod response may facilitate adaptation to climate change in long-distance migratory birds. Proc. R. Soc. Lond. B Biol. Sci. 270, S43–S46.

Crick, H.Q.P., 2004. The impact of climate change on birds. Ibis 146, 48–56.

Crump, M.L., Hensley, F.R., Clark, K.L., 1992. Apparent decline of the golden toad: underground or extinct? Copeia 1992, 413–420.

Daughton, C.G., Ternes, T.A., 1999. Pharmaceuticals and personal care products in the environment: agents of subtle change? Environ. Health Perspect. 107, 907–938.

Davies, S.J.J.F., 1982. Behavioural adaptations of birds to environments where evaporation is high and water is in short supply. Comp. Biochem. Physiol. 71A, 557–566.

Davis, J., Pavlova, A., Thompson, R., Sunnucks, P., 2013. Evolutionary refugia and ecological refuges: key concepts for conserving Australian arid zone freshwater biodiversity under climate change. Glob. Chang. Biol. 19, 1970–1984.

Dean, W.R.J., 2004. Nomadic Desert Birds. Adaptations in Desert Organisms Series, Springer-Verlag, Berlin.

Degani, G., 2016. Cannibalism, among other solutions of adaption, in habitats where food is not available for *Salamandra infraimmaculata* larvae diet in breeding places in xeric habitats. Open J. Anim. Sci. 6, 31–41.

Dehorter, O., Guillemain, M., 2008. Global diversity of freshwater birds (Aves). Hydrobiologia 595, 619–626.

del Hoyo, J., Elliott, A., Sargatal, J., 1992. Handbook of the Birds of the World. Ostrich to Ducks, Vol. 1. Lynx Edicions, Barcelona.

Delibes-Mateos, M., Delibes, M., Ferreras, P., Villafuerte, R., 2008. Key role of European rabbits in the conservation of the Western Mediterranean Basin hotspot. Conserv. Biol. 22, 1106–1117.

Díaz-Paniagua, C., Gómez Rodríguez, C., Portheault, A., de Vries, W., 2005. Los anfibios de Doñana. Ministerio Medio Ambiente, OAPN, Madrid. Serie Técnica.

Dickson, J.G., Williarnson, J.H., Conner, R.N., Ortego, B., 1995. Streamside zones and breeding birds in eastern Texas. Wildl. Soc. Bull. 23, 750–755.

Edwards, D.L., Roberts, J.D., Keogh, J.S., 2007. Impact of Plio-Pleistocene arid cycling on the population history of a southwestern Australian frog. Mol. Ecol. 16, 2782–2796.

Ellery, W.N., Dahlberg, A.C., Strydom, R., Neal, M.J., Jackson, J., 2003. Diversion of water flow from a floodplain wetland stream: an analysis of geomorphological setting and hydrological and ecological consequences. J. Environ. Manag. 68, 51–71.

Ernst, C.H., Barbour, R.W., 1989. Turtles of the World. Smithsonian Institution Press, Washington, DC.

Escoriza, D., Ben Hassine, J., 2015. Niche partitioning at local and regional scale in the North African Salamandridae. J. Herpetol. 49, 276–283.

Escoriza, D., Boix, D., 2014. Reproductive habitat selection in alien and native populations of the genus *Discoglossus*. Acta Oecol. 59, 97–103.

Gehrke, P.C., Harris, J.H., 2001. Regional-scale effects of flow regulation on lowland riverine fish communities in New South Wales, Australia. Regul. Rivers Res. Manag. 17, 369–391.

Gibbons, J.W., Greene, J.L., Congdon, J.D., 1983. Drought-related responses of aquatic turtle populations. J. Herpetol. 17, 242–246.

Gibson, P.P., Olden, J.D., 2014. Ecology, management, and conservation implications of North American beaver (*Castor canadensis*) in dryland streams. Aquat. Conserv. Mar. Freshwat. Ecosyst. 24, 391–409.

Gómez-Rodríguez, C., Díaz-Paniagua, C., Serrano, L., Florencio, M., Portheault, A., 2009. Mediterranean temporary ponds as amphibian breeding habitats: the importance of preserving pond networks. Aquat. Ecol. 43, 1179–1191.

Groom, D.J., Kuchel, L., Richards, J.G., 2013. Metabolic responses of the South American ornate horned frog (*Ceratophrys ornata*) to estivation. Comp. Biochem. Physiol. B: Biochem. Mol. Biol. 164, 2–9.

Hagen, E.M., Sabo, J.L., 2011. A landscape perspective on bat foraging ecology along rivers: does channel confinement and insect availability influence the response of bats to aquatic resources in riverine landscapes? Oecologia 166, 751–760.

Hagen, E.M., Sabo, J.L., 2012. Influence of river drying and insect availability on bat activity along the San Pedro River, Arizona (USA). J. Arid Environ. 84, 1–8.

Hagen, E.M., Sabo, J.L., 2014. Temporal variability in insectivorous bat activity along two desert streams with contrasting patterns of prey availability. J. Arid Environ. 102, 104–112.

Hansson, L.A., Johansson, L., Persson, L., 1987. Effects of fish grazing on nutrient release and succession of primary producers. Limnol. Oceanogr. 32, 723–729.

Harris, G., Thirgood, S., Hopcraft, J.G.C., Cromsigt, J.P.G.M., Berger, J., 2009. Global decline in aggregated migrations of large terrestrial mammals. Endanger. Species Res. 7, 55–76.

Hart, B.T., Bailey, P., Edwards, R., Hortle, K., James, K., McMahon, A., et al., 1991. A review of the salt sensitivity of the Australian freshwater biota. Hydrobiologia 210, 105–144.

Haynes, G., 1991. Mammoths, Mastodonts, and Elephants: Biology, Behavior, and the Fossil Record. Cambridge University Press, New York.

Hernando, M.D., Mezcua, M., Fernández-Alba, A.R., Barceló, D., 2006. Environmental risk assessment of pharmaceutical residues in wastewater effluents, surface waters and sediments. Talanta 69, 334–342.

Hero, J.M., Morrison, C., 2004. Frog declines in Australia: global implications. Herpetol. J. 14, 175–186.

Herrera, C., Pellmyr, O., 2002. Plant Animal Interactions: An Evolutionary Approach. Wiley-Blackwell, Oxford, UK.

Hilty, J.A., Lidicker Jr., W.Z., Merelender, A.M., 2006. Corridor Ecology: The Science and Practice of Linking Landscapes for Biodiversity Conservation. Island Press, Washington, DC.

Holomuzki, J.R., 1995. Oviposition sites and fish-deterrent mechanisms of two stream anurans. Copeia 3, 607–613.

Hughes, G.O., Thuiller, W., Midgley, G.F., Collins, K., 2008. Environmental change hastens the demise of the critically endangered riverine rabbit (*Bunolagus monticulairis*). Biol. Conserv. 141, 23–34.

Jacobson, P.J., 1997. An Ephemeral Perspective on Fluvial Systems: Viewing Ephemeral Rivers in the Context of Current Lotic Ecology. Virginia Polytechnic Institute and State University, Blacksburg, VA. PhD Dissertation.

Jaeger, K.L., Olden, J.D., Pelland, N.A., 2014. Climate change poised to threaten hydrologic connectivity and endemic fishes in dryland streams. Proc. Natl. Acad. Sci. U. S. A. 111, 13894–13899.

Janzen, F.J., 1994. Climate change and temperature dependent sex determination in reptiles. Proc. Natl. Acad. Sci. U. S. A. 91, 7487–7490.

Jones, C.G., Lawton, J.H., Shachak, M., 1994. Organisms as ecosystem engineers. Oikos 69, 373–386.

Kaczensky, P., Sheehy, D.P., Johnson, D.E., Walzer, C., Lhkagvasuren, D., Sheehy, C.M., 2006. Room to Roam? The Threat to Khulan (Wild Ass) From Human Intrusion. East Asia and Pacific Environment and Social Development Department, World Bank, Washington, DC. Mongolia Discussion Papers.

Kingsford, R.T., Thomas, R.F., 1995. The Macquarie Marshes in arid Australia and their waterbirds: a 50-year history of decline. Environ. Manag. 19, 867–878.

Kingsford, R.T., Roshier, D.A., Porter, J.L., 2010. Australian waterbirds: time and space travellers in dynamic desert landscapes. Mar. Freshw. Res. 61, 875–884.

Kok, O.B., Nel, J.A.J., 1996. The Kuiseb River as a linear oasis in the Namib Desert. Afr. J. Ecol. 34, 39–47.

Krueper, D.J., 1993. Conservation Priorities in Naturally Fragmented and Human-Altered Riparian Habitats of the Arid West. U.S. Department of Agriculture, General Technical Report RM-43, Cornell Laboratory of Ornithology, Cornell University, Ithaca, NY.

Lakshminarayanan, N., Karanth, K.K., Goswami, V.R., Vaidyanathan, S., Karanth, K.U., 2016. Determinants of dry season habitat use by Asian elephants in the Western Ghats of India. J. Zool. 298, 169–177.

Lavee, D., Safriel, U.N., Meilijson, I., 1991. For how long do trans-Saharan migrants stop over at an oasis? Ornis Scand. 22, 33–44.

Leigh, C., Boulton, A.J., Courtwright, J.L., Fritz, K., May, C.L., Walker, R.H., et al., 2016. Ecological research and management of intermittent rivers: an historical review and future directions. Freshw. Biol. 61, 1181–1199.

Levick, L., Fonseca, J., Goodrich, D., Hernandez, M., Semmens, D., Stromberg, J., et al., 2008. The Ecological and Hydrological Significance of Ephemeral and Intermittent Streams in the Arid and Semi-Arid American Southwest. EPA/600/R-08/134, ARS/233046, U.S. Environmental Protection Agency and USDA/ARS Southwest Watershed Research Center, Washington, DC.

Ligon, D.B., Peterson, C.C., 2002. Physiological and behavioral variation in estivation among mud turtles (*Kinosternon* spp.). Physiol. Biochem. Zool. 75, 283–293.

Limpus, C.J., 1971. The flatback turtle, *Chelonia depressa* Garman in southeast Queensland, Australia. Herpetologica 27, 431–446.

Lindeque, M., Lindeque, P.M., 1991. Satellite tracking of elephants in northwestern Namibia. Afr. J. Ecol. 29, 196–206.

Luiselli, L., Capizzi, D., Filippi, E., Anibaldi, C., Rugiero, L., Capula, M., 2007. Comparative diets of three populations of an aquatic snake (*Natrix tessellata*, Colubridae) from Mediterranean streams with different hydric regimes. Copeia 2, 426–435.

Maigret, T.A., Cox, J.J., Schneider, D.R., Barton, C.D., Price, S.J., Larkin, J.L., 2014. Effects of timber harvest within streamside management zones on salamander populations in ephemeral streams of southeastern Kentucky. For. Ecol. Manag. 324, 46–51.

Martin, S., 2008. Global diversity of crocodiles (Crocodilia, Reptilia) in freshwater. Hydrobiologia 595, 587–591.

Martin, B.E., Van Devender, T.R., 2002. Seasonal diet changes of Gopherus agassizii (desert tortoise) in desert grassland of southern Arizona and its behavioral implications. Herpetological Nat. Hist. 9, 31–42.

McCarthy, T.S., Ellery, W.N., Bloem, A., 1998. Some observations on the geomorphological impact of hippopotamus (*Hippopotamus amphibius* L.) in the Okavango Delta, Botswana. Afr. J. Ecol. 36, 44–56.

Meyer, J.L., Strayer, D.L., Wallace, J.B., Eggert, S.L., Helfman, G.S., Leonard, N.E., 2007. The contribution of headwater streams to biodiversity in river networks. J. Am. Water Resour. Assoc. 43, 86–103.

Mills, G., Hes, L., 1997. The Complete Book of South African Mammals. Struik Publishers, Cape Town.

Moleón, M., Sánchez-Zapata, J.A., Selva, N., Donázar, J.A., Owen-Smith, N., 2014. Inter-specific interactions linking predation and scavenging in terrestrial vertebrate assemblages. Biol. Rev. 89, 1042–1054.

Mueller, G.A., Carpenter, J., Thornbrugh, D., 2006. Bullfrog tadpole (*Rana catesbeiana*) and red swamp crayfish (*Procambarus clarkii*) predation on early life stages of endangered razorback sucker (*Xyrauchen texanus*). Southwest. Nat. 51, 258–261.

Murphy, M.A., Evans, J.S., Storfer, A., 2010. Quantifying *Bufo boreas* connectivity in Yellowstone National Park with landscape genetics. Ecology 91, 252–261.

Naidoo, R., Chase, M.J., Beytell, P., Du Preez, P., Landen, K., Stuart-Hill, G., et al., 2016. A newly discovered wildlife migration in Namibia and Botswana is the longest in Africa. Oryx 50, 138–146.

Naiman, R.J., Rogers, K.H., 1997. Large animals and system-level characteristics in river corridors. Bioscience 47, 521–529.

Newton, I., 1994. The role of nest sites in limiting the numbers of hole-nesting birds: a review. Biol. Conserv. 70, 265–276.

Nilson, G., Andrén, C., Ioannides, Y., Dimaki, M., 1999. Ecology and conservation of the Milos viper, *Macrovipera schweizeri* (Werner, 1935). Amphibia-Reptilia 20, 355–375.

Nilsson, C., Dynesius, M., 1994. Ecological effects of river regulation on mammals and birds: a review. Regul. Rivers Res. Manag. 9, 45–53.

Ostfeld, R.S., Keesing, F., 2000. Pulsed resources and community dynamics of consumers in terrestrial ecosystems. Trends Ecol. Evol. 15, 232–237.

Owen-Smith, N., 1992. Megaherbivores. The Influence of Very Large Body Size on Ecology. Cambridge University Press, Cambridge.

Parker, M.S., 1994. Feeding ecology of stream-dwelling Pacific giant salamander larvae (*Dicamptodon tenebrosus*). Copeia 1994, 705–718.

Pavey, C.R., Nano, C.E.M., 2009. Bird assemblages of arid Australia: vegetation patterns have a greater effect than disturbance and resource pulses. J. Arid Environ. 73, 634–642.

Pedler, R.D., Ribot, R.F.H., Bennett, A.T.D., 2014. Extreme nomadism in desert waterbirds: flights of the banded stilt. Biol. Lett. 10, 20140547.

Pereira, L.M., Owen-Smith, N., Moleón, M., 2014. Facultative predation and scavenging by mammalian carnivores: seasonal, regional and intra-guild comparisons. Mammal Rev. 44, 44–55.

Petrie, S.A., Rogers, K.H., 2004. Nutrient-reserve dynamics of semiarid-breeding white-faced whistling ducks: a north-temperate contrast. Can. J. Zool. 82, 1082–1090.

Pfennig, D.W., 1999. Cannibalistic tadpoles that pose the greatest threat to kin are most likely to discriminate kin. Proc. R. Soc. Lond. B Biol. Sci. 266, 57–61.

Powell, G.V., 1983. Industrial effluents as a source of mercury contamination in terrestrial riparian vertebrates. Environ. Pollut. B. 5, 51–57.

Price, S.J., Browne, R.A., Dorcas, M.E., 2012. Resistance and resilience of a stream salamander to supraseasonal drought. Herpetologica 68, 312–323.

Puigdefábregas, J., Alonso, J.M., Delgado, L., Domingo, F., Cueto, M., Gutiérrez, L., et al., 1996. Interactions of soil and vegetation along a catena in semiarid Spain. In: Brandt, J., Thornes, J. (Eds.), Mediterranean Desertification and Land Use. John Wiley, Chichester, pp. 137–168.

Rahel, F.J., Olden, J.D., 2008. Assessing the effects of climate change on aquatic invasive species. Conserv. Biol. 22, 521–533.

Ramey, E.M., Ramey, R.R., Brown, L.M., Kelley, S.T., 2013. Desert-dwelling African elephants (*Loxodonta africana*) in Namibia dig wells to purify drinking water. Pachyderm 53, 66–72.

Rees, M., Roe, J.H., Georges, A., 2009. Life in the suburbs: behavior and survival of a freshwater turtle in response to drought and urbanization. Biol. Conserv. 142, 3172–3181.

Robson, B.J., Chester, E.T., Mitchell, B.D., Matthews, T.G., 2013. Disturbance and the role of refuges in mediterranean climate streams. Hydrobiologia 719, 77–91.

Rojas-Ahumada, D.P., Menin, M., 2010. Composition and abundance of anurans in riparian and non-riparian areas in a forest in Central Amazonia, Brazil. South Am. J. Herpetology 5, 157–167.

Rosalino, L.M., Santos-Reis, M., 2008. Fruit consumption by carnivores in Mediterranean Europe. Mammal Rev. 39, 67–78.

Rosenberg, K., Ohmart, R.D., Hunter, W.C., Anderson, B.W., 1991. Birds of the Lower Colorado River Valley. The University of Arizona Press, Tucson.

Rouse, J.D., Bishop, C.A., Struger, J., 1999. Nitrogen pollution: an assessment of its threat to amphibian survival. Environ. Health Perspect. 107, 799–803.

Ruhí, A., Holmes, E.E., Rinne, J.N., Sabo, J.L., 2015. Anomalous droughts, not invasion, decrease persistence of native fishes in a desert river. Glob. Chang. Biol. 21, 1482–1496.

Ruiz-Olmo, J., Delibes, M., 1998. La Nutria en España Ante el Horizonte del Año 2000. SECEM, Málaga.

Ruiz-Olmo, J., Olmo-Vidal, J.M., Mañas, S., Batet, A., 2002. The influence of resource seasonality on the breeding patterns of the Eurasian otter (*Lutra lutra*) in Mediterranean habitats. Can. J. Zool. 80, 2178–2189.

Ruiz-Olmo, J., Jiménez, J., Chacón, W., 2007. The importance of ponds for the otter (*Lutra lutra*) during drought periods in Mediterranean ecosystems: a case study in Bergantes River. Mammalia 71, 16–24.

Säfholm, M., Ribbenstedt, A., Fick, J., Berg, C., 2014. Risks of hormonally active pharmaceuticals to amphibians: a growing concern regarding progestagens. Philos. Trans. R. Soc. B 369, 20130577.

Salvador, A., Ramos, M.A., 1998. Fauna Ibérica. Reptiles. Museo Nacional de Ciencias Naturales-CSIC, Madrid.

Sánchez, M.A., Sánchez-Zapata, J.A., Díez de Revenga, E., 2004. El conejo (*Oryctolagus cuniculus*) en la Región de Murcia. Actas del II Congreso de la Naturaleza de la Región de Murcia, ANSE, Murcia. pp. 169–179.

Sánchez-Montoya, M.M., Moleón, M., Sánchez-Zapata, J.A., Tockner, K., 2016. Dry riverbeds: corridors for terrestrial vertebrates. Ecosphere 7, e01508. http://dx.doi.org/10.1002/ecs2.1508.

Santos, M.J., Matos, H.M., Palomares, F., Santos-Reis, M., 2011. Factors affecting mammalian carnivore use of riparian ecosystems in Mediterranean climates. J. Mammal. 92, 1060–1069.

Seidman, V.M., Zabel, C.J., 2001. Bat activity along intermittent streams in Northwestern California. J. Mammal. 82, 738–747.

Shafroth, P.B., Auble, G.T., Scott, M.L., 1995. Germination and establishment of the native plains cottonwood (*Populus deltoids* Marshall subsp. *monilifera*) and the exotic Russian-olive (*Elaeagnus angustifolia* L.). Conserv. Biol. 9, 1169–1175.

Sheldon, F., Bunn, S.E., Hughes, J.M., Arthington, A.H., Balcombe, S.R., Fellows, C.S., 2010. Ecological roles and threats to aquatic refugia in arid landscapes: dryland river waterholes. Mar. Freshw. Res. 61, 885–895.

Shine, R., Brown, G.P., 2008. Adapting to the unpredictable: reproductive biology of vertebrates in the Australian wet–dry tropics. Philos. Trans. R. Soc. B 363, 363–373.

Shine, T., Böhme, W., Nickel, H., Thies, D.F., Wilms, T., 2001. Rediscovery of relict populations of the Nile crocodile *Crocodylus niloticus* in south-eastern Mauritania, with observations on their natural history. Oryx 35, 260–262.

Simmons, R.E., Barnard, P., Dean, W.R.J., Midgley, G.F., Thuiller, W., Hughes, G., 2004. Climate change and birds: perspectives and prospects from southern Africa. Ostrich 75, 295–308.

Skagen, S.K., Melcher, C.P., Howe, W.H., Knopf, F.L., 1998. Comparative use of riparian corridors and oases by migrating birds in southeast Arizona. Conserv. Biol. 12, 896–909.

Skinner, J.D., Chimimba, C.T., 2005. The Mammals of the Southern African Subregion, third ed. Cambridge University Press, Cambridge.

Smith, M.J., Schreiber, E.S.G., Scroggie, M.P., Kohout, M., Ough, K., Potts, J., et al., 2007. Associations between anuran tadpoles and salinity in a landscape mosaic of wetlands impacted by secondary salinisation. Freshw. Biol. 52, 75–84.

Sogge, M., Sferra, S., Paxton, E., 2008. Saltcedar as habitat for birds: implications to riparian restoration in the southwestern United States. Restor. Ecol. 16, 146–154.

Solís, R., Lobos, G., Walker, S.F., Fisher, M., Bosch, J., 2010. Presence of *Batrachochytrium dendrobatidis* in feral populations of *Xenopus laevis* in Chile. Biol. Invasions 12, 1641–1646.

Souter, N.J., Williams, G.S., 2001. A comparison of macroinvertebrate communities in three South Australian streams with regard to reintroduction of the platypus. Trans. R. Soc. S. Aust. 125, 71–82.

Soykan, C.U., Brand, L.A., Ries, L., Stromberg, J.C., Hass, C., Simmons Jr., D.A., et al., 2012. Multitaxonomic diversity patterns along a desert riparian–upland gradient: is the community more than the sum of its parts? PLoS One 7, e28235.

Sparling, D.W., Fellers, G.M., McConnell, L.L., 2001. Pesticides and amphibian population declines in California, USA. Environ. Toxicol. Chem. 20, 1591–1595.

Stackelberg, P.E., Furlong, E.T., Meyer, M.T., Zaugg, S.D., Henderson, A.K., Reissman, D.B., 2004. Persistence of pharmaceutical compounds and other organic wastewater contaminants in a conventional drinking-water-treatment plant. Sci. Total Environ. 329, 99–113.

Steward, A.L., von Schiller, D., Tockner, K., Marshall, J.C., Bunn, S.E., 2012. When the river runs dry: human and ecological values of dry riverbeds. Front. Ecol. Environ. 10, 202–209.

Stone, P.A., 2001. Movements and demography of the Sonoran mud turtle, *Kinosternon sonoriense*. Southwest. Nat. 46, 41–53.

Stromberg, J.C., Beauchamp, V.B., Dixon, M.D., Lite, S.J., Paradzick, C., 2007. Importance of low-flow and high-flow characteristics to restoration of riparian vegetation along rivers in arid southwestern United States. Freshw. Biol. 52, 651–679.

Stromberg, J.C., McCluney, K.E., Dixon, M.D., Meixner, T., 2013. Dryland riparian ecosystems in the American southwest: sensitivity and resilience to climatic extremes. Ecosystems 16, 411–415.

Thuiller, W., Broennimann, O., Hughes, G., Alkemade, J.R.M., Midgley, G.F., Corsi, F., 2006. Vulnerability of African mammals to anthropogenic climate change under conservative land transformation assumptions. Glob. Chang. Biol. 12, 424–440.

Tieleman, B.I., Williams, J.B., 1999. The role of hyperthermia in the water economy of desert birds. Physiol. Biochem. Zool. 72, 87–100.

Tramer, E.J., Tramer, F.E., 1977. Feeding responses of fall migrants to prolonged inclement weather. Wilson Bull. 89, 166–167.

Tucker, A.D., Limpus, C.J., McCallum, H.I., McDonald, K.R., 1997. Movements and home ranges of *Crocodylus johnstoni* in the Lynd River, Queensland. Wildl. Res. 24, 379–396.

Vanni, M.J., 2002. Nutrient cycling by animals in freshwater ecosystems. Annu. Rev. Ecol. Syst. 33, 341–370.

Vences, M., Köhler, J., 2008. Global diversity of amphibians (Amphibia). Hydrobiologia 595, 569–580.

Veron, G., Patterson, B.D., Reeves, R., 2008. Global diversity of mammals (Mammalia) in freshwater. Hydrobiologia 595, 607–617.

Viljoen, P.J., 1989. Habitat selection and preferred food plants of a desert-dwelling elephant population in the Northern Namib desert, SouthWest Africa/Namibia. Afr. J. Ecol. 27, 227–240.

Vitt, L.J., Caldwell, J.P., 2014. Herpetology: An Introductory Biology of Amphibians and Reptiles. Academic Press, London.

Walker, C., 1996. Signs of the Wild. Struik Publishing, Cape Town.

Walker, H.A., 2008. Floristics and phyiognomy determine migrant land bird response to tamarisk (*Tamarix ramosissima*) invasion in riparian areas. Auk 125, 520–531.

Webb, G., Manolis, C., 1998. Australian Crocodiles: A Natural History. Reed-New Holland, Sydney.

Webb, G.J.W., Manolis, S.C., Buckworth, R., 1983. *Crocodylus johnstoni* in the McKinlay River Area, N.T. VI. Nesting biology. Wildl. Res. 10, 607–637.

Webb, G.J., Choquenot, D., Whitehead, P.J., 1986. Nests, eggs, and embryonic development of *Carettochelys insculpta* (Chelonia: Carettochelidae) from Northern Australia. J. Zool. 1, 521–550.

Weir, J.S., 1972. Spatial distribution of elephants in an African national park in relation to environmental sodium. Oikos 23, 1–3.

Wells, K.D., 2010. The Ecology and Behavior of Amphibians. University of Chicago Press, London.

Welsh Jr., H.H., Hodgson, G.R., Lind, A.J., 2005. Ecogeography of the herpetofauna of a northern California watershed: linking species' patterns to landscape processes. Ecography 28, 521–536.

Western, D., 1975. Water availability and its influence on the structure and dynamics of a savannah large mammal community. Afr. J. Ecol. 13, 265–286.

Whiting, S.D., Whiting, A.U., 2011. Predation by the saltwater crocodile (*Crocodylus porosus*) on sea turtle adults, eggs, and hatchlings. Chelonian Conserv. Biol. 10, 198–205.

Wilkins, R.N., Peterson, N.P., 2000. Factors related to amphibian occurrence and abundance in headwater streams draining second-growth Douglas-fir forests in southwestern Washington. For. Ecol. Manag. 139, 79–91.

Williams, J.B., Tieleman, B.I., 2005. Physiological adaptation in desert birds. Bioscience 55, 416–425.

Wilson, D.E., Mittermeier, R.A., 2009. Handbook of the Mammals of the World. Lynx Edicions, Barcelona.

Wingfield, J.C., Vleck, C.M., Moore, M.C., 1992. Seasonal changes of the adrenocortical response to stress in birds of the Sonoran desert. J. Exp. Zool. 264, 419–428.

Winne, C.T., Willson, J.D., Gibbons, J.W., 2006. Income breeding allows an aquatic snake *Seminatrix pygaea* to reproduce normally following prolonged drought-induced aestivation. J. Anim. Ecol. 75, 1352–1360.

Zale, A., Leslie Jr., V.D.M., Fisher, W.L., Merrifield, S.G., 1989. The Physicochemistry, Flora, and Fauna of Intermittent Prairie Streams: A Review of the Literature. U.S. Fish and Wildlife Service, Washington, DC.

Zeiner, D.C., Laudenslayer, W.F., Mayer, K.E., White, M., 1988–1990. California's Wildlife. Vol. I-III. California Department of Fish and Game, Sacramento.

FOOD WEBS AND TROPHIC INTERACTIONS IN INTERMITTENT RIVERS AND EPHEMERAL STREAMS

4.7

Angus R. McIntosh*, Catherine Leigh[†,‡,§], Kate S. Boersma[¶], Peter A. McHugh,
Catherine Febria*, Emili García-Berthou[††]**

University of Canterbury, Christchurch, New Zealand Irstea, UR MALY, centre de Lyon-Villeurbanne,
Villeurbanne, France[†] CESAB-FRB, Immeuble Henri Poincaré, Aix-en-Provence, France[‡] Griffith University, Nathan,
QLD, Australia[§] University of San Diego, San Diego, CA, United States[¶] Utah State University, Logan, UT,
United States** University of Girona, Girona, Spain[††]*

IN A NUTSHELL

- Contraction of aquatic habitats and expansion of terrestrial habitats associated with cessation of flow and drying in intermittent rivers and ephemeral streams (IRES) change the types, sizes, and abundance of organisms present; the interactions among those organisms; and the overall extent and nature of their trophic interactions
- Drying affects microbial communities and thus detritus processing, heterotrophy can sustain stressed food webs, and autotrophy is important where conditions allow algal buildup
- Spatial and temporal fragmentation of habitats can constrain, compress, and sometimes expand trophic interactions, often substantially increasing interaction strengths as drying progresses and affecting the lateral linking of aquatic and terrestrial food webs
- Overall, drying reduces aquatic food chain length, trophic diversity, and connectance, but also triggers switches toward terrestrial energy pathways with the extent of food web modification linked to regional differences in climate, biota, and riparian conditions

4.7.1 INTRODUCTION

The loss of streamflow, and associated contraction of aquatic ecosystems and expansion of terrestrial ecosystems, inescapably alters the dimensions, conditions, resources, and connections in those habitats (Rolls et al., 2012). This changes the types, sizes, and abundance of organisms present, and the interactions among those organisms, including the extent and nature of their trophic interactions. Thus, the food webs of aquatic and terrestrial systems and their linkage are all affected by the loss of flow and subsequent drying. Here, we review how these linked food webs and trophic interactions are affected by flow cessation and drying, highlighting the mechanisms involved, the emergent patterns within and across river systems, and the insights that can be gleaned both for management and ecological understanding.

BOX 4.7.1 FOOD WEB FUNDAMENTALS IN IRES

Food webs characterize the patterns of energy flow within and between ecosystems by describing the trophic links and interactions among species, including their use of resources (Fig. 4.7.1). At the base of IRES food webs are heterotrophic and autotrophic energy resources. Materials from outside the aquatic ecosystem (i.e., allochthonous sources) including leaves and other organic materials are the major contributors to a detritus pool, which together with the microbes that colonize them (Chapter 4.1), form the heterotrophic energy sources (a). Algae (b) are the major contributors to autotrophic energy pools (Chapter 4.2) and are regarded as an autochthonous energy source because the production is derived within the aquatic ecosystem. Primary consumers (c) of those resources in streams are often classified by the way they feed (Chapter 4.3), and include aquatic invertebrate grazers or scrapers feeding primarily on algae, collector-browsers which feed on a variety of food sources, specialist aquatic detritivores often called shredders which play a key role in breaking down wood and leaf litter (coarse particulate organic matter, CPOM) to finer particles (or fine particulate organic matter, FPOM) which can also be utilized by filter feeders. At higher trophic levels, secondary consumers include a range of predatory aquatic invertebrates (d) and fish (e, Chapter 4.5) in aquatic habitats. Birds, spiders (f), and other predatory taxa (Chapter 4.6) which live in riparian zones can also feed on adult stages of aquatic taxa (dotted arrows), or aquatic stages when they are trapped in pools of drying IRES. After IRES dry, terrestrial detritivores (g) are often important consumers of decaying aquatic organisms and other detritus (Chapter 4.4). Collectively these interactions can be characterized by the length of food chains, a measure of trophic height (h), and the variety of trophic interactions or trophic width (i).

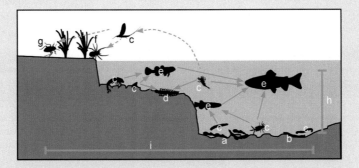

FIG. 4.7.1

Concepts related to IRES food webs and interactions in aquatic (solid arrows) and terrestrial (dashed arrows) environments, including energy sources, primary and secondary consumers, and trophic dimensions. Abbreviations are as follows: a, heterotrophic sources; b, algae; c, aquatic primary consumers; d, predatory aquatic invertebrate; e, fish; f, spider; g, terrestrial detritivore; h, trophic height; i, trophic width. Some silhouettes and original concept developed by Amanda Klemmer, and some original silhouettes also courtesy of the Integration and Application Network, University of Maryland Center for Environmental Science (ian. umces.edu/symbols/).

Food webs (Box 4.7.1) reflect emergent patterns that can be resolved from the feeding interactions of organisms, providing a powerful way of evaluating the structure and functioning of an ecosystem (Thompson et al., 2012). Moreover, because food webs reflect connectivity between organisms and resources (Fig. 4.7.1), they are useful for understanding and summarizing the highly dynamic changes that intermittent rivers and ephemeral streams (IRES) undergo during wetting and drying cycles.

Much of our understanding about how ecosystems maintain structure, are assembled, and function has been based on the patterns that are apparent in food webs. Thus, the process of food web disassembly/assembly (Fig. 4.7.2) associated with drying and rewetting reveals processes and mechanisms relevant to many ecosystems. In particular, theory highlights the importance of habitat size (McCann et al., 2005), the presence of compartments encompassing the resources associated with particular prey

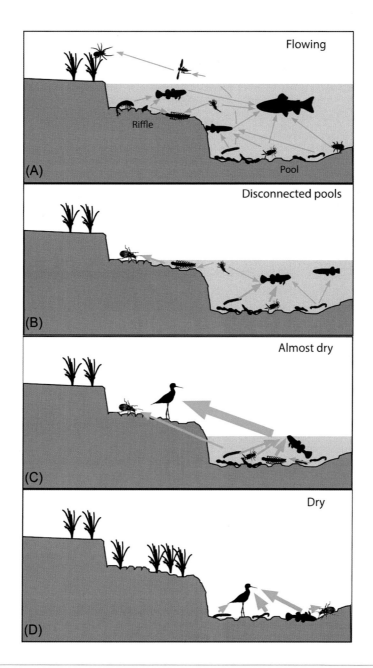

FIG. 4.7.2

Changes in aquatic (*blue arrows*) and terrestrial (*green arrows*) trophic interaction strength (arrow width) associated with flow intermittence (increasing from A to D), illustrating typical changes in interactions and food webs. These involve reductions in body size and richness of consumers, but increased density and interaction strength, associated with shrinking aquatic habitat dimensions over the hydrological cycle of IRES. There is also an associated expansion of terrestrial habitats, with terrestrial predators often feeding on dense aggregations of aquatic prey trapped in fragmented aquatic habitats. Some silhouettes and original concept developed by Amanda Klemmer, and some original silhouettes also courtesy of the Integration and Application Network, University of Maryland Center for Environmental Science (ian.umces.edu/symbols/).

or energy types, the prevalence of particular interactions (or modules) in food webs, and the distributions of both body sizes and interaction strengths (McCann et al., 1998; Thompson et al., 2012). As all of these may be strongly affected by the wetting and drying cycles characteristic of many IRES, much is to be gained from investigating IRES food webs.

The utility of insights derived from studying the food webs of IRES depends, in part, on the consistency with which processes are defined and then used by ecologists and practitioners. Here we are concerned with actual IRES rather than experimental channels, but we briefly summarize insights gained from the latter. We use standard terminology (after Rolls et al., 2012) to refer to IRES throughout, unless giving examples of a particular system, flow regime, or hydrological phase. We first provide an overview of processes affecting IRES food webs in the broadest sense. Then, working across trophic levels, we describe interactions at the base of food webs focusing on how changes in autotrophy and heterotrophy with drying alter resource availability. Next, we consider higher trophic levels, focusing on mechanisms affecting consumer-resource interactions and the spatial dimensions over which changes occur across hydrological phases. We highlight how regional variations in hydrology, vegetation, biodiversity, and aspects of local context affect food webs. Following this, we describe changes in high-level food web properties (e.g., food chain length) across hydrological phases and finish with a look to the future, considering how changes in habitat size, temperature, and water quality might influence food webs.

4.7.2 THE BASE OF IRES FOOD WEBS

The source, quantity, and lability of organic matter can significantly influence trophic interactions from the bottom up. Understanding controls on and the contribution of terrestrial sources of detritus and organic matter processing (i.e., heterotrophy) versus photosynthetically fixed sources of carbon (i.e., autotrophy) is important for understanding food webs in all rivers (Hall and Meyer, 1998). The contribution of detritus to food webs has long been recognized (Hall et al., 2000; Schmid and Schmid-Araya, 2010). Macroinvertebrate shredding is the primary physical breakdown pathway for leaf litter and other coarse organic matter sources (Chapter 4.3) whereas microbial communities release enzymes to enable further breakdown of organic matter compounds (Chapter 3.2). Organic matter from both terrestrial and instream sources make up the majority of invertebrate community diets (Reid et al., 2008; Dekar et al., 2009), increasingly so in times of low flow in some (Closs and Lake, 1994), but not all cases (Álvarez and Pardo, 2009; Vannucchi et al., 2013).

Hydrological extremes alter detrital pools in IRES by altering the quantity and quality of organic matter and the microbial communities responsible for detrital processing (Romaní et al., 2012; Chapter 3.2). During drying, surface water becomes physically contracted (Chapter 2.3) and increasingly disconnected from terrestrial inputs, nutrient processing becomes spatially patchy (Bernal et al., 2013), and food webs become smaller (McHugh et al., 2015). This causes leaf litter breakdown rates to vary as a function of macroinvertebrate shredder density, microbial community and activity rates, and the degree of immersion in water (Bruder et al., 2011; Corti et al., 2011). Microbial community structure itself may drive certain ecosystem functions over others. Sediment experiments suggest that microbial communities responsible for organic matter processing are both altered and reduce their functioning upon drying (Amalfitano et al., 2008; Chapter 3.1). Microbial respiration rates across an intermittent-perennial stream transition indicated carbon stocks to be between four- and eight-fold higher in perennial streams than in either IRES or terrestrial sites (Gerull et al., 2011). Thus, drying likely alters food availability at the food web base. For example, in Mediterranean IRES, the lack of shredders has been associated with the poor quality of organic matter after flow returns, and scrapers or collector-browsers feeding on autochthonous food sources are the dominant consumers in these systems (Álvarez and Pardo, 2009; López-Rodríguez et al., 2012; Vannucchi et al., 2013).

Altered functioning and food web structure may be influenced by shifts in microbial assemblages in longitudinal, vertical, and lateral dimensions during different hydrological phases (Chapter 4.1). Febria et al. (2015) used network path analysis to compare microbial groups in a headwater stream network in Chesapeake Bay, USA, and found perennial streams were dominated by carbon fixers and IRES by nitrogen fixers. The pattern became more enhanced during periods of drying. However, the differences were not attributed to hydrological regime *per se* but to soil moisture, a physical attribute associated with specific streambed microhabitats. Overall evidence suggests that different microbial taxa and functions are associated with IRES food webs compared to perennial streams.

The primary producers in IRES are mostly algae, with macrophytes playing a relatively minor role (Chapter 4.2). Primary production can be considerable in some IRES, such as desert streams. For example, in spatially intermittent Sycamore Creek (Arizona), Fisher et al. (1982) recorded high primary production and rapid recovery of algal communities in response to flooding. The creek's algal assemblage was generally dominated by disturbance-resistant diatoms, but also included macroalgae and cyanobacterial mats (Grimm and Fisher, 1989). Álvarez and Pardo (2009), working in Mediterranean Spain, reported substantial changes in algal abundance measured by chlorophyll pigments over the course of a year. These studies illustrate that primary production can be highly variable in IRES, but is particularly high in warm, open-canopied systems when flows are conducive.

Assessments of how temporal changes in hydrological regime—whether they be short term (e.g., days, seasons) or long term (e.g., years, decades)—show further evidence of shifts between autotrophy and heterotrophy (Fig. 4.7.3). Reid et al. (2008) working in IRES in southeast Australia reported seasonal shifts in basal resources from primarily heterotrophic year-round to autotrophy in some

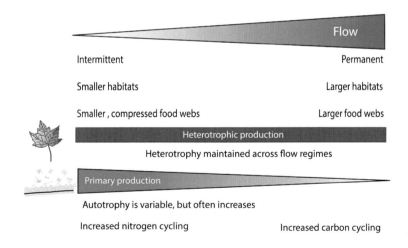

FIG. 4.7.3

Changes in basal food web components across hydrological phases (flow = *blue*) of IRES, indicating constant heterotrophy (*brown*) across flow states, increased autotrophy (*green*) with drying, and changes in the relative increases in nitrogen and carbon cycling. Leaf and algal symbols courtesy of the Integration and Application Network, University of Maryland Center for Environmental Science (ian.umces.edu/symbols/).

streams during periods of disconnection in spring and summer. Across multiple studies, terrestrial carbon pools are a reliable energy source in many IRES (Closs and Lake, 1994; Reid et al., 2008; Dekar et al., 2009; Courtwright and May, 2013), with shifts toward autochthonous production in pools during drying (Fig. 4.7.3). Studies from Mediterranean Europe point toward reliance on autochthonous resources during drying (Álvarez and Pardo, 2009; Vannucchi et al., 2013). Thus, it appears that heterotrophy generally dominates during wet phases, but algal resources and autotrophy can assume prominence during drying/low-flow periods. Further investigation into these patterns at a range of scales is warranted, particularly for determining which carbon source(s) matter most to production at higher levels, whether source shifts are linked to changes in the microbial communities (i.e., structure, functions) and the extent to which ongoing climate warming may impact these food sources.

4.7.3 CONSUMER-RESOURCE DYNAMICS IN IRES

The variable nature of IRES also structures the exchange of matter and energy among trophic levels. Hydrological contraction and fragmentation during drying cause aquatic taxa to concentrate in wetted areas (Fig. 4.7.4). Once fragmentation has occurred, density-dependent processes influence trophic structure and functioning. Densities of predators increase during drying as pools shrink and water quality deteriorates (Dekar and Magoulick, 2007; Dewson et al., 2007; Walters and Post, 2011), and predator-prey ratios typically increase as well (Bogan and Lytle, 2007; McHugh et al., 2015). Strong biotic interactions are usually accompanied by equally strong abiotic stress associated with deteriorating physicochemical conditions (Boulton and Lake, 1990; Drummond et al., 2015; McHugh et al., 2015). Then, upon rewetting, there can be periods when primary producers or primary consumers escape top-down control until herbivores or predators recolonize (Fig. 4.7.4b).

Spatial variation of trophic interactions may be less marked than temporal variability (Closs and Lake, 1994), but the occurrence of strong interactions generally increases with drying, along with changes to abiotic factors that stress aquatic taxa (Stanley et al., 1997). The distribution and abundance of predators depend on the timing and intensity of drying, barriers to movement, and species-specific environmental tolerances and mobility (Walters, 2016). These factors determine the spatial variation in ecological communities, likely causing changes in predation intensity at small spatial scales (Schlosser and Ebel, 1989; Acuña et al., 2005). Pool size and depth are especially important in IRES, determining the density and composition of animal communities (Englund and Krupa, 2000; Magoulick and Kobza, 2003; Wesner, 2013). Abiotic factors may be more important than biotic interactions in shaping fish assemblages of drying stream pools (Capone and Kushlan, 1991; Magoulick, 2000), although interactions can be intense (Power et al., 1985; Williams et al., 2003) and predation by terrestrial vertebrates may be understated (Capone and Kushlan, 1991; Magoulick, 2000).

Hydrological alteration associated with drying may also mediate biotic interactions with nonnative species and affect invasion outcomes. Nonnative fishes often originate in lentic habitats and are less flood-adapted than native species (Propst and Gido, 2004; Kiernan et al., 2012; Murphy et al., 2015), so habitat drying may alter invasion likelihood. However, native IRES fauna is often well adapted to drying (e.g., physiological tolerance, Labbe and Fausch, 2000; Magoulick and Kobza, 2003), so the net results on interactions between native and introduced species are unclear (Propst et al., 2008). Drying has been shown to favor native galaxiids over invasive brown trout in Australia and New Zealand (Closs and Lake, 1996; Leprieur et al., 2006). In other cases, drying may facilitate invasions. Drying has favored

FIG. 4.7.4

See legend on next page.

the displacement of native riparian trees by invasive *Tamarix ramosissima* in southwestern United States (Stromberg et al., 2007). In Arkansas and Missouri, the native crayfish *Orconectes eupunctus* has been replaced by the invasive *Orconectes neglectus* (Larson et al., 2009; but see Magoulick, 2014). In Spain, invasive fishes (e.g., *Lepomis gibbosus*) seem favored by drying compared to endemic cyprinids (Bernardo et al., 2003). Such findings indicate context- or condition-dependency, a pervasive theme in IRES ecology (Walters, 2016), suggesting a match between hydrology and life history mediates native-invader interactions. Thus, understanding the differential response of species to drying not only informs the study of IRES food webs but is also crucial for conservation and mitigating the impacts of climate change and flow alteration (Chapter 5.4).

The large size and limited dispersal ability of many aquatic predators make them especially vulnerable to effects of spatial fragmentation. During drying, they can become confined in pools (Dekar and Magoulick, 2007) and risk extirpation if pools dry completely (Bogan and Lytle, 2011; Christian and Adams, 2014). Upon rewetting, recolonization by large predators is dependent upon dispersal ability (Boersma et al., 2014; Cañedo-Argüelles et al., 2015). Bogan and Lytle (2011), for example, recorded the transition from a perennial but fragmented (isolated pools) state to one of complete drying and noted the loss and prolonged absence of large, dispersal-restricted, predators and their replacement by smaller vagile mesopredators in a multiyear study in Arizona. The variable hydrology of IRES (Chapter 2.2) may increase the vulnerability of IRES fauna to unexpected or prolonged drying and anthropogenic water abstraction (Deacon et al., 2007) and suggests hysteresis effects are likely.

Transitions between lotic and lentic phases generate turnover in aquatic communities, driven frequently by predator assemblages. Bogan and Lytle (2007) highlighted this pattern in Arizonan IRES, where they observed rapid turnover in aquatic invertebrate communities between high- and low-flow seasons. They studied several streams and recorded an increase in the proportion of predators from 24.5% during high flow to 75.2% during low-flow seasons. Large-bodied invertebrate predators (Belostomatidae, Dytiscidae, and Notonectidae) present in low abundances during high flow often become numerically dominant during low flow (Acuña et al., 2005; Bogan and Lytle, 2007).

Even when consumer assemblages remain unaltered between seasons, marked changes in predator diet can occur. In a large floodplain IRES (Cooper Creek, central Australia), Balcombe et al. (2015) found that predatory fishes switched from diets of zooplankton during residency in dry-season pools to a diverse diet of midges, beetle larvae, zooplankton, and terrestrial materials during the wet season spent on the floodplain. Overall, this change led to wider diet breadths for fish collected during wet compared to dry periods (Balcombe et al., 2005). Thus, large changes in trophic interactions can occur across IRES hydrological phases.

FIG. 4.7.4

Variation in resource states and consumer densities across hydrological phases in IRES: (a) algal mats after drying, (b) algal bloom after flow resumption, (c) carcasses of a large number of trout that were trapped in a small pool, (d) flies (a terrestrial invertebrate) on the dried carcass of a fish, (e) tadpoles trapped at high densities in the last remnants of a pool, and (f) *Abedus herberti* (Hemiptera: Belostomatidae) giant water bugs cluster along a 1-m portion of bedrock in a drying pool in Arizona. The giant water bugs are top predators in seasonally fragmenting IRES in southwestern United States and can reach densities of up to $50 \, m^{-2}$ during stream contraction and drying. The adult males in this photo bear eggs on their backs (inset), which will soon hatch and introduce hundreds of additional juvenile predators to the small pool.

Photos: Courtesy A.R. McIntosh (a–d) and K.S. Boersma (e, f).

Transition from wet to dry phases also affects riparian consumers. McCluney and Sabo (2009, 2012) manipulated water availability in the San Pedro River (Arizona) and recorded the response of riparian arthropod food webs. Interaction strengths between predators (wolf spiders: *Hogna antelucana*) and prey (crickets: *Gryllus alogus*) decreased during dry phases relative to wet phases (McCluney and Sabo, 2009). Greenwood and McIntosh (2010) also documented significant drying-associated changes in populations of riparian fishing spider *Dolomedes aquaticus* in New Zealand. *Dolomedes* biomass was lower and individuals were smaller in drying reaches, reflecting their heavy reliance on aquatic prey and vulnerability to desiccation. Lateral interactions between aquatic and terrestrial habitats are thus important determinants of trophic processes in IRES (Section 4.7.4).

While many taxa that inhabit IRES have adapted to withstand seasonal fluctuations in water availability, dispersal-limited predators such as fish and large invertebrates are still vulnerable to complete drying. The giant water bug *Abedus herberti*, the top predator in many fragmented streams in southwestern United States, is exemplary here—a flightless aquatic invertebrate with limited dispersal capacity and high risk of local extinction due to droughts (Bogan and Lytle, 2011). Researchers measured the consequences of such extinctions to food webs by experimentally removing *A. herberti* from some mesocosms but not others (Boersma et al., 2014). Removal caused an increase in the diversity and abundance of mesopredators and unexpected losses in detritivorous taxa. Instream experiments provide additional evidence of the importance of top-down processes in IRES (Williams et al., 2003; Rodríguez-Lozano et al., 2015) and highlight the importance of predatory fish in fragmented pools during drying. Overall though, little is known about how interannual environmental variability modulates predation in IRES food webs, although shifts in timing and magnitude of flow intermittence are expected to alter proximate abiotic drivers.

4.7.4 MULTIDIMENSIONAL INTERACTIONS WITHIN IRES

Materials and organisms cross ecosystem boundaries and transition zones (Strayer et al., 2003). IRES comprise mosaics of terrestrial and aquatic habitat that transition in space and time (Datry et al., 2016), longitudinally along upstream-downstream continua (Leigh et al., 2010a), laterally across river-riparian zone-floodplain continua (Bunn et al., 2006a), and even vertically among channel beds, subsurface sediments, and groundwater compartments (Stubbington et al., 2011) as habitats expand and contract, and dry and wet with changes in streamflow (Stanley et al., 1997; Chapters 2.3 and 4.9). This shifting habitat mosaic not only influences the movements, stability, and persistence of biota inhabiting IRES, but also the spatiotemporal dynamics of trophic links among them (Polis et al., 1996) and the biota and food webs of adjacent aquatic and terrestrial ecosystems (Fig. 4.7.5).

LONGITUDINAL INTERACTIONS

Waters that travel downstream along IRES channels often do so after dry phases when instream aquatic habitat has been absent or restricted to isolated pools and other moist refuges interspersed between exposed, dry riverbed (examples in Chapter 4.4). These events are known as first pulses or advancing wetted fronts and can entrain substantial amounts of organic material: aquatic invertebrates and fish, terrestrial invertebrates, fine detritus, biofilms, leaves, and wood. Travelling down a 7-km dry reach of the intermittent Albarine River, France, one such pulse contained several orders of magnitude more invertebrates than found in peak-flow pulses of perennial rivers (Corti and Datry, 2012). Terrestrial and

FIG. 4.7.5

The four-dimensional nature of food webs in IRES, showing links between aquatic (*blue* boxes) and terrestrial (*green* boxes) compartments along longitudinal (upstream and downstream of the IRES channel), lateral (IRES channel to the riparian zone, floodplain and beyond), and vertical (IRES channel to hyporheic zone and groundwater) dimensions and at different hydrological (temporal) phases: (A) flowing, (B) disconnected pool, (C) dry channel, (D) flooding, and (E) flood recession. *Arrows* show direction of links between compartments; for example, via the passive or active movement of organisms to another compartment or via consumption of organisms by a predator from another compartment (e.g., by migratory bird species in the broader terrestrial zone consuming aquatic prey in IRES channels), and line widths reflect link strength. Upstream and downstream aquatic compartments can include perennial rivers and streams and IRES. The link indicated with "*" is particularly strong when water level is receding. See text for more details.

aquatic invertebrates along with fine suspended organic material were transported downstream with the pulse, entrained for an average of 2 km before being redeposited on the newly rewetted bed. This represents a substantial transfer of energy and associated consumers (Fig. 4.7.5).

Depending on the size of the pulse and subsequent flows, redeposited material can accumulate in-stream, along margins or even above the active channel on banks or floodplain terraces. The importance of these accumulations to adjacent terrestrial ecosystems and downstream aquatic food webs is conjectured as high (Rosado et al., 2015) but remains understudied. Flow pulses may provide peak supplies of nutrients and food resources that stimulate downstream productivity and encourage fish and other mobile consumers to capitalize on the newly available resources (Corti and Datry, 2012; Rosado et al., 2015). The pulsed supplies may also have negative consequences for downstream reaches. Temporary hypoxia may ensue, resulting in fish kills and loss of other aquatic biota with unknown but potentially negative effects on food web structure (e.g., due to declines in prey) and ecosystem processes (e.g., due to unusually high numbers of dead fauna; King et al., 2012). Low-magnitude pulses in dryland IRES may disrupt the production of algae in shallow littoral zones that sustain aquatic food webs through the extended dry phases that occur between flood pulses (Bunn et al., 2006a). Overall, however, the rewetting of dry channels and longitudinal reconnection of isolated aquatic habitats within IRES networks are important for maintaining the integrity of downstream aquatic ecosystems (Bunn et al., 2006a; Leigh et al., 2010a) and supplying downstream communities with labile organic material for assimilation into their food web architecture (Jacobson et al., 2000).

LATERAL INTERACTIONS

As alluded to earlier, rivers, riparian zones, and floodplains are linked laterally via food webs (e.g., Baxter et al., 2005). In IRES, these trophic interactions occur as in perennial rivers; however, spatial and temporal fluctuation in the presence of water may modify the direction, timing, and intensity of energy flows between the adjacent ecosystems.

Given the dynamic nature of IRES hydrology, subsidies of aquatic resources to riparian food webs are spatially variable and temporally pulsed (Fig. 4.7.5). Contracting aquatic habitats offer a temporarily available source of concentrated, trapped, and often highly visible prey that can subsidize riparian food webs (Section 4.7.3). Dekar and Magoulick (2013) demonstrated predation by birds and mammals on fish trapped in artificial pools designed to mimic disconnected aquatic habitats in IRES, with large-bodied individuals being particularly susceptible to predators. Leigh et al. (2013a) found that up to 50% of the vertebrate fauna observed in riparian zones of isolated waterholes along stretches of otherwise dry IRES in Australia consumed aquatic fauna and at a higher rate than in the riparian zone of a nearby perennial channel. Dead aquatic biota accumulated on recently dried IRES channels can also constitute a pulsed resource subsidy to riparian food webs (Boulton and Suter, 1986; Steward, 2012).

The pulsing of aquatic prey availability not only affects lateral food web connections but also population and community dynamics within riparian zones. For example, the low biomass of aquatic insects in drying reaches of a New Zealand IRES was postulated to contribute to the low biomass of *Dolomedes* spiders (Greenwood and McIntosh, 2010; Section 4.7.3). Aquatic subsidies from Sycamore Creek also explain the high abundance and richness of spiders in riparian areas (Sanzone et al., 2003). Conversely, temporal variation in the abundance of riparian consumers potentially influences aquatic prey populations. In the Tagliamento River, a large braided IRES in Italy (Tockner et al., 2003), seasonal variation in the abundance of riparian predatory arthropods coincides with their proportional use of aquatic subsidies (Paetzold et al., 2005). Such predation could represent quantitatively important, pulsed transfers of aquatic secondary production to the riparian food web (Paetzold et al., 2005) while providing a further constraint on aquatic prey abundances during seasonal flow extremes.

Terrestrial invertebrates are also an important food resource for IRES inhabitants. Native brook trout inhabiting forested IRES in Virginia, USA, responded negatively to experimental reductions in terrestrial invertebrate inputs (Courtwright and May, 2013). The experimental reduction was mirrored by a proportional dietary response and fish were unable to make up the difference with more aquatic invertebrates.

Food web connections between adjacent terrestrial and aquatic ecosystems will depend somewhat on the characteristics of the individual ecosystems (e.g., forested vs. grassland) and regional climate and hydrology (Section 4.7.5). Not all IRES food webs rely on terrestrial subsidies (e.g., Reid et al., 2008; Blanchette et al., 2014), and similarly not all adjacent riparian food webs rely foremost on aquatic production (Corti and Datry, 2014; Sections 4.7.2 and 4.7.3). Beyond riparian zones, lateral interactions between IRES channels and adjacent ecosystems may extend to the floodplain (Fig. 4.7.5; e.g., Sheldon et al., 2002; Tockner et al., 2003). Connections between the in-channel and floodplain compartments in these systems are influenced by the lateral advance and retreat of floodwaters, with the pulsed contraction and expansion of aquatic habitats producing fluctuations in the sources of nutrients fuelling production (e.g., Fig. 4.7.4).

During floods, production booms on dryland IRES floodplains, providing an immense food resource for fish and other aquatic consumers (Bunn et al., 2006a) which sustain the aquatic food web well after floodwaters recede and isolated waterholes reform (Balcombe et al., 2015; Fig. 4.7.6). Burford et al. (2008) estimated that 50% of the fish carbon in a waterhole of Cooper Creek, isolated after floodwaters had retreated, was derived from floodplain food sources. Carbon from fish that boomed at high flows but later died in a waterhole fuelled a temporary boom in the waterhole's heterotrophic production. Between floods during extended dry phases, algal production in the littoral zones of isolated waterholes provides a major source of carbon and nitrogen to both primary and secondary aquatic consumers (Bunn et al., 2003, 2006a). The colonization of dry riverbeds by terrestrial plants, and the expansion of terrestrial ecosystems in general, are also likely to have ramifications for aquatic–terrestrial food web dynamics (Fig. 4.7.6).

Lateral connections between IRES and other ecosystems can be spatially extensive, beyond riparian zones and even floodplains, particularly in regions tending toward aridity with infrequent but major flooding (Fig. 4.7.5). For example, waterbirds respond to massive flooding events in dryland IRES to exploit temporarily productive feeding and breeding habitats (Chapter 4.6; Kingsford et al., 1999). Many bird species with wide foraging ranges and long seasonal migration routes observed near IRES waterholes in Australia's wet-dry tropics are known consumers of aquatic fauna (Leigh et al., 2013a). Subsidies from IRES may therefore extend far beyond waterbodies of origin, highlighting the potentially great, but undocumented, importance of IRES at regional scales.

VERTICAL INTERACTIONS

Variable groundwater inputs alter the duration of flow or water permanence in the surface and hyporheic zones of IRES (e.g., Larned et al., 2008; Chapter 2.3), which can influence ecosystem processes affecting overall food web dynamics, including leaf decomposition and nutrient cycling (Fig. 4.7.5; Boulton and Hancock, 2006; Chapters 2.3 and 3.2). Hyporheic sediments in IRES can remain biologically active

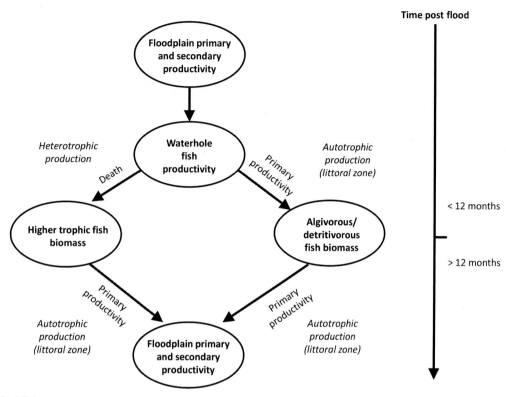

FIG. 4.7.6

Processes driving boom-and-bust fish production in the waterholes of Cooper Creek, which flows through the arid regions of central Australia, over the hydrological cycle starting with flooding of the floodplain and followed by flood recession, early phases of disconnection (all <12 months after flooding), later phases of disconnection and eventual drying (both >12 months after flooding). Floodplain flooding triggers very high productivity leading to very high fish biomass as the waters recede, driven by allochthonous resources (the "boom"). As waterholes dry down and disconnect, there is significant fish mortality (Death) leading to increased heterotrophic production due to the microbial breakdown of the floodplain-derived fish biomass together with increasing autotrophic production from littoral-zone algae. With extended disconnection and severe drying (>12 months), waterholes move into the "bust" phase where the minimal rates of production that prevail are maintained by primary producers.

From Balcombe, S.R., Turschwell, M.P., Arthington, A.H., Fellows, C.S., 2015. Is fish biomass in dryland river waterholes fuelled by benthic primary production after major overland flooding? J. Arid Environ. 116, 71–76, with permission.

throughout periods of surface-flow cessation due to groundwater connection (Leigh et al., 2013a); yet lengthy dry phases put hyporheic inhabitants under increasing water stress (Datry, 2012; Chapter 4.3). In Fuirosos River, a Mediterranean IRES, heterotrophs in biofilms were more resistant to drying than autotrophs, particularly in the hyporheic zone which was buffered against the water stress caused by surficial drying (Timoner et al., 2012). Flow intermittence thus increased the importance of heterotrophic processes and the contribution of the hyporheic zone to overall ecosystem processes. However, there is scant knowledge of how interactions between ground, hyporheic, and surface waters, coupled with flow intermittence, affect the trophic ecology of IRES at higher levels of organization.

4.7.5 EFFECTS OF REGIONAL DIFFERENCES ON FOOD WEBS IN IRES

IRES vary regionally in their hydrology (Chapters 2.2 and 2.3) and biota (Chapters 4.2–4.6); IRES in arid, Mediterranean, polar, tropical, and temperate zones all have distinct climates and flow regimes. This, combined with regionally distinct biota, suggests that IRES food webs should also vary regionally. In Australia's dryland floodplain IRES, the major sources of organic material fuelling higher trophic levels shift from algae in the littoral zones of isolated waterholes during dry phases to material produced on the floodplain during, and shortly after, unpredictable floods (Bunn et al., 2006a; Balcombe et al., 2015; Fig. 4.7.7). In regions with shorter and more predictable dry phases like the Mediterranean and tropics, variation in IRES food web dynamics may be more seasonal. Evidence suggests, however, that high levels of omnivory within food webs of such regions buffers them against fluctuations in flow extremes (e.g., Mediterranean: López-Rodríguez et al., 2012; Sánchez-Carmona et al., 2012; Vannucchi et al., 2013; tropics: Leigh et al., 2010b; Pusey et al., 2010; Blanchette et al., 2014).

Notwithstanding seasonal variation in light, temperature and instream primary production, algae may be a more important basal resource in the food webs of IRES in arid, Mediterranean, and tropical regions (e.g., Álvarez and Pardo, 2009; Pettit et al., 2012; Jardine et al., 2013) than in temperate-zone IRES with well-vegetated riparian zones (e.g., Reid et al., 2008), high litter fall, and comparatively high abundance or diversity of shredders (contrast with arid-zone streams: Bunn et al., 2006b; Fig. 4.7.7). However, riparian communities and leaf-fall timing in relation to hydrological cycles influence the relative importance of autochthonous and allochthonous contributions (Section 4.7.3). In temperate Australia, algae increase in importance as a food resource for aquatic consumers during disconnected pool phases (Reid et al., 2008). In La Solana in Mediterranean Spain, shredders dominate only during periods of major leaf fall, prior to seasonal flow decline (Muñoz, 2003). The dominance of deciduous species in riparian forests of Mediterranean IRES such as La Solana likely also explains the similarities in organic matter dynamics with deciduous forest streams in temperate zones. In Fuirosos River, for example, leaf fall predominantly occurs in autumn and plays an important role in fuelling production in the following season or seasons (e.g., Dekar et al., 2009), in both this (Acuña et al., 2007) and other temperate-zone IRES. In dryland IRES, sparse riparian tree cover limits wood and leaf loads, and low numbers or absence of aquatic invertebrate shredders can limit the aquatic breakdown of leaf litter (Bunn et al., 2006b).

Regional differences in leaf-litter palatability and decomposition rates also play a role in driving regional food web differences. For example, eucalypts are common along many Australian IRES, often dominating the riparian zones of arid-zone IRES. Eucalypts produce litter with relatively low nutritional quality, low palatability to invertebrates, and slow breakdown rates (Francis and Sheldon, 2002). In a South Australian IRES, decomposition of eucalyptus leaves in isolated pools darkened waters and created oxygen-poor conditions lasting up to three months; only two species of larval caddisflies persisted under these conditions (Towns, 1991). Similar changes in water quality, litter decomposition, and invertebrate mortality are observed in IRES in regions where eucalypts are an introduced species (e.g., Canhoto et al., 2013). Considerable regional variation can thus be expected in IRES food webs based on vegetation differences alone.

FIG. 4.7.7

Vegetation in riparian zones of IRES from different regions of the world illustrating differences in density, species, structure, and complexity that likely lead to large variation in influences on food webs: (a) Macadam Creek, during its disconnected waterhole phase in the dry season of Australia's wet-dry tropics; (b) Todd River, Alice Springs in central dryland Australia; (c) Ormiston Creek, central dryland Australia; (d–f) the Santa Maria de Palautordera (d) and Sant Esteve de Palautordera (e and f) reaches of Tordera Stream (Spain) that dry every summer (f) because of water diversion; and (g) the intermittent braided channel of Emanon Creek in the tussock- and shrub-dominated Canterbury high country (~900 m asl, South Island, New Zealand).

Photos: Courtesy C. Leigh (a–c), E. García-Berthou (d), R. Merciai (e, f), and A.R. McIntosh (g).

4.7.6 STRUCTURE OF IRES FOOD WEBS

Despite regional differences in IRES hydrology and energy supply at the base of food webs, consistent patterns emerge in structural properties of food webs. The most obvious changes in drying IRES are the loss of aquatic species and associated changes in community structure; these changes are often cyclic and vary seasonally with flow (Closs and Lake, 1994; Chapters 4.4 and 4.5). This affects energy flow, patterns of food web connectivity and, ultimately, food web dimensions in aquatic ecosystems (Fig. 4.7.1), but also creates opportunities for terrestrial consumers. Changes in terrestrial food web dimensions and aquatic-terrestrial connections (Section 4.7.4) are therefore also likely to occur (Giling et al., 2015).

The loss of aquatic-obligate species associated with flow loss from drying invariably shrinks food web dimensions. For example, in the intermittent Lerderderg River (Victoria, Australia), where detailed food webs were constructed for pools over four time periods spanning an annual hydrological cycle, web size (S, nodes in the food web) varied seasonally from 53 to 24 (Closs and Lake, 1994). Similarly, across the longitudinal perennial-to-drying gradients of IRES in the South Island, New Zealand, S declined from 30 to 15 taxa (McHugh et al., 2015).

A major consequence of changes to web size with drying is reduced trophic area. Trophic area reflects the horizontal (trophic breadth) and vertical (trophic height) dimensions that comprise all feeding interactions and the resource use diversity within a community. Informed by stable carbon-13 and nitrogen-15 isotope ratios, which respectively reflect the horizontal and vertical dimensions (Layman et al., 2007), trophic area signals energy flow through food webs. McHugh et al. (2015) observed a 20-fold decrease in trophic area, primarily driven by reduced trophic height, as habitat size decreased along drying continua. Changes in horizontal food web dimensions also occur, potentially reflecting altered use of basal resources (Section 4.7.4). The dominant change, however, is in trophic height, rather than in the breadth of resources used, perhaps because work to date could not fully resolve basal resource pools.

Reduced trophic height signals a reduction in aquatic food chain length (FCL). FCL reductions have been observed in many drying streams (Fig. 4.7.8), suggesting a consistent response of food webs to drying (Closs and Lake, 1994; Sabo et al., 2010; Woodward et al., 2012; McHugh et al., 2015). In a New Zealand study, FCL differed by approximately two trophic levels between perennial and drying sites. Similarly, FCL was approximately one trophic level longer in perennial rivers compared to IRES in the United States (Sabo et al., 2010). This reduction in FCL is more dramatic than those driven by increasing flood-related disturbance and is also larger than the increase in trophic levels seen along a gradient of increasing stream size (McHugh et al., 2010; Sabo et al., 2010). Interestingly, although seasonal increases in water supply in dynamic systems with extensive changes in the dimensions of aquatic habitat, such as the floodplain IRES of central and northern Australia, can increase consumer dietary breadth (Balcombe et al., 2005), they may not translate into major shifts in FCL (Warfe et al., 2013).

Drying-linked reductions in FCL can be characterized as trophic collapse and are typically associated with the loss of large-bodied top predators, typically fish or predatory invertebrates. Sabo et al. (2010) found top predators in food webs to differ between IRES and perennial streams; in perennial streams, fish were often large and piscivorous whereas top predators in IRES were either invertebrates or smaller insectivorous fish.

Importantly, the shape of the relationship between flow and FCL may be nonlinear in IRES, suggesting there are size thresholds associated with trophic collapse (McHugh et al., 2015). Declines in

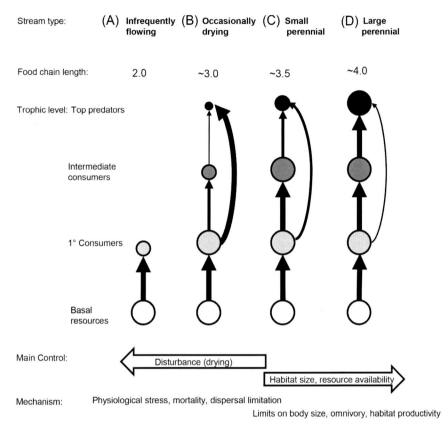

FIG. 4.7.8

Mechanisms influencing food chain length (FCL) in New Zealand IRES, relative to perennially flowing habitats. *Circles* represent food chains of consumers of varying abundance (*circle* size) feeding at progressively higher trophic levels, with arrows indicating relative energy flow (*arrow* size). At the extreme dry end of the IRES spectrum (A), food chains are short because few aquatic taxa can withstand complete desiccation/infrequent rewetting and/or cannot access these sites before they dry again. At the wet end of the IRES spectrum (B), frequent drying disturbance, extreme habitat compression, and/or stress reduce the diversity and size of taxa present, and disproportionately affect higher trophic level taxa. Additionally, because these sites are inherently more variable than perennial sites, top-level taxa are expected to tend toward more trophic omnivory where they also feed on primary consumers more than in more stable, perennial sites resulting in net food chain lengths close to three trophic levels. Food chain lengths in perennial streams (C, D) are expected to vary as a function of habitat size (discharge), resource availability, and disturbance resulting from extreme high flows, with the greatest vertical structuring/FCL observed in large stable systems where most top-level consumers are feeding high up the food web, resulting in food chain lengths closer to four trophic levels.

Based on McHugh, P., Thompson, R.M., Greig, H.S., Warburton, H.J., McIntosh, A.R., 2015. Habitat size influences food web structure in drying streams. Ecography 38, 700–712.

overall habitat size that limit body sizes for high-level consumers (e.g., fish, Walters and Post, 2008) are likely an important driver here, just as they are in perennial streams (McHugh et al., 2010; Sabo et al., 2010). However, in IRES declines are also likely linked to drying-associated stress (Hershkovitz and Gasith, 2013) because FCL-habitat size scaling is steeper in IRES than in perennial streams. McHugh et al. (2015) found FCL to be 0.25–0.50 less in IRES than in perennial reaches of equivalent size. Similarly, Sabo et al. (2010) found that FCL in drying streams in the United States was around two-thirds less compared to perennial streams. Thus, flow cessation initially constrains available space, reducing the size of organisms present (Walters and Post, 2008; Jellyman et al., 2014), and subsequently FCL (e.g., Romanuk et al., 2011). However, severe habitat shrinkage also subjects organisms to harsh conditions associated with, for example, warmer temperatures, reduced oxygen, and lower total resource abundances.

The harshness of intermittence is reflected in decreased energy intake and condition for fish as flow declines (Closs, 1994; Mas-Martí et al., 2010). Larger, longer-lived predatory taxa are often the slowest to recolonize streams upon flow resumption (e.g., Closs and Lake, 1994), consistent with theoretical models linking disturbance and demography to food web structure (e.g., Pimm and Lawton, 1977), although population resistance to drying-related disturbance depends heavily on the magnitude of stress and minimum habitat size (White et al., 2016). Thus, shrinking habitat size and deteriorating conditions force predatory fish to leave or suffer, meaning that top predator loss is an important proximate mechanism (*sensu* Post and Takimoto, 2007) of food web change in IRES. This phenomenon may result from natural or human-driven dewatering associated with abstraction (Chapter 5.1) and has implications for determining environmental flow regimes if damaging effects are to be avoided.

Flow intermittence also affects FCL by reducing the abundance of organisms living at intermediate trophic levels, like predatory invertebrates, or the degree to which top predators feed across trophic levels (i.e., insertion and omnivory, Post and Takimoto, 2007). In contrast to fish, which often respond predictably to low flows and drying (Walters, 2016), invertebrate responses are highly variable (Dewson et al., 2007). Obligate-aquatic invertebrates may respond similarly to fish (i.e., they are quickly lost and slow to recolonize; Closs and Lake, 1994). However, owing to a greater diversity of invertebrate adaptations (Chapters 4.3 and 4.8) for withstanding flow extremes, few generalities can be made regarding which taxa persist at low flow states (Walters, 2011).

While comparative studies examining the effects of extreme flow reduction reveal effects ranging from strong to weak on FCL in aquatic food webs, experiments reveal the mechanisms underlying changes. In general, S, body-size range and overall food web connections decreased in dewatered stream mesocosms in Dorset, UK (Woodward et al., 2012; Ledger et al., 2013), in line with the comparative studies mentioned earlier. However, some network-level properties were unaffected, including connectance, linkage density (i.e., links/species), and interaction diversity. Similarly, Walters and Post (2008, 2011) found that FCL resisted change in response to experimental dewatering, although they noted a response in body size and invertebrate communities to treatments. Thus, while some trophic dimensions of food webs respond considerably to habitat contraction, some structural properties of food webs can be maintained in the face of extreme disturbance.

Given the patterns described earlier, it is important to distinguish between disassembled food webs, those in the process of disassembly and those undergoing repeated disassembly. Webs observed in IRES are often a snapshot, at some stage of disassembly. Their configurations may thus be unstable, perhaps top-heavy or with a high predator-prey ratio (Bogan and Lytle, 2007; McHugh et al., 2015). This could make food webs of drying IRES vulnerable to strong trophic cascades (Bogan and Lytle,

2011). However, the Dorset food webs were exposed to repeated disturbance and had a stable structure, albeit with S considerably reduced. There may be mechanisms which, over time, work to reduce the destabilizing effects of drying, yielding potentially "disturbance-adapted" food webs. Closs and Lake (1994) argued that weak top-down trophic interactions, a heavy reliance on detrital food resources, and cross-ecosystem subsidies were important contributors to the stability of the Lerderderg River food web. Overall, these stabilizing and destabilizing forces require more in-depth study and are particularly relevant in the management of IRES because they will be important in maintaining ecosystem functions in IRES.

Contractions in aquatic food web dimensions offer potential for a corresponding increase in terrestrial food web dimensions with alterations in energy flow toward terrestrial top predators like birds, bats, and spiders (Section 4.7.4). Drying and the subsequent entrapment of aquatic organisms can provide a feeding bonanza for terrestrial predators and decomposers (e.g., Dekar and Magoulick, 2013). Understanding the dynamic temporal changes associated with drying, where potentially large amounts of carbon are transferred across ecosystem boundaries in very short time periods, raises a challenge for food web ecologists. Closs and Lake (1994) regarded their Lerderderg River food web as a "subweb contained within the overall Lerderderg valley food web" (p. 17), and suggested that FCL would inevitably be longer if insectivorous birds were considered.

4.7.7 TROPHIC INTERACTIONS IN IRES IN THE FUTURE

Global change, especially land-use change and climate warming, is expected to increase the spatial and temporal fragmentation of flowing waters (Datry et al., 2014; Chapter 5.1). Given that many of the changes associated with drying increase interaction strengths and produce conditions which potentially destabilize food webs, IRES systems are inherently vulnerable. Some evidence suggests systems adapted to these conditions will remain resilient, but even these systems respond to extended drying, such as during Australia's "millennium drought" (Lake, 2011). Understanding controls on system stability will be particularly important if deleterious consequences are to be avoided, suggesting this is an important area for future research. Moreover, to avoid extinctions, the plight of systems which appear to lack resistance and resilience to drying, such as those associated with the giant water bugs (Section 4.7.3), must be carefully considered.

A critical feature of IRES food webs is their ability to perform ecosystem functions and provide ecosystem services. Although we have a preliminary understanding of food web structure, we have a relatively poor understanding of their functioning. Key aspects include their role in nutrient cycling (Chapter 3.2), the cross-ecosystem energy transfer, and their potential to sustain unique elements of biodiversity.

Large-bodied organisms, including many top predators, are an especially vulnerable group supported by IRES. Studies in both experimental and natural systems indicate large taxa are disproportionately vulnerable to drying (Woodward et al., 2012; Ledger et al., 2013; McHugh et al., 2015). Moreover, because aquatic ectotherms are particularly sensitive to temperature changes, flow loss effects on temperature combined with negative temperature-body size covariance (Daufresne et al., 2009) suggest large bodied taxa in IRES will be further impacted by global climate warming. Thus, maintaining refugia for these organisms will be critical (Magoulick and Kobza, 2003), especially given their potential role in maintaining system stability.

The dynamic nature of IRES habitats and food webs also presents a ripe research opportunity to more effectively incorporate temporal variation into food web theory and network science, as called for in recent reviews (Thompson et al., 2012). Moreover, the spatial interactions of IRES, which occur in multiple dimensions, offer further opportunities to enhance understanding of the changing spatial dimensions of food webs. Answering critical questions concerning trophic interactions in IRES will thus enhance knowledge of more general processes in food web ecology.

REFERENCES

Acuña, V., Muñoz, I., Giorgi, A., Omella, M., Sabater, F., Sabater, S., 2005. Drought and post-drought recovery cycles in an intermittent Mediterranean stream: structural and functional aspects. J. N. Am. Benthol. Soc. 24, 919–933.

Acuña, V., Giorgi, A., Muñoz, I., Sabater, F., Sabater, S., 2007. Meteorological and riparian influences on organic matter dynamics in a forested Mediterranean stream. J. N. Am. Benthol. Soc. 26, 54–69.

Álvarez, M., Pardo, I., 2009. Dynamics in the trophic structure of the macroinvertebrate community in a Mediterranean, temporary stream. Aquat. Sci. 71, 202–213.

Amalfitano, S., Fazi, S., Zoppini, A., Barra Caracciolo, A., Grenni, P., Puddu, A., 2008. Responses of benthic bacteria to experimental drying in sediments from Mediterranean temporary rivers. Microb. Ecol. 55, 270–279.

Balcombe, S.R., Bunn, S., McKenzie-Smith, F.J., Davies, P., 2005. Variability of fish diets between dry and flood periods in an arid zone floodplain river. J. Fish Biol. 67, 1552–1567.

Balcombe, S.R., Turschwell, M.P., Arthington, A.H., Fellows, C.S., 2015. Is fish biomass in dryland river waterholes fuelled by benthic primary production after major overland flooding? J. Arid Environ. 116, 71–76.

Baxter, C.V., Fausch, K.D., Saunders, W.C., 2005. Tangled webs: reciprocal flows of invertebrate prey link streams and riparian zones. Freshw. Biol. 50, 201–220.

Bernal, S., von Schiller, D., Sabater, F., Martí, E., 2013. Hydrological extremes modulate nutrient dynamics in mediterranean climate streams across different scales. Hydrobiologia 719, 31–42.

Bernardo, J.M., Ilhéu, M., Matono, P., Costa, A.M., 2003. Interannual variation of fish assemblage structure in a Mediterranean river: implications of streamflow on the dominance of native or exotic species. River Res. Appl. 19, 521–532.

Blanchette, M.L., Aaron, M.D., Jardine, T.D., Pearson, R.G., 2014. Omnivory and opportunism characterize food webs in a large dry-tropics river system. Freshw. Sci. 33, 142–158.

Boersma, K.S., Bogan, M.T., Henrichs, B.A., Lytle, D.A., 2014. Top predator removals have consistent effects on large species despite high environmental variability. Oikos 123, 807–816.

Bogan, M.T., Lytle, D.A., 2007. Seasonal flow variation allows 'time-sharing' by disparate aquatic insect communities in montane desert streams. Freshw. Biol. 52, 290–304.

Bogan, M.T., Lytle, D.A., 2011. Severe drought drives novel community trajectories in desert stream pools. Freshw. Biol. 56, 2070–2081.

Boulton, A.J., Hancock, P.J., 2006. Rivers as groundwater-dependent ecosystems: a review of degrees of dependency, riverine processes and management implications. Aust. J. Bot. 54, 133–144.

Boulton, A.J., Lake, P.S., 1990. The ecology of two intermittent streams in Victoria, Australia. I. Multivariate analyses of physicochemical features. Freshw. Biol. 24, 123–141.

Boulton, A.J., Suter, P.J., 1986. Ecology of temporary streams—an Australian perspective. In: De Deckker, P., Williams, W.D. (Eds.), Limnology in Australia. CSIRO/Dr W Junk, Melbourne/Dordrecht, pp. 313–329.

Bruder, A., Chauvet, E., Gessner, M.O., 2011. Litter diversity, fungal decomposers and litter decomposition under simulated stream intermittency. Funct. Ecol. 25, 1269–1277.

Bunn, S.E., Davies, P.M., Winning, M., 2003. Sources of organic carbon supporting the food web of an arid zone floodplain river. Freshw. Biol. 48, 619–635.

Bunn, S.E., Thoms, M.C., Hamilton, S.K., Capon, S.J., 2006a. Flow variability in dryland rivers: boom, bust and the bits in between. River Res. Appl. 22, 179–186.

Bunn, S.E., Balcombe, S.R., Davies, P.M., Fellows, C.S., McKenzie-Smith, F.J., 2006b. Aquatic productivity and food webs of desert river ecosystems. In: Kingsford, R.T. (Ed.), Ecology of Desert Rivers. Cambridge University Press, Cambridge, pp. 76–99.

Burford, M.A., Cook, A.J., Fellows, C.S., Balcombe, S.R., Bunn, S.E., 2008. Sources of carbon fuelling production in an arid floodplain river. Mar. Freshw. Res. 59, 224–234.

Cañedo-Argüelles, M., Boersma, K.S., Bogan, M.T., Olden, J.D., Phillipsen, I., Schriever, T.A., et al., 2015. Dispersal strength determines meta-community structure in a dendritic riverine network. J. Biogeogr. 42, 778–790.

Canhoto, C., Calapez, R., Goncalves, A.L., Moreira-Santos, M., 2013. Effects of *Eucalyptus* leachates and oxygen on leaf-litter processing by fungi and stream invertebrates. Freshw. Sci. 32, 411–424.

Capone, T.A., Kushlan, J.A., 1991. Fish community structure in dry-season stream pools. Ecology 12, 983–992.

Christian, J., Adams, G., 2014. Effects of pool isolation on trophic ecology of fishes in a highland stream. J. Fish Biol. 85, 752–772.

Closs, G.P., 1994. Feeding of *Galaxias olidus* Günther (Pisces: Galaxiidae) in an intermittent Australian Stream. Aust. J. Mar. Freshwat. Res. 45, 227–232.

Closs, G.P., Lake, P.S., 1994. Spatial and temporal variation in the structure of an intermittent-stream food web. Ecol. Monogr. 75, 2–21.

Closs, G.P., Lake, P.S., 1996. Drought, differential mortality and the coexistence of a native and an introduced fish species in a south east Australian intermittent stream. Environ. Biol. Fish 47, 17–26.

Corti, R., Datry, T., 2012. Invertebrate and sestonic matter in an advancing wetted front travelling down a dry river bed (Albarine, France). Freshw. Sci. 31, 1187–1201.

Corti, R., Datry, T., 2014. Drying of a temperate, intermittent river has little effect on adjacent riparian arthropod communities. Freshw. Biol. 59, 666–678.

Corti, R., Datry, T., Drummond, L., Larned, S.T., 2011. Natural variation in immersion and emersion affects breakdown and invertebrate colonization of leaf litter in a temporary river. Aquat. Sci. 73, 537–550.

Courtwright, J., May, C.L., 2013. Importance of terrestrial subsidies for native brook trout in Appalachian intermittent streams. Freshw. Biol. 58, 2423–2438.

Datry, T., 2012. Benthic and hyporheic invertebrate assemblages along a flow intermittence gradient: effects of duration of dry events. Freshw. Biol. 57, 563–574.

Datry, T., Larned, S.T., Tockner, K., 2014. Intermittent rivers: a challenge for freshwater ecology. Bioscience 64, 229–235.

Datry, T., Pella, H., Leigh, C., Bonada, N., Hugueny, B., 2016. A landscape approach to advance intermittent river ecology. Freshw. Biol. 61, 1200–1213.

Daufresne, M., Lengfellner, K., Sommer, U., 2009. Global warming benefits the small in aquatic ecosystems. Proc. Natl. Acad. Sci. U. S. A. 106, 12788–12793.

Deacon, J.E., Williams, A.E., Williams, C.D., Williams, J.E., 2007. Fueling population growth in Las Vegas: how large-scale groundwater withdrawal could burn regional biodiversity. Bioscience 57, 688–698.

Dekar, M.P., Magoulick, D.D., 2007. Factors affecting fish assemblage structure during seasonal stream drying. Ecol. Freshw. Fish 16, 335–342.

Dekar, M.P., Magoulick, D.D., 2013. Effects of predators on fish and crayfish survival in intermittent streams. Southeast. Nat. 12, 197–208.

Dekar, M.P., Magoulick, D.D., Huxel, G.R., 2009. Shifts in the trophic base of intermittent stream food webs. Hydrobiologia 635, 263–277.

Dewson, Z.S., James, A.B., Death, R.G., 2007. A review of the consequences of decreased flow for instream habitat and macroinvertebrates. J. N. Am. Benthol. Soc. 26, 401–415.

Drummond, L.R., McIntosh, A.R., Larned, S.T., 2015. Invertebrate community dynamics and insect emergence in response to pool drying in a temporary river. Freshw. Biol. 60, 1596–1612.

Englund, G., Krupa, J.J., 2000. Habitat use by crayfish in stream pools: influence of predators, depth and body size. Freshw. Biol. 43, 75–83.

Febria, C.M., Hosen, J.D., Crump, B.C., Palmer, M.A., Williams, D.D., 2015. Microbial responses to changes in flow status in temporary headwater streams: a cross-system comparison. Front. Microbiol. 6, 522.

Fisher, S.G., Gray, L.J., Grimm, N.B., Busch, D.E., 1982. Temporal succession in a desert stream ecosystem following flash flooding. Ecol. Monogr. 52, 93–110.

Francis, C., Sheldon, F., 2002. River Red Gum (*Eucalyptus camaldulensis* Dehnh.) organic matter as a carbon source in the lower Darling River, Australia. Hydrobiologia 481, 113–124.

Gerull, L., Frossard, A., Gessner, M.O., Mutz, M., 2011. Variability of heterotrophic metabolism in small stream corridors of an early successional watershed. J. Geophys. Res. 116, G02012.

Giling, D.P., Mac Nally, R., Thompson, R.M., 2015. How might cross-system subsidies in riverine networks be affected by altered flow variability? Ecosystems 18, 1151–1164.

Greenwood, M.J., McIntosh, A.R., 2010. Low river flow alters the biomass and population structure of a riparian predatory invertebrate. Freshw. Biol. 55, 2062–2076.

Grimm, N.B., Fisher, S.G., 1989. Stability of periphyton and macroinvertebrates to disturbance by flash floods in a desert stream. J. N. Am. Benthol. Soc. 8, 293–307.

Hall Jr., R.O., Meyer, J.L., 1998. The trophic significance of bacteria in a detritus-based stream food web. Ecology 79, 1995–2012.

Hall Jr., R.O., Wallace, J.B., Eggert, S.L., 2000. Organic matter flow in stream food webs with reduced detrital resource base. Ecology 81, 3445–3463.

Hershkovitz, Y., Gasith, A., 2013. Resistance, resilience, and community dynamics in mediterranean-climate streams. Hydrobiologia 719, 59–75.

Jacobson, P.J., Jacobson, K.M., Angermeier, P.L., Cherry, D.S., 2000. Variation in material transport and water chemistry along a large ephemeral river in the Namib Desert. Freshw. Biol. 44, 481–491.

Jardine, T.D., Hunt, R.J., Faggotter, S.J., Valdez, D., Burford, M.A., Bunn, S.E., 2013. Carbon from periphyton supports fish biomass in waterholes of a wet–dry tropical river. River Res. Appl. 29, 560–573.

Jellyman, P.G., McHugh, P.A., McIntosh, A.R., 2014. Increases in disturbance and reductions in habitat size interact to suppress predator body size. Glob. Chang. Biol. 20, 1550–1558.

Kiernan, J.D., Moyle, P.B., Crain, P.K., 2012. Restoring native fish assemblages to a regulated California stream using the natural flow regime concept. Ecol. Appl. 22, 1472–1482.

King, A.J., Tonkin, Z., Lieshcke, J., 2012. Short-term effects of a prolonged blackwater event on aquatic fauna in the Murray River, Australia: considerations for future events. Mar. Freshw. Res. 63, 576–586.

Kingsford, R.T., Curtin, A.L., Porter, J., 1999. Water flows on Cooper Creek in arid Australia determine 'boom' and 'bust' periods for waterbirds. Biol. Conserv. 88, 231–248.

Labbe, T.R., Fausch, K.D., 2000. Dynamics of intermittent stream habitat regulate persistence of a threatened fish at multiple scales. Ecol. Appl. 10, 1774–1791.

Lake, P.S., 2011. Drought and Aquatic Ecosystems: Effects and Responses. John Wiley and Sons, Chichester.

Larned, S.T., Hicks, D.M., Schmidt, J., Davey, A.J.H., Dey, K., Scarsbrook, M., et al., 2008. The Selwyn River of New Zealand: a benchmark system for alluvial plain rivers. River Res. Appl. 24, 1–21.

Larson, E.R., Magoulick, D.D., Turner, C., Laycock, K.H., 2009. Disturbance and species displacement: different tolerances to stream drying and desiccation in a native and an invasive crayfish. Freshw. Biol. 54, 1899–1908.

Layman, C.A., Arrington, D.A., Montaña, C.G., Post, D.M., 2007. Can stable isotope ratios provide for community-wide measures of trophic structure? Ecology 88, 42–48.

Ledger, M.E., Brown, L.E., Edwards, F.K., Milner, A.M., Woodward, G., 2013. Drought alters the structure and functioning of complex food webs. Nat. Clim. Chang. 3, 223–227.

Leigh, C., Sheldon, F., Kingsford, R.T., Arthington, A.H., 2010a. Sequential floods drive 'booms' and wetland persistence in dryland rivers: a synthesis. Mar. Freshw. Res. 61, 896–908.

Leigh, C., Burford, M.A., Sheldon, F., Bunn, S.E., 2010b. Dynamic stability in dry season food webs within tropical floodplain rivers. Mar. Freshw. Res. 61, 357–368.

Leigh, C., Reis, T.M., Sheldon, F., 2013a. High potential subsidy of dry-season aquatic fauna to consumers in riparian zones of wet–dry tropical rivers. Inland Waters 3, 411–420.

Leprieur, F., Hickey, M.A., Arbuckle, C.J., Closs, G.P., Brosse, S., Townsend, C.R., 2006. Hydrological disturbance benefits a native fish at the expense of an exotic fish. J. Appl. Ecol. 43, 930–939.

López-Rodríguez, M.J., Peralta-Maraver, I., Gaetani, B., Sainz-Cantero, C.E., Fochetti, R., de Figueroa, J.M.T., 2012. Diversity patterns and food web structure in a Mediterranean intermittent stream. Int. Rev. Hydrobiol. 97, 485–496.

Magoulick, D.D., 2000. Spatial and temporal variation in fish assemblages of drying stream pools: the role of abiotic and biotic factors. Aquat. Ecol. 34, 29–41.

Magoulick, D.D., 2014. Impacts of drought and crayfish invasion on stream ecosystem structure and function. River Res. Appl. 30, 1309–1317.

Magoulick, D.D., Kobza, R.M., 2003. The role of refugia for fishes during drought: a review and synthesis. Freshw. Biol. 48, 1186–1198.

Mas-Martí, E., García-Berthou, E., Sabater, S., Tomanova, S., Muñoz, I., 2010. Comparing fish assemblages and trophic ecology of permanent and intermittent reaches in a Mediterranean stream. Hydrobiologia 657, 167–180.

McCann, K., Hastings, A., Huxel, G.R., 1998. Weak trophic interactions and the balance of nature. Nature 395, 794–798.

McCann, K.S., Rasmussen, J.B., Umbanhowar, J., 2005. The dynamics of spatially coupled food webs. Ecol. Lett. 8, 513–523.

McCluney, K.E., Sabo, J.L., 2009. Water availability directly determines per capita consumption at two trophic levels. Ecology 90, 1463–1469.

McCluney, K.E., Sabo, S.L., 2012. River drying lowers the diversity and alters the composition of an assemblage of desert riparian arthropods. Freshw. Biol. 57, 91–103.

McHugh, P., McIntosh, A.R., Jellyman, P., 2010. Dual influences of ecosystem size and disturbance on food chain length in streams. Ecol. Lett. 13, 881–890.

McHugh, P., Thompson, R.M., Greig, H.S., Warburton, H.J., McIntosh, A.R., 2015. Habitat size influences food web structure in drying streams. Ecography 38, 700–712.

Muñoz, I., 2003. Macroinvertebrate community structure in an intermittent and a permanent Mediterranean streams (NE Spain). Limnetica 22, 107–116.

Murphy, C.A., Grenouillet, G., García-Berthou, E., 2015. Natural abiotic factors more than anthropogenic perturbation shape the invasion of Eastern Mosquitofish (*Gambusia holbrooki*). Freshw. Sci. 34, 965–974.

Paetzold, A., Schubert, C.J., Tockner, K., 2005. Aquatic terrestrial linkages along a braided-river: Riparian arthropods feeding on aquatic insects. Ecosystems 8, 748–759.

Pettit, N., Davies, T., Fellman, J., Grierson, P., Warfe, D., Davies, P., 2012. Leaf litter chemistry, decomposition and assimilation by macroinvertebrates in two tropical streams. Hydrobiologia 680, 63–77.

Pimm, S.L., Lawton, J.H., 1977. The numbers of trophic levels in ecological communities. Nature 268, 329–333.

Polis, G.A., Holt, R.D., Menge, B.A., Winemiller, K.O., 1996. Time, space, and life history: influences on food webs. In: Polis, G.A., Winemiller, K.O. (Eds.), Food Webs: Integration of Patterns and Dynamics. Chapman and Hall, New York, pp. 435–460.

Post, D.M., Takimoto, G., 2007. Proximate structural mechanisms for variation in food-chain length. Oikos 116, 775–782.

Power, M.E., Matthews, W.J., Stewart, A.J., 1985. Grazing minnows, piscivorous bass and stream algae: dynamics of a strong interaction. Ecology 66, 1448–1456.

Propst, D.L., Gido, K.B., 2004. Responses of native and nonnative fishes to natural flow regime mimicry in the San Juan River. Trans. Am. Fish. Soc. 133, 922–931.

Propst, D.L., Gido, K.B., Stefferud, J.A., 2008. Natural flow regimes, nonnative fishes, and native fish persistence in arid-land river systems. Ecol. Appl. 18, 1236–1252.

Pusey, B.J., Arthington, A.H., Stewart-Koster, B., Kennard, M.J., Read, M.G., 2010. Widespread omnivory and low temporal and spatial variation in the diet of fishes in a hydrologically variable northern Australian river. J. Fish Biol. 77, 731–753.

Reid, D.J., Quinn, G.P., Lake, P.S., Reich, P., 2008. Terrestrial detritus supports the food webs in lowland intermittent streams of south-eastern Australia: a stable isotope study. Freshw. Biol. 53, 2036–2050.

Rodríguez-Lozano, P., Verkaik, I., Rieradevall, M., Prat, N., 2015. Small but powerful: top predator local extinction affects ecosystem structure and function in an intermittent stream. PLoS One 10, e0117630.

Rolls, R.J., Leigh, C., Sheldon, F., 2012. Mechanistic effects of low-flow hydrology on riverine ecosystems: ecological principles and consequences of alteration. Freshw. Sci. 31, 1163–1186.

Romaní, A.M., Amalfitano, S., Artigas, J., Fazi, S., Sabater, S., Timoner, X., et al., 2012. Microbial biofilm structure and organic matter use in mediterranean streams. Hydrobiologia 719, 43–58.

Romanuk, T.N., Hayward, A., Hutchings, J.A., 2011. Trophic level scales positively with body size in fishes. Glob. Ecol. Biogeogr. 20, 231–240.

Rosado, J., Morais, M., Tockner, K., 2015. Mass dispersal of terrestrial organisms during first flush events in a temporary stream. River Res. Appl. 31, 912–917.

Sabo, J.L., Finlay, J.C., Kennedy, T., Post, D.M., 2010. The role of discharge variation in scaling of drainage area and food chain length in rivers. Science 330, 965–967.

Sánchez-Carmona, R., Encina, L., Rodríguez-Ruiz, A., Rodríguez-Sánchez, M.V., Granado-Lorencio, C., 2012. Food web structure in Mediterranean streams: exploring stabilizing forces in these ecosystems. Aquat. Ecol. 46, 311–324.

Sanzone, D.M., Meyer, J.L., Marti, E., Gardiner, E.P., Tank, J.L., Grimm, N.B., 2003. Carbon and nitrogen transfer from a desert stream to riparian predators. Oecologia 134, 238–250.

Schlosser, I.J., Ebel, K.K., 1989. Effects of flow regime and cyprinid predation on a headwater stream. Ecol. Monogr. 59, 41–57.

Schmid, P., Schmid-Araya, J., 2010. Scale-dependent relations between bacteria, organic matter and invertebrates in a headwater stream. Fundam. Appl. Limnol. 176, 365–375.

Sheldon, F., Boulton, A.J., Puckridge, J.T., 2002. Conservation value of variable connectivity: aquatic invertebrate assemblages of channel and floodplain habitats of a central Australian arid-zone river, Cooper Creek. Biol. Conserv. 103, 13–31.

Stanley, E.H., Fisher, S.G., Grimm, N.B., 1997. Ecosystem expansion and contraction in streams. Bioscience 47, 427–435.

Steward, A., 2012. When the River Runs Dry: The Ecology of Dry River Beds. Griffith University, Brisbane. PhD Thesis.

Strayer, D.L., Power, M.E., Fagan, W.F., Pickett, S.T.A., Belnap, J., 2003. A classification of ecological boundaries. Bioscience 53, 723–729.

Stromberg, J.C., Beauchamp, V.B., Dixon, M.D., Lite, S.J., Paradzick, C., 2007. Importance of low-flow and high-flow characteristics to restoration of riparian vegetation along rivers in arid south-western United States. Freshw. Biol. 52, 651–679.

Stubbington, R., Wood, P.J., Reid, I., 2011. Spatial variability in the hyporheic zone refugium of temporary streams. Aquat. Sci. 73, 499–511.

Thompson, R.M., Brose, U., Dunne, J.A., Hall, R.O., Hladyz, S., Kitching, R.L., et al., 2012. Food webs: reconciling the structure and function of biodiversity. Trends Ecol. Evol. 27, 689–697.

Timoner, X., Acuña, V., von Schiller, D., Sabater, S., 2012. Functional responses of stream biofilms to flow cessation, desiccation and rewetting. Freshw. Biol. 57, 1565–1578.

Tockner, K., Ward, J.V., Arscott, D.B., Edwards, P.J., Kollmann, J., Gurnell, A.M., et al., 2003. The Tagliamento River: a model ecosystem of European importance. Aquat. Sci. 65, 239–253.

Towns, D.R., 1991. Ecology of leptocerid caddisfly larvae in an intermittent South Australian stream receiving eucalyptus litter. Freshw. Biol. 25, 117–129.

Vannucchi, P.E., López-Rodríguez, M.J., de Figueroa, J.M.T., Gaino, E., 2013. Structure and dynamics of a benthic trophic web in a Mediterranean seasonal stream. J. Limnol. 72, 606–615.

Walters, A.W., 2011. Resistance of aquatic insects to a low-flow disturbance: exploring a trait-based approach. J. N. Am. Benthol. Soc. 30, 346–356.

Walters, A.W., 2016. The importance of context dependence for understanding the effects of low-flow events on fish. Freshw. Sci. 35, 216–228.

Walters, A.W., Post, D.M., 2008. An experimental disturbance alters fish size structure but not food chain length in streams. Ecology 89, 3261–3267.

Walters, A.W., Post, D.M., 2011. How low can you go? Impacts of a low-flow disturbance on aquatic insect communities. Ecol. Appl. 21, 163–174.

Warfe, D.M., Jardine, T.D., Pettit, N.E., Hamilton, S.K., Pusey, B.J., Bunn, S.E., et al., 2013. Productivity, disturbance and ecosystem size have no influence on food chain length in seasonally connected rivers. PLoS One 8, e66240.

Wesner, J.S., 2013. Fish predation alters benthic, but not emerging, insects across whole pools of an intermittent stream. Freshw. Sci. 32, 438–449.

White, R.S.A., McHugh, P.A., McIntosh, A.R., 2016. Drought-survival is a threshold function of habitat size and population density in a fish metapopulation. Glob. Chang. Biol. 22, 3341–3348.

Williams, L.R., Taylor, C.M., Warren, M.L., 2003. Influence of fish predation on assemblage structure of macroinvertebrates in an intermittent stream. Trans. Am. Fish. Soc. 132, 120–130.

Woodward, G., Brown, L.E., Edwards, F.K., Hudson, L.N., Milner, A.M., Reuman, R.C., et al., 2012. Climate change impacts in multispecies systems: drought alters food web size structure in a field experiment. Philos. Trans. R. Soc. B 367, 2990–2997.

FURTHER READING

Leigh, C., Stubbington, R., Sheldon, F., Boulton, A.J., 2013b. Hyporheic invertebrates as bioindicators of ecological health in temporary rivers: a meta-analysis. Ecol. Indic. 32, 62–73.

RESISTANCE, RESILIENCE, AND COMMUNITY RECOVERY IN INTERMITTENT RIVERS AND EPHEMERAL STREAMS

Michael T. Bogan*, Edwin T. Chester[†,‡], Thibault Datry[§], Ashley L. Murphy[¶], Belinda J. Robson[†], Albert Ruhi**, Rachel Stubbington[††], James E. Whitney[‡‡]

University of Arizona, Tucson, AZ, United States* Murdoch University, Murdoch, WA, Australia[†] Deakin University, Warrnambool, VIC, Australia[‡] Irstea, UR MALY, centre de Lyon-Villeurbanne, Villeurbanne, France[§] Monash University, Melbourne, VIC, Australia[¶] Arizona State University, Tempe, AZ, United States** Nottingham Trent University, Nottingham, United Kingdom[††] Pittsburg State University, Pittsburg, KS, United States[‡‡]

IN A NUTSHELL

- Intermittent rivers and ephemeral streams (IRES) are temporally and spatially dynamic, experiencing alternating wet and dry phases and supporting both aquatic and terrestrial habitats
- Persistence in these highly variable habitats requires aquatic taxa to be resistant or resilient to disturbances such as drying
- Resistance mechanisms include tolerance of extreme physicochemical conditions and having desiccation-resistant dormant stages; resilience mechanisms require the ability to disperse instream or overland
- Terrestrial taxa inhabiting dry riverbeds must be resistant or resilient to periodic flooding of their habitat
- Resistance and resilience mechanisms interact to facilitate community recovery
- Human disturbance alters recovery pathways and affects long-term persistence of aquatic and terrestrial species in IRES

4.8.1 INTRODUCTION: CHALLENGES OF PERSISTING IN IRES

Intermittent rivers and ephemeral streams (IRES) host complex mosaics of terrestrial and aquatic habitats that vary greatly through time and across landscapes (Chapters 2.2, 2.3, and 4.9). Some IRES flow for as little as several days or weeks each year, while others may flow nearly year-round (Fig. 4.8.1A). Furthermore, flow duration patterns can vary greatly across regions and years (Poff and Ward, 1989). Even within the same stream, adjacent reaches may have perennial flow, flow for several months, or flow only a few days annually (Jaeger and Olden, 2012). This dynamism in the location and duration of flowing, pool and dry habitats poses significant challenges to aquatic and terrestrial species. Despite

Intermittent Rivers and Ephemeral Streams. http://dx.doi.org/10.1016/B978-0-12-803835-2.00013-9

349

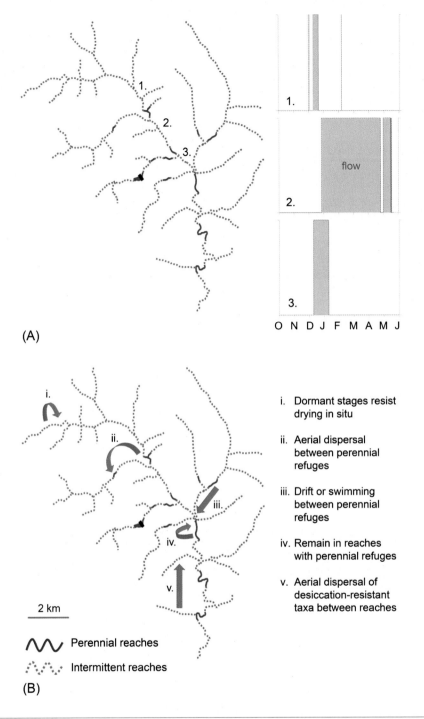

FIG. 4.8.1

Highly variable flow regimes across an IRES network in Mediterranean-climate California and potential dispersal pathways of aquatic organisms. In the top panel (A), three adjacent reaches in Chalone Creek (locations 1–3) exhibited varying periods of flow (shaded *blue*), lasting from a few weeks to several months, during the 2014–2015 wet season. In the bottom panel (B), five of the most common resistance and resilience mechanisms for aquatic species are illustrated. Importantly, these five strategies are not mutually exclusive as some species are capable of using multiple strategies. For example, the caddisfly *Lectrides varians* can use strategies i–v to persist in Australian IRES (Wickson et al., 2012).

these challenges, IRES can support diverse biological communities, including many species of algae, macrophytes, invertebrates, amphibians, reptiles, and fishes (Chapters 4.1–4.6). Although IRES experience numerous types of natural disturbances (e.g., fire, scouring floods), we focus here on the unique challenges that cycles of drying and rewetting present to resident taxa.

Freshwater and riparian taxa employ a wide range of strategies and traits to survive disturbances. Verberk et al. (2008) grouped these into four trait domains: dispersal, reproduction, synchronization, and development. This provides a framework that links the traits we observe to life histories and the phylogenies that constrain them. Traits used by taxa to survive cycles of wetting and drying in IRES exist in all four domains. For example, dispersal is used by many species to escape either wetting or drying (Section 4.8.3). Reproductive strategies used by animals in IRES vary greatly, from the rapid opportunistic production of large numbers of eggs by some desert fish (e.g., Minckley and Marsh, 2009) to long-lived crayfish that only reproduce in years with favorable conditions (e.g., Johnston et al., 2008). Synchronization refers to the timing of life history events such as dispersal, reproduction, and development, and may involve speeding up or slowing down development in response to environmental conditions (Verberk et al., 2008). Flexible development is often assumed to occur in IRES species, but little is known about the outcomes of this flexibility for recovery of stream communities.

Traits in each of these domains may comprise either resistance or resilience strategies for surviving disturbance. Here, we consider resistance and resilience to be distinct components of ecological responses to disturbance. Resistance is defined as the ability of an ecological unit (e.g., population, community) to withstand a disturbance *in situ* whereas resilience is defined as the capacity for an ecological unit to recover following disturbance (Lake, 2000; Nimmo et al., 2015). Aquatic species in many IRES are confronted with periodic loss of water and have various means of resisting, or being resilient to, stream drying (Box 4.8.1; Fig. 4.8.2). Terrestrial taxa must deal with the opposite situation—periodic inundation of their terrestrial habitat. When flow resumes, terrestrial taxa can be flushed downstream, experience high mortality or be forced to disperse laterally into riparian zones (Corti and Datry, 2012; Rosado et al., 2015). If terrestrial taxa cannot resist inundation, then they must crawl or fly overland to recolonize streambeds as flow ceases and aquatic habitat dries.

BOX 4.8.1 RESISTANCE, RESILIENCE, AND RECOVERY PROCESSES IN IRES

Species inhabiting IRES use a variety of *resistance* mechanisms to withstand disturbance *in situ* and *resilience* mechanisms to recover following disturbance (Lake, 2000; Nimmo et al., 2015). For example, many aquatic species persist in drying reaches by tolerating the harsh abiotic conditions in wetted microhabitats that remain after flow ceases (e.g., damp leaf packs, remnant pools; Figure 4.8.2; Chapter 4.3). Other aquatic species enter desiccation-resistant dormant stages during dry periods and reemerge as active individuals when flow returns (Williams, 2006).

In contrast to resistance mechanisms that allow individuals to persist through a disturbance, resilience mechanisms facilitate repopulation of disturbed reaches from elsewhere in the catchment or neighboring catchments. For example, most aquatic insect species have adult stages capable of flight, so adults from perennial streams can fly to, and lay eggs in, neighboring IRES when flow resumes (Figs. 4.8.1B and 4.8.2). Other taxa, such as fish, may swim rapidly through IRES networks during brief periods of flow connectivity (Chapter 4.5), allowing individuals to disperse from distant perennial refuges and recolonize intermittent reaches (Marshall et al., 2016; Datry et al., 2016a; Chapter 4.9).

Both resistance and resilience contribute to *recovery*, considered here to be a return to predisturbance levels of any ecological unit of interest to researchers or managers (e.g., taxonomic diversity, abundance, community composition; Nimmo et al., 2015).

(Continued)

BOX 4.8.1 RESISTANCE, RESILIENCE, AND RECOVERY PROCESSES IN IRES *(Cont'd)*

RESISTANCE
ability to survive flow cessation and persist in a local reach

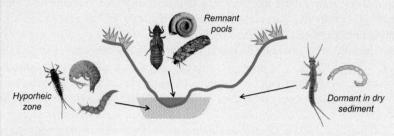

RESILIENCE
ability to recolonize a local reach from elsewhere after flow resumes

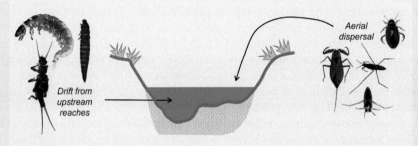

RECOVERY
population or community returns to its *predisturbance* state

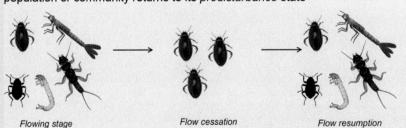

FIG. 4.8.2

Definitions and illustrations of resistance, resilience and recovery as used in this chapter. Although aquatic invertebrates are used here as examples, similar patterns are seen with other aquatic taxa (e.g., algae, fish). Terrestrial communities inhabiting dry riverbeds also exhibit resistance and resilience to inundation, and may demonstrate parallel patterns of community recovery following surface water loss.

Invertebrate drawings: Courtesy P. Fortuño Estrada.

In this chapter, we first explore resistance and resilience mechanisms that aquatic taxa use to persist locally in IRES during flow cessation, including complete stream drying. We then describe how the resistance and resilience mechanisms of individual species scale up to facilitate community recovery to predisturbance levels of diversity, abundance, and composition. Although much remains to be learned about terrestrial invertebrates in dry riverbeds (Chapter 4.4), we summarize what is known about how terrestrial taxa resist or are resilient to inundation in IRES. We also highlight how resistance and resilience mechanisms shape the landscape genetics of IRES taxa. Finally, we discuss how human disturbances can interfere with resistance and resilience pathways and conclude by suggesting several future research directions.

4.8.2 RESISTANCE MECHANISMS OF AQUATIC TAXA

When flow ceases in IRES, aquatic taxa must disperse away from drying reaches (Section 4.8.3) or persist locally. Organisms that can survive dry seasons or years within a given IRES reach are termed "resistant" to the effects of flow cessation. To survive flow cessation, however, organisms must tolerate the often harsh abiotic conditions in remaining wetted habitats or be capable of entering a desiccation-resistant stage. Examples of standing water or damp habitats that can persist after surface flow ceases include remnant pools, crayfish burrows, leaf packs, algal mats, and the hyporheic zone (i.e., saturated sediments below the stream channel) (Chapter 4.3). However, these habitats lack swiftly flowing water and each type presents challenges for aquatic taxa. Even the survival of desiccation-resistant stages is reduced during especially harsh or long dry periods (Storey and Quinn, 2013; Stubbington and Datry, 2013). Three of the most common strategies that aquatic fauna use to resist drying in IRES are surviving in remnant pools, taking refuge locally in the hyporheic zone, and entering desiccation-resistant dormant stages.

REMNANT POOLS

As flow ceases, aquatic organisms in IRES are often concentrated into remnant pools (Boulton and Lake, 1992; Bogan and Lytle, 2007; Chapter 4.3). This concentration effect can result in high densities and diversity of species in relatively small pools (Fig. 4.8.3). In fact, taxa richness in remnant pools during dry periods often equals that of riffles during flowing periods in desert and Mediterranean-climate streams (Bonada et al., 2006; Bogan and Lytle, 2007; Driver and Hoeinghaus, 2016). Remnant pools also can serve important roles beyond local reaches, acting as biodiversity reservoirs for entire drainage networks (Sheldon et al., 2010; Chester and Robson, 2011; Robson et al., 2013). However, remnant pools can also be harsh environments for aquatic organisms. For example, these pools frequently exhibit low dissolved oxygen concentrations and high water temperatures (Boulton et al., 1992). Significant daily fluctuations in these and other water quality variables (Drummond et al., 2015; Datry, 2017) can greatly reduce the survival of aquatic taxa (Driver and Hoeinghaus, 2016). Furthermore, some IRES lack remnant pools, and severe drought may cause formerly perennial pools to dry (Bogan and Lytle, 2011).

Taxa vary greatly in their use of remnant pools to resist flow cessation. Individuals of some algae and diatom species remain "active" in remnant pools and help repopulate intermittent reaches when flow returns (Robson et al., 2008). Aquatic invertebrates frequently use remnant pools to survive dry

FIG. 4.8.3

Refuge pool at Coyote Creek, California, USA and six of the >80 aquatic taxa that inhabited the refuge pool in October 2014: (a) riffle sculpin (*Cottus gulosus*), (b) California floater mussel (*Anodonta californiensis*), (c) giant water bug (*Abedus* sp.), (d) foothill yellow-legged frog (*Rana boylii*), (e) water scavenger beetle (*Berosus* sp.), and (f) darner dragonfly nymph (*Aeshna* sp.).

Photos: Courtesy M. Bogan (a–d, and f) and M. Cover (e).

periods (Boulton and Lake, 2008; Chester and Robson, 2011; Chapter 4.3), especially air-breathing taxa such as adult beetles and true bugs (Bonada et al., 2006; Bogan and Lytle, 2007; Fig. 4.8.3). Pools are especially important for taxa that cannot disperse aerially, such as fish, crustaceans, and amphibians (e.g., Labbe and Fausch, 2000). Many native fishes in IRES are adapted to tolerate the harsh abiotic conditions in remnant pools (Chapter 4.5), including high temperatures, low pH and dissolved oxygen concentrations, and high dissolved organic carbon concentrations (McMaster and Bond, 2008; Driver and Hoeinghaus, 2016). For example, the Australian desert goby can tolerate water temperatures of up to 41°C and salinity levels twice that of seawater (Glover, 1982; Thompson and Withers, 2002). Minnows in Sonoran Desert streams (USA) burrow into wet algal mats as pools dry during daytime heat and then resume activity at night when water levels increase (Minckley and Barber, 1971). Even fish that are traditionally considered sensitive, such as salmonids, can thrive in remnant pools (Wigington et al., 2006; Hwan and Carlson, 2016). While many amphibian species are adapted to exploit seasonal and ephemeral waters (Chapter 4.6), species that cannot estivate in terrestrial habitats require remnant pools to persist in IRES (e.g., yellow-legged frogs in the United States: Fig. 4.8.3; Stebbins, 2003).

HYPORHEIC ZONE

When remnant pools are lost and surface channels dry completely in gravel-bed streams, many taxa passively or actively find refuge in the hyporheic zone (Del Rosario and Resh, 2000; DiStefano et al., 2009; Kawanishi et al., 2013). During dry periods, this zone may offer more stable abiotic conditions than at the surface, and thus host more diverse biotic communities (Stubbington, 2012). In some systems, such as the Eygues River in France, the most common benthic invertebrate taxa are also found in hyporheic zones during brief periods of surface drying, and form the primary source of recolonists when flow returns (Vander Vorste et al., 2016a). Hyporheic refuge use is not limited to invertebrates; some fish persist locally by entering the hyporheic zone when surface water dries (Kawanishi et al., 2013). However, interstitial architecture affects hyporheic refuge use, with fine sediments and small pores excluding large individuals (Gayraud and Philippe, 2001; Descloux et al., 2013; Vadher et al., 2015). Additionally, the physiological condition of individuals can decline with time spent in the hyporheic zone (e.g., fish: Kawanishi et al., 2013). Furthermore, continued drought conditions eventually cause hyporheic water levels to decline. While some species (e.g., crayfish: Flinders and Magoulick, 2003) can actively track falling water tables, doing so may come at a cost. For example, amphipods in IRES can follow declining water tables but their survival may be lower at greater depths (Vander Vorste et al., 2016b).

DESICCATION-RESISTANT STAGES

Having a desiccation-resistant life stage is the only viable resistance strategy when all wetted surface and hyporheic refuges are absent. Primary producers, including chlorophytes and cyanobacteria, commonly use dormant stages to resist drying (Chapter 4.2), becoming active again within hours or days of flow resumption (Robson et al., 2008; Timoner et al., 2014). Aquatic invertebrate taxa may have one or more desiccation-resistant stages within their life history, including resistant eggs, larvae, pupae, or adults (Strachan et al., 2015; Chapters 4.3 and 4.10). "Seedbanks" of active and dormant invertebrates have been widely documented from dry streambeds (Larned et al., 2007; Stubbington and Datry, 2013; Strachan et al., 2015). For example, rehydration experiments conducted in New Zealand and France found that >50% of flowing-phase taxa were found in rehydrated sediments collected from dry gravel-bed streams, suggesting that invertebrate seedbanks can be diverse (Datry et al., 2012; Storey and Quinn, 2013). Larvae of the dobsonfly *Neohermes filicornis* can survive multiple drying and rewetting cycles and may take up to 5 years to reach maturity in IRES with short flow durations (Cover et al., 2015). Additionally, some crayfish and isopod species construct small estivation chambers beneath stones (Fig. 4.8.4). A few vertebrate taxa are also capable of resisting complete stream-drying events. For example, African lungfish can burrow into the mud as streams dry, secrete a cocoon, and lie dormant for up to four years until water returns (Greenwood, 1987). However, reduced streambed moisture can lower the survival rates of active and dormant taxa (Wickson et al., 2012; Stubbington and Datry, 2013; Storey and Quinn, 2013). Furthermore, these desiccation-resistance strategies may not be useful in all IRES. For example, desiccation-resistant invertebrates are rarely found in dry headwater and bedrock streams dominated by coarse substrata or mobile sands (Fritz and Dodds, 2004; Chester and Robson, 2011).

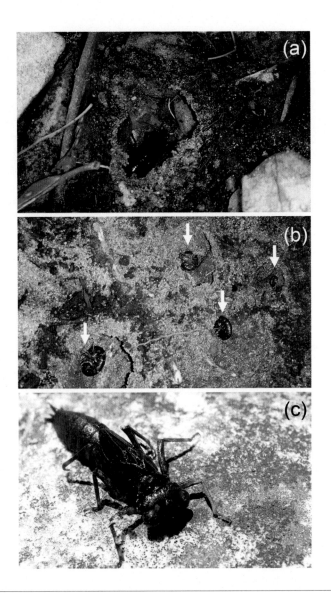

FIG. 4.8.4

Examples of aquatic invertebrates that estivate under stones in dry or damp streambeds to resist drying in some Australian IRES: (a) the crayfish *Geocharax* sp. nov. 1 with stone removed to show estivation chamber, (b) isopods (*Synamphisopus doegi*, as seen below the white arrows) which estivate in damp sand under stones, and (c) a dragonfly nymph (Telephlebiidae) that was found under a large stone.

Photos: Courtesy B. Robson.

4.8.3 RESILIENCE MECHANISMS OF AQUATIC TAXA

If individuals are unable to persist locally when flow ceases, then resilience mechanisms become essential for maintaining populations of aquatic taxa in IRES (Fig. 4.8.2). Resilience via dispersal is frequently cited as the most important means of achieving recovery in IRES (Acuña et al., 2005; Sheldon et al., 2010; Chester and Robson, 2011). The ability to disperse among wetted reaches, including via overland flight, drift dispersal, and swimming, is widely documented among IRES taxa (e.g., Williams and Hynes, 1976; Robson et al., 2008; Lake, 2011). These forms of spatial dispersal allow organisms to recolonize a previously dry reach from elsewhere when flow returns. In fact, many aquatic taxa in IRES can only maintain local populations via spatial dispersal from elsewhere in the catchment or adjacent catchments (e.g., Stanley et al., 1994; Chester et al., 2014). In the following sections, we discuss several commonly used dispersal pathways that confer resilience to IRES populations and explore some of the cues that trigger dispersal away from drying habitats.

DISPERSAL VIA DRIFT

When perennial reaches are located upstream of temporary reaches in IRES, organisms can passively disperse via drift to recolonize downstream reaches. For example, after flow resumes, diatoms rapidly enter the drift or raft on floating detritus, colonizing downstream reaches and influencing assemblage composition (Robson et al., 2008). Drift is especially well studied with regard to aquatic invertebrates (e.g., Larimore et al., 1959; Williams and Hynes, 1976) and certain insect taxa are known to be "drift prone" (e.g., baetid mayflies). Several studies, including those from IRES in tropical Australia and montane temperate North America, have identified drift as being the primary source of invertebrate colonization when flow returns (McArthur and Barnes, 1985; Paltridge et al., 1997). However, in the short term (days), drift in gravel-bed rivers may not contribute as many colonists as emergence from the hyporheic zone (Vander Vorste et al., 2016a). Larval drift can also be an important dispersal mechanism for amphibians, like the fire salamander of Israeli ephemeral streams (Segev and Blaustein, 2014). Drift dispersal of eggs and larvae helps to maintain fish populations in perennial rivers (e.g., Copp et al., 2002; Minckley and Marsh, 2009) and is also likely to promote population resilience in IRES.

DISPERSAL VIA CRAWLING AND SWIMMING

Invertebrates that are not prone to entering the drift, such as heptageniid mayflies, may recolonize IRES by crawling from upstream or downstream perennial reaches when flow resumes (McArthur and Barnes, 1985). In Canadian IRES, midge larvae, nematodes, oligochaetes, ostracods, cyclopoids, and amphipods have been observed crawling and swimming upstream into intermittent reaches from downstream perennial reaches (Williams, 1977). In some desert IRES, flightless giant water bugs (Belostomatidae) crawl and swim through wetted reaches when flow resumes and can even crawl short distances through dry stream channels or terrestrial habitats (Lytle, 1999; Boersma and Lytle, 2014). Although fish are generally limited to instream dispersal pathways, many species disperse quickly through IRES networks when flow returns (Chapter 4.5). In Australia, for example, several fish species swim 70 km or more between isolated waterholes when ephemeral rivers experience flooding (Kerezsy et al., 2013; Marshall et al., 2016). Additionally, fish in IRES may disperse further or more frequently than those from perennial habitats, thus enhancing population resilience following flow resumption, as seen in some Mediterranean streams (Pires et al., 2014). In temperate IRES of the northwestern United States, fish move into intermittent reaches to spawn, and these reaches also act as nurseries for juveniles (Colvin et al., 2009). However, fish species richness decreases with distance from perennial waters, suggesting limitations on dispersal for some species (Colvin et al., 2009).

AERIAL DISPERSAL

Aerial adult life stages of aquatic insects facilitate dispersal both within and between catchments (Figs. 4.8.1B and 4.8.5), providing an important source of colonists in IRES. Adult flight has been cited by some as being the principal recolonization mechanism for many species of aquatic insects (Gray and Fisher, 1981; Abell, 1984). This may especially be the case in dryland regions, where hydrological connections to perennial waters are rare (Davis et al., 2013; Cañedo-Argüelles et al., 2015; Razeng et al., 2016). In fact, genetic studies have demonstrated that entire populations at the reach scale can be the result of just one or two matings of adult aquatic insects (Bunn and Hughes, 1997), highlighting the importance of aerial dispersal in maintaining local populations.

While many aquatic insects (e.g., true flies, mayflies) are relatively weak fliers, others are quite capable of dispersing many kilometers through the air, including beetles, dragonflies, and true bugs (Fig. 4.8.5). Although aerial dispersal abilities vary greatly from species to species, colonization rates of isolated habitats generally decline with distance from source populations as flight capacities are exceeded (Bogan and Boersma, 2012; Chester et al., 2015). Weak dispersers (e.g., flightless giant water bugs, water penny beetles) may be lost from communities when dry period durations and dry reach extents are increased by drought (Bogan and Lytle, 2011; Chester et al., 2015). However, long reaches

FIG. 4.8.5

Overland dispersal of aquatic taxa: (a) water scavenger and predaceous diving beetles (Hydrophilidae and Dytiscidae) escaping a drying stream pool, (b) winter stonefly (*Capnia californica*) emerging into aerial adult stage, (c) flame skimmer dragonfly (*Libellula saturata*) adult, (d) California newt (*Taricha torosa*) dispersing through upland terrestrial habitat, and (e) Sierran treefrog (*Pseudacris sierra*) dispersing along dry stream channel.

Photos: Courtesy M. Bogan.

of dry streambed are not necessarily impediments to aerial dispersal and may even serve as dispersal highways for flying aquatic insect adults (Bogan and Boersma, 2012; Morán-Ordóñez et al., 2015). Many flying insects navigate using visual cues, such as polarized light from water and smooth stones, and ultraviolet light from plants (Collett, 1996; Kriska et al., 2009), which may be restricted to streambeds in arid regions. Dispersal between streams may also be facilitated by concave topographies, such as depressions between sand dunes, which may contain similar visual cues due to water accumulation during rain events (Razeng et al., 2016). During rain events, small, ephemeral pools can form in depressions outside the river network and offer a "stepping-stone" route between streams (Bunn et al., 2006).

ENVIRONMENTAL CUES THAT TRIGGER DISPERSAL

Aquatic taxa use various environmental cues to know when to escape drying reaches or swim through reconnected IRES during flow pulses. Some fish and invertebrates crawl or swim upstream as flow recedes in intermittent reaches (positive rheotaxis: Davey and Kelly, 2007; Lytle et al., 2008). Declining water levels in pools can prompt aquatic insects to crawl out of the water and disperse short distances overland in search of other remnant pools (Fig. 4.8.5; Boersma and Lytle, 2014). Large mixed flocks of aquatic beetle and true bug species have been observed dispersing aerially in dryland regions, often as a result of mass emigration from drying habitats (Kingsley, 1985; Stevens et al., 2007). Indirect drying cues, such as increasing water temperature, may also trigger aerial dispersal of aquatic insects (Velasco and Millan, 1998). Heavy rainfall and associated high-velocity flow conditions can trigger fish to disperse through IRES networks, allowing them to colonize rewetted reaches or move among remnant pools during brief flow events (Labbe and Fausch, 2000; Pires et al., 2014).

4.8.4 AQUATIC COMMUNITY RECOVERY PATTERNS FOLLOWING FLOW RESUMPTION

Aquatic communities in IRES with natural flow regimes tend to recover quickly following flow resumption, but exact timing of recovery varies by taxonomic group and region. Even within a region, recolonization potential is likely to vary with physical connectivity, flow regime, occurrence of perennial refuges, and other environmental factors (Baguette et al., 2013). Despite this variability, biofilms are often the first component of IRES communities to fully recover following flow resumption (Larned et al., 2007; Chapter 4.1). Although full recovery of algal biofilms usually occurs within 12 weeks of flow resumption, the length of the preceding dry period and flow conditions during the previous flow period can alter recovery timing and assemblage trajectories (Robson et al., 2008; Chester and Robson, 2014; Timoner et al., 2014). In some IRES, longer dry periods are associated with higher densities of cyanobacteria on stream stones (Robson et al., 2008). As cyanobacteria are seldom palatable to stream grazers, increases in drying duration may reduce food supplies to algal grazers in IRES (Robson et al., 2008), thereby slowing food web recovery following drying.

Aquatic invertebrate assemblages exhibit great variability in recovery times (Box 4.8.2), but recovery generally occurs within months. Rapid recovery is enabled by both resistance and resilience processes (e.g., Chester and Robson, 2011). Densities can recover in as little as one month, but species richness may take several months longer, especially in isolated sites (Boulton and Lake, 1992; Stanley et al., 1994). In the western United States, for example, IRES invertebrate assemblages are composed

BOX 4.8.2 EXAMPLES OF RECOVERY TRAJECTORIES IN IRES

Aquatic invertebrate communities exhibit a wide range of recovery patterns following flow resumption that are shaped by drying intensity, local habitat conditions, and the spatial configuration of wetted refuges (Chapters 4.3 and 4.9), among other factors. Despite this variability, species richness generally increases through time, with some species returning quickly and others colonizing later. Here, we illustrate how species richness recovers and community composition develops in two systems where these processes have been well documented: temperate alluvial rivers and desert streams.

In alluvial rivers with relatively short dry periods ranging from a few days to a couple of months, numerous species (e.g., mayflies, caddisflies, true flies, amphipods) seek refuge in the hyporheic zone during drying but rapidly return to the surface when flow resumes (Fig. 4.8.6A; Vander Vorste et al., 2016b). Thus, species richness increases quickly in these systems when flow resumes. Additional taxa (e.g., other mayflies and caddisflies, riffle beetles) soon arrive via drift from upstream reaches. Later, aerial colonization occurs via oviposition (e.g., dragonflies and caddisflies), and nonflying taxa (e.g., snails) arrive via crawling or other forms of instream dispersal (Fig. 4.8.6A; Williams, 2006).

In contrast, intermittent desert streams with long dry periods (>6 months/year) and no perennial headwaters tend not to support diverse hyporheic refuges (Bogan et al., 2015). In these systems, species richness recovers slowly. Desiccation-resistant stonefly, blackfly, and midge taxa are the first pioneers following flow resumption (Fig. 4.8.6B; Bogan et al., 2013). After a couple of months, several other taxa may colonize via adult dispersal and oviposition (e.g., mayflies, diving beetles). With longer flow duration, aerial colonization continues by active and passive or phoretic dispersers (e.g., caddisflies and snails, respectively) and a few species may colonize via instream movement from downstream perennial reaches (e.g., adult riffle beetles; Fig. 4.8.6B). Recent work quantifying recovery patterns in little studied IRES (e.g., Neotropical IRES: Datry et al., 2016b) will greatly aid efforts to synthesize recovery patterns across diverse biomes.

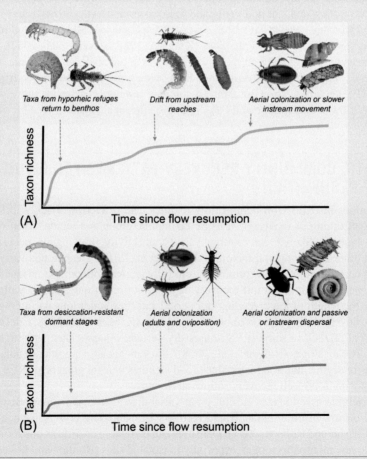

FIG. 4.8.6

Examples of aquatic invertebrate community recovery trajectories in (A) temperate gravel-bed and (B) desert IRES.

Invertebrate drawings: Courtesy P. Fortuño Estrada.

primarily of desiccation-resistant stonefly, blackfly, and midge taxa in the first three months of flow (Bogan et al., 2013). However, longer hydroperiods (e.g., 5–6 months) allow more time for resilient taxa, such as mayflies and dragonflies, to colonize intermittent reaches and promote recovery to pre-drying assemblage composition (Fig. 4.8.6; Bogan et al., 2015). Recovery in highly isolated (e.g., >10 km from perennial refuges) or hydrologically disjunct sites may take multiple years, if it happens at all (Bogan and Lytle, 2011). Because many adult aquatic insects preferentially fly along stream corridors and over short distances (a few kilometers: Briers et al., 2004; Bogan and Boersma, 2012; Chester et al., 2015), long-distance aerial dispersal events that facilitate colonization of isolated IRES may occur infrequently. Some studies report distinct invertebrate assemblages in the first few months after flow resumption, and these compositional differences may be larger in dry than in wet years (Abell, 1984; Acuña et al., 2005). However, full recovery to predrying community composition tends to happen within several months after flow resumption. Studies from both temperate and tropical regions have documented significant recovery in as little as 3 weeks (Datry, 2012; Datry et al., 2016b).

Fish assemblages often take longer to recover from dry periods in IRES because fish species generally rely on hydrological connectivity to disperse (except amphibious species: Clayton, 1993). Dispersal capacity varies greatly among fish species, with some dispersing rarely (Hughes et al., 2012) and other dispersing frequently when the opportunity arises (Bostock et al., 2006). Frequent drying events slow recolonization by most fishes and can result in IRES hosting fish assemblages distinct from those in perennial reaches (Davey and Kelly, 2007). Fish assemblage recovery also depends on the timing of drying; when drying and spawning periods coincide or drying occurs immediately after spawning, recovery can be severely impaired (Detenbeck et al., 1992). Recolonization of previously dry reaches is often a highly ordered process, resulting in temporal nestedness of assemblages in the months following flow resumption (Taylor and Warren, 2001; Whitney et al., 2016). Species that were the most abundant or widely distributed before stream drying, or that have traits such as small body sizes and flexible spawning habits, tend to be first to recolonize rewetted reaches (Detenbeck et al., 1992; Whitney et al., 2016). In contrast, fish species that are larger or take longer to reach maturity may take longer to recover.

Although resilience mechanisms are often reported as being the primary means of community recovery in IRES (e.g., Paltridge et al., 1997; Labbe and Fausch, 2000; Chester and Robson, 2011), both resistance and resilience play a role. Furthermore, the relative importance of their roles varies with local environmental conditions and across biomes. For example, studies of large alluvial rivers have found that hyporheic invertebrates, and resistance in general, contribute greatly to community recovery (e.g., Larned et al., 2007; Datry et al., 2012). In contrast, many bedrock-dominated headwater IRES lack hyporheic communities and rely more on resilience traits for faunal recovery (e.g., Bogan et al., 2015; Chester et al., 2015). Biome and biogeography also play a role, with dryland IRES tending to support more taxa with desiccation-resistance traits than wetter biomes, and thus resistance makes a larger contribution to recovery in those IRES (e.g., Leigh et al., 2016).

4.8.5 DYNAMICS OF TERRESTRIAL COMMUNITIES INHABITING DRY STREAMBEDS

When streambeds dry in IRES (Fig. 4.8.7), new habitat opens up for terrestrial and riparian species and terrestrial successions are initiated (Steward et al., 2011; Corti et al., 2013). When drying commences, riparian invertebrates disperse *en masse* from adjacent riparian zones to benefit from the large quantities

FIG. 4.8.7

IRES transitioning from aquatic to terrestrial habitat. Drying sequences are illustrated for a bedrock and a gravel-bed stream, from high-flow spates (*top*) to declining flows (*middle*), and flow cessation (*bottom*). In the bedrock stream (a–c: Cultivation Creek, Victoria, Australia), remnant pools persist through the dry season. However, in the gravel-bed stream (d–f: Chalone Creek, California, USA), no surface water remains during the dry season.

Photos: Courtesy E. Chester (a–c) and M. Bogan (d–f).

of food left over from the aquatic phase. An initial "clean-up crew" (Abell, 1984) of amphibious and terrestrial organisms consumes the stranded aquatic biota (Fig. 4.8.8), including dead and dying fish, aquatic invertebrates, and macrophytes (Williams, 2006). Some may stay in the dry riverbeds for long periods of time, such as lycosid spiders or carabid beetles (Paetzold et al., 2006; Greenwood and McIntosh, 2010; Datry et al., 2014). Others that require shade or vegetation, such as ants or springtails,

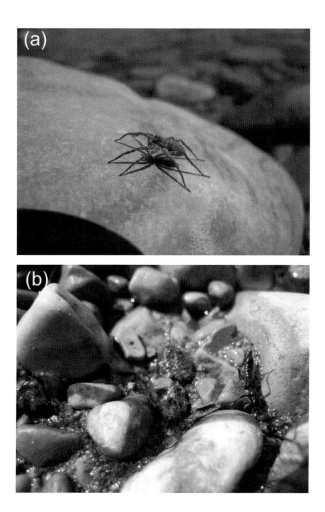

FIG. 4.8.8

Two common arthropods living along intermittent river shorelines and in dry riverbeds after flow ceases are (a) wolf spiders (Lycosidae) and (b) rove beetles (Staphylinidae). The wolf spider *Pardosa wagleri* is from the Asse River, France, and the rove beetle *Paederidus ruficollis* is from a drying reach of the Albarine River, France.

Photos: Courtesy R. Corti.

may use the dry riverbeds only temporarily. Terrestrial communities that develop in dry streambeds can be highly diverse and can differ from adjacent riparian communities (Steward et al., 2011) or just be a subset (Corti and Datry, 2016). As the duration of the dry period increases, streambeds become increasingly terrestrial and fewer aquatic species are able to survive in desiccation-resistant forms (Larned et al., 2007; Datry et al., 2012; Hershkovitz and Gasith, 2013). Despite the recent surge in research regarding the terrestrial phases of IRES, these systems remain poorly understood (Datry et al., 2011, Steward et al., 2012; Chapter 4.4).

During dry phases, IRES can act as dispersal highways, and hence resilience pathways, for terrestrial organisms (Coetzee, 1969; Chapter 4.6). Dry riverbeds contain fewer obstructions (e.g., trees, roads) than adjacent terrestrial habitats and can facilitate an organism's movements on the ground or through the air (Steward et al., 2012; Sánchez-Montoya et al., 2015). In dryland IRES, riparian areas can be more forested than adjacent habitats and this may favor dispersal by terrestrial organisms that are sensitive to desiccation, such as adult aquatic insects (Bogan and Boersma, 2012). In contrast, conditions in dry riverbeds in other regions can be very harsh with strong winds, intense solar radiation, and surface temperatures exceeding 60°C (Steward et al., 2011; Corti et al., 2013). This may limit the daytime movements of some terrestrial organisms along IRES.

When flow resumes, large numbers of terrestrial organisms can be entrained downstream in advancing wetting fronts (Corti and Datry, 2012; Rosado et al., 2015; Fig. 4.8.9), while other terrestrial species escape into the riparian zone (Boumezzough and Musso, 1983; Lambeets et al., 2008). Entrained organisms provide a food source for scavengers and predators from aquatic food webs, but some terrestrial organisms may "surf" these advancing fronts to colonize downstream riparian zones (e.g., some ants, beetles, and spiders: Corti and Datry, 2012; Rosado et al., 2015; Chapter 4.9). This passive dispersal method may be particularly effective for weak dispersers, such as springtails and spiders, and therefore be crucial for maintaining terrestrial invertebrate diversity along IRES corridors (Steward et al., 2012; Rosado et al., 2015).

4.8.6 GENETIC IMPLICATIONS OF DIFFERENT MODES OF RESISTANCE AND RESILIENCE

If resilience is the main recovery pathway in IRES, then most species should be strong dispersers, maintaining large populations with widespread genetic mixing (i.e., panmixia). However, the wide variety of resistance and resilience traits and the combinations of these traits actually observed in IRES show that patterns of population structure among species are far more complex (Murphy et al., 2010; Chester et al., 2015; Phillipsen et al., 2015). Studies of species with different lineages occupying the same landscapes have shown that both phylogenetic and life history differences among IRES species drive population structure (Murphy et al., 2010). Habitat plays a secondary role for species with specific habitat requirements (Chester et al., 2015; Phillipsen et al., 2015). Clearly, not all species are good dispersers, even in streams experiencing frequent disturbances that cause high rates of local extinction. This is an important conceptual advance that has come from research in IRES and emphasizes that present-day habitat conditions (i.e., cycles of drying and rewetting) have not restricted the variety of life histories or population structures in IRES.

Many IRES show dramatic expansions and contractions of habitat, during which some species disperse along river channels (and across floodplains, if present) and others do not (Sheldon et al., 2010). Resilient species that disperse and reproduce rapidly may form large populations with low genetic structure across landscapes (e.g., spangled perch *Leiopotherapon unicolor*, Bostock et al., 2006). In contrast, resistant species that do not disperse widely may show strong geographic structure in genetic patterns, comprising distinct populations that rarely interbreed (e.g., crayfish *Geocharax* sp. nov. 1, Chester et al., 2015; Fig. 4.8.4; Chapter 4.10). We might expect that these two patterns of population structure would dominate IRES species. However, some IRES species possess both resistance and

FIG. 4.8.9

Examples of dry streambed and subsequent flow resumption in four IRES: Bear Gulch, California, USA (a, b), Tea Tree Creek, Victoria, Australia (c, d), John West Fork, California, USA (e, f), and Gila River, New Mexico, USA (g, h).

Photos: Courtesy E. Chester (a–d), M. Bogan (e, f), and J. Whitney (g, h).

resilience traits. For example, the caddisfly *Lectrides varians* has adults capable of dispersal over distances ≥ 100 km (resilience), and larvae capable of estivation (resistance), but shows panmixia (Chester et al., 2015). Hence both resistance and resilience traits contribute to species strategies for surviving the "bust" and exploiting the "boom" (Mossop et al., 2015).

Genetic composition of species in IRES shows remarkable variation and has delivered some surprises in relation to their dispersal strategies. One unexpected pattern is that some species with very poor dispersal maintain either high genetic diversity, high effective population sizes, or both. For example, the Australian amphipod *Wangiannachiltonia guzikae* does not appear able to disperse even over short distances (<1 km), but has higher haplotype diversity and effective population size than other local species (Murphy et al., 2010). Similarly, the flightless giant water bug *Abedus herberti*, which occupies perennial pools of IRES in the southwest United States, maintains genetically distinct local populations, frequently with large effective sizes, that even show behavioral differences (Phillipsen and Lytle, 2013). The factors that enable species with poor dispersal to sustain high genetic diversity and large effective population sizes are unclear, but may include both life history traits (e.g., high fecundity, generalist diet) and intrinsic genetic factors (e.g., high mutation rate) (Murphy et al., 2010). In contrast, other IRES species with poor dispersal are strongly resistant and form small, isolated populations scattered across the landscape. This pattern may be more common in IRES than in perennial river systems and is documented for a variety of fish and invertebrate species (Murphy et al., 2010; Chester et al., 2015; Murphy et al., 2015). These findings demonstrate the importance of protecting isolated perennial waters as both ecological refuges and evolutionary refugia that may act as genetic diversity hotspots for some IRES species (Davis et al., 2013).

4.8.7 HUMAN IMPACTS IMPAIR RESISTANCE AND RESILIENCE PROCESSES

Anthropogenic alterations to IRES can dramatically impair the resistance and resilience processes of resident taxa. For example, water abstraction can increase habitat fragmentation, requiring aquatic taxa to disperse further between remaining wetted reaches. Physical disturbances, like gravel mining and off-road vehicle use (Fig. 4.3.7), may increase habitat fragmentation for terrestrial species inhabiting dry riverbeds and damage the desiccation-resistant resting stages of aquatic invertebrates. In this section, we explore just a few of the hydrological and physical alterations to IRES and how they affect resistance and resilience processes of resident taxa.

HYDROLOGICAL ALTERATIONS

Flow regulation by dams, weirs, and diversions has variable impacts that can either increase or decrease the amount of surface flow in IRES (Fig. 4.8.10; Kingsford et al., 2006; Reich et al., 2010; Chapter 5.1). Reduced population connectivity due to dams and weirs has been widely cited as impairing resilience processes for algae, aquatic invertebrates, and fish (Kingsford et al., 2006; Robson et al., 2008; Minckley and Marsh, 2009). For example, dams on IRES of the Great Plains (USA) blocked instream dispersal of several minnow species and reduced the resilience of their populations, causing extirpations from reaches above dams (Winston et al., 1991). However, taxa with resistance traits (e.g., crayfish) or the ability to disperse aerially (e.g., many aquatic insects) may exhibit stable populations in IRES regulated by weirs and dams, even while native fish are extirpated from the same streams (Chester et al., 2014). Passing "environmental flows" through small weirs in IRES can improve

FIG. 4.8.10

Diversions that reduce the volume and extent of flow in IRES. Granite Reef Dam (a) diverts nearly all of the water in the Salt River (USA) into a canal (upper center) which provides the Phoenix metropolitan area (2012 population: 4,329,534 inhabitants) with irrigation and drinking water. Similarly, but at a much smaller scale, a weir on Number 1 Creek (b) in Victoria, Australia, diverts nearly all flow from the creek, leaving only (c) small remnant pools in the channel downstream.

Photos: Courtesy M. Bogan (a) and E. Chester (b, c).

the dispersal and population connectivity of impacted taxa such as macroinvertebrates (Mackie et al., 2013). Terrestrial taxa that are not resistant to the inundation of dry riverbeds, or those that can only disperse through stream channels during dry periods, may benefit from reduced flow durations caused by water abstraction. However, other research suggests that stream drying reduces the diversity of terrestrial invertebrate taxa along riverbeds (McCluney and Sabo, 2012).

In contrast, artificially enhanced flows in IRES may increase the connectivity of formerly isolated reaches, which could increase the dispersal potential and resilience of populations. This alteration may benefit native species that prefer flowing water and are prone to dispersal via drift, such as some mayflies and caddisflies (Reich et al., 2010). However, artificially perennial flow can also facilitate the dispersal of nonnative fish species through IRES networks (Kingsford et al., 2006). Further research is needed to fully understand how artificially perennial flow regimes affect resistance and resilience processes in IRES.

The amount and duration of surface flow in IRES can be dramatically reduced by groundwater abstraction (Falke et al., 2011; Perkin et al., 2015). This abstraction can impair resistance processes in IRES by reducing the occurrence of remnant pools available for organisms to survive dry periods. For example, if groundwater pumping in the Arikaree River catchment (USA) continues to increase at current rates, only 36% of the already small number of remnant perennial pools are predicted to remain by 2045 (Falke et al., 2011). As well as reducing habitat in which aquatic individuals can resist dry periods, the loss of perennial pools will also impair resilience processes. The few remaining pools in the Arikaree River by 2045 will be concentrated in the lower reaches of the catchment, increasing the distance aquatic species will have to disperse during periods of flow.

PHYSICAL ALTERATIONS

Given the lower perceived value of IRES by most people, the amount of anthropogenic modification of these systems may be much higher than in perennial systems (Chapters 5.1 and 5.2). Dry streambeds are used as transportation corridors (e.g., seasonal roads and highways) in many regions, while off-road vehicles use is high in other regions. However, the physical alterations most likely to impair resistance and resilience processes in IRES are riparian vegetation clearing and riverbed gravel mining.

Riparian vegetation is commonly cleared along IRES in agricultural and urban areas. This clearing significantly reduces allochthonous organic matter biomass in IRES channels, depriving detritivores of essential food resources (Reid et al., 2008a,b). Such a loss of plant biomass is likely to reduce population resilience from stream drying, since these food resources are less available to colonists arriving from perennial refuges. However, this may not always be the case. For example, in southern Australia, large beds of submerged macrophytes often develop in agricultural IRES after flow resumes (Paice et al., 2016a). Carbon from these macrophyte stands can contribute substantially to aquatic food webs and may partially compensate for the loss of allochthonous resources (Paice et al., 2016b).

Riparian vegetation clearing may also reduce the resistance of aquatic taxa in dry riverbeds. Increased streambed temperature due to reduced canopy cover may affect the survival of desiccation-resistant invertebrates in dry streambeds. For example, temperatures in New Zealand streambeds climbed from ~18°C in forested reaches to ~40°C in cleared pasture reaches, and the abundance of aquatic invertebrates emerging from dormancy decreased by as much as 90% under higher temperature regimes (Storey and Quinn, 2013). However, in warmer biomes, even streambeds with intact riparian canopy can reach temperatures exceeding 60°C when dry (Steward et al., 2011), so dormant invertebrates in those IRES must tolerate higher temperatures.

Wholesale removal of dry riverbed sediments via gravel mining is a widespread disturbance to aquatic and riparian taxa, especially in dryland IRES like those of the southwestern United States (Bull and Scott, 1974). Gravel mining not only impacts the ability of fish and invertebrates to use and recolonize habitats (Brown et al., 1998) but can also remove or destroy desiccation-resistant resting stages in dry sediment. The mining process itself, with large machinery driven through the streambed and damaging riparian vegetation, clearly would cause direct mortality of terrestrial taxa inhabiting dry streambeds. Furthermore, when extensive reaches are mined for gravel, in-channel habitat connectivity would be reduced, thus decreasing the resilience of both riparian and aquatic taxa. Our understanding of human impacts on dry streambeds, including those not considered here, such as biological pollution and the introduction of nonnative species, is still in its infancy. Each of these topics deserves further research attention.

4.8.8 KNOWLEDGE GAPS AND FUTURE RESEARCH DIRECTIONS

Although significant advances have been made in understanding resistance, resilience, and recovery processes in IRES, many unanswered questions remain. We know relatively little about the life history of most IRES species or the flexibility of different forms of dormancy in response to repeated wetting and drying cycles (Robson et al., 2011; Strachan et al., 2015; Stubbington et al., 2016). There have been few multiyear studies in IRES, limiting inferences about population structure and synchronization of life stages and recovery processes, among other themes. These are important because if we are to predict and manage the impacts of climate change and other disturbances in IRES, we must understand life histories, population dynamics, and the factors influencing them. Additionally, there is disparity in research focus among taxonomic groups. For example, aquatic invertebrates are relatively well studied, but primary producers and microbial taxa have received far less attention despite their supporting role in food webs and influence on recovery trajectories.

While the variety of refuges used by aquatic species to resist flow cessation in IRES is now quite well known, their spatial distribution and the interannual variability in their frequency, extent, and role in community recovery have not been well described. There is, for example, ongoing debate about the importance of the invertebrate seedbank in recovery of IRES assemblages, largely because rivers in different biomes appear to differ greatly in the abundance and richness of species emerging from river sediments (Chester and Robson, 2011; Stubbington and Datry, 2013). In addition to desiccation-resistant individuals, perennial refuges (e.g., remnant pools) are commonly cited as important sources of colonists to intermittent reaches when flow returns. But exactly how the spatial configuration of refuges, and community structure within those refuges, influences recolonization sequences in IRES following flow resumption has rarely been studied.

Resilience pathways can be difficult to study because it is hard to trace the movement of individuals, especially small taxa such as invertebrates and diatoms. Population genetics has helped greatly with understanding effective dispersal that leads to reproduction, but it cannot trace other movements that animals make (e.g., adult insects seeking refuge in caves or hollow trees when streams are dry). Many more measurements of actual movements by organisms are needed to fully understand resilience and recovery. Furthermore, new models for population structures unique to IRES may augment the models of population structure developed through studies of genetic patterns in perennial streams. Additional studies of multiple species with different lineages and life history strategies are needed to explore population structures across IRES.

Finally, alterations to flow regimes caused by climate change (Chapter 2.2) and prolonged drying exacerbated by water withdrawals will affect both terrestrial and aquatic dynamics, but more research is required to understand how these changes will alter IRES ecosystems. These data will be especially important in guiding conservation and management decisions. For example, what are the best approaches for conserving resistant versus resilient species in IRES? When should managers focus on protecting connectivity among habitat patches versus preserving specific habitat patches that harbor endemic species (i.e., ecological refuges and evolutionary refugia, Davis et al., 2013)? Is assisted dispersal a viable strategy for conserving natural communities in IRES with artificially extended dry periods or complete drying (Lawler and Olden, 2011)? These will be pressing questions to address as flow regimes change in IRES because of increased human water demands and altered precipitation patterns in a globally changing climate.

REFERENCES

Abell, D.L., 1984. Benthic invertebrates of some California intermittent streams. In: Jain, S., Moyle, P.B. (Eds.), Vernal Pools and Intermittent Streams. University of California, Davis, CA, pp. 46–60.

Acuña, V., Munoz, I., Giorgi, A., Omella, M., Sabater, F., Sabater, S., 2005. Drought and postdrought recovery cycles in an intermittent Mediterranean stream: structural and functional aspects. J. N. Am. Benthol. Soc. 24, 919–933.

Baguette, M., Blanchet, S., Legrand, D., Stevens, V.M., Turlure, C., 2013. Individual dispersal, landscape connectivity and ecological networks. Biol. Rev. 88, 310–326.

Boersma, K.S., Lytle, D.A., 2014. Overland dispersal and drought-escape behavior in a flightless aquatic insect, *Abedus herberti* (Hemiptera: Belostomatidae). Southwest. Nat. 59, 301–302.

Bogan, M.T., Boersma, K.S., 2012. Aerial dispersal of aquatic invertebrates along and away from arid-land streams. Freshwater Sci. 31, 1131–1144.

Bogan, M.T., Lytle, D.A., 2007. Seasonal flow variation allows 'time-sharing' by disparate aquatic insect communities in montane desert streams. Freshw. Biol. 52, 290–304.

Bogan, M.T., Lytle, D.A., 2011. Severe drought drives novel community trajectories in desert stream pools. Freshw. Biol. 56, 2070–2081.

Bogan, M.T., Boersma, K.S., Lytle, D.A., 2013. Flow intermittency alters longitudinal patterns of invertebrate diversity and assemblage composition in an arid-land stream network. Freshw. Biol. 58, 1016–1028.

Bogan, M.T., Boersma, K.S., Lytle, D.A., 2015. Resistance and resilience of invertebrate communities to seasonal and supraseasonal drought in arid-land headwater streams. Freshw. Biol. 60, 2547–2558.

Bonada, N., Rieradevall, M., Prat, N., Resh, V.H., 2006. Benthic macroinvertebrate assemblages and macrohabitat connectivity in Mediterranean-climate streams of northern California. J. N. Am. Benthol. Soc. 25, 32–43.

Bostock, B.M., Adams, M., Laurenson, L.J.B., Austin, C.M., 2006. The molecular systematics of *Leiopotherapon unicolor* (Gunther, 1859): testing for cryptic speciation in Australia's most widespread freshwater fish. Biol. J. Linn. Soc. 87, 537–552.

Boulton, A.J., Lake, P.S., 1992. The ecology of two intermittent streams in Victoria, Australia. III. Temporal changes in faunal composition. Freshw. Biol. 27, 123–138.

Boulton, A.J., Lake, P.S., 2008. Effects of drought on stream insects and its ecological consequences. In: Lancaster, J., Briers, R.A. (Eds.), Aquatic Insects: Challenges to Populations. CABI International, Wallingford, pp. 91–102.

Boulton, A.J., Peterson, C.G., Grimm, N.B., Fisher, S.G., 1992. Stability of an aquatic macroinvertebrate community in a multiyear hydrologic disturbance regime. Ecology 73, 2192–2207.

Boumezzough, A., Musso, J.-J., 1983. Etude des communautés animales ripicoles du bassin de la rivière Aille (Var-France). I. Aspects biologiques et éco-éthologiques. Ecologia Mediterranea 9, 31–56.

Briers, R.A., Gee, J.H.R., Cariss, H.M., Geoghegan, R., 2004. Inter-population dispersal by adult stoneflies detected by stable isotope enrichment. Freshw. Biol. 49, 425–431.

Brown, A.V., Lyttle, M.M., Brown, K.B., 1998. Impacts of gravel mining on gravel bed streams. Trans. Am. Fish. Soc. 127, 979–994.

Bull, W.B., Scott, K.M., 1974. Impact of mining gravel from urban stream beds in the southwestern United States. Geology 2, 171–174.

Bunn, S.E., Hughes, J.M., 1997. Dispersal and recruitment in streams: evidence from genetic studies. J. N. Am. Benthol. Soc. 16, 338–346.

Bunn, S.E., Thoms, M.C., Hamilton, S.K., Capon, S.J., 2006. Flow variability in dryland rivers: boom, bust and the bits in between. River Res. Appl. 22, 179–186.

Cañedo-Argüelles, M., Boersma, K.S., Bogan, M.T., Olden, J.D., Phillipsen, I.C., Schriever, T.A., et al., 2015. Dispersal strength determines meta-community structure in a dendritic riverine network. J. Biogeogr. 42, 778–790.

Chester, E.T., Robson, B.J., 2011. Drought refuges, spatial scale and recolonisation by invertebrates in non-perennial streams. Freshw. Biol. 56, 2094–2104.

Chester, E.T., Robson, B.J., 2014. Do recolonization processes in intermittent streams have sustained effects on benthic algal density and assemblage composition? Mar. Freshw. Res. 65, 784–790.

Chester, E.T., Matthews, T.G., Howson, T.J., Johnston, K., Mackie, J.K., Strachan, S.R., et al., 2014. Constraints upon the response of fish and crayfish to environmental flow releases in a regulated headwater stream network. PLoS One 9, e91925.

Chester, E.T., Miller, A., Valenzuela, I., Wickson, S., Robson, B.J., 2015. Drought survival strategies, dispersal potential and persistence of invertebrate species in an intermittent stream landscape. Freshw. Biol. 60, 2066–2083.

Clayton, D.A., 1993. Mudskippers. Oceanogr. Mar. Biol. Annu. Rev. 31, 507–577.

Coetzee, C.G., 1969. The distribution of mammals in the Namib Desert and adjoining inland escarpment. Sci. Pap. Namib Desert Res. Station 40, 23–36.

Collett, T., 1996. Insect navigation en route to the goal: multiple strategies for the use of landmarks. J. Exp. Biol. 199, 227–235.

Colvin, R., Giannico, G.R., Li, J., Boyer, K.L., Gerth, W.J., 2009. Fish use of intermittent watercourses draining agricultural lands in the Upper Willamette River Valley, Oregon. Trans. Am. Fish. Soc. 138, 1302–1313.

Copp, G.H., Faulkner, H., Doherty, S., Watkins, M.S., Majecki, J., 2002. Diel drift behaviour of fish eggs and larvae, in particular barbel, *Barbus barbus* (L.), in an English chalk stream. Fish. Manag. Ecol. 9, 95–103.

Corti, R., Datry, T., 2012. Invertebrates and sestonic matter in an advancing wetted front travelling down a dry river bed (Albarine, France). Freshwater Sci. 31, 1187–1201.

Corti, R., Datry, T., 2016. Terrestrial and aquatic invertebrates in the riverbed of an intermittent river: parallels and contrasts in community organisation. Freshw. Biol. 61, 1308–1320.

Corti, R., Larned, S.T., Datry, T., 2013. A comparison of pitfall-trap and quadrat methods for sampling ground-dwelling invertebrates in dry riverbeds. Hydrobiologia 717, 13–26.

Cover, M.R., Seo, J.H., Resh, V.H., 2015. Life history, burrowing behavior, and distribution of *Neohermes filicornis* (Megaloptera: Corydalidae), a long-lived aquatic insect in intermittent streams. West. North Am. Nat. 75, 474–490.

Datry, T., 2012. Benthic and hyporheic invertebrate assemblages along a flow intermittence gradient: effect of duration of dry events. Freshw. Biol. 57, 565–574.

Datry, T., 2017. Ecological effects of flow intermittence in gravel bed rivers. In: Tsutsumi, D., Laronne, J. (Eds.), Gravel-Bed Rivers: Processes and Disasters. Wiley, Chichester, pp. 261–298.

Datry, T., Arscott, D.B., Sabater, S., 2011. Recent perspectives on temporary river ecology. Aquat. Sci. 73, 453–457.

Datry, T., Corti, R., Philippe, M., 2012. Spatial and temporal aquatic-terrestrial transitions in the temporary Albarine River, France: responses of invertebrates to experimental rewetting. Freshw. Biol. 57, 716–727.

Datry, T., Larned, S.T., Fritz, K.M., Bogan, M.T., Wood, P.J., Meyer, E.I., et al., 2014. Broad-scale patterns of invertebrate richness and community composition in temporary rivers: effects of flow intermittence. Ecography 37, 94–104.

Datry, T., Bonada, N., Heino, J., 2016a. Towards understanding the organisation of metacommunities in highly dynamic ecological systems. Oikos 125, 149–159.

Datry, T., Moya, N., Zubieta, J., Oberdorff, T., 2016b. Determinants of local and regional communities in intermittent and perennial headwaters of the Bolivian Amazon. Freshw. Biol. 61, 1335–1349.

Davey, A.J.H., Kelly, D.J., 2007. Fish community responses to drying disturbances in an intermittent stream: a landscape perspective. Freshw. Biol. 52, 1719–1733.

Davis, J., Pavlova, A., Thompson, R., Sunnucks, P., 2013. Evolutionary refugia and ecological refuges: key concepts for conserving Australian arid zone freshwater biodiversity under climate change. Glob. Chang. Biol. 19, 1970–1984.

Del Rosario, R.B., Resh, V.H., 2000. Invertebrates in intermittent and perennial streams: is the hyporheic zone a refuge from drying? J. N. Am. Benthol. Soc. 19, 680–696.

Descloux, S., Datry, T., Marmonier, P., 2013. Benthic and hyporheic invertebrate assemblages along a gradient of increasing streambed colmation by fine sediment. Aquat. Sci. 75, 493–507.

Detenbeck, N.E., Devore, P.W., Niemi, G.J., Lima, A., 1992. Recovery of temperate-stream fish communities from disturbance: a review of case studies and synthesis of theory. Environ. Manag. 16, 33–53.

Distefano, R.J., Magoulick, D.D., Imhoff, E.M., Larson, E.R., 2009. Imperiled crayfishes use hyporheic zone during seasonal drying of an intermittent stream. J. N. Am. Benthol. Soc. 28, 142–152.

Driver, L.J., Hoeinghaus, D.J., 2016. Spatiotemporal dynamics of intermittent stream fish metacommunities in response to prolonged drought and reconnectivity. Mar. Freshw. Res. 67, 1667–1679.

Drummond, L.R., McIntosh, A.R., Larned, S.T., 2015. Invertebrate community dynamics and insect emergence in response to pool drying in a temporary river. Freshw. Biol. 60, 1596–1612.

Falke, J.A., Fausch, K.D., Magelky, R., Aldred, A., Durnford, D., Riley, L.K., et al., 2011. The role of groundwater pumping and drought in shaping ecological futures for stream fishes in a dryland river basin of the western Great Plains. Ecohydrology 4, 682–697.

Flinders, C.A., Magoulick, D.D., 2003. Effects of stream permanence on crayfish community structure. Am. Midl. Nat. 149, 134–147.

Fritz, K.M., Dodds, W.K., 2004. Resistance and resilience of macroinvertebrate assemblages to drying and flood in a tallgrass prairie stream. Hydrobiologia 527, 99–112.

Gayraud, S., Philippe, M., 2001. Does subsurface interstitial space influence general features and morphological traits of the benthic macroinvertebrate community in streams? Arch. Hydrobiol. 151, 667–686.

Glover, C.J.M., 1982. Adaptations of fishes in arid Australia. In: Barker, W.R., Greenslade, P.J.M. (Eds.), Evolution of the Flora and Fauna of Arid Australia. Peacock Publications, Adelaide, pp. 241–246.

Gray, L.J., Fisher, S.G., 1981. Postflood recolonization pathways of macroinvertebrates in a lowland Sonoran Desert stream. Am. Midl. Nat. 106, 249–257.

Greenwood, P.H., 1987. The natural history of African lungfishes. In: Bemis, W.E., Burggren, W.W., Kemp, N.E. (Eds.), The Biology and Evolution of Lungfishes. Alan R. Liss, New York, pp. 163–179.

Greenwood, M.J., McIntosh, A.R., 2010. Low river flow alters the biomass and population structure of a riparian predatory invertebrate. Freshw. Biol. 55, 2062–2076.

Hershkovitz, Y., Gasith, A., 2013. Resistance, resilience, and community dynamics in Mediterranean-climate streams. Hydrobiologia 719, 59–75.

Hughes, J.M., Real, K.M., Marshall, J.C., Schmidt, D.J., 2012. Extreme genetic structure in a small-bodied freshwater fish, the purple-spotted gudgeon, *Mogurnda adspersa* (Eleotridae). PLoS One 7, e40546.

Hwan, J.L., Carlson, S.M., 2016. Fragmentation of an intermittent stream during seasonal drought: intra-annual and interannual patterns and biological consequences. River Res. Appl. 32, 856–870.

Jaeger, K.L., Olden, J.D., 2012. Electrical resistance sensor arrays as a means to quantify longitudinal connectivity of rivers. River Res. Appl. 28, 1843–1852.

Johnston, K., Robson, B.J., Austin, C.M., 2008. Population structure and life history characteristics of *Euastacus bispinosus* and *Cherax destructor* (Parastacidae) in the Grampians National Park, Australia. Freshwater Crayfish 16, 165–173.

Kawanishi, R., Inoue, M., Dohi, R., Fujii, A., Miyake, Y., 2013. The role of the hyporheic zone for a benthic fish in an intermittent river: a refuge, not a graveyard. Aquat. Sci. 75, 425–431.

Kerezsy, A., Balcombe, S.R., Tischler, M., Arthington, A.H., 2013. Fish movement strategies in an ephemeral river in the Simpson Desert, Australia. Austral Ecol. 38, 798–808.

Kingsford, R.T., Lemly, A.D., Thompson, J.R., 2006. Impacts of dams, river management and diversions on desert rivers. In: Kingsford, R.T. (Ed.), Ecology of Desert Rivers. Cambridge University Press, Cambridge, pp. 203–247.

Kingsley, K.J., 1985. *Eretes sticticus* (L.) (Coleoptera: Dytiscidae): life history observations and an account of a remarkable event of synchronous emigration from a temporary desert pond. Coleopt. Bull. 39, 7–10.

Kriska, G., Bernáth, B., Farkas, R., Horváth, G., 2009. Degrees of polarization of reflected light eliciting polarotaxis in dragonflies (Odonata), mayflies (Ephemeroptera) and tabanid flies (Tabanidae). J. Insect Physiol. 55, 1167–1173.

Labbe, T.R., Fausch, K.D., 2000. Dynamics of intermittent stream habitat regulate persistence of a threatened fish at multiple scales. Ecol. Appl. 10, 1774–1791.

Lake, P.S., 2000. Disturbance, patchiness, and diversity in streams. J. N. Am. Benthol. Soc. 19, 575–592.

Lake, P.S., 2011. Drought and Aquatic Ecosystems: Effects and Responses. Wiley Blackwell, Chichester.

Lambeets, K., Vandegehuchte, M.L., Maelfait, J.P., Bonte, D., 2008. Understanding the impact of flooding on trait-displacements and shifts in assemblage structure of predatory arthropods on river banks. J. Anim. Ecol. 77, 1162–1174.

Larimore, R.W., Childers, W.F., Heckrotte, C., 1959. Destruction and re-establishment of stream fish and invertebrates affected by drought. Trans. Am. Fish. Soc. 88, 261–285.

Larned, S.T., Datry, T., Robinson, C.T., 2007. Invertebrate and microbial responses to inundation in an ephemeral river reach in New Zealand: effects of preceding dry periods. Aquat. Sci. 69, 554–567.

Lawler, J.J., Olden, J.D., 2011. Reframing the debate over assisted colonization. Front. Ecol. Environ. 9, 569–574.

Leigh, C., Bonada, N., Boulton, A.J., Hugueny, B., Larned, S.T., Vander Vorste, R., et al., 2016. Invertebrate assemblage responses and the dual roles of resistance and resilience to drying in intermittent streams. Aquat. Sci. 78, 291–301.

Lytle, D.A., 1999. Use of rainfall cues by *Abedus herberti* (Hemiptera: Belostomatidae): a mechanism for avoiding flash floods. J. Insect Behav. 12, 1–12.

Lytle, D.A., McMullen, L.E., Olden, J.D., 2008. Drought-escape behaviors of aquatic insects may be adaptations to highly variable flow regimes characteristic of desert rivers. Southwest. Nat. 53, 399–402.

Mackie, J.K., Chester, E.T., Matthews, T.G., Robson, B.J., 2013. Macroinvertebrate response to environmental flows in headwater streams in western Victoria, Australia. Ecol. Eng. 53, 100–105.

Marshall, J.C., Menke, N., Crook, D.A., Lobegeiger, J.S., Balcombe, S.R., Huey, J.A., et al., 2016. Go with the flow: the movement behaviour of fish from isolated waterhole refugia during connecting flow events in an intermittent dryland river. Freshw. Biol. 61, 1242–1258.

McArthur, J.V., Barnes, J.R., 1985. Patterns of macroinvertebrate colonization in an intermittent Rocky Mountain stream in Utah. Great Basin Nat. 45, 117–123.

McCluney, K.E., Sabo, J.L., 2012. River drying lowers the diversity and alters the composition of an assemblage of desert riparian arthropods. Freshw. Biol. 57, 91–103.

McMaster, D., Bond, N., 2008. A field and experimental study on the tolerances of fish to *Eucalyptus camaldulensis* leachate and low dissolved oxygen concentrations. Mar. Freshw. Res. 59, 177–185.

Minckley, W.L., Barber, W.E., 1971. Some aspects of the biology of the longfin dace, a cyprinid fish characteristic of streams of the Sonoran Desert. Southwest. Nat. 15, 459–464.

Minckley, W.L., Marsh, P.C., 2009. Inland Fishes of the Greater Southwest: Chronicle of a Vanishing Biota. The University of Arizona Press, Tucson.

Morán-Ordóñez, A., Pavlova, A., Pinder, A.M., Sim, L., Sunnucks, P., Thompson, R.M., et al., 2015. Aquatic communities in arid landscapes: local conditions, dispersal traits and landscape configuration determine local biodiversity. Divers. Distrib. 21, 1230–1241.

Mossop, K.D., Adams, M., Unmack, P.J., Smith Date, K.L., Wong, B.B.M., Chapple, D.G., 2015. Dispersal in the desert: ephemeral water drives connectivity and phylogeography of an arid-adapted fish. J. Biogeogr. 42, 2374–2388.

Murphy, N.P., Guzik, M.T., Worthington-Wilmer, J., 2010. The influence of landscape on population structure of four invertebrates in groundwater springs. Freshw. Biol. 55, 2499–2509.

Murphy, A.L., Pavlova, A., Thompson, R., Davis, J., Sunnucks, P., 2015. Swimming through sand: connectivity of aquatic fauna in deserts. Ecol. Evol. 5, 5252–5264.

Nimmo, D.G., Mac Nally, R., Cunningham, S.C., Haslem, A., Bennett, A.F., 2015. Vive la résistance: reviving resistance for 21st century conservation. Trends Ecol. Evol. 30, 516–523.

Paetzold, A., Bernet, J.F., Tockner, K., 2006. Consumer-specific responses to riverine subsidy pulses in a riparian arthropod assemblage. Freshw. Biol. 51, 1103–1115.

Paice, R.L., Chambers, J.M., Robson, B.J., 2016a. Native submerged macrophyte distribution in seasonally-flowing, south-western Australian streams in relation to stream condition. Aquat. Sci. http://dx.doi.org/10.1007/s00027-016-0488-x.

Paice, R.L., Chambers, J.M., Robson, B.J., 2016b. Potential of submerged macrophytes to support food webs in lowland agricultural streams. Mar. Freshw. Res. http://dx.doi.org/10.1071/MF15391.

Paltridge, R.M., Dostine, P.L., Humphrey, C.L., Boulton, A.J., 1997. Macroinvertebrate recolonization after re-wetting of a tropical seasonally-flowing stream (Magela Creek, Northern Territory, Australia). Mar. Freshw. Res. 48, 633–645.

Perkin, J.S., Gido, K.B., Costigan, K.H., Daniels, M.D., Johnson, E.R., 2015. Fragmentation and drying ratchet down Great Plains stream fish diversity. Aquat. Conserv. Mar. Freshwat. Ecosyst. 25, 639–655.

Phillipsen, I.C., Lytle, D.A., 2013. Aquatic insects in a sea of desert: population genetic structure is shaped by limited dispersal in a naturally fragmented landscape. Ecography 36, 731–743.

Phillipsen, I.C., Kirk, E.H., Bogan, M.T., Mims, M.C., Olden, J.D., Lytle, D.A., 2015. Dispersal ability and habitat requirements determine landscape-level genetic patterns in desert aquatic insects. Mol. Ecol. 24, 54–69.

Pires, D.F., Beja, P., Magalhaes, M.F., 2014. Out of pools: movement patterns of Mediterranean stream fish in relation to dry season refugia. River Res. Appl. 30, 1269–1280.

Poff, N.L., Ward, J.V., 1989. Implications of streamflow variability and predictability for lotic community structure: a regional analysis of streamflow patterns. Can. J. Fish. Aquat. Sci. 46, 1805–1818.

Razeng, E., Morán-Ordóñez, A., Brim Box, J., Thompson, R., Davis, J., Sunnucks, P., 2016. A potential role for overland dispersal in shaping aquatic invertebrate communities in arid regions. Freshw. Biol. 61, 745–757.

Reich, P., McMaster, D., Bond, N., Metzeling, L., Lake, P.S., 2010. Examining the ecological consequences of restoring flow intermittency to artificially perennial lowland streams: patterns and predictions from the Broken-Boosey Creek system in Northern Victoria, Australia. River Res. Appl. 26, 529–545.

Reid, D.J., Lake, P.S., Quinn, G.P., Reich, P., 2008a. Association of reduced riparian vegetation cover in agricultural landscapes with coarse detritus dynamics in lowland streams. Mar. Freshw. Res. 59, 998–1014.

Reid, D.J., Quinn, G.P., Lake, P.S., Reich, P., 2008b. Terrestrial detritus supports the food webs in lowland intermittent streams of south-eastern Australia: a stable isotope study. Freshw. Biol. 53, 2036–2050.

Robson, B.J., Matthews, T.G., Lind, P.R., Thomas, N.A., 2008. Pathways for algal recolonization in seasonally-flowing streams. Freshw. Biol. 53, 2385–2401.

Robson, B.J., Chester, E.T., Austin, C.M., 2011. Why life history information matters: drought refuges and macroinvertebrate persistence in non-perennial streams subject to a drier climate. Mar. Freshw. Res. 62, 801–810.

Robson, B.J., Chester, E.T., Mitchell, B.D., Matthews, T.G., 2013. Disturbance and the role of refuges in Mediterranean climate streams. Hydrobiologia 719, 77–91.

Rosado, J., Morais, M., Tockner, K., 2015. Mass dispersal of terrestrial organisms during first flush events in a temporary stream. River Res. Appl. 31, 912–917.

Sánchez-Montoya, M.D.M., Von Schiller, D., Ruhí, A., Pechar, G.S., Proia, L., Miñano, J., et al., 2015. Responses of ground-dwelling arthropods to surface flow drying in channels and adjacent habitats along Mediterranean streams. Ecohydrology 9, 1376–1387.

Segev, O., Blaustein, L., 2014. Influence of water velocity and predation risk on fire salamander (*Salamandra infraimmaculata*) larval drift among temporary pools in ephemeral streams. Freshwater Sci. 33, 950–957.

Sheldon, F., Bunn, S.E., Hughes, J.M., Arthington, A.H., Balcombe, S.R., Fellows, C.S., 2010. Ecological roles and threats to aquatic refugia in arid landscapes: dryland river waterholes. Mar. Freshw. Res. 61, 885–895.

Stanley, E.H., Buschman, D.L., Boulton, A.J., Grimm, N.B., Fisher, S.G., 1994. Invertebrate resistance and resilience to intermittency in a desert stream. Am. Midl. Nat. 131, 288–300.

Stebbins, R.C., 2003. A Field Guide to Western Reptiles and Amphibians, third ed. Houghton Mifflin Harcourt, Boston, MA.

Stevens, L.E., Polhemus, J.T., Durfree, R.S., Olson, C.A., 2007. Large mixed-species disperal flights of predatory and scavenging aquatic Heteroptera and Coleoptera, northern Arizona, USA. West. North Am. Nat. 67, 587–592.

Steward, A.L., Marshall, J.C., Sheldon, F., Harch, B., Choy, S., Bunn, S.E., et al., 2011. Terrestrial invertebrates of dry river beds are not simply subsets of riparian assemblages. Aquat. Sci. 73, 551–566.

Steward, A.L., Von Schiller, D., Tockner, K., Marshall, J.C., Bunn, S.E., 2012. When the river runs dry: human and ecological values of dry riverbeds. Front. Ecol. Environ. 10, 202–209.

Storey, R.G., Quinn, J.M., 2013. Survival of aquatic invertebrates in dry bed sediments of intermittent streams: temperature tolerances and implications for riparian management. Freshwater Sci. 32, 250–266.

Strachan, S.R., Chester, E.T., Robson, B.J., 2015. Freshwater invertebrate life history strategies for surviving desiccation. Springer Sci. Rev. 3, 57–75.

Stubbington, R., 2012. The hyporheic zone as an invertebrate refuge: a review of variability in space, time, taxa and behaviour. Mar. Freshw. Res. 63, 293–311.

Stubbington, R., Datry, T., 2013. The macroinvertebrate seedbank promotes community persistence in temporary rivers across climate zones. Freshw. Biol. 58, 1202–1220.

Stubbington, R., Gunn, J., Little, S., Worrall, T.P., Wood, P.J., 2016. Macroinvertebrate seedbank composition in relation to antecedent duration of drying and multiple wet-dry cycles in a temporary stream. Freshw. Biol. 61, 1293–1307.

Taylor, C.M., Warren, M.L., 2001. Dynamics in species composition of stream fish assemblages: environmental variability and nested subsets. Ecology 82, 2320–2330.

Thompson, G.G., Withers, P.C., 2002. Aerial and aquatic respiration of the Australian desert goby, *Chlamydogobius eremius*. Comp. Biochem. Physiol. A Mol. Integr. Physiol. 131, 871–879.

Timoner, X., Acuña, V., Frampton, L., Pollard, P., Sabater, S., Bunn, S.E., 2014. Biofilm functional responses to the rehydration of a dry intermittent stream. Hydrobiologia 727, 185–195.

Vadher, A.N., Stubbington, R., Wood, P.J., 2015. Fine sediment reduces vertical migrations of *Gammarus pulex* (Crustacea: Amphipoda) in response to surface water loss. Hydrobiologia 753, 61–71.

Vander Vorste, R., Malard, F., Datry, T., 2016a. Is drift the primary process promoting the resilience of river invertebrate communities? A manipulative field experiment in an intermittent alluvial river. Freshw. Biol. 61, 1276–1292.

Vander Vorste, R., Mermillod-Blondin, F., Hervant, F., Mons, R., Forcellini, M., Datry, T., 2016b. Increased depth to the water table during river drying decreases the resilience of *Gammarus pulex* and alters ecosystem function. Ecohydrology 9, 1177–1186.

Velasco, J., Millan, A., 1998. Insect dispersal in a drying desert stream: effects of temperature and water loss. Southwest. Nat. 43, 80–87.

Verberk, W.C.E.P., Siepel, H., Esselink, H., 2008. Life history strategies in freshwater macroinvertebrates. Freshw. Biol. 53, 1722–1738.

Whitney, J.E., Gido, K.B., Martin, E.C., Hase, K.J., 2016. The first to arrive and the last to leave: colonisation and extinction dynamics of common and rare fishes in intermittent prairie streams. Freshw. Biol. 61, 1321–1334.

Wickson, S.J., Chester, E.T., Robson, B.J., 2012. Aestivation provides flexible mechanisms for survival of stream drying in a larval trichopteran (Leptoceridae). Mar. Freshw. Res. 63, 821–826.

Wigington Jr., P.J., Ebersole, J.L., Colvin, M.E., Leibowitz, S.G., Miller, B., Hansen, B., et al., 2006. Coho salmon dependence on intermittent streams. Front. Ecol. Environ. 4, 513–518.

Williams, D.D., 1977. Movements of benthos during the recolonization of temporary streams. Oikos 29, 306–312.

Williams, D.D., 2006. The Biology of Temporary Waters. Oxford University Press, Oxford.

Williams, D.D., Hynes, H.B., 1976. The recolonization mechanisms of stream benthos. Oikos 27, 265–272.

Winston, M.R., Taylor, C.M., Pigg, J., 1991. Upstream extirpation of four minnow species due to damming of a prairie stream. Trans. Am. Fish. Soc. 120, 98–105.

FURTHER READING

Stubbington, R., Boulton, A.J., Little, S., Wood, P.J., 2015. Changes in invertebrate assemblage composition in benthic and hyporheic zones during a severe supraseasonal drought. Freshwater Sci. 34, 344–354.

HABITAT FRAGMENTATION AND METAPOPULATION, METACOMMUNITY, AND METAECOSYSTEM DYNAMICS IN INTERMITTENT RIVERS AND EPHEMERAL STREAMS

4.9

Thibault Datry*,†, Roland Corti*,†, Jani Heino‡, Bernard Hugueny§, Robert J. Rolls¶, Albert Ruhí**

Irstea, UR MALY, centre de Lyon-Villeurbanne, Villeurbanne, France* Leibniz-Institute of Freshwater Ecology and Inland Fisheries (IGB), Berlin, Germany† Finnish Environment Institute, Oulu, Finland‡ UMR "BOREA" CNRS, DMPA, Museum National d'Histoire Naturelle, Paris, France§ Institute for Applied Ecology, University of Canberra, ACT, Australia¶ Arizona State University, Tempe, AZ, United States**

IN A NUTSHELL

- Intermittent rivers and ephemeral streams (IRES) are dynamic mosaics of lotic (flowing), lentic (nonflowing), and terrestrial habitats
- Dynamic changes in hydrological connectivity shape population, community, and ecosystem patterns and underlying processes in IRES
- Fragmentation can lead to thresholds and tipping points in metacommunity and metapopulation dynamics
- The dynamic changes in connectivity make IRES ideal arenas for exploring temporal dynamics of metapopulations, metacommunities, and metaecosystems from novel perspectives
- Conceptual developments in the dynamics of metapopulations, metacommunities, and metaecosystems have implications for IRES management and conservation (e.g., to define reference conditions and predict restoration trajectories)

4.9.1 INTRODUCTION

Habitat fragmentation is defined as the process during which a large expanse of habitat is transformed into a number of smaller patches of smaller total area isolated from each other by a matrix of habitats unlike the original (Fahrig, 2003). As a result, habitat fragmentation leads to both habitat loss and habitat disintegration, both of which affect biodiversity (Benton et al., 2003; Fahrig, 2003; Haddad et al., 2015). For many species, populations scattered in space are prone to extinction

(Fahrig and Merriam, 1994) if the networks of patches are not sufficiently connected by dispersal (Hanski, 1999; Bowne and Bowers, 2004; Van Dyck and Baguette, 2005). This connection depends on the availability of dispersing individuals and the ease with which these individuals can move across the landscape. This ease of movement is often termed "landscape connectivity" and is a central concept in conservation biology that is of paramount importance for population persistence, patterns of biodiversity, and functioning of ecosystems across landscapes (Fahrig and Merriam, 1994; Kindlmann and Burel, 2008).

Landscape features and geomorphological constraints dictate where and how far individuals can move in all ecosystems thus affecting species coexistence and local community structure (Salomon et al., 2006; Altermatt et al., 2011). In river systems, the dendritic structure of drainage networks results in isolated exterior branches (i.e., the headwaters) converging to form the mainstream channel (Fagan, 2002; Carrara et al., 2012; Chapter 2.1). Moreover, the unidirectional downstream movement of water down river networks carries materials and promotes dispersal of individuals in the downstream direction. Therefore, landscape connectivity and its effects on populations and communities in dendritic networks require a reconsideration of the notion of patch isolation, fragmentation, and connectivity (Fagan, 2002).

Ecological processes operating over multiple spatial and temporal scales determine patterns of population viability and distribution, biodiversity, and ecosystem functioning (Loreau et al., 2003). Specifically, processes structuring populations, communities, and ecosystems occur both at local and large scales, whereby local-scale processes are influenced by processes occurring over much larger scales (Hubbell, 2001; Leibold et al., 2004; Logue et al., 2011). These metasystem concepts (i.e., metapopulations, metacommunities, and metaecosystems) recognize that local discrete populations, communities, and ecosystems form networks connected by gene flow, dispersal, and flows of material and energy. Associated with the recent developments in spatial ecology (Massol et al., 2011), this concept is highly relevant to explore how populations, communities, and ecosystems are organized in intermittent rivers and ephemeral streams (IRES). Stream systems possess several physical characteristics that differentiate them from patch-like systems such as sets of ponds, lakes, or meadows (e.g., Fagan, 2002). For instance, the dendritic structure of drainage networks directs the dispersal of organisms, restricting most dispersal within the stream corridors (Tonkin et al., 2014). Also, the unidirectional flow of water means that passive drift dispersal travels primarily downstream. Such restrictions dictate that dispersal is mostly directional in stream systems, although overland dispersal may also contribute to species distributions across riverscapes. In IRES, the different contributions of spatial and temporal dispersal may complicate assumptions about the importance of directional dispersal for metapopulation and metacommunity dynamics. Therefore, IRES are excellent model systems in which to apply metasystem approaches (Datry et al., 2016a).

In this chapter, we address the effects of riverine network fragmentation by flow intermittence on metapopulation, metacommunity, and metaecosystem dynamics. Considering IRES as coupled aquatic-terrestrial systems, we explore how spatiotemporal patterns of flowing, nonflowing, and dry phases influence the local and regional processes involved in these dynamics. We first present the essential concepts of metasystem and spatial ecology and their relevance to the study of IRES. Then, we review and illustrate metapopulation, metacommunity, and metaecosystem dynamics in IRES, identifying promising research areas. We conclude with a synthesis of research gaps and an illustration of the value of metasystem concepts for the management and restoration of IRES.

4.9.2 CONNECTIVITY, FLUXES, AND DISPERSAL IN RIVER NETWORKS FRAGMENTED BY FLOW INTERMITTENCE

IRES AS DYNAMIC AND SHIFTING HABITAT MOSAICS

IRES are shifting habitat mosaics of lotic (flowing), lentic (nonflowing), and terrestrial habitats whose extent, spatial arrangement, and connectivity constantly vary, following cycles of expansion and contraction (Larned et al., 2010; Datry et al., 2016a; Fig. 4.9.1). Associated with temporal changes in the composition of the mosaic, the longitudinal, lateral, and vertical connectivity between river channels, floodplains, and groundwaters varies greatly (Chapter 2.3). During flow recession, lotic habitats contract and decline in depth and area; lateral aquatic habitats (e.g., floodplains, backwaters, and anabranches) become disconnected from active channels (Boulton, 2003; Chapter 2.3). As contraction continues and flow ceases, the emersion of riffles and other high topographic points along river channels creates longitudinal sequences of disconnected pools (lentic habitats), although these are sometimes connected longitudinally by hyporheic flow (Boulton, 2003). These pools can dry completely due to drainage, seepage, and evaporation, creating terrestrial habitats along dry riverbeds and leading to vertical disconnections with the groundwaters (Chapter 2.3). In some systems, however, disconnected

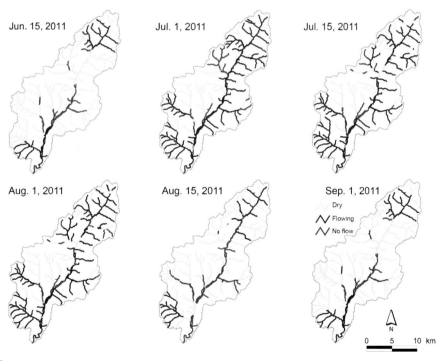

FIG. 4.9.1

Example of a dynamic shifting habitat mosaic of lotic (*blue lines*), lentic (*red lines*), and terrestrial (*yellow lines*) patches in an IRES, the Tude catchment in France during summer, 2011.

pools can persist for months, years, or remain permanent (Hamilton et al., 2005). Flow resumption can take the form of sudden rewetting events when driven by high discharge following rainfall (e.g., Corti and Datry, 2012), instigating a rapid reversal of the drying sequence from terrestrial to lotic habitats. Such events can strongly contribute to groundwater recharge (Chapter 2.3). Rewetting can also be more gradual in large river networks when rainfall is localized to upstream catchments or when driven by rising groundwater levels (Larned et al., 2010). These temporal changes in the mosaic's composition translate into temporal variations in connectivity (Fig. 4.9.1).

FLOW INTERMITTENCE AND CONNECTIVITY

Connectivity can be considered along longitudinal, lateral, vertical, and temporal dimensions in riverine ecosystems (Ward, 1989; Chapter 2.3). Dispersal of organisms and material between habitats occurs in all four dimensions (Fullerton et al., 2010). For example, different stream sites are connected by the longitudinal flow of water. The degree of hydrological connectivity is determined by the presence and amount of water, as well as slope and other geomorphological characteristics. Different sites are also connected by transport of material, such as fine and coarse organic material, usually coming from upstream (Vannote et al., 1980). These materials can be exchanged between terrestrial and stream ecosystems as reciprocal subsidies between streams and riparian zones, thus enhancing connectivity between distinct ecosystems (Soininen et al., 2015) and representing the lateral dimension (Ward, 1989). Stream sites are also connected by the dispersal of organisms, which occurs via hydrological connectivity, active swimming or flight, or passively via aquatic drift or wind (Bilton et al., 2001). These biological dispersal events can be considered as occurring in longitudinal, lateral, or vertical dimensions (Ward, 1989). Vertical connectivity also relates to exchanges of material and organisms between groundwater and surface-water environments. Finally, connectivity typically varies in time, representing the temporal dimension (Ward, 1989).

FLOW INTERMITTENCE AND DISPERSAL

Dispersal is a very important process which affects the distribution of individuals, populations, and species as well as the structure of local communities and metacommunities (Leibold et al., 2004). In IRES, dispersal occurs in both time and in the three spatial dimensions described earlier. *Spatial dispersal* is the movements of individuals between localities and occurs via water, air, or land. In water, organisms crawl or swim actively, drift passively, or move along with other aquatic organisms in the longitudinal dimension (Bilton et al., 2001; Bohonak and Jenkins, 2003; Chapter 4.8). Fish, shrimps, and amphipods are typical examples of active-swimming dispersers in IRES, whereas insect and fish larvae often disperse longitudinally via active or passive drift. Over land, dispersal occurs by flying stages of insects (often adult), passive dispersal by wind, or the use of host organisms as vectors (Bilton et al., 2001, Chapter 4.8). In IRES, an important component of dispersal also incorporates the lateral movements of riparian terrestrial organisms to dry river beds (Corti and Datry, 2016; Chapter 4.4). In particular, these movements may follow diurnal cycles to cope with changes in environmental condition (Corti and Datry, 2016). Additionally, rewetting events can act as mass dispersal events in all three spatial dimensions (Corti and Datry, 2012; Rosado et al., 2015, Chapter 4.3).

The vertical dimension of dispersal of organisms in IRES is probably the least understood but may be highly significant for population and community dynamics (Datry et al., 2016a, Chapter 2.2).

During drying phases, several groups of aquatic organisms are able to actively move into the saturated sediments underneath and lateral to the river channel, defined as the hyporheic zone (e.g., Vander Vorste et al., 2016a,b). In IRES that maintain connectivity with the hyporheic zone throughout the dry period, the primary source of colonists from the hyporheic zone can greatly outnumber those from longitudinal drift upon rewetting (Vander Vorste et al., 2016b). Conversely, the hyporheic zone of some IRES can dry up a few hours or days after the surface water recedes in the riverbed (e.g., Datry et al., 2012) limiting its role in promoting population and community persistence in IRES.

However, the persistence of desiccation-resistant forms in the dry sediments is another way by which populations and communities can be maintained in IRES, and this is termed *temporal dispersal* (Stubbington and Datry, 2013; Datry et al., 2016a). Temporal dispersal constitutes "traveling in time," whereby the resting stages of organisms in the sediments can withstand unsuitable conditions and rapidly reappear when conditions become suitable (Bohonak and Jenkins, 2003; Stubbington and Datry, 2013; Chapters 4.8 and 4.10). In addition, the hyporheic zone can be a refuge during these unsuitable periods and also promote temporal dispersal of aquatic organisms (Stubbington, 2012; Vander Vorste et al., 2016b).

In the following sections, we explore the effects of the temporal variations of the habitat mosaic composition in terms of lotic, lentic, and terrestrial habitat on biodiversity dynamics and ecosystem processes in IRES. Specifically, we explore how variation in connectivity can influence dynamics at the population, community, and ecosystem scales according to the metasystem perspective.

4.9.3 METAPOPULATION DYNAMICS IN IRES
CONCEPTUAL BACKGROUND OF METAPOPULATION IDEAS

In a given landscape, a set of local populations connected through dispersal and gene flows is referred to as a metapopulation (Levins, 1969). For organisms that disperse in water, flow cessation and drying of IRES fragments patches of habitat (as water bodies) with little or no movement possible among patches. In IRES, species persistence is governed by the ability of species to survive in isolated water bodies during periods of no flow and colonize newly created habitats during rewatering (Chapter 2.2; Datry et al., 2016b). The interplay between within-patch persistence and dispersal among patches suggests that the metapopulation concept is an effective framework to address applied and fundamental issues of population and community dynamics in IRES (Datry et al., 2016a).

Levins (1969) proposed the first modeling framework of the regional persistence of a metapopulation, defined as a set of local populations connected by regional dispersal. Levins' (1969) model is simple and assumes an infinite number of identical favorable patches within a matrix of inhospitable habitat. Each patch is described by two states: occupied or unoccupied, depending on whether or not the focal species inhabits the patch. An occupied patch may become empty due to population extinction, and an empty patch may become occupied after having been colonized from an occupied patch. Levins (1969) further assumed that colonization rate depends on the proportion of patches occupied. The important message conveyed by his model is that a metapopulation may persist in a landscape only if the colonization rate is higher than extinction rate. If the landscape is modified in a way that extinction rate increases (e.g., patches reduced in size) and/or colonization rate decreases (e.g., more barriers to dispersal), then such changes may drive the entire metapopulation extinct. This reasoning suggests that habitat fragmentation, destruction, or conversion may result in species extinction even if some amount of suitable habitat persists.

In the 1990s, approaches emerged that intended to apply the Levins (1969) framework to real meta-populations within a spatially explicit framework. These approaches also boosted the discipline of metapopulation ecology by providing promising tools to deal with habitat fragmentation, which is a major concern for conservation biologists. For instance, Hanski (1994) introduced the incidence function model (IFM) that could be parameterized simply by fitting a statistical model to data that are widely available (e.g., snapshots of patch occupancies in a landscape). The IFM assumes that extinction rate per patch is an inverse function of its size and that the probability for a patch to be colonized from a second one is a decreasing function of the distance separating them. In contrast with Levins' (1969) initial approach, the IFM is spatially explicit (dispersal occurs between patches with known spatial coordinates), it accounts for heterogeneity among patches in extinction rate through patch size variability, and it deals with a finite number of patches. As in the Levins' model, the regional persistence of a species depends on landscape features. In parallel to these simplifying approaches, more elaborate models have been proposed to generalize the IFM by allowing it to deal with multiple-patch occupancy snapshots of the same landscape taken at different times, leading to more accurate estimates of extinction and extinction rates (Etienne et al., 2004).

METAPOPULATIONS IN IRES: WHY SO FEW MODELS?

Nevertheless, the potential to apply the framework suggested by Levins (1969) to highly dynamic IRES metapopulations is limited. First, IRES are dendritic structures so that aquatic dispersal is constrained, which contrasts with the two-dimensional dispersal assumed in Levins-type models. This problem obviously matters with strictly aquatic organisms or aquatic phases of amphibiotic organisms, but it may also apply to adult insects with a flight stage that disperse preferentially along the river network (e.g., Bogan and Boersma, 2012). Experimental studies have shown that, for the same landscape, very different metapopulation dynamics are observed whether or not dendritic dispersal is chosen over a two-dimensional (2D) dispersal (Carrara et al., 2012). Contrasting metapopulation dynamics are relevant for species such as many aquatic insects that disperse both in the water as larvae and overland as flying adults; the wrong choice of the dominant dispersal mode may lead to the wrong conclusions. When aquatic dispersal is passive (e.g., drift) and dependent on water flow, it is necessarily asymmetric, from upstream to downstream. Asymmetry could also occur in species that disperse actively but that show a preference for upstream (e.g., rainbow trout *Oncorhynchus mykiss*, Hwan and Carlson, 2015) or downstream movements (e.g., Davey and Kelly, 2007).

HOW TO DEAL WITH DISPERSAL IN DENDRITIC NETWORKS?

Asymmetric dispersal potentially has consequences for metapopulation persistence (e.g., Altermatt et al., 2011). Both dendritic and asymmetric dispersal can be dealt with by using spatially explicit Levins-like models such as IFM. The main difficulty arises when several dispersal modes can be used by the same species (e.g., dendritic-aquatic and 2D overland). Ecologists are therefore challenged in accurately estimating the various model parameters when fitting empirical data. Another major difficulty in applying Levins-like models in IRES is the existence of desiccation-resistant stages (Stubbington and Datry, 2013; Chapter 4.8) and/or the possibility of sheltering in micro-refuges or hyporheic zones (Vander Vorste et al., 2016a; Chapter 4.8).

The occurrence of resistant life history stages means that a population is not locally extinct even if no functioning individuals are found in a habitat (e.g., during periods of no water), yet the population can produce dispersing individuals. A third state would need to be added to account for local population presence in cryptic stages and not contributing to dispersal. Additionally, individuals in dormant stages or in "micro-shelters" are generally difficult to sample, which further complicates estimation of extinction and colonization rates in the field. Insights about the main dispersal mode of a species can be obtained by looking at its regional genetic structure (Chaput-Bardy et al., 2008; Chester et al., 2014) or by using manipulative experiments (Vander Vorste et al., 2016b). For instance, Chester et al. (2014) found that some insect species had a genetic signature implying dendritic dispersal whereas others had a signature suggesting overland dispersal within the same river basin.

HOW TO DEAL WITH DYNAMIC AND SHIFTING HABITAT MOSAICS?

Another challenging feature of IRES is their dynamic character in space and time; the number and quality of patches vary in space and time along with dispersal routes connecting them. The first consequence of flow cessation is the gradual loss of lotic habitats from the riverscape, leaving only isolated lentic habitats separated by dry patches (Chapter 2.3). After their isolation, pools or waterholes typically experience changes in their physicochemistry such as increased water temperature and conductivity and decreased dissolved oxygen (Boulton, 2003; Datry, 2017; Chapter 3.1). Additionally, a decrease in the number of patches reduces the number of colonization sources and increases dispersal distances.

Such dynamism of the IRES landscape in terms of quality and number of patches was not covered by the original Levins-like metapopulation model but subsequent models allow for patches that disappear and recur at constant rates within the landscape. A comparison of pulsed dispersal (prior to patch destruction) versus continuous dispersal strategies found that pulsed dispersal increased metapopulation persistence (Reigada et al., 2015). Pulsed dispersal may explain why some species of fishes are more mobile during the onset of the dry period (Matthews and Marsh-Matthews, 2003). Additionally, De Woody et al. (2005) showed that metapopulation persistence is favored when both population extinction and patch destruction probabilities are inversely related to patch size. In IRES, shallow pools and waterholes generally dry up more quickly than large ones, although there are exceptions (e.g., Beesley and Prince, 2010; Chapter 2.2). Interestingly, this model offers great potential for use in real situations because it can deal with spatially explicit landscapes (i.e., dendritic features, directionally biased dispersal). However, there are still too few theoretical works dealing with dynamic landscapes to draw general conclusions or generate hypotheses that could help attain a better theoretical understanding of IRES metapopulations (Datry et al., 2016a).

EXISTING METAPOPULATION STUDIES IN IRES

Modeling real metapopulations in IRES can be extremely challenging if the focal species has stages that (1) can be active or estivating and (2) can disperse through aquatic pathways as larvae and by flying overland as adults (e.g., the caddisfly *Lectrides varians*, Chester et al., 2014). This implies that many parameters should be measured in the field (Lowe, 2002). At the opposite end of the spectrum, the most model-friendly organisms are lentic fishes because they can disperse only via aquatic habitats and few can persist without surface water. Consequently, extinction and colonization rates can be handled realistically with a few parameters (Falke et al., 2012) and the dynamics of the landscape can be readily described by the number, size, location, and longevity of pools or waterholes (Datry et al., 2016b). Moreover, fishes with limited dispersal capacities are also more likely to have difficulties in persisting as a metapopulation in IRES than most invertebrates. For this reason, fishes frequently gain attention from conservation biologists (Labbe and Fausch, 2000; Jaeger et al., 2014), and the few quantitative studies that have addressed metapopulation issues in IRES focus on fishes.

The first step in conducting a quantitative metapopulation study in IRES is to estimate extinction and colonization rates, while accounting for patch creation (e.g., rewetting) and destruction (e.g., pool drying) if the landscape is dynamic. Such approaches are still rare (e.g., Falke et al., 2012) but are expected to be useful in comparing metapopulation dynamics among species, among years, or in the same study among streams differing in their level of flow intermittence.

Quantifying both landscape and occupancy dynamics is also the first step toward more comprehensive modeling in which the persistence of a metapopulation is explicitly described to quantify the impact of different natural or anthropogenic environmental changes. Few studies have incorporated these latter features. In an IRES in the Murray-Darling Basin in Australia, a spatially explicit metapopulation model for golden perch (*Macquaria ambigua*), present in isolated waterholes during the dry season, estimated metapopulation persistence under different scenarios including climate changes and human water extraction (Bond et al., 2015). One interesting result was the high modeled vulnerability of the metapopulation when surface water extraction preferentially targets large waterholes because this mirrors expectations from the simpler and more general model of De Woody et al. (2005). According to this model, metapopulation persistence is lowered when the probability of patch destruction (e.g., waterhole drying due to surface water extraction) is correlated with the size of the patch (e.g., large waterholes). Conversely, the metapopulation may cope better with a scenario of increasing aridity because, in this case, the probability of drying is inversely related to waterhole size. This study illustrates how different landscape dynamics may have contrasting consequences on metapopulation persistence. Both population and landscape dynamics need to be accounted for to provide insightful guidelines for metapopulation management.

Many studies solely consider landscape changes under different scenarios to assess impacts on a few species of interest, such as mapping aquatic refuges and their isolation under different hydrological regimes. Modeling the dynamics of the landscape is a necessary but not sufficient step and is improved by better integration of climatic, hydrological, and human extraction models (e.g., Falke et al., 2012); the availability of physical-chemical sensors allowing for data logging (e.g., Larned et al., 2011); and the use of citizen scientists in river monitoring to describe dynamics of habitat mosaics in IRES (e.g., Datry et al., 2016b).

WHERE TO GO FROM HERE?

Metapopulation theory suggests that a species will become extinct in a landscape before the suitable habitat (e.g., refuges) is entirely gone. Consequently, the next step is to build on our increasing ability to model landscapes and riverscapes in space and time to add true metapopulation dynamics and to accept "the ephemeral nature of local populations while planning and monitoring for regional persistence" (Labbe and Fausch, 2000). As demonstrated by Falke et al. (2012) and Bond et al. (2015), landscape and occupancy dynamics could and should be estimated jointly to derive a reliable understanding of the fate of a metapopulation and to assess its long-term persistence when faced with different management options. However, this is mainly true for fishes. For species with more complex life cycles combining active and dormant stages and/or different dispersal modes, the parameterization of a metapopulation model is so challenging that it could prevent empirical field-based approaches. The relevance of metapopulation ecology to management of environmental scenarios has been improved by the development of general and simple models that are easily parameterized with few field data. Such a framework is largely lacking for metapopulations embedded within a dynamic landscape, such as IRES. This stresses the need for more empirical and theoretical works to identify which questions could be answered using general and simple models and which ones require more specific and data-demanding models.

4.9.4 METACOMMUNITY DYNAMICS IN IRES
CONCEPTUAL BACKGROUND OF METACOMMUNITY IDEAS

Metacommunity ecology examines how local dynamics of ecological communities are associated with regional dispersal of species (Leibold et al., 2004). Consistent with metapopulation theory, dispersal is an integral part of the four main metacommunity perspectives. First, in species sorting, sufficient dispersal is necessary so that species are able to track variation in abiotic and biotic environmental conditions among sites. Hence, with sufficient dispersal, all species in the regional species pool (i.e., a metacommunity) can potentially reach all sites, and a subset of species is filtered to occur at a site according to its environmental conditions. Second, with mass effects, high dispersal rates homogenize, at least to some extent, the community structure of adjacent sites even though they may possess different environmental conditions. Mass effects through high dispersal rates may also decouple local communities from purely environmental control.

Third, in neutral dynamics, random speciation, extinction, immigration, and emigration processes structure communities of functionally equivalent species (Hubbell, 2001). Community structure should be different between distant sites, with community similarity decaying along spatial gradients irrespective of any environmental influences. Fourth, in patch dynamics, a colonization-competition trade-off dictates that some species are good dispersers but poor competitors, whereas other species are poor dispersers but good competitors. Again, dispersal limitation plays an important role in patch dynamics. While these four perspectives are useful in metacommunity studies, they are rarely mutually exclusive and many, if not most, natural freshwater metacommunities show signs of multiple perspectives (Heino et al., 2015a,b).

HOW TO STUDY METACOMMUNITIES IN IRES?

Metacommunity organization has been studied using both observational and manipulative approaches. Observational approaches have analyzed field data on species distributions using multivariate analytical methods to infer the relative roles of different assembly processes. Commonly used multivariate analytical approaches include using spatial eigenfunction analysis in association with constrained ordination to tease apart environmental and spatial (e.g., dispersal) effects. Another approach has been to use distance-based methods, such as Mantel tests and regression of distance matrices to associate biological and environmental distance matrices (distance-decay relationships, Legendre and Legendre, 2012). While the advantages and disadvantages of these two descriptive approaches have been discussed intensively (Legendre et al., 2005; Tuomisto and Ruokolainen, 2006), they may be seen as complementary. The ordination-based approach may be better in describing community-environment relationships, whereas the distance-based approach may be better in examining true effects of distance on the decay (i.e., change) of community similarity along geographical gradients.

Experimental approaches include the use of meso- and microcosms (Altermatt et al., 2013), where local environmental conditions, species interactions, and dispersal rates can be manipulated. Although experimental approaches are highly efficient when focusing on the mechanisms affecting metacommunity organization, they may only be conducted at unrealistic scales that are too small to validly extrapolate to natural situations. On the other hand, observational field-based studies are plagued by difficulties in associating patterns with mechanisms, typically because of missing abiotic and biotic variables as well as spatially autocorrelated environmental variables (Jacobson and Peres-Neto, 2010; Heino et al., 2015a,b). The latter problem largely precludes meaningful comparisons of environmental and spatial effects on local community structure in field-based studies.

SPECIES TURNOVER AND NESTEDNESS IN IRES METACOMMUNITIES

The terms "nestedness" and "turnover" are directly related to interpreting and explaining variation in community composition among sites (i.e., beta diversity, Anderson et al., 2011) and are strongly related to metacommunity organization. Variation in community composition among multiple locations varies due to the replacement of species among sites (e.g., sites support unique species), species loss (e.g., one site supports a subset of species present at another), or a combination of both processes (Baselga, 2010; Legendre, 2014; Fig. 4.9.2). These concepts are useful for the development of basic theory of biodiversity by illustrating how local communities are assembled, and also for applied conservation to identify which communities (locations) support most of regional biodiversity.

Increasing duration or prevalence of flow cessation can filter species out and cause local communities from IRES to be nested within perennial ones (Larned et al., 2010; Chapter 4.8). This may explain why in many situations alpha diversity of aquatic species declines with increasing duration of flow cessation (e.g., Datry et al., 2014a; Table 4.9.1). Habitat fragmentation is also linked to nestedness,

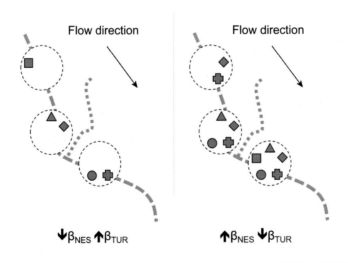

FIG. 4.9.2

Turnover (β_{TUR}) refers to the replacement of species among samples, whereas nestedness (β_{RIC}) indicates the degree to which species-poor communities are a subset of relatively richer communities. Different-shaped symbols represent the different species in each local community distributed along the intermittent river network (*dashed circle*).

Table 4.9.1 General effects of flow intermittence on biodiversity in IRES across different spatial components of species diversity

Component of diversity[a]	Potential response to flow intermittence	Organism group	Examples
Alpha diversity (number of species present in a location)	Declines with increasing duration or frequency of drying	Riparian and floodplain plants	Stromberg et al. (2005)
	Unimodal peak in alpha diversity at intermittently inundated locations compared to permanently or ephemerally inundated sites	Riparian plants	Katz et al. (2012)
	Declines with decreased habitat persistence	Fish, macroinvertebrates	Datry et al. (2007), Beesley and Prince (2010)
	Declines with increasing habitat fragmentation and isolation	Bacteria, plankton, macroinvertebrates	Fazi et al. (2013)
	Negatively associated with habitat size (minimum discharge)	Macroinvertebrates, fish	McHugh et al. (2015)
Beta diversity (spatial variation in species composition among locations)	Change in community composition along a gradient of hydrological intermittence (drying, discharge permanence)	Riparian plants, fish, macroinvertebrates, algae	Datry et al. (2014a)
	Communities become more spatially homogenous during periods of low discharge and during and immediately after flooding, and show greatest variation when hydrological conditions are between extremes of flooding and drying	Macroinvertebrates, fish	Fernandes et al. (2009), and Buendia et al. (2014)
	Spatial variation in composition along intermittence gradients is driven by loss (nestedness) or turnover (replacement) of species due to increasing fragmentation or environmental harshness	Macroinvertebrates, fish	Bogan et al. (2013), Miyazono and Taylor (2015), Datry et al. (2014a, 2016a), Ruhí et al. (2015)
Gamma diversity (number of species present in a region)	Less well studied in IRES; however, evidence indicates that intermittent river reaches contribute to gamma diversity for an entire river network	Macroinvertebrates	Larned et al. (2010)

[a]*Definitions based on terminology of Whittaker (1960) and Anderson et al. (2011).*

whereby communities in sites that experience prolonged fragmentation are often highly nested subsets of connected habitats (Miyazono and Taylor, 2015; Datry et al., 2014a). In some rare cases, compositional differences between intermittent and perennial sites can be driven by species turnover, where communities in intermittent sites support taxa that are specialized to occur and dominate in such sites (Chapters 4.1–4.5). Turnover among contrasting flow regimes has been observed for some invertebrate and diatom communities (Bogan et al., 2013; Tornés and Ruhí, 2013). As the prevalence of turnover and nestedness varies among organism groups and regions, it may be hypothesized that the degree to which they relate to spatial variation in composition depends on a combination of dispersal mode, regional species pool selection, and long-term hydrological history. However, this hypothesis requires further development and testing (Tonkin et al., 2015; Datry et al., 2016c; Leigh and Datry, 2017).

METACOMMUNITIES IN IRES

The alternation of contrasting hydrological phases in IRES generates dynamic variation in the respective metacommunity assembly processes, which may lead to dynamic community patterns in the network (Cañedo-Argüelles et al., 2015; Datry et al., 2016b,c). For example, shifts from flowing to nonflowing conditions are likely to cause a rapid increase in the importance of species sorting, including adaptions to environmental conditions that occur during flow cessation, biotic interactions within contracting pools, and predation by terrestrial organisms (Fig. 4.9.3). Subsequently, the relative importance of dispersal in accounting for community structure increases with the arrival of large specialist predators with strong flying abilities, such as dragonflies (Odonata), diving beetles (Coleoptera), and some true bugs (Heteroptera) (Bogan and Boersma, 2012; Bonada et al., 2006; Chapter 4.8; Fig. 4.9.3). Rewetting of habitats can promote colonization via two processes: dispersal of organisms from refuges (Datry et al., 2014a; Chapter 4.8) and emergence of desiccation-resistant forms from rewetted sediments (temporal dispersal, Stubbington and Datry, 2013; Chapter 4.8).

The relative importance of these two recolonization processes likely depends on two factors. The first is the duration of substrate drying, where viability of desiccation-resistant forms can decrease with increasing drying duration. The second is the geographic distance from refuges supporting source of colonists, where recolonization via dispersal is more rapid when refuges are close compared to locations separated by long distances. After these first phases of recolonization involving dispersal, species sorting becomes gradually more important (Fig. 4.9.3). Similar temporal dynamics can be predicted for semiaquatic and terrestrial communities in IRES, although the restructuring processes involved during each phase have been poorly explored thus far.

Consequently, the structure of a metacommunity may reflect the temporal variability in the relative roles of community assembly processes. The most obvious example is probably the respective proportions of lotic, lentic, and terrestrial species at a given location. Community structure may vary sharply during the different hydrological phases, with notable dominance by lotic species during flowing phases, by lentic species during nonflowing phases, and by terrestrial species during dry phases (Bonada et al., 2006; Corti and Datry, 2016). Many metrics of community structure, such as alpha diversity, species relative abundances, or the proportion of predatory species may undergo abrupt changes that include "step changes" in response to changes in hydrological conditions (Boulton, 2003). The temporal variability in community assembly processes may also lead to predictable spatial patterns of metacommunity organization. During phases dominated by dispersal, nestedness may be observed, particularly for weak to moderate dispersers (Larned et al., 2010; Datry

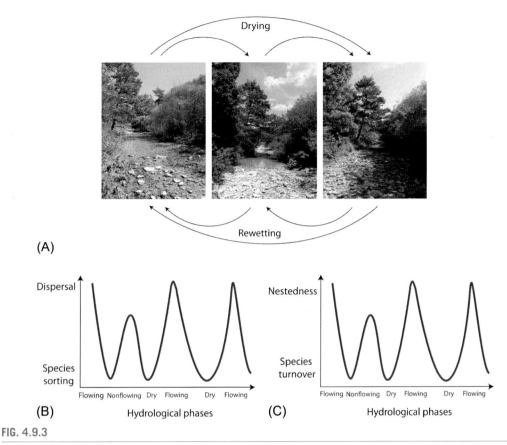

FIG. 4.9.3

Alternating cycles of flowing, nonflowing, and dry phases in IRES (A) provide the dynamic environment for two hypothetical examples of how these hydrological phases and shifts influence the respective importance of species sorting and dispersal on metacommunity organization (B) and potential changes in spatial organization of communities (C). Panels B and C are derived from Datry et al. (2016a).

Photos: Courtesy B. Launay.

et al., 2014a; Datry et al., 2016c; Fig. 4.9.3). In contrast, when species sorting dominates, species turnover may be observed more commonly (Datry et al., 2016b; Tonkin et al., 2015; Fig. 4.9.3).

Metacommunity patterns in IRES are likely to alternate in response to alternating flowing, nonflowing, and dry phases. Consequently, snapshot spatial views of metacommunities are likely to provide incomplete understanding of which community assembly processes are structuring local community composition. Although the temporal dynamics of metacommunities have not been explored in IRES to date, evidence from other ecosystems supports these predictions. For example, recurrent hurricanes temporarily reduce the nestedness of gastropod assemblages in tropical wet forests (Bloch et al., 2007). In seasonal floodplains, fish metacommunity structure varies between the initial and late phases of the flooding in response to a shift in the importance of connectivity (cf. dispersal) versus local environmental conditions (cf. species sorting) in structuring local communities (Fernandes et al., 2009). In perennial river systems, different models of metacommunity structure apply along gradients of flooding disturbance (Campbell et al., 2015) and environmental harshness (Datry et al., 2016c). In clusters of constructed ponds, environmental filtering by drought increased nested structures in invertebrate metacommunities (Ruhí et al., 2013). Finally, in temporary wetlands, the transitions from terrestrial to aquatic phases modify community assembly processes, and the importance of species sorting decreases with inundation time (Vanschoenwinkel et al., 2009).

Compared to other temporary systems, such as rock pools, vernal pools, wetlands, or intertidal zones, the temporal variability in the relative roles of community assembly processes associated with flow cessation and drying may differ spatially in IRES due to the dendritic nature of river networks and directionally biased dispersal (Brown and Swan, 2010; Altermatt et al., 2011; Datry et al., 2016a). Contrary to the paradigm that headwater communities are driven by species sorting whereas those of lowland downstream reaches are driven by dispersal due to convergence of all channels and downstream water flow (Brown and Swan, 2010), more complex patterns probably occur in IRES. This is because of distinctive configurations of flow cessation and channel drying. For example, headwater communities of a network whose drying reaches are upstream may be shaped more by dispersal limitation than by species sorting, since the sources of colonists are downstream or in either the saturated sediments or dry underlying sediments (Chapter 4.8; Datry et al., 2016a). In turn, metacommunities of IRES are likely to show patterns that vary spatially depending on the location of drying events in the network. For example, nestedness related to dispersal may predominate in mid-reaches or downstream drying systems (e.g., Datry et al., 2014a,b), whereas turnover related to species sorting should be more common in systems with intermittent headwaters or drying out completely (e.g., Fagan, 2002). Although these ideas remain speculative, the emergence of metacommunity studies in IRES allows for their further testing.

BIODIVERSITY PARTITIONING IN IRES

Biodiversity has multiple spatial components (Whittaker, 1960), and different components of biodiversity vary along hydrological gradients in IRES (Table 4.9.1; Box 4.9.1). Local species richness (i.e., number of taxa within a location) is termed alpha (α) diversity, the overall number of species in a region

BOX 4.9.1 PARTITIONING OF SPECIES DIVERSITY IN IRES

The combination of flow directionality, associated drift, and network topology means that an upstream local assemblage may share more species with a downstream local assemblage than with a local community of a different headwater site (Fig. 4.9.4). This is why, in the absence of drying, the turnover component of beta diversity may largely explain differences in composition among headwaters, and differences in the richness component are more important between headwater and downstream (main stem) communities. Under this scenario, both components contribute to overall beta diversity in the metacommunity (1A and 2A in Fig. 4.9.4). The unevenness in the distribution of dissimilarity across the dendritic network is because if drying starts in the headwaters (1B in Fig. 4.9.4), the metacommunity may be homogenized via an increase in the nestedness component (all headwaters lose idiosyncratic, tolerant species). In contrast, mainstem drying (2B in Fig. 4.9.4) may reduce the relative importance of this component and increase that of turnover (in 2B, the mainstem has lost tolerant species that were contributing to nestedness in 1B). Finally, when low- or nonflowing conditions affect the whole river network (1C and 2C in Fig. 4.9.4), only tolerant species persist. At this point, dissimilarity across local communities is determined by the tolerant species that constituted the local communities before drought (Ruhí et al., 2013). Complementary patterns in alpha, beta, and gamma diversities of terrestrial faunas colonizing the dry river beds could be expected but have not yet been described.

BOX 4.9.1 PARTITIONING OF SPECIES DIVERSITY IN IRES *(Cont'd)*

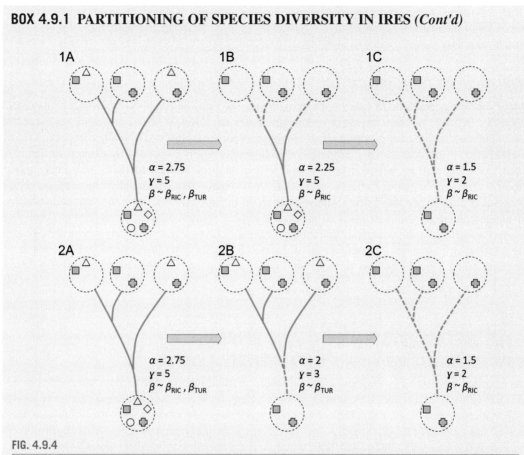

FIG. 4.9.4

Predicted patterns of alpha diversity (α), beta diversity (β), and its components (richness: β_{RIC} turnover: β_{TUR}), and gamma diversity (γ) overtime. The first row represents drying (*dashed blue line*) starting in the headwaters; the second row represents drying starting in the main stem. Dashed circles represent local communities containing species (various symbols) that are either tolerant (shaded symbols) or sensitive (unshaded symbols) to low-flow or no-flow conditions.

is termed gamma diversity (γ), and variation in the composition of local communities within regional units (i.e., across locations) or within a location or groups of locations (i.e., over time) is termed beta (β) diversity (Anderson et al., 2011). Compared to both alpha and gamma diversities, beta diversity is conceptually more complex because there are multiple aspects of communities that can be used to describe variation among locations. For example, variation in beta diversity can be explained by variation in richness differences, in the identity of the species, or in their relative abundances (see Anderson et al., 2011; Box 4.9.1). Accordingly, this topic has attracted much interest from both fundamental and applied perspectives (Baselga, 2010; Legendre and Cáceres, 2013; Legendre, 2014). Importantly, each component of diversity is relevant to aspects of spatial scaling (e.g., grain size and spatial extent) that is often determined by researchers on the basis of the basic biology of study taxa, scale of environmental variables of interest to patterns of diversity, and geographic variables (e.g., drainage divide, biogeographical pool).

WHERE TO GO FROM HERE?

Metacommunity research in IRES is progressing rapidly but the temporal dynamics of metacommunities in IRES and other dynamic systems remain unexplored. Whereas dispersal processes may be of paramount relevance during high-flow phases, species sorting may be more important during nonflowing or dry phases. This temporal variability is certainly enhanced by discrete and punctuated biotic interactions between aquatic and terrestrial communities in IRES and other coupled aquatic-terrestrial systems. In the case of dendritic systems with directionally biased dispersal, the location and spatial extent of disturbances, such as drying, may interact with the temporal variations in community assembly processes to produce complex spatiotemporal variability in local community structure. In the context of increasing extreme climatic events and ecosystem disturbances, understanding how metacommunities are organized in highly dynamic systems is becoming a key research topic. Translating this research into efficient management guidelines is urgently needed (Section 4.9.6).

4.9.5 METAECOSYSTEM DYNAMICS IN IRES
CONCEPTUAL BACKGROUND OF METAECOSYSTEM IDEAS

As a natural extension of metapopulation and metacommunity theory, metaecosystems represent a set of local ecosystems connected by spatial flows of energy, materials, and organisms across ecosystem boundaries (Loreau et al., 2003; Table 4.9.1). The metaecosystem concept recognizes that ecosystems act as sinks and sources of subsidies and organisms, such that different ecosystems fulfill different functions (Loreau et al., 2003). A striking example is the lateral reciprocal subsidies between streams and riparian zones (Polis et al., 1997). The concept was also extended to dendritic ecological networks such as river networks across which carbon is being continuously supplied, processed, and stored in habitats ranging from headwaters, river floodplains, and the hyporheic zone to terminal lakes, estuaries, and oceans (Vannote et al., 1980; Battin et al., 2009).

IRES AS METAECOSYSTEMS: INSIGHTS FROM ORGANIC MATTER DYNAMICS

IRES are ideal systems to develop further the metaecosystem perspective of river networks owing to their dynamic mosaic of flowing, nonflowing, and dry habitats (Section 4.9.2). We contend that the dynamic mosaic functions as a set of local ecosystems contributing differentially to the transport, deposition, and processing of material. In the absence of empirical synthesis studies, we illustrate this metaecosystem perspective of IRES using examples derived from field studies on coarse particulate organic matter (CPOM), but the same insights are also applicable to dissolved organic carbon, nutrients, and autotrophic carbon.

In comparison to flowing habitats, the longitudinal and lateral transport of CPOM is halted in nonflowing and dry habitats (Boulton, 1991; Langhans et al., 2008; Corti et al., 2011) and CPOM

FIG. 4.9.5

Accumulation of CPOM in the dry channels (a, b) and disconnected pools (c, d) of IRES. Terrestrial vegetation encroaches onto the alluvial sediment when surface water dwindles (e, f). The sites are as follows: Demnitzer River, Germany (a), Albarine River, France (b, c, e, and f), and the Eygue River, France (c).

Photos: Courtesy R. Corti.

from riparian zones accumulates and vegetation develops on dry sediments (Boulton, 1991; Corti et al., 2011; Fig. 4.9.5). The types of processes involved in CPOM decomposition and their primary agents also differ within the dynamic mosaic (Chapters 3.1 and 3.2). In flowing and nonflowing habitats, decomposition rate decreases with the duration of past and current drying events, due to the loss of aquatic fungi and shredders (Bruder et al., 2011; Corti et al., 2011; Datry et al., 2011).

In dry habitats, the decomposition rate is reduced to almost zero due to the low abundance of terrestrial shredders, reduced microbial activity, absence of mechanical breakdown, and low leaching rates (Maamri et al., 2001; Corti et al., 2011). However, both dry and nonflowing periods can induce substantial variability in organic matter chemistry, bioavailability, and decomposition rates (Dieter et al., 2013; Chapter 3.2), due to the effects of leaching during rain events, photodegradation by ultraviolet light (Austin and Vivanco, 2006; Fellman et al., 2013), fermentation and accumulation of toxic compounds in anoxic pools (Dieter et al., 2013; Canhoto et al., 2013), and nutrient uptake by microbes and invertebrates in standing pools and in the hyporheic zone (Corti et al., 2011; Febria et al., 2012).

Alternation of flowing, nonflowing, and dry phases controls the flow of material within the habitat mosaic and connects the set of ecosystems (Larned et al., 2010; Datry et al., 2014b). Parallel to the predictions of the River Continuum Concept (Vannote et al., 1980), CPOM in IRES networks is progressively transported to be stored and processed in retention sites downstream; the river network functions as a longitudinal biogeochemical reactor (Battin et al., 2009, Fig. 4.9.6). However, unlike perennial networks, IRES function as a punctuated longitudinal biogeochemical reactor (Larned et al., 2010). While flow is absent, transport of CPOM among patches is inhibited and ecosystems are isolated (Fig. 4.9.6). When flow resumes, CPOM is transported and stored downstream, especially when floodwaters recede (Corti and Datry, 2012; Rosado et al., 2015). The quantity of transported CPOM can be highly variable, depending on climate and seasonal variations in leaf litter inputs (Corti and Datry, 2012). Consequently, the frequency, timing, and magnitude of rewetting events control the degree of connectivity among ecosystems; frequent, long, and intensive rewetting events increase the exchange of CPOM among habitats.

The spatial organization of the habitat mosaic controls the transport, storage, and processing of CPOM, ultimately influencing its quantity, quality, and bioavailability downstream (Box 4.9.2). Downstream waters are expected to receive more material when dry habitats are located upstream, compared to perennial networks and networks with only downstream dry habitats. This is because of the combined effect of lower rates of decomposition and increased accumulation in dry patches (Box 4.9.2). In the temperate Albarine River in France, for instance, the concentration of transported CPOM at the leading edge of a rewetting front advancing over a dry section exceeded $600 \, g \, m^{-3}$ in autumn, and was 50-fold higher than the concentration transported in the upstream perennial section (Corti and Datry, 2012). Large, sudden pulses of CPOM transport and storage can produce "hot moments" of biogeochemical transfer and transformation (Datry et al., 2014b), reducing the quantity of processed material received downstream. However, the processing of the massive input of terrestrial labile compounds can also cause dissolved oxygen to plummet, killing invertebrates and fishes (Hladyz et al., 2011; Chapter 3.2) and exporting unprocessed material. The bioavailability of the CPOM reaching downstream habitats also depends on chemical transformations occurring upstream or on floodplain habitats. For instance, the decomposition of leaves previously affected by solar radiation in upstream dry habitats can be reduced by 33% (Dieter et al., 2013). In addition, the mixing at confluences of CPOM previously processed in different habitats could generate either positive or negative effects on decomposition (Gartner and Cardon, 2004; Hättenschwiler and Gasser, 2005). Ultimately, the effects of the different processes involved in the processing, storage, and transport of CPOM along IRES are likely to propagate down the river network and influence the annual CPOM budget.

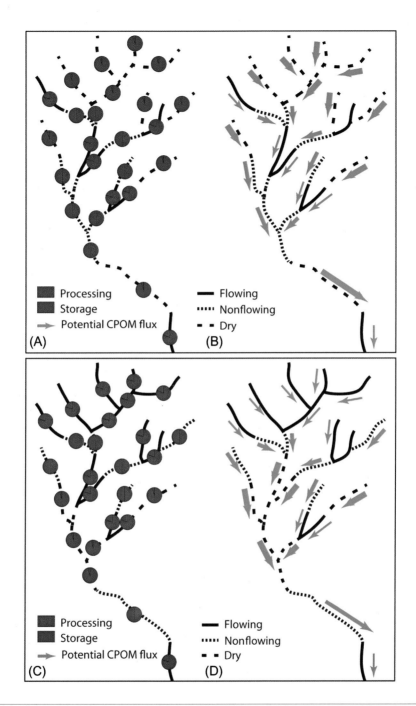

FIG. 4.9.6

Two theoretical networks of IRES (A, C) composed of flowing, nonflowing, and dry habitats. The contribution by each habitat to the processing and storage of CPOM differs among habitats; flowing habitats have higher decomposition rates whereas dry habitats have higher accumulation rates (B, D). The potential contribution of habitats to CPOM fluxes also varies with habitat and is greatest in dry habitats. In addition, the spatial organization of flowing, nonflowing, and dry habitats across the network determines the potential fluxes of CPOM among the habitats of the mosaic (see Box 4.9.2).

BOX 4.9.2 COARSE PARTICULATE ORGANIC MATTER DYNAMICS IN IRES WITH CONTRASTING SPATIAL PATTERNS OF FLOW INTERMITTENCE

Integrating the roles and contributions of IRES to the dynamics of coarse particulate organic matter (CPOM) at the scale of entire river networks (from headwaters to oceans or terminal lakes) could greatly alter our current perception of the role of rivers in the global carbon cycle (Larned et al., 2010; Datry et al., 2014b). We contend that current global estimates of carbon (C) fluxes might be inaccurate because the role of IRES in the storage, processing, and release of organic material has not yet been considered.

The spatial arrangement of flowing, nonflowing, and dry habitats within IRES networks is likely to induce pronounced differences in C fluxes compared to networks entirely composed of perennially flow habitats (perennial-river network). In a perennial-river network (Fig. 4.9.7A), CPOM is progressively stored and processed during its transport downstream and, with distance downstream, CPOM quantity increases due to the increase in retention structures. In a river network with upstream IRES (Fig. 4.9.7B), dry and nonflowing habitats temporarily reduce the quantity of CPOM delivered

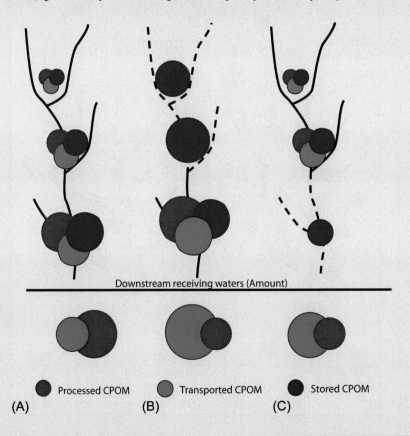

FIG. 4.9.7

Schematic representation of three geomorphically identical river networks with different spatial flow patterns (A-C) showing potential variation in the quantity of processed (*red circles*), transported (*green circles*), and stored (*blue circles*) CPOM, and the annual budget of CPOM in downstream receiving waters. Circle diameter is proportional to the quantity of CPOM; dashed lines represent intermittent reaches and solid lines represent perennial reaches.

BOX 4.9.2 COARSE PARTICULATE ORGANIC MATTER DYNAMICS IN IRES WITH CONTRASTING SPATIAL PATTERNS OF FLOW INTERMITTENCE *(Cont'd)*

downstream and potentially decrease how much CPOM is processed within the network. Consequently, the annual amount of processed CPOM in downstream receiving waters is expected to be lower compared to a perennial-river network whereas the annual amount of unprocessed transported CPOM might be greater, due to sudden huge pulses of CPOM carried during rewetting events coupled with low decomposition rates in intermittent habitats.

In a river network with downstream IRES, smaller quantities of CPOM are expected to accumulate in dry habitats due to lower inputs of CPOM from riparian zones in larger streams. Consequently, the amount of material reaching downstream waters is greater than in perennial-river networks but lower than in networks with upstream IRES (Fig. 4.9.7C). To determine how biogeochemical differences between IRES and perennial-river networks might challenge current estimates of C fluxes in river networks, large-scale measurements of CPOM deposited along riverbeds, their changes over time, and estimates of the biodegradability of the organic matter are needed. These measurements must be supplemented with concurrent data on the environmental drivers controlling these processes (e.g., riparian cover, climate, substrates type, drying duration, and frequency) and will yield crucial insights into global dynamics of C and their likely changes in response to climate change and other factors affecting flow regimes of rivers worldwide.

WHERE TO GO FROM HERE?

Increasing knowledge about transport, accumulation, and processing rates during the different hydrological phases combined with spatial and metamodeling approaches is a promising avenue to model CPOM dynamics in IRES. Incorporating IRES into global carbon models and flux estimates is clearly important but will be challenging. Metasystem approaches can also be used to include adjacent systems (e.g., riparian zones, groundwaters) when exploring biodiversity and ecosystem processes in IRES.

4.9.6 RESEARCH AND MANAGEMENT PERSPECTIVES

One theme in this chapter is that the metasystem approach opens up numerous perspectives for both basic and applied IRES ecology. Although there are many questions to explore, we list some of the main points and challenges that we believe should be addressed using metasystem approaches to improve our understanding of the processes structuring ecosystems across spatial scales in IRES.

Metapopulation modeling in IRES is challenging, yet adapting approaches developed in more static systems would increase our ability to predict species distributions in these and other dynamic ecological systems. Increased ability to predict metapopulation persistence in IRES would be highly valuable for species conservation issues in the face of global changes. For example, accounting for desiccation-resistant stages in metapopulation models will represent an important advance.

Metacommunity organization in IRES should continue to examine the link between community variation across space and underlying mechanisms driving differences in community composition. One promising approach is to incorporate spatial and temporal variation simultaneously to reveal the essence of dynamics in IRES: how spatial variation in community structure changes through time in response to variation in the hydrological regimes of IRES over multiple spatial and temporal dimensions.

Modeling CPOM dynamics in IRES is a conceptually challenging task due to the strong influence of fragmented spatial mosaics and temporal hydrological pulses. Incorporating IRES into global carbon models and flux estimates will refine our understanding of the role played by river networks in global biogeochemical cycles and draw attention to the significant role probably played by IRES in these cycles.

More generally, it is essential to test whether *water* intermittence and *flow* intermittence produce similar relationships across different scale-dependent components of biodiversity (alpha, beta, gamma). Are the effects of flow intermittence across multiple spatial and temporal scales similar in IRES with and without hyporheic zones? Answer to these questions are important for management of flow regimes and restoration plans because they would inform guidelines for water policy decisions if the specific roles of water are identified as driving changes in biodiversity. Additionally, identifying how trends in biodiversity across different spatial components are influenced by flow intermittence in IRES would emphasize the value of biodiversity in IRES. This will also contribute to the knowledge of global biodiversity change, possibly significantly modifying general trends currently reported in the literature (e.g., McGill et al., 2015).

4.9.7 CONCLUSIONS

IRES are complex and dynamic systems. The structure and dynamics of metapopulations, metacommunities, and metaecosystems are inherently related to flow intermittence and to the associated patterns of hydrological connectivity (Chapter 2.3). The study of metapopulations in IRES is challenging due to the varying habitat template and the multiple dispersal pathways of organisms; accordingly, it requires modifying existing conceptual models and empirical approaches. Metacommunity studies have revealed that alpha, beta, and gamma diversities may be predictably associated with the degree of flow intermittence. Although the relevance of dispersal processes (and thus nested metacommunity structures) may be paramount during high-flow phases, species sorting (and thus metacommunity structures dominated by turnover) may be more important during nonflowing or dry phases. These ideas deserve further testing and serve as a suitable starting point to better understand the drivers of beta diversity in IRES. Finally, metaecosystem research in IRES has mainly focused on carbon dynamics, whereby the processed, transported, and stored portions of CPOM vary with changes in flow intermittence. There is now a need for further studies on other biogeochemical fluxes and their interactions with intermittence in different types of IRES.

Metasystem concepts are also very relevant for the management and conservation of IRES because this approach recognizes that sites are not isolated but connected by individuals, materials, and energy. If among-site dynamics are not well understood and explicitly considered, we may not be able to correctly appraise the resilience of these ecosystems, hindering optimized conservation strategies for species and communities. The intensification of the global hydrological cycle, combined with increasing exploitation of freshwater resources by humans, is expanding the extent and duration of hydrological droughts across many parts of the globe. Thus, we need to predict how alterations in patterns of flow intermittence patterns induced by these global changes will affect riverine metapopulations, metacommunities, and metaecosystems if we are to successfully adapt management and conservation strategies for IRES (Chapters 5.4 and 5.5).

REFERENCES

Altermatt, F., Schreiber, S., Holyoak, M., 2011. Interactive effects of disturbance and dispersal directionality on species richness and composition in metacommunities. Ecology 92, 859–870.
Altermatt, F., Seymour, M., Martinez, N., 2013. River network properties shape α-diversity and community similarity of aquatic insect communities across major drainage basins. J. Biogeogr. 12, 2249–2260.

Anderson, M.J., Crist, T.O., Chase, J.M., Vellend, M., Inouye, B.D., Freestone, A.L., et al., 2011. Navigating the multiple meanings of β diversity: a roadmap for the practicing ecologist. Ecol. Lett. 14, 19–28.

Austin, A.T., Vivanco, L., 2006. Plant litter decomposition in a semi-arid ecosystem controlled by photodegradation. Nature 442, 555–558.

Baselga, A., 2010. Partitioning the turnover and nestedness components of beta diversity. Glob. Ecol. Biogeogr. 19, 134–143.

Battin, T.J., Luyssaert, S., Kaplan, L.A., Aufdenkampe, A.K., Richter, A., Tranvik, L.J., 2009. The boundless carbon cycle. Nat. Geosci. 2, 598–600.

Beesley, L.S., Prince, J., 2010. Fish community structure in an intermittent river: the importance of environmental stability, landscape factors and within-pool habitat descriptors. Mar. Freshw. Res. 61, 605–614.

Benton, T.G., Vickery, J.A., Wilson, J.D., 2003. Farmland biodiversity: is habitat heterogeneity the key? Trends Ecol. Evol. 18, 182–188.

Bilton, D.T., Freeland, J.R., Okamura, B., 2001. Dispersal in freshwater invertebrates. Annu. Rev. Ecol. Syst. 32, 159–181.

Bloch, C.P., Higgins, C.L., Willig, M.R., 2007. Effects of large-scale disturbance on metacommunity structure of terrestrial gastropods: temporal trends in nestedness. Oikos 116, 395–406.

Bogan, M.T., Boersma, K.S., 2012. Aerial dispersal of aquatic invertebrates along and away from arid-land streams. Freshwater Sci. 31, 1131–1144.

Bogan, M.T., Boersma, K.S., Lytle, D.A., 2013. Flow intermittency alters longitudinal patterns of invertebrate diversity and assemblage composition in an arid-land stream network. Freshw. Biol. 58, 1016–1028.

Bohonak, A.J., Jenkins, D.G., 2003. Ecological and evolutionary significance of dispersal by freshwater invertebrates. Ecol. Lett. 6, 783–796.

Bonada, N., Rieradevall, M., Prat, N., Resh, V.H., 2006. Benthic macroinvertebrate assemblages and microhabitat connectivity in Mediterranean-climate streams in northern California. J. N. Am. Benthol. Soc. 25, 32–43.

Bond, N.R., Balcombe, S.R., Crook, D.A., Marshall, J.C., Menke, N., Lobegeiger, J.S., 2015. Fish population persistence in hydrologically variable landscapes. Ecol. Appl. 25, 901–913.

Boulton, A.J., 1991. Eucalypt leaf decomposition in an intermittent stream in south-eastern Australia. Hydrobiologia 211, 123–136.

Boulton, A.J., 2003. Parallels and contrasts in the effects of drought on stream macroinvertebrate assemblages. Freshw. Biol. 48, 1173–1185.

Bowne, D.R., Bowers, M.A., 2004. Interpatch movements in spatially structured populations: a literature review. Landsc. Ecol. 19, 1–20.

Brown, B.L., Swan, C.M., 2010. Dendritic network structure constrains metacommunity properties in riverine ecosystems. J. Anim. Ecol. 79, 571–580.

Bruder, A., Chauvet, E., Gessner, M.O., 2011. Litter diversity, fungal decomposers and litter decomposition under simulated stream intermittency. Funct. Ecol. 25, 1269–1277.

Buendia, C., Gibbins, C.N., Vericat, D., Batalla, R.J., 2014. Effects of flow and fine sediment dynamics on the turnover of stream invertebrate assemblages. Ecohydrology 7, 1105–1123.

Campbell, R.E., Winterbourn, M.J., Cochrane, T.A., McIntosh, A.R., 2015. Flow-related disturbance creates a gradient of metacommunity types within stream networks. Landsc. Ecol. 30, 667–680.

Cañedo-Argüelles, M., Boersma, K.S., Bogan, M.T., Olden, J.D., Phillipsen, I., Schriever, T.A., et al., 2015. Dispersal strength determines meta-community structure in a dendritic riverine network. J. Biogeo. 42, 778–790.

Canhoto, C., Calapez, R., Gonçalves, A.L., Moreira-Santos, M., 2013. Effects of *Eucalyptus* leachates and oxygen on leaf-litter processing by fungi and stream invertebrates. Freshwater Sci. 32, 411–424.

Carrara, F., Altermatt, F., Rodriguez-Iturbe, I., Rinaldo, A., 2012. Dendritic connectivity controls biodiversity patterns in experimental metacommunities. Proc. Natl. Acad. Sci. U. S. A. 109, 5761–5766.

Chaput-Bardy, A., Lemaire, C., Picard, D., Secondi, J., 2008. In-stream and overland dispersal across a river network influences gene flow in a freshwater insect, *Calopteryx splendens*. Mol. Ecol. 17, 3496–3505.

Chester, E.T., Matthews, T.G., Howson, T.J., Johnston, K., Mackie, J.K., Strachan, S.R., et al., 2014. Constraints upon the response of fish and crayfish to environmental flow releases in a regulated headwater stream network. PLoS One 9, e91925.

Corti, R., Datry, T., 2012. Invertebrates and sestonic matter in an advancing wetted front travelling down a dry river bed (Albarine, France). Freshwater Sci. 31, 1187–1201.

Corti, R., Datry, T., 2016. Terrestrial and aquatic invertebrates in the riverbed of an intermittent river: parallels and contrasts in community organisation. Freshw. Biol. 61, 1308–1320.

Corti, R., Datry, T., Drummond, L., Larned, S.T., 2011. Natural variation in immersion and emersion affects breakdown and invertebrate colonization of leaf litter in a temporary river. Aquat. Sci. 73, 537–550.

Datry, T., Corti, R., Philippe, M., 2012. Spatial and temporal aquatic–terrestrial transitions in the temporary Albarine River, France: responses of invertebrates to experimental rewetting. Freshw. Biol. 57, 716–727.

Datry, T., Larned, S.T., Scarsbrook, M.R., 2007. Responses of hyporheic invertebrate assemblages to large-scale variation in flow permanence and surface-subsurface exchange. Freshw. Biol. 52, 1452–1462.

Datry, T., Larned, S.T., Fritz, K.M., Bogan, M.T., Wood, P.J., Meyer, E.I., et al., 2014a. Broad-scale patterns of invertebrate richness and community composition in temporary rivers: effects of flow intermittence. Ecography 37, 94–104.

Datry, T., Larned, S.T., Tockner, K., 2014b. Intermittent rivers: a challenge for freshwater ecology. Bioscience 64, 229–235.

Datry, T., Bonada, N., Heino, J., 2016a. Towards understanding metacommunity organisation in highly dynamic systems. Oikos 125, 149–159.

Datry, T., Pella, H., Leigh, C., Bonada, N., Hugueny, B., 2016b. A landscape approach to advance intermittent river ecology. Freshw. Biol. 61, 1200–1213.

Datry, T., Melo, A.S., Moya, N.B., Zubieta, J., De la Barra, E., Oberdorff, T., 2016c. Metacommunity patterns across three Neotropical catchments with varying environmental harshness. Freshw. Biol. 61, 277–293.

Datry, T., 2017. In: Tsutsumi, D., Laronne, J. (Eds.), Ecological effects of flow intermittence in gravel bed rivers. Gravel-Bed Rivers: Processes and Disasters, Wiley, pp. 261–298.

Davey, A.J.H., Kelly, D.J., 2007. Fish community responses to drying disturbances in an intermittent stream: a landscape perspective. Freshw. Biol. 52, 1719–1733.

De Woody, Y.D., Feng, Z., Swihart, R.K., 2005. Merging spatial and temporal structure within a metapopulation model. Am. Nat. 166, 42–55.

Dieter, D., Frindte, K., Krüger, A., Wurzbacher, C., 2013. Preconditioning of leaves by solar radiation and anoxia affects microbial colonisation and rate of leaf mass loss in an intermittent stream. Freshw. Biol. 58, 1918–1931.

Etienne, R.S., 2004. On optimal choices in increase of patch area and reduction of interpatch distance for metapopulation persistence. Ecol. Model. 179(1), 77–90.

Fagan, W.F., 2002. Connectivity, fragmentation, and extinction risk in dendritic metapopulations. Ecology 83, 3243–3249.

Fazi, S., Vazquez, E., Casamayor, E.O., Amalfitano, S., Butturini, A., 2013. Stream hydrological fragmentation drives bacterioplankton community composition. PLoS One 8, e64109.

Fahrig, L., 2003. Effects of habitat fragmentation on biodiversity. Annu. Rev. Ecol. Evol. Syst. 34, 487–515.

Fahrig, L., Merriam, G., 1994. Conservation of fragmented populations. Conserv. Biol. 8, 50–59.

Falke, J.A., Bailey, L.L., Fausch, K.D., Bestgen, K.R., 2012. Colonization and extinction in dynamic habitats: an occupancy approach for a Great Plains stream fish assemblage. Ecology 93, 858–867.

Febria, C.M., Beddoes, P., Fulthorpe, R.R., Williams, D.D., 2012. Bacterial community dynamics in the hyporheic zone of an intermittent stream. ISME J. 6, 1078–1088.

Fellman, J.B., Petrone, K.C., Grierson, P.F., 2013. Leaf litter age, chemical quality, and photodegradation control the fate of leachate dissolved organic matter in a dryland river. J. Arid Environ. 89, 30–37.

Fernandes, R., Gomes, L.C., Pelicice, F.M., Agostinho, A.A., 2009. Temporal organization of fish assemblages in floodplain lagoons: the role of hydrological connectivity. Environ. Biol. Fish 85, 99–108.

Fullerton, A.H., Burnett, K.M., Steel, E.A., Flitcroft, R.L., Pess, G.R., Feist, B.E., et al., 2010. Hydrological connectivity for riverine fish: measurement challenges and research opportunities. Freshw. Biol. 55, 2215–2237.

Gartner, T.B., Cardon, Z.G., 2004. Decomposition dynamics in mixed-species leaf litter. Oikos 104, 230–246.

Haddad, N.M., Brudvig, L.A., Clobert, J., Davies, K.F., Gonzalez, A., Holt, R.D., et al., 2015. Habitat fragmentation and its lasting impact on Earth's ecosystems. Sci. Adv. 1, e1500052.

Hamilton, S.K., Bunn, S.E., Thoms, M.C., Marshall, J.C., 2005. Persistence of aquatic refugia between flow pulses in a dryland river system (Cooper Creek, Australia). Limnol. Oceanogr. 50, 743–754.

Hanski, I., 1994. A practical model of metapopulation dynamics. J. Anim. Ecol. 151–162.

Hanski, I., 1999. Habitat connectivity, habitat continuity, and metapopulations in dynamic landscapes. Oikos 87, 209–219.

Hättenschwiler, S., Gasser, P., 2005. Soil animals alter plant litter diversity effects on decomposition. Proc. Natl. Acad. Sci. U. S. A. 102, 1519–1524.

Heino, J., Melo, A.A., Siqueira, T., Soininen, J., Valanko, S., Bini, L.M., 2015a. Metacommunity organisation, spatial extent and dispersal in aquatic systems: patterns, processes and prospects. Freshw. Biol. 60, 845–869.

Heino, J., Melo, A.S., Bini, L.M., 2015b. Reconceptualising the beta diversity-environmental heterogeneity relationship in running water systems. Freshw. Biol. 60, 223–235.

Hladyz, S., Cook, R., Petrie, R., Nielsen, D., 2011. Influence of substratum on the variability of benthic biofilm stable isotope signatures: implications for energy flow to a primary consumer. Hydrobiologia 664, 135–146.

Hubbell, S.P., 2001. The Unified Neutral Theory of Biodiversity and Biogeography (MPB-32). Princeton University Press, Princeton, NJ.

Hwan, J.L., Carlson, S.M., 2015. Fragmentation of an intermittent stream during seasonal drought: intra-annual and interannual patterns and biological consequences. River Res. Appl. 32, 856–870.

Jacobson, B., Peres-Neto, P.R., 2010. Quantifying and disentangling dispersal in metacommunities: how close have we come? How far is there to go? Landsc. Ecol. 25, 495–507.

Jaeger, K.L., Olden, J.D., Pelland, N.A., 2014. Climate change poised to threaten hydrologic connectivity and endemic fishes in dryland streams. Proc. Natl. Acad. Sci. U. S. A. 111, 13894–13899.

Katz, G.L., Denslow, M.W., Stromberg, J.C., 2012. The Goldilocks effect: intermittent streams sustain more plant species than those with perennial or ephemeral flow. Freshw. Biol. 57, 467–480.

Kindlmann, P., Burel, F., 2008. Connectivity measures: a review. Landsc. Ecol. 23, 879–890.

Labbe, T.R., Fausch, K.D., 2000. Dynamics of intermittent stream habitat regulate persistence of a threatened fish at multiple scales. Ecol. Appl. 10, 1774–1791.

Langhans, S.D., Tiegs, S.D., Gessner, M.O., Tockner, K., 2008. Leaf-decomposition heterogeneity across a riverine floodplain mosaic. Aquat. Sci. 70, 337–346.

Larned, S.T., Datry, T., Arscott, D.B., Tockner, K., 2010. Emerging concepts in temporary-river ecology. Freshw. Biol. 55, 717–738.

Larned, S.T., Schmidt, J., Datry, T., Konrad, C.P., Dumas, J.K., Diettrich, J.C., 2011. Longitudinal river ecohydrology: flow variation down the lengths of alluvial rivers. Ecohydrology 4, 532–548.

Legendre, P., 2014. Interpreting the replacement and richness difference components of beta diversity. Glob. Ecol. Biogeogr. 23, 1324–1334.

Legendre, P., Legendre, L., 2012. Numerical Ecology, third English ed. Elsevier, Amsterdam.

Legendre, P., Cáceres, M., 2013. Beta diversity as the variance of community data: dissimilarity coefficients and partitioning. Ecol. Lett. 16, 951–963.

Legendre, P., Borcard, D., Peres-Neto, P.R., 2005. Analyzing beta diversity: partitioning the spatial variation of community composition data. Ecol. Monogr. 75, 435–450.

Leibold, M.A., Holyoak, M., Mouquet, N., Amarasekare, P., Chase, J.M., Hoopes, M.F., et al., 2004. The metacommunity concept: a framework for multi-scale community ecology. Ecol. Lett. 7, 601–613.

Leigh, C.L., Datry, T., 2017. Drying as a primary hydrological determinant of biodiversity in river systems: a broad-scale analysis. Ecography 40(4), 487–499.

Levins, R., 1969. Some demographic and genetic consequences of environmental heterogeneity for biological control. Bull. Entomol. Soc. Am. 15, 237–240.

Logue, J.B., Mouquet, N., Peter, H., Hillebrand, H., Metacommunity Working Group, 2011. Empirical approaches to metacommunities: a review and comparison with theory. Trends Ecol. Evol. 26, 482–491.

Loreau, M., Mouquet, N., Holt, R.D., 2003. Meta-ecosystems: a theoretical framework for a spatial ecosystem ecology. Ecol. Lett. 6, 673–679.

Lowe, W.H., 2002. Landscape-scale spatial population dynamics in human-impacted stream systems. Environ. Manag. 30, 225–233.

Maamri, A., Bärlocher, F., Pattee, E., Chergui, H., 2001. Fungal and bacterial colonisation of Salix pedicellata leaves decaying in permanent and intermittent streams in eastern Morocco. Int. Rev. Hydrobiol. 86, 337–348.

Massol, F., Gravel, D., Mouquet, N., Cadotte, M.W., Fukami, T., Leibold, M.A., 2011. Linking community and ecosystem dynamics through spatial ecology. Ecol. Lett. 14, 313–323.

Matthews, W.J., Marsh-Matthews, E., 2003. Effects of drought on fish across axes of space, time and ecological complexity. Freshw. Biol. 48, 1232–1253.

McGill, B.J., Dornelas, M., Gotelli, N.J., Magurran, A.E., 2015. Fifteen forms of biodiversity trend in the Anthropocene. Trends Ecol. Evol. 30, 104–113.

McHugh, P.A., Thompson, R.M., Greig, H.S., Warburton, H.J., McIntosh, A.R., 2015. Habitat size influences food web structure in drying streams. Ecography 38, 700–712.

Miyazono, S., Taylor, C.M., 2015. Fish species incidence patterns in naturally fragmented Chihuahuan Desert streams. Ecol. Freshw. Fish 25, 545–552.

Polis, G., Anderson, W., Holt, R., 1997. Toward an integration of landscape and food web ecology: the dynamics of spatially subsidized food webs. Annu. Rev. Ecol. Syst. 28, 289–316.

Reigada, C., Schreiber, S.J., Altermatt, F., Holyoak, M., 2015. Metapopulation dynamics on ephemeral patches. Am. Nat. 185, 183–195.

Rosado, J., Morais, M., Tockner, K., 2015. Mass dispersal of terrestrial organisms during first flush events in a temporary stream. River Res. Appl. 31, 912–917.

Ruhí, A., Boix, D., Gascón, S., Sala, J., Quintana, X.D., 2013. Nestedness and successional trajectories of macroinvertebrate assemblages in man-made wetlands. Oecologia 171, 545–556.

Ruhí, A., Holmes, E.E., Rinne, J.N., Sabo, J.L., 2015. Anomalous droughts, not invasion, decrease persistence of native fishes in a desert river. Global Change Biol. 21, 1482–1496.

Salomon, A.K., Ruesink, J.L., DeWreede, R.E., 2006. Population viability, ecological processes and biodiversity: valuing sites for reserve selection. Biol. Conserv. 128, 79–92.

Soininen, J., Bartels, P., Heino, J., Luoto, M., Hillebrand, H., 2015. Toward more integrated ecosystem research in aquatic and terrestrial environments. Bioscience 65, 174–182.

Stromberg, J.C., Bagstad, K.J., Leenhouts, J.M., Lite, S.J., Makings, E., 2005. Effects of stream flow intermittency on riparian vegetation of a semiarid region river (San Pedro River, Arizona). River Res. Appl. 21, 925–938.

Stubbington, R., 2012. The hyporheic zone as an invertebrate refuge: a review of variability in space, time, taxa and behaviour. Mar. Freshw. Res. 63, 293–311.

Stubbington, R., Datry, T., 2013. The macroinvertebrate seedbank promotes community persistence in temporary rivers across climate zones. Freshw. Biol. 58, 1202–1220.

Tonkin, J.D., Stoll, S., Sundermann, A., Haase, P., 2014. Dispersal distance and the pool of taxa, but not barriers, determine the colonisation of restored river reaches by benthic invertebrates. Freshw. Biol. 59, 1843–1855.

Tonkin, J.D., Stoll, S., Jähnig, S.C., Haase, P., 2015. Contrasting metacommunity structure and beta diversity in an aquatic-floodplain system. Oikos 125, 686–687.

Tornés, E., Ruhí, A., 2013. Flow intermittency decreases nestedness and specialisation of diatom communities in Mediterranean rivers. Freshw. Biol. 58, 2555–2566.

Tuomisto, H., Ruokolainen, K., 2006. Analyzing or explaining beta diversity? Understanding the targets of different methods of analysis. Ecology 87, 2697–2708.

Van Dyck, H., Baguette, M., 2005. Dispersal behaviour in fragmented landscapes: routine or special movements? Basic Appl. Ecol. 6, 535–545.

Vander Vorste, R., Corti, R., Sagouis, A., Datry, T., 2016a. Invertebrate communities in gravel-bed, braided rivers are highly resilient to flow intermittence. Freshwater Sci. 35, 164–177.

Vander Vorste, R., Malard, F., Datry, T., 2016b. Is drift the primary process promoting the resilience of river invertebrate communities? A manipulative field experiment in an alluvial, intermittent river. Freshw. Biol. 61, 1276–1292.

Vannote, R.L., Minshall, G.W., Cummins, K.W., Sedell, J.R., Cushing, C.E., 1980. The river continuum concept. Can. J. Fish. Aquat. Sci. 37, 130–137.

Vanschoenwinkel, B., Hulsmans, A.N.N., De Roeck, E., De Vries, C., Seaman, M., Brendonck, L., 2009. Community structure in temporary freshwater pools: disentangling the effects of habitat size and hydroregime. Freshw. Biol. 54, 1487–1500.

Ward, J.V., 1989. The four-dimensional nature of lotic ecosystems. J. N. Am. Benthol. Soc. 8, 2–8.

Whittaker, R.H., 1960. Vegetation of the Siskiyou Mountains, Oregon and California. Ecol. Monogr. 30, 279–338.

FURTHER READING

Albanese, B., Angermeier, P.L., Peterson, J.T., 2009. Does mobility explain variation in colonisation and population recovery among stream fishes? Freshw. Biol. 54, 1444–1460.

Barton, P.S., Cunningham, S.A., Manning, A.D., Gibb, H., Lindenmayer, D.B., Didham, R.K., 2013. The spatial scaling of beta diversity. Glob. Ecol. Biogeogr. 22, 639–647.

Lecerf, A., Risnoveanu, G., Popescu, C., Gessner, M.O., Chauvet, E., 2007. Decomposition of diverse litter mixtures in streams. Ecology 88, 219–227.

Malmquist, B., 2002. Aquatic invertebrates in riverine landscapes. Freshw. Biol. 47, 679–694.

GENETIC, EVOLUTIONARY, AND BIOGEOGRAPHICAL PROCESSES IN INTERMITTENT RIVERS AND EPHEMERAL STREAMS

4.10

**Núria Bonada*, Stephanie M. Carlson[†], Thibault Datry[‡], Debra S. Finn[§],
Catherine Leigh[‡,¶,**], David A. Lytle[§], Michael T. Monaghan[††], Pablo A. Tedesco[‡‡]**

*Universitat de Barcelona (UB), Barcelona, Spain[*] University of California, Berkeley, CA, United States[†] Irstea, UR MALY,
centre de Lyon-Villeurbanne, Villeurbanne, France[‡] Missouri State University, Springfield, MO, United States[§]
CESAB-FRB, Immeuble Henri Poincaré, Aix-en-Provence, France[¶] Griffith University, Nathan, QLD, Australia[**]
Leibniz Institute of Freshwater Ecology and Inland Fisheries (IGB), Berlin, Germany[††] Université Paul Sabatier Toulouse,
Toulouse, France[‡‡]*

IN A NUTSHELL

- Intermittent rivers and ephemeral streams (IRES) are ideal ecosystems in which to study how ecological and evolutionary processes interact across levels of organization, from genes to ecosystems
- Recurrent disturbance and succession processes are strong drivers of evolution in IRES and occur at varying spatial and temporal scales
- Species inhabiting IRES exhibit many adaptations that are subject to trade-offs related to flow intermittence
- Flow intermittence affects genetic and species diversity, speciation, and extinction, all of which affect ecosystem functions

4.10.1 INTRODUCTION

Despite occupying only 0.8% of the Earth's surface, freshwater ecosystems contain nearly 6% of all described species (Dudgeon et al., 2006). Understanding this diversity requires consideration of how ecology and evolution interact. Ecologists first recognized the importance of this interaction in the mid-20th century. Hutchinson's (1965) "Ecological theatre and evolutionary play" emphasized that evolutionary processes play out within a specific ecological context. Margalef (1959) also related ecology to evolution within the framework of ecological succession. He considered Hutchinson's ecological context to be highly dynamic because ecosystems are subject to disturbances that cause succession which then guide evolutionary processes. Margalef (1959) concluded that disturbances are key elements promoting the evolution of species.

Intermittent Rivers and Ephemeral Streams. http://dx.doi.org/10.1016/B978-0-12-803835-2.00015-2

Given that intermittent rivers and ephemeral streams (IRES) are characterized by flow cessation and, often, drying events that cause disturbance (here defined as events that reset biological communities; Williams and Hynes, 1977; Boulton, 2003), drying can be considered a selection pressure that gives rise to adaptations that allow organisms to cope with drying. IRES consist of habitat mosaics that disconnect and reconnect in both time and space (Datry et al., 2016a, Chapter 2.3). This dynamic landscape exerts strong controls on the population and community structure of aquatic organisms and on ecosystem functions (EFs). Therefore, IRES are ideal ecosystems in which to study how ecological and evolutionary processes interact.

Evolutionary ecology is the study of how ecological and evolutionary processes interact, with consequences across organizational levels from the individual to the biosphere (Fig. 4.10.1). Following,

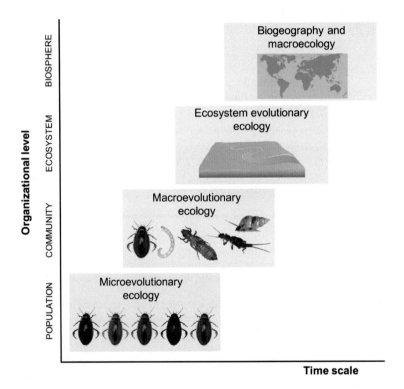

FIG. 4.10.1

Interactions between evolution and ecology have relevance for IRES across spatial and temporal scales and organizational levels. Microevolutionary ecology focuses on individual variability, here depicted with different colors of diving beetles (Dytiscidae). Macroevolutionary ecology focuses on species variability, here depicted with different invertebrate taxa (left to right: the diving beetle Dytiscidae, the midge Chironomidae, the dragonfly Gomphidae, the stonefly Nemouridae, and the snail Hydrobiidae; aquatic invertebrate drawings by Pau Fortuño). Ecosystem evolutionary ecology focuses on processes acting at the ecosystem scale, here represented with a meandering reach (symbol "ian-symbol-river-3d-meandering-with-sandy-banks" designed by Tracey Saxby, courtesy of the Integration and Application Network, University of Maryland Center for Environmental Science (ian.umces.edu/imagelibrary/). Biogeography and macroecology focus on patterns that range from large regions to the biosphere.

we address genetic, evolutionary, and biogeographical/macroecological processes in IRES by analyzing the link between ecology and evolution at different levels. At the population level, we consider how IRES may influence selection, adaptation, and potential trade-offs, and how IRES spatiotemporal dynamics influence genetic diversity. At the community level, we consider how IRES promote niche differentiation and diversification, how trait variability can modify EFs, and how ecosystem characteristics influence adaptive evolution. At the biogeographical/macroecological level, we discuss the relative importance of local and regional factors and how speciation, extinction, and immigration can determine global diversity patterns. Our aim is to stimulate future research in IRES by adopting an evolutionary ecology approach. An overall goal is to emphasize that IRES have much to contribute to the field of evolutionary ecology.

4.10.2 MICROEVOLUTIONARY ECOLOGY PROCESSES IN IRES

Microevolutionary ecology focuses on individual- and population-level changes that result in adaptations to the environment. A centerpiece of microevolutionary ecology is the phenotype, or an observable trait, that directly or indirectly affects individual fitness. Traits are subject to intrinsic trade-offs because natural selection cannot optimize all traits simultaneously because resources are limited. Natural selection, mutation, genetic drift, gene flow, and sexual selection are the mechanisms of evolution, and along with fitness, adaptation, and phenotypic plasticity, are key concepts that apply to IRES (Box 4.10.1 and Section 4.10.2). Fig. 4.10.2 highlights open questions and hypotheses concerning the importance of each evolutionary process in each hydrological phase.

CHARACTERISTIC TRAITS TO COPE WITH LIFE IN IRES

Individuals respond to changing environmental conditions by dispersing, withstanding the event, or expressing phenotypic plasticity, all of which play a role in biotic responses to dry-phase disturbances. Drying can be a predictable (e.g., seasonal) or nonpredictable disturbance in the annual cycle of IRES. Although any repeated disturbance can lead to adaptive evolution, theory suggests that trait evolution will be particularly rapid when disturbances are frequent and/or predictable (Lytle et al., 2008).

Organisms can adapt to drying via behavioral, life history, and morphological changes (Lytle and Poff, 2004; Lytle, 2008) (Fig. 4.10.3; Chapters 4.7 and 4.8). Behavioral adaptations utilize environmental cues correlated with drying (e.g., increasing pool temperatures, decreasing water levels; Velasco and Millan, 1998) as a signal to escape by dispersing within a river or leaving the river entirely. Dispersal within the river network is important for highly mobile taxa such as fishes (Kerezsy et al., 2013; Chapter 4.5) and some larger-bodied invertebrates (Boersma and Lytle, 2014; Chapter 4.3). Dispersal out of the river is especially important for mobile, air-breathing taxa such as adults of aquatic Coleoptera and Hemiptera. Life history adaptations are expected in taxa that cannot respond immediately to drying with behavioral avoidance or dispersal, but that may be able to anticipate drying from seasonal cues. For these organisms, phenology is closely tied to recurring seasonal drying (e.g., synchronous emergence; Cover et al., 2015).

Another example is diapause, common in arthropods but also occurring in some fishes such as Cyprinodontiformes (Chapter 4.5). Like dispersal that spreads mortality risk through space, diapause can be thought of as a risk-spreading strategy that operates through time. Resting stages can occur at all life stages and persist for varying lengths of time from single dry seasons (e.g., the stonefly *Mesocapnia arizonensis*

BOX 4.10.1 DEFINITIONS OF KEY EVOLUTIONARY FORCES AND TERMS, EMPHASIZING THEIR RELEVANCE IN STUDIES OF IRES

One theme of this chapter is the implications of evolutionary processes in IRES. Central among these processes is *natural selection*, which winnows within-population trait variation while simultaneously adapting populations to their local environments. In IRES, natural selection is primarily imposed by the river's flow regime, particularly the predictability, timing, and duration of high, low, and zero flow events. *Genetic drift* is a population-level process that results from random fluctuations in gene frequencies, particularly in small populations such as isolated headwater populations or populations in disconnected dryland habitats. Small isolated populations are also vulnerable to inbreeding depression and extinction. *Dispersal* (*gene flow*) can "rescue" such populations via the influx of genetic variation (genetic rescue) or through the numerical addition of migrants (demographic rescue) (reviewed in Carlson et al., 2014). In IRES, dispersal among populations is an important mechanism by which many organisms escape drought conditions by moving to perennial refuges. Once flow resumes, dispersal from refuges to recently rewetted areas facilitates population recovery. *Mutation* is the generation of novel genetic variation *in situ*, which creates fodder for selection. Beyond these key evolutionary processes, we also define various terms from evolutionary ecology that appear throughout this chapter.

	Definition	Relevance in IRES
Absolute Fitness (population level)	A population-level parameter reflecting the average number of surviving offspring produced per capita per time period (i.e., the λ growth parameter).	Population models can be used to estimate the growth parameter over successive cycles of drying and rewetting phases in IRES.
Adaptation	A trait that confers increased fitness in an environment. An adaptation is an evolutionary response to selection.	Adaptations in IRES include resistance (e.g., diapause) and resilience traits (e.g., dispersal) that allow organisms to cope with specific drying and/or flowing conditions.
Dispersal	Involves the movement of individuals from one area (region, population) to another. In population genetics, migration indicates gene flow that results from movement between populations.	In IRES, dispersal to perennial refuges is a critical mechanism by which many organisms avoid desiccation during dry phases. Once flow resumes, dispersal then allows recolonization of rewetted habitats.
Fitness (Individual)	An individual's reproductive success, often estimated via fitness components such as survival, fecundity, or mating success. Individuals with higher fitness relative to others will be favored by natural selection.	Fitness estimates in IRES can vary greatly depending on the hydrological phase (dry vs. wet) and must be considered in study designs.
Genetic Drift	A population-level phenomenon that results from the random fluctuations in allele frequency owing to small population size.	Genetic drift is strongest in small isolated populations. Thus, drift may be pronounced during drying in IRES when populations are highly fragmented.
Isolation-by-Distance (IBD)	A pattern of increasing genetic distance among populations with increasing geographic distance	An example of IBD in IRES is provided in Section 4.10.2 (Fig. 4.10.6).
Mutation	A physical/chemical change in DNA that occurs at the individual level. Mutations generate variation on which natural selection can act.	Mutation in IRES populations is one means of increasing genetic diversity, thereby creating variation on which selection can act. Mutation rates increase with stress, which may be caused by abrupt changes in environmental conditions (e.g., sudden loss of water or flash flooding in IRES).

BOX 4.10.1 DEFINITIONS OF KEY EVOLUTIONARY FORCES AND TERMS, EMPHASIZING THEIR RELEVANCE IN STUDIES OF IRES *(Cont'd)*

Natural Selection	The differential survival or reproductive success of individuals with different phenotypes. Natural selection acts on individuals, but it is measured as a change in the phenotype distribution within a generation. Selection can occur at different stages in an individual's life cycle.	In IRES, natural selection is imposed by the flow regime, particularly the timing and duration of high, low, and zero flow events.
Neutral Molecular Markers	Genetic markers for which mutational changes do not have any direct effect on fitness (i.e., selectively neutral).	Neutral molecular markers are often used— including in studies of IRES—to study dispersal and gene flow among populations.
Phenotypic Plasticity	The ability to express alternate phenotypes under different environments.	Plasticity may be common in IRES, where individuals adjust their growth and/ or development in response to changing environmental conditions.
Sexual Selection	A special case of natural selection in which the traits under selection influence an individual's ability to acquire mates and reproduce.	Similar to natural selection, in IRES sexual selection is imposed by the flow regime, particularly the timing and duration of flow (and zero flow) events and how these factors influence mate acquisition and reproduction.

Gray, 1981) to multiple dry years (e.g., some microcrustaceans and rotifers, Jenkins and Boulton, 2007; Chapters 4.3 and 4.8). Morphological adaptations to drying include diapausing eggs, osmoregulatory structures, thickened cuticles, and other anatomical structures that confer desiccation resistance (Chapter 4.8). Fish species that estivate during the dry season can undergo metabolic suppression (African lungfish *Protopterus aethiopicus*, Delaney et al., 1974), possess an elongate body form for burrowing or maneuvering through small spaces (loach *Cobitis shikokuensis*, Kawanishi et al., 2013), or secrete mucus (West Australian salamanderfish *Lepidogalaxias salamandroides*, Allen and Berra, 1989; Chapter 4.5). In some cases, suites of traits may be favored in IRES and might produce a variety of adaptive landscapes (i.e., a variety of phenotypes or genotypes and their corresponding fitness options).

While many resistance and resilience traits are presumed to be adaptations to disturbance in IRES (Chapter 4.8), empirical studies of selection and adaptation in IRES remain rare. Selection (natural or sexual) is often studied in nature by examining trait shifts before and after some episode of selection. Understanding the causes of selection requires understanding the environmental circumstances that give rise to variation in fitness among individuals with different traits (Wade and Kalisz, 1990). An example from IRES is the caddisfly *Mesophylax aspersus* which avoids the dry season by estivating in nearby caves while eggs mature in the adult females. Salavert et al. (2011) found that natural selection favored large-bodied females during estivation, presumably because they had the energy reserves to succor their eggs and survive the estivation period.

Comparative studies provide important information and hypotheses, but conclusive studies of adaptation traditionally rely on common environment or reciprocal transplant experiments (Reznick and Travis, 1996). Lytle et al. (2008) quantified the flood-escape behavior of multiple populations of the

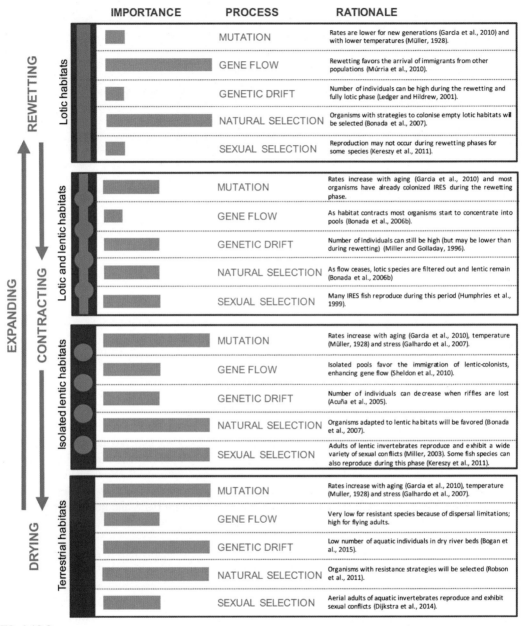

FIG. 4.10.2

Hypotheses about the relative importance (represented by the width of the gray bars) of the different evolutionary processes in each hydrological phase of an IRES, including our underlying rationale. Hydrological phases are presented from the rewetting phase where lotic habitats dominate to the contracting phase to the drying phase where terrestrial habitats dominate. Our rationale is focused on aquatic organisms; however, the importance of processes during the rewetting phase would be similar to those for terrestrial organisms during the drying phase. Blue rectangles indicate lotic habitats, blue circles indicate lentic habitats, and brown rectangles indicate terrestrial habitats. Arrows indicate contracting and expanding paths within wet and dry phases in IRES.

FIG. 4.10.3

Examples of IRES species with different traits to cope with drying: (a) the caddisfly *Limnephilus guadarramicus* has a stick case that increases resistance to drying; (b) some individuals, such as Gammaridae amphipod crustaceans (the brown patches indicated by an arrow), can move upstream or downstream through the aquatic habitat to find perennial refuges; (c) several species of aquatic beetles, such as *Hydrobius convexus*, can inhabit isolated pools during flow contraction by breathing aerial oxygen; (d) the caddisfly *Mesophylax aspersus* has diapausing adults that estivate in caves during dry phases; (e) the oleander *Nerium oleander* has leaves with stomata concentrated in crypts with trichomas that reduce transpiration; (f) many mollusks, such as *Bithynia tentaculata*, have an operculum that allow them to resist desiccation; (g) some amphibian species have tadpoles with life cycles synchronized with drying whereas others, such as those in this photo, are not; (h) many microbes and algae in biofilms can desiccate to withstand drying, reactivating upon rewetting; (i) the killifish *Nothobranchius furzeri* estivates during the dry season.

Photos: Courtesy A. Rúfusová (a), N. Bonada (b,g and h), J. Arribas (c), T. Pérez-Fernández (d), T. Herrera (e), T. Machacek (f) and A. Furness (i).

flightless water bug *Abedus herberti* across a gradient of disturbance predictability, defined as the ability of a cue (rainfall) to predict a disturbance (flash flood). Common environment experiments, where populations from a range of environments were subject to the same conditions, revealed that individuals from predictable environments responded to rainfall by moving out of the water quickly, whereas individuals from less predictable environments did not. The study provides compelling evidence for adaptive divergence among populations in flood-escape behavior that correlates with the environmental predictability at their home sites. It also indicates that drying is not the only evolutionary driver in IRES and, due to cotolerance, many traits selected by other disturbances such as floods can favor the presence of some taxa over others (Vander Vorste et al., 2016).

Aquatic organisms also can alter life history strategies or behaviors to match a range of drying conditions through plasticity (Lytle, 2008). Insects in IRES benefit from plastic strategies that allow different trajectories of growth and development, contingent on whether or not a dry phase occurs (Johansson et al., 2001; Wissinger et al., 2003; De Block and Stoks, 2004, 2005). These strategies require sufficient proximate cues to signal whether or not a dry phase will occur. For populations studied during drying events, phenotypic plasticity does appear to be common; individuals adjust their rates of growth, development, or both to compensate for different environments. For instance, experiments with alpine caddisflies (Shama and Robinson, 2006) and damselflies (De Block and Stoks, 2004) found that plasticity, rather than local adaptation, accounted for most observed differences between populations from perennial rivers and IRES. Thus, it is possible that IRES may select for traits that facilitate phenotypic plasticity across environments that dry, rather than favoring traits focused solely on surviving dry conditions (Vander Vorste et al., 2016). The prevalence of phenotypic plasticity across taxa will depend largely on whether there exists a trade-off between plasticity itself versus specialization in environments prone to drying. Phenotypic plasticity has been associated with epigenetic mechanisms (i.e., a change in gene expression without a change in genotype). How environmental cues influence heritable epigenetic mechanisms is still an open question in many fields of biology (Duncan et al., 2014), and the environmental cues provided by drying offer a good opportunity to further investigate epigenetic mechanisms in IRES species.

POTENTIAL TRADE-OFFS IN IRES: TO STAY OR TO LEAVE?

Dispersing or remaining can be viewed as alternative strategies for spreading risk through space and time (Den Boer, 1968), but most studies focus on one, not both (Bohonak and Jenkins, 2003), limiting our ability to examine trade-offs between them. A trade-off should lead to negative covariation between the two strategies (e.g., organisms with complex life histories may disperse through space less than organisms with more simple life histories, Fig. 4.10.4), as predicted from theory. Although much of the existing empirical literature focuses on plants with dormancy capability (Venable and Lawlor, 1980), aquatic invertebrates with desiccation-resistant life stages hold considerable potential as an empirical study system for exploring such trade-offs (see also Bohonak and Jenkins, 2003). How the two strategies interact to influence the probability of persistence is an open question (Buoro and Carlson, 2014) with much relevance for the study of IRES.

GENETIC STRUCTURE IN IRES

Whether a species is connected or isolated depends on its dispersal ability and the spatial characteristics of flow intermittence and hydrological connectivity (Zickovich and Bohonak, 2007; Sheldon et al., 2010; Chester et al., 2015; Box 4.10.2 and Fig. 4.10.5). Phillipsen et al. (2015) studied the

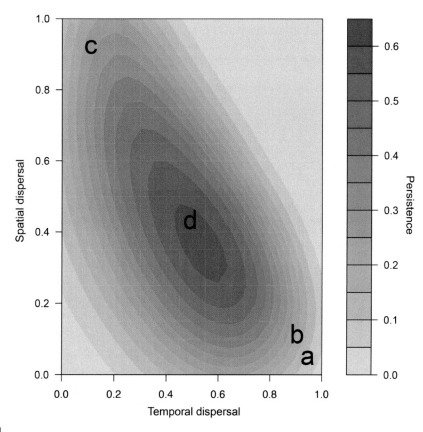

FIG. 4.10.4

Hypothesized relationship between persistence probabilities at a site as a function of spatial and temporal dispersal strategies. Persistence probabilities were generated from simulations using a positive effect of both dispersal strategies on persistence, but a negative effect of interactions between the two (modified from Buoro and Carlson, 2014). Overlain on the figure are hypotheses about where certain species of IRES specialists fall along the curve: (a) the stonefly *Mesocapnia arizonenis* (Michael T. Bogan, personal communication) has a complex life history including dormancy, suggesting high potential for temporal dispersal, but males are effectively flightless, suggesting little opportunity for spatial dispersal at the species level; (b) the dobsonfly *Neohermes filicornis* (Matthew R. Cover, personal communication) also has a complex life history, including dormancy, but little is known about the spatial dispersal capabilities of this species; (c) the beetle *Boreonectes aequinoctialis* has high spatial dispersal capabilities (inferred from Phillipsen et al., 2015), (d) the caddisfly *Mesophylax aspersus* has adults that disperse several kilometers during the dry season and estivate in caves until flow resumes (Salavert et al., 2008), suggesting intermediate temporal and spatial dispersal in this species. Whether the observed differences lead to similar persistence probabilities remains unknown.

BOX 4.10.2 GENERAL MODELS OF GENE FLOW IN STREAMS

Death Valley Model (DVM) (Fig. 4.10.5A). Local populations exhibit little or no gene flow and, in some cases, evolve into species flocks of incipiently diversifying species. The model is named for its original application to describe genetic patterns among highly isolated spring populations in arid regions such as the Mojave Desert of the southwestern United States (Meffe and Vrijenhoek, 1988). The DVM might be highly applicable for low-dispersal species in IRES stream networks with high flow intermittence, where aquatic connectivity is frequently severed.

Headwater Model (HWM) (Fig. 4.10.5B). Headwater specialist taxa show more genetic affinity across short terrestrial distances than within the stream network itself. The HWM was originally described for species occupying the upper limits of predominantly intermittent stream networks and that have some ability to disperse terrestrially (Finn et al., 2007). The HWM is likely to fit many headwater taxa in IRES, especially aquatic insects that disperse terrestrially, because drought often disrupts aquatic connectivity of intervening reaches (Chester et al., 2015; Cañedo-Argüelles, et al., 2015; Razeng et al., 2016).

Stream Hierarchy Model (SHM) (Fig. 4.10.5C). Genetic connectivity is nested within the hierarchical structure of river stream networks (Meffe and Vrijenhoek, 1988). This is the pattern to be expected in species that occur throughout stream networks and that rarely or never disperse terrestrially, such as many fishes and some flightless invertebrates. In IRES, the SHM would be expected in systems with low flow intermittence, where surface flows connect local populations.

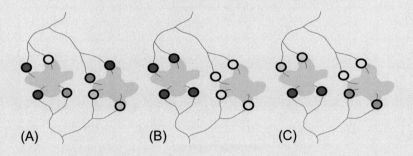

FIG. 4.10.5

General models of gene flow in streams (adapted from Hughes et al., 2009; Crook et al., 2015): the Death Valley Model (A), the Headwater Model (B), and the Stream Hierarchy Model (C).

Modified from Crook, D.A., Lowe, W.H., Allendorf, F.W., Erös, T., Finn, D.S., Gillanders, B.M., et al., 2015. Human effects on ecological connectivity in aquatic ecosystems: integrating scientific approaches to support management and mitigation, Sci. Total Environ. 534, 52–64, with permission from Elsevier.

population genetics of three species of aquatic insect inhabiting an IRES network in arid southeastern Arizona. The first, the giant water bug *A. herberti*, is restricted to perennial stream reaches and exhibited extreme between-population differentiation that was uncorrelated with geographic distance (Fig. 4.10.6). Significant divergence occurred between populations in the same stream, separated by less than 1 km of intermittent habitat, a pattern perhaps common in IRES species that require aquatic habitat (Fig. 4.10.2). The second, the stonefly *M. arizonensis*, which is only found in IRES reaches that sometimes lack surface water, exhibited an isolation-by-distance pattern, likely due to the ability of individuals to fly within and among streams. The water beetle *Boreonectes aequinoctialis* showed a panmictic population genetic pattern due to its ability to fly long distances and utilize nearly any surface water as habitat (Fig. 4.10.6). These different dispersal abilities of aquatic organisms and the

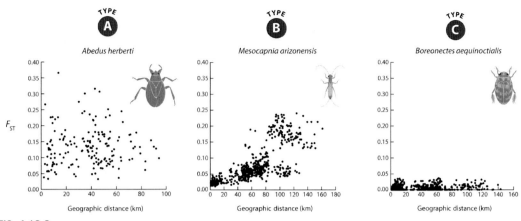

FIG. 4.10.6

The spatial structure of IRES produces distinctive population patterns, depending on the dispersal ability of the organism. F_{ST} is a measure of pairwise genetic difference between populations (i.e., higher the F_{ST}, the greater the distinction between populations). Type A shows a flightless perennial-water specialist, the giant water bug *Abedus herberti*. Populations are strongly differentiated due to genetic drift. Type B shows an intermittent-habitat specialist, the stonefly *Mesocapnia arizonensis*, that disperses short distances aerially. Populations exhibit an isolation-by-distance pattern where genetic distance increases with geographic distance. Type C shows a strong-flying habitat generalist found in IRES, the beetle *Boreonectes aequinoctialis*. Genetic structure is panmictic (i.e., random mating), with no apparent spatial pattern.

From Phillipsen, I.C., Kirk, E.H., Bogan, M.T., Mims, M.C., Olden, J.D., Lytle, D.A., 2015. Dispersal ability and habitat requirements determine landscape-level genetic patterns in desert aquatic insects. Mol. Ecol. 24, 54–69.

network configuration of flow intermittence influence the potential for metapopulation dynamics in IRES (Múrria et al., 2010; Datry et al., 2016a; Chapter 4.9).

4.10.3 MACROEVOLUTIONARY ECOLOGY PROCESSES IN IRES

Macroevolution focuses on the species and lineage levels, and aims to understand community ecology from an evolutionary perspective (Fig. 4.10.1). Many studies on IRES report that some species are IRES specialists whereas others are generalists (Leigh et al., 2016a; Chapter 4.3). However, most community studies of IRES lack an evolutionary perspective and, at best, only describe the biological traits of the species within a community (Bonada et al., 2007; Datry et al., 2014). Linking community ecology to evolutionary biology should provide a more theoretical and fundamental basis for understanding species diversification and assemblages. In this vein, Vellend (2010) suggested that similar rules apply to both population genetics and community ecology. For example, natural selection favors individuals with the highest fitness, while environmental filtering favors species with traits that better match the environment. Genetic drift in population genetics can be considered analogous to ecological drift in community ecology. Much like mutation is the ultimate source of genetic variation, speciation is the ultimate source of species diversity. Finally, the ecological equivalent to gene flow in population ecology is dispersal in community ecology. All of these concepts are also fundamental to understanding communities within IRES.

Empirical and theoretical evidence suggests that local communities in streams are structured predominantly by a dynamic balance between the environmental characteristics of the local habitat (species sorting) and regional-scale dispersal and postdisturbance recolonization (mass effects) (Thompson and Townsend, 2006; Brown et al., 2011). Deterministic biotic interactions are thought to be less important drivers of community structure. In IRES, the repeated pattern of drying and rewetting suggests that recolonization by aquatic-obligate species is an essential process in structuring local communities, and that mass effects might be even more important in IRES than species sorting during particular hydrological phases (Datry et al., 2016b).

Aquatic communities in IRES, particularly those occupying the "harsh" end of the intermittence gradient (i.e., those spending a greater proportion of time without than with surface flow), tend to comprise predictable subsets of the regional metacommunity that have high-dispersal ability (Datry et al., 2014; Chapter 4.9). This pattern suggests a sink-type nature of IRES, in which the local communities rely on connectivity with communities in more stable streams that act as recolonization sources (Fig. 4.10.7). Therefore, well-connected metacommunities can be said to confer resilience. However,

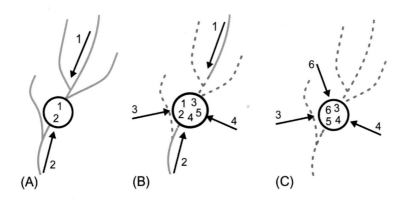

FIG. 4.10.7

Trait variability of colonizing individuals/species in a perennial stream (A) and two IRES networks with different degrees of flow permanence (B and C). Circles represent particular sites and the *solid blue* and *dashed orange* lines represent perennial and intermittent river sections, respectively. Arrows represent the origin of colonizing individuals/species, and numbers represent the genetic configuration of colonizing organisms. In the perennial network (A), individuals/species of a particular site have the trait configuration of the individuals/species arriving from upstream and downstream permanent sites. Arrival of individuals/species from nearby sites is likely to be limited (i.e., habitat is already occupied). In the IRES network with intermediate levels of flow permanence (B), there is an increase of trait variability in populations/communities following flow resumption because individual/species arrive from upstream, downstream, and nearby locations. In the IRES network with low levels of flow permanence (C), trait variability decreases during flow contraction but these systems can still have relatively high levels of trait variability due to immigration of individuals/species with different trait values arriving randomly from nearby locations. In both IRES examples (B and C), the number 5 refers to characteristic IRES individuals/species with resistance forms that are present year-round.

growing evidence suggests that a rarer component of IRES communities is resistant to the inherently harsh wet-dry alternations (Chapter 4.8) and remains in place (Fig. 4.10.7). This component includes the estivating *M. arizonensis* described earlier (Phillipsen et al., 2015). The "low-disperser" component of the metacommunity that includes *M. arizonensis* shows a pattern of distance decay of regional community similarity (Cañedo-Argüelles et al., 2015). This spatial pattern is analogous to the pattern of isolation-by-distance among populations of *M. arizonensis* shown in Fig. 4.10.6. At the community level, this pattern reflects the neutral effects of dispersal limitation across the landscape (e.g., Thompson and Townsend, 2006) and suggests that evolution *in situ* could be more likely in low-dispersal species of the full assemblage than in high-dispersal species.

Resistance is important for macroevolutionary processes such as speciation because diversification is expected to proceed more rapidly in less dispersive lineages that are more likely to remain in a local habitat. Research comparing lotic and lentic environments has found lentic species to have larger ranges (Hof et al., 2006; Abellán and Ribera, 2011), steeper species accumulation curves (Ribera et al., 2003), and lower population differentiation (Hjalmarsson et al., 2015; Marten et al., 2006) than lotic species. The hypothesis is that the comparatively short geological life span of standing water bodies requires lentic lineages to be more dispersive than lotic lineages (Ribera, 2008). Perhaps IRES and lentic systems act similarly, albeit at shorter time scales for IRES. The only explicit test of habitat stability on diversification in IRES found no significant link to habitat (Dijkstra et al., 2014), but this hypothesis warrants further investigation.

In the absence of significant spatial dispersal, species that are resistant to drying have the potential to diversify in allopatry (i.e., while separated geographically). Allopatric diversification can be driven by different selective pressures in different localities (adaptive divergence, Lytle et al., 2008; Watanabe et al., 2014), by neutral genetic drift in isolation or by a combination of the two. Furthermore, neutral drift leading to diversification can result from spatial or temporal isolation. Genetic divergence at putatively neutral markers (e.g., microsatellites) can also occur when dispersal between localities is low enough to permit populations to diverge, even without strong selective forces (Finn et al., 2006, 2007). A developing idea in stream ecology that is likely to be highly relevant to IRES is that even species with moderate dispersal ability that have populations with minimal spatial isolation can evolve in divergent directions if environmental conditions vary among localities (Watanabe et al., 2014). One important example is when conditions lead to significant differences in the timing of key life history events. For example, mayfly populations occupying nearby alpine streams fed by either glacier meltwater or groundwater reach maturity at significantly different times of the season due to the extremely different thermal and hydrological regimes of these two stream types (Finn et al., 2014). Boumans (2014) also found strong evidence for divergence/ speciation in stoneflies occupying sites with different hydrological regimes that influenced the timing of adult emergence. Such evolutionary diversification via separation in time rather than space has been termed "temporal allopatry" (Boumans, 2014) or temporal "reproductive isolation" (Finn et al., 2014). Although not widely studied in aquatic insects (see Dijkstra et al., 2014), temporal variation in the durations of drying in IRES may explain closely related species that are distributed longitudinally along river systems (Mey, 2003) or the co-occurrence of distinct lineages within recognized species in sympatry (Leys et al., 2014). Because IRES produce brief windows of opportunity for life history events, and because the timing is likely to vary, IRES provide multiple opportunities to investigate the degree to which such temporal allopatry might be driving evolutionary diversification and, eventually, speciation.

4.10.4 EVOLUTIONARY PROCESSES AT THE ECOSYSTEM LEVEL IN IRES

Ecosystem ecologists study processes that occur at the ecosystem level, such as fluxes of matter (including organisms) and energy. They consider individuals and species as part of the ecosystem, but only recently have started to analyze how evolutionary differences among them (i.e., genetic and species diversity, respectively) determine EFs (Matthews et al., 2011). Each individual and species plays a particular functional role in an ecosystem, and even closely related individuals or species can have functional differences. For instance, several closely related species of the caddisfly genus *Drusus* have different feeding habits (grazers, shredders, or predators) acquired through evolution, thus contributing differently to the organic matter processing of river ecosystems (Pauls et al., 2008). Understanding how levels of genetic and species diversity affect EFs is where evolution and ecosystem ecology meet. Obviously, EFs not only depend on the genetic and species diversity in the ecosystem but also on present environmental factors such as temperature and light. Disentangling evolutionary and environmental effects involved in EFs helps to tease apart purely evolutionary effects from environmentally related effects.

BIOLOGICAL TRAITS IN IRES AND THEIR RELATION TO ECOSYSTEM FUNCTIONS

The functional role of an individual or species in an ecosystem is expressed through their biological traits (Violle et al., 2007). To understand evolutionary effects at the ecosystem level in IRES, a comprehensive knowledge of traits related to EFs is needed (Table 4.10.1). For aquatic organisms inhabiting IRES, traits such as feeding habits, body size, life cycle characteristics, and behavior have been related to EFs (Bonada et al., 2006a). Feeding habits influence organic matter processing, whereas body size and life cycle duration are related to ecosystem productivity. Especially relevant in IRES are life cycle characteristics, dispersal abilities, or strategies to withstand drying, and these are indirectly related to ecosystem resilience or resistance (i.e., the capacity of an ecosystem to respond to a disturbance, Datry et al., 2014; Chapter 4.8).

Traits related to EFs can be considered neutral or nonneutral in terms of their influence on ecosystem performance. Neutral traits would not change EFs under specific environmental conditions, whereas nonneutral traits would be favored by natural selection. Following flow resumption in IRES, individuals and species with traits that favor colonization (e.g., higher dispersal abilities, short life cycles) would be positively selected, enhancing ecosystem resilience. Other traits would be neutral for resilience following flow resumption. For instance, being a shredder or not would not specifically affect an organism's resilience to drought.

BIODIVERSITY-ECOSYSTEM FUNCTIONS RELATIONSHIPS IN IRES

Disturbance strongly influences genetic and species diversity (Banks et al., 2013). Given the constant presence of human disturbances, there is growing concern about the functional consequences of biodiversity loss. Scientists are now striving to understand the relationships between biodiversity and EFs. Overall, there is a positive relationship between biodiversity and ecosystem performance (see Srivastava and Vellend, 2005; Fig. 4.10.8A) and this relationship applies to species and genetic diversity (Whitham et al., 2006). The biodiversity-EF relationship is hypothesized to be highly dynamic in space and time in IRES. Flow intermittence decreases species (Datry et al., 2014; Fig. 4.10.8B) and

Table 4.10.1 Biological traits of aquatic organisms that relate to ecosystem functions

Biological trait	Related ecosystem function	Examples
Resources supply	Organic matter production	Algae, macrophytes, and riparian vegetation produce organic matter through photosynthesis
	Organic matter processing	Shredders process organic matter accumulated in river beds
	Carbon sequestration	Autotrophic organisms (algae, macrophytes, and riparian vegetation) capture and store carbon dioxide
	Carbon emission	Heterotrophic organisms (microbes, invertebrates, vertebrates) release carbon dioxide through their metabolism
	Nutrient cycling	Through their metabolism, fish excrete nitrogen and phosphorus
	Food web structure	Feeding habits of species determine food web structure
	Food web stability	Diversity of feeding habits of species determines food web stability
Body size	Biomass	Large fish can provide larger amounts of food resources for top predators
	Productivity	Large species have been traditionally associated with lower productivity values
	Stability	Large consumers can increase ecosystem stability
Dispersal ability	Resilience	Species with high dispersal ability will rapidly colonize new empty habitats
	Connectivity	Species with high dispersal ability will increase population connectivity and increase the probability of persistence
Life history	Production	Species with short life cycles and several generations per year increase production
	Resilience	Species with short life cycles and several generations per year increase ecosystem resilience
Behavior characteristics	Resilience	Species with large numbers of offspring increase ecosystem resilience
	Engineering	Burrowing organisms promote vertical connectivity and nutrient cycling
	Pollination	Plant species with higher ability to mobilize gametes increase pollination
	Retention of solids	Species that build cases with mineral substrates contribute to retention of solids

genetic diversity (Meffe and Vrijenhoek, 1988; but see Zickovich and Bohonak, 2007). According to the biodiversity-EF relationship, low diversity in IRES will reduce EFs (Fig. 4.10.8A and B). However, the few studies exploring this relationship in IRES have found substantial functional redundancy (Schriever et al., 2015; Vander Vorste et al., 2016), indicating that even if species diversity decreases, EF can remain relatively stable.

The genetic diversity-EF relationship has been little studied in freshwaters. The few existing examples from terrestrial and marine ecosystems (e.g., Reusch et al., 2005; Kettenring et al., 2014) indicate that genetic diversity increases EFs. We propose that during nonflowing phases, only individuals with

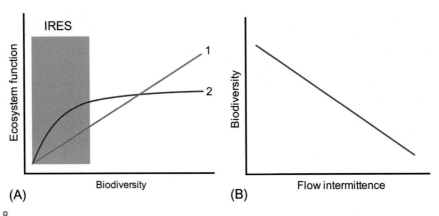

FIG. 4.10.8

The relationship between biodiversity and ecosystem function (A) according to two general models (1: linear and 2: log-linear) (Srivastava and Vellend, 2005). IRES (represented here with a gray box) are expected to have lower ecosystem function than perennial rivers because of the negative relationship between biodiversity and flow intermittence (B) (Datry et al., 2014).

resistance to drying will sustain EFs (Roger et al., 2012). However, these individuals are expected to have lower genetic variability, with genotypes highly adapted to the local environment and with high levels of isolation because of their limited dispersal. Several fish species from isolated populations in desert IRES, for instance, show low genetic diversity within populations and high genetic divergence among populations (Meffe and Vrijenhoek, 1988).

IRES are ecosystems with highly dynamic metapopulations and metacommunities (Chapter 4.9) where local adaptation to flow intermittence (the resting individuals or species) and dispersal (the vagile individuals or species) can regulate EFs. Colonization sources in IRES can be very diverse (Datry et al., 2016b). Following flow resumption or during the nonflowing phase, individuals recolonize from upstream or downstream perennial reaches or from other reaches in nearby catchments (Datry et al., 2016b; Chapters 4.8 and 4.9). Alternatively, they may display a high degree of within-reach affinity. Many of the vagile individuals will randomly arrive from all these sources, but only those with favorable traits will succeed. These surviving individuals will potentially increase the individual trait variability and thus enhance the suite of EFs (Fig. 4.10.7). More research is needed to more fully understand how genetic diversity in these highly dynamic systems affects EFs (Chapter 4.9).

RECURRENT ECOLOGICAL SUCCESSION IN IRES AS A FRAMEWORK FOR FAST EVOLUTION

Flow resumption in IRES transforms terrestrial habitats to aquatic ones, flow cessation extinguishes lotic habitats, and drying transforms aquatic habitats to terrestrial ones. All these events open new spaces for succession to proceed and for evolution to act. Several sets of ecosystem properties change as ecological succession proceeds. Complexity, biomass, trophic levels, specialization, and coevolution increase, whereas the production:biomass ratio and irreversibility decrease (Margalef, 1959; Fig. 4.10.9). These changes are thought to be universal rules of succession and imply that conditions for natural

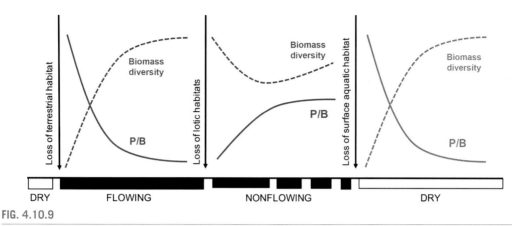

FIG. 4.10.9

Application of Margalef's (1959, 1986) ideas on the relationship between evolution and ecological succession in IRES. The x-axis symbolizes the changes in hydrological phases over time. Vertical arrows symbolize critical moments for biota and ecosystem functions. Blue lines correspond to aquatic organisms and orange lines to terrestrial organisms. Following flow resumption, the biomass and diversity of aquatic organisms increase whereas productivity (P/B: Production/Biomass) decreases. With the loss of lotic habitats, the biomass, diversity and productivity of aquatic organisms is also expected to decrease until communities collapse due to the loss of surface aquatic habitat. Finally, the colonization of dry riverbeds results in an increase of the biomass and diversity of terrestrial organisms and a decrease in their productivity.

selection to act should also change as succession proceeds (Margalef, 1959). Only species adapted to the next successional stage of these universal changes pass through the environmental filter and are favored by natural selection. Ecosystems are recurrently subject to disturbances and adaptive evolution should therefore be guided by successional rules (Margalef, 1986).

Recurrent ecological succession can lead to rapid evolution in IRES. Evidence supporting this idea comes from the numerous adaptations that exist in both aquatic and terrestrial species of IRES to cope with flow cessation and resumption conditions. The loss of riffles in IRES represents a tipping point for aquatic communities (Chapters 2.2, 2.3, and 4.9). Lotic species disappear, reducing biomass and diversity but lentic species that slowly increase in biomass, diversity, and productivity are favored (Fig. 4.10.9). These include species with few offspring, low-dispersal abilities, or long life cycles (Margalef, 1959). Many dragonflies, beetles, and true bugs inhabiting IRES when riffles are lost have some of these characteristics (Bonada et al., 2007; Chapter 4.3). As habitat contraction proceeds, pools eventually disappear and a newly available habitat—the terrestrial habitat—appears. Aquatic species disperse to more suitable habitats, become locally extinct, or become part of the "seed bank" (Chapter 4.3), whereas terrestrial species progressively colonize the dry riverbed (Steward et al., 2012; Chapter 4.4). In turn, terrestrial species present at these initial stages would be, again, opportunistic according to the succession rules, peaking in productivity and slowly increasing in biomass and diversity with time (Fig. 4.10.9).

Flow resumption can also promote rapid evolution in terrestrial and aquatic species. Numerous terrestrial invertebrates may be entrained by flow; some of these die, while others escape from inundation by displaying different behaviors (Corti and Datry, 2012; Chapter 4.3). In contrast, aquatic species with numerous offspring, high-dispersal abilities, short life cycles, and successful competition relative to congeneric species are favored. The evolutionary rate of these aquatic species (i.e., changes across many generations) is expected to be fast and species that are not opportunistic would most likely not be present (Margalef, 1959). However, the presence of resting species in IRES that survive the dry period in the "seed bank" (Chapter 4.3) may also contribute to succession and affect the evolution of colonizing species by adding an extra constraint (i.e., biotic interactions). Traits that favor colonization and reduce biotic interactions with already existing species would be favored by natural selection. Species possessing these favored traits will increase productivity at the beginning of the flow resumption period, whereas biomass and diversity will slowly increase with time (Fig. 4.10.9).

4.10.5 GLOBAL BIOGEOGRAPHICAL AND MACROECOLOGICAL PATTERNS OF AQUATIC ORGANISMS IN IRES

IRES have received little attention in biogeographical and macroecological studies of aquatic organisms (but see Leigh and Datry, 2016). This is surprising considering the global ubiquity of IRES (Chapter 1) and the increase in their ecological investigation in the last decade (Datry et al., 2011; Leigh et al., 2016a). A precise global inventory of IRES and their characteristics (i.e., duration, frequency, and magnitude of drying) is needed for a macroecological evaluation of the diversity they sustain. As more temporal and spatial water flow data become available and hydrological models more accurate (Chapter 2.2), this research gap should be filled.

Global biogeographical and macroecological studies of freshwater diversity have included organisms adapted to both permanent and intermittent freshwater habitats. The characteristics of IRES certainly contribute to current diversity patterns across different regions. Flow-related environmental variables in global or continental freshwater biodiversity assessments give, at best, a broad perspective of the importance of the hydrological regime for freshwater diversity (e.g., Tisseuil et al., 2013), basically spreading a message of "less flow-less diversity." This certainly applies to IRES aquatic diversity which is negatively related to flow intermittence (Datry et al., 2014).

LOCAL AND REGIONAL FACTORS DETERMINING DIVERSITY IN IRES

The study of biogeography and macroecology considers large-scale drivers of diversity as affected by factors of habitat-size (i.e., the species-area effect), resource availability (i.e., the species-energy effect), and historical contingencies (e.g., impacts of major past climate changes like the Pleistocene glaciations). Given the strong influence of flow intermittence on diversity and nestedness of aquatic communities, local factors such as the distribution of aquatic refuges across river landscapes or the dispersal abilities of IRES species (Chapters 4.8 and 4.9) may be important in IRES with high levels of intermittence (Leigh and Datry, 2016), whereas area, climate, and latitude may be more important in IRES with low levels of flow intermittence and with surface water most of the year. Razeng et al. (2016) found that the community composition of aquatic invertebrates from central Australia, a region where IRES dominate the landscape, was influenced primarily by topographic connectivity via overland dispersal pathways. However, predictions may differ between strictly and nonstrictly aquatic organisms because they have different dispersal capacities (Fig. 4.10.10). For strictly aquatic organisms (e.g., fish, some aquatic invertebrates) that have a single pathway to colonize IRES from perennial stretches or refuges, local factors driving diversity may be more relevant than for nonstrictly aquatic

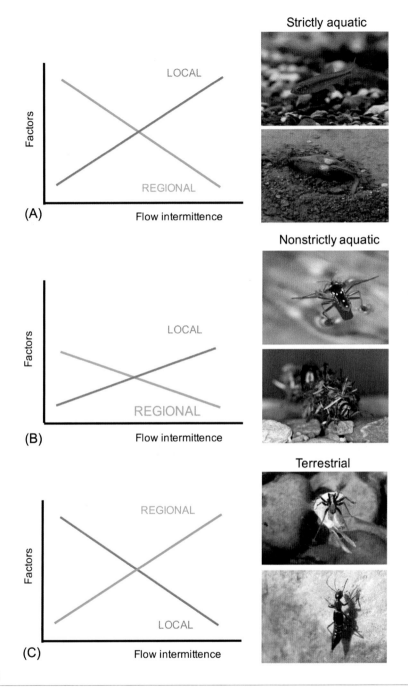

FIG. 4.10.10

Hypothesized assumptions about the relative importance of local (in orange) and regional (in green) factors in explaining diversity in IRES of organisms that are: (A) strictly aquatic such as the fish *Squalius malacitanus* and the mollusk *Unio* sp., (B) not strictly aquatic such as the veliid bug *Velia caprai caprai* and the caddisfly *Limnephilus guadarramicus*, and (C) terrestrial such as the wolf spider *Pardosa saltans* and the rove beetle *Paederidus ruficollis*.

Photos: Courtesy T. Herrera - the fish, mollusk and veliid bug; J. Ortiz (Associació CEN) - the caddisfly; and R. Corti - the spider and rove beetle.

organisms (e.g., those invertebrates that can fly) at similar levels of intermittence (Cañedo-Argüelles et al., 2015; Datry et al., 2016b). For terrestrial organisms (i.e., those colonizing dry river beds), regional factors may become more important as flow intermittence increases in duration and riverbeds remain dry for longer periods of time (Fig. 4.10.10).

SPECIATION, EXTINCTION, AND IMMIGRATION IN IRES

IRES often comprise the headwaters of river networks (Benstead and Leigh, 2012) where local diversity is usually low (Finn et al., 2011). Compared to terrestrial landscapes, where individuals move through several dispersal routes, the movements of aquatic organisms within drainage basins (especially for strictly aquatic organisms) are restricted along the aquatic branches of dendritic river networks (Benda et al., 2004). In IRES, the naturally higher isolation of headwaters combined with their intermittence should have important consequences for the mechanisms driving speciation, extinction, and immigration (Fig. 4.10.11). As a consequence, and because isolation and fragmentation greatly influence genetic divergence of aquatic organisms (Burridge et al., 2006; see Section 4.10.2), different levels of intermittence are likely to affect these mechanisms and contribute to large-scale diversity gradients

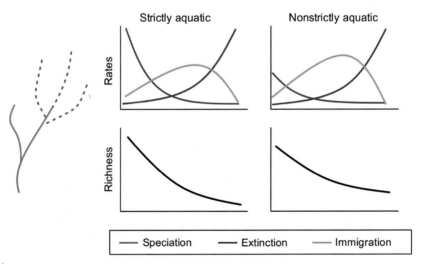

FIG. 4.10.11

Assumptions about evolutionary processes shaping diversity patterns in headwater streams along a gradient of flow intermittence (horizontal axes) for strictly and nonstrictly aquatic organisms, assuming perennial flow in downstream sections provides a constant source of organisms. The decreasing pattern of richness along this flow-intermittence gradient would be mainly driven by speciation at low levels of intermittence, then giving way to communities supported by immigration from downstream permanent reaches before finally being outweighed by extinction rates at high levels of intermittence. Because nonstrictly aquatic organisms implicitly have greater dispersal capacities, speciation rates would be lower and immigration rates greater than for strictly aquatic organisms at similar levels of intermittence, resulting in a lower slope of the decreasing richness pattern with flow intermittence. Blue and orange lines in the network represent perennial and intermittent river sections, respectively.

(Fig. 4.10.11). For instance, low levels of intermittence in headwaters should enhance speciation processes by long-term hydrological stability and isolation-by-distance mechanisms. As the intensity of intermittence increases, immigration should increase. Empirical evidence suggests that there is a higher number of aerial dispersal organisms and/or the role of overland dispersal in shaping aquatic communities increases in importance during the isolated-pools phase (Bonada et al., 2007; Razeng et al., 2016). Finally, with increasing flow intermittence, extinction processes should be more important and immigration should decrease as populations cannot persist, leaving no time for genetic differentiation and speciation. Of course, the importance of these processes could vary, resulting in different diversity patterns, if drying occurs in the lower reaches of the network or across the whole network.

Indirect evidence of differential evolutionary mechanisms in IRES shaping large-scale diversity patterns comes from global analyses of freshwater diversity. For instance, rivers located in Mediterranean and arid regions, where IRES are abundant, have high levels of endemism and beta-diversity (Leprieur et al., 2011). Macroinvertebrates, fish, and riparian communities of Mediterranean rivers harbor greater regional diversity, beta diversity, rarity, and endemicity levels than temperate rivers (Bonada and Resh, 2013). Furthermore, Tedesco et al. (2012) suggest that the evolutionary processes shaping endemism in Mediterranean, semiarid, and arid regions differ from the mechanisms shaping the typically high levels of endemism found in tropical drainages. The diversity found at Mediterranean and arid latitudes may result from anagenetic speciation (i.e., the gradual transformation of a single species from ancestral species) through isolation and species range contractions. The intermediate position of these climatic regions between temperate and tropical climates makes them more sensitive to cycles of expansion and contraction at both longer and shorter geological time scales. For instance, global biome reconstruction for the Middle Pliocene (c. 3.6–2.6 Ma) suggests that tropical savannahs and woodland expanded in Africa and Australia at the expense of deserts (Salzmann et al., 2008), and pollen data suggests that the mid-Holocene (6000 years BP) climate was generally wetter than today in southern Europe and northern Africa (Wu et al., 2007). These historical climatic changes have certainly enhanced range expansions followed by contractions and isolation of previously connected populations, producing high levels of endemism and beta-diversity in IRES through local extinctions and speciation.

GLOBAL TRAIT PATTERNS IN IRES

IRES exhibit complex spatiotemporal flow patterns (Datry et al., 2016a; Chapters 2.2 and 2.3). Because differences in drying duration, predictability, and frequency among regions and within catchments have not yet been compiled globally, it is difficult to outline the "big picture" of how their aquatic assemblages have formed and which trait adaptations and strategies have been shaped and selected. Even if continental scale classifications of hydrological regimes exist (Chapter 2.3), the use of different methods impedes intercontinental comparisons of taxonomic and trait diversity from IRES. Although not all latitudes were included, Datry et al. (2014) adopted a global approach and compared 14 rivers from multiple biogeographic regions spanning a wide range of flow intermittence and spatial arrangements of perennial and IRES reaches. They found widespread congruence in the responses of invertebrate communities to intermittence in terms of richness and persistence strategies (resistance being less common than resilience). However, there was little or no influence of the spatial arrangement of perennial and IRES reaches, suggesting a common response of organisms regardless of the biogeographic species pool.

Resilience and resistance strategies are also common attributes found among biota from Mediterranean climate regions, with resilience being found more frequently than resistance (Hershkovitz and Gasith, 2013). Functional similarities (i.e., adaptive biological traits) among these regions have been widely recognized for a large variety of individual species and biological traits (Bonada and Resh, 2013), as have the dynamics of recovery following disturbance (a measure of the resilience of these ecosystems, Carmel and Flather, 2004). This overall functional convergence of individual species suggests that ecological and evolutionary processes driving patterns of trait adaptations to drying in IRES are deterministic and influenced by large-scale climatic and geomorphological factors shaping IRES globally, rather than being dependent on historical contingencies.

For a comprehensive evaluation of trait adaptations to intermittence, a wider comparison must include IRES from all climatic regions, spanning varying levels of magnitude, frequency, seasonal timing, predictability, and duration of dry periods. A recent analysis of datasets from temperate, Mediterranean, and arid regions examining the responses of aquatic macroinvertebrate assemblages to drying in relation to traits of resistance and resilience found that resistance strategies appear more prevalent in arid climates (Leigh et al., 2016b). However, nearby perennial sites are probably important sources of taxa with resilience strategies that can colonize IRES when flow returns (Chapter 4.8).

4.10.6 IRES ARE IDEAL SETTINGS FOR THE STUDY OF ECOLOGY AND EVOLUTION: CONCLUSIONS AND FUTURE RESEARCH

Evolutionary processes have been largely neglected in studies of IRES, despite the fact that these systems provide an ideal setting for studying the link between ecology and evolution. We base this conclusion on six main points. First, alternating wet and dry phases exert strong selection pressures in IRES that result in a myriad of adaptations. Second, IRES impose continual trade-offs for individuals, including how to respond to approaching dry phases, but also challenges for species stability and how to respond to disturbances. Third, the recurrent extinction and recolonization processes that affect IRES allow us to study how nonneutral and neutral processes influence genetic patterns and, similarly, how environmental sorting and dispersal influence the composition of local communities. Fourth, the high spatial and temporal habitat variability of IRES may lead to rapid niche differentiation and thus species diversification. Fifth, the drying period may impose changes in the relative roles of local and regional factors controlling biological communities and the relative importance of speciation, extinction, and immigration. Sixth, habitat contraction might change the relative effects of different evolutionary processes acting at population and community levels, which may ultimately influence EFs.

We hope our chapter stimulates future research on how ecology and evolution interact in IRES. In particular, studies should seek to understand (1) how evolutionary processes change in space and time; (2) the adaptive significance of resistance and resilience traits common in IRES biota; (3) how speciation, migration, and extinction processes vary across IRES networks; and (4) how evolutionary processes influence ecosystem-level processes in IRES. Future studies should not constrain themselves to single levels of organization (Fig. 4.10.1), but should focus on how evolutionary processes acting at one level influence those at other levels. The application of more powerful molecular techniques that are now available for the study of evolutionary ecology will help to elucidate some of the open questions presented here (Pauls et al., 2014). A more comprehensive view of the relationship between ecology and evolution in IRES will help us to predict how populations, communities, and ecosystems will adapt to increasing flow intermittence resulting from global change and other anthropogenic causes.

REFERENCES

Abellán, P., Ribera, I., 2011. Geographic location and phylogeny are the main determinants of the size of the geographical range in aquatic beetles. BMC Evol. Biol. 11, 344.

Allen, G.R., Berra, T.M., 1989. Life history aspects of the West Australian Salamanderfish, *Lepidogalaxias salamandroides* Mees. Rec. West. Aust. Museum 14, 253–267.

Banks, S.C., Cary, G.J., Smith, A.L., Davies, I.D., Driscoll, D.A., Gill, A.M., et al., 2013. How does ecological disturbance influence genetic diversity? Trends Ecol. Evol. 28, 670–679.

Benda, L., Poff, N.L., Miller, D., Dunne, T., Reeves, G., Pess, G., et al., 2004. The network dynamics hypothesis: how channel networks structure riverine habitats. Bioscience 54, 413–427.

Benstead, J.P., Leigh, D.S., 2012. An expanded role for river networks. Nat. Geosci. 5, 678–679.

Boersma, K.S., Lytle, D.A., 2014. Overland dispersal and drought-escape behavior in a flightless aquatic insect, *Abedus herberti* (Hemiptera: Belostomatidae). Southwest. Nat. 59, 301–302.

Bogan, M.T., Boersma, K.S., Lytle, D.A., 2015. Resistance and resilience of invertebrate communities to seasonal and supraseasonal drought in arid-land headwater streams. Freshw. Biol. 60, 2547–2558.

Bohonak, A.J., Jenkins, D.G., 2003. Ecological and evolutionary significance of dispersal by freshwater invertebrates. Ecol. Lett. 6, 783–796.

Bonada, N., Resh, V.H., 2013. Mediterranean-climate streams and rivers: geographically separated but ecologically comparable freshwater systems. Hydrobiologia 719, 1–29.

Bonada, N., Prat, N., Resh, V.H., Statzner, B., 2006a. Developments in aquatic insect biomonitoring: a comparative analysis of recent approaches. Annu. Rev. Entomol. 51, 495–523.

Bonada, N., Rieradevall, M., Prat, N., 2007. Macroinvertebrate community structure and biological traits related to flow permanence in a Mediterranean river network. Hydrobiologia 589, 91–106.

Boulton, A.J., 2003. Parallels and contrasts in the effects of drought on stream macroinvertebrate assemblages. Freshw. Biol. 48, 1173–1185.

Boumans, L., 2014. Speciation, sexual communication and reproductive barriers in northern hemisphere stoneflies (Plecoptera, Arctoperlaria). Akademika forlag, Oslo. PhD thesis.

Brown, B.L., Swan, C.M., Auerbach, D.A., Grant, E.H.C., Hitt, N.P., Maloney, K.O., et al., 2011. Metacommunity theory as a multispecies, multiscale framework for studying the influence of river network structure on riverine communities and ecosystems. J. N. Am. Benthol. Soc. 30, 310–327.

Buoro, M., Carlson, S.M., 2014. Life-history syndromes: integrating dispersal through space and time. Ecol. Lett. 17, 756–767.

Burridge, C.P., Craw, D., Waters, J.M., 2006. River capture, range expansion, and cladogenesis: the genetic signature of freshwater vicariance. Evolution 60, 1038–1049.

Cañedo-Argüelles, M., Boersma, K.S., Bogan, M.T., Olden, J.D., Phillipsen, I., Schriever, T.A., et al., 2015. Dispersal strength determines meta-community structure in a dendritic riverine network. J. Biogeogr. 42, 778–790.

Carlson, S.M., Cunningham, C.J., Westley, P.A.H., 2014. Evolutionary rescue in a changing world. Trends Ecol. Evol. 29, 521–530.

Carmel, Y., Flather, C.H., 2004. Comparing landscape scale vegetation dynamics following recent disturbance in climatically similar sites in California and the Mediterranean basin. Landsc. Ecol. 19, 573–590.

Chester, E.T., Miller, A.D., Valenzuela, I., Wickson, S.J., Robson, B.J., 2015. Drought survival strategies, dispersal potential and persistence of invertebrate species in an intermittent stream landscape. Freshw. Biol. 60, 2066–2083.

Corti, R., Datry, T., 2012. Invertebrates and sestonic matter in an advancing wetted front travelling down a dry river bed (Albarine, France). Freshwater Sci. 31, 1187–1201.

Cover, M.R., Seo, J.H., Resh, V.H., 2015. Life history, burrowing behavior, and distribution of *Neohermes filicornis* (Megaloptera: Corydalidae), a long-lived aquatic insect in intermittent streams. West. North Am. Nat. 75, 474–490.

Crook, D.A., Lowe, W.H., Allendorf, F.W., Erös, T., Finn, D.S., Gillanders, B.M., et al., 2015. Human effects on ecological connectivity in aquatic ecosystems: integrating scientific approaches to support management and mitigation. Sci. Total Environ. 534, 52–64.

Datry, T., Corti, R., Claret, C., Philippe, M., 2011. Flow intermittence controls leaf litter breakdown in a French temporary alluvial river: the "drying memory". Aquat. Sci. 73, 471–483.

Datry, T., Larned, S.T., Fritz, K.M., Bogan, M.T., Wood, P.J., Meyer, E.I., et al., 2014. Broad-scale patterns of invertebrate richness and community composition in temporary rivers: effects of flow intermittence. Ecography 37, 94–104.

Datry, T., Pella, H., Leigh, C., Bonada, N., Hugueny, B., 2016a. A landscape approach to advance intermittent river ecology. Freshw. Biol. 61, 1200–1213.

Datry, T., Bonada, N., Heino, J., 2016b. Towards understanding the organisation of metacommunities in highly dynamic ecological systems. Oikos 125, 149–159.

De Block, M., Stoks, R., 2004. Cannibalism-mediated life history plasticity to combined time and food stress. Oikos 106, 587–597.

De Block, M., Stoks, R., 2005. Pond drying and hatching date shape the tradeoff between age and size at emergence in a damselfly. Oikos 108, 485–494.

Delaney, R.G., Lahiri, S., Fishman, A.P., 1974. Aestivation of the African lungfish *Protopterus aethiopicus*: cardiovascular and respiratory functions. J. Exp. Biol. 61, 111–128.

Den Boer, P.J., 1968. Spreading of risk and stabilization of animal numbers. Acta Biotheor. 17, 165–194.

Dijkstra, K.-D.B., Monaghan, M.T., Pauls, S.U., 2014. Freshwater biodiversity and aquatic insect diversification. Annu. Rev. Entomol. 59, 143–163.

Dudgeon, D., Arthington, A., Gessner, M.O., Kawabata, Z.-I., Knowler, D.J., Lévêque, C., et al., 2006. Freshwater biodiversity: importance, threats, status and conservation challenges. Biol. Rev. 81, 163–182.

Duncan, E.J., Gluckman, P.D., Dearden, P.K., 2014. Epigenetics, plasticity, and evolution: how do we link epigenetic change to phenotype? J. Exp. Zool. B Mol. Dev. Evol. 322B, 208–220.

Finn, D.S., Theobald, D.M., Black, W.C.I.V., Poff, N.L., 2006. Spatial population genetic structure and limited dispersal in a Rocky Mountain alpine stream insect. Mol. Ecol. 15, 3553–3566.

Finn, D.S., Blouin, M.S., Lytle, D.A., 2007. Population genetic structure reveals terrestrial affinities for a headwater stream insect. Freshw. Biol. 52, 1881–1897.

Finn, D.S., Bonada, N., Múrria, C., Hughes, J.M., 2011. Small but mighty: headwaters are vital to stream network biodiversity at two levels of organization. J. N. Am. Benthol. Soc. 30, 963–980.

Finn, D.S., Zamora-Muñoz, C., Múrria, C., Sáinz-Bariáin, M., Alba-Tercedor, J., 2014. Evidence from recently deglaciated mountain ranges that *Baetis alpinus* (Ephemeroptera) could lose significant genetic diversity as alpine glaciers disappear. Freshwater Sci. 33, 207–216.

Gray, L., 1981. Species composition and life histories of aquatic insects in a lowland Sonoran Desert stream. Am. Midl. Nat. 106, 229–242.

Hershkovitz, Y., Gasith, A., 2013. Resistance, resilience, and community dynamics in mediterranean-climate streams. Hydrobiologia 719, 59–75.

Hjalmarsson, A., Bergsten, J., Monaghan, M.T., 2015. Dispersal is linked to habitat use in 59 species of water beetles (Coleoptera: Adephaga) on Madagascar. Ecography 38, 732–739.

Hof, C., Brändle, M., Brandl, R., 2006. Lentic odonates have larger and more northern ranges than lotic species. J. Biogeogr. 33, 63–70.

Hughes, J.M., Schmidt, D.J., Finn, D.S., 2009. Genes in streams: using DNA to understand the movement of freshwater fauna and their riverine habitat. Bioscience 59, 573–583.

Hutchinson, G.E., 1965. The Ecological Theatre and the Evolutionary Play. Yale University Press, New Haven, CT.

Jenkins, K.M., Boulton, A.J., 2007. Detecting impacts and setting restoration targets in arid-zone rivers: aquatic micro-invertebrate responses to reduced floodplain inundation. J. Appl. Ecol. 44, 823–832.

Johansson, F., Stoks, R., Rowe, L., De Block, M., 2001. Life history plasticity in a damselfly: effects of combined time and biotic constraints. Ecology 82, 1857–1869.

Kawanishi, R., Inoue, M., Dohi, R., Fujii, A., Miyake, Y., 2013. The role of the hyporheic zone for a benthic fish in an intermittent river: a refuge, not a graveyard. Aquat. Sci. 75, 425–431.

Kerezsy, A., Balcombe, S.R., Tischler, M., Arthington, A.H., 2013. Fish movement strategies in an ephemeral river in the Simpson Desert, Australia. Austral Ecol. 38, 798–808.

Kettenring, K.M., Mercer, K.L., Reinhart Adams, C., Hines, J., 2014. Application of genetic diversity-ecosystem function research to ecological restoration. J. Appl. Ecol. 51, 339–348.

Leigh, C., Datry, T., 2016. Drying as a primary hydrological determinant of biodiversity in river systems: a broad-scale analysis. Ecography 39, 1–13.

Leigh, C., Bonada, N., Boulton, A.J., Hugueny, B., Larned, S.T., Vorste, R.V., et al., 2016a. Invertebrate assemblage responses and the dual roles of resistance and resilience to drying in intermittent rivers. Aquat. Sci. 78, 291–301.

Leigh, C., Boulton, A.J., Courtwright, J.L., Fritz, K., May, C.L., Walker, R.H., et al., 2016b. Ecological research and management of intermittent rivers: an historical review and future directions. Freshw. Biol. 61, 1181–1199.

Leprieur, F., Tedesco, P.A., Hugueny, B., Beauchard, O., Dürr, H.H., Brosse, S., et al., 2011. Partitioning global patterns of freshwater fish beta diversity reveals contrasting signatures of past climate changes. Ecol. Lett. 14, 325–334.

Leys, M., Keller, I., Räsänen, K., Gattolliat, J.-L., Robinson, C.T., 2014. Distribution and population genetic variation of cryptic species of the Alpine mayfly *Baetis alpinus* (Ephemeroptera: Baetidae) in the Central Alps. BMC Evol. Biol. 16, 77.

Lytle, D.A., 2008. Life-history and behavioural adaptations to flow regime in aquatic insects. In: Lancaster, J., Briers, R. (Eds.), Aquatic Insects: Challenges to Populations. CAB International, Wallingford, pp. 122–138.

Lytle, D., Poff, N.L., 2004. Adaptation to natural flow regimes. Trends Ecol. Evol. 19, 94–100.

Lytle, D.A., Bogan, M.T., Finn, D.S., 2008. Evolution of aquatic insect behaviours across a gradient of disturbance predictability. Proc. R. Soc. B Biol. Sci. 275, 453–462.

Margalef, R., 1959. Ecología, biogeografía y evolución. Revista de la Universidad de Madrid 8, 221–273.

Margalef, R., 1986. Sucesión y evolución: su proyección biogeográfica. Paleontologia i evolució 20, 7–26.

Marten, A., Brändle, M., Brandl, R., 2006. Habitat type predicts genetic population differentiation in freshwater invertebrates. Mol. Ecol. 15, 2643–2651.

Matthews, B., Narwani, A., Hausch, S., Nonaka, E., Peter, H., Yamamichi, M., et al., 2011. Toward an integration of evolutionary biology and ecosystem science. Ecol. Lett. 14, 690–701.

Meffe, G.K., Vrijenhoek, R.C., 1988. Conservation genetics in the management of desert fishes. Conserv. Biol. 2, 157–169.

Mey, W., 2003. Insular radiation of the genus *Hydropsyche* (Insecta, Trichoptera: Hydropsychidae) Pictet, 1834 in the Philippines and its implications for the biogeography of Southeast Asia. J. Biogeogr. 30, 227–236.

Múrria, C., Bonada, N., Ribera, C., Prat, N., 2010. Homage to the Virgin of Ecology, or why an aquatic insect unadapted to desiccation may maintain populations in very small, temporary Mediterranean streams. Hydrobiologia 653, 179–190.

Pauls, S.U., Graf, W., Haase, P., Lumbsch, H.T., Waringer, J., 2008. Grazers, shredders and filtering carnivores—the evolution of feeding ecology in Drusinae (Trichoptera: Limnephilidae): insight from a molecular phylogeny. Mol. Phylogenet. Evol. 46, 776–791.

Pauls, S.U., Alp, M., Bálint, M., Bernabò, P., Ciampor, F., Ciamporová-Zatovicovňa, Z., et al., 2014. Integrating molecular tools into freshwater ecology: developments and opportunities. Freshw. Biol. 59, 1559–1576.

Phillipsen, I.C., Kirk, E.H., Bogan, M.T., Mims, M.C., Olden, J.D., Lytle, D.A., 2015. Dispersal ability and habitat requirements determine landscape-level genetic patterns in desert aquatic insects. Mol. Ecol. 24, 54–69.

Razeng, E., Morán-Ordóñez, A., Brim Box, J., Thompson, R., Davis, J., Sunnucks, P., 2016. A potential role for overland dispersal in shaping aquatic invertebrate communities in arid regions. Freshw. Biol. 61, 745–757.

Reusch, T.B.H., Ehlers, A., Hämmerli, A., Worm, B., 2005. Ecosystem recovery after climatic extremes enhanced by genotypic diversity. Proc. Natl. Acad. Sci. U. S. A. 102, 2826–2831.

Reznick, D., Travis, J., 1996. The empirical study of adaptation in natural populations. In: Rose, M.R., Lauder, G.V. (Eds.), Adaptation. Academic Press, San Diego, CA, pp. 243–290.

Ribera, I., 2008. Habitat constraints and the generation of diversity in freshwater macroinvertebrates. In: Lancaster, J., Briers, R. (Eds.), Aquatic Insects: Challenges to Populations. CAB International, Wallingford, pp. 289–331.

Ribera, I., Foster, G.N., Vogler, A.P., 2003. Does habitat use explain large scale species richness patterns of aquatic beetles in Europe? Ecography 26, 145–152.

Roger, F., Godhe, A., Gamfeldt, L., 2012. Genetic diversity and ecosystem functioning in the face of multiple stressors. PLoS One 7, e45007.

Salavert, V., Zamora-Muñoz, C., Ruiz-Rodríguez, M., Fernández-Cortés, A., Soler, J.J., 2008. Climatic conditions, diapause and migration in a troglophile caddisfly. Freshw. Biol. 53, 1606–1617.

Salavert, V., Zamora-Muñoz, C., Ruiz-Rodríguez, M., Soler, J.J., 2011. Female-biased size dimorphism in a diapausing caddisfly, *Mesophylax aspersus*: effect of fecundity and natural and sexual selection. Econ. Entomol. 36, 389–395.

Salzmann, U., Haywood, A.M., Lunt, D.J., Valdes, P.J., Hill, D.J., 2008. A new global biome reconstruction and data-model comparison for the Middle Pliocene. Glob. Ecol. Biogeogr. 17, 432–447.

Schriever, T.A., Bogan, M.T., Boersma, K.S., Cañedo-Argüelles, M., Jaeger, K.L., Olden, J.D., et al., 2015. Hydrology shapes taxonomic and functional structure of desert stream invertebrate communities. Freshwater Sci. 34, 399–409.

Shama, L.N.S., Robinson, C.T., 2006. Sex-specific life-history responses to seasonal time constraints in an alpine caddisfly. Evol. Ecol. Res. 8, 169–180.

Sheldon, F., Bunn, S.E., Hughes, J.M., Arthington, A.H., Balcombe, S.R., Fellows, C.S., 2010. Ecological roles and threats to aquatic refugia in arid landscapes: dryland river waterholes. Mar. Freshw. Res. 61, 885–895.

Srivastava, D.S., Vellend, M., 2005. Biodiversity-ecosystem function research: is it relevant to conservation? Annu. Rev. Ecol. Evol. Syst. 36, 267–294.

Steward, A.L., von Schiller, D., Tockner, K., Marshall, J.C., Bunn, S.E., 2012. When the river runs dry: human and ecological values of dry riverbeds. Front. Ecol. Environ. 10, 202–209.

Tedesco, P.A., Leprieur, F., Hugueny, B., Brosse, S., Dürr, H.H., Beauchard, O., et al., 2012. Patterns and processes of global riverine fish endemism. Glob. Ecol. Biogeogr. 21, 977–987.

Thompson, R., Townsend, C., 2006. A truce with neutral theory: local deterministic factors, species traits and dispersal limitation together determine patterns of diversity in stream invertebrates. J. Anim. Ecol. 75, 476–484.

Tisseuil, C., Cornu, J.-F., Beauchard, O., Brosse, S., Darwall, W., Holland, R., et al., 2013. Global diversity patterns and cross-taxa convergence in freshwater systems. J. Anim. Ecol. 82, 365–376.

Vander Vorste, R., Corti, R., Sagouis, A., Datry, T., 2016. Invertebrate communities in gravel-bed, braided rivers are highly resilient to flow intermittence. Freshwater Sci. 35, 164–177.

Velasco, J., Millan, A., 1998. Insect dispersal in a drying desert stream: effects of temperature and water loss. Southwest. Nat. 43, 80–87.

Vellend, M., 2010. Conceptual synthesis in community ecology. Q. Rev. Biol. 85, 183–206.

Venable, D.L., Lawlor, L., 1980. Delayed germination and dispersal in desert annuals: escape in space and time. Oecologia 46, 272–282.

Violle, C., Navas, M.-L., Vile, D., Kazakou, E., Furtunel, C., Hummel, I., et al., 2007. Let the concept of trait be functional! Oikos 116, 882–892.

Wade, M.J., Kalisz, S., 1990. The causes of natural selection. Evolution 44, 1947–1955.

Watanabe, K., Kazama, S., Omura, T., Monaghan, M.T., 2014. Adaptive genetic divergence along narrow environmental gradients in four stream insects. PLoS One 9, e93055.

Whitham, T.G., Bailey, J.K., Schweitzer, J.A., Shuster, S.M., Bangert, R.K., LeRoy, C.J., et al., 2006. A framework for community and ecosystem genetics: from genes to ecosystems. Nat. Rev. Genet. 7, 510–523.

Williams, D.D., Hynes, H.B.N., 1977. Ecology of temporary streams: general remarks on temporary streams. Internationale Revue der gesamten Hydrobiologie und Hydrographie 62, 53–61.

Wissinger, S., Brown, W., Jannot, J., 2003. Caddisfly life histories along permanence gradients in high-altitude wetlands in Colorado (USA). Freshw. Biol. 48, 255–270.

Wu, H., Guiot, J., Brewer, S., Guo, Z., 2007. Climatic changes in Eurasia and Africa at the last glacial maximum and mid-Holocene: reconstruction from pollen data using inverse vegetation modelling. Clim. Dyn. 29, 211–229.

Zickovich, J.M., Bohonak, A.J., 2007. Dispersal ability and genetic structure in aquatic invertebrates: a comparative study in southern California streams and reservoirs. Freshw. Biol. 52, 1982–1996.

FURTHER READING

Acuña, V., Muñoz, I., Giorgi, A., Omella, M., Sabater, F., Sabater, S., 2005. Drought and postdrought recovery cycles in an intermittent Mediterranean stream: structural and functional aspects. J. N. Am. Benthol. Soc. 24, 919–933.

Bogan, M.T., Lytle, D.A., 2007. Seasonal flow variation allows 'time-sharing' by disparate aquatic insect communities in montane desert streams. Freshw. Biol. 52, 290–304.

Bonada, N., Rieradevall, M., Prat, N., Resh, V.H., 2006b. Benthic macroinvertebrate assemblages and macrohabitat connectivity in Mediterranean-climate streams of northern California. J. N. Am. Benthol. Soc. 25, 32–43.

Cardinale, B.J., Duffy, J.E., Gonzalez, A., Hooper, D.U., Perrings, C., Venail, P., et al., 2012. Biodiversity loss and its impact on humanity. Nature 486, 59–67.

Galhardo, R.S., Hastings, P.J., Mosenberg, S.M., 2007. Mutation as a stress response and the regulation of evolvability. Crit. Rev. Biochem. Mol. Biol. 42, 399–435.

Garcia, A.M., Calder, R.B., Dollé, M.E.T., Lundell, M., Kapahi, P., Vijg, J., 2010. Age- and temperature-dependent somatic mutation accumulation in *Drosophila melanogaster*. PLoS Genet. 6, e1000950.

Humphries, P., King, A.J., Koehn, J.D., 1999. Fish, flows and flood plains: links between freshwater fishes and their environment in the Murray-Darling river system, Australia. Environ. Biol. Fish 56, 129–151.

Hutchison, D.W., Templeton, A.R., 1999. Correlation of pairwise genetic and geographic distance measures: inferring the relative influences of gene flow and drift on the distribution of genetic variability. Evolution 53, 1898–1914.

Kerezsy, A., Balcombe, S.R., Arthington, A.H., Bunn, S.E., 2011. Continuous recruitment underpins fish persistence in the arid rivers of far-western Queensland, Australia. Mar. Freshw. Res. 62, 1178–1190.

Ledger, M., Hildrew, A., 2001. Recolonization by the benthos of an acid stream following a drought. Arch. Hydrobiol. 152, 1–17.

Ledger, M.E., Brown, L.E., Edwards, F.K., Milner, A.M., Woodward, G., 2012. Drought alters the structure and functioning of complex food webs. Nat. Clim. Chang. 3, 223–227.

Miller, K.B., 2003. The phylogeny of diving beetles (Coleoptera: Dytiscidae) and the evolution of sexual conflict. Biol. J. Linn. Soc. 79, 359–388.

Miller, A.M., Golladay, S.W., 1996. Effects of spates and drying on macroinvertebrate assemblages of an intermittent and a perennial prairie stream. J. N. Am. Benthol. Soc. 15, 670–689.

Müller, H.J., 1928. The measurement of the gene mutation rate in *Drosophila*, its high variability, and its dependence upon temperature. Genetics 13, 279–357.

Ricklefs, R.E., 2004. A comprehensive framework for global patterns in biodiversity. Ecol. Lett. 7, 1–15.

Rodríguez-Lozano, P., Verkaik, I., Rieradevall, M., Prat, N., 2015. Small but powerful: top predator local extinction affects ecosystem structure and function in an intermittent stream. PLoS One 10, e0117630.

Romaní, A.M., Amalfitano, S., Artigas, J., Fazi, S., Sabater, S., Timoner, X., et al., 2013. Microbial biofilm structure and organic matter use in mediterranean streams. Hydrobiologia 719, 43–58.

Tronstad, L., Tronstad, B.P., Benke, A.C., 2007. Aerial colonization and growth: rapid invertebrate responses to temporary aquatic habitats in a river floodplain. J. N. Am. Benthol. Soc. 26, 460–471.

ANTHROPOGENIC THREATS TO INTERMITTENT RIVERS AND EPHEMERAL STREAMS

Ming-Chih Chiu[*,†], Catherine Leigh[‡,§,¶], Raphael Mazor[**], Núria Cid[††], Vincent Resh[*]

*University of California, Berkeley, CA, United States[*] National Chung Hsing University, Taichung City, Taiwan[†]
Irstea, UR MALY, centre de Lyon-Villeurbanne, Villeurbanne, France[‡] CESAB-FRB, Immeuble Henri Poincaré,
Aix-en-Provence, France[§] Griffith University, Nathan, QLD, Australia[¶] Southern California Coastal Water Research
Project, Costa Mesa, CA, United States[**] Universitat de Barcelona (UB), Barcelona, Spain[††]*

IN A NUTSHELL

- In intermittent rivers and ephemeral streams (IRES), significant anthropogenic threats result from alterations in natural flow regimes, changes in geomorphological features, and degraded water quality. These changes may favor invasion by alien species, a major ecological problem in IRES.
- Anthropogenic threats in IRES often exceed the thresholds of native species' adaptations and can change community structure and ecological processes. Impacts may cascade to downstream waterbodies and adjacent riparian wetlands.
- Hydrology, geomorphology, and water quality impacts often occur concurrently (e.g., from mining activities) and result in multiple stressors in IRES, which can increase their vulnerability to alien species.
- IRES are under-appreciated in terms of the ecosystem services they provide, resulting in a lack of biological monitoring systems which consequently hinders effective management strategies.

5.1.1 INTRODUCTION: THE MAIN THREATS TO IRES

Intermittent rivers and ephemeral streams (IRES) occur worldwide and make important contributions to freshwater biodiversity and biogeochemical cycles at local, catchment, regional, and global scales (Chapters in Sections 1–4; Datry et al., 2014). IRES in all parts of the world are subject to many of the same human-induced threats as perennial rivers and streams. However, most IRES have far less legal protection than their perennial counterparts (Nikolaidis et al., 2013; Acuña et al., 2014; Chapter 5.3), reflecting the low value that society places on their ecological attributes and ecosystem services (Chapter 5.2). Given this lack of appreciation, IRES are probably under greater threat of degradation than perennial systems. They can serve as sites for the dumping of trash and dredging of sediment or as conduits for waste water disposal and road traffic (Fig. 5.1.1). Furthermore, the natural variability of their flow regimes and hydrological phases, which can result in extended and unpredictable periods of zero flow (Chapters 2.2 and 2.3), means that these activities, along with anthropogenic changes

FIG. 5.1.1

See legend on opposite page.

to their hydrology induced by artificial dewatering or augmented flows (Fig. 5.1.1), may be wrongly dismissed as unproblematic.

Flow intermittence and the shifting aquatic-terrestrial habitat dynamics of IRES (Chapter 4.9) present challenges to both the aquatic and terrestrial biota inhabiting these systems. The biota must either withstand the episodic unfavorable conditions such as dry habitat for aquatic biota or flooded habitat for terrestrial biota in situ (i.e., exhibit resistance) or escape them to return later when conditions again become favorable (i.e., exhibit resilience). Many taxa that inhabit IRES thus possess adaptive traits of resistance and/or resilience to flow extremes, including drying and flooding (e.g., Bêche et al., 2006; Leigh et al., 2016a; Chapter 4.8). However, anthropogenic activities that alter the natural flow regimes, geomorphology, and water quality of IRES, and even the composition of biotic communities (e.g., via introductions of alien species or clearing of vegetation) may push these ecosystems and biota beyond their adaptation thresholds.

In this chapter, we review anthropogenic threats to IRES and their biota, broadly classified into the following (Fig. 5.1.2): hydrological alterations, including withdrawals or additions of water that change inundation durations and the dynamic interplay of aquatic vs terrestrial habitat (as well as altering other ecologically relevant aspects of the flow regime (Poff et al., 2010)); physical and chemical alterations, including sedimentation, sediment mining, and water pollution that degrade aquatic and terrestrial habitats in IRES; and biological alterations, in particular the introduction or invasion of species that can threaten native communities. We discuss these threats and their individual and interactive effects on IRES, both those that undergo natural flow intermittence and those once-perennial rivers that now undergo anthropogenically induced intermittence and their adjacent ecosystems. Furthermore, we consider how climate change may interact with these threats or even induce hydrological, physical, chemical, and biological alterations directly (Box 5.1.1). Finally, we consider future needs and challenges for research and management to effectively prevent, reduce, or mitigate the impacts of anthropogenic alterations to IRES ecosystems.

5.1.2 HYDROLOGICAL ALTERATIONS IN IRES

Flow regulation, surface and groundwater abstraction, and water diversions for flood control, irrigation, and other purposes have altered the flow regimes of many rivers worldwide (Vörösmarty and Sahagian, 2000). Anthropogenic hydrological alterations in rivers are ubiquitous (Dudgeon et al., 2006), especially in regions where flow regimes are highly seasonal or variable, and aim to meet human demands for

FIG. 5.1.1
___
Examples of anthropogenic impacts that affect or create IRES: (a) an artificial IRES created by a small hydropower station in Riu Freser, Girona, Catalonia, Spain; (b) eutrophic conditions of San Juan Creek, an IRES in Orange County, California; (c) a heavily modified IRES (rambla) used for vehicular transport in Río Seco, Málaga, Spain; (d) mining activities within and surrounding Sandy Creek, an IRES in northeast Australia; (e) an IRES below a dam built by an aggregate-extraction company in Barranc de l'Estany, Tarragona, Spain; (f) invasion of *Lantana camara* throughout the riparian zone of a forested headwater IRES in Brisbane, Queensland, Australia, traversed by a well-used bush-walking track; (g) nonnative Mexican fan palms (*Washingtonia robusta*) are frequent invaders in IRES experiencing hydrological alteration in southern California: a young sapling has taken root in Jeronimo Creek (Orange County, California), adjacent to a stormwater outfall that has perennialized the creek's lower portions.

Photos: Courtesy P. Bonada (a), R. Mazor (b and g), N. Bonada (c), F. Barron (d), N. Cid/TRivers project (e) and C. Leigh (f).

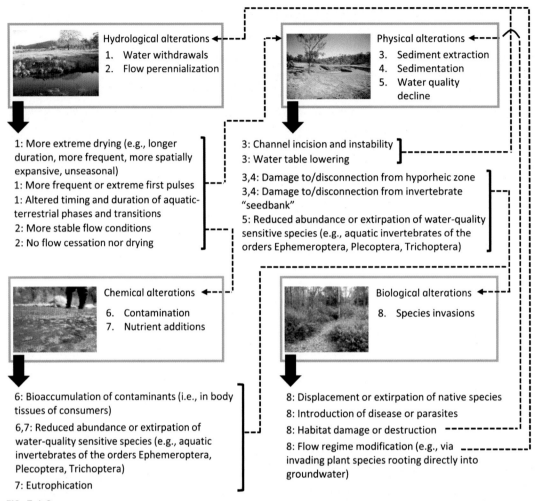

FIG. 5.1.2

The prevailing anthropogenic threats to IRES (blue-lined boxes) and interactions among them: hydrological (water withdrawals; perennialization), physical (sediment extraction; sedimentation; physical water quality decline such as hypoxia), chemical (contamination by heavy metals and other toxic substances; nutrient additions), and biological (species invasions) alterations. Thick arrows indicate potential consequences of threats, as numbered within boxes. Thin broken lines with arrowheads show where potential consequences of threats may exacerbate existing states or produce novel changes in the hydrological, physical, chemical, or biological conditions of IRES. See text for more details.

Photos: Courtesy R. Rolls (hydrological alterations; water abstraction from a small IRES in northern New South Wales, Australia), F. Barron (physical alterations; sediment mining in northeast Australian IRES), C. Leigh (biological alterations; Lantana camara invasion of an eastern Australian IRES), R. Mazor (chemical alterations; eutrophication of an IRES in western United States).

BOX 5.1.1 CLIMATE CHANGE AND IRES

IRES are especially sensitive to climate change because the durations of their wet and dry phases, and the thermal thresholds of the organisms present, are strongly affected by precipitation patterns and temperature maxima (Brooks, 2009). Ensuing hydrological changes such as increased flow intermittence (e.g., Döll and Schmied, 2012; van Vliet et al., 2013) could have similar ecological impacts (e.g., species extinctions, altered biogeochemical processes; Jaeger et al., 2014) on IRES to those of direct anthropogenic manipulations of flow involving water removal (e.g., Dewson et al., 2007).

In the dry phase of IRES, climate change may result in local extinctions, species replacement, and alteration of metapopulation dynamics (Chapter 4.9) by increasing flow intermittence and reducing habitat connectivity and availability (Bond et al., 2010; Jackson and Sax, 2010), as well as leading to changes in accumulations of organic matter on dry streambeds (Boulton and Lake, 1992; Section 5.1.5). Biogeochemical processes will be influenced by extended drying (Chapter 3.2) and the increased possibility of extreme flooding from increased precipitation (Acuña and Tockner, 2010).

Increased water temperature could have similar effects to the impacts of many anthropogenic activities (Section 5.1.3). In addition, future warming may increase instream metabolism of organic carbon in IRES, but these effects are reported to be much smaller than those caused by flow regime alterations (Acuña and Tockner, 2010). Considerable uncertainty still surrounds the likely impacts of current threats superimposed on the effects of climate change in IRES, especially where stressors are likely to combine in a nonadditive way and provide novel conditions that exceed the tolerances of native biota.

regular, reliable, and safe water supply and storage (Grey and Sadoff, 2007). Unfortunately, this means that IRES are often under threat of or subject to anthropogenic flow alteration. IRES are renowned for their flow variability (Davies et al., 1994; Puckridge et al., 1998; Poff et al., 2006; Chapter 2.2), not least because of their continual alternations between the two extremes of flow—flooding and drying—that create complex mosaics of dry channels and lentic waters interspersed among flowing reaches and floodplains (e.g., Gallart et al., 2012; Datry et al., 2016; Chapters 1 and 2.3). The importance of this natural flow variability to the ecology of river systems, including IRES, was recognized many years ago (e.g., Walker et al., 1995; Poff and Ward, 1989). Accordingly, scientists raised concerns about the threats posed by anthropogenic hydrological alterations to the ecological integrity of IRES (e.g., Walker et al., 1997; Boulton et al., 2000); these are concerns that remain to this day as discussed later.

HYDROLOGICAL ALTERATIONS ASSOCIATED WITH ANTHROPOGENIC WATER WITHDRAWALS

Water removed from rivers and aquifers by abstraction, retention, or diversion for exploitation by humans (hereafter, "water withdrawals") can alter the hydraulics and hydrology of rivers. In perennial streams and rivers, water withdrawals may cause flow cessation and drying when and where these events would not otherwise occur (e.g., exposing dry channel beds in reaches otherwise permanently watered, Benejam et al., 2010). This may also occur as a result of water withdrawals from natural IRES, along with prolonged periods of drying (King et al., 2015; Table 5.1.1). Changes in the timing, location, duration, and frequency of flow cessation and drying in natural IRES or novel flow cessation and drying in once-perennial rivers may place biota at risk of local or regional extinction (Box 5.1.1). For example, in the once-perennial Tordera River in Spain, from which 34% of its water is abstracted for human use, the channel is dry 97% of the time when it would otherwise flow and native fish species have consequently declined in number (Benejam et al., 2010). Water abstraction from IRES for irrigation is also linked to the proliferation of nonnative fish species (Godinho et al., 2014), which may then

Table 5.1.1 Hydrological alterations resulting from anthropogenic water withdrawals, specifically those inducing flow cessation or drying (From King, A.J., Townsend, S.A., Douglas, M.M., Kennard, M.J., 2015. Implications of water extraction on the low-flow hydrology and ecology of tropical savannah rivers: an appraisal for northern Australia. Freshw. Sci. 34, 741–758.), with examples of observed biotic changes that may ensue

Natural flow regime	Altered flow regime	Hydrological alteration and habitat changes	Biotic changes
Intermittent/ephemeral	Intermittent/ephemeral	Lower volumes during period of flow	Changes as per those documented for perennial rivers, below
		Earlier, longer and more spatially extensive periods of flow cessation or drying, and possible novel flow cessation or drying	Rapid recovery of macroinvertebrate communities to pre-drying composition if source(s) of colonists remain intact (Vander Vorste et al., 2016)
		Shorter persistence of isolated pool habitats	Potential collapse of freshwater food webs (Chapter 4.7) and slowed recovery of biotic communities following flow resumption
Perennial	Intermittent/ephemeral	Lower flow volumes	Increase in density of macroinvertebrates associated with decreased habitat area (Dewson et al., 2007)
		Novel flow cessation, isolated pool habitats and/or drying	Decline in proportion of intolerant fish species (e.g., those showing restricted distributions or decrease in abundances in response to anthropogenic disturbances) and benthic fish species (Benejam et al., 2010);
			Potentially no change in macroinvertebrate diversity but change in composition, e.g., proportional increase in Chironomidae taxa (Hille et al., 2014);
			Populations of fishes and other obligate lotic species eliminated during drying
		Novel flow resumption and post-drying period	Rapid recovery of macroinvertebrate communities to pre-drying composition if source(s) of colonists remain intact (Dudgeon, 1992);
			Partial or slow recovery of fish populations if source(s) of colonists remain intact (Skoulikidis et al., 2011)

place additional pressure on the native taxa (Section 5.1.4). On a regional scale, for example, the IUCN considers water abstraction one of the greatest threats to fishes of the Mediterranean basin (Smith and Darwall, 2006), a well-recognized hot spot of freshwater biodiversity (Myers et al., 2000) where many rivers undergo natural or anthropogenic flow intermittence.

The effects of anthropogenic water withdrawal on river habitat, water quality, and biota appear to mimic those described for meteorological drought (Lake, 2011) as well as natural drying (Boulton, 2003); aquatic habitats contract and become fragmented, along with declines in water quality and changes to food resource availability (Rolls et al., 2012; Section 5.1.3). Consequently, the survival, reproduction, and movement of water-dependent species are typically reduced. With complete loss of surface water, aquatic biota are eliminated from surface habitat, but they can recolonize following flow resumption (Kawanishi et al., 2013; Vander Vorste et al., 2016).

However, the rate and success of recolonization depends on the proximity and availability of aquatic refuges, the intensity and duration of drying, the traits of the taxa involved (Chapter 4.8) and the cause of intermittence (natural vs anthropogenic). In once-perennial rivers, for example, recovery of fish communities following extensive channel desiccation resulting from water abstraction may take several years of "normal" streamflow (Skoulikidis et al., 2011), whereas macroinvertebrate communities may recover rapidly after flow resumption as long as upstream sources of colonists remain intact (Dudgeon, 1992; Skoulikidis et al., 2011). However, when water quality declines (e.g., increased conductivity and temperature, decreased dissolved oxygen; Section 5.1.3) in tandem with water volume, the recovery of macroinvertebrate communities may be delayed and there may be extreme changes in community composition (Box 5.1.1). For example, temperatures exceeding 30°C, together with water abstractions and diversions for irrigation extracting over 90% of original discharge volumes and causing instream drying, shifted macroinvertebrate community composition from dominance by detritus-collecting Ephemeroptera, Plecoptera, and Trichoptera taxa to dominance by predatory insects, noninsect taxa, and algae-scraping beetles (Miller et al., 2007). In summary, the persistence of aquatic biota is threatened by persistent and intense water withdrawals that induce or prolong river drying.

HYDROLOGICAL ALTERATIONS THAT INDUCE PERENNIAL FLOW IN NATURAL IRES

Flow regulation and the addition of water to IRES can induce perennial flow in natural IRES, a process we refer to as "perennialization" and for which the causes are manifold. For example, IRES channels can be used to convey water for livestock and irrigation or other uses, particularly during dry seasons when flow would naturally cease (Gasith and Resh, 1999; Bond et al., 2010). In urban areas, impervious surfaces and landscape irrigation with imported water (Roy et al., 2009; Wenger et al., 2009) or releases of wastewater effluent (Bischel et al., 2013; Lawrence et al., 2014) can cause IRES to flow continuously. Nutrient-rich wastewater inputs to ephemeral streams of Israel and Palestine, for example, have created continuous flow and caused the rapid development of riparian vegetation (Hassan and Egozi, 2001). Removal of riparian vegetation and consequent reduction in evapotranspiration may actually induce perennialization (Ingebo, 1971; Salemi et al., 2012). For example, Hibbs et al. (2012) observed that perennialization in southern California resulted from a combination of factors, including vegetation removal, importation of water, and increased imperviousness, that together incised stream channels below the groundwater table.

Although perennialization of individual reaches has been documented for some IRES (e.g., Bond et al., 2010; Hibbs et al., 2012), the spatial extent of this phenomenon across river networks and at the global level is unknown and remains poorly understood. Stream surveys at the local level, however, have found that systems in urban and agricultural areas have higher proportions of perennial stream length than expected, relative to those in comparable natural areas, a difference attributed to anthropogenically augmented flows from discharge and runoff (SWAMP, 2011; Mazor, 2015a).

When considered as "streamflow augmentation," perennialization is commonly associated with benefits such as increased water quality, habitat area for aquatic and riparian biota, and esthetic and recreational amenity (Brooks et al., 2006). Calera Creek, an urban IRES in California that has been perennialized by tertiary-treated effluent discharged in unnatural diurnal pulses, now has "improved" water quality relative to that of its highly degraded, nonperennialized reaches upstream (Halaburka et al., 2013). Other benefits include a drastic reduction in pest mosquito populations and an overall increase in benthic macroinvertebrate abundance relative to the degraded, nonperennialized reach; however, diversity has remained relatively low (Halaburka et al., 2013). The augmentation project also involved creation of an urban park and extensive aquatic habitat restoration. Four hectares of wetland and about 20 ha of stream bank and buffer areas were revegetated, habitat was recreated for the endangered San Francisco garter snake (*Thamnophis sirtalis tetrataenia*) and the threatened Californian red-legged frog (*Rana draytonii*), and the formerly diverted and channelized lower creek was returned to its historic flowpath. Thus, the physical, biological, and social benefits of the project cannot be attributed exclusively to perennialization and research is needed to identify the ecological effects ("positive" or "negative") of perennialization on this and other IRES.

Legitimate concerns have been raised about perennialization of IRES (Brooks et al., 2006). For example, both public and ecosystem health may be at risk depending on the quality of the water perennializing the IRES (Section 5.1.3), which might contain pharmaceutical products, trace contaminants or elevated nutrient levels, or be released at temperatures dramatically different to that of the receiving waterbody (e.g., treated effluent has been recorded at temperatures up to 10°C above that of receiving stream water; Plumlee et al., 2012). Restoring a natural hydrograph (i.e., restoring periods of flow cessation and drying) to perennialized reaches of IRES may even facilitate the recovery of native assemblages. For example, mathematical modeling of fish distributions in the Broken-Boosey Creek system, a southeast Australian IRES which has been perennialized by flow regulation for over 100 years, predicts that flow restoration will facilitate the reestablishment of populations of at least two native fish species that have long been eliminated from the system (Bond et al., 2010).

5.1.3 PHYSICAL AND CHEMICAL ALTERATIONS IN IRES
EXPLOITATION OF IRES AND DRY RIVERBEDS BY HUMANS

Dry riverbeds of IRES are exploited by humans for numerous uses such as recreation and resource extraction (Steward et al., 2012). The sediment of dry channels is exposed and, depending on the IRES location and surrounding infrastructure, constitutes a readily accessible commodity for mining. This is often done using heavy machinery such as bulldozers which extract sand and gravel directly from dry riverbeds, a process known as dry-pit mining (Kondolf, 1994). The physical and ecological effects of such intense activity are diverse: riparian and instream vegetation is damaged and sediment deficits can cause lateral channel instability, bed armoring, and progressive incision both upstream and downstream (e.g., headcutting) that may then alter the frequency of floodplain inundation, lower the water table, and even destroy infrastructure such as bridges (Rinaldi et al., 2005). In the Upper Ewaso Ngiro North River Basin of Kenya, for example, sediment-mining activities are estimated to have lowered the ephemeral riverbed by up to 2m (Gichuki, 2004).

Removal of the overlying layer of moist and even dry sediment during dry-pit mining is likely to negatively affect the recolonization processes that occur in IRES upon flow resumption and habitat rewetting (Chapter 4.8). When waters recede in anthropogenically unmodified IRES, moist sediments in the surface and potentially the subsurface (e.g., the hyporheic zone) are used as refuges by many animals, including invertebrates and, in rare cases, fish that can then recolonize the surface habitat when flow returns (Kawanishi et al., 2013; Vander Vorste et al., 2016). These moist zones provide a potential refuge that promotes the survival of biota lacking life history adaptations to withstand complete drying (Stubbington, 2012; Chapter 4.3). The macroinvertebrate "seedbank" within sediments of dry riverbeds can also contribute to the recolonization and persistence of aquatic communities in many IRES (Storey and Quinn, 2008; Datry, 2012; Stubbington and Datry, 2013).

Dry riverbeds can also provide a fertile but often scarce source of land for short-term agricultural activity (e.g., Hans et al., 1999) or use as roads (Fig. 5.1.1), which can increase both the mobilization and compaction of river sediments (e.g., by releasing fine sediments that clog substrate interstices). This forms a physical barrier that restricts movement of biota into the hyporheic zone or "seedbank," limiting the functions of these refuges during surface water loss (Vadher et al., 2015). This refuge function is also threatened by livestock grazing, which in IRES and perennial rivers alike can cause bank erosion, increase soil compaction and nutrient concentrations, and remove or damage riparian vegetation (Belsky et al., 1999; Herbst et al., 2012). Even light trampling by livestock can cause compaction and destroy habitat features of dry riverbeds such as biological or salt crusts and mud cracks or curls (e.g., Steward et al., 2012; CWMW, 2015; Chapter 4.4) that support unique assemblages of terrestrial invertebrates, identified as distinct from those inhabiting adjacent upland areas (Steward et al., 2011; Chapter 4.4).

Such effects are not limited to livestock; activities of feral animals may have similar impacts on IRES sediments and habitat. For example, Doupe et al. (2009) documented the degradation of ephemeral wetlands on the floodplain of Laura River, an IRES in tropical north Australia, by feral pigs (*Sus scrofa*) whose waste excretions and foraging activities combined to destroy macrophyte communities, upheave sediments, increase turbidity by several orders of magnitude and produce anoxic conditions, pH imbalances, and nutrient enrichment that persisted for months.

IRES AND WATER QUALITY DEGRADATION

Inputs of contaminated water, excess nutrients, and sediment from anthropogenic activities threaten the water quality of rivers the world over, including IRES. In IRES, inputs that occur during flow cessation or drying are not diluted by river discharge and usually remain *in situ* until flow resumes. Alternatively, if inputs occur during flow, the contaminants, nutrients, or sediments may be transported and deposited in downstream accumulations when flow ceases. In either case, flow intermittence may prolong the exposure of riverine and riparian biota to contaminants, nutrients, and sediments for longer and at higher concentrations than might occur if flow was continuous (e.g., Resh and Jackson, 1993). In the unregulated Guadiamar River, an IRES in southwest Spain that dries completely in several reaches during summer, a major spill from a zinc and silver mine during the spring of 1998 resulted in 5 million cubic meters of toxic waste entering the system (Sola et al., 2004). Even after 2 years of restoration activities involving removal of metal-polluted sediment, fewer than half the macroinvertebrate taxa present in surrounding unaffected reaches were found in the impacted reach, whose sediments still had concentrations of heavy metals up to 120-fold higher than control concentrations (Sola et al., 2004). Metal concentrations in *Hydropsyche* caddisflies collected from the impacted reach were up to 35 times higher than those at control sites (Sola et al., 2004).

Inputs of excess nutrients and pharmaceuticals to IRES from wastewater effluents can also lead to water quality impairments such as hypoxia or elevated pH, cause eutrophication, and expose biota directly to harmful chemicals (Dodds and Welch, 2000; Camargo and Alonso, 2006). For example, the upper Arc River in Provence, France, which naturally ceases flow in several reaches during summer, receives inputs of treated and sometimes untreated sewage from wastewater treatment plants.

Therefore, its flow during this season is predominantly effluent (Comoretto and Chiron, 2005) and is also when the highest concentrations of the pharmaceuticals carbamazepine ($1.15\,\mu g\,L^{-1}$) and bezafibrate ($0.78\,\mu g\,L^{-1}$) are observed, exposing multiple generations of the river's aquatic inhabitants to these pollutants (Comoretto and Chiron, 2005).

Although few studies provide direct comparisons, the hydrology of IRES probably render them more sensitive to nutrients and other pollutants than comparably sized perennial rivers. Indeed, the risks of eutrophication are greater in the low-velocity, warm conditions that are typical of IRES during low-flow conditions and flow cessation when pools of surface water become isolated, as seen for example in the Evrotas River in Greece (Tzoraki et al., 2014). In IRES of southern California, indicators of eutrophication such as streambed-smothering growth of filamentous green *Cladophora* sp. and large fluctuations in pH and dissolved oxygen have been observed, even in the IRES with moderate nutrient concentrations (e.g., $<1\,mg\,L^{-1}$ total nitrogen; Mazor et al., 2012; Mazor et al., 2014). These IRES also receive poor biological integrity scores, suggesting that small additions of nutrients may have disproportionately large impacts on the water quality and biota of IRES (Mazor et al., 2012; Mazor et al., 2014). Similar sensitivity has been noted in the intermittent Fuirosos River in Spain, where moderate nutrient inputs (i.e., twofold and threefold increases in nitrogen and phosphorus, respectively, above ambient concentrations) shifted the taxonomic and trophic structure of the biota by producing conditions favoring organisms such as grazing snails that could take advantage of the resultant increases in algal resources and fine particulate organic matter (Sabater et al., 2011).

5.1.4 BIOLOGICAL ALTERATIONS IN IRES: INVASIVE SPECIES

Invasion of fresh waters by alien species (Fig. 5.1.1) is one of the five major threats to freshwater biodiversity (Dudgeon et al., 2006) and this is no less true for the biodiversity of IRES. Invasive species can prey upon or compete with native biota for resources and destroy native species' habitat. Because native biota can be constrained to low population densities under harsh conditions (Chesson and Huntly, 1997), such as can occur in IRES during flow cessation and drying (e.g., Fritz and Dodds, 2005; Chapter 3.1), they may be highly susceptible to extinction from predation by or competition with invasive species. For example, the recruitment and persistence of low-density populations of the native Sonora mud turtle (*Kinosternon sonoriense*) which inhabits ephemeral Arizonan IRES are threatened by the invasion of the alien crayfish (*Orconectes virilis*) because it preys on the hatchlings of this vulnerable species (Hensley et al., 2010).

Invasive species also threaten IRES via processes or behaviors that modify flow regimes or habitat. For example, the deep-rooted riparian tamarisk (*Tamarix* spp.), an alien species in the United States and Australia, grows quickly and can tolerate long periods of drought and inundation, making it well suited to colonization of riverbanks, especially along IRES (Tickner et al., 2001; Stromberg et al., 2007). Deep alluvial groundwater below IRES favors tamarisk over the shallower-rooted, native riparian plants (e.g., Stromberg et al., 2007). Because tamarisk can root directly into groundwater, it can contribute to flow cessation and river drying through significant alteration of groundwater hydrology. Once dense stands are established, their trapping and stabilization of sediment can alter fluvial geomorphology and instream habitat (Tickner et al., 2001). Feral animals and livestock that concentrate around permanent pools in IRES channels following flow cessation can trample shallow habitat, contaminate waters with feces and urine, and increase water turbidity, potentially shifting primary producer communities from dominance by aquatic macrophytes to dominance by cyanobacteria (Pettit et al., 2012).

Alien species can also introduce parasites into IRES, threatening native fauna previously unexposed (Clavero et al., 2015). For example, Stone et al. (2007) found that nonnative fishes provided a mechanism for the dispersal of alien aquatic species in IRES; species such as common carp (*Cyprinus carpio*) that hosted the parasitic Asian tapeworm (*Bothriocephalus acheilognathi*) were potential invaders of habitat along the Little Colorado River, Arizona, having survived in upstream pools over the dry period and moved downstream with subsequent freshets. Indeed, some nonnative species appear to be favored by or can at least tolerate IRES flow regimes. For example, extended drying in Californian IRES facilitates establishment and success of the invasive green sunfish (*Lepomis cyanellus*) which also tolerate the extreme high flows that occur in these systems (Bêche et al., 2009; Fig. 5.1.3).

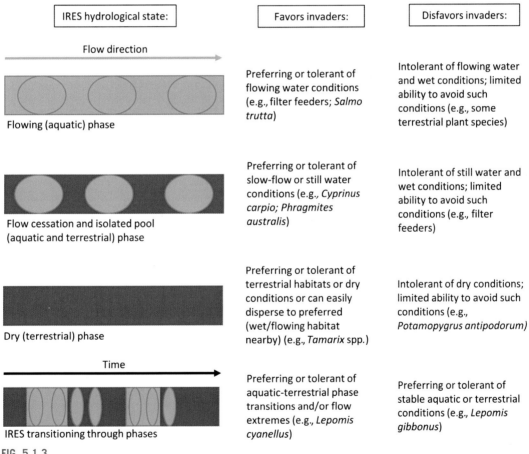

FIG. 5.1.3

Hydrological conditions of IRES favoring or disfavoring the invasion of alien species as dependent on species tolerances, intolerances or dispersal abilities, with some examples provided (for more detail, see text and Closs and Lake, 1996; Tickner et al., 2001; Koehn, 2004; Stone et al., 2007; Abramson, 2009; Arce et al., 2009; Bêche et al., 2009; Clavero et al., 2015).

Conversely, the natural flow variability of IRES may offer some protection against invasion by alien species (Costelloe et al., 2010; Fig. 5.1.3), because many alien species depend on permanent aquatic habitats. In an Australian IRES, the Lerderderg River, nonnative *Salmo trutta* had high levels of mortality during an extended dry period whereas native *Galaxias olidus* survived in isolated pools, suggesting that the nonnative species was constrained by its inability to tolerate the naturally harsh conditions that occurred during summer flow cessation and subsequent habitat fragmentation (Closs and Lake, 1996). Likewise, unregulated upstream reaches of Moroccan IRES are presently relatively free of pumpkinseed sunfish (*Lepomis gibbosus*) and other nonnative fishes, but these species appear to have successfully colonized the flow-regulated reaches downstream and consequently lowered the abundance of native fish (Clavero et al., 2015). In California, many of the most widespread invasive animals (e.g., bullfrogs *Lithobates catesbeinus*, New Zealand mudsnails *Potamopyrgus antipodarum*, mosquito fish *Gambusia affinis*) cannot tolerate lengthy dry periods. For example, in Malibu Creek in California, mudsnails have invaded nearly all perennial reaches, which means that intermittent reaches (and, in some cases, upstream perennial reaches) now remain the only uninfested portions of the watershed (Abramson, 2009). Thus, intermittent reaches may provide critical refugia from alien species, and protecting these refugia may be essential to sustaining the overall biodiversity of the watershed.

5.1.5 CASCADING EFFECTS OF HYDROLOGICAL, PHYSICAL, CHEMICAL, AND BIOLOGICAL ALTERATIONS IN IRES ON ADJACENT ECOSYSTEMS

The natural flow regimes and the presence and prevalence of IRES can strongly influence downstream ecosystems by altering the timing, quality, and quantity of sediment, water, and nutrients delivered to them (Chapters 3.1 and 3.2). During drying, large volumes of riparian leaf litter and sediments may accumulate, and terrestrial vegetation gradually replaces aquatic plants and algae in the dry riverbeds (Steward et al., 2012; Datry et al., 2014). When stream flow resumes, these materials are released and transported downstream by advancing wetted fronts (Corti and Datry, 2012) which rewet dry riverbeds and entrain terrestrial and aquatic invertebrates, suspended sediments, and both fine and particulate organic matter. For example, in the Albarine River in France, invertebrates in an advancing wetted front were found to be mainly of terrestrial origin, and entrained aquatic taxa represented ~15% of the benthic taxa downstream (Corti and Datry, 2012).

Hydrological alterations can influence these material fluxes and cause cascading ecological effects on adjacent ecosystems, including downstream aquatic and terrestrial habitats. For example, hypoxic blackwater events in IRES can occur when large amounts of organic material accumulated during natural or anthropogenic drying are mobilized and metabolized following flow resumption. These blackwater events cause dramatic declines in dissolved oxygen, killing fish and zooplankton in downstream waters (King et al., 2012; Small et al., 2014). The likelihood of these events can depend on the amount and type of accumulated carbon in the dry channel, the duration and spatial extent of the preceding zero-flow period, and the temperature and timing of rewetting flow (Acuña and Tockner, 2010; Hladyz et al., 2011; Datry et al., 2014). Prolonged drying (i.e., beyond that which would occur naturally) caused by either climate change (Box 5.1.1) or anthropogenic activities such as water withdrawals can result in greater quantities of organic material accumulating. Anthropogenic activities can also add directly to these accumulations (e.g., forestry slash entering streams, leaf fall after intentional fire).

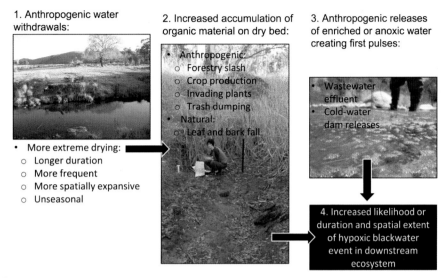

1. Anthropogenic water withdrawals:

- More extreme drying:
  o Longer duration
  o More frequent
  o More spatially expansive
  o Unseasonal

2. Increased accumulation of organic material on dry bed:

- Anthropogenic:
  o Forestry slash
  o Crop production
  o Invading plants
  o Trash dumping
- Natural:
  o Leaf and bark fall

3. Anthropogenic releases of enriched or anoxic water creating first pulses:

- Wastewater effluent
- Cold-water dam releases

4. Increased likelihood or duration and spatial extent of hypoxic blackwater event in downstream ecosystem

FIG. 5.1.4

Cascading and interactive effects of anthropogenic hydrological and chemical alterations in IRES (as depicted in Fig. 5.1.2) and potential consequences for downstream ecosystems.

Photos: Courtesy R. Rolls (anthropogenic water withdrawals from a small IRES in northern New South Wales, Australia), F. Sheldon (Catherine Leigh observing the organic matter accumulated on the dry bed of an IRES near Adelaide, Australia), R. Mazor (anthropogenic eutrophication of an IRES in California, United States). See text for more details.

Consequently, the likelihood of a hypoxic event occurring upon flow resumption increases. Higher magnitude and/or unseasonal flushing flows from effluent inputs or dam releases onto dry riverbeds could further increase the potential occurrence, extent, and duration of such events in IRES and other ecosystems downstream (Fig. 5.1.4).

Extended drying or perennialization in IRES may also have cascading effects on riparian ecosystems. Many riparian animals such as arthropods, bats, lizards, and birds rely on consumption of aquatic biota from IRES, particularly during nonflow periods when aquatic biota are concentrated in isolated pools (e.g., Leigh et al., 2013a). Hydrological alterations that cause these pools to dry early (Table 5.1.1) could thus reduce the longevity of aquatic food supply for consumers inhabiting the riparian zones of IRES. Vegetation in riparian zones may also be affected by hydrological alterations in IRES because water availability is a key driver of distribution patterns of riparian plant species (Stromberg and Merritt, 2016). For example, normalized difference vegetation index scores have shown that water loads from anthropogenic sources in southern California have not only perennialized IRES but also supplemented runoff to riparian habitats during naturally dry phases (Cervantes, 2013).

Anthropogenic alterations of flow regimes, water quality, and sediment dynamics in IRES could have cascading effects not only on neighboring (perennial, intermittent, or ephemeral) reaches and riparian zones but also on hyporheic zones, as discussed in Section 5.1.3. Beyond the hyporheic zone, first pulses in IRES can remobilize sediments with adsorbed pollutants and carry them downstream or deposit them on riverbanks and floodplains (Obermann et al., 2009). For example, in the Vène River in southern France, which receives inputs of wastewater effluent into its intermittent reaches, accumulation of anthropogenic pollutants (organotins, fecal indicator bacteria, and nutrients) in a downstream lagoon that accounts for 10% of French shellfish production occurs predominantly via sediment resuspension and pollutant remobilization during high flows (Chahinian et al., 2012). With runoff and sediment mobilization occurring only 10–15 s after rainfall starts, dry riverbed sediments are also an important source of nutrients and bacterial contamination in waters downstream from IRES, particularly in agricultural watersheds. This runoff poses risks not only to water quality and river biota but also to human health (Frey et al., 2015)

In terms of biological alterations, alien species that occupy IRES can also invade adjacent ecosystems. For example, in the Little Colorado River in Arizona, nonnative fishes such as *C. carpio* are restricted to isolated pools after flows cease, but they invade downstream perennial reaches during flash floods, threatening local native fish populations (e.g., endangered humpback chub, *Gila cypha*) through resource competition and infection by introduced parasites such as *B. acheilognathi* as discussed in Section 5.1.4 (Stone et al., 2007).

The anthropogenic hydrological, physical, chemical, and biological alterations that threaten IRES also threaten multiple adjacent ecosystems, often in complex and interactive ways (Fig. 5.1.4). Although the threats to IRES are, at some level, similar to those facing or that can cascade into perennial streams and rivers, there are some important differences. IRES are subjected to certain threats rarely experienced by perennial rivers (e.g., exploitation of the dry riverbed) and respond to other threats more acutely (e.g., sensitivity to nutrients and pulsed contaminants). IRES contribute significantly to regional biodiversity (e.g., their unique terrestrial invertebrate fauna) and play outsize roles in certain ecosystem services such as processing of nutrients and organic matter (Chapter 3.2). Effective management and condition assessment of aquatic resources in IRES are incomplete without consideration of these differences.

5.1.6 THE CHALLENGE OF ASSESSING THE BIOLOGICAL AND ECOLOGICAL CONDITION OF IRES

The management of threats and mitigation of their impacts on IRES requires the implementation of integrated assessments within monitoring programs that explicitly consider hydrological variability (Bond and Cottingham, 2008; Prat et al., 2014; Box 5.1.2; Chapter 5.5). Biological and ecological assessments are widely used in streams and rivers worldwide (Carter et al., 2006) but are still in development for IRES (Nikolaidis et al., 2013; Mazor et al., 2014; Leigh et al., 2016b). Many regional surveys of stream condition explicitly or effectively exclude IRES (e.g., Hall et al., 1998; Sheldon, 2005; Dobbie and Negus, 2013), often because information on the spatial and temporal hydrological conditions of the river is missing (i.e., when, where, and for how long does the IRES flow?). This makes it difficult or impossible for managers to use existing standard methods developed for perennial streams and rivers to assess IRES.

In regions where IRES prevail, biological and water samples may not be collected if the target reach is dry at the time of sampling. Furthermore, if samples are collected during flow or when disconnected pools are present, biological quality may not effectively represent the river's ecological condition (Sheldon, 2005; Munné and Prat, 2011; Mazor et al., 2014). For instance, biotic indices from "reference" and "least

BOX 5.1.2 IMPROVING BIOASSESSMENT OF IRES: TWO EXAMPLES

The MIRAGE toolbox (Prat et al., 2014), now incorporated into the open access software TREHS (Temporary Rivers Ecological and Hydrological Status), is a new approach developed to improve assessment of IRES in the Mediterranean basin. The toolbox allows users to classify a river's flow regime as "perennial," "intermittent-pools," "intermittent-dry," and "ephemeral/episodic" and provides information on the frequency of different aquatic states (e.g., flowing, isolated pools, dry) occurring throughout the year (Gallart et al., 2012; Table 5.1.2). It integrates data from several sources, including flow data from gauging stations, modeling, interviews, and aerial photographs (Gallart et al., 2016) and simultaneously conducts a hydrological assessment to ascertain whether flow cessation is due to natural or anthropogenic causes. This information then facilitates decisions on the potential application of existing standard methods of ecological assessment for each river (Table 5.1.2).

Demonstration of the validity of bioassessment indices in IRES is crucial to including them in management programs. For example, a recently promulgated "total maximum daily load" in Malibu Creek, California, demonstrated that poor bioassessment scores were associated with excess nutrients and sedimentation, not with the intermittent flow regimes prevalent in the watershed (USEPA, 2013). In arid southern California, an ongoing probabilistic regional stream monitoring program recently expanded its target population to include IRES (Mazor, 2015b) and will assess them with benthic macroinvertebrate and algal indices, along with riparian wetland assessment indicators (e.g., CWMW, 2015). Indicators to assess highly ephemeral portions are in development, as are predictive models to estimate flow regimes at ungauged sites. Data from these regional assessments will be used to include IRES in management programs intended to protect high-quality streams, prevent degradation relative to baseline conditions, and prioritize degraded streams for improvement.

Table 5.1.2 Potential use of existing standard methods to assess biological and ecological condition of natural IRES

Natural flow regime as defined within the MIRAGE toolbox	Can current standard biotic indices developed for perennial streams and rivers be used
Intermittent-pools: Flow cessation occurs during the dry season maintaining disconnected pools	Yes. Sampling to be conducted during flow. Sampling calendar needs to be adapted for each station and conducted in those months in which the probability of finding flow is the highest
Intermittent-dry: Flow cessation occurs during the dry season and the IRES dries completely at its surface	No. Biological indices based on aquatic fauna need to be adapted and sampling needs to be conducted during flow. Development of new methods, not based on aquatic fauna, could be an option (Chapter 4.4)
Ephemeral/episodic: Flow and disconnected pools are occasional	No. Development of new methods, not based on aquatic fauna, is required

disturbed" sites (used as standards against which the conditions of other streams are assessed) in the Mediterranean basin were so variable, even when they were only sampled while flowing, that defining environmental targets for this region's IRES became extremely difficult (Sanchez-Montoya et al., 2010; Feio et al., 2014). This difficulty is partly due to the responses of the aquatic biota inhabiting IRES to flow cessation and drying, particularly invertebrates which are commonly used as biological indicators. As discussed earlier, when surface-water pools disconnect following flow cessation, the number of invertebrate taxa usually declines (Boulton, 2003) and those adapted to flow cessation (e.g., many Odonata, Coleoptera and Heteroptera: OCH) become predominant (Bonada et al., 2006; Cid et al., 2016). This reduced taxonomic richness, together with the low values that OCH taxa often score in many biological assessment schemes and indices (Chessman, 2003; Alba Tercedor et al., 2004), essentially forces managers to assess IRES only during times of flow (Bonada et al., 2006; Buffagni et al., 2010).

The use of terrestrial or hyporheic invertebrates as alternative indicators of biological condition during dry or zero-flow periods is now being explored for IRES (e.g., Steward et al., 2012; Corti et al., 2013; Leigh et al., 2013b) and concerted efforts are leading to improvements or new practices and tools for assessing the ecological condition of IRES (Box 5.1.2). In California, for example, assessment indices based on benthic macroinvertebrates (e.g., Ode et al., 2005) or algae (e.g., Fetscher et al., 2014) have been calibrated with nominally perennial reference streams. Most of these streams exhibit dynamic flow regimes and many are actually IRES, particularly in dry years. Consequently, the indices perform well in some IRES and show potential for setting management targets for biological integrity (Mazor et al., 2014).

5.1.7 CONCLUSIONS

IRES commonly occur throughout the world, but the ecosystem services that they provide are often unappreciated (Chapter 5.2). In addition to the same hydrological, physiochemical, and biological impacts that adversely influence perennial systems, IRES are subject to specific types of impact when they cease flow and dry (e.g., dry-pit mining). Climate change may have direct and indirect effects that are analogous to anthropogenic activities, such as water withdrawals or perennialization. Inadequate public awareness and scientific understanding have hampered development of management techniques to effectively prevent, reduce, or mitigate the impacts of anthropogenic alterations on IRES ecosystems.

Increased demands for water, and consequent impacts on IRES, will be unavoidable in the future. Moreover, threats to ecosystem services (e.g., carbon storage or release, biodiversity) of IRES are poorly documented. Consequently, effective conservation of these systems will require information about additive or synergistic effects of the multiple threats that these unique systems face. Future efforts to better understand the socioeconomic value of ecosystem services may increase public awareness of the importance of IRES (Chapter 5.2). Likewise, legislative mandates may be a key component of both conservation efforts and the basis for effective management of IRES (Chapters 5.3–5.5).

ACKNOWLEDGMENTS

MCC was supported by grants from the Ministry of Science and Technology, Taiwan (Project ID: 104-2917-I-564-009). NC was supported by the EU Project LIFE+ TRIVERS (LIFE13 ENV/ES/000341).

REFERENCES

Abramson, M., 2009. Tracking the invasion of the New Zealand mudsnail, *Potamopyrgus antipodarum*, in the Santa Monica Mountains. Urban Coast 1, 21–27.

Acuña, V., Tockner, K., 2010. The effects of alterations in temperature and flow regime on organic carbon dynamics in Mediterranean river networks. Glob. Chang. Biol. 16, 2638–2650.

Acuña, V., Datry, T., Marshall, J., Barceló, D., Dahm, C.N., Ginebreda, A., et al., 2014. Why should we care about temporary waterways? Science 343, 1080–1081.

Alba Tercedor, J., Jáimez-Cuéllar, P., Álvarez, M., Avilés, J., Bonada i Caparrós, N., Casas, J., et al., 2004. Caracterización del estado ecológico de los ríos mediterráneos ibéricos mediante el índice IBMWP (antes BMWP'). Limnetica 21, 175–185.

Arce, M.I., Gomez, R., Vidal-Abarca, M.D., Suarez, M.L., 2009. Effects of *Phragmites australis* growth on nitrogen retention in a temporal stream. Limnetica 28, 229–241.

Bêche, L.A., McElravy, E.P., Resh, V.H., 2006. Long-term seasonal variation in the biological traits of benthic-macroinvertebrates in two Mediterranean-climate streams in California, U.S.A. Freshw. Biol. 51, 56–75.

Bêche, L.A., Connors, P.G., Resh, V.H., Merenlender, A.M., 2009. Resilience of fishes and invertebrates to prolonged drought in two California streams. Ecography 32, 778–788.

Belsky, A.J., Matzke, A., Uselman, S., 1999. Survey of livestock influences on stream and riparian ecosystems in the western United States. J. Soil Water Conserv. 54, 419–431.

Benejam, L., Angermeier, P.L., Munné, A., García-Berthou, E., 2010. Assessing effects of water abstraction on fish assemblages in Mediterranean streams. Freshw. Biol. 55, 628–642.

Bischel, H.N., Lawrence, J.E., Halaburka, B.J., Plumlee, M.H., Bawazir, A.S., King, J.P., et al., 2013. Renewing urban streams with recycled water for streamflow augmentation: hydrologic, water quality, and ecosystem services management. Environ. Eng. Sci. 30, 455–479.

Bonada, N., Rieradevall, M., Prat, N., Resh, V.H., 2006. Benthic macroinvertebrate assemblages and macrohabitat connectivity in Mediterranean-climate streams of northern California. J. N. Am. Benthol. Soc. 25, 32–43.

Bond, N., Cottingham, P., 2008. Ecology and hydrology of temporary streams: implications for sustainable water management. eWater Technical Report, Canberra. http://ewater.org.au/uploads/files/Bond_Cottingham-2008-Temporary_Streams.pdf Accessed 16 August 2016.

Bond, N., McMaster, D., Reich, P., Thomson, J.R., Lake, P.S., 2010. Modelling the impacts of flow regulation on fish distributions in naturally intermittent lowland streams: an approach for predicting restoration responses. Freshw. Biol. 55, 1997–2010.

Boulton, A.J., 2003. Parallels and contrasts in the effects of drought on stream macroinvertebrate assemblages. Freshw. Biol. 48, 1173–1185.

Boulton, A.J., Lake, P.S., 1992. The ecology of two intermittent streams in Victoria, Australia. III. Temporal changes in faunal composition. Freshw. Biol. 27, 123–138.

Boulton, A.J., Sheldon, F., Thoms, M.C., Stanley, E.H., 2000. Problems and constraints in managing rivers with variable flow regimes. In: Boon, P.J., Davies, B.R., Petts, G.E. (Eds.), Global Perspectives on River Conservation: Science, Policy and Practice. John Wiley and Sons, London, pp. 415–430.

Brooks, R.T., 2009. Potential impacts of global climate change on the hydrology and ecology of ephemeral freshwater systems of the forests of the northeastern United States. Clim. Chang. 95, 469–483.

Brooks, B.W., Riley, T.M., Taylor, R.D., 2006. Water quality of effluent-dominated ecosystems: ecotoxicological, hydrological, and management considerations. Hydrobiologia 556, 365–379.

Buffagni, A., Erba, S., Armanini, D.G., 2010. The lentic-lotic character of Mediterranean rivers and its importance to aquatic invertebrate communities. Aquat. Sci. 72, 45–60.

Camargo, J.A., Alonso, A., 2006. Ecological and toxicological effects of inorganic nitrogen pollution in aquatic ecosystems: a global assessment. Environ. Int. 32, 831–849.

Carter, J.M., Myers, M.J., Hannaford, M., Resh, V.H., 2006. Macroinvertebrates as biotic indicators of environmental quality. In: Hauer, F.R., Lamberti, G.A. (Eds.), Methods in Stream Ecology, second ed. Academic Press, San Diego, CA, pp. 805–833.

Cervantes, L., 2013. The Effects of Dry-Season Urban Runoff on Normalized Differential Vegetation Index by Riparian Vegetation in San Diego County, California. California State University, San Marcos, CA. Masters Dissertation.

Chahinian, N., Bancon-Montigny, C., Caro, A., Got, P., Perrin, J.L., Rosain, D., et al., 2012. The role of river sediments in contamination storage downstream of a waste water treatment plant in low flow conditions: organotins, faecal indicator bacteria and nutrients. Estuar. Coast. Shelf Sci. 114, 70–81.

Chessman, B.C., 2003. New sensitivity grades for Australian river macroinvertebrates. Mar. Freshw. Res. 54, 95–103.

Chesson, P., Huntly, N., 1997. The roles of harsh and fluctuating conditions in the dynamics of ecological communities. Am. Nat. 150, 519–553.

Cid, N., Verkaik, I., Garcia-Roger, E.M., Rieradevall, M., Bonada, N., Sanchez-Montoya, M.M., et al., 2016. A biological tool to assess flow connectivity in reference temporary streams from the Mediterranean Basin. Sci. Total Environ. 540, 178–190.

Clavero, M., Esquivias, J., Qninba, A., Riesco, M., Calzada, J., Ribeiro, F., et al., 2015. Fish invading deserts: non-native species in arid Moroccan rivers. Aquat. Conserv. Mar. Freshwat. Ecosyst. 25, 49–60.

Closs, G.P., Lake, P.S., 1996. Drought, differential mortality and the coexistence of a native and an introduced fish species in a south east Australian intermittent stream. Environ. Biol. Fish 47, 17–26.

Comoretto, L., Chiron, S., 2005. Comparing pharmaceutical and pesticide loads into a small Mediterranean river. Sci. Total Environ. 349, 201–210.

Corti, R., Datry, T., 2012. Invertebrates and sestonic matter in an advancing wetted front travelling down a dry river bed (Albarine, France). Freshwater Sci. 31, 1187–1201.

Corti, R., Larned, S.T., Datry, T., 2013. A comparison of pitfall-trap and quadrat methods for sampling ground-dwelling invertebrates in dry riverbeds. Hydrobiologia 717, 13–26.

Costelloe, J.F., Reid, J.R.W., Pritchard, J.C., Puckridge, J.T., Bailey, V.E., Hudson, P.J., 2010. Are alien fish disadvantaged by extremely variable flow regimes in arid-zone rivers? Mar. Freshw. Res. 61, 857–863.

CWMW (California Wetlands Monitoring Workgroup), 2015. California Rapid Assessment Method (CRAM), Episodic Riverine User's Manual and Field Book, version 1.0. http://www.cramwetlands.org/sites/default/files/CRAM_Fieldbook_EpisodicRiverine_v1.0.pdf. Accessed 5 November 2016.

Datry, T., 2012. Benthic and hyporheic invertebrate assemblages along a flow intermittence gradient: effects of duration of dry events. Freshw. Biol. 57, 563–574.

Datry, T., Larned, S.T., Tockner, K., 2014. Intermittent rivers: a challenge for freshwater ecology. Bioscience 64, 229–235.

Datry, T., Pella, H., Leigh, C., Bonada, N., Hugueny, B., 2016. A landscape approach to advance intermittent river ecology. Freshw. Biol. 61, 1200–1213.

Davies, B.R., Thoms, M.C., Walker, K.F., O'Keefe, J.H., Gore, J.A., 1994. Dryland rivers: their ecology, conservation and management. In: Calow, P., Petts, G.E. (Eds.), The Rivers Handbook: Hydrological and Ecological Principles. Blackwell Scientific Publishing, Oxford, pp. 484–511.

Dewson, Z.S., James, A.B.W., Death, R.G., 2007. A review of the consequences of decreased flow for instream habitat and macroinvertebrates. J. N. Am. Benthol. Soc. 26, 401–415.

Dobbie, M.J., Negus, P., 2013. Addressing statistical and operational challenges in designing large-scale stream condition surveys. Environ. Monit. Assess. 185, 7231–7243.

Dodds, W.K., Welch, E.B., 2000. Establishing nutrient criteria in streams. J. N. Am. Benthol. Soc. 19, 186–196.

Döll, P., Schmied, H.M., 2012. How is the impact of climate change on river flow regimes related to the impact on mean annual runoff? A global-scale analysis. Environ. Res. Lett. 7. 014037.

Doupe, R.G., Schaffer, J., Knott, M.J., Dicky, P.W., 2009. A description of freshwater turtle habitat destruction by feral pigs in tropical north-eastern Australia. Herpetol. Conserv. Biol. 4, 331–339.

Dudgeon, D., 1992. Effects of water transfer on aquatic insects in a stream in Hong Kong. Regul. Rivers Res. Manag. 7, 369–377.

Dudgeon, D., Arthington, A.H., Gessner, M.O., Kawabata, Z.I., Knowler, D.J., Leveque, C., et al., 2006. Freshwater biodiversity: importance, threats, status and conservation challenges. Biol. Rev. 81, 163–182.

Feio, M.J., Ferreira, J., Buffagni, A., Erba, S., Dorflinger, G., Ferreol, M., et al., 2014. Comparability of ecological quality boundaries in the Mediterranean basin using freshwater benthic invertebrates. Statistical options and implications. Sci. Total Environ. 476, 777–784.

Fetscher, A.E., Stancheva, R., Kociolek, J.P., Sheath, R.G., Stein, E.D., Mazor, R.D., et al., 2014. Development and comparison of stream indices of biotic integrity using diatoms vs. non-diatom algae vs. a combination. J. Appl. Phycol. 26, 433–450.

Frey, S.K., Gottschall, N., Wilkes, G., Gregoire, D.S., Topp, E., Pintar, K.D.M., et al., 2015. Rainfall-induced runoff from exposed streambed sediments: an important source of water pollution. J. Environ. Qual. 44, 236–247.

Fritz, K.M., Dodds, W.K., 2005. Harshness: characterisation of intermittent stream habitat over space and time. Mar. Freshw. Res. 56, 13–23.

Gallart, F., Prat, N., Garcia-Roger, E.M., Latron, J., Rieradevall, M., Llorens, P., et al., 2012. A novel approach to analysing the regimes of temporary streams in relation to their controls on the composition and structure of aquatic biota. Hydrol. Earth Syst. Sci. 16, 3165–3182.

Gallart, F., Llorens, P., Latron, J., Cid, N., Rieradevall, M., Prat, N., 2016. Validating alternative methodologies to estimate the regime of temporary rivers when flow data are unavailable. Sci. Total Environ. 565, 1001–1010.

Gasith, A., Resh, V.H., 1999. Streams in Mediterranean climate regions: abiotic influences and biotic responses to predictable seasonal events. Annu. Rev. Ecol. Syst. 30, 51–81.

Gichuki, F.N., 2004. Managing the externalities of declining dry season river flow: a case study from the Ewaso Ngiro North River Basin, Kenya. Water Resour. Res. 40. W08S03.

Godinho, F.N., Pinheiro, P.J., Oliveira, J.M., Azedo, R., 2014. Responses of intermittent stream fish assemblages to irrigation development. River Res. Appl. 30, 1248–1256.

Grey, D., Sadoff, C.W., 2007. Sink or swim? Water security for growth and development. Water Policy 9, 545–571.

Halaburka, B.J., Lawrence, J.E., Bischel, H.N., Hsiao, J., Plumlee, M.H., Resh, V.H., et al., 2013. Economic and ecological costs and benefits of streamflow augmentation using recycled water in a California coastal stream. Environ. Sci. Technol. 47, 10735–10743.

Hall, R., Husby, P., Wolinsky, G., Hansen, O., Mares, M., 1998. Site access and sample frame issues for R-EMAP Central Valley, California, stream assessment. Environ. Monit. Assess. 51, 357–367.

Hans, R.K., Farooq, M., Babu, G.S., Srivastava, S.P., Joshi, P.C., Viswanathan, P.N., 1999. Agricultural produce in the dry bed of the River Ganga in Kanpur, India—a new source of pesticide contamination in human diets. Food Chem. Toxicol. 37, 847–852.

Hassan, M.A., Egozi, R., 2001. Impact of wastewater discharge on the channel morphology of ephemeral streams. Earth Surf. Process. Landf. 26, 1285–1302.

Hensley, F.R., Jones, T.R., Maxwell, M.S., Adams, L.J., Nedella, N.S., 2010. Demography, terrestrial behavior, and growth of Sonora mud turtles (*Kinosternon sonoriense*) in an extreme habitat. Herpetol. Monogr. 24, 174–193.

Herbst, D.B., Bogan, M.T., Roll, S.K., Safford, H.D., 2012. Effects of livestock exclusion on in-stream habitat and benthic invertebrate assemblages in montane streams. Freshw. Biol. 57, 204–217.

Hibbs, B.J., Hu, W.N., Ridgway, R., 2012. Origin of stream flows at the wildlands-urban interface, Santa Monica Mountains, California, USA. Environ. Eng. Geosci. 18, 51–64.

Hille, S., Kristensen, E.A., Graeber, D., Riis, T., Jorgensen, N.K., Baattrup-Pedersen, A., 2014. Fast reaction of macroinvertebrate communities to stagnation and drought in streams with contrasting nutrient availability. Freshwater Sci. 33, 847–859.

Hladyz, S., Watkins, S.C., Whitworth, K.L., Baldwin, D.S., 2011. Flows and hypoxic blackwater events in managed ephemeral river channels. J. Hydrol. 401, 117–125.

Ingebo, P.A., 1971. Suppression of channel-side chaparral cover increases streamflow. J. Soil Water Conserv. 26, 79–81.

Jackson, S.T., Sax, D.F., 2010. Balancing biodiversity in a changing environment: extinction debt, immigration credit and species turnover. Trends Ecol. Evol. 25, 153–160.

Jaeger, K.L., Olden, J.D., Pelland, N.A., 2014. Climate change poised to threaten hydrologic connectivity and endemic fishes in dryland streams. Proc. Natl. Acad. Sci. 111, 13894–13899.

Kawanishi, R., Inoue, M., Dohi, R., Fujii, A., Miyake, Y., 2013. The role of the hyporheic zone for a benthic fish in an intermittent river: a refuge, not a graveyard. Aquat. Sci. 75, 425–431.

King, A.J., Tonkin, Z., Lieshcke, J., 2012. Short-term effects of a prolonged blackwater event on aquatic fauna in the Murray River, Australia: considerations for future events. Mar. Freshw. Res. 63, 576–586.

King, A.J., Townsend, S.A., Douglas, M.M., Kennard, M.J., 2015. Implications of water extraction on the low-flow hydrology and ecology of tropical savannah rivers: an appraisal for northern Australia. Freshw. Sci. 34, 741–758.

Koehn, J.D., 2004. Carp (*Cyprinus carpio*) as a powerful invader in Australian waterways. Freshw. Biol. 49, 882–894.

Kondolf, G.M., 1994. Geomorphic and environmental-effects of instream gravel mining. Landsc. Urban Plan. 28, 225–243.

Lake, P.S., 2011. Drought and Aquatic Ecosystems: Effects and Responses. Wiley-Blackwell, Chichester.

Lawrence, J.E., Pavia, C.P.W., Kaing, S., Bischel, H.N., Luthy, R.G., Resh, V.H., 2014. Recycled water for augmenting urban streams in mediterranean-climate regions: a potential approach for riparian ecosystem enhancement. Hydrol. Sci. J. 59, 488–501.

Leigh, C., Reis, T.M., Sheldon, F., 2013a. High potential subsidy of dry-season aquatic fauna to consumers in riparian zones of wet-dry tropical rivers. Inland Waters 3, 411–420.

Leigh, C., Stubbington, R., Sheldon, F., Boulton, A.J., 2013b. Hyporheic invertebrates as bioindicators of ecological health in temporary rivers: a meta-analysis. Ecol. Indic. 32, 62–73.

Leigh, C., Bonada, N., Boulton, A.J., Hugueny, B., Larned, S.T., Vorste, R.V., et al., 2016a. Invertebrate assemblage responses and the dual roles of resistance and resilience to drying in intermittent rivers. Aquat. Sci. 78, 291–301.

Leigh, C., Boulton, A.J., Courtwright, J.L., Fritz, K., May, C.L., Walker, R.H., et al., 2016b. Ecological research and management of intermittent rivers: an historical review and future directions. Freshw. Biol. 61, 1181–1199.

Mazor, R.D., 2015a. Bioassessment of perennial Streams in Southern California: a report on the first five years of the stormwater monitoring coalition's regional stream survey. Technical Report #844. Southern California Coastal Water Research Project, Costa Mesa, CA.

Mazor, R.D., 2015b. bioassessment survey of the stormwater monitoring coalition: workplan for years 2015 through 2019, Version 1.0. Technical Report #849. Southern California Coastal Water Research Project, Costa Mesa, CA.

Mazor, R.D., Stein, E.D., Ode, P.R., Schiff, K., 2012. Final Report on bioassessment in nonperennial streams. Technical Report #695. Southern California Coastal Water Research Project, Costa Mesa, CA.

Mazor, R.D., Stein, E.D., Ode, P.R., Schiff, K., 2014. Integrating intermittent streams into watershed assessments: applicability of an index of biotic integrity. Freshwater Sci. 33, 459–474.

Miller, S.W., Wooster, D., Li, J., 2007. Resistance and resilience of macroinvertebrates to irrigation water withdrawals. Freshw. Biol. 52, 2494–2510.

Munné, A., Prat, N., 2011. Effects of Mediterranean climate annual variability on stream biological quality assessment using macroinvertebrate communities. Ecol. Indic. 11, 651–662.

Myers, N., Mittermeier, R.A., Mittermeier, C.G., da Fonseca, G.A.B., Kent, J., 2000. Biodiversity hotspots for conservation priorities. Nature 403, 853–858.

Nikolaidis, N.P., Demetropoulou, L., Froebrich, J., Jacobs, C., Gallart, F., Prat, N., et al., 2013. Towards sustainable management of Mediterranean river basins: policy recommendations on management aspects of temporary streams. Water Policy 15, 830–849.

Obermann, M., Rosenwinkel, K.H., Tournoud, M.G., 2009. Investigation of first flushes in a medium-sized mediterranean catchment. J. Hydrol. 373, 405–415.

Ode, P.R., Rehn, A.C., May, J.T., 2005. A quantitative tool for assessing the integrity of southern coastal California streams. Environ. Manag. 35, 493–504.

Pettit, N.E., Jardine, T.D., Hamilton, S.K., Sinnamon, V., Valdez, D., Davies, P.M., et al., 2012. Seasonal changes in water quality and macrophytes and the impact of cattle on tropical floodplain waterholes. Mar. Freshw. Res. 63, 788–800.

Plumlee, M.H., Gurr, C.J., Reinhard, M., 2012. Recycled water for stream flow augmentation: benefits, challenges, and the presence of wastewater-derived organic compounds. Sci. Total Environ. 438, 541–548.

Poff, N.L., Ward, J.V., 1989. Implications of streamflow variability and predictability for lotic community structure - a regional analysis of streamflow patterns. Can. J. Fish. Aquat. Sci. 46, 1805–1818.

Poff, N.L., Olden, J.D., Pepin, D.M., Bledsoe, B.P., 2006. Placing global stream flow variability in geographic and geomorphic contexts. River Res. Appl. 22, 149–166.

Poff, N.L., Richter, B.D., Arthington, A.H., Bunn, S.E., Naiman, R.J., Kendy, E., et al., 2010. The ecological limits of hydrologic alteration (ELOHA): a new framework for developing regional environmental flow standards. Freshw. Biol. 55, 147–170.

Prat, N., Gallart, F., Von Schiller, D., Polesello, S., Garcia-Roger, E.M., Latron, J., et al., 2014. The MIRAGE toolbox: an integrated assessment tool for temporary streams. River Res. Appl. 30, 1318–1334.

Puckridge, J.T., Sheldon, F., Walker, K.F., Boulton, A.J., 1998. Flow variability and the ecology of large rivers. Mar. Freshw. Res. 49, 55–72.

Resh, V.H., Jackson, J.K., 1993. Rapid assessment approaches to biomonitoring using benthic macroinvertebrates. In: Rosenberg, D.M., Resh, V.H. (Eds.), Freshwater Biomonitoring and Benthic Macroinvertebrates. Chapman and Hall, New York, pp. 195–223.

Rinaldi, M., Wyzga, B., Surian, N., 2005. Sediment mining in alluvial channels: physical effects and management perspectives. River Res. Appl. 21, 805–828.

Rolls, R.J., Leigh, C., Sheldon, F., 2012. Mechanistic effects of low-flow hydrology on riverine ecosystems: ecological principles and consequences of alteration. Freshwater Sci. 31, 1163–1186.

Roy, A.H., Dybas, A.L., Fritz, K.M., Lubbers, H.R., 2009. Urbanization affects the extent and hydrologic permanence of headwater streams in a midwestern US metropolitan area. J. N. Am. Benthol. Soc. 28, 911–928.

Sabater, S., Artigas, J., Gaudes, A., Muñoz, I., Urrea, G., Romani, A.M., 2011. Long-term moderate nutrient inputs enhance autotrophy in a forested Mediterranean stream. Freshw. Biol. 56, 1266–1280.

Salemi, L.F., Groppo, J.D., Trevisan, R., de Moraes, J.M., Lima, W.D., Martinelli, L.A., 2012. Riparian vegetation and water yield: a synthesis. J. Hydrol. 454, 195–202.

Sanchez-Montoya, M.M., Vidal-Abarca, M.R., Suarez, M.L., 2010. Comparing the sensitivity of diverse macroinvertebrate metrics to a multiple stressor gradient in Mediterranean streams and its influence on the assessment of ecological status. Ecol. Indic. 10, 896–904.

Sheldon, F., 2005. Incorporating natural variability into the assessment of ecological health in Australian dryland rivers. Hydrobiologia 552, 45–56.

Skoulikidis, N.T., Vardakas, L., Karaouzas, I., Economou, A.N., Dimitriou, E., Zogaris, S., 2011. Assessing water stress in Mediterranean lotic systems: insights from an artificially intermittent river in Greece. Aquat. Sci. 73, 581–597.

Small, K., Kopf, R.K., Watts, R.J., Howitt, J., 2014. Hypoxia, blackwater and fish kills: experimental lethal oxygen thresholds in juvenile predatory lowland river fishes. PLoS One 9. e94524.

Smith, K.G., Darwall, W.R., 2006. The Status and Distribution of Freshwater Fish Endemic to the Mediterranean Basin. IUCN, Gland, Switzerland and Cambridge.

Sola, C., Burgos, M., Plazuelo, A., Toja, J., Plans, M., Prat, N., 2004. Heavy metal bioaccumulation and macroinvertebrate community changes in a Mediterranean stream affected by acid mine drainage and an accidental spill (Guadiamar River, SW Spain). Sci. Total Environ. 333, 109–126.

Steward, A.L., Marshall, J.C., Sheldon, F., Harch, B., Choy, S., Bunn, S.E., et al., 2011. Terrestrial invertebrates of dry river beds are not simply subsets of riparian assemblages. Aquat. Sci. 73, 551–566.

Steward, A.L., von Schiller, D., Tockner, K., Marshall, J.C., Bunn, S.E., 2012. When the river runs dry: human and ecological values of dry riverbeds. Front. Ecol. Environ. 10, 202–209.

Stone, D.M., Van Haverbeke, D.R., Ward, D.L., Hunt, T.A., 2007. Dispersal of nonnative fishes and parasites in the intermittent Little Colorado River, Arizona. Southwest. Nat. 52, 130–137.

Storey, R.G., Quinn, J.M., 2008. Composition and temporal changes in macroinvertebrate communities of intermittent streams in Hawke's Bay, New Zealand. N. Z. J. Mar. Freshw. Res. 42, 109–125.

Stromberg, J.C., Merritt, D.M., 2016. Riparian plant guilds of ephemeral, intermittent and perennial rivers. Freshw. Biol. 61, 1259–1275.

Stromberg, J.C., Lite, S.J., Marler, R., Paradzick, C., Shafroth, P.B., Shorrock, D., et al., 2007. Altered stream-flow regimes and invasive plant species: the *Tamarix* case. Glob. Ecol. Biogeogr. 16, 381–393.

Stubbington, R., 2012. The hyporheic zone as an invertebrate refuge: a review of variability in space, time, taxa and behaviour. Mar. Freshw. Res. 63, 293–311.

Stubbington, R., Datry, T., 2013. The macroinvertebrate seedbank promotes community persistence in temporary rivers across climate zones. Freshw. Biol. 58, 1202–1220.

SWAMP (Surface Water Ambient Monitoring Program), 2011. Extent of California's Perennial and Non-perennial Streams. State Water Resources Control Board, Sacramento, CA.

Tickner, D.P., Angold, P.G., Gurnell, A.M., Mountford, J.O., 2001. Riparian plant invasions: hydrogeomorphological control and ecological impacts. Prog. Phys. Geogr. 25, 22–52.

Tzoraki, O., Nikolaidis, N.P., Cooper, D., Kassotaki, E., 2014. Nutrient mitigation in a temporary river basin. Environ. Monit. Assess. 186, 2243–2257.

USEPA (United States Environmental Protection Agency), 2013. Malibu Creek and Lagoon TMDLs for Sedimentation and Nutrients to address Benthic Community Impairments. US EPA, Pacific Southwest Region, San Francisco, CA.

Vadher, A.N., Stubbington, R., Wood, P.J., 2015. Fine sediment reduces vertical migrations of *Gammarus pulex* (Crustacea: Amphipoda) in response to surface water loss. Hydrobiologia 753, 61–71.

van Vliet, M.T.H., Franssen, W.H.P., Yearsley, J.R., Ludwig, F., Haddeland, I., Lettenmaier, D.P., et al., 2013. Global river discharge and water temperature under climate change. Glob. Environ. Chang.-Hum. Policy Dimens. 23, 450–464.

Vander Vorste, R., Malard, F., Datry, T., 2016. Is drift the primary process promoting the resilience of river invertebrate communities? A manipulative field experiment in an intermittent alluvial river. Freshw. Biol. 61, 1276–1292.

Vörösmarty, C.J., Sahagian, D., 2000. Anthropogenic disturbance of the terrestrial water cycle. Bioscience 50, 753–765.

Walker, K.F., Sheldon, F., Puckridge, J.T., 1995. A perspective on dryland river ecosystems. Regul. Rivers Res. Manag. 11, 85–104.

Walker, K.F., Puckridge, J.T., Blanch, S.J., 1997. Irrigation development on Cooper Creek, central Australia—prospects for a regulated economy in a boom-and-bust ecology. Aquat. Conserv. Mar. Freshwat. Ecosyst. 7, 63–73.

Wenger, S.J., Roy, A.H., Jackson, C.R., Bernhardt, E.S., Carter, T.L., Filoso, S., et al., 2009. Twenty-six key research questions in urban stream ecology: an assessment of the state of the science. J. N. Am. Benthol. Soc. 28, 1080–1098.

ECOSYSTEM SERVICES, VALUES, AND SOCIETAL PERCEPTIONS OF INTERMITTENT RIVERS AND EPHEMERAL STREAMS

5.2

Phoebe Koundouri[*,†,‡], **Andrew J. Boulton**[§], **Thibault Datry**[¶], **Ioannis Souliotis**[‡,‖]

Athens University of Economics and Business, Athens, Greece[*] *London School of Economics, London, United Kingdom*[†] *International Centre for Research on the Environment and the Economy, Athens, Greece*[‡] *University of New England, Armidale, NSW, Australia*[§] *Irstea, UR MALY, centre de Lyon-Villeurbanne, Villeurbanne, France*[¶] *Imperial College London, London, United Kingdom*[‖]

IN A NUTSHELL

– Intermittent rivers and ephemeral streams (IRES) contribute multiple ecosystem services (ES) that are not as widely recognized or appreciated compared with those of perennially flowing waters.

– Different ES are provided by IRES at different stages of their flow regime although the inherent uncertainty of, for example, flow may influence the way people value such services.

– Also undervalued is the spatial arrangement of ES provision by a river network that comprises perennial and IRES; the suite of ES in a region is probably more relevant than those provided by individual IRES alone.

– Economic valuation should consider the use and nonuse values of different IRES and during different flow phases.

– Strategies to enhance society's appreciation of the ES provided by IRES should translate into improved legislative protection, including restoration of IRES and their ES.

5.2.1 INTRODUCTION

Ecosystem services (ES) are the benefits that people obtain that are directly attributable to the ecological functioning of ecosystems (de Groot et al., 2002). Considering ecosystems from this perspective helps practitioners understand human relationships with nature, set management priorities, and formulate environmental policies (Carpenter et al., 2009; Daily et al., 2009; Seidl, 2014). There are many ways of classifying ES (Box 5.2.1), such as the classification by the Millennium Ecosystem Assessment (MEA, 2005a) that recognizes four broad types of ES: provisioning, regulating, supporting, and cultural services. Although this classification has several serious drawbacks (Box 5.2.1), it has been widely used for the last decade and distinguishes the direct and indirect links of ES that subsequently affect their valuation. For example, provisioning services have direct links to human needs

BOX 5.2.1 CLASSIFYING ES IN FRESHWATERS

One of the most widely used classifications of ES is the one prepared by the Millennium Ecosystem Assessment (MEA 2005a,b) that allocates ES according to their functional significance into provisioning, regulating, cultural, and supporting services. In this classification, provisioning services are those that lead to products from ecosystems, and include food, fiber, and water. Regulating services provide benefits by regulating ecosystem processes such as water purification as well as ameliorating climatic extremes and disease. Cultural services are nonmaterial benefits from ecosystems and include educational, spiritual, recreational, and inspirational benefits. All three of these broad groups of services are underpinned by supporting services such as nutrient cycling and primary production (MEA, 2005a). Although this classification is widely used, it has been criticized for its ambiguity, inability to explicitly match services with people, and its double-counting of some services (Fisher et al., 2009; Landers and Nahlick, 2013). The double-counting is particularly problematic because "intermediate services" are mixed with "final services." For example, in assessing the ES from a river supporting edible herbivorous fish, the MEA (2005a) classification would distinguish a supporting service (primary production) from a provisioning service (fish) even though the supply of fish results from primary production.

These drawbacks to the MEA classification and similar ones such as the "hydrological services" framework by Brauman et al. (2007) have led to more recent schemes that consider only the end-products as ES and omit the intermediate services which are not directly consumed. One of these more recent schemes is the comprehensive classification developed for the US Environmental Protection Agency by Landers and Nahlik (2013) that identifies 352 "Final Ecosystem Goods and Services" (FEGS). These FEGS are provided by 15 environmental subclasses (e.g., rivers and streams, wetlands, lakes, and ponds) and are used by 38 subcategories of beneficiaries such as irrigators, aquaculturalists, and researchers. Flows of ES from natural systems to socioeconomic ones are mapped by matching users with FEGS in each environmental subclass in a matrix. Application of this two-way matrix approach highlights a key point often missed in discussions about ES: benefits must be known to flow from ecosystems to end-users before an ecosystem function is considered to provide an ES (Fisher et al., 2009; Schwerdtner Máñez et al., 2014). Unfortunately, our knowledge about the relationships between ecosystem functions, their linkages with benefits for humans, and their exploitation and valuation by different users is often limited, and this is especially true in IRES.

for nutrition, shelter, or safety and are relatively easy to quantify economically (Costanza et al., 2014), whereas regulating and supporting services usually have more complex links with human needs and are less easy to quantify and value. Although most cultural services are readily linked to human values and are often used to raise public support for protecting ecosystems (Gobster et al., 2007; Daniel et al., 2012), many of these ES cannot be readily quantified in monetary terms. Nonetheless, there is growing demand for their explicit incorporation into ecosystem management and environmental policy agendas (Carpenter et al., 2009; Mace, 2014).

The current paradigm underpinning ecosystem management aims at ensuring sustainable provision of ES to society while maintaining the integrity, ecological function, and biodiversity of natural and, increasingly, novel ecosystems (Hobbs et al., 2014; Mace, 2014). This paradigm has evolved through several conceptual developments over the past two decades. After the importance of ES for sustaining human well-being was initially articulated and started to become popular (Daily, 1997), the next major conceptual development was the classification of ES into groups (e.g., MEA, 2005a; reviewed in Box 5.2.1). These classifications helped practitioners organize the diversity of different ES and communicate their relationships to resource managers, politicians, and the general public. This heralded the next conceptual development which entailed economic evaluations of each ES using various systematic approaches. One example is The Economics of Ecosystems and Biodiversity (TEEB, 2010) that extended the seminal work by Costanza et al. (1997) to value the economic significance of different ES by quantifying their monetary value in different ecosystems. The latest conceptual development has been the integration of ecological understanding and economic valuation of ES into a unified perspective that is rapidly gaining importance at local, regional, and global policy levels. Recent initiatives include the establishment of the Intergovernmental Science-Policy Platform on Biodiversity and Ecosystem Services (IPBES) and the incorporation of ES in the 2020 targets set by the 10th Conference of Parties to the Convention on Biological Diversity (Carpenter et al., 2009; Larigauderie and Mooney, 2010; Matzdorf and Meyer, 2014).

In aquatic ecosystems, these conceptual developments of the ES paradigm have been evident in the increased focus on ES in strategies for management and conservation over the last two decades (Green et al. 2015; Boulton et al., 2016). All rivers and streams provide multiple ES, ranging from the supply of water for household, agricultural, and industrial uses through to the mitigation of flood and drought damage, and the provision of esthetic and recreational values (Brauman et al., 2007; Palmer et al., 2009). However, research on the provision and valuation of ES has focused almost solely on perennial rivers and streams; in stark contrast, ES and their values in intermittent rivers and ephemeral streams (hereafter, IRES) have been largely overlooked (Boulton, 2014). This oversight is surprising considering the prevalence of IRES on Earth (Chapter 1) and the physical, functional, and biological links between IRES and perennial waterways (Acuña et al., 2014; Dahm et al., 2015). Moreover, the global distribution of IRES is expanding owing to drying climates and increasing human demands for fresh water, and many once-perennial rivers are now intermittent (Larned et al., 2010; Gleick and Palaniappan, 2010; Jaeger et al., 2014). Such changes in flow regimes are likely to alter the provision of ES at local and landscape scales, especially when the biodiversity and ecosystem processes that underpin various ES are altered by increased frequency and duration of intermittence (e.g., Chapters 3.2, 4.3, and 4.10).

In the first part of this chapter, we explore the types of ES likely to be provided by IRES and how these ES might differ over time when conditions fluctuate between flowing and nonflowing phases, including loss of surface water. The second part of this chapter deals with different techniques to evaluate ES in IRES based on monetary and nonmonetary values. This valuation step is a crucial part of the ES paradigm (Haines-Young and Potschin, 2010). However, deciding which technique to use is challenging because all valuation techniques are social constructs and their "currency" is entirely dictated by human value systems and social attitudes (Larson et al., 2013; Sukhdev et al., 2014). Ignorance of the values of ES provided by IRES may be largely responsible for the widespread environmental degradation of these ecosystems (Boulton, 2014), and we urge systematic efforts to evaluate the ES of IRES globally. Our chapter concludes with speculation about the potential benefits of considering IRES within the framework of the current ES paradigm and how this perspective might guide wiser management and conservation of these undervalued ecosystems.

5.2.2 WHAT ECOSYSTEM SERVICES ARE PROVIDED BY IRES?

Although the concept of valuing natural ecosystems for their ES has been well established for over two decades (e.g., Costanza et al., 1997; Daily, 1997), there does not appear to have ever been a concerted effort to explicitly list and assess all of the ES of IRES as there has been for perennial streams and rivers. Indeed, despite considerable published research on ES, especially during the last decade, it seems that the application of the ES paradigm to IRES has been largely overlooked. For example, a literature search (Web of Science, October 15, 2015) on the topic words "ecosystem service* AND river*" yielded over 2000 titles whereas a search on "ecosystem service* AND intermittent river*" yielded only 16 titles. A closer look at these 16 publications indicates that most references to ES are tangential and none of these publications includes an attempt to list the ES of IRES or compare them with those in perennial rivers and streams.

In addition to societal ignorance leading to undervaluation of IRES (Section 5.2.4), there are other probable reasons why a list of ES explicitly for IRES has not yet been attempted. First, identifying and allocating ES is complicated by major inconsistencies in how different ES have been defined in the literature (see review in Nahlik et al., 2012). Although the four broad ES posited in the highly influential MEA (2005a) report are heuristically relevant, there is still no accepted approach for consistently identifying individual goods and services across studies of different ecosystems (Landers and Nahlik, 2013; La Notte et al., 2015). This lack of unification currently confounds the use of a universal and complete "checklist" of well-defined specific ES to apply to IRES. Instead, different countries have adopted different classifications (e.g., the UK National Ecosystem Assessment (UK NEA, 2011); the Final Ecosystem Goods and Services Classification System (Landers and Nahlik, 2013) in the United States; the Spanish National Ecosystem Assessment (SNEA, 2014)). Second, even where specific ES can be identified, establishing clear linkages between ecosystem functions and human well-being remains elusive (Ringold et al., 2013) because of our incomplete understanding of the mechanisms and processes that provide ES in many natural ecosystems. This is especially true in IRES where considerable uncertainty surrounds the influence of flow intermittence on many of the ecosystem processes that underpin ES (e.g., Chapters 3.2, 4.9, and 4.10). In addition, there is limited understanding of the interactions among these processes before and after flow ceases in IRES which further confounds assessment of the linkages between ecosystem functions and the provision of different ES at different stages of the flow regime.

Given these problems, we have adopted a very conservative approach to list the likely ES of IRES provided at different phases of their flow regime. This conservative approach acknowledges the many knowledge gaps about IRES ecosystem processes that currently hamper unequivocal allocation of ES and avoids the detailed subdivision of individual ES evident in other approaches (e.g., Landers and Nahlick, 2013). Therefore, despite its limitations (Box 5.2.1), we adopted the MEA's (2005a) heuristic classification of provisioning, regulating, supporting, and cultural services to categorize the ES provided by or derived from "wetlands." The MEA report defined wetlands in their broadest sense according to the Ramsar Convention on Wetlands which includes all perennial and nonperennial flowing and nonflowing waters. For the analyses in this chapter, we used the list tabulated in MEA (2005b) because this report focuses specifically on aquatic ES. The MEA (2005b) list omits some ES (e.g., transportation, use in mining) and suffers the broader constraints associated with the MEA (2005a) classification (Box 5.2.1). However, the list has the major advantages that (1) it covers the primary ES, (2) is a peer-reviewed and widely accepted table, and (3) was developed by an international team of respected scientists explicitly for application to wetlands defined in the broadest sense to include IRES.

To encompass the influence of flow intermittence and surface drying in IRES, we extended the MEA's (2005b) table of ES to identify which services continue to be provided when flow ceases and when surface water disappears, including some hypotheses about how different ES are altered and whether additional ES might result (Table 5.2.1). Our table provides the basis for exploring the landscape-level provision of ES that result from the "spatial mosaic" of flowing, nonflowing (pool), and surface-dry channels typical of IRES at different stages of the flow regime and of linked perennial-intermittent river networks and their adjacent riparian zones, floodplains, and alluvial groundwaters.

Table 5.2.1 ES provided by or derived from IRES

Ecosystem service	Examples	Provision according to flow phase		
		Flowing	Pools	Dry
Provisioning				
Fresh water	Surface water for domestic, industrial, and agricultural use	+	+ (but may be water quality issues)	Lost (or relies on access to subsurface water)
Food	Production of fish, wild game, fruits, and grains	+	+	Reduced or altered (some vegetation may derive water from groundwater)
Fiber and fuel	Production of logs, fuelwood, peat, and fodder	+	+	Reduced or altered (some vegetation may derive water from groundwater)
Biochemical	Extraction of medicines and other materials from biota	+	+ (if available in lentic biota)	Altered (may be derived from terrestrial biota)
Genetic materials	Genes for resistance to plant pathogens, ornamental species, etc.	+	+ (if available in lentic biota)	Lost (unless from biota that can use groundwater)
Regulating				
Climate regulation	Source of and sink for greenhouse gases; influence local and regional temperature, precipitation, and other climatic processes	+ (flow pulses may affect efflux of greenhouse gases)	+ (sediment OM important for sequestering carbon)	Altered (heat released or stored by dry channel may alter local air temperatures and humidity)
Water regulation (hydrological flows)	Groundwater recharge/discharge	+	Reduced (groundwater recharge may be reduced through loss of advection)	Lost (from surface sources)
Water purification and waste treatment	Retention, recovery, and removal of excess nutrients and other pollutants	+	Reduced or altered (loss of flow removes physical component of nutrient removal)	Reduced or altered (loss of water alters biogeochemical processes, likely reducing some purification processes)
Erosion regulation	Retention of soils and sediments	− (flow usually removes soils and sediments)	+	+ (dry channel may be sink for sediments carried by wind and other erosion)

Continued

Table 5.2.1 ES provided by or derived from IRES—cont'd

Ecosystem service	Examples	Provision according to flow phase		
		Flowing	**Pools**	**Dry**
Natural hazard regulation	Flood control, storm protection	+	+	+ (dry channel plays important role as sink for floodwaters; recharge for alluvial aquifers)
Supporting				
Soil formation	Sediment retention and accumulation of organic matter (OM)	− (flow usually removes sediments and OM)	+	+ (dry channel may be a sink for sediments and OM, especially as OM decomposition is slowed by drying (Chapter 3.2))
Nutrient cycling	Storage, recycling, processing, and acquisition of nutrients	+ (flow promotes nutrient spiraling; flow pulses affect cycling and storage (Chapter 3.2))	+	Reduced or altered (nutrient processing is slowed by drying; storage and microbial uptake altered in dry sediments)
Cultural				
Spiritual and inspirational	Many religions attach spiritual and religious values to aspects of wetlands; source of inspiration	+	+	Reduced or altered (values associated with water are lost but replaced by values associated with dry channels and gorges)
Recreational	Opportunities for recreational activities	+ (water-sports, fishing, etc.)	+ (water-sports, fishing, etc.)	Altered (walking, riding, etc., in dry channel)
Esthetic	Many people find beauty or esthetic value in aspects of wetland ecosystems (ecotourism)	+	+	Reduced (most people prefer channels containing water; however, dry gorges also attract tourists)
Educational	Opportunities for formal and informal education and training	+	+	+ (dry channels as "terra incognita")

The ES, arranged according to the four broad categories proposed by MEA (2005a), are ones listed by MEA (2005b) as provided by or derived from wetlands. For many of them, their provision (+ = provided; − = absent) varies with flow phase at the surface (defined as "flowing," "pools," and "dry"). Specific examples are discussed further in the text.

Assessments of ES must be considered from a broad perspective that explicitly acknowledges spatial variation in hydrological connectivity (Chapter 2.3) within individual IRES as well as across multiple river networks, both perennial and nonperennial.

Provisioning services from surface waters are typically lost or reduced by drying in IRES (Table 5.2.1) although alluvial water may be pumped from shallow aquifers below and along the channel. Obviously, where surface water is directly required for an ES, the service is lost when the IRES dries. However, because water quality often deteriorates when flow ceases (Chapter 3.1), some provisioning services are reduced during the nonflowing pool phase as well, even though surface water remains. It is likely that this collective loss or diminution of provisioning services underpins the societal undervaluation of IRES (Section 5.2.4) because the provisioning ES, particularly fresh water for domestic, agricultural, and industrial uses, are the most obvious ones to the public. This differential public valuation of provisioning services over regulating and supporting ES is also true for perennial streams and rivers (e.g., Gutiérrez and Alonso, 2013). Variations in hydrological phases not only alter the ecosystem processes that give rise to ES, but they can also change the accessibility to the goods derived from provisioning services. For example, in IRES, sediment extraction (Chapter 5.1) is easiest when flow has ceased and surface water is absent but flow is required to replenish mined stores via sedimentation and other geomorphological processes (Chapter 2.1).

Regulating services are lost, altered, or promoted by drying (Table 5.2.1). Recharge of shallow alluvial groundwater ceases when surface water disappears. However, the dry channel has an enhanced capacity to act as a sink for floodwaters (Fig. 5.2.1a and b) and sediments, helping to regulate the effects of erosion and natural hazards such as flooding. The pulsed flows characteristic of most IRES result in wide variation in some of the regulating ES provided during different flow phases. A promising area for future research on ES in IRES would be to assess how pulsed flow and intermittence affect the dynamics of greenhouse gases associated with aspects of local and global climate change. It is likely that the alternating redox conditions and other biogeochemical changes wrought by flow pulses (Chapter 3.1) mediate other crucial ES beyond climate regulation and the supporting service of nutrient cycling.

Provision of many supporting and cultural services also varies according to flow phase. Most are promoted during the flowing and pool phases but reduced or altered during dry phases (Table 5.2.1). For example, nutrient spiraling occurs during the flowing phase but when flow ceases, the downstream transport of nutrients in surface water ceases. When the channel dries, many of the physical and microbially mediated processes involved in nutrient cycling cease or become much slower (Amalfitano et al., 2008; Arce et al., 2014). Dry gorges of IRES are often major tourist attractions and the diverse other cultural services provided by dry channels are becoming better documented (Steward et al., 2012). When flowing, many IRES hold significant cultural values, especially in semiarid and arid regions (e.g., for indigenous peoples in central and northern Australia, Finn and Jackson, 2011).

There are two important points to consider when comparing the ES of IRES with those of perennial streams and rivers. The first is how the spatial arrangement of IRES and permanently flowing channels in a river network interact to affect the provision of ES by the different systems. The provision of many ES, especially those within the regulating and supporting categories (MEA, 2005a), is probably enhanced in both types of systems when intermittent and perennial sections intergrade. This enhancement arises from the diversity of different environmental conditions provided by the "spatial mosaic" at the landscape and catchment scale. For example, the ES of flood- and erosion-regulation (Table 5.2.1) are likely to be optimized where perennial sections can feed into IRES whose dry channels serve to buffer erosive effects and maximize bed surface area for recharge of shallow alluvial aquifers with excess floodwater (Chapter 2.3).

FIG. 5.2.1

Some examples of IRES at different hydrological phases and performing different ES. An IRES in the Flinders Ranges, South Australia, carrying floodwaters (a); dry phase in Chaki Mayu, an IRES from the tropical Piedmont of Bolivia whose dry channel can serve as a sink for nutrients (Table 5.2.1) (b); pool phases of an IRES in the karstic west coast of the South Island of New Zealand where nutrient and organic matter cycling occur (c and d).

Photos: Courtesy A. Boulton (a) and T. Datry (b–d).

The spatial arrangement of linked intermittent and perennial stream reaches also favors the life cycles of some biota. Juvenile coho salmon (*Oncorhynchus kisutch*) with access to an intermittent tributary in an Oregon river grew faster in winter than fish restricted to the perennial channel, and this tributary also harbored some of the highest densities of spawning salmon in November-December (Wigington et al., 2006). The importance of the spatial arrangement of these linkages to the provision of different ES also extends to the adjacent riparian zone and underlying alluvial groundwater of IRES. Currently, we have scant knowledge of how ES provision is governed by the spatial arrangement, connectivity, and edge dynamics of "patches" of channels with flow durations ranging from perennial to ephemeral, and this would be another promising area for future research. There is increasing interest in developing catchment-scale restoration strategies in, for example, semiarid river basins (Trabucchi et al., 2014) that recognize the spatial arrangement of river network components when establishing approaches to enhance the delivery of ES at this scale and that prioritize patches (e.g., subcatchments) for restoration according to their potential to deliver the optimum combination of key ES for the entire basin.

This leads to the second important point to consider: ES do not operate in isolation but in suites of "bundles" of co-occurring processes. These bundles of ES result in collective outcomes, requiring the negotiation of trade-offs among the desired benefits for different stakeholders (Raudsepp-Hearne et al., 2010). When comparing the provision of ES between perennial and IRES, it is more logical to compare bundles of ES than to address them individually. This is especially true for management of sociological ecosystems because optimizing one ES without affecting the provision of others is unlikely and unrealistic (Seppelt et al., 2011; Berry et al., 2015). Palmer et al. (2014) provide a typical example of this trade-off of ES bundles that occurs during restoration strategies targeting incised channels of low-order perennial, intermittent or ephemeral stream reaches in urban areas. In an effort to maximize the bundle of ES including the reduction of bank erosion and promotion of the retention of nutrients and suspended sediments, the channel is converted into a stormwater management structure designed to reduce peak flows and enhance hydraulic retention. Although this design modifies the hydrological responses during some storm events and has potential to achieve the ES of sediment retention, there is no consistent pattern of nitrogen retention or removal that would lead to net annual benefits (Palmer et al., 2014).

Finally, flow regimes and ecological conditions have been so altered in various IRES in regions such as California and the Iberian Peninsula (Arthington et al., 2014) that they now support novel ecosystems. These novel ecosystems often harbor new combinations of species and, potentially, different ecological processes from those in natural IRES, with likely implications for the provision of ES. Where alterations to flow regime, water quality, and biota are so severe that restoration back to near-natural conditions is no longer feasible, there may be impairment or even loss of particular ES. This must be assessed when undertaking the sorts of "reconciliation ecology" advocated by Moyle (2014) for restoring ES in severely altered IRES. A promising future research direction is to ascertain how the proliferation of novel ecosystems will influence provision of different ES in river networks, including hybrid systems of naturally and artificially intermittent reaches. It is possible that these novel ecosystems may have intermittent flow yet not provide the same sorts of ES as natural IRES because of crucial differences in biodiversity or the ecosystem processes that underpin the provision of various ES. Furthermore, the ES of novel ecosystems may differ from natural ecosystems in their monetary and nonmonetary values.

5.2.3 VALUING THE ES OF IRES

Putting values on ES is a challenging and controversial exercise, and economists have often been criticized for attempting to put a price tag on nature (Silvertown, 2015). However, agencies responsible for managing or conserving natural resources must decide how to allocate scarce resources of money and time (Boulton et al., 2016), often involving trade-offs as discussed earlier. As these are economic decisions based either explicitly or implicitly on society's values, some form of economic valuation is needed to justify priorities and strategies for rational management and protection programs. Economic valuation provides monetary measures of the value of ES relative to other goods and services on which individuals spend their disposal income (Farber et al., 2006).

The economic value of benefits of ES can be classified into "use" and "nonuse" values (Pearce et al., 2006). Use values are those that are derived from the actual direct (e.g., water provisioning) and indirect use (e.g., fishing and bird-watching) of an ES. Nonuse values are those that do not actually

Table 5.2.2 Relationship between the MEA (2005a) categories of ES (Table 5.2.1) of IRES and their use and nonuse economic values

Category of ecosystem service	Types of values				
	Use values		Nonuse values		
	Direct	Indirect	Option	Existence	Bequest
Provisioning	✓		✓		✓
Regulating		✓	✓		✓
Supporting		✓	✓		✓
Cultural	✓		✓	✓	✓

Cultural services encompass existence value, reflecting the benefit that humans enjoy from knowing that a resource exists. Box 5.2.2 describes the different types of economic values. Ticks indicate the type of values that may relate to the four ES.

consume an ES (e.g., esthetic value), and can be classified into option, existence, and bequest values (Table 5.2.2). Option value refers to the utility placed on maintaining or preserving a good, even if there is no likelihood of using it in the future (Pearce et al., 2006). Existence values relate to the benefit that individuals obtain from just knowing that an ecosystem and its ES exist. Finally, bequest values refer to the utility that individuals attach to ensuring that future generations will also enjoy benefits from ES.

Economic benefits from ES do not come without cost. Maintaining the level of provision of such services involves opportunity costs that concern the alternative uses of IRES. When policy options are being assessed, both costs and benefits should be considered in a cost-benefit analysis (Currie et al., 2009). Costs can be considered as "negative benefits." From this perspective, lower levels of an ES are seen as yielding lower economic benefits or imposing financial costs to society for maintaining higher levels of that ES.

Table 5.2.2 shows how each ES relates to at least three different categories of economic values. For example, although direct use values are only associated with provisioning and cultural services of IRES, humans may place nonuse values (e.g., option and bequest values) on the same ES because these benefits are also obtained. However, supporting ES are only linked to nonuse values because humans do not actively consume such services. The same could be claimed for regulating services; however, as these include flood and climate change regulation, indirect use values are also potentially relevant. Option and bequest values can be attributed to all services provided by IRES because these ES represent potential demand for which full information might not currently be available. If these values were better known and publicized, individuals might be more inclined to protect IRES in light of the wide range of known and anticipated ES instead of causing irreversible changes through activities such as, for example, mining (Chapter 5.1) to exploit a short-term and transient ES, an activity that is likely to impair provision of many other ES in the future.

Economic value does not relate to the stock of ES that is available, but to increases or decreases in the level of provisioning of ES that cause changes in the magnitude of socioeconomic benefits. Fluctuation in the level of provided benefits will create variation in the value derived from ES. However, the total value of environmental goods cannot only be attributed to ES, but also to capital made by humans because, in some cases, environmental goods are the result of combining natural and manufactured capital such as infrastructure (Bateman et al., 2011). Following this argument, although we can assign economic values to various ES, it should not be regarded that if an ES ceases to exist, individual users' utility decreases to zero (as might be the case when the flow of IRES decreases or stops). Such a belief could lead to overstating the values of ES and cast doubt on the accuracy of economic assessment. Therefore, before attempting to estimate the economic values of IRES, it is important to identify all relevant ES. This is because the type and magnitude of use and nonuse values differ among different IRES. For example, IRES far away from residential areas would be expected to generate lower direct use values compared to those closer to where individuals reside, due to higher costs of directly using the resources (e.g., water abstraction) when greater distances are involved for access to the ES. As IRES are often more abundant in less populated areas (e.g., arid and semiarid zones, Ortega et al., 2013), it is rational to assume that nonuse values are of high importance for such ecosystems. Although most individuals may live far away from IRES, they are still likely to attribute bequest and option values to them.

The identification of the relevant types of values helps to obtain robust results and defines the choice of the most relevant economic valuation technique (Table 5.2.3). This is because not all economic valuation techniques are able to elicit both use and nonuse values (Garrod and Willis 1999; Pearce et al., 2006). On one hand, *revealed preference techniques* (RP), a family of techniques that utilize the relationship between individuals' behavior and ES by using information taken from surrogate markets, can only estimate the use values of a resource (e.g., estimates of the recreational value of IRES based on the price of properties in the vicinity), but not the nonuse values attached to it. On the other hand, *stated preference techniques* (SP) are able to elicit both use and nonuse values, obtained by giving questionnaires to a representative sample of stakeholders. For this reason, SP techniques such as contingent valuation and choice modeling are more appropriate to estimate the total economic value of IRES than RP.

Table 5.2.3 The three main families of economic valuation techniques, their main application and notes on their relevance for valuing ES of IRES

Technique	Examples	Main application	Relevance for valuing ES of IRES
Revealed preference (RP) techniques (e.g., Bark-Hodgins, 2005)	Hedonic pricing Travel cost	Able to value changes in the status of IRES that have already taken place	Market data are required. These techniques are not able to estimate the value of supporting services (e.g., soil formation, nutrient cycling)
Stated preference (SP) techniques (e.g., Koundouri et al., 2014b)	Contingent valuation – Choice modeling – Choice experiment – Choice ranking	Able to estimate changes in the status of IRES that have taken or hypothetically could take place	These are more costly and time- and data-intensive measures but are able to monetize the benefits of all ES of IRES due to their survey format
Value transfer techniques (e.g., Koundouri et al., 2016)	Unit transfer Function transfer Meta-analysis	Can be used similarly to any of the previously mentioned. Initial studies might involve either RP or SP or both techniques	These techniques are easier to implement and are able to value all ES of IRES if these have previously been valued for other areas

Revealed preference techniques use market to value changes in IRES that have already occurred. On the other hand, stated preference techniques are based on data from surveys and can be used for ex ante (hypothetical changes) and ex post assessments. Value transfers can value both actual and hypothetical changes in IRES.

As IRES nonuse values are probably highly significant (discussed in detail later), successful planning of policies (Chapter 5.3) should be based on information about both use and nonuse values. One implication of failing to do so could be that use values would be favored over nonuse values, leading to inadequate economic valuation and, in worst-case scenarios, seriously compromising the future integrity of IRES and their capacity to provide other ES. Besides RP and SP techniques, *value transfer techniques* are also commonly used (Table 5.2.3). These belong to a different family of techniques that use results from earlier primary studies (either RP or SP) in other areas similar to the area under investigation. To account for socioeconomic differences, several adjustments (e.g., income differences, levels of prices, currency) are needed. Finally, other techniques exist, such as *averting behavior* based on the use of market prices to estimate the cost of investments to prevent an unwanted environmental change from occurring and *replacement cost* which is the estimated cost to replace an ES. Although several examples of the last two techniques are provided later, this chapter mainly focuses on the first three families of techniques: RP, SP, and value transfer techniques.

Aside from provisioning services, economic values associated with the other types of ES can be monetized using a number of techniques. Fig. 5.2.2 presents the associations between different types of values, ES and valuation techniques. Special attention is needed for valuing regulating and supporting

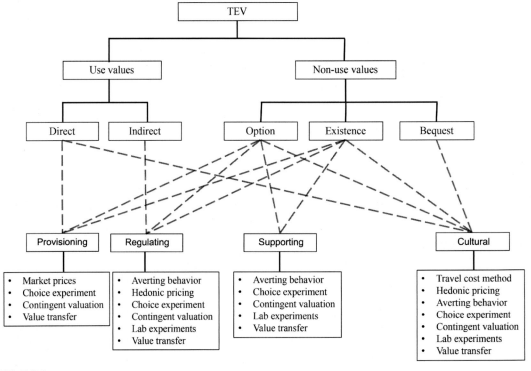

FIG. 5.2.2

Total economic value of ecosystems (TEV). The dashed lines denote the connections between the categories of economic value and the four main types of ES proposed by MEA (2005a). Under each ES category, a box presents examples of the many economic techniques that could be used to estimate economic value.

services. These types of ES are seldom adequately understood and, as a result, barely taken into account when policies are designed (Emerton et al., 2002). Generally, they are underestimated (Wood et al., 2010) and, as a consequence of this underestimation, other ES that depend on regulating and supporting processes might be threatened.

Given variations in the current management practices and the natural variability in morphological, chemical, and ecological status of IRES, economic valuation should be concerned with the impact of these changes on human welfare due to changes in provision of ES (Pagiola et al., 2004). When the ES of interest are related only to use values, RP techniques (such those mentioned earlier) are appropriate (Fig. 5.2.2). For example, Acuña et al. (2013) estimated the value of several ES in IRES using information from actual markets. These authors used the market price of brown trout (*Salmo trutta*) for estimating the value of the provision of fish and the mean market price of fishing permits was used to infer the value of opportunities of recreation. Additionally, the cost of replacing ES with technology (replacement cost technique) was used to estimate the benefits of water purification and the avoided cost technique (based on the investment cost to avoid a negative impact) was used to calculate the value of erosion control. Although several techniques were used, none of them was able to account for the nonuse values associated with four streams in the Añarbe reservoir in Spain. Further to the previously mentioned array of techniques, there are integrated models such the InVEST model (an open-source suite of tools used to map ES). These have been used to investigate the changes in the provision of ES and subsequent changes in value and land uses (e.g., Nelson et al., 2009; Sánchez-Canales et al., 2012; Bangash et al., 2013); however, in most applications, only use values are considered.

Hedonic pricing is an RP technique that has been used extensively. This technique enables the user to trace the footprint of the value of ES by observing a surrogate market. For instance, in two areas that have similar socioeconomic characteristics and differ only in the ES provided by the IRES, the value attributed by individuals to higher provision of ES can be assessed by comparing the prices of the properties in the two areas while keeping all other variables constant. Colby and Wishart (2002) used this technique to estimate the value of the Tanque Verde Wash in northeast Tuscany, which relates to ES associated with the scenic view, wildlife, and buffer from noise and pollution. House prices drop 0.45% for each 1% increase in distance from Tanque Verde, illustrating the spatial patterns of values of these ES. Another hedonic pricing study of ES in IRES (Bark-Hodgins, 2005) indicated that home-buyers assigned a higher value to the parts of the river where water flowed perennially or intermittently than ephemerally, due to the increased vegetation in the perennial and intermittent reaches. A final example is presented in Box 5.2.2 outlining the techniques used to estimate the economic value of several services provided by the Asopos river basin (Koundouri and Papandreou, 2014).

One key issue is the distinction between the total economic value of IRES that dry to pools and those that dry completely (broadly, the distinction between "intermittent" and "ephemeral", Chapter 1). To the authors' knowledge, there has not been a study that explicitly compares the economic values of the ES of these two types of IRES. However, the existence of pools secures higher levels of certain ES and their economic values than those present when the river is completely dry and some ES are even lost (Table 5.2.1). It is logical to assume that when accounting for nonuse values, the economic value of an IRES is higher during its pool phase than the dry phase, but further research is required on the economic importance of nonuse values during the dry periods. Transitioning between wet and dry phases may rejuvenate some ES that are less prevalent in perennial streams (e.g., some of the biogeochemical processes favored by wetting and drying, Chapter 3.2), and this has implications for the choice of valuation techniques that might correspond to the values associated with the different phases of IRES.

BOX 5.2.2 VALUATION OF THE ECONOMIC BENEFITS OF ES OF THE ASOPOS RIVER BASIN IN GREECE

Several techniques were used by Koundouri and Papandreou (2014) to estimate the value of different environmental goods in the Asopos River basin in Greece. The Asopos River is a typical Mediterranean IRES. It has a total length of 57 km^2 and an annual runoff of 73 hm^3. Approximately 224 industrial units (e.g., metallurgical and animal-feed industries) operate in the area (Koundouri et al., 2014a).

The study of this river basin included a Choice Experiment (CE) to estimate the economic value of higher water quality in terms of (i) environmental conditions (ecological status); (ii) impact on the local economy (tourism/recreation), local production, and cost of living of households; and (iii) impact on human health described through different water uses. CE is founded on the theory of Lancaster (1966) which proposes that each good can be described in terms of its characteristics. These characteristics are identified through stakeholder focus groups, experts' opinions, and the literature. Therefore, this technique is able to elicit preferences for use and nonuse values of an IRES.

In the case of the Asopos river, after selecting relevant attributes, defining attribute levels and choosing the design of the survey, personal interviews were carried out with local and nonlocal residents. Respondents were asked to rank alternative scenarios. Each hypothetical scenario included a combination of different attributes and suggested a price that indicated the cost that would have to be borne by the respondents. For example, respondents could choose as their least preferred, an option with no improvement of the ecological status; negative impact on the local economy; and no access to water for drinking, cooking, and irrigation at a given price over an alternative with good ecological status, positive impact on the local economy, and access to water for drinking, cooking, and irrigation at a higher price. Comparing the willingness to pay (WTP) between local residents and a sample from the population in Athens (based on the fact that Athenians either visit the basin often or own summer houses in the area), improvements in water available for all uses (drinking, cooking, agriculture) were valued at €475,000 for the local population (23,000 households) and €45,144,000 for the Athenian population (1,800,000 households) per year, whereas moderate improvements would result in values of €882,200 and €72,500,000 respectively (Koundouri et al., 2014b).

The same study in the Asopos river basin also used the value transfer technique to make an economic estimation of the benefits of mitigating industrial pollution. The value for improvements from "bad" to "very good" ecological status (Directive, 2000/60/EC) was €116.94 and from "bad" to "good" was €88.28 (per household per year in 2005 prices). Experimental techniques, such as laboratory experiments are useful to elicit WTP to avoid potential risks in relation to the use of environmental resources. To investigate whether consumers were willing to pay a price premium to avoid health risks from poor water quality in the Asopos river basin, a laboratory experiment was conducted by Drichoutis et al. (2014). A well-structured case scenario was presented to the respondents who were directed to choose their preferred option. A sample of consumers was asked to take part in three hypothetical and three actual auctions for two prizes. The subjects were given a kilogram of potatoes from a river basin with the characteristics of the Asopos case study and offered the opportunity to exchange their endowment with a kilogram of potatoes from a river basin with good water ecological status that would not entail any health risks. The results showed that when people perceived the basin to be in good ecological status and free of health risks, they were willing to bid more (60 cents per kilogram of potatoes).

If valuation techniques focus on provisioning services (e.g., fresh water) that are lost when the river dries, then techniques heavily based on market data will give a different and potentially less accurate perspective than SP that account for both use and nonuse values (Table 5.2.3). Despite this, the accuracy of SP has often been criticized due to their susceptibility to biases.

Finally, when the goods of various ES of IRES are traded in a market and thus their price is indicative of their value, all valuation techniques can be used for the monetization of relevant benefits from provisioning, regulating, and most cultural services (Koundouri et al., 2016; Table 5.2.4). Exceptions may be the supporting services and some benefits accruing from cultural services. For the estimation of generated economic value, RP should be avoided since the market fails to reveal the demand for such services because they are provided for free by IRES and therefore people are unlikely to state how much they would be willing to pay for them. Consequently, the price that would allow the internalization of

Table 5.2.4 Four types of ES (from MEA, 2005), their goods and benefits, and the nonmarket techniques used to quantify their economic values

Ecosystem services	Benefits (goods and services)	Nonmarket techniques for the quantification of economic values
Provisioning	These ES are related to goods that are of direct use for humans. Examples include water from an IRES that is used in agriculture as production input	Revealed preference Stated preference Value transfer
Regulating	These ES can be thought as benefits that control natural phenomena that help sustain or improve human life	Revealed preference Stated preference Value transfer
Supporting	These ES are vital for other ES, including ES in other ecosystems. For example, supporting ES in IRES can benefit riparian ecosystem services	Stated preference Value transfer
Cultural	These ES are nontangible goods that benefit human well-being	Revealed preference Stated preference Value transfer

Three families of nonmarket techniques (Table 5.2.3) can be used to estimate the benefits of ES in IRES.

externalities cannot be determined. To tackle this problem, SP and value transfer techniques should be used to avoid disregarding components of nonuse values. On the other hand, when the biggest share of the total economic value of the ES is attributed to use values, RP might be the best alternative. This argument is reinforced by the fact that the economic values of IRES are subject to great uncertainty arising from both uncertainty related to the timing and size of the flows as well as uncertainty and biases embedded in SP techniques. These associations between the four groups of ES proposed by MEA (2005a) and the families of economic valuation techniques are summarized in Table 5.2.4.

5.2.4 SOCIETAL PERCEPTIONS OF IRES

Societal decision-making and patterns of behavior are underpinned by perceptions (recognition and awareness of a state) and attitudes (evaluations of an object or outcome) (Kaiser et al., 1999). As IRES are readily identifiable parts of the landscape, it is reasonable to assume that most humans would recognize these ecosystems, even when dry, if the channels are well defined. When the waterways are flowing, people may not be aware that the waterways are IRES. However, as most IRES cease flow or are dry for over half the time, perceptions of IRES in the landscape can be assumed to be the norm for local landholders and most occasional visitors.

In contrast, attitudes toward IRES and their values are likely to be more diverse. Given the cryptic nature of many of the ES provided by these systems (Section 5.2.2), it would be predicted that most humans would undervalue IRES. This is especially true given the loss or reduction of provisioning services such as fresh water and food from surface waters when IRES dry (Table 5.2.1). In some areas, there is even an aversion to IRES because they are perceived as dangerous. For example, in many Mediterranean regions, the risk of flash-flooding within the channels and floodplains of ephemeral streams in peri-urban areas is mapped (e.g., Camarasa-Belmonte et al., 2011), and this is likely to influence land values and occupation patterns.

Consequently, legislation and incentives for management, protection, and restoration of IRES (Chapters 5.3–5.5) are strongly influenced by attitudes and, to a lesser degree, perceptions of these types of waterways. Surprisingly, few studies have been done to quantify human attitudes about IRES—a contrast to the rich literature on societal views on, for example, stream riparian buffer zones (e.g., Ryan et al., 2003; Kenwick et al., 2009). The most comprehensive study of human attitudes to IRES is by Armstrong et al. (2012) who interviewed landowners living along either perennial rivers or IRES in a small Pennsylvanian catchment about how "important" the stream was to them, how often it flowed ("always," "most of the time," "sometimes," "rarely"), together with other sociodemographic questions about age, education, length of ownership, and length of residence. Respondents were typically postmiddle-aged (mean = 62 years old) and had owned their riparian property for an average of 27 years. Those living along streams that flowed either always or most of the time rated the stream as significantly more important to them compared with the ratings of importance given by respondents living along IRES that flowed either "sometimes" or "rarely." Further data analysis indicated that the higher value given to perennial streams than IRES reflected a perception that water quality was better in the permanently flowing waterways (Armstrong et al., 2012), implying that land-owners' value judgments were based on provisioning ES for fresh water as well as perhaps cultural ES of recreation. The conclusions by Armstrong et al. (2012: p. 857) are blunt: "Landowner perceptions and attitudes reveal a disproportionate lack of concern towards ephemeral or intermittent streams."

Another line of evidence that IRES are less valued than perennial waterways is reflected by case studies of substantial demographic shifts when flow regimes have been changed. One example is a case study from the Água Limpa stream basin in the Jequitinhonha Valley of Minas Gerais, Brazil (Nogueira de Andrade and Leite, 2013). Over the last 50 years, stream flow (particularly base-flow during low rainfall periods) has declined associated with changes in land-use and management. Nogueira de Andrade and Leite (2013) claim that this increase in stream intermittence has been responsible for an exodus of rural residents to other parts of the country, resulting in marked socioeconomic changes in the region as rural percentages fell from 72% in the 1970s to only 53% in 1991.

One of the implications of societal attitudes and perceptions of IRES being lower than those of perennial waterways is the far greater possibility that channels of IRES will be used as convenient dumping grounds for rubbish (Fig. 5.2.3) and subjected to other intentional pollution and physical degradation (Chapter 5.1) less likely in the more-valued perennial waterways. A second implication is that the development of legislative protection for IRES will lag behind that for perennial waters (Acuña et al., 2014; Chapter 5.3). A third is the lower priority likely to be given to efforts to conserve or restore IRES, even by local landowners. Until there is a wider understanding of the ES provided by IRES leading to a change in public attitudes to these ecosystems, society will continue to undervalue IRES and our activities will continue to compromise many of the ES, especially the regulating and supporting ones listed in Table 5.2.1. Interestingly, these attitudes and perceptions of IRES closely resemble those reported from studies of perceived values of ES from terrestrial ecosystems in semiarid regions (e.g., Castro et al., 2014; Iniesta-Arandia et al., 2014). For example, in these terrestrial ecosystems, maintenance of water flow was considered to be the most important ES in both sociocultural and economic dimensions for all stakeholders.

FIG. 5.2.3

This dry channel of an ephemeral stream on a grazing property in Victoria, Australia, is used as a dumping ground for refuse despite the fact that the channel runs into a perennial waterway that provides a watering point for the farm's stock.

Photo: Courtesy A. Boulton.

5.2.5 CONCLUSIONS AND PROGNOSIS

During the flowing phase, IRES likely provide very similar ES to those found in perennial rivers. Many of these provisioning, regulating, supporting, and cultural services persist when flow ceases and, in some cases, when surface water dries in the channel. The main difference between these ecosystems is that perennial rivers typically provide most of their ES at a more or less constant rate whereas in IRES, the marked changes in the flow regime and water permanence result in variable rates and provision of many ES. This variability and sometimes complete loss of ES, such as fresh surface water, as well as the current lack of understanding of supporting and regulating services leads to society valuing IRES less than nearby perennial rivers. Another reason why IRES are undervalued is that risk-averse individuals and management agencies perceive flash-flooding as being dangerous which impacts the land uses of areas within and near such ecosystems.

Economic techniques such as RP and SP are able to elicit societal preferences and attitudes towards IRES and to express the monetary value that individuals place on them. What is important is that different techniques should be used for the economic valuation of different ES of IRES. One of our main conclusions is that very few studies of ES and their values have been undertaken in IRES, corroborating the perception that IRES have been given little attention despite their ubiquity (Chapter 1). However, water scarcity caused by climate change and other stressors, such as overextraction of water (Chapter 5.1), will probably increase the prevalence and relevance of such ecosystems and their consideration by policy makers (Chapter 5.3). In particular, climate change gives rise to synergistic effects of multiple stressors, especially during periods of water shortage (Navarro-Ortega et al., 2015), an issue that is increasingly relevant in arid and semiarid areas. The significance of studying such poorly known ecosystems lies in the fact that the exploitation of use values risks jeopardizing or even losing nonuse values that could be relatively more important.

Faced with threats of water shortage through climate change and increasing human exploitation, coupled with the widespread misperceptions about IRES, one role of science should be to address such complex socioecological systems (Folke, 2006) and communicate results to relevant stakeholders who benefit from use and nonuse values of the diverse ES provided by intact IRES. Additionally, there should be more focus on the intangible benefits and ES of IRES, such as the existence value, because these values potentially comprise the biggest share of the total economic value of environmental resources (Johnston et al., 2003). Finally, scientific results should be integrated into policies in order to design measures that can efficiently raise awareness of the ES and economic values of IRES and, at the same time, create incentives for stakeholders to use these ecosystems in a more sustainable way. Educational programs on IRES and their importance for human welfare could help improve individuals' understanding and preferences in a way that protection of IRES could be better secured by targeted legislation and legal frameworks (Chapter 5.3).

ACKNOWLEDGMENTS

We are grateful to two referees (Drs Frank Jensen and Maria Rosario Vidal-Abarca) and the Handling Editor (Dr Núria Bonada) for their very helpful comments which substantially improved the text.

REFERENCES

Acuña, V., Díez, J.R., Flores, L., Meleason, M., Elosegi, A., 2013. Does it make economic sense to restore rivers for their ecosystem services? J. Appl. Ecol. 50, 988–997.

Acuña, V., Datry, T., Marshall, J., Barceló, D., Dahm, C.N., Ginebreda, A., et al., 2014. Why should we care about temporary waterways? Science 343, 1080–1082.

Amalfitano, S., Fazi, S., Zoppini, A., Caracciolo, A.B., Grenni, P., Puddu, A., 2008. Responses of benthic bacteria to experimental drying in sediments from Mediterranean temporary rivers. Microb. Ecol. 55, 270–279.

Arce, M.I., Sanchez-Montoya, M.D., Vidal-Abarca, M.R., Suarez, M.L., Gomez, R., 2014. Implications of flow intermittency on sediment nitrogen availability and processing rates in a Mediterranean headwater stream. Aquat. Sci. 76, 173–186.

Armstrong, A., Stedman, R.C., Bishop, J.A., Sullivan, P.J., 2012. What's a stream without water? Disproportionality in headwater regions impacting water quality. Environ. Manag. 50, 849–860.

Arthington, A.H., Bernado, J.M., Ilhéu, M., 2014. Temporary rivers: linking ecohydrology, ecological quality and reconciliation ecology. River Res. Appl. 30, 1209–1215.

Bangash, R.F., Passuello, A., Sanchez-Canales, M., Terrado, M., López, A., Elorza, J., et al., 2013. Ecosystem services in Mediterranean river basin: climate change impact on water provisioning and erosion control. Sci. Total Environ. 458, 246–255.

Bark-Hodgins, R.H., 2005. Do homebuyers care about the 'quality' of natural habitats? Doctoral dissertation, School of Life Sciences, Arizona State University, Phoenix, AZ.

Bateman, I.J., Mace, G.M., Fezzi, C., Atkinson, G., Turner, K., 2011. Economic analysis for ecosystem service assessments. Environ. Resour. Econ. 48, 177–218.

Berry, P.M., Brown, S., Chen, M., Kontogianni, A., Rowlands, O., Simpson, G., et al., 2015. Cross-sectoral interactions of adaptation and mitigation measures. Clim. Chang. 128, 381–393.

Boulton, A.J., 2014. Conservation of ephemeral streams and their ecosystem services: what are we missing? Aquat. Conserv. Mar. Freshwat. Ecosyst. 24, 733–738.

Boulton, A.J., Ekebom, J., Gíslason, G.M., 2016. Integrating ecosystem services into conservation strategies for freshwater and marine habitats: a review. Aquat. Conserv. Mar. Freshwat. Ecosyst. 26, 963–985.

Brauman, K.A., Daily, G.C., Duarte, T.K., Mooney, H.A., 2007. The nature and value of ecosystem services: an overview highlighting hydrologic services. Annu. Rev. Environ. Resour. 32, 67–98.

Camarasa-Belmonte, A.M., López-García, M.J., Soriano-García, J., 2011. Mapping temporally-variable exposure to flooding in small Mediterranean basins using land-use indicators. Appl. Geogr. 31, 136–145.

Carpenter, S.R., Mooney, H.A., Agard, J., Capistrano, D., DeFries, R.S., Díaz, S., et al., 2009. Science for managing ecosystem services: Beyond the Millennium Ecosystem Assessment. Proc. Natl. Acad. Sci. 106, 1305–1312.

Castro, A.J., Verburg, P.H., Martín-López, B., Garcia-Llorente, M., Cabello, J., Vaugh, C.V., et al., 2014. Ecosystem service trade-offs from supply to social demand: a landscape-scale spatial analysis. Landsc. Urban Plan. 132, 102–110.

Colby, B.G., Wishart, S., 2002. Riparian Areas Generate Property Value Premium for Landowners. Agricultural and Resource Economics, College of Agriculture and Life Sciences, University of Arizona, Tucson, AZ.

Costanza, R., d'Arge, R., de Groot, R.S., Farber, S., Grasso, M., Hannon, B., et al., 1997. The value of the world's ecosystem services and natural capital. Nature 387, 253–260.

Costanza, R., de Groot, R., Sutton, P., van der Ploeg, S., Anderson, S.J., Kubiszewski, I., et al., 2014. Changes in the global value of ecosystem services. Glob. Environ. Chang. 26, 152–158.

Currie, B., Milton, S.J., Steenkamp, J.C., 2009. Cost–benefit analysis of alien vegetation clearing for water yield and tourism in a mountain catchment in the Western Cape of South Africa. Ecol. Econ. 68, 2574–2579.

Dahm, C.N., Candelaria-Ley, R.I., Reale, C.S., Reale, J.K., Van Horn, D.J., 2015. Extreme water quality degradation following a catastrophic forest fire. Freshw. Biol. 60, 2584–2599.

Daily, G.C., 1997. Nature's Services: Societal Dependence on Natural Ecosystems. Island Press, Washington, DC.

Daily, G.C., Polasky, S., Goldstein, J., Kareiva, P.M., Mooney, H.A., Pejchar, L., et al., 2009. Ecosystem services in decision making: time to deliver. Front. Ecol. Environ. 7, 21–28.

Daniel, T.C., Muhar, A., Arnberger, A., Aznar, O., Boyd, J.W., Chan, K.M., et al., 2012. Contributions of cultural services to the ecosystem services agenda. Proc. Natl. Acad. Sci. 109, 8812–8819.

De Groot, R.S., Wilson, M.A., Boumans, R.M.J., 2002. A typology for the classification, description and valuation of ecosystem functions, goods and services. Ecol. Econ. 41, 393–408.

Directive 2000/60. EC of the European Parliament and of the Council establishing a framework for the Community action in the field of water policy. Off. J. L 327, 1–73. http://eur-lex.europa.eu/legal-content/En/TXT/?uri=CELEX:32000L0060 Accessed 4 November 2016.

Drichoutis, A., Koundouri, P., Remoundou, K., 2014. A laboratory experiment for the estimation of health risks: policy recommendations. In: Koundouri, P., Papandreou, N.A. (Eds.), Water Resources Management Sustaining Socio-Economic Welfare. Springer, Netherlands, pp. 129–137.

Emerton, L., Seilava, R., Pearith, H., 2002. Bokor, Kirirom, Kep and Ream National Parks, Cambodia: Case studies of Economic and Development Linkages Field Study Report, Review of Protected Areas and their Role in the Socio-economic Development of the Four Countries of the Lower Mekong Region. International Centre for Environmental Management, Brisbane and IUCN - The World Conservation Union Regional Environmental Economics Programme, Karachi.

Farber, S., Costanza, R., Childers, D.L., Erickson, J., Gross, K., Grove, M., et al., 2006. Linking ecology and economics for ecosystem management. Bioscience 56, 121–133.

Finn, M., Jackson, S., 2011. Protecting indigenous values in water management: a challenge to conventional environmental flow assessments. Ecosystems 14, 1232–1248.

Fisher, B., Turner, R.K., Morling, P., 2009. Defining and classifying ecosystem services for decision making. Ecol. Econ. 68, 643–653.

Folke, C., 2006. Resilience: the emergence of a perspective for social–ecological systems analyses. Glob. Environ. Chang. 16, 253–267.

Garrod, G., Willis, K.G., 1999. Economic Valuation of the Environment: Methods and Case Studies. John Wiley and Sons, Chichester.

Gleick, P.H., Palaniappan, M., 2010. Peak water limits to freshwater withdrawal and use. Proc. Natl. Acad. Sci. 107, 11155–11162.

Gobster, P.H., Nassauer, J.I., Daniel, T.C., Fry, G., 2007. The shared landscape: what does aesthetics have to do with ecology? Landsc. Ecol. 22, 959–972.

Green, P.A., Vörösmarty, C.J., Harrison, I., Farrell, T., Sáenzc, L., Fekete, B.M., 2015. Freshwater ecosystem services supporting humans: pivoting from water crisis to water solutions. Glob. Environ. Chang. 34, 108–118.

Gutiérrez, M.R.V., Alonso, M.L.S., 2013. Which are, what is their status and what can we expect from ecosystem services provided by Spanish rivers and riparian areas? Biodivers. Conserv. 22, 2469–2503.

Haines-Young, R., Potschin, M., 2010. The links between biodiversity, ecosystem services and human well-being. In: Raffaelli, D.G., Frid, C.L.J. (Eds.), Ecosystem Ecology: A New Synthesis. Cambridge University Press, Cambridge, pp. 110–139.

Hobbs, R.J., Higgs, E., Hall, C.M., Bridgewater, P., Chapin, F.S., Ellis, E.C., et al., 2014. Managing the whole landscape: historical, hybrid, and novel ecosystems. Front. Ecol. Environ. 12, 557–564.

Iniesta-Arandia, I., García-Llorente, M., Aguilera, P.A., Montes, C., Martín-López, B., 2014. Socio-cultural valuation of ecosystem services: uncovering the links between values, drivers of change, and human well-being. Ecol. Econ. 108, 36–48.

Jaeger, K.L., Olden, J.D., Pelland, N.A., 2014. Climate change poised to threaten hydrologic connectivity and endemic fishes in dryland streams. Proc. Natl. Acad. Sci. 111, 13894–13899.

Johnston, R.J., Besedin, E.Y., Wardwell, R.F., 2003. Modeling relationships between use and nonuse values for surface water quality: a meta-analysis. Water Resour. Res. 39, 221–248.

Kaiser, F.G., Wolfing, S., Fuhrer, U., 1999. Environmental attitude and ecological behaviour. J. Environ. Psychol. 19, 1–19.

Kenwick, R.A., Shammin, R., Sullivan, W.C., 2009. Preferences for riparian buffers. Landsc. Urban Plan. 91, 88–96.

Koundouri, P., Papandreou, N.A., 2014. Water Resources Management Sustaining Socio-economic Welfare: The Implementation of the European Water Framework Directive in Asopos River Basin in Greece. vol. 7. Springer, Netherlands.

Koundouri, P., Papandreou, N., Stithou, M., Mousoulides, A., Anastasiou, Y., Mousoulidou, M., et al., 2014a. The economic characterization of Asopos River Basin. In: Koundouri, P., Papandreou, N.A. (Eds.), Water Resources Management Sustaining Socio-Economic Welfare. Springer, Netherlands, pp. 49–70.

Koundouri, P., Scarpa, R., Stithou, M., 2014b. A choice experiment for the estimation of the economic value of the river ecosystem: management policies for sustaining NATURA (2000) species and the coastal environment. In: Koundouri, P., Papandreou, N.A. (Eds.), Water Resources Management Sustaining Socio-Economic Welfare. Springer, Netherlands, pp. 101–112.

Koundouri, P., Rault, P.K., Pergamalis, V., Skianis, V., Souliotis, I., 2016. Development of an integrated methodology for the sustainable environmental and socio-economic management of river ecosystems. Sci. Total Environ. 540, 90–100.

La Notte, A., Liquete, C., Grizzetti, B., Maes, J., Egoh, B., Paracchini, M.L., 2015. An ecological-economic approach to the valuation of ecosystem services to support biodiversity policy. A case study for nitrogen retention by Mediterranean rivers and lakes. Ecol. Indic. 48, 292–302.

Lancaster, K.J., 1966. A new approach to consumer theory. J. Polit. Econ. 74, 132–157.

Landers, D.H., Nahlik, A.M., 2013. Final Ecosystem Goods and Services Classification System (FEGS-CS)EPA/600/R-13/ORD-004914. Office of Research and Development, U.S. Environmental Protection Agency, Washington, DC.

Larigauderie, A., Mooney, H.A., 2010. The Intergovernmental science-policy Platform on Biodiversity and Ecosystem Services: moving a step closer to an IPCC-like mechanism for biodiversity. Curr. Opin. Environ. Sustain. 2, 9–14.

Larned, S.T., Datry, T., Arscott, D.B., Tockner, K., 2010. Emerging concepts in temporary-river ecology. Freshw. Biol. 55, 717–738.

Larson, S., De Freitas, D.M., Hicks, C.C., 2013. Sense of place as a determinant of people's attitudes towards the environment: implications for natural resources management and planning in the Great Barrier Reef, Australia. J. Environ. Manag. 117, 226–234.

Mace, G.M., 2014. Whose conservation? Science 345, 1558–1560.

Matzdorf, B., Meyer, C., 2014. The relevance of the ecosystem services framework for developed countries' environmental policies: a comparative case study of the US and EU. Land Use Policy 38, 509–521.

Millennium Ecosystem Assessment, 2005a. Ecosystems and Human Well-being: Synthesis. World Resources Institute, Washington, DC.

Millennium Ecosystem Assessment, 2005b. Ecosystems and Human Well-being: Wetlands and Water Synthesis. World Resources Institute, Washington, DC.

Moyle, P.B., 2014. Novel aquatic ecosystems: the new reality for streams in California and other Mediterranean climate regions. River Res. Appl. 30, 1335–1344.

Nahlik, A.M., Kentula, M.E., Fennessy, M.S., Landers, D.H., 2012. Where is the consensus? A proposed foundation for moving ecosystem service concepts into practice. Ecol. Econ. 77, 27–35.

Navarro-Ortega, A., Acuña, V., Bellin, A., Burek, P., Cassiani, G., Choukr-Allah, R., et al., 2015. Managing the effects of multiple stressors on aquatic ecosystems under water scarcity. The GLOBAQUA project. Sci. Total Environ. 503, 3–9.

Nelson, E., Mendoza, G., Regetz, J., Polasky, S., Tallis, H., Cameron, D., et al., 2009. Modeling multiple ecosystem services, biodiversity conservation, commodity production, and trade-offs at landscape scales. Front. Ecol. Environ. 7, 4–11.

Nogueira de Andrade, L., Leite, M.G.P., 2013. An analysis of the human activities impact on water quantity in the Jequitinhonha Valley, MG/Brazil. Manag. Environ. Qual. 24, 383–393.

Ortega, J.A., Razola, L., Garzón, G., 2013. Recent human impacts and change in dynamics and morphology of ephemeral rivers. Nat. Haz. Hydrol. Earth Syst. Sci. Discuss. 1, 917–956.

Pagiola, S., Von Ritter, K., Bishop, J., 2004. Assessing the economic value of ecosystem conservation. Environment Department Papers, Environmental Economics Number 101, World Bank, Washington, DC. https://openknowledge.worldbank.org/handle/10986/18391 Accessed 4 November 2016.

Palmer, M.A., Lettenmaier, D.P., Poff, N.L., Postel, S.L., Richter, B., Warner, R., 2009. Climate change and river ecosystems: protection and adaptation options. Environ. Manag. 44, 1053–1068.

Palmer, M.A., Filoso, S., Fanelli, R.M., 2014. From ecosystems to ecosystem services: stream restoration as ecological engineering. Ecol. Eng. 65, 62–70.

Pearce, D., Atkinson, G., Mourato, S., 2006. Cost-benefit analysis and the environment: recent developments. Organisation for Co-operation and Development (OECD), Paris.

Raudsepp-Hearne, C., Peterson, G.D., Bennett, E.M., 2010. Ecosystem service bundles for analyzing tradeoffs in diverse landscapes. Proc. Natl. Acad. Sci. 107, 5242–5247.

Ringold, P.L., Boyd, J.W., Landers, D.H., Weber, M.A., 2013. What data should we collect? A framework for identifying indicators of ecosystem contributions to human well-being. Front. Ecol. Environ. 11, 98–105.

Ryan, R.L., Erikson, D.L., DeYoung, R., 2003. Farmers' motivations for adopting conservation practices along a riparian zone in a Midwestern agricultural watershed. J. Environ. Plan. Manag. 46, 19–37.

Sánchez-Canales, M., Benito, A.L., Passuello, A., Terrado, M., Ziv, G., Acuña, V., et al., 2012. Sensitivity analysis of ecosystem service valuation in a Mediterranean watershed. Sci. Total Environ. 440, 40–153.

Schwerdtner Máñez, K., Krause, G., Ring, I., Glaser, M., 2014. The Gordian knot of mangrove conservation: disentangling the role of scale, services and benefits. Glob. Environ. Chang. 28, 120–128.

Seidl, R., 2014. The shape of ecosystem management to come: anticipating risks and fostering resilience. Bioscience 64, 1159–1169.

Seppelt, R., Dormann, C.F., Eppink, F.V., Lautenbach, S., Schmidt, S., 2011. A quantitative review of ecosystem service studies: approaches, shortcomings and the road ahead. J. Appl. Ecol. 48, 630–636.

Silvertown, J., 2015. Have ecosystem services been oversold? Trends Ecol. Evol. 30, 641–648.

SNEA (Spanish National Ecosystem Assessment), 2014. Ecosystems and biodiversity for human wellbeing. Synthesis of the key findings. Biodiversity Foundation of the Spanish Ministry of Agriculture, Food and Environment, Madrid.

Steward, A.L., von Schiller, D., Tockner, K., Marshall, J.C., Bunn, S.E., 2012. When the river runs dry: human and ecological values of dry rivers. Front. Ecol. Environ. 10, 202–209.

Sukhdev, P., Wittmer, H., Miller, D., 2014. The economics of ecosystems and biodiversity (TEEB): challenges and responses. In: Helm, D., Hepburn, C. (Eds.), Nature in the Balance: The Economics of Biodiversity. Oxford University Press, Oxford, pp. 135–152.

TEEB, 2010. The Economics of Ecosystems and Biodiversity Ecological and Economic Foundations. In: Kumar, P. (Ed.), Earthscan, London, Washington.

Trabucchi, M., O'Farrell, P.J., Notivol, E., Comín, F.A., 2014. Mapping ecological processes and ecosystem services for prioritizing restoration efforts in a semi-arid Mediterranean river basin. Environ. Manag. 53, 1132–1145.

UK National Ecosystem Assessment (UK NEA), 2011. The UK national ecosystem assessment: synthesis of the key findings. http://uknea.unep-wcmc.org/ Accessed 4 November 2016.

Wigington, P.J., Ebersole, J.L., Colvin, M.E., Miller, B., Hansen, B., Lavigne, H., et al., 2006. Coho salmon dependence on intermittent streams. Front. Ecol. Environ. 4, 514–519.

Wood, M.D., Kumar, P., Negandhi, D., Verma, M., 2010. Guidance manual for the valuation of regulating services. UNEP, University of Liverpool, Liverpool. www.unep.org/pdf/Guidance_Manual_for_the_Regulating_Services.pdf Accessed 4 November 2016.

FURTHER READING

Apostolopoulos, K., 2010. Assessment of the Economic Importance of Water Use in the Industrial and Agricultural Areas of Asopos RB. Department of International and European Studies, Athens University of Economics and Business, Athens. http://www.aueb.gr/users/resees/en/econcharen.html (In Greek, Accessed 29.October 16).

Cummings, R.G., Taylor, L.O., 1999. Unbiased value estimates for environmental goods: a cheap talk design for the contingent valuation method. Am. Econ. Rev. 89, 649–665.

Fisher, B., Turner, R.K., 2008. Ecosystem services: classification for valuation. Biol. Conserv. 141, 1167–1169.

Kareiva, P., Watts, S., McDonald, R., Boucher, T., 2007. Domesticated nature: shaping landscapes and ecosystems for human welfare. Science 316, 1866–1869.

Maes, J., Paracchini, M.L., Zulian, G., Dunbar, M.B., Alkemade, R., 2012. Synergies and trade-offs between ecosystem service supply, biodiversity, and habitat conservation status in Europe. Biol. Conserv. 155, 1–12.

Ostrom, E., 2009. A general framework for analyzing sustainability. Science 325, 419.

GOVERNANCE, LEGISLATION, AND PROTECTION OF INTERMITTENT RIVERS AND EPHEMERAL STREAMS

5.3

Ken Fritz*, Núria Cid[†], Brad Autrey*

USEPA/ORD/NERL, Cincinnati, OH, United States[*] Universitat de Barcelona (UB), Barcelona, Spain[†]

IN A NUTSHELL

- Effective water governance requires systems-thinking and therefore recognizes intermittent rivers and ephemeral streams (IRES) as key components to river networks.
- Most water governance systems have focused more attention on perennial waterways than IRES.
- Although many existing water governance systems do not address IRES, these waterways have had a rich history of water law.
- Resetting expectations of decision makers and developing fresh approaches are needed to better integrate IRES into water governance systems to enhance the preparedness for large-scale changes in water quality and quantity.

5.3.1 INTRODUCTION

Since the dawn of civilization, water has presented society with two primary challenges: the conveyance to move water where it was needed and control to keep water from where it would cause damage. Both of these activities have construction and maintenance costs. Planning and supervising such activities were key administrative roles of many early governments and associated water law systems (Dellapenna, 2014). While many civilizations developed in valleys of perennial rivers, the water is not equally distributed. Some regions suffer from recurring droughts while others suffer from recurring floods. However, as described in Chapters 1 and 2.2, intermittent rivers and ephemeral streams (IRES) that periodically cease flowing occur in varying abundance around the globe and are connected to and influence adjacent land and downstream waters. For instance, a third (71) of the 237 nations recognized by the United Nations have at least 25% of their land within arid desert (Bw), arid steppe (Bs), and Mediterranean (Cs) Köppen climate classes, climates most likely to have IRES (Fig. 5.3.1A). Of those 71 nations, most are in Africa, Asia, and Europe (Fig. 5.3.1B), but those in the Americas (i.e., Argentina, Chile, Mexico, and the United States) and Oceania represent high proportions of the areas of those world regions.

Intermittent Rivers and Ephemeral Streams. http://dx.doi.org/10.1016/B978-0-12-803835-2.00019-X

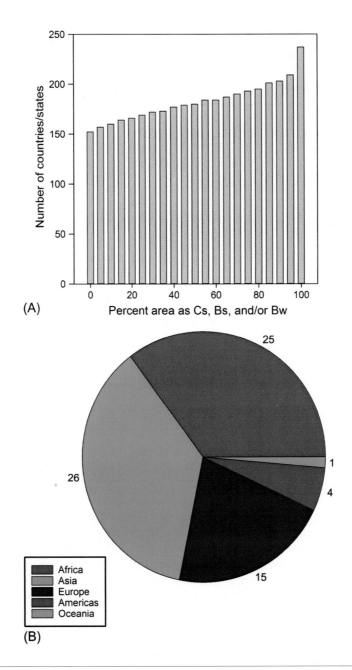

FIG. 5.3.1

Representation of arid (Bs and Bw) and Mediterranean (Cs) Köppen classes as a surrogate for IRES abundance as (A) cumulative distribution across nations by area, and (B) world regional proportions of nations with at least one-fourth of their area represented by Bs, Bw, and Cs classes.

We begin this chapter by briefly reviewing the history of water laws and their legacy in existing water governance systems. We then present a short characterization of several water governance systems, delve into specific areas where IRES are currently addressed and discuss how flow intermittence has complicated water governance and led to several legal and policy debates. Before closing, we identify where the growing knowledge about IRES can be interjected into policies that support and encourage the sustainable management of IRES.

5.3.2 BRIEF HISTORY OF WATER LAW AROUND THE WORLD

Existing water resource administration and policies reflect social history and environmental factors. All contemporary water law around the world is derived from one or more historical legal frameworks of customary law, Roman law, Islamic law, or Hindu-Buddhist law. The earliest hydraulic civilizations developed around major river valleys like the Nile, Indus, and Huang Ho. The core of Mesopotamia was the Tigris and Euphrates Rivers that provided irrigation water and, at times, transportation in the hot dry climate. The Hammurabi Code, among the earliest known documented water regulations, includes four articles on water control (Caponera and Nanni, 2007).

Customary water law reflects the traditional uses of water and determines water rights and their administration. Historically, these were uncodified and therefore not well defined. In arid North Africa and the Middle East where IRES are common, the local administration of customary water law combined with Islamic law is still prevalent (Caponera and Nanni, 2007). Fundamental to Islamic law is the idea that water should be available to all members of Moslem society. Early Hindu water law, documented in the Code of Manu, also prohibits the private ownership of water. Under Islamic law, domestic, livestock, and irrigation uses are considered substantially more important than environmental uses (Caponera and Nanni, 2007). Nomadic societies following the Islamic system have a first-come, first-served approach to water rights but those with precedence cannot refuse water to others when there is a surplus. The administration of floodwater irrigation from IRES (*wadis*) is traditionally divided into upstream, midstream, and downstream zones with each having separate water masters (*shaikh*) to manage the water resource. Under this system, the nearest to the water (i.e., the riparian landowner) gets first access, priority is based on first-established water rights among landowners, and higher elevation landowners take water before lower elevation landowners. In North Africa and the Middle East, water resource administration has largely become more centralized and adopted hybrid Islamic and European water legal systems.

In Roman law (507 BC–AD 27), the legal status of water followed land status such that where land was public, all associated water bodies were considered public. Strips of land and perennial rivers (*flumina*) were often used as property borders and so were considered public. However, the smaller IRES (*flumina torrentia*) were considered private because their streambeds were deemed too unreliable for property boundaries (Caponera and Nanni, 2007). Regulation of land use and its alteration of runoff, and therefore the formation or alteration of ephemeral channels, was addressed in Roman law (*actio aquae fluviae arcendae*) which stated that downstream landowners could take upstream landowners to court if they believed their property was impaired by upgradient changes to natural rainwater drainage (Caponera and Nanni, 2007). By the end of the Roman Empire (AD 286–565), IRES became more widely recognized as public waters.

During the Middle Ages, Roman water law merged with German traditional law which focused on water use. For instance, the *Lex Visigothorum* distinguished major rivers and immediate tributaries

FIG. 5.3.2

Map showing pathways (arrows) of historical influence for major legal systems (ovals) on water law.

Modified from Radosevich, G.E., Giner Boria, V., Daines, D.P., Skogerboe, G.V., Vlachos, E.C., 1976. International Conference on Global Water Law Systems—Summary Report. AID Publication-US Agency for International Development (USA), Number PN-AAC-592, Colorado State University, Fort Collins.

from minor watercourses. Where feudalism became the dominant social system in Western Europe, the nobility held land and water rights in exchange for legal and military obligations. From the 11th-century Feudal revolution in France until 1910, only navigable and floatable rivers were considered public waterways. By 1992, the legal definition of public waters in France expanded to include all surface and ground waters (Caponera and Nanni, 2007). While these and other examples show the imprints of past governments on existing water laws and governance systems (Fig. 5.3.2), we describe later how these historical policies imported to different regions have become increasingly inappropriate as changes in climate and population size have affected water scarcity (Falkenmark et al., 2014).

5.3.3 IRES RECOGNITION IN WATER AND ENVIRONMENTAL LAW
INTERNATIONAL TREATIES: THE RAMSAR CONVENTION

The Ramsar Convention was held in 1971 in Ramsar, Iran, to provide a framework for national action and international cooperation for the conservation and wise use of wetlands and their resources. The Ramsar treaty broadly defines wetlands as:

"… areas of marsh, fen, peatland or water, whether natural or artificial, permanent or temporary, with water that is static or flowing, fresh, brackish or salt, including areas of marine water the depth of which at low tide does not exceed six metres." (Ramsar Convention, 1971, Article 1.1)

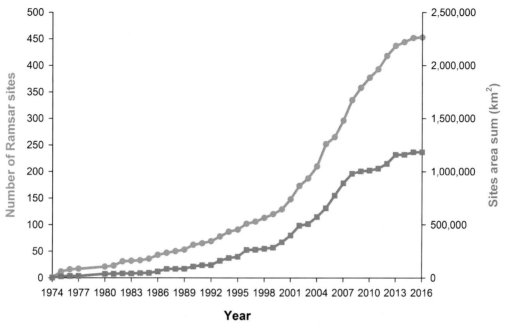

FIG. 5.3.3

Cumulative annual number (orange) and area (gray) of Ramsar sites that contain the wetland type "seasonal/intermittent/irregular rivers/streams/creeks."

Data from Ramsar Sites Information Service, 2016. https://rsis.ramsar.org/ Accessed 20 April 2016.

In 1990, the creation of a wetland classification system was recommended in the Fourth Meeting of the Conference of the Contracting Parties (Matteus, 1993). The classification system includes "Seasonal/intermittent/irregular rivers/streams/creeks" among the listed inland wetland types. According to the Ramsar database, 457 Ramsar sites have been classified into this type, mostly in Africa (Figs. 5.3.3 and 5.3.4). However, this wetland habitat lacks a specific definition and most Ramsar sites under this classification may also include lakes, meadows, coastal lagoons, wetlands, and oases. This classification was adopted by the International Union for Conservation of Nature (IUCN) and is currently used to describe the major habitats in which taxa included in the Red List occur (http://www.iucnredlist.org/technical-documents/classification-schemes/habitats-classification-scheme-ver3). Regarding the management implications of the Ramsar treaty, the Convention's Contracting Parties have assumed a wide range of related obligations. For example, in the European Union (EU), most Ramsar sites are covered by the European Habitats Directive (Council Directive 92/43/EEC) and Water Framework Directive (European Commission, 2000). A second example is the Etosha Pan, Lake Oponono and Cuvelai drainage, a Ramsar site in Namibia designated as National Park in 1995 (Box 5.3.1).

UNITED STATES

The basic tenets of United States surface water regulations are founded in British Common Law which arose primarily from court or magistrate decisions to resolve disputes among parties. These rules were used in the United States as a matter of tradition and many have been formally codified by statute over the years.

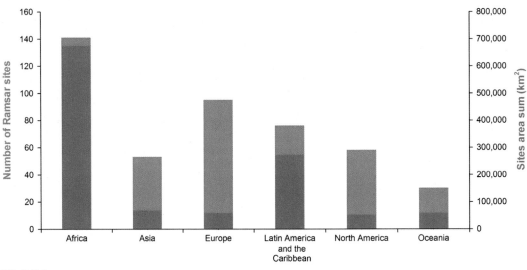

FIG. 5.3.4

Number (orange) and area (gray) of Ramsar sites by world region that contain the wetland type "seasonal/intermittent/irregular rivers/streams/creeks."

Data from Ramsar Sites Information Service, 2016. https://rsis.ramsar.org/ Accessed 20 April 2016.

Two different doctrines of water law, each reflecting the relative scarcity of water of the region that adopted it, have been produced in the United States: Riparian Rights and Prior Appropriation. Under the Riparian Rights doctrine, common in the more mesic eastern United States, if land abuts a body of water, the land owner may use water from it so long as the water use is reasonable and for the land with the water rights. Water is not properly owned until extracted and using water does not grant future water ownership. However, for the more arid western United States, the tradition of Prior Appropriation is more common. Under Prior Appropriation, parties who first put the water to beneficial use have the rights to it. Water usage under this first-in-time, first-in-right doctrine does not need to be associated with the land from which it was taken. This doctrine could have some bearing on IRES since those with first rights to the water might be in a position to put great pressure on the resource (e.g., Knox, 2001).

BOX 5.3.1 IRES UNDER THE RAMSAR CONVENTION—ETOSHA PAN, LAKE OPONONO, AND CUVELAI DRAINAGE

Etosha Pan, Lake Oponono, and Cuvelai drainage is an African Ramsar site in Namibia where the wetland type "seasonal/intermittent/irregular rivers/streams/creeks" is predominant (Ramsar Sites Information Service, 2016). The Cuvelia system terminates at Etosha Pan (6000 km^2) which is fed by numerous ephemeral rivers and is seasonally connected to Lake Oponono (70 km north of the pan) by the Ekuma River. The system supports 45% of Namibia's human population, who exist by fishing and subsistence farming on the floodplains and seasonal wetlands. The site supports populations of several large mammals, such as springbok (*Antidorcas marsupialis*) and gemsbok (*Oryx gazelle*) (Fig. 5.3.5a), and endangered or rare species such as black rhinoceros (*Diceros bicornis*), African elephant (*Loxodonta africana*, Fig. 5.3.5b), and roan antelope (*Hippotragus equinus*). In exceptional rainy seasons, Etosha Pan also serves as a breeding ground for waterbirds, including flamingos (*Phoenicopterus ruber* and *Phoeniconaias minor*).

BOX 5.3.1 IRES UNDER THE RAMSAR CONVENTION—ETOSHA PAN, LAKE OPONONO, AND CUVELAI DRAINAGE *(Cont'd)*

FIG. 5.3.5

Large mammals in Etosha National Park: (a) springbok (*Antidorcas marsupialis*) and gemsbok (*Oryx gazelle*) and (b) African elephant (*Loxodonta africana*).

Photos: Courtesy M.M. Sánchez-Montoya.

The 1948 Federal Water Pollution Control Act was enacted to give power to the states to control water pollution. The 1972 amendment to the Federal Water Pollution Control Act became known as the Clean Water Act (CWA). The US Environmental Protection Agency (EPA) and the US Army Corps of Engineers (ACE) coadminister the CWA and its goal is to "restore and maintain the chemical, physical, and biological integrity of the Nation's waters." Because of ambiguity in the statute, an area of concern that arose was how to determine the boundaries of federal jurisdiction over water bodies under the CWA (e.g., Broderick, 2005).

There has been a series of high-level court cases addressing legal questions about the extent of CWA jurisdiction. In the most recent case (Rapanos v. United States, 547 U.S. 715, 2006), the Supreme Court handed down a complex splintered decision. While all justices agreed "waters of the United States" extend beyond traditionally navigable waters and their adjacent wetlands, two different standards for determining CWA jurisdiction emerged for waters not considered traditionally navigable: Justice Kennedy's "significant nexus" standard and the plurality's "relatively permanent" standard. The plurality advocated limiting CWA jurisdiction to wetlands that are adjacent to and have continuous surface connections with relatively permanent bodies of water connected to traditional navigable waters. This standard explicitly excludes from CWA jurisdiction IRES that do not have at least seasonal flow. Under the significant nexus standard, a water body has CWA jurisdiction if it "either alone or in combination with similarly situated lands in the region" significantly affects the chemical, physical, and biological integrity of waters more readily understood as navigable. The dissenting opinion concluded that the agencies' interpretation of "waters of the United States" to include all tributaries and their adjacent wetlands was a reasonable interpretation of the CWA. The decision left the United States without a majority opinion to definitively differentiate between waters that fall under CWA jurisdiction and those that do not. After the decision, joint agency guidance stated that jurisdiction under the CWA exists if either the plurality's or Justice Kennedy's standard is satisfied.

In 2015, the EPA and ACE published and put into effect a rule (Clean Water Rule) to clarify what are considered jurisdictional waters under the CWA. The rule recognizes legal uncertainty surrounding CWA jurisdiction and seeks to more precisely define and protect tributaries that impact the health of downstream waters. The rule establishes that all tributaries of traditional navigable waters, interstate waters, or territorial seas, regardless of flow duration or size, having physical features of channelized flowing water (i.e., bed, bank, and "ordinary high water mark") are categorically jurisdictional under CWA. In developing this rule, the EPA and ACE used scientific evidence, including a report (USEPA, 2015) that summarizes >1200 peer-reviewed studies and concludes that small streams, regardless of their flow permanence, and wetlands are important to the integrity of larger, downstream water bodies. However, at the time of writing (October 2016), the rule has been suspended by the courts pending the outcome of litigation, and agencies are using their prior regulations and guidance documents for jurisdictional determinations.

According to the CWA, states, territories, and authorized tribes (hereafter "states") must develop, revise, and adopt water quality standards that must be approved by the EPA. State water quality standards inform water quality goals for waterbodies or waterbody segments, permitted effluent limits and antidegradation policy. States may develop water quality standards that are more stringent than required by EPA regulation and may have state definitions of regulated waters that are broader than federal CWA jurisdiction. The primary components of state water quality standards are the establishment of waterbody-designated uses and setting criteria to protect those designated uses. Generally, designated uses describe recreational (e.g., fishing, swimming), ecological (e.g., propagation of aquatic life), public drinking water supply, agricultural, industrial, or navigational purposes.

IRES are recognized within state water quality standards in two ways. The first is whether or not a state defines IRES in their water quality standards. Among the 56 US states and territories, 17 define ephemeral waterways and 20 define intermittent waterways in water quality standards (Table 5.3.1). Most states differentiate perennial streams from IRES by the presence of year-round surface flow and distinguish ephemeral and intermittent flow based on their flow source or bed elevation relative to the groundwater table. However, specific state definitions vary. South Carolina

Table 5.3.1 US states and territories with definitions for intermittent and ephemeral waterways and associated designated uses in their water quality standards

State	Defined		Designated uses	
	Intermittent	Ephemeral	Intermittent	Ephemeral
Alabama	X		X	
Alaska				
Arizona	X	X	X	X
Arkansas			X[a]	
California	X[b]	X[b]		
Colorado		X		
Connecticut				
Delaware	X	X		
Florida				
Georgia				
Hawaii	X[c]			
Idaho	X		X	
Illinois				
Indiana				
Iowa			X	X
Kansas				
Kentucky	X			
Louisiana	X			
Maine				
Maryland	X			
Massachusetts				
Michigan				
Minnesota			X	
Mississippi		X		X
Missouri			X	X
Montana	X	X		
Nebraska				
Nevada				
New Hampshire				
New Jersey	X			
New Mexico	X	X	X	X
New York				
North Carolina	X	X		
North Dakota				
Ohio				
Oklahoma		X	X	X
Oregon				

Continued

Table 5.3.1 US states and territories with definitions for intermittent and ephemeral waterways and associated designated uses in their water quality standards—cont'd

State	Defined		Designated uses	
	Intermittent	**Ephemeral**	**Intermittent**	**Ephemeral**
Pennsylvania				
Rhode Island				
South Carolina	X	X		
South Dakota				
Tennessee		X[d]		X
Texas	X		X	
Utah				
Vermont				
Virginia				
Washington				
West Virginia	X	X[d]		
Wisconsin	X[e]	X	X	
Wyoming	X	X	X	X
American Samoa				
District of Columbia				
Guam				
Northern Mariana Is.				
Puerto Rico		X[f]		
Virgin Islands				

[a] Uses the term "seasonal" to describe intermittent.
[b] California has nine Regional Water Quality Control Boards that each develops separate water quality standards. Two regions define "intermittent" and "ephemeral" in their water quality standards.
[c] Defines perennial streams as those that have only continuous flow and those that have both continuous and intermittent reaches.
[d] Describes ephemeral streams as wet-weather conveyances or streams.
[e] Describes intermittent streams as noncontinuous streams.
[f] Uses the term "intermittent streams" to describe ephemeral streams.

specifies that ephemeral streams flow for ≤ 29 consecutive days. Wisconsin defines "noncontinuous" streams as those with natural average 7-day low discharge ($7Q_{10}$) of <0.1 cubic feet per second. Alabama, in part, defines intermittent streams based on recent topographic map designations. Hawaii considers streams that are "interrupted" because of "downward seepage" or anthropogenic water diversions to be perennial.

Not all states with definitions for IRES associate them with designated uses (Table 5.3.1). Some states (e.g., Alabama, New Mexico, Oklahoma, Wisconsin) describe IRES as limited resource or marginal surface waters, having minimum water quality standards when and where water is present. Other state water quality standards, like those for North Dakota, include in their descriptions of lower-tiered (based on associated water quality criteria) designated-use classes (i.e., agricultural and industrial use), streams that "generally have low average flows with prolonged periods of no flow." Louisiana distinguishes only IRES with perennial pools as supporting recreational uses and habitat for "desirable species of aquatic biota." The designated uses assigned to IRES in other states, including Arizona and Missouri, vary depending upon site-specific information such as the aquatic life present (e.g., coldwater vs warmwater), presence of perennial pools, or effluent-dependent flow. There are also state standards like those for California and Idaho that assign the same designated uses to IRES and perennial streams. Similarly, some territories (e.g., Guam) and tribal authorities (e.g., Makah Indian Nation) explicitly recognize IRES in their water quality standards. Lastly, some states such as New Jersey tie IRES status to the water quality in the perennial segment immediately downstream.

The other way IRES are included in state water quality standards is in antidegradation policies. States can downgrade a designated use to one with lower water quality standards through a use-attainability analysis. This analysis must provide sufficient evidence that a designated use is unattainable because of natural conditions, or the removal of human-induced conditions would cause more damage or result in significant and widespread socioeconomic impact. Among the natural causes identified by many states is a waterway having naturally ephemeral or intermittent flow that cannot be compensated for by sufficient effluent without violating water conservation requirements. Such management decisions are likely to become commonplace in regions expected to experience flow-regime shifts caused by climatic and land-use changes (Chapter 5.1).

The CWA allows states to adjust their water quality standards to reflect improvements in water quality criteria and monitoring to be more protective than the federal regulations. Although climate varies substantially across states and territories (e.g., precipitation <250 mm $year^{-1}$ in Nevada and >3000 mm $year^{-1}$ in American Samoa), there is no apparent pattern in the legislative recognition of IRES related to the resulting likely prevalence of IRES. There also appears to be no lingering association with the origin of the state water rights and the extent to which state water quality standards define or recognize designated uses for IRES.

EUROPEAN UNION

In 1992, the Habitats Directive (HD; Council Directive 92/43/EEC) was approved in the EU to meet the obligations under the 1979 Convention on the Conservation of European Wildlife and Natural Habitats. Within the HD, *natural habitat types of Community interest* are habitats in danger of disappearance from their natural range, have a small natural range, or are outstanding examples of one or more of the nine EU biogeographical regions: Alpine, Atlantic, Black Sea, Boreal, Continental, Macaronesian, Mediterranean, Pannonian, and Steppic. The HD Annex I requires these to be designated conservation areas and so a European ecological network known as Natura 2000 was created.

In the HD, two habitat-type definitions refer solely to IRES: (1) intermittently flowing Mediterranean rivers of the Paspalo-Agrostidion and (2) riparian formations on intermittent Mediterranean watercourses with *Rhododendron ponticum, Salix,* and other species. The first habitat type is running waters with interrupted flow and either a dry bed or pools for part of the year with characteristic wetland plant communities (i.e., *Polygonum amphibium, Ranunculus fluitans, Potamogeton natans, Potamogeton nodosus, Potamogeton pectinatus*). This habitat is mainly in Portugal, Spain, France, Italy, Greece, and Cyprus (Fig. 5.3.6A). The second habitat's description only refers to the plant communities present and is restricted to small areas in Spain and Portugal (Fig. 5.3.6B).

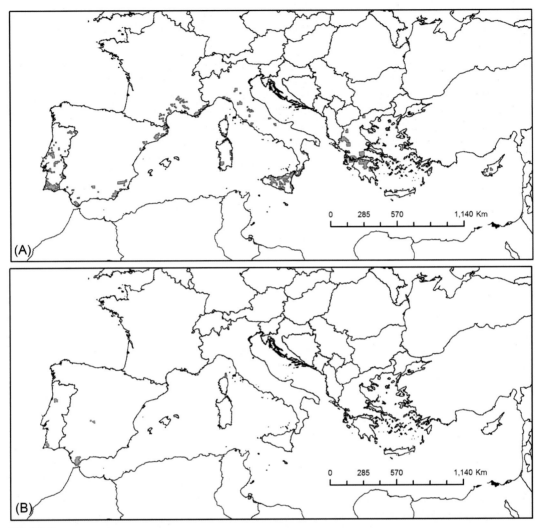

FIG. 5.3.6

Distribution of the IRES habitats described in the EU Habitats Directive Annex I: (A) Intermittently flowing Mediterranean rivers of the *Paspalo-Agrostidion*, and (B) Riparian formations on intermittent Mediterranean watercourses with *Rhododendron ponticum*, *Salix,* and other species.

Data from European Environmental Agency.

Two other habitat types include IRES in their definition: (1) Southern riparian galleries and thickets (*Nerio-Tamaricetea* and *Securinegion tinctoriae*) which are widely distributed across the Mediterranean and Canary Islands (Fig. 5.3.7A), and (2) *Platanus orientalis* and *Liquidambar orientalis* woods (*Plantanion orientalis*) which are restricted to the Eastern Mediterranean (Fig. 5.3.7B). EU Member States are required to monitor the designated natural habitat types and report on their conservation status every six years to the European Commission (EC) (Table 5.3.2).

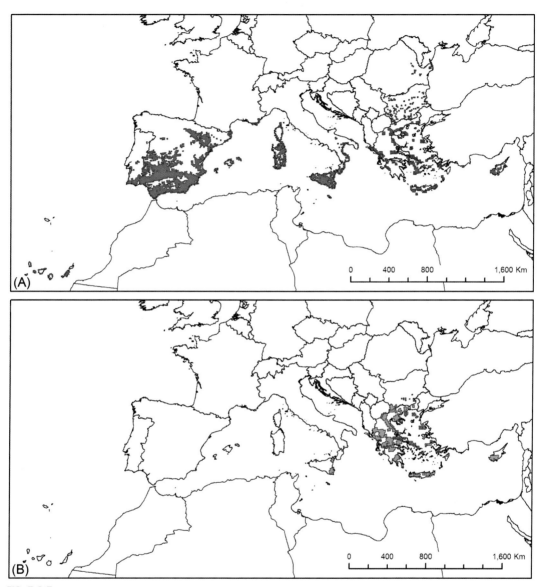

FIG. 5.3.7

Distribution of the IRES habitats that included other water body types (e.g., perennial streams, springs) described in the Annex I of the EU Habitats Directive Annex I: (A) Southern riparian galleries and thickets (*Nerio-Tamaricetea* and *Securinegion tinctoriae*), and (B) *Platanus orientalis* and *Liquidambar orientalis* woods (*Plantanion orientalis*).

Data from European Environmental Agency.

Table 5.3.2 Summary of the report under Article 17 of the EU Habitats Directive for IRES habitats, 2007–12 (European Environment Agency, 2016)

	Intermittently flowing Mediterranean rivers of the Paspalo-Agrostidion	Riparian formations on intermittent Mediterranean watercourses
Conservation status	Cyprus: Favorable France: Unfavorable Spain: Unfavorable Portugal: Favorable Italy: Favorable Greece: Favorable	Spain: Unfavorable Portugal: Favorable
Main pressures and threats	Changes in water bodies' conditions Forestry activities Mining and quarrying Roads, railroads, and paths Soil pollution and solid waste (excluding discharges) Other changes to ecosystems	Invasive alien species Abiotic changes (climate change)
Conservation measures	Establish protected areas/sites Legal protection of habitats and species Wetland-related measures Urban and industrial waste management	Managing water abstraction
Area of the habitat covered by the Natura 2000 network	Cyprus: 100% France: 100% Spain: 80% Portugal: Not reported Italy: 84% Greece: Not reported	Spain: 100% Portugal: Not reported

The EU Water Framework Directive (WFD) (EC, 2000) underpins European policy for safeguarding aquatic ecosystems. It adopts hydrological and ecological criteria, representing a significant change in water resources management for most Member States. The WFD incorporates the river-basin scale as the management unit, considering different ecosystem types in a catchment (e.g., rivers, lakes and ground waters, estuarine, and coastal waters) but does not specifically mention IRES. To determine ecological status, water body types must be defined by Member States. The identification and delimitation of surface water body types is a crucial step for implementing river basin management plans (RBMP) under the WFD. Therefore, the protection of IRES under the WFD depends upon whether the Member States recognize IRES among the water body types within their jurisdiction. Nevertheless, efforts were made to include IRES in an intercalibration (IC) exercise required by the WFD to ensure comparability of assessment methods and ecological status classes developed by each Member State (European Commission, 2011; Section 5.3.4).

For adequate WFD implementation, Member States must adapt or create regulations. For instance, Spain's WFD implementation is conducted through River Basin management legislation (i.e.,

Instrucción de Planificación Hidrológica (IPH) (ORDEN ARM/2656/2008)). The IPH defines 32 river typologies into which each surface water body is classified and for which different typology-based ecological status classes exist. No typologies are defined by flow permanence and therefore none refers specifically to IRES, although many are likely naturally intermittent or ephemeral. This complicates the ecological status assessment of those IRES designated as water bodies (Section 5.3.4, Chapter 5.1). The Spanish IPH states that only streams with mean natural flows $>0.1\,m^3\,s^{-1}$ and with catchment areas $>10\,km^2$ must be assessed for ecological status. Consequently, most IRES are excluded by biomonitoring programs due to their small catchment size and low mean annual flows (Box 5.3.2), despite their substantial aggregate contribution to larger waters downstream.

BOX 5.3.2 GAPS AND COMPLEMENTARITY BETWEEN THE EU HD AND WFD: THE LAS BÁRDENAS REALES (SPAIN) EXAMPLE

Bárdenas Reales is a semiarid area in northern Spain (Ebro Basin) cataloged as a Biosphere Reserve by United Nations Educational Scientific and Cultural Organization. It is also part of the Natura 2000 network because it contains natural habitat types of Community interest listed in the HD. For instance, Southern riparian galleries and thickets (*Nerio-Tamaricetea* and *Securinegion tinctoriae*), where IRES (Fig. 5.3.8) may be included, occupy >576 ha. Nevertheless, IRES draining Bárdenas Reales are not monitored under the Water Framework Directive (WFD) because these streams are too small to be designated as WFD water bodies (not mapped as blue lines, e.g., Fig. 5.3.9).

FIG. 5.3.8

IRES from Las Bárdenas Reales (Navarra, Spain).

Photo: Courtesy N. Bonada.

(Continued)

BOX 5.3.2 GAPS AND COMPLEMENTARITY BETWEEN THE EU HD AND WFD: THE LAS BÁRDENAS REALES (SPAIN) EXAMPLE *(Cont'd)*

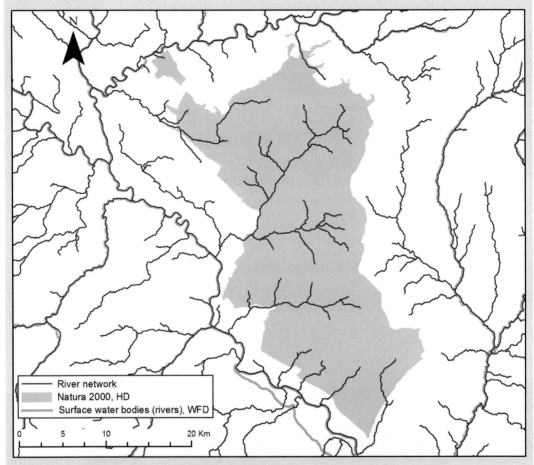

FIG. 5.3.9

River network, surface water bodies considered under the WFD (blue lines) and Natura 2000 protected area (green shading) in Las Bárdenas Reales (Navarra, Spain).

Data from European Environmental Agency.

Notwithstanding, the IPH defines river types according to level of intermittence for the characterization of environmental flows. The resulting four classes—perennial, temporary or seasonal, intermittent or strongly seasonal, and ephemeral (Table 5.3.3)—are based on rainfall-runoff models. In this context, the IPH describes criteria for characterizing natural flow regimes that are used to determine the degree to which a river is naturally or anthropogenically an IRES.

Table 5.3.3 Definition of the different flow regimes by Spanish implementation of the WFD (i.e., IPH)

Natural flow regime	Definition
Perennial	Watercourses in which, under natural flow regimes, flow is always present
Temporary or seasonal	Watercourses that, under natural flow regimes, present marked seasonality with reduced flow or dry riverbeds in summer and that flow for an average period of 300 days year^{-1}
Intermittent or strongly seasonal	Watercourses that, under natural flow regimes, are highly intermittent and flow for an average period of 100–300 days year^{-1}
Ephemeral	Watercourses that, under natural flow regimes, only flow sporadically, mainly during storm episodes, and flow for an average period of <100 days year^{-1}

This classification is only used for environmental flows characterization.

Table 5.3.4 IRES classification by the Italian implementation of the WFD

Stream type	Description
Temporary	Watercourse that can dry out completely and/or at some stretches
• Intermittent	Water is present more than 8 months a year. It may dry out in some river stretches and/or several times a year
• Ephemeral	Water is present <8 months a year. Disconnected pools may remain
• Episodic	Water only present after heavy rains, once every 5 years

The Italian WFD implementation (Decreto M. 131-2008) includes an IRES classification (Table 5.3.4) stating that all water bodies with a catchment area >10 km^2 must be included in the RBMP, except episodic streams which carry water less than once every 5 years. Thus, Italian episodic streams are excluded from WFD regulations without any assessment of their environmental role.

The EU Floods Directive (EC, 2007) establishes a framework for the assessment and management of flood risks, aimed at reducing adverse consequences on human health, environment, cultural heritage, and economic activity in coordination with the WFD. Article 2 defines a flood as *"the temporary covering by water of land not normally covered by water. This shall include floods from rivers, mountain torrents, Mediterranean ephemeral water courses, and floods from the sea in coastal areas, and may exclude floods from sewerage systems."* This is the only part of the Floods Directive that mentions IRES.

LATIN AMERICA AND THE CARIBBEAN

In this region, IRES have been legally defined only in Argentina and Brazil (Table 5.3.5). The 2001 water legal code of the Corrientes Province, Argentina, contains three articles on IRES. These three articles define IRES, prohibit activities that could impede their flow, and state that regulations governing human activities in riparian areas and floodplains also apply for IRES.

In Brazil, legislation (Resolução n° 141) was developed to establish criteria and guidelines for implementing water allocations and classifying water bodies according to their predominant uses.

Table 5.3.5 List of IRES designations in water law of Latin American and Caribbean countries

Name/definition	Country	Legal text	Direct management implications mentioned
Quebradas Waters flowing discontinuously	Chile	Codigo de Aguas (1981)	No. Only mentioned in watershed definition (Article 3). Indicates a need to establish a monitoring network in each basin and the need for establishing minimum ecological flows, but no special mention to IRES
Surface waters flowing discontinuously	Honduras	Decreto N° 181/09—Ley general de aguas (2009)	No. Only mentioned in surface waters definition (Article 6) and in public water domain definition (Article 25)
Surface waters with intermittent water discharge	Colombia	Decreto N° 1.640—Reglamenta los instrumentos para la planificación, ordenación y manejo de las cuencas hidrográficas y acuíferos (2012)	No. Only mentioned in watershed definition (Article 3)
Intermittent surface waters	Costa Rica	Ley de Aguas (1942)	No. Only mentioned in watershed definition (Article 1)
Discontinuous running waters	El Salvador	Ley General de Aguas (2012) *Draft*	No. Only mentioned in continental waters (Article 5) and public water domain (Article 9) definitions
Gullies	Jamaica	Natural Resources Conservation (Wastewater and sludge) regulations (2013)	Yes. Method to calculate discharge fees. Surface waters (including dry gullies) are listed
Intermittent surface waters	Paraguay	Ley n° 37 de los recursos hídricos del Paraguay (2007)	No. Only mentioned in stream channels definition (Article 5)
Discontinuous running waters	Ecuador	Ley orgánica de recursos hídricos, usos y aprovechamiento del agua (2014)	No. Only mentioned in stream channels definition (Article 5)
Intermittent running waters	Bolivia	Decreto Supremo N° 28.590—Reglamento Ambiental Minero para el aprovechamiento de áridos en cauces de ríos y afluentes (2006)	No. Only mentioned in river definition (Article 5)
Discontinuous waters	Bolivia	Ley de Aguas (1906)	No. Only mentioned in public waters definition (Article 4)
Intermittent stream channels with intermittent water discharge	Panamá	Ley 44 de 5 de agosto de 2002 que establece el Régimen Administrativo especial para el manejo, protección y conservación de las cuencas hidrográficas de la República de Panamá	No. Only mentioned in watershed definition (Article 2)

Table 5.3.5 List of IRES designations in water law of Latin American and Caribbean countries —cont'd

Name/definition	Country	Legal text	Direct management implications mentioned
Intermittent streams	Mexico	Ley de Aquas Nacionales 1992	No. Only mentioned in river definition (Article 3)
Intermittent rivers	Argentina (Corrientes Province)	Decreto Ley N° 191/01 Código de Aguas de la Província de Corrientes (2001)	Yes. Three articles of the law are dedicated to IRES. The law prohibits activities that could impede the free flow of these water bodies and states that the riparian legal code (Ley N° 5.588) has to be implemented also in IRES
Discontinuous running waters/intermittent rivers	Venezuela	Ley de Agua (2006)	No. Only mentioned in surface waters (Article 2) and public water domain (Article 6) definitions
Intermittent, ephemeral and perennialized rivers	Brazil	Resolução n° 141, de 10 de julho de 2012	Yes. Establishes criteria for the use of water resources and prohibits the release of nontreated effluent in these water bodies

This includes definitions for intermittent, ephemeral, and perennialized rivers (i.e., IRES that have perennial flow due to human intervention, Chapter 5.1). Although the resolution mainly focuses on water use, it also directs the regulation of untreated effluents and considers the monitoring of IRES. Other Brazilian legislation (Resolução n° 357) states that environmental targets for IRES may vary throughout the year and that special conditions may be defined for the discharge of treated effluents into dry riverbeds. Although there has been an effort to incorporate biomonitoring for water bodies in Brazil, it is not required for IRES or any other water body type.

In most Latin American and Caribbean countries for which water law information is available, IRES are only considered in the definitions of waters. A lack of adequate definitions describing waterway

BOX 5.3.3 DISPUTE OVER AN IRES WATER RESOURCE—THE SILALA RIVER

Since 1997, Bolivia and Chile have had an ongoing dispute over the usage rights to the Silala River, which is among the few flowing waterways in the Atacama Desert. Bolivia claims that the river only flows to Chile because it was modified but Chile claims that the river is a naturally perennial, transboundary river and that under international law, Chile does not have to pay Bolivia for the river's water. In 1908, Bolivia granted a concession to a Chilean mining company to construct canals in the Bolivian part of the Silala River and use the water for steam engines that transported minerals to the Pacific Coast. In 1997, Bolivia revoked the concession stating that Chile had long ago used the water for different purposes than was originally agreed. Bolivia has sued Chile for more than a billion US dollars for historic water uses and demands that any negotiations should include Chile granting Bolivia permanent access to the Pacific Ocean. Under international water law, surface runoff flowing in marginally or undefined channels across international waters does not apply. If the river was, in fact, intermittent then Bolivia, under liability rules, would not be held responsible for not supplying water to Chile as violations can only be attributed to human actions (Mulligan and Eckstein, 2011).

hydrological regimes may contribute to water resource disputes (Box 5.3.3). Water laws include regional terminology for IRES, such as *quebradas* in the Atacama Desert of northern Chile or *gullies* in Jamaica. Some countries such as Chile recognize the need for a monitoring network in each basin and minimum ecological flows but do not specify IRES. Conversely, Chilean law states that ecological flows may not be applied during drought periods. In Jamaica, regulations include a method to calculate discharge fees with a specific weighting factor applied for different water body types, including dry *gullies*.

AUSTRALIA AND NEW ZEALAND

Water governance in Australia is largely through state and territory legislation. An exception is the Commonwealth Water Act of 2007 which established an independent authority for the sustainable and integrated management of the Murray-Darling River, Australia's largest river system (>1 million km^2) which drains four states and a territory. Australia is among the driest nations in the world and a high proportion of its river network length is IRES (Finlayson and McMahon, 1988; Kennard et al., 2010). Much of the state and territorial legislation for water resource management recognizes regulated watercourses to include those that are perennially and intermittently flowing (Table 5.3.6), but not all state and territory agencies have water quality objectives in place for IRES (Table 5.3.7).

New Zealand is a unitary state government but its key water legislation, the Resource Management Act (RMA) of 1991, sets the responsibilities between local governments (i.e., city and district councils) and the central government. The central government is responsible for guiding and directing the local governments through setting the National Policy Statement for Freshwater Management, National Coastal Policy Statement, drinking-water legislation, national environmental standards, and other regulations. The local governments are responsible for freshwater planning and managing water quality and quantity at the community and regional scales (Richmond et al., 2004). The RMA defines "river"

Table 5.3.6 Summary of Australian federal, state and territory legislative interpretations of IRES

Geographic scope	Legal text	Legal definitions
Murray-Darling River basin	Commonwealth Water Act 2007	Watercourse: "a river, creek, stream or other natural channel in which water flows (whether continuously or intermittently)."
Australian Capital Territory	ACT Water Resources Act 2007	Waterway: "a river, creek, stream or other natural channel in which water flows (whether continuously or intermittently)."
		The waterway includes the bed and banks that water normally flows over and between, respectively, but does not include land that is covered from time to time by floodwaters from the waterway
New South Wales	NSW Water Management Act 2000	River: "any watercourse, whether perennial or intermittent" and any other watercourse from which a river flows into or from
New South Wales	Protection of the Environment Operations Act 1997	Pollution of waters is relevant to any waters or "dry bed of any waters, or in any drain, channel or gutter" that receives or passes rainwater, floodwater, or any unpolluted water

Table 5.3.6 Summary of Australian federal, state and territory legislative interpretations of IRES—cont'd

Geographic scope	Legal text	Legal definitions
Northern Territory	Northern Territory Water Act 1983	Waterway: "a river, creek, stream or watercourse; a natural channel in which water flows, whether or not the flow is continuous…land which is intermittently covered by water from a waterway…but does not include any artificial channel or work which diverts water away from such a waterway." Bed and banks of a waterway: "the land over which normally flows, or which is normally covered by, the water of the waterway, whether permanently or intermittently, but does not include land from time to time temporarily covered by the flood waters of the waterway and abutting on or adjacent to its bed and banks"
Queensland	Water Act 2000	Watercourse: rivers, creeks, or other streams "in which water flows permanently or intermittently" but excludes "drainage feature." Drainage features: gullies, drains, drainage depressions, or other erosion features formed by or acts to concentrate overland flow for short durations immediately after rainfall events and does not have the sufficient continuous flow to establish a riverine environment
South Australia	South Australia Natural Resources Management Act 2004	Watercourse: "a river, creek or natural watercourse…in which water is contained or flows whether permanently or from time to time"
Tasmania	Tasmania Water Management Act 1999	Watercourse—"a river, creek or other natural stream of water flowing in a defined channel, or between banks, notwithstanding that the flow may be intermittent or seasonal or the banks not clearly or sharply defined and includes…the floodplain of such stream but does not include…a drain or drainage depression in the contours on the land which only serves to relieve upper land of excess water in times of major precipitation." Depending upon context, reference to watercourse can be "the bed and banks of the watercourse or the water for the time being within the bed and banks of the watercourse." Channel: an entire or part of "a drain, gutter or pipe."
Victoria	Water Act 1989	Waterway: "a river, creek, stream or watercourse or natural channel in which water regularly flows whether or not the flow is continuous."
Western Australia	Rights in Water and Irrigation Act 1914	Watercourse: "any river, creek, stream or brook in which water flows…a flow or collection of water comes within that definition even though it is only intermittent or occasional." Watercourse bed: "the land over which normally flows, or which is normally covered by, the water thereof, whether permanently or intermittently, but does not include land from time to time temporarily covered by the flood waters of such watercourse or wetland and abutting on or adjacent to such bed."
Western Australia	Water Resources Management Bill 2009	Watercourse: "a channel of water that flows permanently or intermittently, and includes…the bed that the water in the watercourse normally flows over or is covered by; and…the banks that the water in the watercourse normally flows between or is contained by."

Table 5.3.7 Examples of water quality guidelines for IRES in Australia and New Zealand

Geographic scope	Water quality guidelines set for IRES
Australia and New Zealand	Not developed for intermittent or ephemeral systems
Queensland	Some developed for ephemeral streams
Victoria	Not developed for intermittent or ephemeral systems
Western Australia	Does not differentiate intermittent or ephemeral systems
Auckland Region, New Zealand	Under development for intermittent streams
Wellington Region, New Zealand	Not developed for intermittent or ephemeral systems

as "a continually or intermittently flowing body of fresh water; and includes a stream and modified watercourse." These RMA "rivers" do not include artificial watercourses such as irrigation canals. Some regional councils are developing water quality objectives specifically for intermittent streams (Table 5.3.7).

CANADA

Water governance at the federal level in Canada consists of jurisdiction for fisheries, navigation, federal lands, and the management of boundary waters shared with the United States. The primary federal water legislation is the Canada Water Act administered by the Minister of the Environment but there are other legislation and government agencies involved in the management and protection of water bodies and aquatic life (e.g., Navigation Protection Act administered by the Minister of Transport, Fisheries Act administered by the Minister of Fisheries and Oceans). IRES are not defined in any Canadian Federal legislation but do occur in Canada (Buttle et al., 2012). Canadian provinces have laws governing water quality and quantity. For instance, Ontario's primary water legislation is the Ontario Water Resources Act of 1990. This act prohibits the discharge of pollutants in or near waters and regulates discharge of municipal waste into waters. The definition of waters under this legislation includes perennial and intermittent watercourses and groundwater. British Columbia's legal definition for streams includes natural watercourses, whether or not they have been modified and whether or not they usually contain water, including ice. Under Manitoba's Water Protection Act, IRES have separate water quality objectives from perennial streams. The objectives vary depending on whether discharge exceeds or is $<0.003\,\mathrm{m^3s^{-1}}$. However, IRES are not recognized by all Canadian provinces (e.g., Saskatchewan Water Security Act of 2005).

SOUTH AFRICA

South Africa's average annual rainfall is only about $500\,\mathrm{mm\,year^{-1}}$, so it is considered a water-scarce country. South African water resources are governed by two national laws: the Water Services Act (1997) which pertains to how municipalities should provide drinking water and wastewater services, and the National Water Act (1998) which deals with source water protection, development, and management. The National Water Act preamble recognizes water scarcity and the resulting need to integrate water management in a sustainable manner. The definition of watercourses includes rivers in "which water flows regularly or intermittently." Although about two-thirds of rivers and streams in

South Africa are IRES (Davies and Day, 1998), they represent only about 3% of records in the South Africa national bioassessment database (Watson and Dallas, 2013). A national assessment of South African river (1:500,000 scale coverage) status for conservation purposes determined that rivers likely to have intermittent or ephemeral flow were proportionally less threatened than their perennial counterparts (Nel et al., 2007). Existing bioassessment protocols developed for assessing the condition of perennial water resources in South Africa yield equivocal results when applied to IRES (Watson and Dallas, 2013), and there are ongoing efforts to develop protocols and the water quantity and quality required for maintaining ecological conditions for South African IRES (e.g., Rossouw et al., 2005; Seaman et al., 2010).

INDIA

India has a wide range of hydroclimatic regions with varying human densities. Water is generally copious but IRES are common in the more arid regions of Rajasthan, Haryana, and Gujarat in the northwest and the semiarid interior portions of Karnataka and Andhra Pradesh (e.g., Ramachandra et al., 2014). Groundwater depletion is a major problem in arid regions, so management to enhance aquifer recharge through IRES and rainwater-harvesting techniques is common (e.g., Glendenning and Vervoort, 2010). The Indian Easements Act of 1882, based largely upon British Common Law, states that landowners have rights to draw from and discharge into water passing in streams, which can be perennial or intermittent, at a justifiable or reasonable level. The proliferation of domestic and industry waste entering streams and rivers in India led to the Water Prevention and Control of Pollution Act of 1974, the primary central government legislation for the management of Indian water quality. Streams defined in this Act include watercourses whether "flowing or for the time being dry." A regional law, the Punjab Land Preservation Act of 1900 (aka the Chos Act), gives the Punjab and Haryana state governments jurisdictional control over the beds of *chos* or ephemeral stream channels. In contrast to many of the legal definitions pertaining to water protection, this law was enacted to protect land prone to soil erosion. Historically, the hills of this region were covered with acacia and pine but overharvesting of trees and overgrazing in the mid-19th century led to land degradation (Behari, 1967; Bowen, 1985). This law regulates, restricts, and prohibits certain land practices (e.g., clear-cutting, burning, herding) and includes local government's jurisdiction and control over the *chos* to regulate the flow of water and reclaim or protect land adjacent to *chos*.

ISRAEL AND PALESTINE

The Israeli Water Law in 1959 first codified water law in the region (Tal and Rabbo, 2010). Water sources under this law include springs, streams, rivers, lakes, and water reservoirs, whether above or below ground, having natural or regulated flow, or flowing or standing perennially or intermittently. The original Water Law focused on economic objectives for the public good, namely, domestic needs, agriculture, industry, and public services. Environmental protection was not considered a legitimate use of water (Tal and Katz, 2012). In addition to the social and military conflicts in the region and scarcity of freshwater, the heavy use of the limited freshwater resources without strong environmental protection led to degradation of natural waterways. Recent desalination developments, land-use changes, and water resource policies have sparked interest in natural waterway rehabilitation in Israel and Palestine. While there is a long way to go, adoption of the Water Law by the Palestinian Water Authority in 2002 and amendments to the Water Law in 2004, which included environmental flow requirements and the expansion of wastewater systems to collect, treat, and reuse water, are positive signs for improvement (Tal and Rabbo, 2010).

5.3.4 FLOW INTERMITTENCE COMPLICATING THE GOVERNANCE OF WATER USE AND ENVIRONMENTAL PROTECTION

While the degree to which governments legally recognize IRES varies, almost every governance system lags behind in implementing water resource management for IRES relative to that of perennial waterways. Already-stretched resources available to monitor and characterize perennial waters in many nations compete with significant channel lengths of IRES yet to be assessed, and probably explain part of this disparity. However, IRES also have some inherent properties that complicate the governance of water use and have contributed to this lag.

First, many IRES lack water much of the time. Most legislation for protecting and restoring water bodies primarily focuses on water quality and suitability for public health and aquatic life. Other elements, such as habitat degradation, are present in many governance systems and their goals but are often secondary to water quality or intended to protect minimum flows in perennial rivers. An exception is IRES in Queensland, Australia, where laws include thresholds set to limit the duration of no-flow conditions to protect perennial refugia and riparian vegetation (Acuña et al., 2014).

Water quality and quantity are inextricably tied to one another; this is fundamental in most laws for surface water protection (e.g., European Commission, 2012). Water governance systems that separately govern water quantity and quality (e.g., some parts of western United States) complicate water resource management through practices such as water withdrawals and interbasin transfers that have serious implications for IRES. A major purpose for surface water laws is to prevent municipal or industrial effluents from degrading the chemical, physical, and biological integrity of receiving waters. Water quality criteria or standards set for permitted effluent discharges take into account the magnitude (i.e., pollutant concentration) as well as the discharge duration and frequency relative to those of the receiving water. The recurrent loss of flow and associated physicochemical changes of IRES (Chapters 3.1 and 3.2) mean that these waters sometimes lack the dilution capacity present in perennial waterways (Chapter 5.1). This creates situations where IRES become effluent-dominated or effluent-dependent streams. The novel characteristics (e.g., stable perennial flow, predominance of introduced species) of such streams are sometimes valued and maintained, often at the risk of losing natural characteristics and benefits provided by IRES (Moyle, 2014; Luthy et al., 2015).

Second, predicting when water and aquatic life are present in IRES is often challenging. This makes evaluating the water quality and aquatic life status in IRES more difficult than in their perennial counterparts (Chapter 5.1). Assessments of perennial waters, particularly from their biota, tend to occur when flooding is less likely but this period often coincides with when IRES are dry or reduced to isolated pools. Having separate index periods for perennial and IRES segments within river networks may be less desirable as a sampling strategy but is feasible (e.g., Prat et al., 2014). Distinguishing when an ungauged IRES reach last dried or where it lies in its "boom-bust cycle" at the time of sampling are also

---

BOX 5.3.4 IRES AND THE WFD INTERCALIBRATION PROCESS

The WFD requires intercalibration (IC) to ensure comparability of assessment methods and ecological status classes developed by Member States (European Commission, 2011). Because different Member States share similar water body types, experts from different countries formed Geographical Intercalibration Groups (GIGs). The Mediterranean GIG (Cyprus, France, Greece, Italy, Malta, Portugal, Slovenia, and Spain) was the only one that included IRES. Five river typologies based on catchment area, altitude, geology, and flow regime were defined for this GIG, including one for IRES. This IRES type was described as lowland temporary streams, with small catchments ($10-100\,km^2$) and at altitudes $<300\,msl$.

   Crucial to the IC process is reference condition comparability, which is the basis for establishing quality class boundaries (European Commission, 2003). The IC exercise identified differences between IRES and perennial rivers for abiotic and biotic conditions (Feio et al., 2014a). For instance, a lower threshold value for reference condition based upon dissolved oxygen saturation was identified for temporary rivers (60%) compared to that of perennial streams (74%). Due to this difference, IRES communities were treated as a separate group to develop ecological quality classes for macroinvertebrates and phytobenthos (Almeida et al., 2014; Feio et al., 2014a). Nevertheless, as ecological quality boundaries using IRES macroinvertebrates were not sufficiently comparable across the Mediterranean region, each Member State must independently determine ecological status for IRES within their jurisdiction (Feio et al., 2014b; Reyjol et al., 2014).

---

major challenges. Also, not recognizing the degree to which a reach was isolated during the dry phase relative to perennial refugia may complicate data interpretation. A more comprehensive assessment of the status of IRES and perennial rivers would be gained by developing approaches for data integration across space and time.

   Third, surface-water governance systems rely on water body classification to set assessment expectations and account for variation. Many programs use the reference condition approach (Stoddard et al., 2006) to set benchmarks for water quality objectives or goals for assessment and restoration. Reference condition is not a single value; rather, it is a distribution that represents sampling error and natural variability. The spatial and temporal variation in hydrology that drives concomitant variation in many traditionally measured physicochemical and biological properties is commonly higher in IRES than in perennial waterways (Boulton et al., 2000; Sánchez-Montoya et al., 2011). Therefore, distinguishing between reference conditions and varying degrees of impairment (Davies and Jackson, 2006) for IRES will require careful typology and/or characterization of covariates such as flow duration or refuge availability that influence assessment metrics and their intercalibration (Box 5.3.4).

   Lastly, the recurrent cessation and resumption of flow distinguishes IRES from perennial waterways in various landscapes. Under most governance systems, the legal definition of a waterway includes some way to distinguish a public waterway from the adjacent private land. For instance, "ordinary high water mark," a legal construct adopted from English common law, is used in the United States to demarcate the lateral jurisdictional limits of nontidal waters in the absence of adjacent wetlands. The regulatory definition (33 CFR 328.3) of ordinary high water mark is "that line on the shore established by the fluctuations of water and indicated by physical characteristics such as a clear, natural line impressed on the bank, shelving, changes in the character of soil, destruction of terrestrial vegetation, the presence of litter and debris, or other appropriate means that consider the characteristics of the surrounding areas." Such physical characteristics along stream channels reflect not only the erosion and deposition processes that occur during a range of flow magnitudes but also counteracting terrestrial processes such as riparian vegetation growth and soil formation (Wolman and Gerson, 1978). Continuous flow and inundation limits the extent to which terrestrial processes can encroach into the active channel of perennial waterways. The longer an IRES is dry, the more time is available for high water marks to be obscured by terrestrialization. However, terrestrial processes will encroach more slowly in arid than mesic climates. Documenting the legal boundaries between public waterways and adjacent land can be complicated for many IRES, especially those in arid regions, which differ from the single-threaded, perennial rivers in humid and temperate regions from which the legal construct of "ordinary high water mark" originated (Lichvar and McColley, 2008).

5.3.5 A WAY FORWARD FOR SUSTAINABLE GOVERNANCE, LEGISLATION, AND PROTECTION OF IRES

Addressing conflicting water uses in governance systems is exacerbated where water is scarce. In the past, historical practices, culture, and religious beliefs have been more important than science when setting water policies. However, science-based governance strategies will be critical for society to solve future water quality and supply challenges (Zimmerman et al., 2008; Chapter 5.5). Rather than be reactive to increasing water scarcity, governance systems need to build in adaptive capacities at national and local scales so they can nimbly implement or adjust mitigation actions or incentives (Wilhite et al., 2005; Engle, 2013). Steps toward incorporating systems-thinking to make water governance adaptive include recognizing IRES as key components of natural river networks and not as "stressed" or damaged versions of perennial streams.

There is growing recognition by many governments of the current extent of IRES and their future potential expansion. Three key steps to breaking the mold set by many governance systems focused mainly on perennial waterways are as follows: (1) breaking from stationarity in time and space (i.e., notion that flow fluctuates within "an unchanging envelope of variability" (NRC, 2011)), (2) embracing systems-thinking, and (3) shifting perceptions and expectations.

Novel monitoring and modeling approaches not restrained by stationarity are needed. For example, Sheldon (2005) advocates a "trend approach" to better isolate the natural variation inherent to IRES from anthropogenic impairment over time. In this approach, condition determinations or effectiveness of a management practice is evaluated by how indicators respond or change over time since flow resumption or flooding. The traditional single-point assessment over time is likely to be misinterpreted when it lacks the context of expected natural trajectories of IRES.

Stationarity also includes connectivity over space. For example, incorporating recent developments in hydrological modeling of flow intermittence (e.g., Snelder et al., 2013) and in characterizing hydrological connectivity to downstream waters (e.g., Caruso, 2015) should enhance on-the-ground assessments of beneficial use impairments. IRES vary spatially in surface water connectivity (Chapter 2.3), which is critical for the movement of aquatic life and transport of materials. Developments in technology (e.g., Hooshyar et al., 2015) and approaches (Datry et al., 2016; Gallart et al., 2016; Chapters 2.2 and 2.3) to describe surface-water fragmentation and how it changes with expansion and contraction can inform assessments and management decisions. Traditional indicators in river assessments may be less sensitive to stressors in IRES or simply not applicable (Chapter 5.1). Better conceptual understanding of the effects of antecedent conditions, such as durations of dry or pooled phases, on water quality and aquatic life in IRES will enable us to set more realistic expectations for condition assessments. Indicators developed for IRES assessment during the dry phase, such as emergence from the propagule bank (Angeler and García, 2005), pollutant bioavailability (White et al., 1998), and gas fluxes (Gallo et al., 2014), may overcome the difficulties of sampling during the sometimes short and less-predictable flow phase.

Except those in endorheic basins that terminate in dune fields or playas, most IRES are periodically connected to perennial waters (rivers, lakes, wetlands, or oceans) via surface connections and vertically to aquifers and the atmosphere via groundwater–surface water exchange and evaporation, respectively. There is a growing understanding of the critical contributions of groundwater to the spatiotemporal dynamics of flow and water quality in IRES (e.g., Fleckenstein et al., 2004; Boulton and Hancock, 2006). Policies and programs addressing surface and ground water separately need to be integrated at levels ranging from local catchment groups to international agreements for transboundary waters (Schlager and Heikkila, 2011). Piecemeal approaches to managing waterways as components (e.g., surface and ground water) hamper the ability to see the cumulative impacts of policies and management practices manifested across river networks. Because of where many IRES are positioned in catchments and because their dry beds are commonly unregulated and treated as land rather than waterways, stronger laws to protect IRES will better integrate land (especially riparian habitats and floodplains) and water resource management. Government agencies working on connections or the interface between land and water (e.g., nonpoint source pollutants, riparian buffers) should discuss how their associated regulations can be in accord rather than in conflict, removing confusion and burden for the regulated community.

Inherent to the historical importation of water governance doctrines are the associated societal perceptions and expectations of water resources that may not align with the climate and landscape (Boulton et al., 2000). Providing science that informs policy and educates the public about IRES is key to shifting perceptions and expectations. These shifts can be tempered through historical and local insight gained from management approaches used by traditional and indigenous cultures that derived benefits from IRES (e.g., Norton et al., 2002; Barmuta, 2003). International collaborations among researchers, managers, and policy makers have made strong progress (e.g., Tal et al., 2010; Nikolaidis et al., 2013) but more research on ecosystem goods and services provided or mediated by IRES is needed. This would not only better inform appropriate designation of beneficial uses and mitigation valuations but, perhaps more importantly, inform decision-makers about the societal benefits of IRES. Characterizing how ecosystem goods and services shift during different hydrological phases will provide insight on how to best manage IRES for the production of multiple and interacting benefits (Chapter 5.2).

5.3.6 CONCLUSIONS

Examples of water governance systems around the world show that the level of IRES management varies. While this chapter is not a full account, our examples show that water governance systems have generally paid less attention to IRES than to perennial waterways. Expectations and implementation of water governance systems for IRES have largely been focused on water quality and aquatic biota rather than the conditions and uses provided by IRES while they are dry or isolated pools. There is a growing recognition of IRES by water governance systems because of their connections to the status of perennial water bodies and the direct benefits they provide to society. Effective water governance requires systems-thinking, which includes the protection and restoration of IRES

(Chapter 5.4) and their longitudinal, lateral, and vertical connections. Advances in the sustainable management of IRES will be made when their characteristic spatiotemporal variation is embraced rather than evaded.

ACKNOWLEDGMENTS

We thank Laurie Alexander, Mike Dunbar, Francesc Gallart, Rose Kwok, Jon Marshall, Antoni Munné, Megan Niesen, and Narcís Prat for their advice and comments. NC was supported by the EU Project LIFE+ TRIVERS (LIFE13 ENV/ES/000341). The views expressed in this article are those of the authors and do not necessarily represent the views or policies of the US Environmental Protection Agency.

REFERENCES

Acuña, V., Datry, T., Marshall, J., Barceló, D., Dahm, C.N., Ginebreda, A., et al., 2014. Why should we care about temporary waterways? Science 343, 1080–1081.

Almeida, S.F.P., Elias, C., Ferreira, J., Tornés, E., Puccinelli, C., Delmas, F., et al., 2014. Water quality assessment of rivers using diatom metrics across Mediterranean Europe: a methods intercalibration exercise. Sci. Total Environ. 476–477, 768–776.

Angeler, D.G., García, G., 2005. Using emergence from soil propagule banks as indicators of ecological integrity in wetlands: advantages and limitations. J. N. Am. Benthol. Soc. 24, 740–752.

Barmuta, L.A., 2003. Imperilled rivers of Australia: challenges for assessment and conservation. Aquat. Ecosyst. Health Manag. 6, 55–68.

Behari, M., 1967. Chos, their harmful effects and control in the Hoshiarpur Division of Punjab. Indian Forester 93, 228–238.

Boulton, A.J., Hancock, P.J., 2006. Rivers as groundwater-dependent ecosystems: a review of degrees of dependency, riverine processes and management implications. Aust. J. Bot. 54, 133–144.

Boulton, A.J., Sheldon, F., Thoms, M.C., Stanley, E.H., 2000. Problems and constraints in managing rivers with variable flow regimes. In: Boon, P.J., Davies, B.R., Petts, G.E. (Eds.), Global Perspectives on River Conservation: Science, Policy and Practice. Wiley, Chichester, pp. 415–430.

Bowen, R., 1985. Hydrogeology of the Bist Doab and adjacent areas, Punjab, India. Nord. Hydrol. 16, 33–44.

Broderick, G.T., 2005. From migratory birds to migratory molecules: the continuing battle over the scope of federal jurisdiction under the Clean Water Act. Columbia J. Environ. Law 30, 473–523.

Buttle, J.M., Boon, S., Peters, D.L., Spence, C., van Meerveld, H.J.I., Whitfield, P.H., 2012. An overview of temporary stream hydrology in Canada. Can. Water Res. J. 37, 279–310.

Caponera, D.A., Nanni, M., 2007. Principles of Water Law and Administration: National and International, second ed. Taylor and Francis, London.

Caruso, B.S., 2015. A hydrological connectivity index for jurisdictional analysis of headwater streams in a montane watershed. Environ. Monit. Assess. 187, 635.

Council Directive 92/43/EEC, 1992. 21 May 1992 on the conservation of natural habitats and of wild fauna and flora. Off. J. Eur. Communities 35, 7–50.

Datry, T., Pella, H., Leigh, C., Bonada, N., Hugueny, B., 2016. A landscape approach to advance intermittent river ecology. Freshw. Biol. 61, 1200–1213.

Davies, B., Day, J., 1998. Vanishing Waters. University of Cape Town Press, Cape Town.

Davies, S.P., Jackson, S.K., 2006. The biological condition gradient: a descriptive model for interpreting change in aquatic ecosystems. Ecol. Appl. 16, 1251–1266.

Dellapenna, J.W., 2014. Patterns in water law. In: Bhaduri, A., Bogardi, J., Leentvaar, J., Marx, S. (Eds.), The Global Water System in the Anthropocene. Springer Water, Heidelberg, pp. 401–413.

Engle, N.L., 2013. The role of drought preparedness in building and mobilizing adaptive capacity in states and their community water systems. Climate Change 118, 291–306.

European Commission, 2000. Directive 2000/60/EC of the European Parliament and of the Council of 23 October 2000 establishing a framework for Community action in the field of water policy. Off. J. Eur. Communities and L 327, 1.

European Commission, 2003. Common Implementation Strategy for the Water Framework Directive (2000/60/EC). Working Group REFCOND, Guidance Document Number 10: Rivers and Lakes—Typology, Reference Conditions and Classification Systems. Office for Official Publications of the European Communities, Luxembourg.

European Commission, 2007. Directive 2007/60/EC of the European Parliament and of the Council of 23 October 2007 on the assessment and management of flood risks. Off. J. Eur. Communities, L 288, 27.

European Commission, 2011. Common Implementation Strategy for the Water Framework Directive (2000/60/EC). Guidance Document Number 14: Guidance Document on the Intercalibration Process 2008-2011. Office for Official Publications of the European Communities, Luxembourg.

European Commission, 2012. A Blueprint to Safeguard Europe's Water Resources. Communication from the Commission to the European Parliament, the Council, the European Economic and Social Committee and the Committee of the Regions. http://www.eea.europa.eu/policy-documents/a-blueprint-to-safeguard-europes Accessed 5 November 2016.

European Environment Agency, 2016. European Topic Centre on Biological Diversity. Report under the Article 17 of the Habitats Directive Period 2007-2012 http://art17.eionet.europa.eu/article17/reports2012/habitat/summary/ Accessed 20 April 16.

Falkenmark, M., Jägerskoga, A., Schneidera, K., 2014. Overcoming the land–water disconnect in water-scarce regions: time for IWRM to go contemporary. Int. J. Water Resour. Dev. 30, 391–408.

Feio, M.J., Aguiar, F.C., Almeida, S.F.P., Ferreira, J., Ferreira, M.T., Elias, C., et al., 2014a. Least disturbed condition for European Mediterranean rivers. Sci. Total Environ. 476–477, 745–756.

Feio, M.J., Ferreira, J., Buffagni, A., Erba, S., Dörflinger, G., Ferréol, M., et al., 2014b. Comparability of ecological quality boundaries in the Mediterranean basin using freshwater benthic invertebrates. Statistical options and implications. Sci. Total Environ. 476–477, 777–784.

Finlayson, B.L., McMahon, T.A., 1988. Australia v. the world: a comparative analysis of streamflow characteristics. In: Warner, R.F. (Ed.), Fluvial Geomorphology of Australia. Academic Press, Sydney, pp. 17–40.

Fleckenstein, J., Anderson, M., Fogg, G., Mount, J., 2004. Managing surface water-groundwater to restore fall flows in the Cosumnes River. J. Water Resour. Plan. Manag. 130, 301–310.

Gallart, F., Llorens, P., Latron, J., Cid, N., Rieradevall, M., Prat, N., 2016. Validating alternative methodologies to estimate the regime of temporary rivers when flow data are unavailable. Sci. Total Environ. 565, 1001–1010.

Gallo, E.L., Lohse, K.A., Ferlin, C.M., Meixner, T., Brooks, P.D., 2014. Physical and biological controls on trace gas fluxes in semi-arid urban ephemeral waterways. Biogeochemistry 121, 189–207.

Glendenning, C.J., Vervoort, R.W., 2010. Hydrological impacts of rainwater harvesting (RWH) in a case study catchment: the Arvari River, Rajasthan, India. part 1: field-scale impacts. Agric. Water Manag. 98, 331–342.

Hooshyar, M., Kim, S., Wang, D., Medeiros, S.C., 2015. Wet channel network extraction by integrating LiDAR intensity and elevation data. Water Resour. Res. 51. WR018021.

Kennard, M.J., Pusey, B.J., Olden, J.D., Mackay, S.J., Stein, J.L., Marsh, N., 2010. Classification of natural flow regimes in Australia to support environmental flow management. Freshw. Biol. 55, 171–193.

Knox, K.W., 2001. The La Plata River Compact: administration of an ephemeral river in the arid southwest. Univ. Denver Water Law Rev. 5, 104–120.

Lichvar, R.W., McColley, S.M., 2008. A field guide to the identification of the ordinary high water (OHWM) in the Arid West Region of the Western United States. ERDC/CRREL Technical Report, TR-08-12, US Army Corps of Engineers, Engineer Research and Development Center, Cold Regions Research and Engineering Laboratory, Hanover.

Luthy, R.G., Sedlak, D.L., Plumlee, M.H., Austin, D., Resh, V.H., 2015. Wastewater-effluent-dominated streams as ecosystem-management tools in a drier climate. Front. Ecol. Environ. 13, 477–485.

Matteus, G.V.T., 1993. The Ramsar Convention on Wetlands: Its History and Development. Ramsar Convention Bureau, Gland.

Ministerio de Medio Ambiente, y Medio Rural y Marino, 2008. ORDEN ARM/2656/2008, de 10 de septiembre, por la que se aprueba la instrucción de planificación hidrológica. http://www.boe.es/diario_boe/txt.php?id=BOE-A-2008-15340 Accessed 5 November 2016.

Moyle, P.B., 2014. Novel aquatic ecosystems: the new reality for streams in California and other Mediterranean climate regions. River Res. Appl. 30, 1335–1344.

Mulligan, B.M., Eckstein, G.E., 2011. The Silala/Siloli watershed: dispute over the most vulnerable basin in South America. Water Resour. Dev. 27, 595–606.

Nel, J., Roux, D.J., Maree, G., Kleynhans, C.J., Moolman, J., Reyers, B., et al., 2007. Rivers in peril inside and outside protected areas: a systematic approach to conservation assessment of river ecosystems. Divers. Distrib. 13, 341–352.

Nikolaidis, N.P., Demetropoulou, L., Froebrich, J., Jacobs, C., Gallart, F., Prat, N., et al., 2013. Towards sustainable management of Mediterranean river basins: policy recommendations on management aspects of temporary streams. Water Policy 15, 830–849.

Norton, J.B., Bowannie, F., Peynestsa, P., Quandelacy, W., Siebert, S.F., 2002. Native American methods for conservation and restoration of semiarid ephemeral streams. J. Soil Water Conserv. 57, 250–258.

NRC (National Research Council), 2011. Global Change and Extreme Hydrology: Testing Conventional Wisdom. National Academies Press, Washington, DC.

Prat, N., Gallart, F., Von Schiller, D., Polesello, S., García-Roger, E.M., Latron, J., et al., 2014. The MIRAGE toolbox: an integrated assessment tool for temporary streams. River Res. Appl. 30, 1318–1334.

Ramachandra, T.V., Vinay, S., Bharath, H.A., Bharath, S., Shashishankar, A., 2014. Environmental flow assessment in rivers originating at the Western Ghats. In: Proceedings of Lake 2014: Conference on Conservation and Sustainable Management of Wetland Ecosystems in Western Ghats, Indian Institute of Science, Bangalore. http://wgbis.ces.iisc.ernet.in/energy/lake2014/proceedings.php Accessed 22 November 2016.

Ramsar Convention, 1971. Convention on Wetlands of International Importance especially as Waterfowl Habitat. Ramsar (Iran), 2 February 1971. UN Treaty Series No. 14583. As amended by the Paris Protocol, 3 December 1982, and Regina Amendments, 28 May 1987. http://www.ramsar.org/about-the-ramsar-convention Accessed 22 November 2016.

Ramsar Sites Information Service, 2016. https://rsis.ramsar.org/ Accessed 20 April 2016.

Reyjol, Y., Argillier, C., Bonne, W., Borja, A., Buijse, A.D., Cardoso, A.C., et al., 2014. Assessing the ecological status in the context of the European Water Framework Directive: where do we go now? Sci. Total Environ. 497–498, 332–344.

Richmond, C., Froude, V., Fenemor, A., Zuur, B., 2004. Management and conservation of natural waters. In: Harding, J., Mosley, P., Pearson, C., Sorrell, B. (Eds.), Freshwaters of New Zealand. New Zealand Hydrological Society and New Zealand Limnological Society, Christchurch, pp. 44.1–44.19.

Rossouw, L., Avenant, M.F., Seaman, M.T., King, J.M., Barker, C.H., du Preez, P.J., et al., 2005. Environmental water requirements in non-perennial systems. Water Research Commission Report No. 1414/1/-05, Water Research Commission, Pretoria.

Sánchez-Montoya, M.M., Gómez, R., Suárez, M.L., Vidal-Abarca, M.R., 2011. Ecological assessment of Mediterranean streams and the special case of temporary streams. In: Elliot, H.S., Martin, L.E. (Eds.), River Ecosystems: Dynamics, Management and Conservation. Nova Science Publishers, New York, pp. 109–148.

Schlager, E., Heikkila, T., 2011. Left high and dry? Climate change, common-pool resource theory, and the adaptability of western water compacts. Public Adm. Rev. 71, 461–470.

Seaman, M.T., Avenant, M.F., Watson, M., King, J., Armour, J., Barker, C.H., et al., 2010. Developing a method for determining the environmental water requirements for non-perennial systems: WRC Report No TT459/10. Water Research Commission, Pretoria.

Sheldon, F., 2005. Incorporating natural variability into the assessment of ecological health in Australian dryland rivers. Hydrobiologia 552, 45–56.

Snelder, T.H., Datry, T., Lamouroux, N., Larned, S.T., Sauquet, E., Pella, H., et al., 2013. Regionalization of patterns of flow intermittence from gauging station records. Hydrol. Earth Syst. Sci. 15, 2685–2699.

Stoddard, J.L., Larsen, D.P., Hawkins, C.P., Johnson, R.K., Norris, R.H., 2006. Setting expectations for the ecological condition of streams: the concept of reference condition. Ecol. Appl. 16, 1267–1276.

Tal, A., Katz, D., 2012. Rehabilitating Israel's streams and rivers. Int. J. River Basin Manag. 10, 317–330.

Tal, A., Rabbo, A.A., 2010. Water Wisdom: Preparing the Groundwork for Cooperative and Sustainable Water Management in the Middle East. Rutgers University Press, New Brunswick.

Tal, A., Al Khateeb, N., Nagouker, N., Akerman, H., Diabat, M., Nassar, A., et al., 2010. Chemical and biological monitoring in ephemeral and intermittent streams: a study of two transboundary Palestinian-Israeli watersheds. Int. J. River Basin Manage. 8, 185–205.

USEPA, 2015. Connectivity of streams and wetlands to downstream waters: a review and synthesis of the scientific evidence (Final report). EPA/600/R-14/475F. US Environmental Protection Agency, Washington, DC.

Watson, M., Dallas, H.F., 2013. Bioassessment in ephemeral rivers: constraints and challenges in applying macroinvertebrate sampling protocols. Afr. J. Aquat. Sci. 38, 35–51.

White, J.C., Quinones-Rivera, A., Alexander, M., 1998. Effect of wetting and drying on the bioavailability of organic compounds sequestered in soil. Environ. Toxicol. Chem. 17, 2378–2382.

Wilhite, D.A., Hayes, M.J., Knutson, C.L., 2005. Drought preparedness planning: building institutional capacity. In: Wilhite, D.A. (Ed.), Drought and Water Crises: Science, Technology, and Management Issues. CRC Press, Boca Raton, FL, pp. 93–135.

Wolman, M.G., Gerson, R., 1978. Relative scales of time and effectiveness of climate in watershed geomorphology. Earth Surf. Process. 3, 189–208.

Zimmerman, J.B., Mihelcic, J.R., Smith, J., 2008. Global stressors on water quality and quantity. Environ. Sci. Technol. 42, 4247–4254.

FURTHER READING

Environmental Law Institute, 2012. The Clean Water Act Jurisdictional Handbook, second ed. Environmental Law Institute, Washington, DC.

ISPRA (L'istituto Superiore per la Protezione e la Ricerca Ambientale), 2014. Linee guida per la valutazione della componente macrobentonica fluviale ai sensi del DM 260/2010. http://www.isprambiente.gov.it/it/pubblicazioni/manuali-e-linee-guida/linee-guida-per-la-valutazione-della-componente-macrobentonica-fluviale-ai-sensi-del-dm-260-2010 Accessed 22 November 2016.

Ministero dell'Ambiente e della tutela del territorio e del mare, 2008. Decreto Ministero, Ambiente 16/06/2008 n° 131, G.U. 11/08/2008. http://www.gazzettaufficiale.it/atto/serie_generale/caricaDettaglioAtto/originario?atto.dataPubblicazioneGazzetta=2008-08-11&atto.codiceRedazionale=008G0147&elenco30giorni=false Accessed 22 November 2016.

Radosevich, G.E., Giner Boria, V., Daines, D.P., Skogerboe, G.V., Vlachos, E.C., 1976. International Conference on Global Water Law Systems—Summary Report. AID Publication-US Agency for International Development (USA), Number PN-AAC-592, Colorado State University, Fort Collins.

RESTORATION ECOLOGY OF INTERMITTENT RIVERS AND EPHEMERAL STREAMS

5.4

Philip S. Lake*, Nick Bond[†], Paul Reich[‡]

Monash University, Clayton, VIC, Australia[*] Murray-Darling Freshwater Research Centre, La Trobe University, Wodonga, VIC, Australia[†] Arthur Rylah Institute for Environmental Research, Heidelberg, VIC, Australia[‡]

IN A NUTSHELL

- Many intermittent rivers and ephemeral streams (IRES) have been degraded by past and current human disturbances acting on channels and catchments. Ecological restoration seeks to halt the degradation and to build viable self-sustaining systems.
- Well-documented restoration projects focused on IRES are few compared with those on perennial streams.
- Ecological restoration of IRES should initially be planned at the catchment level before the reach level. Restoration should proceed in progressive steps, starting with assessing the disturbance regime, implementing restoration measures and ending with full analysis and reporting on the project outcomes.
- The major form of IRES restoration has focused on revegetating riparian zones. Restoration of stream channels has involved increasing connectivity to allow high flows (floods) to occur unhindered and to allow biota to move. Reinstallation of habitat structure and replenishment of drought refuges have also been restoration targets.
- Ecological restoration of degraded IRES demands urgent attention and action to offset further climate change-induced impacts.

5.4.1 INTRODUCTION

Ecological restoration attempts to return degraded ecological entities (populations, communities, ecosystems) toward their original state or at least to a valued and enduring "self-sustaining" state that displays a number of distinctive attributes relative to degraded ecosystems. Restoration can be guided by concepts, ideas, and empirical results from the discipline of restoration ecology, which in turn relies on principles from ecological theory (Palmer et al., 2006; Lake et al., 2007). Restoration ecology is a relatively new field and while ecological concepts can be, and have been, successfully applied, progress is largely by empirical means (i.e., monitoring, analyzing, and reporting of both successful and unsuccessful restoration projects) and revision of approaches (learning by doing).

In Europe, North America, and Australia, site-based river restoration projects are a major activity for natural resource management groups. However, robust evidence to gauge the outcomes of these projects is scarce (Bernhardt et al., 2005; Brooks and Lake, 2007) despite improvements in the frequency of monitoring (Boulton et al., 2013). Criteria advocated by Hobbs and Norton (1996), Palmer et al. (2005), and Lake et al. (2007) provide a useful guide to maximize the value of restoration projects

(Table 5.4.1). These criteria apply equally to IRES as they do to perennial systems, although available data to guide the assessment may be more difficult to find. Overall, there is a paucity of literature that explicitly addresses the restoration and management of IRES (Fig. 5.4.1).

We begin with a brief outline of the distinctive features of IRES from a restoration perspective. We then summarize the general principles and approaches now adopted in stream restoration, focusing particularly on the importance of understanding the broader landscape processes (both natural and anthropogenic) influencing a stream when assessing and prioritizing restoration actions. After considering the specific types of disturbances and associated restoration actions that are most relevant to IRES,

Table 5.4.1 Restoration steps for IRES, associated issues, and management emphases

Restoration step	Issue	Management emphases
System understanding	Poor characterization of hydrology (particularly at zero flow)	Installation of depth recorders, mapping of deep pools, longitudinal surveys of channel
	Poor characterization of system condition due to dynamism with respect to hydrological state	Requires long-term data collected from degraded and unimpacted systems
	Incomplete understanding of how system originally worked due to limited unimpacted analogs	Compile and synthesize relevant technical literature; develop testable hypotheses about how the system works
Degrading forces	Direct impacts on aquatic habitat and fauna during dry phase	Fencing of channel to prevent livestock and vehicle access, removal of barriers, limit water extraction during zero-flow periods
	Impacts of catchment degradation, exacerbated by ephemerality and hydrological flashiness	Sediment and nutrient management, revegetation, riparian buffers
	Impacts of catchment storage on hydrology (timing and duration of zero-flow periods and surface water persistence)	Farm dam/pond regulation, low-flow by-passes
	Strong interaction between geomorphology and hydrology	Protect and restore deep pools, prevent livestock access, manage sediment sources by revegetation, consider allowing channel migration
	Limited information about site history	Obtain oral histories, historical land surveys, and records
Objective setting	Strong dynamism of biota and ecological functions reflect hydrological phase	Set objectives that reflect hydrological dynamism of system (dynamic targets at relevant spatial and temporal scales)
	Strong but gradual cumulative landscape and downstream benefits of site-scale restoration	Set objectives that recognize downstream and catchment benefits
Management delivery	Relatively more stream length to manage/restore	Management should be planned and undertaken at large scales to maximize cumulative responses and ensure key ecological processes (e.g., connectivity) are restored or maintained
Monitoring	Dynamism of biota and function reflect hydrological phase	Indicators should be chosen with respect to hydrological phase: flowing, nonflowing, and dry
	Responses to restoration may be enhanced, delayed, or modified by hydrological sequences	Long-term monitoring should be anticipated to ensure hydrological phases and sequences

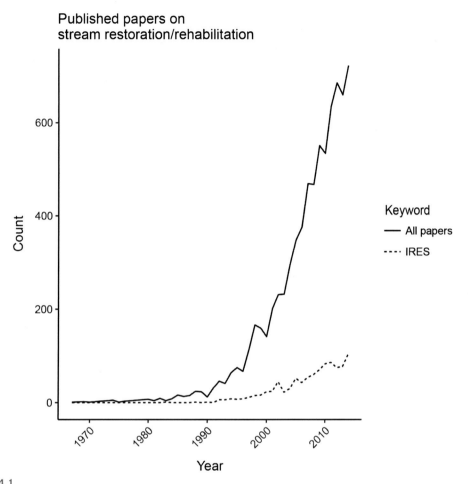

FIG. 5.4.1

Temporal trends in the number of papers published on stream restoration (solid line) and those that refer explicitly to hydrology and IRES-relevant terms more specifically in the title, abstract, and keywords (broken line).

we conclude with suggestions of things to consider now and into the future. Throughout the chapter, we aim to emphasize the importance of using system understanding to guide restoration actions and being aware of potential bottlenecks that may affect system recovery by considering catchment-scale processes potentially influencing restoration trajectories.

5.4.2 IRES AND THEIR CATCHMENTS

Intermittent streams have regular flow periods of differing magnitudes and predictability whereas ephemeral streams have only occasional episodic flows that may include extreme floods. In intermittent streams, flows are either highly variable in volume and timing (e.g., many arid and semiarid regions, Levick et al., 2008) or predictably seasonal (e.g., Mediterranean climate regions, Cooper et al., 2013). Many IRES occur in the headwaters (stream orders 1–3) of perennial systems and can be the crucial water-gathering streams of downstream rivers. For example, in the well-watered Triassic Basins Ecological Region of North Carolina, Russell et al. (2015) located 105 intermittent stream origins in comparison to 52 perennial stream origins, with 35% of stream length being intermittent. The need to conserve headwater streams is widely recognized (Lowe and Likens, 2005), mainly for the supply of good quality water. Water supply areas often provide the protection for headwater streams that is often lacking in areas cleared for farming.

Flow in streams can occur along three spatial dimensions—longitudinal, lateral, and vertical—and along the critical dimension of time (Ward, 1989; Chapter 2.3). In IRES, the duration of flow is critical. When flows occur, avenues of connectivity of biota, sediments, organic matter, nutrients, and trace elements are created, and their strength, directionality, and duration fluctuate with changes in flow (Larned et al., 2010).

In arid and semiarid regions, ephemeral streams can dominate (Levick et al., 2008). These streams are characterized by long dry periods interrupted by sudden episodic floods when heavy rain falls on their dry, impervious catchments. When flow occurs in ephemeral streams, the major axis of flow is longitudinal. If the channel is shallow and relatively unrestricted, there may be channel widening (Friedman and Lee, 2002). Vertical flow to the ground water is usually very limited, although it may occur when major floods occur in quick succession (Baillie et al., 2007). Connectivity involving biota and ecological processes is generally limited in ephemeral streams, and periodic opportunities for dispersal during floods can be critical to biotic persistence (Chapter 4.5). The condition of the catchment largely controls the amount and type of materials exported downstream during flash floods (Steward et al., 2012). Floods may carry large amounts of sediments, nutrients, and organic matter and, in the process, reshape channel morphology (Bull, 1997; Levick et al., 2008; Chapter 2.1). These flows can have considerable transmission losses as they move downstream and may even cease flowing downstream (Chapter 2.3).

Intermittent streams usually have clearly delimited channels and flows span a range of regimes from irregular to regularly seasonal. Like ephemeral streams, intermittent ones when they flow have longitudinal connectivity, but they differ in having both lateral and vertical connectivity. As flows build, water may move sideways into the parafluvial margins; further increases in flow may move water out of the channel and into and over the riparian zone. With water in the channel, the hyporheic zone can become activated, linking surface water with the groundwater table. The hyporheic zone with its specialized biota and range of biogeochemical processes can exert a strong influence on the ecological dynamics of the channel biota (Boulton et al., 1998).

Riparian zones in ephemeral streams in arid and semiarid regions may be limited in species composition, cover, and spatial extent. Nevertheless, in arid areas riparian habitats can have a strong influence on the distribution and abundance of biota (Chapter 4.6) and on ecological processes (Shaw and Cooper, 2008). As channel flow and groundwater availability become more substantial in intermittent streams, riparian zones become more robust and productive. Even though they may represent only a small percentage of the catchment area, they may contain high levels of biodiversity compared to the catchment as a whole.

5.4.3 STREAM RESTORATION BACKGROUND

The fields of both restoration ecology and stream restoration have evolved considerably over the last few decades, with a considerable shift in emphasis away from thinking about restoration at individual sites to thinking about individual sites within a broader landscape/catchment context (e.g., Fig. 5.4.2;

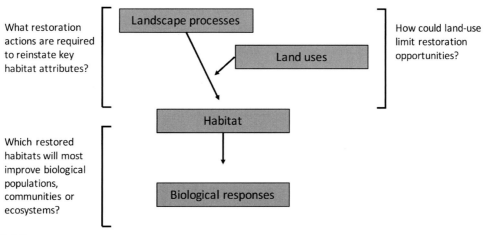

What restoration actions are required to reinstate key habitat attributes?

How could land-use limit restoration opportunities?

Landscape processes

Land uses

Habitat

Which restored habitats will most improve biological populations, communities or ecosystems?

Biological responses

FIG. 5.4.2

Diagram highlighting the role of a landscape perspective in deriving system understanding and planning restoration activities.

Modified from Beechie, T.J, Pess, G., Roni, P., Giannico, G., 2008. Setting river restoration priorities: A review of approaches and a general protocol for identifying and prioritizing actions. N. Am. J. Fish. Manag. 28, 891–905.

Bond and Lake, 2003a; Beechie et al., 2008, 2010). To a large degree, this shift in emphasis has arisen from the recognized failings from approaching restoration within a narrow "site-specific" perspective. A good example of this shift comes in managing restoration of salmonid habitat in streams of the Pacific North West impacted by logging. Here, there has been a gradual evolution from a focus on recreating local habitat-bearing log-jams to restoring connectivity between high-quality habitats and the processes that support natural habitat regeneration, and finally to considering off-shore drivers affecting the population dynamics of diadromous fishes (e.g., Roni et al., 2002; Beechie et al., 2008, 2010). Such shifts in thinking are equally relevant to IRES where connectivity with perennial streams can be critical, and many of the degrading forces and factors that can constrain recovery occur at landscape scales.

Arguably, one of the most important steps in carrying out successful restoration is to first compile a detailed understanding of the stream and its catchment, including the physical and biological processes that operated in the past, operate currently, and are necessary to support a more desirable state (Table 5.4.1; Fig. 5.4.2; Beechie et al., 2008, 2010). This involves gathering historical and contemporary information on climate; geology; catchment geomorphology; vegetation cover; land use; surface and groundwater hydrology and water quality; and the nature, localities, and historical sequences of human settlement and land-use change (e.g., Beller et al., 2016). Stream characteristics to assess include channel morphology and substrata, extent and condition of riparian zones, hydrology (both volumes and timing of flowing and nonflowing periods), and human interventions (e.g., dams, weirs, channel works, bridges).

A vital but often neglected step is to understand the historical sequence of disturbances leading to degradation (e.g., Mika et al., 2010; Beller et al., 2016). This may include the interactions between natural disturbances and the degradation from both legacy (e.g., gold dredging, bushfires) (Bain et al., 2012) and contemporary disturbances (e.g., grazing, water extraction, invasive species). For example, the creation of arroyos (eroded gulches with steep sides) in IRES of southwestern United States largely occurred from the 1880s to the 1920s with the advent of widespread livestock grazing and frequent floods (Turner et al., 2003; Webb and Leake, 2006). With the cutting of the arroyos, groundwater levels subsequently dropped, threatening the integrity of the riparian vegetation. Attempts must be made to determine the forms of degradation created by the various anthropogenic and natural disturbances and their interactions. In many cases, this may be difficult, largely due to poor historical records. Nevertheless, the historical and current regime of disturbances should be estimated, if not understood, to determine appropriate restoration strategies (Roni et al., 2002). If this is not attempted, then further disturbances or mixtures of disturbances may compromise the restoration effort. This step and subsequent steps in the restoration sequence are summarized in Table 5.4.1.

With an understanding, however incomplete, of the stream and its catchment and of the disturbance regime, an attempt should then be made to devise an overall vision or guiding image of the restored stream and its catchment (Palmer et al., 2005; Mika et al., 2010). Such a vision of the restored condition is essential to guide site selection and types of restorative interventions to achieve project objectives.

The selection of the location and types of interventions that are required depends largely on spatio-temporal scaling. If the restorative intervention is carried out at a large spatial extent (e.g., catchment), then the overall effects will take a long time to eventuate. In contrast, restoration efforts at small spatial extents (e.g., isolated sites) may be relatively rapid to implement and the planned effects may emerge in the short term. If the aim of restoration is to remove a single structure (e.g., dam, barrier) or remove point pollution sources, then the intervention is local, albeit with large-scale positive downstream effects along the stream. However, in many cases the disturbances creating the degradation may affect the whole catchment, even though restoration activities may be occurring at more local scales (Bond and Lake, 2003a; Kondolf and Podolak, 2014). This spatial mismatch may mean that only modest results can be expected from restoration and that prior planning must account for the influence of catchment and landscape processes in deciding what, where, and when to restore (Beechie et al., 2008). This is especially true in the case of catchment disturbances affecting IRES because of the often diffuse and dynamic nature of the terrestrial-aquatic boundary (Chapters 2.3 and 4.9).

After selection of the area or areas of a catchment to be restored, quantifiable goals/targets need to be set. One goal may be to attain the reference condition determined by the conditions in relatively undisturbed streams of the same stream type. For IRES, there may be considerably less published information or primary data to help define targets, due to the historical tendency to focus research on perennial streams in many parts of the world (although there are regional exceptions, Chapter 1). While historical data may be obtained on geomorphology and catchment land use, data describing the hydrology, biota, and ecosystem processes may be limited. The lack of hydrological data for IRES is particularly notable. This occurs because far fewer gauges have historically been located on IRES compared with perennial streams (Chapter 2.2) because IRES often contribute less to catchment water yields. Furthermore, even with gauging data, when cease-to-flow periods occur, it is difficult to determine hydrological attributes, such as the nature of the groundwater system and the distribution and quantity of persistent wetted sections of stream, which provide critical refuge habitats (Jaeger and Olden, 2012). As a result, there is often very poor information on the current and historical flow regimes and the characteristics of IRES to use in understanding their ecology and guiding objective setting. This is further compounded by inadequate descriptions in published ecological work of key hydrological parameters (Fig. 5.4.1), making it difficult to compare across studies.

Despite these challenges, we implore practitioners to try and piece together this historical perspective as there is an obvious and vital role for historical ecology in guiding restoration interventions (Wohl, 2005; Beller et al., 2016). Nontechnical accounts can be valuable, such as the notes from early explorers and from newspaper clippings, both of which can help create a picture of the historical landscape or mention biota with known habitat requirements that may no longer be present. The latter is particularly true for species targeted by recreational anglers and hunters. Photographs, especially "repeats" from the same place, can be valuable for assessing geomorphological and vegetation changes (Turner et al., 2003). With this knowledge from both scientists and informed locals, a plan of expectations could be compiled for the various components and goals set for both flowing and dry periods.

In many cases, due to the strength and forms of legacy and current disturbances, a vision based on reference conditions is infeasible. Instead, it may be necessary to accept irreversible changes and plan for restoration to achieve a novel state (Hobbs et al., 2009). Planning for such a state should be ecologically based and not simply a managerial compromise. In IRES, major components to consider in planning include the riparian vegetation (Stromberg et al., 2009) and fish (Moyle, 2014), the latter often being underappreciated as inhabitants of IRES (Chapter 4.5).

Finally, to assess progress, indicators must be selected. Ideally, they are selected and monitored long before the restorative intervention to provide before-and-after signals of change. Ideal indicators should be inexpensive, easy to measure, and capable of providing a reliable indication of change due to the restoration measure or measures. The feasibility of each of these characteristics will vary by region and with the objectives being measured. It is also important to distinguish between the indicators of intervention measures (e.g., environmental flows) and the indicators responding to the intervention. Some indicators, such as riparian vegetation and channel form, may be sampled at regular intervals while others may need to be monitored on particular occasions (event-based) such as when there is water in the stream (e.g., water quality, aquatic biota; Table 5.4.1, Chapter 5.1).

5.4.4 DEGRADATION AND RESTORATION IN A CATCHMENT CONTEXT

Currently, there is a strong push to examine degradation and recovery processes, and hence restoration activities, from a landscape or whole-of-catchment perspective (Fig. 5.4.2). A major challenge in assessing IRES is discerning where their channels and catchment areas begin, which in their uppermost reaches are often poorly defined. In addition, groundwater features such as recharge areas and aquifers are rarely well mapped, even though they may be critical to streamflow. Consequently, many IRES headwaters are poorly protected by legislation due to inadequate mapping and, in some cases, a lack of concern for IRES (e.g., Svec et al., 2005; Fritz et al., 2013; Chapter 5.2). For example, ephemeral streams are not considered in the US Clean Water Act (O'Connor et al., 2014). Given this limited protection and as their catchments are exploited by humans for a range of uses (Chapter 5.1), the spatial extent and strength of anthropogenic disturbances affecting IRES can be formidable (e.g., Levick et al., 2008; Cooper et al., 2013). Most of the specific threats overlap with those that affect perennial waterways, but the magnitude of impact in many cases can be more severe. Anthropogenic drivers of degradation include livestock grazing, land clearing, mining, timber harvesting, groundwater withdrawal, streamflow diversion, channelization, urbanization, cropping, roads and road construction, off-road vehicle use, camping, hiking, and vegetation conversion (Levick et al., 2008).

Catchment-wide disturbances can be climatic (e.g., floods, droughts, heat waves) or anthropogenic (e.g., land clearing, urbanization, invasive species); in many cases, mixtures of these disturbing forces interact and co-occur (Cooper et al., 2013). Wild fires can damage vegetation cover, including riparian vegetation (Verkaik et al., 2015; Douglas et al., 2015), triggering erosion that can leave a considerable legacy of degradation in affected streams (Reich and Lake, 2015; Verkaik et al., 2015). Similarly, past grazing and floods have left the legacy of arroyos in streams of southwestern United States (Turner et al., 2003, Webb and Leake, 2006). There can also be large-scale disturbances emanating from specific sites that exert strong degrading effects downstream. Examples include point-source pollution from mining operations with widespread toxic effects downstream (Lottermoser, 2003) and dams below which flow may be strongly regulated and/or reduced (Kondolf et al., 2013). In many cases, flow regulation co-occurs with agriculture and pollution (e.g., Cooper et al., 2013; Theodoropoulos et al., 2015), requiring effective restoration to address multiple sources of disturbance. At the subcatchment scale, these may include disturbances such as heavy sedimentation from upstream sources, disruption of riparian zones, stream channelization, and salinization (Cooper et al., 2013).

In urbanized landscapes, the piping and filling of streams (i.e., stream burial) is an extreme form of modification that often affects IRES (Roy et al., 2009). For example, in Baltimore City, Maryland, 66% of all stream reaches have been buried (Elmore and Kaushal, 2008). These low-order IRES are typically areas of high biological activity in undisturbed systems (Meyer and Wallace, 2001). Piping of headwaters causes downstream impacts via increased flow velocities, altered carbon and nutrient inputs, and amplified nitrogen transport; these effects can be exacerbated by increased climatic variability (Kaushal et al., 2008). In the United States, there is particular interest in understanding how headwater degradation affects downstream waters because of important ramifications for regulation under the US Clean Water Act (Leibowitz et al., 2008; Chapter 5.3).

Extraction of groundwater by bores, wells, and pumping from the channel can lower the groundwater table, leading to a loss of flow and increases in cease-to-flow events (Boulton and Hancock, 2006; Wohl et al., 2009). In addition, increasing catchment plant cover (e.g., from grassland to tree plantations, Adelana et al., 2015) can lower the groundwater table and decrease streamflow. With time, reducing groundwater extraction from catchments can elevate the groundwater table and restore streamflow and riparian vegetation (Katz et al., 2009). Another form of a severe catchment-scale disturbance affecting IRES is dryland salinity caused by large-scale clearing of native vegetation. In southwestern Western Australia, for example, the replacement of deep-rooted native vegetation with shallow-rooted crops has led to rising saline groundwater tables, which can then salinize the IRES (Halse et al., 2003). In both the cases of catchment disturbance described earlier, effects on IRES may be amplified by the periodic lack of flow. Thus, the accumulation of pollutants during cease-to-flow periods may generate significantly higher pollutant concentrations in IRES than those in perennial streams, and the periodic lack of connectivity can mean that most aquatic fauna cannot escape to less impacted sites.

Disturbances acting at these catchment scales can be extremely difficult to reverse, in part because of the sheer scale of the impacts and in part because there may be strong positive feedbacks reinforcing the degraded state. For example, to address catchment salinization, it is estimated that 40%–50% of a catchment must be revegetated. In itself, this is an ambitious target for large catchments, but can be almost impossible to achieve in low rainfall areas where the soils are heavily degraded (Ghassemi et al., 1995). In other cases, channel incision can greatly increase the shear stress associated with even small floods, exacerbating the problem of trying to restore natural channel morphology. However, there may be isolated sources of degradation such as gully erosion contributing sediment to streams that can be efficiently targeted by restoration activities to reduce a catchment-scale impact (Hermoso et al., 2015). A consistent theme for ameliorating catchment-scale disturbances (as for others discussed below) is thus the need to undertake careful planning to decide on the most effective and efficient means of achieving restoration goals (Beechie et al., 2008).

5.4.5 HYDROLOGICAL IMPACTS AND THEIR RESTORATION
ALTERED FLOODING REGIMES

As with perennial streams, floods are a powerful shaping force in IRES. Large floods can reshape the channel, destroying and building new stream habitat, rejuvenating riparian vegetation, replenishing groundwater levels, and distributing plant propagules (e.g., Stromberg, 2001; Stromberg et al., 2007; Levick et al., 2008; Kondolf et al., 2013). Infrequent floods trigger riparian tree (e.g., cottonwood *Populus* spp.) recruitment in ephemeral sand-bed streams (Friedman and Lee, 2002), often with decadal intervals between floods and the accompanying recruitment. Similarly, large floods down IRES in the arid and semiarid lands of Australia promote recruitment of river red gum (*Eucalyptus camaldulensis*). For river red gums to remain healthy, flooding is needed every 3–4 years (Roberts and Marston, 2011; Catelotti et al., 2015). In both cases, recruitment can be greatly reduced by insufficient soil wetting, low groundwater levels, and grazing livestock. In IRES, the use of environmental flows released from dams to mimic natural floods could be a powerful restoration measure to restore riparian vegetation, but well-documented cases remain to be found.

The impacts of such extreme events, particularly floods, droughts, and wildfires, may be greatly magnified by human degradation of catchment condition. Extreme floods aided the creation of the arroyos in southwestern United States, cutting deep channels and removing riparian vegetation (Webb and Leake, 2006). Land uses such as intensive agriculture can exacerbate the impacts of floods. For example, a flash flood down a catchment which had been cleared and the channel straightened produced very heavy sediment loads and damaging erosive changes in channel configuration whereas in another catchment where intensive agriculture was declining and reforestation was extensive, the same flood event produced much lower sediment loads (Ortega et al., 2014). However, there was still severe erosion at sites with active channelization works that produced increased velocities and erosive energy.

LOSS OF FLOWS

Urbanization can threaten riparian zones by channelization (divorcing the stream from its riparian zone) and by clearing or fragmenting the riparian zone for the construction of buildings and infrastructure (Goodwin et al., 1997; Groffman et al., 2003). A further threat from urbanization and agricultural activities is the diversion of surface flows, reduced infiltration from impervious surfaces, and the extraction of groundwater, all of which may lower groundwater tables below the root zones of riparian plants (Stromberg et al., 1996, 2005; Webb and Leake, 2006) and contribute to increased flash flooding. The latter threat may favor the invasion of exotic plants (from ground cover to trees) which may supplant the native flora through competition and/or by taking advantage of the lowered groundwater tables (Lite and Stromberg, 2005; Richardson et al., 2007; Stromberg et al., 2007).

INCREASED PERIODS OF ZERO FLOWS

As dry periods are intrinsic to IRES, much of the biota has strategies to survive such events (Chapter 4.8); these strategies may even allow survival through severe supraseasonal droughts (Lake, 2011). From a restoration management point of view, both the timing of flows and spatiotemporal pattern of channel drying are critical. There are at least three distinctive spatial patterns of stream drying: the upstream channel is perennial whereas the downstream channel is intermittent and dries out either completely or to pools, the upstream channels are intermittent while downstream channels remain perennial, or the stream from its source to downstream reaches goes from being perennial to intermittent along its length (Lake, 2003; Chapter 2.3). Each of these patterns has different implications for the ability of individual taxa to access suitable refuge habitats. During nonflowing periods, headwater refuges may benefit very different taxa to those occurring downstream in refuge pools of lowland rivers (Chapter 4.8).

Frequent or prolonged drying of the channel may allow dry streambeds to be used by terrestrial animals to move, forage, or shelter, and colonized by terrestrial plants (Steward et al., 2012). Dry streambeds are also used by humans as a transport corridor and as venues for cultural activities. Changes in patterns of flow intermittence can arise from a number of causes. Farm dams or ponds, mainly for supplying water for livestock, are common on many dryland catchments. However, they can reduce stream flow and greatly extend the natural cease-to-flow period by acting as additional storage and evaporation basins (Callow and Smettem, 2009; Nathan and Lowe, 2012), with particularly damaging effects during and after droughts (e.g., O'Connor, 2001). The impact of farm dams on streamflow is especially critical during low rainfall seasons, capturing up to 30% of net runoff and greatly prolonging nonflowing periods, an effect that will be exacerbated by climate change in many areas (Fig. 5.4.3).

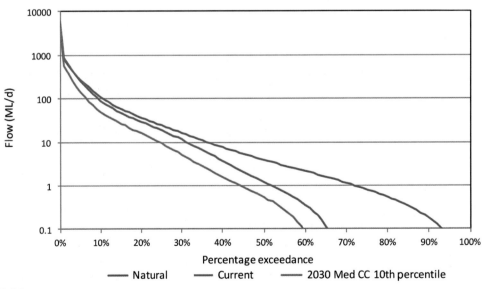

FIG. 5.4.3

Flow duration curves for Castle Ck, a second-order IRES in southeastern Australia. Individual curves represent the natural (blue line), current (green line), and projected future flows (orange line), highlighting the relative influences associated with farm dams (current) and climate change impacts (2030 med) on flow permanence.

BOX 5.4.1 ARTIFICIALLY INCREASED FLOW PERMANENCE IN IRES: EFFECTS AND RESTORATION CONSTRAINTS

The effects of artificially increased flow permanence, particularly when perennial flows replace previously intermittent ones, largely fall into three categories: (1) changes in the composition and abundance of flow-dependent biota; (2) associated changes in susceptibility to invasion by nonnative species; and (3) changes in water chemistry and biogeochemical cycling, especially at the soil-water interface. The lack of direct case studies means that these effects are simply inferred from comparisons between naturally perennial and naturally or artificially intermittent reaches. For example, most studies (Fritz and Dodds, 2005; Wood et al., 2005) report higher diversity and abundance of aquatic invertebrates in perennial streams. Similar trends have also been reported for fish (Rolls et al., 2012), and in most cases, there is some degree of nestedness (i.e., the fauna of intermittent sites is largely a subset of that from perennial sites (Dodds et al., 2004)).

It would be wrong, however, to conclude that perennialization leads to *positive* outcomes for the biota. For example, many plants and invertebrates have life history strategies that involve dormant seeds or eggs which hatch only after desiccation and rewetting (Boulton, 1989; Brock et al., 2003; Jenkins and Boulton, 2003; Chapter 4.8). Studies of wetlands subjected to increased hydroperiods suggest that increasing flow permanence can act as an major disturbance in these systems (Golladay et al., 1997; Brock et al., 2003) by disrupting life cycles and modifying refuge habitats. In a restoration context, Bond et al. (2010) concluded that restoration of an intermittent flow regime would preferentially suit native fishes over introduced species that proliferated under artificially perennial flows. Conversely, naturally intermittent flows have been shown to facilitate persistence of native fishes in main stem and floodplain wetland habitats, respectively, due to the physiological constraints (e.g., temperature and DO tolerance) of some invasive species (McNeil and Closs, 2007). There are very few examples where naturally intermittent flows have been actively restored. Instead, there may be strong social pressures and, in some cases, conservation targets that promote establishing artificially perennial flows even where these preclude the reinstatement of the locally native ecosystems adapted to intermittent flows (Bateman et al., 2015).

ANTIDROUGHT

While increased intermittence from water extraction is perhaps the most widely considered form of hydrological disturbance affecting IRES (Chapter 5.1), the inverse situation may also arise when streams have higher levels of flows and flow permanence than would occur naturally (so-called antidrought; McMahon and Finlayson, 2003). Such inverse flow patterns are most likely to result from river regulation and the delivery of high flows for human use during the natural low-flow season, a common pattern in irrigation regions around the world (McMahon and Finlayson, 2003; Magilligan and Nislow, 2005; Magilligan et al., 2013). While some hydrological analyses demonstrate that irrigation releases can lead to increases in flow permanence and decreases in flow variability (Fernández et al., 2012; Alexandre et al., 2013), there are relatively few studies that specifically document the effects of increased flow permanence, particularly artificially induced perennial flows (perennialization, Chapter 5.1) on the structure and function of IRES. In part, this may reflect the underreporting of the effects of flow alteration and stream restoration on IRES; seldom have naturally intermittent flows been intentionally restored (Box 5.4.1).

5.4.6 RIPARIAN ZONE DEGRADATION AND RESTORATION

Riparian zones and their vegetation support a wide range of important ecological functions (Ewel et al., 2001). These include bank stabilization, provision of living and dead organic matter (both dissolved and particulate), and of habitat (including wood) for both aquatic and terrestrial biota, capture of sediment, retention and processing of nutrients, and moderation of extreme temperatures by shading (Ewel et al., 2001; Davies, 2010; Burt et al., 2013; Reich et al., 2016). As IRES mainly occur in arid and semiarid regions, the presence of surface water and groundwater fosters the development of distinct riparian zones replete with herbs, shrubs, and trees (Chapter 4.2), which provide key habitat and resources to a rich fauna, from insects to mammals. This fauna can be quite different from and richer than the fauna in the surrounding resource-poor hinterland (Sabo et al., 2005; Johnson et al., 2007).

Given this importance, it is not surprising that restoration of IRES has concentrated on riparian zones rather than the much more difficult task of restoring catchments. Indeed, restoring riparian zones may be a way to quarantine streams from the impacts of disturbances in their catchments. Many disturbances such as the clearance of floodplains, livestock grazing, and the spread of exotic plants impact on riparian zones of IRES and may operate synergistically. The nature and variety of disturbances largely sets the agenda for restoration projects. Often, catchment clearing will decrease the width and tree density of riparian zones as well as fragmenting their continuity (Goodwin et al., 1997; Stella et al., 2013; Beller et al., 2016). If livestock are allowed access to the riparian zones and streambeds, vegetation can be lost through grazing, further compromising bank stability, and aquatic habitats may be destroyed by trampling and addition of excess nutrients by defecation (Fig. 5.4.4; Line et al., 2000; Steward et al., 2012; Reich et al., 2016). Furthermore, maintaining access for stock to drink from remnant pools is often seen as a factor motivating farmers to keep riparian zones along IRES unfenced.

Invasion of riparian zones of IRES by highly competitive plants, including trees and grasses, can be a major problem (Richardson et al., 2007). In southwestern arid United States, *Tamarix* (tamarisk, saltcedar) and *Elaeagnus angustifolia* (Russian olive) are widespread and highly successful invaders of riparian zones along both intermittent and perennial streams (Friedman et al., 2005; Shafroth et al., 2005; Stromberg et al., 2009). Tamarisks readily invade riparian zones of IRES and there have been many projects involving millions of dollars to eradicate this plant (Shafroth et al., 2005; Shafroth and Briggs, 2008; Sher and Quigley, 2013). Tamarisk invasion and spread are substantially aided by a variety of anthropogenic disturbances, ranging from riparian vegetation clearing to regulated river flows. With its capacity to survive dry periods (including droughts), its high tolerance of soil salinity and its capacity to reach deep sources of groundwater, tamarisk may be a symptom of river degradation. Nevertheless, it also offers a means to prevent riparian zone erosion and provide shelter for terrestrial wildlife (Shafroth et al., 2005; Stromberg et al., 2007, 2009; Shafroth and Briggs, 2008; Sher and Quigley, 2013).

FIG. 5.4.4

A dry IRES in southeastern Australia showing the effects of stock access and overgrazing on bankside vegetation and morphology. Note the pugging of sediments in the channel from stock access during a moist phase.

Photo: Courtesy P. Reich.

Many IRES, some of considerable size, occur in arid to semiarid rangelands where production of livestock (especially sheep and cattle) is the major economic activity. In such areas, riparian zones offer shade, shelter from the heat, and access to water and forage (Kauffman and Krueger, 1984). Livestock having access to riparian zones can degrade water quality; alter stream channel morphology and riparian soils; and reduce instream, streambank, and riparian vegetation (Kauffman and Krueger, 1984; Goodwin et al., 1997; Belsky et al., 1999; Reich et al., 2016). Fencing off the riparian zone of streams to prevent livestock access can help restore both the stream and the riparian zone (Fig. 5.4.5; Kauffman and Krueger, 1984; Line et al., 2000; Kueper et al., 2003; Miller et al., 2010). In implementing fencing, it is important for the riparian zone to approach its pregrazing width (Hansen et al., 2015).

While there have been many livestock exclusion projects along IRES, published studies of riparian restoration by livestock exclusion are few (Reich et al., 2016). In a survey of fenced-off riparian zones along intermittent and perennial creeks, Burger et al. (2010) found that soil carbon concentrations increased with restoration and that soil nitrate and plant-available phosphorus were influenced by the concentrations in adjacent farmland. Sunohara et al. (2012) in a 5-year study on an IRES compared a 356-m length of riparian zone (3–5 m wide) with restricted cattle access with a 348-m riparian section downstream with cattle access. Excluding cattle from the riparian zone and channel of the IRES reduced mean loads of nutrients, total coliform bacteria, *Escherichia coli*, and *Enterococcus* spp. at both low and high flow conditions. Despite the short lengths of stream involved, these results indicate that intact riparian zones and restricted livestock access can reduce stream loads of nutrients and pathogenic bacteria even under the fluctuating flow conditions of IRES.

While fencing to exclude livestock and allowing natural regeneration of riparian vegetation is seen as an extremely cost-effective restoration strategy, several studies have demonstrated that the depletion of the original seed-bank coupled with the predominance of propagules of invasive plants, especially introduced pasture and weed species, may render this approach largely ineffective (Williams et al., 2008; O'Donnell et al., 2016). For example, in southeastern Australia, fencing off riparian zones and excluding livestock grazing along IRES resulted in an introduced pasture grass (*Phalaris aquatica*) rapidly coming to dominate the ground cover (Williams et al., 2008; Reich et al., 2016). Active restoration

FIG. 5.4.5

A drying IRES in southeastern Australia that has been fenced from stock and replanted. Note replantings in the background and significant growth of fringing vegetation.

Photo: Courtesy P. Reich.

involving physical and/or chemical removal of invaders and boosting native plant recruitment by adding seeds and/or seedlings may be necessary for effective recovery (Richardson et al., 2007). Many of the invading plants are nitrophilous and thus lowering the soil nitrogen levels may favor native plants. One approach that has been trialed effectively is the addition of freely available carbon (e.g., sugar) to riparian soils which increases nitrogen uptake by bacteria, thereby impeding the growth of invasive plants (e.g., Prober et al., 2005; Cole et al., 2016).

Riparian restoration of IRES may also restore terrestrial faunal assemblages. The restoration of bird assemblages appears to rely on successful restoration of native vegetation, together with the size of the restored area. Four years after the removal of cattle along perennial and intermittent sections of the San Pedro River in the San Pedro Riparian National Conservation Area, there was a substantial increase in riparian vegetation and abundance and species diversity of birds (Kueper et al., 2003). In contrast, after 8 years of riparian vegetation restoration, involving the removal of livestock and replanting along IRES in southern Australia (Fig. 5.4.6), recovery of the bird fauna was difficult to detect (Hale et al., 2015). These studies illustrate that the time for successful riparian restoration varies considerably, depending on the initial conditions, target indicators, and restoration goals.

FIG. 5.4.6

Aerial view of a site with riparian fencing and replanting along the banks of an IRES in southeastern Australia (see Reich et al. (2016) for details).

Photo: Courtesy P. Reich.

5.4.7 RESTORATION OF INSTREAM HABITAT AND REFUGES

For many degraded streams, restoration of instream habitat is one of the most commonly employed restoration strategies (Bernhardt et al., 2005, 2007). In a recent survey of restoration methods employed globally, "channel hydromorphic" and "instream hydromorphic" approaches accounted for 70% of restoration projects (Palmer et al., 2014). Such habitat restoration activities can occur at a number of scales.

At a local scale, habitat restoration most commonly focuses on the addition of habitat structure (e.g., rock, large woody debris) to provide cover and help restore local geomorphic diversity. For example in streams degraded by sedimentation, channelization, and removal of fallen timber, the addition of instream structures can increase hydraulic diversity and begin the process of restoring channel morphology over relatively short timescales. This action can thus complement the processes expected to unfold over decades following riparian restoration. Such methods have been shown to be effective in IRES but their effectiveness may wane during severe dry periods. In a survey of two sand-slugged IRES, Bond and Lake (2003b) reported that fish were mostly found around structurally complex habitat, notably that created by large wood. In an attempt to restore habitat and both fish and invertebrate assemblages, wood structures (railway sleepers) were positioned at nine sites in each creek devoid of existing habitat structure. Fish surveys occurred before and after wood installation. In the short term, there were increases in abundance, notably of the galaxiid fish *Galaxias olidus* at the wood structures (Bond and Lake, 2005). However, during a period of extreme drought, fish disappeared from these restored reaches, but persisted within remnant refugial pools (Bond, 2012). As sand sedimentation has greatly reduced the availability of these pools, and as the drought was particularly severe (10 years), subsequent recolonization was predicted to be slow (Bond, 2012).

At larger spatial scales, there have been efforts to restore the reach-scale channel form by constructing stable, possibly meandering, channels (Kondolf, 2012; Kondolf et al., 2013). However, given the tendency for periodic large floods in many IRES, such "stable" channels are prone to destruction. Instead, restoration should promote channels capable of evolving in response to these episodic floods (Kondolf et al., 2013; Ortega et al., 2014; Beagle et al., 2016). Restoration of this flood-dependent nature involves restoring zones both longitudinally and laterally where the river in flood "can flood, erode, deposit, and shift course without conflicting with human uses" (Kondolf et al., 2013). Such a zone has been called the channel migration zone (Beagle et al., 2016), "territorio fluvial" (Ollero and Ibisate, 2012) or "espace de liberté" (Kondolf, 2012). As Kondolf (2012, p. 239) says, "The most effective approach to restoring rivers will often be for us to stand aside, and give the river its space." Large episodic floods in IRES are a vivid example of the power of extreme events (Reich and Lake, 2015), although their restorative powers may be limited in cleared catchments.

5.4.8 RESTORING REFUGES AND COLONIZATION PATHWAYS

IRES are, by their very nature, highly variable systems with disturbances at both ends of the flow regime ranging from seasonal and supraseasonal droughts to floods (Chapters 2.2 and 2.3). Although native biota are typically adapted to persist over multiple generations in these environments, the ability to move between breeding, residential, and refuge habitat areas can be critical (Bond and Lake, 2003a). Thus, as well as restoring permanent refuges, it is necessary to restore subsequent postdisturbance recolonization pathways (connectivity). If not, then restoration is unlikely to succeed (Bond et al., 2008; Robson et al., 2013).

Refuge habitats are used by a wide spectrum of organisms from attached algae to micro- and macroinvertebrates and fish (Boulton and Lake, 2008; Robson et al., 2008, 2013). In IRES restoration, provision of persistent and drought-resistant refuge habitats is a critical challenge (Bond et al., 2008). Restoration of refuge habitats can be compromised in sand-bed IRES by the mobility of the streambed on the falling limb of the hydrograph, which infills pools prior to the loss of flow.

Restoring connectivity as a means of facilitating recovery following drying events can also be critical. For example, Dexter et al. (2014) demonstrated the need for hydrological connectivity to allow fish (*Galaxias olidus*) to access breeding sites within headwater reaches of a lowland stream network. This only occurs during nondrought years, when flows persist throughout the breeding season. In very dry years, access to breeding habitats is likely blocked (Dexter et al., 2014). Given that *G. olidus* live for only 3–4 years, relatively frequent connections are critical to their population persistence. Furthermore, to achieve hydrological connectivity, removal of artificial barriers may be an important restoration activity. In IRES, the predominance of low flows can mean that even small barriers, such as road pipes and culverts, can prevent recolonization following the recurrence of stream flows.

One aspect of connectivity that warrants consideration is the connection between multiple small-scale restoration projects distributed throughout the catchment (Kondolf et al., 2008). If restoration can only proceed at the local scale and projects are incrementally begun, then at the catchment level, a program of restoration should be devised that maximizes ecological benefits as sites are sequentially added. This raises the question: if restoration projects are at small spatial extents, are there ways by which their spatial distribution can be implemented to provide progressively greater restoration benefits at larger spatial extents? This question appears to be unresolved, and yet it remains pertinent to IRES restoration (Table 5.4.1).

5.4.9 MEASURING RESTORATION TRAJECTORY

The selection of indicators to assess the success of restoration projects can be a challenge, especially in IRES. To identify salient indicators for loss of connectivity and habitat fragmentation in an IRES, the Odeouca in Portugal, prior to the completion of a large dam, Hughes et al. (2010) assessed a range of potential indicators from macrophytes to birds. The most appropriate indicators of future flow changes were benthic macroinvertebrates and endemic fish. In many streams, changes in flow from perennial to intermittent and vice versa rely on the groundwater levels. One example is the San Pedro River in Arizona. Changes in groundwater level in this IRES are reflected in riparian vegetation composition with herbs reacting more rapidly than trees to groundwater changes (Stromberg et al., 2006, 2013). In perennial sections, native trees are dominant but with declining groundwater levels, *Tamarix*, tolerant of the low groundwater levels, may invade.

The importance of monitoring ecosystem processes in restoration has become increasingly recognized (Dolédec and Bonada, 2013; Boudell et al., 2015) along with the monitoring of ecosystem services (Palmer et al., 2014). The selection of indicators should consider the prevailing hydrological context (flowing, nonflowing and dry) as well as the timing (i.e., season). Understanding how each indicator responds to antecedent or prevailing hydrological states (e.g., time since last state, duration of current state, Table 5.4.1) is also likely to be important (Chapter 5.1).

5.4.10 THINGS TO CONSIDER FOR NOW AND FOR THE FUTURE

IRES and their catchments experience a wide variety of human disturbances varying in spatiotemporal extent and severity. As many IRES occur in dryland regions and their catchments are often of comparatively low economic value with low human population density, concern for their protection and conservation has lagged behind that directed toward more productive and populated areas. Consequently, human disturbances through land clearing, livestock grazing, low-yield grain production, and mining may be extensive and, in turn, impact severely on the morphology and ecology of dryland IRES.

Compared with restoration of perennial streams, concern for restoration of IRES appears limited, possibly due to their low perceived economic and social values (Chapter 5.2). A further difficulty in attempting to restore IRES lies in their very nature with variable wet and dry periods. This makes setting instream restoration targets an awkward exercise due to the very dynamic ecological changes which occur both when water in the channel is flowing, standing, or absent. Perhaps this explains the finding that, judging from the literature, the most common form of restoration in IRES is riparian zone revegetation.

Successful restoration of IRES is most likely to be achieved by adopting a catchment-wide approach. This involves identifying and assessing the spatial extent of legacy and current disturbances, and implementing interventions to curb or eliminate the disturbances. With lower human populations and less intensive forms of exploitation, it may be less challenging to execute restoration measures at the catchment scale in IRES than in perennial streams.

Successful restoration over a range of spatial extents is tightly linked to the length of time required to reach the set targets. Some targets, such as ground cover in riparian restoration, may be reached in a relatively short period while others such as tree maturity and channel reshaping may take many years, if not decades. Most stream restoration projects take substantial time yet funding and monitoring are rarely supported up to completion. This problem of limited funding is common and may be more severe for IRES restoration as many of these streams are not highly valued (Chapter 5.2) and restoration is slow, being governed by the periodicity of flow.

Most restoration measures can be described as "reactive"—reacting to legacy and current disturbances (Palmer et al., 2008, Bain et al., 2012). However, restoration can be planned and implemented in a proactive way (Palmer et al., 2008), which involves implementing measures now which may curb the forces of future disturbances. Given current and projected declines in rainfall and runoff due to climate change in dryland and Mediterranean regions (Jaeger et al., 2014; Pumo et al., 2016), IRES in these regions can also be expected to flow less often, with reduced flows and less groundwater recharge (Meixner et al., 2016). Thus, restoration could involve proactive measures such as the implementation of environmental flows (Palmer et al., 2009), increasing the effectiveness of refuges (e.g., pools) in streams, reducing water capture and extraction in catchments (e.g., decommissioning farm dams), increasing connectivity (e.g., dam removal), and restoring riparian vegetation (Davies, 2010; Stromberg et al., 2013).

Restoration goals in IRES usually focus on the flowing phase. However, nonflowing or dry streambeds are an essential component in the ecological dynamics of IRES (Steward et al., 2012). They can be degraded by human activities such as rubbish dumping, livestock access and grazing, gravel/sand extraction, and input of wastewater and sewage. Thus, in many cases it is essential that restoration actions in IRES consider both the wet (flowing and nonflowing) and dry streambed conditions.

5.4.11 CONCLUSIONS

A great number and variety of IRES with a wide range of flow patterns occur globally. Judging from the literature, restoration activities appear much rarer in IRES than in perennial streams and their catchments. From local sites to catchments, IRES host a wide range of human disturbances operating over a broad range of spatiotemporal scales and which appear to be less well understood than those operating in perennial systems. Ecological restoration of IRES, as for all restoration projects, needs to be carefully planned with a schedule of successive steps essential to produce ecologically beneficial outcomes.

Ideally, restoration of IRES should be planned and implemented at the catchment scale. However, this rarely occurs. There have been two forms of restoration at this scale: altering catchment land management and large-scale channel restoration. The major form of IRES restoration reported in the literature focuses on riparian zone restoration, with the major activity being revegetation. To be successful, revegetation takes time (years-decades), a requirement which is seldom funded and achieved. At local scales such as sites and reaches, adding habitat structure and bolstering dry-period refuges have been the major restoration activities. Achieving the latter requires careful attention to stream hydrology, including groundwater interactions as well as geomorphic attributes. Where restoration occurs via multiple small-scale restoration projects, such projects must be planned in a way that maximizes the benefits at the catchment scale. This requires prioritizing the types and location of various restoration activities (Beechie et al., 2008).

Finally, there remains a strong need for further monitoring of well-designed restoration projects focused on IRES. Not only must monitoring treat these as long-term projects (rather than as the short-term inadequate and incomplete projects that now prevail), monitoring should occur only where restoration activities have been well designed and planned with specific hypotheses and timelines regarding the expected biophysical response trajectories.

ACKNOWLEDGMENTS

We acknowledge the helpful comments provided by two anonymous reviewers and thank Andrew Boulton for his generous support and insight.

REFERENCES

Adelana, M.S., Dresel, P.E., Hehmeijer, P., Zydor, H., Webb, J.A., Reynolds, M., et al., 2015. A comparison of streamflow, salt and water balances in adjacent farmland and forest catchments in south-western Victoria, Australia. Hydrol. Process. 29, 1630–1643.

Alexandre, C.M., Ferrera, T.F., Almeida, P.R., 2013. Fish assemblages in non-regulated and regulated rivers from permanent to temporary Iberian systems. River Res. Appl. 29, 1042–1058.

Baillie, M.N., Hagan, J.F., Erwurgel, B.E., Wahi, A.K., Eastoe, J., 2007. Quantifying water sources to a semiarid riparian ecosystem, San Pedro River, Arizona. J. Geophys. Res. 112. GO3S02.

Bain, D.J., Green, M.B., Campbell, J.L., Chamblee, J.F., Chaoka, S., Fraterrigo, J.M., et al., 2012. Legacy effects in material flux: structural catchment changes predate long-term studies. Bioscience 62, 575–584.

Bateman, H.L., Stromberg, J.C., Banville, M.J., Makings, E., Scott, B.D., Suchy, A., et al., 2015. Novel water sources restore plant and animal communities along an urban river. Ecohydrology 8, 792–811.

Beagle, J.R., Kondolf, G.M., Adams, R.M., Marcus, L., 2016. Anticipatory management for instream habitat: application to Carneros Creek, California. River Res. Appl. 32, 280–294.

Beechie, T.J., Pess, G., Roni, P., Giannico, G., 2008. Setting river restoration priorities: a review of approaches and a general protocol for identifying and prioritizing actions. N. Am. J. Fish Manag. 28, 891–905.

Beechie, T.J., Sear, D., Olden, J.D., Pess, G.R., Buffington, J.M., Moir, H., et al., 2010. Process-based principles for restoring river ecosystems. Bioscience 60, 209–222.

Beller, E.E., Downs, P.W., Grossinger, R.M., Orr, B.K., Salomon, M.N., 2016. From past patterns to future potential: using historical ecology to inform river restoration on an intermittent California river. Landsc. Ecol. 31, 581–600.

Belsky, A.J., Matzke, A., Uselman, S., 1999. Survey of livestock influences on stream and riparian ecosystems in western United States. J. Soil Water Conserv. 54, 419–431.

Bernhardt, E.S., Palmer, M.A., Allan, J.D., Alexander, G., Brooks, S., Carr, J., et al., 2005. Synthesizing U.S. river restoration efforts. Science 308, 636–637.

Bernhardt, E.S., Sudduth, E.B., Palmer, M.A., Allan, J.D., Meyer, J.L., Alexander, G., et al., 2007. Restoring rivers one reach at a time: results from a survey of U.S. river restoration practitioners. Restor. Ecol. 15, 482–493.

Bond, N., 2012. Fish Responses to Low Flows in Lowland Streams: A Summary of Findings From the Granite Creeks System, Victoria. National Water Commission, Canberra.

Bond, N.R., Lake, P.S., 2003a. Local habitat restoration in streams: constraints on the effectiveness of restoration for stream biota. Ecol. Manag. Restor. 4, 193–198.

Bond, N.R., Lake, P.S., 2003b. Characterising fish-habitat associations in streams as the first step in ecological restoration. Austral Ecol. 28, 611–621.

Bond, N.R., Lake, P.S., 2005. Ecological restoration and large-scale ecological disturbance: the effects of drought on the response by fish to a habitat restoration experiment. Restor. Ecol. 13, 39–48.

Bond, N.R., Lake, P.S., Arthington, A.H., 2008. The impacts of drought on freshwater ecosystems: an Australian perspective. Hydrobiologia 600, 3–16.

Bond, N.R., McMaster, D., Reich, P., Thomson, J.R., Lake, P.S., 2010. Modelling the impacts of flow regulation on fish distributions in naturally intermittent lowland streams: an approach for predicting restoration responses. Freshw. Biol. 55, 1997–2010.

Boudell, J.A., Dixon, M.D., Rood, S.B., Stromberg, J.C., 2015. Restoring functional riparian ecosystems. Ecohydrology 8, 747–752.

Boulton, A.J., 1989. Over-summering refuges of aquatic macroinvertebrates in two intermittent streams in central Victoria. Trans. R. Soc. S. Aust. 113, 23–34.

Boulton, A.J., Lake, P.S., 2008. Effects of drought on stream insects and its ecological consequences. In: Lancaster, J., Briers, R.A. (Eds.), Aquatic Insects: Challenges to Populations. CAB International, Wallingford, CT, pp. 81–102.

Boulton, A.J., Hancock, P.J., 2006. Rivers as groundwater-dependent ecosystems: A review of degrees of dependency, riverine processes and management implications. Aust. J. Bot. 54, 133–144.

Boulton, A.J., Findlay, S., Marmonier, P., Stanley, E.H., Valett, H.M., 1998. The functional significance of the hyporheic zones in streams and rivers. Annu. Rev. Ecol. Syst. 29, 59–81.

Boulton, A.J., Dahm, C., Correa, L., Kingsford, R., Negishi, J., Nakamura, F., et al., 2013. Good news: progress in successful river conservation and restoration. In: Elosegi, A., Sabater, S. (Eds.), River Conservation: Challenges and Opportunities. Fundación BBVA, Bilbao, pp. 331–357.

Brock, M.A., Nielsen, D.L., Shiel, R.J., Green, J.D., Langley, J.D., 2003. Drought and aquatic community resilience: the role of eggs and seeds in sediments of temporary wetlands. Freshw. Biol. 48, 1207–1218.

Brooks, S.S., Lake, P.S., 2007. River restoration in Victoria, Australia: change is in the wind, and none too soon. Restor. Ecol. 15, 584–591.

Bull, W.B., 1997. Discontinuous ephemeral streams. Geomorphology 19, 227–276.

Burger, B., Reich, P., Cavagnaro, T.R., 2010. Trajectories of change: Riparian vegetation and soil conditions following livestock removal and replanting. Austral Ecol. 35, 980–987.

Burt, T., Pinay, G., Grimm, N., Harms, T., 2013. Between the land and the river: river conservation and the riparian zone. In: Elosegi, A., Sabater, S. (Eds.), River Conservation: Challenges and Opportunities. Fundación BBVA, Bilbao, pp. 217–240.

Callow, J.N., Smettem, K.R.J., 2009. The effect of farm dams and constructed banks on hydrologic connectivity and runoff estimation in agricultural landscapes. Environ. Model. Softw. 24, 959–968.

Catelotti, K., Kingsford, R.T., Bino, G.R., Bacon, P., 2015. Inundation requirements for persistence and recovery of river red gum (Eucalyptus camaldulensis) in semi-arid Australia. Biol.Conserv. 184, 346–356.

Cole, I.A., Prober, S., Lunt, I., Koen, T.B., 2016. Nutrient versus seed bank depletion approaches to controlling exotic annuals in threatened Box Gum woodlands. Austral Ecol. 41, 40–52.

Cooper, S.D., Lake, P.S., Sabater, S., Melack, J.M., Sabo, J.L., 2013. The effects of land uses on streams and rivers in Mediterranean climates. Hydrobiologia 719, 383–425.

Davies, P.M., 2010. Climate change implications for river restoration in global biodiversity hotspots. Restor. Ecol. 18, 261–268.

Dexter, T., Bond, N., Hale, R., Reich, P., 2014. Dispersal and recruitment of fish in an intermittent stream network. Austral Ecol. 39, 225–235.

Dodds, W.K., Gido, K.B., Whiles, M.R., Fritz, K.M., Matthews, W.J., 2004. Life on the edge: the ecology of Great Plains prairie streams. Bioscience 54, 205–216.

Dolédec, S., Bonada, N., 2013. So what? Implications of the loss of biodiversity for ecosystem functioning. In: Elosegi, A., Sabater, S. (Eds.), River Conservation: Challenges and Opportunities. Fundación BBVA, Bilbao, pp. 169–192.

Douglas, M.M., Setterfield, S.A., McGuinness, K., Lake, P.S., 2015. The impact of fire on riparian vegetation in Australia's tropical savanna. Freshw. Sci. 34, 1351–1365.

Elmore, A.J., Kaushal, S.S., 2008. Disappearing headwaters: patterns of stream burial due to urbanization. Front. Ecol. Evol. 6, 308–312.

Ewel, K.C., Cressa, C., Kneib, R.T., Lake, P.S., Levin, L.A., Palmer, M.A., et al., 2001. Managing critical transition zones. Ecosystems 4, 452–460.

Ferńandez, J.A., Martinez, C., Magdaleno, F., 2012. Application of indicators of hydrologic alterations to the designation of heavily modified waterbodies in Spain. Environ. Sci. Technol. 16, 31–43.

Friedman, J.M., Lee, V.J., 2002. Extreme floods, channel change, and riparian forests along ephemeral streams. Ecol. Monogr. 72, 409–425.

Friedman, J.M., Auble, G.T., Shafroth, P.B., Scott, M.L., Merigliano, M.F., Freehling, M.D., et al., 2005. Dominance of non-native trees in western USA. Biol. Invasions 7, 747–751.

Fritz, K.M., Dodds, W.K., 2005. Harshness: characterisation of intermittent stream habitat over space and time. Mar. Freshw. Res. 56, 13–23.

Fritz, K.M., Hagenbuch, E., D'Amico, E., Reif, M., Wiginton, P.J., Leibowitz, S.G., et al., 2013. Comparing the extent and permanence of headwater streams from two field surveys to values from hydrographic databases and maps. J. Am. Water Resour. Assoc. 49, 867–882.

Ghassemi, F., Jakeman, A.J., Nix, H.A., 1995. Salinisation of Land and Water Resources: Human Causes, Extent, Management and Case Studies. University of New South Wales Press, Sydney.

Golladay, S.W., Taylor, B.W., Palik, B.J., 1997. Invertebrate communities of forested limesink wetlands in southwest Georgia, USA: habitat use and influence of extended inundation. Wetlands 17, 383–393.

Goodwin, C.N., Hawkins, C.P., Kershner, J.L., 1997. Riparian restoration in the western United States: overview and perspective. Restor. Ecol. 5, 4–14.

Groffman, P.M., Bain, D.J., Band, I.E., Belt, K.T., Brush, G.S., Grove, J.M., et al., 2003. Down by the riverside: urban riparian ecology. Front. Ecol. Environ. 1, 315–321.

Hale, R., Reich, P., Johnson, M., Hansen, B.D., Lake, P.S., Thomson, J.R., et al., 2015. Bird responses to riparian management of degraded lowland streams in southeastern Australia. Restor. Ecol. 23, 104–112.

Halse, S.A., Ruprecht, J.K., Pinder, A.M., 2003. Salinisation and prospects for biodiversity in rivers and wetlands of south-west Western Australia. Aust. J. Bot. 51, 673–688.

Hansen, B.D., Reich, P., Cavagnaro, T.R., Lake, P.S., 2015. Challenges in applying scientific evidence to width recommendations for riparian management in agricultural Australia. Ecol. Manag. Restor. 16, 50–57.

Hermoso, V., Pantos, F., Olley, J., Linke, S., Mugodo, J., Lea, P., 2015. Prioritising catchment rehabilitation for multi-objective management: an application from SE-Queensland, Australia. Ecol. Model. 316, 168–175.

Hobbs, R.J., Norton, D.A., 1996. Towards a conceptual framework for restoration ecology. Restor. Ecol. 4, 93–110.

Hobbs, R.J., Higgs, E., Harris, J.A., 2009. Novel ecosystems: implications for conservation and restoration. Trends Ecol. Evol. 24, 599–605.

Hughes, S.J., Santos, J., Ferreira, T., Mendes, A., 2010. Evaluating the response of biological assemblages as potential indicators for restoration measures in an intermittent Mediterranean river. Environ. Manag. 46, 285–301.

Jaeger, K.L., Olden, J.D., 2012. Electrical resistance sensor arrays as a means to quantify longitudinal connectivity of rivers. River Res. Appl. 28, 1843–1852.

Jaeger, K.L., Olden, J.D., Pelland, N.A., 2014. Climate change poised to threaten hydrologic connectivity and endemic fishes in dryland streams. Proc. Natl. Acad. Sci. 111, 13894–13899.

Jenkins, K.M., Boulton, A.J., 2003. Connectivity in a dryland river: short-term aquatic microinvertebrate recruitment following floodplain inundation. Ecology 84, 2708–2723.

Johnson, M., Reich, P., Mac Nally, R., 2007. Bird assemblages of a fragmented agricultural landscape and the relative importance of vegetation structure and landscape pattern. Wildl. Res. 34, 185–193.

Katz, G.L., Stromberg, J.C., Denslow, M.W., 2009. Streamside herbaceous vegetation response to hydrologic restoration on the San Pedro River, Arizona. Ecohydrology 2, 213–225.

Kauffman, J.B., Krueger, W.C., 1984. Livestock impacts on riparian ecosystems and streamside management implications. J. Range Manag. 37, 430–438.

Kaushal, S.S., Groffman, P.M., Band, L.E., Shields, C.A., Morgan, R.P., Palmer, M.A., et al., 2008. Interaction between urbanization and climate variability amplifies watershed nitrate export in Maryland. Environ. Sci. Technol. 42, 5872–5878.

Kondolf, G.M., 2012. The *éspace de liberté* and restoration of fluvial processing: when can the river restore itself and when must we intervene? In: Boon, P.J., Raven, P.J. (Eds.), River Conservation and Management. John Wiley and Sons, Chichester, pp. 225–241.

Kondolf, G.M., Podolak, K., 2014. Space and time scales in human-landscape systems. Environ. Manag. 53, 76–87.

Kondolf, G.M., Angermeier, P., Cummins, K., Dunne, T., Healey, M., Kimmerer, W., et al., 2008. Prioritizing river restoration: projecting cumulative benefits of multiple projects: an example from the Sacramento-San Jaoquin River system in California. Environ. Manag. 42, 933–945.

Kondolf, G.M., Podolak, K., Grantham, T.E., 2013. Restoring Mediterranean-climate rivers. Hydrobiologia 719, 527–545.

Kueper, D., Bart, J., Rich, T.D., 2003. Response of vegetation and breeding birds to the removal of cattle on the San Pedro River, Arizona (USA). Conserv. Biol. 17, 607–615.

Lake, P.S., 2003. Ecological effects of perturbation by drought in flowing waters. Freshwat. Biol. 48, 1161–1172.

Lake, P.S., 2011. Drought and Aquatic Ecosystems: Effects and Responses. Wiley-Blackwell, Chichester.

Lake, P.S., Bond, N., Reich, P., 2007. Linking ecological theory with stream restoration. Freshw. Biol. 52, 597–615.

Larned, S.T., Datry, T., Arscott, D.B., Tockner, K., 2010. Emerging concepts in temporary-river ecology. Freshw. Biol. 55, 717–738.

Leibowitz, S.G., Wigington, P.J., Downs, M.C., Downing, D.M., 2008. Non-navigable streams and adjacent wetlands: addressing science needs following the Supreme Court's Rapanos decision. Front. Ecol. Evol. 6, 364–371.

Levick, L., Fonseca, J., Goodrich, D., Hernandez, M., Semmens, D., Stromberg, J., et al., 2008. The ecological and hydrological significance of ephemeral and intermittent streams in the arid and semi-arid American Southwest. US Environmental Protection Agency and USDA/ARS Southwest Watershed Research Center, EPA/600/R-08/134 ARS/233046, Washington, DC.

Line, D.E., Harman, W.A., Jennings, G.D., Thompson, E.J., Osmond, D.L., 2000. Nonpoint-source pollutant load reductions associated with livestock exclusion. J. Environ. Qual. 29, 1882–1890.

Lite, S.J., Stromberg, J.C., 2005. Surface water and ground-water thresholds for maintaining Populus-Salix forests, San Pedro River, Arizona. Biol. Conserv. 125, 153–167.

Lottermoser, B.G., 2003. Mine Wastes: Characterization, Treatment and Environmental Impacts. Springer-Verlag, Berlin.

Lowe, W.H., Likens, G.E., 2005. Moving headwater streams to the head of the class. Bioscience 55, 196–197.

Magilligan, F.J., Nislow, K.H., 2005. Changes in hydrologic regime by dams. Geomorphology 71, 61–78.

Magilligan, F.J., Nislow, K.H., Renshaw, C.E., 2013. Flow regulation by dams. In: Wohl, E. (Ed.), Treatise on Geomorphology. Fluvial Geomorphology, 9. Academic Press, San Diego, CA, pp. 794–808.

McMahon, T.J., Finlayson, B.H., 2003. Droughts and anti-droughts: the low flow hydrology of Australian rivers. Freshw. Biol. 48, 1147–1160.

McNeil, D.G., Closs, G.P., 2007. Behavioural responses of a south-east Australian floodplain fish community to gradual hypoxia. Freshw. Biol. 52, 412–420.

Meixner, T., Manning, A.H., Stonestrom, D.A., Allen, D.M., Ajami, H., Blasch, K.W., et al., 2016. Implications of projected climate change for groundwater recharge in the western United States. J. Hydrol. 534, 124–138.

Meyer, J.L., Wallace, J.B., 2001. Lost linkages and lotic ecology: rediscovering small streams. In: Press, M.C., Huntly, N.J., Levin, S. (Eds.), Ecology: Achievement and Challenge. Blackwell Science, Maiden, pp. 295–317.

Mika, S., Hoyle, J., Kyle, G., Howell, T., Wolfenden, B., Ryder, D., et al., 2010. Inside the "black box" of river restoration: using catchment history to identify disturbances and response mechanisms to set targets for process-based restoration. Ecol. Soc. 15, 8.

Miller, J.J., Chanasyk, D.S., Curtis, T., Willms, W.D., 2010. Influence of streambank fencing on the environmental quality of cattle-excluded pastures. J. Environ. Qual. 39, 991–1000.

Moyle, P.B., 2014. Novel aquatic ecosystems: the new reality for streams in California and other Mediterranean climate regions. River Res. Appl. 30, 1335–1344.

Nathan, R., Lowe, L., 2012. The hydrologic impacts of small dams. Aust. J. Water Res. 16, 75–83.

O'Connor, T.G., 2001. Effect of a small catchment dam on downstream vegetation of a seasonal river in semi-arid African savanna. J. Appl. Ecol. 38, 1314–1325.

O'Connor, B.L., Hamada, Y., Bowen, E.E., Grippo, M.A., Hartmann, H.N., Patton, T.L., et al., 2014. Quantifying the sensitivity of ephemeral streams to land disturbance in arid ecosystems at the watershed scale. Environ. Monit. Assess. 186, 7075–7095.

O'Donnell, J., Fryirs, K.A., Leishman, M.R., 2016. Seed banks as a source of vegetation regeneration to support the recovery of degraded rivers: a comparison of river reaches of varying condition. Sci. Total Environ. 542, 591–602.

Ollero, A., Ibisate, A., 2012. Space for the river: a flood management tool. In: Wong, T.S.W. (Ed.), Flood Risk and Flood Management. Nova Science, Hauppage, pp. 199–218.

Ortega, J.A., Razola, I., Garzon, G., 2014. Recent human impacts and change in dynamics and morphology of ephemeral rivers. Nat. Hazards Earth Syst. Sci. 14, 713–730.

Palmer, M.A., Bernhardt, E.S., Allan, J.D., Lake, P.S., Alexander, G., Brooks, S., et al., 2005. Standards for ecologically successful river restoration. J. Appl. Ecol. 42, 208–217.

Palmer, M.A., Falk, D.A., Zedler, J.B., 2006. Ecological theory and restoration ecology. In: Palmer, M.A., Falk, D.A., Zedler, J.B. (Eds.), Foundations of Restoration Ecology. Island Press, Washington, DC, pp. 1–10.

Palmer, M.A., Reidy, C.A., Nilsson, C., Flörke, M., Alcamo, J., Lake, P.S., et al., 2008. Climate change and the world's river basins: anticipating management options. Front. Ecol. Environ. 6, 81–89.

Palmer, M.A., Lettenmaier, D.P., Poff, N.L., Postel, S.L., Richter, B., Warner, R., 2009. Climate change and river ecosystems: protection and adaptation options. Environ. Manag. 44, 1053–1068.

Palmer, M.A., Filoso, S., Fanelli, R.M., 2014. From ecosystems to ecosystem services: stream restoration as ecological engineering. Ecol. Eng. 65, 62–70.

Prober, S.M., Thiele, K.R., Lunt, I.D., Koen, T.B., 2005. Restoring ecological function in temperate grassy woodlands: manipulating soil nutrients, exotic annuals and native perennial grasses through carbon supplements and spring burns. J. Appl. Ecol. 42, 1073–1085.

Pumo, D., Caracciolo, D., Viola, F., Noto, L.V., 2016. Climate change effects on the hydrological regime of small non-perennial river basins. Sci. Total Environ. 542, 76–92.

Reich, P., Lake, P.S., 2015. Extreme hydrological events and the ecological restoration of flowing waters. Freshw. Biol. 60, 2639–2652.

Reich, P., Williams, L., Cavagnaro, T., Lake, P.S., 2016. The ongoing challenge of restoring Australia's riparian zones. In: Capon, S., James, C., Reid, M. (Eds.), Vegetation of Australian Riverine Landscapes. CSIRO Publishing, Collingwood, VIC, pp. 343–364.

Richardson, D.M., Holmes, P.M., Esler, K.J., Galatowitsch, S.M., Stromberg, J.C., Kirkman, S.P., et al., 2007. Riparian vegetation: degradation, alien plant invasions, and restoration prospects. Divers. Distrib. 13, 126–139.

Roberts, J., Marston, F., 2011. Water Regime for Wetland and Floodplain Plants. A Source Book for the Murray-Darling Basin. National Water Commission, Canberra.

Robson, B.J., Chester, E.T., Mitchell, B.D., Matthews, T.G., 2008. Identification and Management of Refuges for Aquatic Organisms: Waterlines Report Number 11. National Water Commission, Canberra.

Robson, B.J., Chester, E.T., Mitchell, B.D., Matthews, T.G., 2013. Disturbance and the role of refuges in Mediterranean climate streams. Hydrobiologia 719, 77–91.

Rolls, R.J., Leigh, C., Sheldon, F., 2012. Mechanistic effects of low-flow hydrology on riverine ecosystems: ecological principles and consequences of alteration. Freshw. Sci. 31, 1163–1186.

Roni, P., Beechie, T.J., Bilby, R.E., Leonetti, F.E., Pollock, M.M., Pess, G.R., 2002. A review of restoration techniques and a hierarchical strategy for prioritizing restoration in Pacific Northwest watersheds. N. Am. J. Fish Manag. 22, 1–20.

Roy, A.H., Dybas, A.L., Fritz, K.M., Lubbers, H.Y., 2009. Urbanization affects the extent and hydrologic permanence of headwater streams in a midwestern US metropolitan area. J. N. Am. Benthol. Soc. 28, 911–928.

Russell, P.P., Gale, S.M., Muñoz, B., Dorney, J.R., Rubino, M.J., 2015. A spatially explicit model for mapping headwater streams. J. Am. Water Resour. Assoc. 51, 226–239.

Sabo, J.L., Sponseller, R., Dixon, M., Gade, K., Harma, T., Heffernan, J., et al., 2005. Riparian zones increase regional species richness by harbouring different, not more species. Ecology 86, 56–62.

Shafroth, P.B., Briggs, M.K., 2008. Restoration ecology and invasive riparian plants: an introduction to the Special Section on *Tamarix* spp. in western North America. Restor. Ecol. 16, 94–96.

Shafroth, P.B., Cleverley, J.R., Dudley, T.L., Taylor, J.P., Van Piper III, C., Weeks, E.P., et al., 2005. Control of *Tamarix* in the western United States: implications for water salvage, wildlife use, and riparian restoration. Environ. Manag. 35, 231–246.

Shaw, J.R., Cooper, D.J., 2008. Linkages among watersheds, stream reaches and riparian vegetation. J. Hydrol. 350, 68–82.

Sher, A., Quigley, M.F. (Eds.), 2013. *Tamarix:* A Case Study of Ecological Change in the American West. Oxford University Press, New York.

Stella, J.C., Rodríguez-González, P., Dufour, S., Bendix, J., 2013. Riparian vegetation research in Mediterranean-climate regions: common patterns, ecological processes and considerations for management. Hydrobiologia 719, 291–315.

Steward, A.L., Von Schiller, D., Tockner, K., Marshall, J., Bunn, S.E., 2012. When the river runs dry: human and ecological values of dry river beds. Front. Ecol. Environ. 10, 202–209.

Stromberg, J.C., 2001. Restoration of riparian vegetation in south-western United States: importance of flow regimes and fluvial dynamism. J. Arid Environ. 49, 17–34.

Stromberg, J.C., Tiller, R., Richter, B., 1996. Effects of groundwater decline on riparian vegetation of semiarid regions: the San Pedro River, AZ. Ecol. Appl. 6, 113–131.

Stromberg, J.C., Bagstad, K.J., Leenhouts, J.M., Lite, S.J., Makings, E., 2005. Effect of stream flow intermittency on riparian vegetation of a semiarid region river (San Pedro River, Arizona). River Res. Appl. 21, 925–938.

Stromberg, J.C., Lite, S.J., Rychener, T.J., Levick, L.R., Dixon, M.D., Watts, J.M., 2006. Status of the riparian ecosystem in the upper San Pedro River, Arizona: application of an assessment model. Environ. Monit. Assess. 115, 145–173.

Stromberg, J.C., Beauchamp, V.B., Dixon, M.D., Lite, S.J., Paradzick, C., 2007. Importance of low-flow and high-flow characteristics to restoration of riparian vegetation along rivers in arid south-western United States. Freshw. Biol. 52, 651–679.

Stromberg, J.C., Chew, M.K., Nagler, P.L., Glenn, E.P., 2009. Changing perceptions of change: the role of scientists in *Tamarix* and river management. Restor. Ecol. 17, 177–186.

Stromberg, J.C., McCluney, K.E., Dixon, M.D., Meixner, T., 2013. Dryland riparian ecosystems in the American Southwest: sensitivity and resilience to climatic extremes. Ecosystems 16, 411–415.

Sunohara, M.D., Topp, E., Wilkes, G., Gottschall, N., Neumann, N., Jones, T.H., et al., 2012. Impact of riparian zone protection from cattle on nutrient, bacteria, F-5 coliphage, *Cryptosporidium* and *Giardia* loading of an intermittent stream. J. Environ. Qual. 41, 1301–1314.

Svec, J.R., Kolka, R.K., Stringer, J.W., 2005. Defining perennial, intermittent and ephemeral channels in eastern Kentucky: application to forestry best management practices. For. Ecol. Manag. 214, 170–182.

Theodoropoulos, C., Aspiridis, D., Iliopoulou-Georgudaki, J., 2015. The influence of land use on freshwater macroinvertebrates in a regulated and temporary Mediterranean river network. Hydrobiologia 751, 201–213.

Turner, R.M., Webb, R.H., Bowers, J.E., Hastings, J.R., 2003. The Changing Mile Revisited. University of Arizona Press, Tucson, AZ.

Verkaik, I., Vila-Escalé, M., Rieradevall, M., Baxter, C.V., Lake, P.S., Minshall, G.W., et al., 2015. Stream macroinvertebrate community responses to fire: are they the same in different fire-prone biogeographic regions? Freshw. Sci. 34, 1527–1541.

Ward, J.V., 1989. The four-dimensional nature of lotic ecosystems. J. N. Am. Benthol. Soc. 8, 2–8.

Webb, R.H., Leake, S.A., 2006. Ground-water surface-water interactions and long-term change in riverine riparian vegetation in the southwestern United States. J. Hydrol. 320, 302–323.

Williams, L., Reich, P., Capon, S.J., Raulings, E., 2008. Soil seed banks of degraded riparian zones in southeastern Australia and their potential contribution to the restoration of understory vegetation. River Res. Appl. 24, 1002–1017.

Wohl, E., 2005. Compromised rivers: understanding historical human impacts on rivers in the context of restoration. Ecol. Soc. 10, 2.

Wohl, E., Egenhoff, D., Larkin, K., 2009. Vanishing riverscapes: a review of the historical channel changes on the western Great Plains. In: James, L.A., Rathbun, S.L., Whittecar, K. (Eds.), Management and Restoration of Fluvial Systems with Broad Historical Changes and Human Impacts. Geological Society of America, Boulder, CO, pp. 131–142.

Wood, P.J., Gunn, J., Smith, H., Abas-Kutty, A., 2005. Flow permanence and macroinvertebrate community diversity within groundwater dominated headwater streams and springs. Hydrobiologia 545, 55–64.

FURTHER READING

Berg, M.D., Popescu, S.C., Wilcox, B.P., Angerer, J.P., Rhodes, E.C., McAlister, J., et al., 2015. Small farm ponds: overlooked features with important impacts on watershed sediment transport. J. Am. Water Resour. Assoc. 52, 67–76.

Davis, J., Finlayson, B., 2000. Sand slugs and stream degradation: the case of the Granite Creeks, north-east Victoria. Technical Report 7/2000. Cooperative Research Centre for Freshwater Ecology, Canberra.

Downes, B.J., Lake, P.S., Glaister, A., Bond, N.R., 2006. Effects of sand sedimentation on the macroinvertebrate fauna of lowland streams: are the effects consistent? Freshw. Biol. 51, 144–160.

Lake, P.S., 2000. Disturbance, patchiness and diversity in streams. J. N. Am. Benthol. Soc. 19, 573–592.

Palmer, M.A., McDonough, T., 2013. Ecological restoration to conserve and recover river ecosystem services. In: Elosegi, A., Sabater, S. (Eds.), River Conservation: Challenges and Opportunities. Fundación BBVA, Bilbao, pp. 279–300.

Robson, B.J., Mitchell, B.D., 2010. Metastability in a river subject to multiple disturbances may constrain restoration options. Mar. Freshw. Res. 61, 778–785.

Society for Ecological Restoration International Science and Policy Working Group, 2004. The SER International Primer on Ecological Restoration. www.ser.org Accessed 7 November 2016.

Vigiak, O., Newman, L.T.H., Whitford, J., Roberts, A.M., Rattray, D., Melland, A.R., 2011. Integrating farming systems and landscape processes to assess management impacts on suspended sediment loads. Environ. Model. Softw. 26, 144–162.

Waters, T.F., 1995. Sediment in Streams: Sources, Biological Effects and Control. American Fisheries Society, Bethesda, MD.

White, J.M., Stromberg, J.C., 2011. Resilience, restoration, and riparian ecosystems: case study of a dryland urban river. Restor. Ecol. 19, 101–111.

Williams, L.J., Cavagnaro, T.R., Reich, P., Lake P.S., 2009. Phalaris in Australian ecosystems: Understanding its biology and ecology as the first step towards effective management. Report to Goulburn-Broken Catchment Management Authority. Australian Centre for Biodiversity, Monash University, Victoria.

Wolman, M.G., Gerson, R., 1978. Relative scales of time and effectiveness of climate in geomorphic processes. J. Geol. 68, 54–74.

STRATEGIC ADAPTIVE MANAGEMENT (SAM) OF INTERMITTENT RIVERS AND EPHEMERAL STREAMS

Richard T. Kingsford*, Dirk J. Roux[†,‡], Craig A. McLoughlin[§], John Conallin[¶], Vol Norris[‖]

UNSW Australia, Sydney, NSW, Australia[*] South African National Parks, George, South Africa[†] Nelson Mandela Metropolitan University, George, South Africa[‡] Resilient Learning Consultancy, Armidale, NSW, Australia[§] UNESCO-IHE Institute for Water Education, Delft, Netherlands[¶] Desert Channels Group, Longreach, QLD, Australia[‖]

IN A NUTSHELL

- Linking science to value-driven objectives must 'avoid the tail wagging the dog.' Strategic Adaptive Management aims to explicitly link a vision to fine-scale measurable objectives, providing direction for scientific investment.
- Key indicators must be identified that explain enough of the variation and are sensitive to management intervention. Indicators need to be sensitive to management intervention, otherwise they are not helpful for guiding or assessing efficacy of management.
- Transparent management is essential. Explicit identification of clear measurable objectives linked to a vision is critical, followed by transparent decision-making around management scenarios informed by monitoring.
- Conceptual modeling, experimentation, and monitoring are all essential for the process of learning by doing. A key objective for understanding and managing intermittent rivers and ephemeral streams is to identify cause-and-effect processes for indicators which can then inform management. Conceptual modeling, experimentation, and monitoring allow for probing and building this understanding.

5.5.1 INTRODUCTION

Intermittent rivers and ephemeral streams (IRES) stop flowing temporally and/or spatially, and make up more than 50% of the global river network (Acuña et al., 2014; Datry et al., 2014). In the arid areas comprising half of the world's land area (Middleton and Thomas, 1997), evaporation generally exceeds precipitation (e.g., Cooper Creek, Hamilton et al., 2005) and IRES are dominant (Kingsford and Thompson, 2006). IRES have characteristically high temporal and spatial variability in their flow and inundation regimes (Puckridge et al., 1998; McMahon et al., 2007; Datry et al., 2016; Chapters 1, 2.2, and 2.3), although some have seasonal pulses that depend on climatic drivers (Young and Kingsford, 2006). For example, a regular seasonal pulse of water flows from catchments in the Angola highlands to the Cubango-Okavango River and the Okavango Swamp, where inundation is spatially and temporally variable (McCarthy et al., 2000). Even in more mesic regions of the world, IRES are common (Datry et al., 2014), and regulation and abstraction from rivers (Nilsson et al., 2005) may further increase flow intermittence (Chapter 5.1). In this chapter, we focus on rivers or parts of rivers where intermittence is a major component of their flow and flooding regimes. However, we also discuss their dependent wetlands, often floodplains, supporting high biodiversity, complex ecological processes, and important ecosystem services (Kingsford et al., 2016).

The management of IRES is inevitably complex, involving social, economic, and environmental dimensions or drivers. Policy mandates of different governing bodies often overlap and are not clearly defined, either geographically or even thematically. Most governments have environment agencies responsible for the management and protection of biodiversity, including protected areas which may be wetland systems, yet have little or no control over water supply managed by a separate government water agency (Chapter 5.3). For example, South African National Parks manages protected areas but the Department of Water and Sanitation manages the rivers that flow through these protected areas. This differentiation can be perverse for biodiversity conservation. On the Darling River system at Menindee Lakes in Australia, about 28% of Kinchega National Park is wetland but its management only passes over to the environment agency (New South Wales National Parks and Wildlife) when the lakes are dry; otherwise, the water agency (New South Wales Office of Water) is the manager.

People want many things from rivers, including water for drinking, industry and agriculture, generation of hydroelectricity, and cultural ecosystem services such as recreation and conservation (Chapter 5.2). Many ecosystem services (e.g., water quality, nutrient management, fishing, pastoralism) are significant economically (Costanza et al., 1997; Chapter 5.2). Governments and their communities often depend on these services and yet undervalue them. River management is particularly challenging when rivers cross international borders where one country's goals can differ from another's. For example, the Colorado River starts in the United States and reaches Mexico where it now seldom flows (Postel, 2000). Upstream users control the river and benefit at the potential cost of downstream users, including the environment.

There are ways to integrate social, political, economic, and environmental complexities across IRES. Explicit identification of agreed goals, driven by a coherent vision collectively crafted by stakeholders, provides direction. This approach is particularly well developed and applied to rivers flowing through Kruger National Park in South Africa (Rogers and Bestbier, 1997; McLoughlin et al., 2011a) and increasingly considered in Australia (Kingsford et al., 2011a). In this chapter, we focus on four case studies of IRES and their associated wetlands: two from South Africa and two from Australia. Our aim is to show the SAM approach in disparate IRES with widely differing spatial frameworks and to illustrate the importance of idiosyncrasy which captures values of the stakeholders and government agencies. We also discuss SAM implementation across IRES of the world, integrating the issues covered in Chapters 5.1–5.4 in an adaptive management context.

5.5.2 THE MANAGEMENT CHALLENGE
ECOLOGICAL CONTEXT

Catchment processes often occur over long distances, involving many ecological and social processes, overlapping policy mandates, and diverse user expectations and values. Lateral, longitudinal, and vertical connectivity (Chapter 2.3) make river management complex, particularly as the drivers to be managed are often beyond local control. Changes to flow regimes through direct or indirect anthropogenic interventions (e.g., dams, invasive species) affect resilience, degrading dependent organisms (Chapter 5.1; Poff et al., 2007; Olden and Naiman, 2010; Belmar et al., 2013; Catelotti et al., 2015). Competition for fresh water between anthropogenic and ecological uses of water can be fierce (Kingsford et al., 2011b; Pollard and Du Toit, 2011), challenging decision-making. Different stakeholders have differing values and objectives, often competing, interacting with scales of social-ecological complexity (Chapter 5.3).

The river catchment is fundamental to structuring management. Spatial geomorphological structure and associated hydrology (Chapters 2.1–2.3) intersect with the socioeconomic spatial structure of human needs. This structure includes large dams to control water for the generation of hydroelectricity or store water for irrigation, drinking supply, industry, and flood mitigation. Sometimes, reliant human communities seldom interact with the river except to access its water. For example, irrigated agricultural areas divert river water to crops, often carrying it considerable distances from upstream biodiverse floodplains (Kingsford, 2015). Management complexity also varies with level of river development, with increasing management focusing on reducing impacts on ecosystem services and ecosystems.

SOCIAL CONTEXT

Management requires a strong focus, clear processes, learning, and demonstration of effectiveness to adequately realize expectations. Communities and governments often fail to deliver on goals for IRES, reflecting the highly contested space, lack of knowledge, and poor processes (Larned et al., 2010; Acuña et al., 2014). Management effectiveness is limited by incoherent articulation of goals and objectives, insufficient transformation of these into transparent processes for action, lack of learning, and deficient monitoring to determine evidence.

Engagement of stakeholders (community, industry, government, scientific) in colearning (Fig. 5.5.1) is essential, even though their goals may differ. Ubiquitous 'silos' of practices within river management, policy, and science make this difficult. Management and science have a long and often

FIG. 5.5.1

Stakeholders collaborate to craft a joint vision and identify objectives, drivers, and thresholds of potential concern at workshops, such as this one for the Crocodile River in Kruger National Park.

Photo: Courtesy C. McLoughlin.

uneasy history of integration (Roux et al., 2006). Managers sometimes do not adequately understand the importance of data, measurement of variability, and sampling rigor. Scientists seldom fully appreciate the pressure on river managers to make rapid decisions at large scales where provision of timely and understandable information is critical. There is a mismatch between the timescales to deliver a peer-reviewed paper and decide about delivery of environmental flows. Even managers have competing objectives (Knight et al., 2011). For example, a river manager from a water agency provides water for all uses (e.g., irrigation, ecosystems) whereas a conservation manager focuses on ecological, cultural, and recreational values.

These competing objectives are often hardwired in legislation, with few clear mechanisms to overcome the challenges of integration. A common result is a decline in biodiversity while river management fails to deliver on sustainability objectives. To address this, SAM offers considerable promise. Although still in its infancy (albeit well established in Kruger National Park for about 20 years, Kingsford et al., 2011a; Roux and Foxcroft, 2011), SAM can 'professionalize' river management and integrate across different cultures and stakeholders to achieve agreed goals and objectives. Environmental management seldom adopts explicit transparent management practices, instead relying primarily on idiosyncratic management. Institutions and individuals often lack the ability or motivation to alter current command and control practices (Holling and Meffe, 1996), incurring great cost and poor outcomes.

5.5.3 STRATEGIC ADAPTIVE MANAGEMENT (SAM)
DEFINITION

Adaptive management has acquired different meanings, from rigorous experimentation (Walters and Holling, 1990) to simple adaptability (Kingsford et al., 2011a). Neither serves adaptive management well. The former is a straitjacket on the reality of managing large complex social-ecological systems with few opportunities for experimentation while the latter allows managers to define mere changes in decisions as adaptive management.

Adaptive management should include governance, planning, implementation of decisions, and monitoring and evaluation of subsequent outcomes. Importantly, SAM includes iterative colearning and knowledge coproduction by managers, stakeholders, and scientists and ultimately triple loop learning (Armitage et al., 2008, 2009), driven by a "strategic" focus on the future desired state. Single loop learning is what to do (the rules), double loop learning is how to do (insights), and triple loop learning is why to do (principles) (Flood and Romm, 1996).

FRAMEWORK

There are four broad steps, beginning with setting the desired future condition (Fig. 5.5.2). Stakeholders for IRES include managers (conservation and river), scientists, and users of water. They jointly articulate their collective vision (Figs. 5.5.1 and 5.5.2, Box 5.5.1), which captures the desired state succinctly, becoming the 'sounding board' for objectives and actions. It may change with specification of fine-scale objectives, if a key value is missed. Hierarchical visions may be appropriate in catchments, with a basin-scale vision capturing broad ecological and social values that increase in focus and rigor with fine scales of governance (McLoughlin and Thoms, 2015), culminating in a specific vision for a target

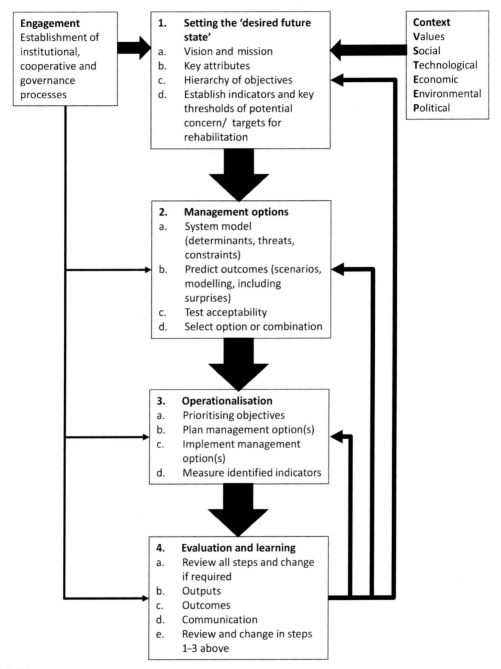

FIG. 5.5.2

Strategic Adaptive Management (SAM) framework which can be applied to the management of IRES and their social, economic, and ecological values. It involves four key steps which encourage colearning and management to deliver the best environmental outcomes, supported by rigorous science.

BOX 5.5.1 STAKEHOLDER ENGAGEMENT: THE SHARED JOURNEY OF LEARNING TOGETHER

Knight et al. (2011) succinctly capture the importance of social engagement: "effective conservation planning is a social process informed by science, not a scientific process which engages society." Stakeholders need to be at the center of managing natural resources in social-ecological systems, underlining the principles of engagement (Arnstein, 1969; Ostrom, 1990). Effective stakeholder engagement relies on building trust and ownership among stakeholders, valuing all opinions and including different forms of information (local and formal scientific). Two-way communication pathways are established using information and processes understandable by all stakeholders. Local community members and champions within government make the process happen, facilitating participation and communication among stakeholders, especially among their own groups. Strategic Adaptive Management allows stakeholders to feel valued and part of the decision-making, through the V-STEEP (values, social, technical, economic, environmental, and political, Fig. 5.5.2) context setting, and culminating in an agreed vision before breaking down the vision into agreed-upon measurable endpoints. Allowing stakeholders to remain engaged and be part of decision-making through SAM builds trust and ownership. SAM incorporates principle-based stakeholder engagement, centered on inclusiveness, transparency, and trust.

Eight-point check list for stakeholder engagement (Moore, 2013):

1. Expect to make mistakes; acknowledge them, learn from them, try to forgive them
2. Reveal relevant information, records, relationships,
3. Communicate immediately when there is a problem; don't let it fester
4. Believe what I am saying; don't assume I am lying
5. Don't set me up
6. Don't bash me in the media
7. Keep the door open
8. Be able to differ with me in a respectful way, to disengage without malice.

---

aquatic ecosystem. Setting the desired state usually occurs in workshops (Fig. 5.5.1), moving from the vision to more specific objectives. The context of values and social, technological, economic, environmental, and political factors defines the boundaries (VSTEEP, Fig. 5.5.2). For highly contentious river management, this process may require independent facilitation.

Agreement on 5-15 key attributes (e.g., endangered fish species, riparian vegetation) of the IRES focuses management and ultimately monitoring (Kingsford and Biggs, 2012; Walker and Salt, 2012). These are designed to capture much of the variability of the system and the issues that matter to people, necessarily reflecting place and knowledge: the requisite simplicity of the social-ecological system (Stirzaker et al., 2010). Broad-scale (e.g., jurisdictional) approaches to adaptive management can fail if they do not adequately reflect these key attributes. Following the vision, the hierarchy of broad value-laden objectives ("inverted tree") is developed (Rogers and Biggs, 1999; Kingsford and Biggs, 2012). This is seldom adequately achieved in river management and usually stops at a generic level with no specificity for clear management direction or scientific investment (Pollard and Du Toit, 2007; Kingsford et al., 2011a). Specification is the engagement point for science, where clarity links to measurement through adopting measurable indicators useful for management and reporting outcomes. These indicators allow identification of thresholds of potential concern (TPCs, Box 5.5.2, Fig. 5.5.3), beyond natural stochastic variation (Biggs et al., 2011), identified quantitatively using historical data to derive confidence levels (e.g., 95% confidence limits). For some highly degraded social-ecological systems such as the Macquarie Marshes (see later), these indicators may be targets for rehabilitation (Kingsford et al., 2011a). Conceptual models are important because they help portray how the system

BOX 5.5.2 THRESHOLDS OF POTENTIAL CONCERN

Thresholds of Potential Concern (TPCs) are explicit, measurable endpoints of the Strategic Adaptive Management (SAM) objectives hierarchy. Using the best available information to determine TPCs, SAM monitors trends and then mandates reflection on collaboratively identified goals, guiding mutually agreed action (Pollard and Du Toit, 2011). These TPCs are typically "decision thresholds," an optimization of ecological (scientific/model-based) and utility (value/objectives-based) thresholds (Martin et al., 2009), rather than specific predicted ecosystem thresholds (Biggs et al., 2011). TPCs need to be developed collaboratively between scientists, managers, and stakeholders, within the context of understanding priority drivers within a system. They need to be measurable response indicators of change related to these drivers and sensitive to management interventions. The natural variability (resilience) of the response indicators is recognized by incorporating upper and lower levels (thresholds) of acceptable change (Fig. 5.5.3).

Importantly, construction of TPCs does not require complete understanding about the system: a continuum from empirically developed, or informed by expert opinion, or intelligent early guesswork, or from a conceptual understanding of the system (Biggs et al., 2011). TPCs are hypotheses of acceptable change and open to challenge and refinement, forming an inductive approach to adaptive management (Biggs and Rogers, 2003). Collectively, TPCs define the measurable component ('tent-boundary') of a "desired future state" under SAM. Management aims to keep the system within this tent boundary; if the system moves outside this boundary, it becomes a "target" for rehabilitation (Fig. 5.5.3, cf. Biggs et al., 2011).

TPCs are the early warning system (red flag) before actual threshold (often theoretical) boundaries are reached. If and when TPCs are exceeded (detected by monitoring and/or modeling), the scientists, managers, and stakeholders collectively examine why before deciding the management response or revising the TPCs. Monitoring is essential for such feedback and so must be practicable, fast, affordable, and effective (McLoughlin et al., 2011a). TPCs can be refined with new knowledge. Development, use, and auditing of TPCs in SAM are essential feedback mechanisms for management, within an iterative and adaptive process (Rogers and Biggs, 1999).

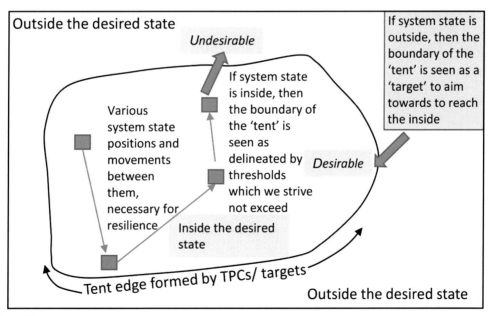

FIG. 5.5.3

Diagram showing how a 'tent boundary' defines the difference between natural bounds of variability and intervention, identified using Thresholds of Potential Concern (after Biggs et al., 2011).

works, including the potential social and ecological drivers for management (e.g., Florida Everglades, Ogden et al., 2005). The drivers can ultimately be modeled quantitatively for scenario planning and exploring management options (Fig. 5.5.2).

Operationalization (step 2) involves actions while evaluation and learning (step 4) audit implementation and ensure that learning is occurring through feedback, a key stumbling block in implementation (Stirzaker et al., 2011). Achievement of objectives needs to be evaluated including whether or not they are achievable (Pollard and Du Toit, 2007; Roux and Foxcroft, 2011). For example, there may be insufficient environmental flows to the Macquarie Marshes to restore river red gum forests (Catelotti et al., 2015).

MAKING IT WORK

The SAM approach is generic, implementable on any IRES and capable of incorporating different spatial scales, social-ecological complexity, and development (Kingsford et al., 2011a). It is not dependent on significant resources, targeting management issues prioritized with available resources. We show how the process works in four case studies from South Africa and Australia. The most mature example is from Kruger National Park while the other three (Tankwa Karoo National Park, Macquarie Marshes, and Edward-Wakool system) are at relatively early stages. We then show how SAM could be implemented in other areas where IRES are common.

5.5.4 CASE STUDIES—STATUS, SUCCESSES, CHALLENGES, AND THE FUTURE

CROCODILE RIVER, KRUGER NATIONAL PARK (SOUTH AFRICA)

Status

The Crocodile River traverses Kruger National Park (KNP), northeastern South Africa. Regulation by Kwena Dam occurs in the upper catchment (Fig. 5.5.4A), providing water for irrigation. Natural sediment scouring and transport and frequencies of small floods have declined, potentially increasing intermittence. Elevated sedimentation smothers bedrock (Rountree and Rogers, 2004), negatively impacting biodiversity (van Coller et al., 2000), such as fish (e.g., *Chiloglanis paratus*) and woody riparian vegetation (e.g., *Breonadia salicina*). Aquatic and terrestrial biodiversity of KNP depends on its rivers (Table 5.5.1), which have a steep and stable macro-channel bank where riparian vegetation differentially distributes along the elevation-inundation gradient (O'Keeffe and Rogers, 2003). The highly variable flow regime (including tributary IRES) and geology govern patterns of sediment erosion, transportation, and deposition (Rountree and Rogers, 2004), producing diverse alluvial- and bedrock-controlled geomorphic features and their habitats (Van Niekerk et al., 1995; Heritage et al., 1999) for biota (Russell, 1997a; van Coller et al., 2000). KNP's rivers have the highest fish species diversities in South Africa, with many species critically dependent on the fast-flowing well-oxygenated bedrock habitats (Russell, 1997b).

KNP's well-developed history of SAM (Pollard and Du Toit, 2007, SANParks, 2008; Roux and Foxcroft, 2011) includes a vision, hierarchy of objectives for river management, and related TPCs (Table 5.5.1, Box 5.5.2) focused on maintenance of biodiversity in all its natural facets and fluxes (Rogers and Bestbier, 1997). Biodiversity-related TPCs include variables related to geomorphology,

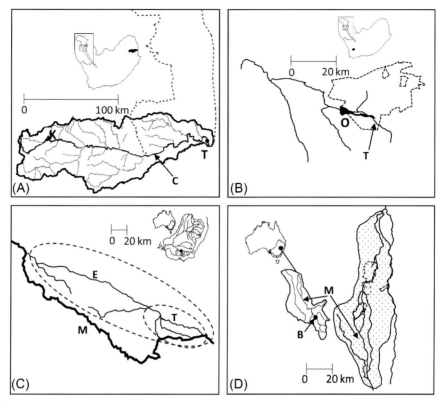

FIG. 5.5.4

Locations of the four IRES case studies: (A) Crocodile River with its catchment in South Africa (inset shows South Africa and filled catchment), within the southern region of Kruger National Park (dashed boundary), showing Kwena Dam (K), the Crocodile River (C) and Ten Bosch gauge (T); (B) Tankwa Karoo National Park (dashed boundary) in South Africa (inset shows South Africa and filled catchment), showing Oudebaaskraal Dam (O) and Tankwa River (T); (C) Edward-Wakool system (large dashed ellipse) in the Murray-Darling Basin, southeastern Australia (inset, filled circle), with the River Murray (M), Edward River (E) and Tuppal (T) and Jimaringle-Cockrans Creeks system (small dashed ellipse); (D) the Macquarie Marshes Nature Reserve (dashed boundaries) within the Macquarie Marshes (stippled) within the Macquarie-Bogan catchment in the Murray-Darling Basin in southeastern Australia, showing the Macquarie River (M) and the largest dam (Burrendong Dam, B).

fish, macroinvertebrates, riparian vegetation (McLoughlin et al., 2011a), and alien vegetation (Foxcroft, 2009). These TPCs reflect biodiversity change from human-induced pressures, predominantly upstream and outside the protected area, which alter river flow and water quality, increasing sedimentation rates. Objectives of KNP's SAM plan vary among its major rivers. The Crocodile River forms KNP's southern boundary (Fig. 5.5.4A), receiving flow and sediment from IRES inside and outside the protected area (Rountree and Rogers, 2004). Management of environmental flows in the Crocodile River is critical for alleviating unnatural sedimentation (Fig. 5.5.5, Table 5.5.1).

Table 5.5.1 Characteristics of the four case studies reviewed for applicability and implementation of the Strategic Adaptive Management approach (see Fig. 5.5.4 for locations of case studies)

Case study	Catchment size	Hydrological description	Ecological values	Government agencies	Stakeholders	References
Kruger National Park: Crocodile River	Total: 10,420 km²; (Kruger: 2390 km²)	River length: 320 km; (Kruger: 115 km) Rainfall: 400 mm y⁻¹; (Kruger: 1200 mm y⁻¹ on Drakensberg Escarpment) Mean annual runoff: 73-1444 × 10⁶ m³ y⁻¹ Mean monthly flow: 93.5 × 10⁶ m³); CV monthly flow: 116.1 One major dam (Kwena) in the upper catchment (158 × 10⁶ m³) supplying irrigated agriculture, producing abnormally high and stable winter flows.	Large contributor to biodiversity within Kruger National Park. High structural and biotic diversity - due to variable flow regime and mixed bedrock-alluvial character of river. Variety of habitat types available for different assemblages of riparian vegetation, fishes, macroinvertebrates, and waterbirds. Also critical for terrestrial biodiversity.	Dept. of Water Affairs and Sanitation; Inkomati-Usuthu Catchment Management Agency; Dept. of Agriculture; Dept. of Land Affairs; Dept. of Environmental Affairs; Mbombela Municipality; South African National Parks; Mpumalanga Tourism and Parks Agency.	Crocodile River Main Irrigation Board; Kruger National Park; Komati Basin Water Authority; Association for Water and Rural Development, consultants and researchers.	van Niekerk et al., 1995; van Coller et al., 2000; O'Keeffe and Rogers, 2003; Pollard and Du Toit, 2011
Tankwa Karoo National Park: Tankwa River	River length: 142.4 km (inside park: 25.6 km, 16%); catchment area: 644 725 ha (inside park: 124 050 ha, 19.2%) Mean annual rainfall: 100 mm Oudebaaskraal Dam built in 1969 within the park (34 × 10⁶ m³)	The Tankwa River catchment is characterized by sparse vegetation with succulents on shallow soils and grass, and ephemerals on sandy alluvium. The river experiences occasional floods during heavy rainstorms and the sandy riverbed is often devoid of surface water	South African National Parks	Neighboring farmers, tourism sector, visitors (local, national, international)	Rubin, 1998

System	Area	Rivers/Hydrology	Wetlands/Ecology	Water Agencies/Organizations	Communities/Stakeholders	References
Edward-Wakool System (Tuppal, Jimaringle-Cockrans Creeks)	~1000 km^2	Major permanent rivers include the Wakool River (~365 km) and the Edward River (~400 km); a total >2000 km of rivers and creeks, many intermittent	Over 500 wetlands (Ramsar-listed Werai, Millewa and Koondrook-Perricoota forests), 12 listed threatened species and one endangered ecological community	NSW Water Agencies (NSW Office of Environment and Heritage, New South Wales Office of Water) Australian Government (Murray-Darling Basin Authority, Commonwealth Environmental Water Holder) Local Land Services (catchment management agency)	Landholders and rural communities, Indigenous communities	Green, 2001; Baumgartner et al., 2014
Macquarie Marshes—Macquarie River	74,000 km^2	River length: 960 km Annual rainfall: 300-500 mm (Macquarie Marshes)—1200 mm (Great Dividing Range) Range of annual runoff: 73-1444 × 10^6 m^3; Mean daily flow: 1873 ML (Warren Weir) There are two major dams (Burrendong—1 190 110 ML; Windamere—353 000 ML), three smaller dams and other weirs along the river in the upper catchment, supplying irrigated agriculture.	The Macquarie Marshes is now about 200 000 ha with current flows although it likely flooded at least twice as great an area in the past. It is a complex mosaic of diverse vegetation communities which support significant waterbird communities, particularly breeding colonial waterbirds as well as diverse reptile, frog, and mammal communities.	New South Wales Office of Environment and Heritage (OEH)—two functional areas, management of the Macquarie Marshes reserves and environmental flows, Commonwealth Environmental Water Office, New South Wales New South Wales Department of Primary Industries Water, WaterNSW, New South Wales Department of Primary Industries (Fisheries)	Macquarie Food and Fibre (Irrigation industry), Floodplain graziers, catchment management organization.	Kingsford, 1999; Kingsford and Johnson, 1998; Kingsford and Thomas, 1995; Kingsford and Auld, 2005; Rayner et al., 2009; Ralph and Hesse, 2010; Ren et al., 2010; Ren and Kingsford, 2011; Thomas et al., 2011; Steinfeld and Kingsford, 2012; Bino et al., 2014b; Ocock et al., 2014; Bino et al., 2015; Catelotti et al., 2015; Thomas et al., 2015

FIG. 5.5.5

Researchers measuring bed-rock changes in the Crocodile River in Kruger National Park to inform thresholds of potential concern that provide triggers for management.

Photo: Courtesy C. McLoughlin.

Despite increased environmental flows for the Crocodile River, river monitoring and reporting systems of SAM were not adequately implemented until the mid-2000s, reducing adaptive feedbacks and capacity for evaluation and learning. The Water Research Commission explored deployment of SAM adaptive feedback for evaluation and learning (McLoughlin et al., 2011b), coinciding with implementation of environmental flows of the Ecological Reserve (Pollard and Du Toit, 2011). This is the quantity and quality of water required to protect ecological functions for humans, which must be determined and maintained in natural water resources, defined under the *South African National Water Act* (1998) (O'Keeffe and Rogers, 2003). Determination of the Ecological Reserve for the Crocodile River (Department of Water Affairs, 2010) promoted implementation of SAM. Further, the Inkomati-Usuthu Catchment Management Agency adopted SAM to manage the river's fresh water resources (Jackson, 2015). In 2010, the Crocodile River Operations Committee (CROCOC) comprising catchment stakeholders was initiated to oversee management of the Ecological Reserve and allocation of urban, rural (irrigation 70% of available allocations), and industrial water (Department of Water Affairs, 2010).

Successes

SAM is reasonably mature in the management of the Crocodile River compared to other IRES in South Africa and elsewhere (Kingsford et al., 2011a; Table 5.5.1). Critically, operational monitoring and reporting was introduced in 2011: the Crocodile River's Rapid Response System, with rapid (daily or weekly) communication among stakeholders of CROCOC (Table 5.5.1). The system has different low-flow 'worry-levels,' related to the Ecological Reserve, measured at Ten Bosch flow gauging weir (Fig. 5.5.4A) with each linked to a management response (e.g., water releases or restrictions, McLoughlin et al., 2011a). Adaptive feedbacks ensure weekly calculated benchmarks are achieved.

Monthly reports track successes and failures (e.g., % time river flows in each 'worry-level'), and consequent adaptations and improvements as well as informal gatherings of stakeholders. Success relies heavily on the Water Resources Planning and Operations Manager of the Inkomati-Usuthu Catchment Management Agency and the KNP river technician championing stakeholder interactions and feedbacks among stakeholders (single loop learning; McLoughlin, 2016).

TPC operationalization is also successful, incorporating adaptive feedbacks (McLoughlin, 2016). Tabled TPCs (exceeded or close to being exceeded) are essential for adaptive evaluation and learning, changing management. For example, weekly calculation replaced monthly assessment of the low-flow Ecological Reserve, avoiding the risk of exceeding fish TPCs (single loop learning; McLoughlin, 2016). TPCs can be revised. For example, the original geomorphology indicators and TPCs (Rogers and Bestbier, 1997) were replaced with ones improving measurement of bedrock influence in the river (McLoughlin et al., 2011a), relative to ecological categories (Kleynhans and Louw, 2007) of the Ecological Reserve, and this enhanced subsequent decision-making (double loop learning; McLoughlin, 2016).

Challenges

Delivery of the low-flow Ecological Reserve, a component of the flow regime, may be at the expense of auditing and tabling of the biodiversity-related TPCs (e.g., fish, macroinvertebrate, and geomorphology TPCs), skewing reporting toward this abiotic measure, rather than the outcome of ecological integrity. Assessing whether the Ecological Reserve delivers a particular ecological value is still lacking as is assessing its impacts on other users in this highly water-contested river. Return flows from irrigation farming exacerbate pollution in the river, a focus for water quality monitoring, using TPCs, for potential mitigation.

Further, there are no high-flow or flooding requirements as specified in the Ecological Reserve study (Department of Water Affairs, 2010). Lack of adequate dam design for releasing water for small flooding events challenges flow manipulation. Finally, ecological flow management does not extend into neighboring Mozambique, although an international agreement stipulates a minimum flow requirement.

Future

CROCOC stakeholders continue their learning, delivering the low-flow Ecological Reserve, even where water access is highly contested. TPCs need to be institutionalized (McLoughlin, 2016), so that learning responds to auditing. CROCOC stakeholders also negotiate options and governance processes (triple loop learning), reflecting improvements in knowledge of inputs, outputs, assumptions, and hypotheses (Huxham and Vangen, 2000; Pahl-Wostl et al., 2007). This contributes to implementation of an adequate Ecological Reserve, essential for maintaining biodiversity in KNP.

TANKWA KAROO NATIONAL PARK (SOUTH AFRICA)

Status

Tankwa Karoo National Park (146 747 ha, Fig. 5.5.4B, Table 5.5.1) was proclaimed in 1986 and conserves rich floral diversity of the arid Succulent Karoo Biome. SAM underpins park management (SANParks, 2014), including a vision and objectives hierarchy (Fig. 5.5.2) focused on biodiversity, cultural heritage, and tourism objectives. A 'water in the landscape' subobjective aims to understand the roles of surface and groundwater as drivers of ecological functions. The Tankwa River intermittently flows through the Park. It is, a main tributary of the Doring River, one of South Africa's 19 flagship free-flowing rivers (Figs. 5.5.4B and 5.5.6, Table 5.5.1). Ironically, Oudebaaskraal Dam in the Park regulates flows by capturing floods and interrupts sediment transport, affecting economic productivity

FIG. 5.5.6

The highly intermittent Tankwa River flows through Tankwa Karoo National Park where it is impounded, altering its flow regime downstream and creating conflicting objectives for conservation.

Photo: Courtesy D. Roux.

(e.g., grape farming, tourism, water supply) and ecological function (e.g., estuarine health, endemic fish species) but no longer serves its original function for irrigation (Nel et al., 2006, 2011).

Successes

Adaptive planning introduced new ideas about freshwater biodiversity, contrasting local utility value and water-related ecosystem services benefitting users outside park boundaries. It provided a focus for freshwater conservation that was not previously established, particularly as a focus for management and policy.

Challenges

Objective-setting and involvement of stakeholders lacked a catchment-scale focus, limiting the objectives to the protected area and neighbors. Upstream and downstream influences were undefined constraints on achieving unarticulated objectives and outcomes, including water-related ecosystem services to downstream users (e.g., grape farming). The current objectives hierarchy, developed by scientists and managers, highlighted the value of freshwater ecosystems for terrestrial animals (game) and consequent tourism but freshwater conservation objectives remain relatively poorly articulated with little development of TPCs. Freshwater monitoring is not within the skillset of current rangers or a priority to resource.

Future

Removal of Oudebaaskraal Dam would restore hydrological connectivity and downstream ecological processes of the Tankwa River, potentially benefitting conservation of two endangered fish species, the ecological integrity of the Doring River and Olifants estuary, and downstream water users.

The dam provides novel habitat for waterbirds and tourism (angling and kayaking). Scientists and managers discussed the merits of this action, and potentially conflicting objectives between biodiversity and tourism. Neighboring farmers believe that the dam contributes to recharge of groundwater, vital to this arid landscape, but this remains untested. The dam wall could be modified for fish migrations and releases of environmental flows, using funds from recreational and nature-based tourism activities. SAM provides a coherent framework for linking flow regimes to multiple objectives within and outside the boundaries of the national park, involving diverse and dispersed stakeholders in the park's management.

EDWARD-WAKOOL RIVER SYSTEM—TUPPAL CREEK (AUSTRALIA)

Status

The Edward-Wakool System (Fig. 5.5.4C) includes the Ramsar-listed Werai, Millewa, and Koondrook-Perricoota Forests, supporting diverse vegetation communities (part of the Murray River endangered ecological community) and threatened and iconic species (Table 5.5.1). It also has diverse indigenous values. Large upstream dams and extractions regulate its rivers and streams for agriculture but the riparian zone and instream habitat remain in good condition. Intermittent Tuppal and Jimaringle-Cockrans Creeks (Figs. 5.5.4 and 5.5.7) used to flow for a few months each year, leaving residual pools (Baldwin, 2007) but now require management of environmental water, guided by objectives, defined in strategic plans and consistent with high-level objectives of the Murray-Darling Basin Plan.

Water management is complicated, with 'real-time' management responsibilities for environmental water delivery by Commonwealth and state governments (Table 5.5.1). There are also influential stakeholders (fisher and river associations, irrigation, Table 5.5.1). Traditional mistrust and perceived mismanagement led to significant conflict between some stakeholders and governments, exacerbated by the Murray-Darling Basin Plan, leading to perceptions of winners and losers. Environmental water is delivered each year in both systems, guided by participation and stakeholder-derived objectives that are supported by inclusive monitoring and understood by stakeholders.

FIG. 5.5.7

Tuppal and Jimaringle-Cockrans Creeks are heavily modified river systems, affected by upstream dams, diversions, and adjacent agricultural development. These rivers are intermittent, existing sometimes as a series of pools (A) that are joined during high flows (B).

Photos: Courtesy J. Campbell.

Successes

A SAM process began in the Edward-Wakool System in early 2010, contrasting with most management approaches across the Murray-Darling Basin. Building trust and ownership reduced conflict and associated transaction costs. A vision and objectives engaged government and local stakeholders. This refined environmental flow management. In the Tuppal-Jimaringle system, although there was no formal objectives' hierarchy, the agreed vision influenced objectives, increasing trust and ownership by stakeholders, who then encouraged government implementation. Real-time monitoring (water quality, frogs, vegetation, fish) linked directly to objectives. Implementation was phased to initial-season flows and late-season replenishment flows. Landholders reported flow fronts, water levels, and fauna (e.g., frogs, birds, water rats *Hydromys chrysogaster*) when environmental flows reached their farms. Simple water-quality and water-depth TPCs were set for refuge pools, informing the timing of replenishment flows; this process reflected the requisite simplicity (Stirzaker et al., 2010). Evaluation, communication, and adaptation of delivery were informed by real-time monitoring. Meetings evaluated whether objectives were met each year, recommending future environmental flow management, supported by champions within government and community, improving collaboration and trust.

Challenges

There is no institutionalized adaptive management and that renders this collaborative approach vulnerable. Stakeholders expect commitment to this nascent program but there is little agreement on long-term management objectives or measurable targets for long-term social-ecological outcomes (Stirzaker et al., 2010). Current objectives remain simple, aimed at short-term water quality and vegetation improvements, without adequately including more complex flow-ecological and long-term objectives relevant to the Murray-Darling Basin Plan. For example, there is a lack of understanding of how groundwater interacts with the creek systems and impacts on saline intrusion into the creeks. Finally, government turnover is high, undermining trust and ownership, challenging effective management.

Future

Participatory decision-making and adaptive management principles should support long-term management objectives for environmental water delivery. These need to complement riparian fencing and control of invasive fauna and flora and be integrated into a hierarchy of objectives, recognizing overlapping responsibilities and opportunities for management synergies and priority setting. Complexity of management will increase with understanding of needs for a future desired state which must be supported by participatory decision-making, informed by monitoring. SAM helped bring stakeholders together but more elements are needed to capture relevant objectives of the Murray-Darling Basin Plan.

MACQUARIE MARSHES (AUSTRALIA)

Status

The Macquarie Marshes (Figs. 5.5.4D and 5.5.8) are supplied by the intermittent Macquarie River (Table 5.5.1), which can dry up completely in its distributary streams but also inundates large areas at high flows (Kingsford and Thomas, 1995; Ren and Kingsford, 2011; Thomas et al., 2011, 2015). Flooding is critical for heterogeneous wetland plant communities (Bino et al., 2015), including large areas of river red gum *Eucalyptus camaldulensis* forests, woodlands, and reed beds (Fig. 5.5.8), and stimulating significant breeding of colonial waterbirds (Kingsford and Johnson, 1998). Large dams regulate flows, predominantly for irrigated agriculture upstream of the wetland (Kingsford and Thomas,

FIG. 5.5.8

The Ramsar-listed Macquarie Marshes, supplied by the Macquarie River, are highly intermittent and affected by upstream river regulation and diversions, requiring restoration with environmental flows.

Photo: Courtesy R.T. Kingsford.

1995). About a tenth of the 200,000-ha Macquarie Marshes floodplain includes disjoint areas of the Macquarie Marshes Nature Reserve (and the State Conservation area of Pillicawarrina) which are listed as a Ramsar site of international importance, with privately owned areas.

Regulation has detrimentally affected waterbird and vegetation communities (Kingsford and Johnson, 1998; Thomas et al., 2010; Bino et al., 2015; Catelotti et al., 2015), decreasing river flows (Kingsford and Thomas, 1995) and reducing frequency and extent of inundation and increasing periods of intermittence (Ren et al., 2010; Ren and Kingsford, 2011; Thomas et al., 2011). In 2009, the Ramsar Convention Secretariat was notified under Article 3.2 that there was a likely change in ecological character. State and Australian Governments have bought water, previously supplied for irrigation, as "environmental" water. When the dams are full, more than 300,000 ML of annual environmental water are available, owned by the Australian (Commonwealth Environmental Water Holder) and New South Wales (NSW Office of Environmental and Heritage) Governments. This flow is used primarily to deliver small floods and also contribute to large natural floods. Management by the NSW Office of Environmental and Heritage is informed by an environmental water advisory group including environmental, irrigation, and floodplain grazing interests and government (fisheries, water, environment, Fazey et al., 2006; http://www.environment.nsw.gov.au/environmentalwater/macquarie-ewag.htm). Government responsibility overlaps in the complex management of the Macquarie Marshes.

Successes

Ecological values and processes and drivers are well understood (NSW DECCW, 2010). Scientific effort has focused on understanding links between ecological communities and abiotic drivers, including increasing intermittence and other anthropogenic impacts (Table 5.5.1). Area of inundation is a key surrogate of environmental flow outcomes (Thomas et al., 2011, 2015). There are broad environmental objectives (called targets) for the Murray-Darling Basin Plan (Murray-Darling Basin Authority, 2014), applicable to the Macquarie River and providing the basis for river-specific low-level objectives which could be integrated into a hierarchy of objectives developed by the New South Wales Office of Environment and Heritage (Kingsford et al., 2011a; Bino et al., 2014a). This process could drive development of indicators and targets for rehabilitation.

Challenges

The New South Wales Office of Environment and Heritage is the key management agency, responsible for the Nature Reserve and environmental flows. In 2015, it developed a position statement committing to adaptive management (http://www.environment.nsw.gov.au/research/adaptive-management.htm), with all the essential elements of adaptive management but not fully implemented for the Macquarie Marshes. Linkages between scientific effort and active management remain weak. The agency's main method of delivery of management of the Nature Reserve is primarily through the Macquarie Marshes Plan of Management, completed in 1993, currently more than two decades out of date, reflecting the complexity of management. Management has remained primarily responsive to the functions and the structure of the New South Wales Office of Environment and Heritage. Responsibilities for management of the protected area and environmental flows are in separate parts of the organization, although there is considerable communication. Additional complexity comes with involvement of other government agencies and stakeholders (Table 5.5.1). Environmental flow management primarily focuses on water delivery, informed by advice from the reference group (Fazey et al., 2006) and basin-scale objectives. There is no integration through an adaptive management planning framework.

The hierarchy of objectives requires integration and development, linked to TPCs or restoration and appropriate monitoring data. Importantly, objectives for key ecosystem functions are also needed. There is little rigorous monitoring to inform objectives, apart from explicit measurement of inundation as a surrogate of wetland persistence (Thomas et al., 2011, 2015). Integration of different management responsibilities (e.g., management of floodplain earthworks, Steinfeld and Kingsford, 2012) is part of different processes such as floodplain management, http://www.water.nsw.gov.au/water-management/floodplain-management).

Future

There are considerable opportunities to implement SAM in the Macquarie Marshes, improving transparency and accountability, and linking to science (Lindenmayer and Likens, 2010; Lindenmayer et al., 2011). There is a need to articulate a clear vision of the desired state to guide objective setting. This requires active implementation of the organization's commitment to adaptive management, an embracing of the importance of SAM, and involvement of stakeholders in developing the appropriate vision and objectives hierarchy. Internal and external champions are essential. Otherwise, the current management framework is vulnerable because of the lack of integration and clarity about key objectives and their outcomes.

5.5.5 MAKING IT WORK—A SHARED JOURNEY

SAM of IRES remains challenging across all complex social-ecological dimensions. Generalities emerge from South Africa and Australia (Table 5.5.1), through commonalities and differences (Table 5.5.2). Scales of the social-ecological systems are starkly contrasted, from the small (Tankwa Karoo National

Table 5.5.2 Generalities (commonalities, differences, critical elements) among attributes for the four case studies and their implications for implementation of Strategic Adaptive Management

Generalities	Attributes	Implications
Commonalities	Flow intermittence	All rivers have aspects of intermittence which need to be accounted for in their management to ensuring that key attributes are protected by flow regime management
	Biotic and abiotic drivers	The four IRES varied in their levels of flow intermittence but the biotic (e.g., fish communities, food webs) and abiotic drivers (climate, geomorphology) were broadly similar
	Environmental and cultural values	There are high environmental and cultural values which were readily identifiable and articulated by a broad range of stakeholders within a vision or objectives hierarchy
	Environmental protection goals	Based on these values, it was possible to articulate specific environmental protection goals
	Anthropogenic drivers	Alterations to flow regimes and invasive species were strong direct and indirect drivers
	Strategic Adaptive Management	All case studies have attempted some form of SAM although progress has varied considerably dependent on which critical elements are missing
Differences	Spatial scale	Wide variation from relatively small (Tankwa Karoo National Park) to extensive (Macquarie Marshes, Crocodile River)
	Biotic and abiotic drivers	Biotic communities varied in terms of species and processes while characteristics of abiotic drivers also varied (e.g., level of intermittence)
	Characteristic environmental values	These are necessarily idiosyncratic reflecting the different values of the systems and the different plants, animal, and other biotic communities
	Anthropogenic drivers	Detailed characteristics of types of anthropogenic drivers varied in terms of impacts on the river. For example, this varied from a dam with no extraction (Tankwa Karoo National Park) to dams with extraction (e.g., Macquarie Marshes) altering flow regimes
	Development stage	Rivers varied from highly developed (e.g., Crocodile River and Macquarie Marshes) to free flowing
	Stage of Strategic Adaptive Management development	Differed widely from well developed for Kruger National Park to first steps in development of a broad vision and objectives for the Tankwa Karoo National Park
	Legislation, policy, and management frameworks	Differences occurred between the two countries but also within jurisdictions and were interrelated with level of water resource development
	Resourcing	There was considerable variation in amount of resourcing available for implementation of SAM, reflecting institutional support, but also differences in ecosystem values (e.g., Kruger National Park contrasted to Tankwa Karoo National Park)
Critical elements	Vision	A shared vision is essential, built among stakeholders and reflecting vital attributes of the IRES and its environmental, cultural, and socioeconomic values

Continued

Table 5.5.2 Generalities (commonalities, differences, critical elements) among attributes for the four case studies and their implications for implementation of Strategic Adaptive Management—cont'd

Generalities	Attributes	Implications
	Objectives hierarchy	This allowed fine-scale development of indicators and thresholds of potential concern and potential targets for rehabilitation which could be connected to monitoring.
	Collaboration among policy-makers, managers, and scientists	There is a need for shared development of a vision and objectives hierarchy linked to specific responsive indicators which can then report on the desired state of the ecosystem
	Institutional support	Government and stakeholder support for the process is critical, given the time and energy required for full development
	Capacity	It is critical to provide access to key people to develop SAM for a social-ecological system with resourcing of time, personnel, and reporting essential
	Champions	Improved development depends heavily on individuals who can promote the process within organizations and facilitate social learning processes across organizations/stakeholders
	Monitoring of key indicators	Learning and altering management depend heavily on feedback loops that provide information on outcomes from indicators, stimulating increased management attention or a change in management, including revision of endpoint objectives if required.
	Transparent reporting	SAM provides a transparent framework for reporting on management linked to key values of a social-ecological ecosystem but it also requires development of a range of reporting systems (e.g., indicators, objectives)

Park) to the large (Macquarie Marshes, Crocodile River, Table 5.5.2). Despite this, similar biotic and abiotic (e.g., precipitation, evaporation) factors drive flow regimes, biodiversity, and food webs, although characteristics of each system's environmental values and processes vary (Table 5.5.1). Anthropogenic drivers are also similar, affecting potential intervention management. This necessarily demands different visions, stakeholders, and management objectives, reflecting development stage and locality.

People and their institutions have attempted SAM to varying degrees in all our IRES. There are clear opportunities and challenges in implementation (Fig. 5.5.9), with maturity of SAM primarily reflecting time, commitment to and institutionalization of the process, and how opportunities have overcome challenges (Fig. 5.5.9). SAM negotiates the challenges (Fig. 5.5.9) which thwart implementation of other frameworks. South Africa has a well-developed culture of SAM compared to Australia, reflected in the successes and challenges in the respective systems. Resourcing for SAM inevitably reflects conservation importance (Table 5.5.2), illustrated by contrasts between the Crocodile River of KNP and the Tankwa Karoo National Parks. Implementation of SAM also depends heavily on legislative policy and management frameworks (Table 5.5.2), reflected in personnel and other resources available for SAM (Table 5.5.1).

Critical elements for successful implementation emerge (Table 5.5.2, Fig. 5.5.9). The development of a shared vision among stakeholders is particularly important, reflecting the values that underpin the future desired state. Subsequently, the objectives hierarchy winnows this vision into clear endpoints for management and scientific effort. Willing collaboration among different stakeholders (Fig. 5.5.1) including government staff from different agencies and scientists is particularly important (Table 5.5.2). Institutional support is essential, providing policy drive and capacity (Table 5.5.2, Fig. 5.5.9) for workshops and articulation of the key elements of SAM (Figs. 5.5.2 and 5.5.9). Champions within institutions and stakeholders are especially important, driving organizational inertia or intercepting traditional command and control processes (Fig. 5.5.9). Rigorous testing of assumptions about management outcomes and system behavior can create discomfort, requiring conflict to be approached with integrity (Box 5.5.1).

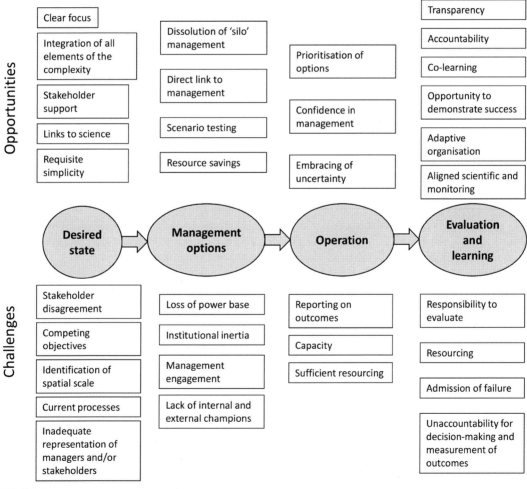

FIG. 5.5.9

Synthesis of the opportunities and challenges applicable to the four steps of Strategic Adaptive Management and affecting its implementation.

Fine-scale objectives provide the scientific reporting framework for responsive indicators that reflect the system's values. This becomes the opportunity, not only for retrospective analysis of the effects of past management, but also development of TPCs or targets for rehabilitation for management options. Increasing certainty of predictions as a result of improved understanding of cause-and-effect relationships can help in identifying the most responsive management drivers. Transparent reporting is essential to allow for others to learn and track management outcomes and progress (Table 5.5.2).

5.5.6 GLOBAL APPLICATIONS

SAM offers considerable promise for management of the world's IRES, systems covering over half of the Earth's land area (Middleton and Thomas, 1997; Datry et al., 2014; Chapter 1). Many are actively or passively managed, affecting biological communities (Boix et al., 2010; Munné and Prat, 2011; Acuña et al., 2014; Chapters 5.1 and 5.4). The need for management will only grow, as increasing pressure on the world's freshwater resources and climate change potentially increase intermittence in many parts of the world (Pahl-Wostl, 2007; Datry et al., 2014; Chapter 2.3). This management includes active management of environmental flows for restoration (Arthington, 2012; Chapter 5.4). Despite this growing need, adaptive management of rivers remains poorly developed.

There is a long history of inadequate implementation of adaptive management (Walters, 1997; Gunderson and Light, 2006), often primarily reflecting insufficient incorporation of planning, collaboration, modeling, and evaluation or learning (Schreiber et al., 2004; Kingsford et al., 2011a). The importance of social learning is critical (Pahl-Wostl et al., 2007), including understanding and open sharing, diversification of financial resources, cross-sectoral analysis, purposeful learning, and self-organization (Pahl-Wostl, 2007), all essential components of SAM. Conceptual models, also called pressure-state-impact-response models, are critical for implementing adaptive management (Pahl-Wostl, 2007), promoting common understanding and a platform for predicting management scenarios, increasingly valuable for management of rivers and their biota (Gunderson and Light, 2006). Development of objectives is also critical, underpinned by an understanding of ecological status. For example, the European Water Framework Directive requires the assessment of the ecological status of water bodies, including IRES (Acuña et al., 2014; Prat et al., 2014; Chapter 5.3), allowing development of environmental objectives such as quantification of targets of environmental flow varying across different streams (Nikolaidis et al., 2013).

SAM is globally applicable to IRES, providing aspects missing in much of current management. There are clear policy drivers around the world for managing IRES effectively, including the European Water Framework Directive and large-scale restoration projects such as the Florida Everglades and the Murray-Darling Basin. SAM provides a holistic framework of management, which includes engagement of stakeholders, specification of a desired state, and learning by doing (Figs. 5.5.2 and 5.5.8). It has been primarily developed in the world's most variable IRES in South Africa and Australia (Puckridge et al., 1998). Articulation of the desired state, involving the full range of stakeholders (Figs. 5.5.2 and 5.5.8), is often poorly developed in management and culminates in development of a hierarchy of objectives linked to measurement of management outcomes. For example, criticism of the management of the Florida Everglades focuses on the absence of critical aspects of governance in a system where management responds primarily through technical fixes and government mandates because vested interests inhibit resolution (Gunderson and Light, 2006). Increasingly, there is recognition that adaptive governance, focusing on social learning and openness (Folke et al., 2005), is critical for implementation of adaptive management for rivers. SAM, with more than two decades of development, offers hope for providing a generic framework for managing rivers in increasingly challenging social and ecological environments.

5.5.7 CONCLUSIONS

SAM challenges current organizational cultures and management processes (Stirzaker et al., 2011) but offers a comprehensive structure, incorporating different dimensions of complex social-ecological systems and their responsible natural resource management agencies. Importantly, SAM highlights the interdependencies between science and management and facilitates bidirectional feedbacks (Biggs and Rogers, 2003), providing direction for understanding cause-and-effect relationships. Our focus on implementation of SAM in IRES provides good examples of the complex dimensions of spatial scale, collaborating communities and institutional challenges. Our case studies all represent adaptive management journeys that involve shared understanding among scientists, policy-makers, managers, and relevant stakeholders. The framework is globally applicable.

Connecting the social and ecological dimensions sometimes languishes because of inadequate institutional support or absence of internal champions. Given its usefulness and transparency, the framework needs to continue to be developed across the world, improving investment of significant resources in river management and supported by objective measures of effectiveness. If more widely adopted, it can empower communities to track achievement of environmental outcomes as opposed to outputs.

SAM is essential for adequate management of IRES around the world. It can be resource-intensive but even few resources will suffice to focus on the importance and value of targeted management (e.g., Tankwa Karoo National Park). Currently, resourcing for river management is delivered to agencies and individuals but seldom integrated with a largely undirected scientific effort. Management of IRES needs to be improved, integrating these processes for more effective and transparent delivery of intended outcomes to reach a desired state. SAM shows considerable promise for this process, evidenced by the successes in the IRES described in this chapter, and given there are no obvious alternatives that are working.

ACKNOWLEDGMENTS

We thank Jessica Rooke for assistance with the figures and Rachael Thomas for her contribution to the Macquarie Marshes case study. We also thank the NSW Office of Environment and Heritage, Murray Local Land Services, Australian Commonwealth Environmental Water Office, and community members for providing information for the Edward-Wakool case study.

REFERENCES

Acuña, V., Datry, T., Marshall, J., Barcelo, D., Dahm, C.N., Ginebreda, A., et al., 2014. Why should we care about temporary waterways? Science 343, 1080–1081.

Armitage, D., Marschke, M., Plummer, R., 2008. Adaptive co-management and the paradox of learning. Glob. Environ. Chang. 18, 86–98.

Armitage, D.R., Plummer, R., Berkes, F., Arthur, R.I., Charles, A.T., Davidson-Hunt, I.J., et al., 2009. Adaptive co-management for social-ecological complexity. Front. Ecol. Environ. 7, 95–102.

Arnstein, S.R., 1969. A ladder of citizen participation. J. Am. Inst. Plann. 35, 216–224.

Arthington, A.H., 2012. Environmental Flows: Saving Rivers in the Third Millennium. University of California Press, Berkeley, CA.

Baldwin, D.S., 2007. A Biophysical ('Aquatic Habitat') Assessment of the Natural and Current Condition of Tuppal Creek. Murray-Darling Freshwater Research Centre, Wodonga, VIC.

Baumgartner, L.J., Conallin, J., Wooden, I., Campbell, B., Gee, R., Robinson, W.A., et al., 2014. Using flow guilds of freshwater fish in an adaptive management framework to simplify environmental flow delivery for semi-arid riverine systems. Fish Fish. 15, 410–427.

Belmar, O., Bruno, D., Martinez-Capel, F., Barquin, J., Velasco, J., 2013. Effects of flow regime alteration on fluvial habitats and riparian quality in a semiarid Mediterranean basin. Ecol. Indic. 30, 52–64.

Biggs, H., Rogers, K.H., 2003. An adaptive system to link science, monitoring and management in practice. In: Du Toit, J.T., Rogers, K.H., Biggs, H.C. (Eds.), The Kruger Experience: Ecology and Management of Savanna Heterogeneity. Island Press, Washington, DC, pp. 59–80.

Biggs, H., Ferreira, S., Ronaldson, S.F., Grant-Biggs, R., 2011. Taking stock after a decade: does the 'thresholds of potential concern' concept need a socio-ecological revamp? Koedoe 53, 60–68.

Bino, G., Kingsford, R.T., Jenkins, K., 2014a. Climate adaptation and adaptive management planning for the Macquarie Marshes: a wetland of international importance. In: Palutikof, J.P., Boulter, S.L., Barnett, J., Rissik, D. (Eds.), Applied Studies in Climate Adaptation. John Wiley and Sons, Chichester, pp. 95–106.

Bino, G., Steinfeld, C., Kingsford, R.T., 2014b. Maximizing colonial waterbirds' breeding events using identified ecological thresholds. and environmental flow management. Ecol. Appl. 24, 142–157.

Bino, G., Sisson, S.A., Kingsford, R.T., Thomas, R.F., Bowen, S., 2015. Developing state and transition models of floodplain vegetation dynamics as a tool for conservation decision-making: a case study of the Macquarie Marshes Ramsar wetland. J. Appl. Ecol. 52, 654–664.

Boix, D., García-Berthou, E., Gascón, S., Benejam, L., Tornés, E., Sala, J., et al., 2010. Response of community structure to sustained drought in Mediterranean rivers. J. Hydrol. 383, 135–146.

Catelotti, K., Kingsford, R.T., Bino, G., Bacon, P., 2015. Inundation requirements for persistence and recovery of river red gums (*Eucalyptus camaldulensis*) in semi-arid Australia. Biol. Conserv. 184, 346–356.

Costanza, R., d'Arge, R., de Groot, R., Farber, S., Grasso, M., Hannon, B., et al., 1997. The value of the world's ecosystem services and natural capital. Nature 387, 253–260.

Datry, T., Larned, S.T., Tockner, K., 2014. Intermittent rivers: a challenge for freshwater ecology. Bioscience 64, 229–235.

Datry, T., Pella, H., Leigh, C., Bonada, N., Hugueny, B., 2016. A landscape approach to advance intermittent river ecology. Freshw. Biol. 61, 1200–1213.

Department of Water Affairs, 2010. Comprehensive Reserve Determination Study for Selected Water Resources (Rivers, Groundwater and Wetlands) in the Inkomati Water Management Area: DWA Report Number 26/8/3/10/12/015. Department of Water Affairs, Pretoria.

Fazey, I.R.A., Proust, K., Newell, B., Johnson, B., Fazey, J.A., 2006. Eliciting the implicit knowledge and perceptions of on-ground conservation managers of the Macquarie Marshes. Ecol. Soc. 11, 25.

Flood, R.L., Romm, N.R., 1996. Diversity Management: Triple Loop Learning. John Wiley, Chichester.

Folke, C., Hahn, T., Olsson, P., Norberg, J., 2005. Adaptive governance of social-ecological systems. Annu. Rev. Environ. Resour. 30, 441–473.

Foxcroft, L.C., 2009. Developing thresholds of potential concern for invasive alien species: hypotheses and concepts. Koedoe 51. http://dx.doi.org/10.4102/koedoe.v51i1.157.

Green, D., 2001. The Edward-Wakool System River Regulation and Environmental Flows: Report for Department of Land and Water Conservation. NSW Department of Land and Water Conservation, Murray Region, Deniliquin NSW.

Gunderson, L., Light, S.S., 2006. Adaptive management and adaptive governance in the Everglades ecosystem. Policy. Sci. 39, 323–334.

Hamilton, S.K., Bunn, S.E., Thoms, M.C., Marshall, J.C., 2005. Persistence of aquatic refugia between flow pulses in a dryland river system (Cooper Creek, Australia). Limnol. Oceanogr. 50, 743–754.

Heritage, G.L., Van Niekerk, A.W., Moon, B.P., 1999. Geomorphology of the Sabie River, South Africa: an incised bedrock-influenced channel. In: Miller, A.J., Gupta, A. (Eds.), Varieties of Fluvial Form. John Wiley and Sons, Chichester, pp. 53–79.

Holling, C.S., Meffe, G.K., 1996. Command and control and the pathology of natural resource management. Conserv. Biol. 10, 328–337.

Huxham, C., Vangen, S., 2000. Leadership in the shaping and implementation of collaboration agendas: how things happen in a (not quite) joined-up world. Acad. Manag. J. 43, 1159–1175.

Jackson, B., 2015. An Adaptive Operational Water Resources Management Framework for the Crocodile River Catchment. University of KwaZulu-Natal, South Africa. Masters Thesis.

Kingsford, R.T., 1999. Social and economic costs and benefits of taking water from our rivers: the Macquarie Marshes as a test case. In: Robertson, A.I., Watts, R.J. (Eds.), Preserving Rural Australia: Issues and Solutions. CSIRO Publishing, Collingwood, pp. 125–143.

Kingsford, R.T., 2015. Conservation of floodplain wetlands-out of sight, out of mind. Aquat. Conserv. Mar. Freshwat. Ecosyst. 25, 727–732.

Kingsford, R.T., Auld, K.M., 2005. Waterbird breeding and environmental flow management in the Macquarie Marshes, arid Australia. River Res. Appl. 21, 187–200.

Kingsford, R.T., Biggs, H.C., 2012. Strategic Adaptive Management Guidelines for Effective Conservation of Freshwater Ecosystems in and Around Protected Areas of the World. IUCN WCPA Freshwater Taskforce, Australian Wetlands and Rivers Centre, Sydney.

Kingsford, R.T., Johnson, W.J., 1998. The impact of water diversions on colonially nesting waterbirds in the Macquarie Marshes in arid Australia. Colon. Waterbirds 21, 159–170.

Kingsford, R.T., Thomas, R.F., 1995. The Macquarie Marshes in arid Australia and their waterbirds: a 50-year history of decline. Environ. Manag. 19, 867–878.

Kingsford, R.T., Thompson, J.R., 2006. Desert rivers of the world—an introduction. In: Kingsford, R.T. (Ed.), Ecology of Desert Rivers. Cambridge University Press, Cambridge, pp. 3–10.

Kingsford, R.T., Biggs, H.C., Pollard, S.R., 2011a. Strategic Adaptive Management in freshwater protected areas and their rivers. Biol. Conserv. 144, 1194–1203.

Kingsford, R.T., Walker, K.F., Lester, R.E., Young, W.J., Fairweather, P.G., Sammut, J., et al., 2011b. A Ramsar wetland in crisis - the Coorong, Lower Lakes and Murray Mouth, Australia. Mar. Freshw. Res. 62, 255–265.

Kingsford, R.T., Basset, A., Jackson, L., 2016. Wetlands: conservation's poor cousin. Aquat. Conserv. Mar. Freshwat. Ecosyst. 26, 892–916.

Kleynhans, C.J., Louw, M.D., 2007. Module A: EcoClassification and Ecostatus Determination in River EcoClassification: Manual for EcoStatus Determination (version 2): Joint Water Research Commission and Department of Water Affairs and Forestry report. WRC Report Number TT 333/08. Water Research Commission, Pretoria.

Knight, A.T., Cowling, R.M., Boshoff, A.F., Wilson, S.L., Pierce, S.M., 2011. Walking in STEP: lessons for linking spatial prioritisations to implementation strategies. Biol. Conserv. 144, 202–211.

Larned, S.T., Datry, T., Arscott, D.B., Tockner, K., 2010. Emerging concepts in temporary-river ecology. Freshw. Biol. 55, 717–738.

Lindenmayer, D.B., Likens, G.E., 2010. The science and application of ecological monitoring. Biol. Conserv. 143, 1317–1328.

Lindenmayer, D.B., Gibbons, P., Bourke, M.A.X., Burgman, M., Dickman, C.R., Ferrier, S., et al., 2011. Improving biodiversity monitoring. Austral Ecol. 37, 285–294.

Martin, J., Runge, M.C., Nichols, J.D., Lubow, B.C., Kendall, W.L., 2009. Structured decision making as a conceptual framework to identify thresholds for conservation and management. Ecol. Appl. 19, 1079–1090.

McCarthy, T.S., Cooper, G.R.J., Tyson, P.D., Ellery, W.N., 2000. Seasonal flooding in the Okavango Delta, Botswana - recent history and future prospects. S. Afr. J. Sci. 96, 25–33.

McLoughlin, C.A., 2016. Reflexive Learning in the Practice of Adaptive Freshwater Management. PhD Thesis, University of New England, Armidale, New South Wales, Australia.

McLoughlin, C.A., Thoms, M.C., 2015. Integrative learning for practicing adaptive resource management. Ecol. Soc. 20, 34.

McLoughlin, C.A., Deacon, A., Sithole, H., Gyedu-Ababio, T., 2011a. History, rationale, and lessons learned: thresholds of potential concern in Kruger National Park river adaptive management. Koedoe 53. http://dx.doi.org/10.4102/koedoe.v53i2.996.

McLoughlin, C.A., MacKenzie, J., Rountree, M., Grant, R., 2011b. Implementation of Strategic Adaptive Management for Freshwater Protection Under the South African National Water Policy: WRC Research Report Number 1797/1/11. Water Research Commission, Pretoria.

McMahon, T.A., Vogel, R.M., Peel, M.C., Pegram, G.G.S., 2007. Global streamflows—part 1: characteristics of annual streamflows. J. Hydrol. 347, 243–259.

Middleton, N.J., Thomas, D.S.G., 1997. World Atlas of Desertification. UNEP/ Edward Arnold, London.

Moore, L., 2013. Common Ground on Hostile Turf: Stories from an Environmental Mediator. Island Press, Washington, DC.

Munné, A., Prat, N., 2011. Effects of Mediterranean climate annual variability on stream biological quality assessment using macroinvertebrate communities. Ecol. Indic. 11, 651–662.

Murray-Darling Basin Authority, 2014. Basin-Wide Environmental Watering Strategy. Murray-Darling Basin Authority, Canberra.

Nel, J., Driver, A., Strydom, W., Maherry, A., Petersen, C., Hill, L., et al., 2011. Atlas of Freshwater Ecosystem Priority Areas in South Africa: Maps to Support Sustainable Development of Water Resources: Report No TT 500/11. Water Research Commission, Pretoria.

Nel, J.L., Belcher, A., Impson, N.D., Kotze, I.M., Paxton, B., Schonegevel, L.Y., et al., 2006. Conservation Assessment of Freshwater Biodiversity in the Olifants/Doorn Water Management Area: Final report. CSIR Report Number CSIR/NRE/ECO/ER/2006/0182/C. Council for Scientific and Industrial Research, Stellenbosch.

Nikolaidis, N.P., Demetropoulou, L., Froebrich, J., Jacobs, C., Gallart, F., Prat, N., et al., 2013. Towards sustainable management of Mediterranean river basins: policy recommendations on management aspects of temporary streams. Water Policy 15, 830–849.

Nilsson, C., Reidy, C.A., Dynesius, M., Revenga, C., 2005. Fragmentation and flow regulation of the world's large river systems. Science 308, 405–408.

NSW DECCW (New South Wales Department of Environment, Climate Change and Water), 2010. Macquarie Marshes Adaptive Environmental Plan. New South Wales Department of Environment, Climate Change and Water, Sydney. http://www.environment.nsw.gov.au/resources/environmentalwater/100224-aemp-macquarie-marsh.pdf (accessed 01.12.16).

NSW NPWS (New South Wales National Parks and Wildlife Service), 1993a. Macquarie Marshes Nature Reserve Plan of Management. New South Wales National Parks and Wildlife Service, Sydney. <http://www.environment.nsw.gov.au/resources/parks/pomfinalmacquariemarshes.pdf > (accessed 01.12.16.).

O'Keeffe, J., Rogers, K.H., 2003. Heterogeneity and management of the Lowveld rivers. In: Du Toit, J.T., Rogers, K.H., Biggs, H.C. (Eds.), The Kruger Experience: Ecology and Management of Savanna Heterogeneity. Island Press, Washington, DC, pp. 447–468.

Ocock, J.F., Kingsford, R.T., Penman, T.D., Rowley, J.J.L., 2014. Frogs during the flood: differential behaviours of two amphibian species in a dryland floodplain wetland. Austral Ecol. 39, 929–940.

Ogden, J.C., Davis, S.M., Jacobs, K.J., Barnes, T., Fling, H.E., 2005. The use of conceptual ecological models to guide ecosystem restoration in South Florida. Wetlands 25, 795–809.

Olden, J.D., Naiman, R.J., 2010. Incorporating thermal regimes into environmental flows assessments: modifying dam operations to restore freshwater ecosystem integrity. Freshw. Biol. 55, 86–107.

Ostrom, E., 1990. Governing the Commons: The Evolution of Institutions for Collective Action. Cambridge University Press, Cambridge.

Pahl-Wostl, C., 2007. Transitions towards adaptive management of water facing climate and global change. Water Resour. Manag. 21, 49–62.

Pahl-Wostl, C., Craps, M., Dewulf, A., Mostert, E., Tabara, D., Taillieu, T., 2007. Social learning and water resources management. Ecol. Soc. 12, 5.

Poff, N.L., Olden, J.D., Merritt, D.M., Pepin, D.M., 2007. Homogenization of regional river dynamics by dams and global biodiversity implications. Proc. Natl. Acad. Sci. 104, 5732–5737.

Pollard, S., Du Toit, D., 2007. Guidelines for Strategic Adaptive Management: Experiences From Managing the Rivers of the Kruger National Park. Scientific Services, Skukuza.

Pollard, S., Du Toit, D., 2011. Towards adaptive integrated water resources management in southern Africa: the role of self-organisation and multi-scale feedbacks for learning and responsiveness in the Letaba and Crocodile catchments. Water Resour. Manag. 25, 4019–4035.

Postel, S.L., 2000. Entering an era of water scarcity: the challenges ahead. Ecol. Appl. 10, 941–948.

Prat, N., Gallart, F., Von Schiller, D., Polesello, S., García-Roger, E., Latron, J., et al., 2014. The MIRAGE toolbox: an integrated assessment tool for temporary streams. River Res. Appl. 30, 1318–1334.

Puckridge, J.T., Sheldon, F., Walker, K.F., Boulton, A.J., 1998. Flow variability and the ecology of arid zone rivers. Mar. Freshw. Res. 49, 55–72.

Ralph, T.J., Hesse, P.P., 2010. Downstream hydrogeomorphic changes along the Macquarie River, southeastern Australia, leading to channel breakdown and floodplain wetlands. Geomorphology 118, 48–64.

Rayner, T.S., Jenkins, K.M., Kingsford, R.T., 2009. Small environmental flows, drought and the role of refugia for freshwater fish in the Macquarie Marshes, arid Australia. Ecohydrology 2, 440–453.

Ren, S., Kingsford, R., 2011. Statistically integrated flow and flood modelling compared to hydrologically integrated quantity and quality model for annual flows in the regulated Macquarie River in arid Australia. Environ. Manag. 48, 177–188.

Ren, S.Q., Kingsford, R.T., Thomas, R.F., 2010. Modelling flow to and inundation of the Macquarie Marshes in arid Australia. Environmetrics 21, 549–561.

Rogers, K., Bestbier, R., 1997. Development of a Protocol for the Definition of the Desired State of Riverine Systems in South Africa. Department of Environmental Affairs and Tourism, Pretoria.

Rogers, K., Biggs, H., 1999. Integrating indicators, endpoints and value systems in strategic management of the rivers of the Kruger National Park. Freshw. Biol. 41, 439–451.

Rountree, M., Rogers, K., 2004. Channel pattern changes in the mixed bedrock/alluvial Sabie River, South Africa: response to and recovery from large infrequent floods. In: de Jalon, D.G., Vizcaino, P. (Eds.), Proceedings of the Fifth International Symposium on Ecohydraulics. IAHR, Madrid, pp. 318–324.

Roux, D.J., Foxcroft, L.C., 2011. The development and application of strategic adaptive management within South African National Parks. Koedoe 53. http://dx.doi.org/10.4102/koedoe.v53i2.1049.

Roux, D.J., Rogers, K.H., Biggs, H.C., Ashton, P.J., Sergeant, A., 2006. Bridging the science-management divide: moving from unidirectional knowledge transfer to knowledge interfacing and sharing. Ecol. Soc. 11, 4.

Rubin, F., 1998. The physical environment and major plant communities of the Tankwa-Karoo National Park. Koedoe 41, 61–94.

Russell, I., 1997a. Spatial variation in the structure of fish assemblages in the Vaalbos National Park, South Africa. Koedoe 40, 113–123.

Russell, I.S., 1997b. Monitoring the Conservation Status and Diversity of Fish Assemblages in the Major Rivers of the Kruger National Park. University of the Witwatersrand, Johannesburg. PhD Thesis.

SANParks, 2008. Kruger National Park management plan. South African National Parks, https://www.sanparks.org/assets/docs/conservation/park_man/knp-management-plan1.pdf (accessed 01.12.16.).

SANParks, 2014. Tankwa Karoo National Park—park management plan. South African National Parks, https://www.sanparks.org/assets/docs/conservation/park_man/tankwa_approved_plans.pdf (accessed 01.12.16.).

Schreiber, E.S.G., Bearlin, A.R., Nicol, S.J., Todd, C.R., 2004. Adaptive management: a synthesis of current understanding and effective application. Ecol. Manag. Restor. 5, 177–182.

Steinfeld, C.M.M., Kingsford, R.T., 2012. Disconnecting the floodplain: earthworks and their ecological effect on a dryland floodplain in the Murray–Darling Basin, Australia. River Res. Appl. 29, 206–218.

Stirzaker, R., Biggs, H., Roux, D., Cilliers, P., 2010. Requisite simplicities to help negotiate complex problems. Ambio 39, 600–607.

Stirzaker, R.J., Roux, D.J., Biggs, H.C., 2011. Learning to bridge the gap between adaptive management and organisational culture. Koedoe 53, 28–33.

Thomas, R., Bowen, S., Simpson, S., Cox, S., Sims, N., Hunter, S., et al., 2010. Inundation response of vegetation communities of the Macquarie Marshes in semi-arid Australia. In: Saintilan, N., Overton, I. (Eds.), Ecosystem Response Modelling in the Murray-Darling Basin. CSIRO Publishing, Collingwood, pp. 137–150.

Thomas, R.F., Kingsford, R.T., Lu, Y., Hunter, S.J., 2011. Landsat mapping of annual inundation (1979–2006) of the Macquarie Marshes in semi-arid Australia. Int. J. Remote Sens. 32, 4545–4569.

Thomas, R.F., Kingsford, R.T., Lu, Y., Cox, S.J., Sims, N.C., Hunter, S., 2015. Mapping inundation in the heterogeneous floodplain wetlands of the Macquarie Marshes, using Landsat Thematic Mapper. J. Hydrol. 524, 194–213.

van Coller, A.L., Rogers, K.H., Heritage, G.L., 2000. Riparian vegetation-environment relationships: complementarity of gradients versus patch hierarchy approaches. J. Veg. Sci. 11, 337–350.

Van Niekerk, A., Heritage, G., Moon, B., 1995. River classification for management: the geomorphology of the Sabie River in the eastern Transvaal. S. Afr. Geogr. J. 77, 68–76.

Walker, B., Salt, D., 2012. Resilience Thinking: Sustaining Ecosystems and People in a Changing World. Island Press, Washington, DC.

Walters, C., 1997. Challenges in adaptive management of riparian and coastal ecosystems. Conserv. Ecol. 1, 1. http://www.consecol.org/vol1/iss2/art1/ (accessed 3.12.16).

Walters, C.J., Holling, C.S., 1990. Large-scale management experiments and learning by doing. Ecology 71, 2060–2068.

Young, W.J., Kingsford, R.T., 2006. Flow variability in large unregulated dryland rivers. In: Kingsford, R.T. (Ed.), Ecology of Desert Rivers. Cambridge University Press, Cambridge, pp. 11–46.

FURTHER READING

NSW NPWS (New South Wales National Parks and Wildlife Service), 1993b. Macquarie Marshes Nature Reserve Plan of Management. New South Wales National Parks and Wildlife Service, Sydney. <http://www.environment.nsw.gov.au/resources/parks/pomfinalmacquariemarshes.pdf > (accessed 01.12.16.).

Rogers, K.H., O'Keeffe, J., 2003. River heterogeneity: ecosystem structure, function and management. In: Du Toit, J.T., Rogers, K.H., Biggs, H.C. (Eds.), The Kruger Experience: Ecology and Management of Savanna Heterogeneity. Island Press, Washington DC, pp. 189–218.

CONCLUSIONS: RECENT ADVANCES AND FUTURE PROSPECTS IN THE ECOLOGY AND MANAGEMENT OF INTERMITTENT RIVERS AND EPHEMERAL STREAMS

Thibault Datry*, Núria Bonada[†], Andrew J. Boulton[‡]

Irstea, UR MALY, centre de Lyon-Villeurbanne, Villeurbanne, France[*] Universitat de Barcelona (UB), Barcelona, Spain[†] University of New England, Armidale, NSW, Australia[‡]

IN A NUTSHELL

- The 20 chapters in this book synthesize the scattered information on intermittent rivers and ephemeral streams (IRES), exploring the effects of flow intermittence on geomorphology, hydrology, biogeochemistry, and ecology in these ecosystems.
- Here, we distil this information into fourteen themes, most of which emphasize how intermittence and pulsed flows govern almost every physical, chemical, and ecological aspect of IRES and must be explicitly considered in conservation, restoration, and ecosystem management.
- Despite substantial advances in scientific knowledge of IRES in the last two decades, major methodological, geographical/disciplinary, and conceptual gaps remain.
- IRES are widespread yet undervalued ecosystems that deserve greater recognition of their services and more effective legislative protection and management in the face of global climate change and increasing human demands for water.

6.1 INTRODUCTION

One of the primary intentions of this book was to collate and review the scientific literature on intermittent rivers and ephemeral streams (IRES) across the world, especially the results of the prolific research published during the last decade (Datry et al., 2011; Leigh et al., 2016a). This collation enabled identification of broad themes or generalizations about these poorly understood ecosystems to complement findings from a concurrent international research program ("Intermittent River Biodiversity and Synthesis," IRBAS, http://irbas.cesab.org/) whose focus was to compile and synthesize the scattered

Intermittent Rivers and Ephemeral Streams. http://dx.doi.org/10.1016/B978-0-12-803835-2.00031-0

knowledge on the biodiversity of IRES (Datry et al., 2014a,b; Leigh et al., 2016b). A second major intention of this book was to review how management and restoration of IRES currently use scientific information, and whether emergent themes and generalizations from recent research could be better integrated into strategies to manage IRES.

For the book, experts were invited to collaborate with international colleagues to write chapters reviewing relevant literature in subtopics selected to span the diverse but related disciplines of geomorphology, hydrology, biogeochemistry, ecology, and management of IRES. Authors of all chapters were asked to adopt a global perspective and suggest fruitful research directions or promising management options for IRES. Drawing on the results of these endeavors, this final chapter derives fourteen themes about the ecology and management of IRES and posits exciting avenues for future work.

We commence by presenting the 14 themes, many of which relate to the implications of the characteristic feature of all IRES: flow intermittence. Although most of these themes can be stated as broad generalizations, some are posed as general hypotheses that deserve testing. Nine of the themes cover the physical, chemical, and biological aspects of IRES while the remaining five themes relate to the management of IRES, again focusing on the implications of flow intermittence. Many gaps in our knowledge remain, and we conclude by identifying some of the common ones, arranged in three broad categories: methodological, geographical/disciplinary, and conceptual gaps and perspectives.

6.2 FOURTEEN THEMES IN THE ECOLOGY AND MANAGEMENT OF IRES

The 20 chapters in the four main sections in this book complement each other, providing the context for the integrative themes outlined later. The geographic setting (e.g., climate, latitude, and altitude) interacts with geomorphology, catchment, and hydrogeology features across the world (Theme 1) to collectively drive hydrological connectivity and flow regime (Themes 2 and 3), especially intermittence in IRES (Fig. 6.1). These govern water quality and physicochemistry (Theme 4) that, together with the direct effects of flow intermittence on hydrological connectivity and habitat availability, influence biodiversity and ecological processes such as trophic interactions and colonization dynamics (Themes 5–9). Societal values of IRES are strongly influenced by the ecosystem services provided by biodiversity and ecological processes, and often dictate the management strategies used to protect and restore IRES in the face of diverse human impacts (Themes 10–11; Fig. 6.1). As we come to better understand the ecosystem services provided by IRES, legislation, policy, and management strategies that protect and restore IRES (Themes 12–14) will become more prevalent, capitalizing on the recent advances in knowledge about how these ecosystems function and their reliance on intermittence.

THEME 1

IRES are the most globally widespread flowing-water ecosystem and are expanding further through human intervention.

IRES occur on all continents, including Antarctica. They predominate in arid, semiarid, Mediterranean, and dry-subhumid regions but are also common in temperate and well-watered areas where they often comprise the majority (up to 70%) of headwater streams (Fritz et al., 2013; Datry et al., 2014a,b). A trend of climatic drying in many parts of the globe, coupled with rapidly increasing human demands as the planet's population escalates, is causing an overall increase in the occurrence and spatial and temporal

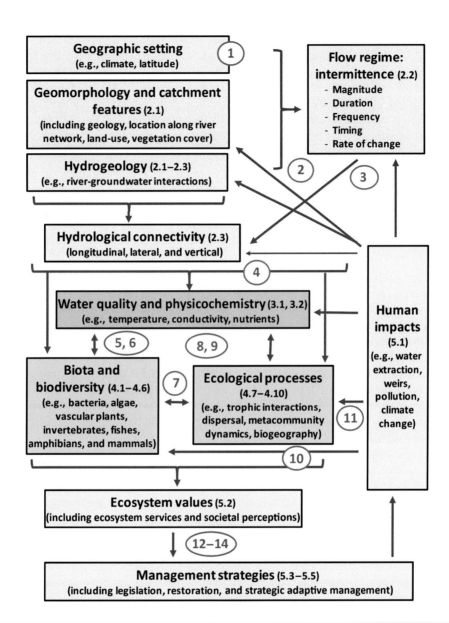

FIG. 6.1

Inter-related aspects of IRES, color-coded to match the sections and chapters of the book where they are described: *blue*—Section 2: geomorphology and hydrology; *orange*—Section 3: physicochemistry and biogeochemical processes; *green*—Section 4: biota, biodiversity and ecological processes; *yellow*—Section 5: management issues and strategies. *Circled numbers* represent elements related to the numbered themes discussed in the text; *arrows* indicate contextual relationships among topics.

extent of IRES worldwide (Döll and Schmied, 2012). At the same time, flows in many natural IRES are becoming more intermittent, and there are longer periods when channels have zero flow or are dry. Controlled releases from dams; discharge of agricultural, industrial, and urban effluents; and interbasin transfers have reduced intermittence of many IRES, even turning some of them perennial (Chapter 5.1). However, these activities are unlikely to offset the increasing global expansion of intermittence and IRES. Therefore in this Anthropocene era, IRES are arguably the most widespread flowing-water ecosystem in the world and are expanding further through human intervention and climate change (Chapter 1).

THEME 2

The global diversity of geomorphic, climatic, and hydrological settings of IRES generates extreme spatial and temporal variability in flow regime and hydrological connectivity along longitudinal, lateral, and vertical dimensions, often challenging our capacity to generalize about the effects of flow intermittence on biota and ecological processes in IRES.

IRES occur in all climates and across a wide range of tectonic, lithological, and physiographic settings, leading to a variety of flow regimes. Valley-floor (i.e., channel and floodplain) processes and morphologies in IRES systems are extremely diverse and extend along a spectrum from distinctly different to overlapping with perennial systems. Sediment (Chapter 2.1) and hydrological (Chapter 2.2) regimes are the major drivers of IRES valley-floor processes and forms and hydrological connectivity, providing a complex and dynamic physical template that influences most ecological processes and organisms in IRES (Chapter 2.3). Valley-floor morphology in IRES systems may persist for years to decades or longer, but also can be highly transient, adjusting with every major flow event (Chapter 2.1). Diversity of geomorphology and sediment regimes both within and between the different geomorphological zones (upland, piedmont, lowland, and floodout, Tooth and Nanson, 2011) promotes ecological processes and patterns in IRES that can be very distinct from perennial river systems. Similarly, the flow regimes (quantity, timing, predictability, and variability in flow) of IRES are typically more variable than in nearby equivalent-sized perennial rivers and streams (Chapter 2.2).

Hydrological connectivity mediated by either flowing or nonflowing water extends along three spatial dimensions (longitudinal, lateral, and vertical, Chapter 2.3) and varies over the fourth dimension of time (Ward, 1989). Flow intermittence disrupts this connectivity, operating through complex hydrological transitions (e.g., between flowing and nonflowing phases). Concurrent interactions and physical discontinuities in hydrological connectivity along these three dimensions produce complex mosaics of physicochemical patches at different scales whose boundaries fluctuate over time in response to the flow regime (Chapter 2.3). This complex patchiness underpins the characteristic physical, chemical, and biological diversity at multiple scales along longitudinal, lateral, and vertical hydrological dimensions in IRES, and often challenges our capacity to derive consistent generalizations about the effects of flow intermittence on biota and ecological processes in IRES. It probably also constrains the applicability to IRES of river ecosystem models developed for perennial rivers and used to guide management and restoration activities (see Theme 13).

THEME 3

Current approaches for summarizing flow intermittence and its variability in IRES are seldom adequate for deriving unambiguous ecohydrological associations or for detecting subtleties in flow regime that should be protected or restored in IRES.

Flow regimes of IRES have been primarily characterized using data from gaging stations, usually only sparsely situated along most IRES (Chapter 2.2). These flow data are typically summarized as hydrological metrics (e.g., variance in frequency, duration, timing, and rate of onset of intermittence) that have been used to classify flow regimes of many of the world's rivers (e.g., Kennard et al., 2010; Snelder et al., 2013) and establish flow-ecology relationships. However, associations between these hydrological metrics in IRES and ecological variables are seldom robust or consistent (e.g., Snelder et al., 2013; Leigh et al., 2016b), often owing to the metrics' inabilities to capture fine-scale subtleties of the flow regime or other hydrological processes such as surface water-groundwater interactions in IRES (Chapter 2.3). Other issues arise because of temporal and spatial mismatches between flow gage data and ecological data, as well as the fact that flow gauge data usually lump nonflowing and dry conditions together as "zero-flow" conditions. The current era of dramatic changes to local and global hydrological cycles is one where IRES are increasingly prevalent in many regions of the globe. Many of these additional IRES are "novel" ecosystems (*sensu* Hobbs et al., 2014) with altered flow regimes of unprecedented or increased intermittence artificially created by human activities (Chapter 5.1). For adequate management or restoration of the flow regimes of natural and modified IRES, hydrological metrics that effectively summarize relevant features of historical, current, and likely future flow regimes of IRES are needed. Alternatively, or as a supplement, we may need other approaches to measuring or modeling flow regimes (Chapter 2.2) and ecohydrological relationships, and this is a promising direction for future work (Section 6.1).

THEME 4

Flow intermittence typically causes spatiotemporal variability in water physicochemistry in IRES to exceed ranges in nearby perennial rivers and streams, requiring modifications of water quality standards for IRES that also encompass nonflowing conditions.

In most IRES, extremes and spatiotemporal variability of almost every component of water physicochemistry (e.g., water temperature, electrical conductivity, pH, dissolved oxygen, and nutrient concentrations) exceed ranges in nearby perennial rivers and streams. This is usually as a direct result of flow cessation and drying (e.g., von Schiller et al., 2011; Arce et al., 2014) but is also affected by flooding, substrate type, vegetation inputs, groundwater interactions, and human activities (Chapters 3.1, 3.2, and 5.1). Fluctuations in flow govern the biogeochemical heartbeat of IRES (Chapter 3.2) which translates into pulses of biogeochemical "hot spots" and "hot moments" (*sensu* McClain et al., 2003) that generate transient and dynamic patches with highly variable water physicochemistry. At some stages of the flow regime (e.g., immediately after flow cessation or in the final stages of surface water drying), there are frequently sharp changes in concentrations of dissolved materials that coincide with diel extremes in physical factors such as water temperature. Not only do these changes affect the biota and ecological processes in IRES, but they also confound the use of many conventional water quality standards and triggers that are typically developed using data from perennial rivers and streams. Modified water quality standards and guidelines are needed that explicitly acknowledge the inherently greater physicochemical variability of IRES and that are tailored to compensate for the spatiotemporal fluctuations in water quality associated with natural intermittence including during nonflowing periods.

THEME 5

Microbial diversity, function, and trophic linkages in IRES are strongly influenced by intermittence and drying, in turn affecting the rates of biogeochemical processes such as nutrient cycling and organic matter breakdown.

Heterotrophic microbial diversity and composition in IRES are controlled by habitat and substrate availability during the different hydrological phases. Prokaryotes, fungi, and protozoans play crucial roles in ecosystem functions such as nutrient and carbon cycling, particle retention, and detrital processing (Leff et al., 2016), and form a key trophic component of instream and semiaquatic food webs (Chapters 3.2, 4.1, and 4.7). They have different physiological and behavioral strategies to resist drying or recover swiftly upon rewetting (Romaní and Sabater, 1997; Timoner et al., 2014a). Many also can tolerate harsh conditions (e.g., anoxia in drying pools), making them especially relevant contributors to ecosystem functions when other organisms are excluded or inhibited. These adaptations to intermittence and fluctuating physicochemistry (Theme 4) cause "waves" of microbial functions and biodiversity to covary with hydrological phases, potentially affecting ecosystem functioning and higher trophic levels (Chapter 4.1). Strategies to manage, monitor, or restore IRES (Chapter 5.4) rely on improved understanding of the effects of flow intermittence on microbial communities and functions because microbes mediate important ecosystem services such as cycling of nitrogen and carbon.

THEME 6

Flow intermittence governs provision of nutrients and water required by algae and vascular plants for autotrophic production, and the pulse of primary production that occurs immediately after rewetting often supplies high-quality organic matter to aquatic and terrestrial consumers.

All IRES support primary producers including cyanobacteria, algae, aquatic macrophytes, and riparian and terrestrial plants. Their abundance and community composition in IRES depend on the degree of flow intermittence and the duration and timing of the different hydrological phases. Flow cessation and drying deprive primary producers of crucial nutrients and water so that amounts and rates of primary production are governed by intermittence. Many primary producers in IRES have morphological and physiological adaptations to cope with drying such as desiccation-resistant stages, thickened cell walls, and flexible life cycle strategies to capitalize on optimum periods for growth and reproduction (Brock et al., 2003; Timoner et al., 2014b). Upon rewetting and flow resumption, most aquatic primary producers recover swiftly and there is a pulse of production of high-quality organic matter that is available for consumption by aquatic and terrestrial consumers (Chapter 4.2). These pulses of production associated with flow intermittence and wetting-drying cycles underpin episodic aquatic-terrestrial subsidies of energy, especially in IRES in arid and semiarid areas where stream channels and their associated riparian zones and floodplains are hot spots of regional production when water is available (Fig. 6.2).

THEME 7

Flow intermittence affects the abundance, assemblage composition, and activity of the diverse heterotrophs (aquatic, semiaquatic, and terrestrial invertebrates and vertebrates) that occupy IRES during different hydrological phases and play fundamental roles in mediating ecological processes and generating ecosystem services.

FIG. 6.2

Instream primary producers in this South Australian IRES (Enorama Creek, Flinders Ranges) include green algae (*Spirogyra* and *Cladophora*) along the edge of a groundwater-fed pool scoured at the base of another primary producer, a river red gum (*Eucalyptus camaldulensis*) which has trapped a small wall of organic matter swept down in a previous flood. Small herbs, protected at the base of the debris dam, also contribute primary production but will be inundated when flow resumes.

Photo: Courtesy A. Boulton.

The stream channel and riparian zone support diverse communities of heterotrophs during each hydrological phase in IRES. Aquatic vertebrates and invertebrates dominate instream habitats of IRES during flowing and nonflowing phases (Chapters 4.3 and 4.5), while semiaquatic and terrestrial organisms dominate the channels during dry phases and occupy exposed bed sediments, riparian zones, and floodplain habitats during all phases (Chapters 4.4 and 4.6). Aquatic invertebrate communities, which include some specialist taxa confined to IRES, have diverse adaptations to survive conditions in dwindling pools and dry bed sediments of IRES channels during nonflowing phases (Chapter 4.3). Similarly, fishes and other aquatic vertebrates in IRES use physiological and/or behavioral strategies to cope with fluctuating flow and, in some cases, drying (e.g., Kerezsy et al., 2013; Whitney et al., 2015). A diverse and often abundant terrestrial and semiaquatic invertebrate (Chapter 4.4) and vertebrate (Chapter 4.6) fauna inhabits IRES, often feeding on aquatic biota trapped in drying pools and along receding stream margins (Steward, 2012). Although little is known about the species composition and ecological roles of semiaquatic and riparian heterotrophs of IRES, it appears that they play the same fundamental roles in many ecological processes (e.g., organic matter cycling, landscape engineering, seed dispersal, and provision of trophic subsidies, Ellery et al., 2003; Corti and Datry, 2016; Chapter 4.7) as aquatic organisms in IRES (Chapters 4.3, 4.7, and 4.8). These ecological processes underpin vital ecosystem services (Chapter 5.2) that will be impaired or lost if anthropogenic activities alter conditions in IRES (e.g., flow regimes, channel complexity) in such a way as to deplete communities of resident aquatic, semiaquatic, and riparian heterotrophs.

THEME 8

The alternating expansion and contraction (including fragmentation) of aquatic and terrestrial habitats in IRES modifies the intensity of trophic interactions among species, changes the relative contributions of autotrophy and heterotrophy to food webs, and governs the relative dominance of terrestrial and aquatic subsidies over time in IRES.

The sequence of flowing, nonflowing, and dry phases in IRES not only alters community composition but also their biotic interactions and food webs. For example, flow cessation and drying alter proportions of microbes, autotrophs, and detritivorous heterotrophs (Chapters 4.1–4.6) sometimes but not always favoring heterotrophy over autotrophy (Chapter 4.7), a trend typically reversed upon rewetting and flow resumption. Thus, fluctuations in hydrological connectivity over space and time in IRES (Chapter 2.3) constrain, intensify, or expand trophic interactions among species, often due to disproportionate effects of drying on top predators (McHugh et al., 2015). On one hand, flow intermittence usually shortens aquatic food chain length and trophic diversity (Closs and Lake, 1994; Sabo et al., 2010), often resulting in a collapse of aquatic food webs. On the other hand, it can trigger transitions toward terrestrial energy pathways and expand IRES food webs into and beyond the riparian zone. Flow resumption reverses these processes, although the extent of this reversal is context-specific, largely governed by antecedent conditions of hydrological connectivity along longitudinal, lateral, and vertical dimensions in each IRES.

THEME 9

Diverse modes of dispersal, recolonization, and desiccation resistance of aquatic, semiaquatic, and terrestrial biota promote high beta diversity and temporal variability in metapopulations, metacommunities, and metaecosystems in IRES.

Although alpha diversity of aquatic assemblages in IRES is generally lower than in nearby perennial reaches or rivers (Datry et al., 2011, 2014a; Chapters 4.1–4.6), beta diversity is often higher. This contrast in biodiversity reflects the prevalence of diverse strategies of resistance and resilience to drying, including the ability to persist in dry sediments (Stubbington and Datry, 2013, Chapters 4.8 and 4.10) or to swiftly recolonize from nearby refuges (Bonada et al., 2006; Bogan et al., 2015). Further, the shifting spatiotemporal mosaic of lotic, lentic, and dry habitats at the catchment scale favors the concurrent presence of multiple taxa and successional stages, increasing beta diversity in space and time (Larned et al., 2010; Chapter 4.8). The diversity of modes of dispersal, resistance, and resilience to flooding, rewetting, intermittence, and drying in IRES also increases the temporal variability of metapopulations, metacommunities, and metaecosystems, and IRES are ideal arenas for exploring these temporal dynamics (Chapter 4.9).

THEME 10

Even while dry or not flowing, IRES provide diverse ecosystem services whose provision is poorly understood and likely to be impaired by human activities that alter flow regimes and catchment conditions of IRES.

Ecosystem services are the benefits that people obtain that are directly attributable to the ecological functioning of ecosystems (de Groot et al., 2002). The ecosystem services provided by IRES are not as widely recognized or appreciated compared with those from perennially flowing waters, and this may largely explain why IRES are undervalued in monetary and nonmonetary terms (Chapter 5.2). Many ecological functions in IRES continue during nonflowing and dry phases (Chapters 3.2 and 4.1), providing regulating and supporting services such as erosion control and buffering catchment runoff (Chapter 2.1). However, there is far less known about the provision of ecosystem services by IRES than by perennial streams and rivers (Acuña et al., 2014), and consequently less impetus to protect the ecological functions that underpin these services (Boulton, 2014), especially when the channel dries (Fig. 6.3). IRES experience the same human activities that adversely influence perennial waters (Chapter 5.1) but the effects of these activities on ecosystem service provision in IRES are unclear, especially the effects of reduced intermittence when wastewater is released into the channel for long periods.

THEME 11

The pulsed flows in IRES, coupled with alternating hydrological fragmentation and reconnection along longitudinal, lateral, and vertical dimensions, exacerbate the impacts of human activities such as eutrophication, salinization, and water pollution because of fluctuations in water volume and associated physicochemistry.

Flow intermittence contributes to the extremes in flow regimes of IRES (Chapter 2.2) that disrupt hydrological connectivity along longitudinal, lateral, and vertical dimensions (Costigan et al., 2015), resulting in wide fluctuations in water physicochemistry (Chapters 3.1 and 3.2) and microbial activity (Chapter 4.1). The pulsed flows and concurrent spatiotemporal fluctuations in drying and rewetting naturally generate extremes in water quality (e.g., dissolved oxygen, electrical conductivity), especially in dwindling pools. These natural extremes already pose physiological challenges to resident aquatic biota (e.g., Bruton, 1979; Chester et al., 2015). However, when superimposed on the effects of pulsed concentrations of nutrients, salt, and other aquatic pollutants from human activities and intensified by flow cessation and drying (Chapter 5.1), these extended extremes overwhelm all but the hardiest organisms. Thus, the pulsed flow regimes and hydrological fragmentation characteristic of all IRES exacerbate the impacts of human activities such as eutrophication, salinization, and other water pollution and must be considered when setting water quality guidelines for IRES (Theme 4).

THEME 12

The importance of IRES and their diverse contributions to entire river networks is gradually being recognized by targeted legislation, facilitating protection and improvement of water quality and quantity in IRES.

FIG. 6.3

IRES are especially vulnerable to direct human impacts when the channel dries, often being used as sediment mining sites (a, Jankho Khala River, Cochabamba, Bolivia), sites to grow marijuana (b, Beni M'Hamed River, Rif, Morocco), open sewers (c, Seco River, Chimore, Bolivia), convenient gravel roads (d, Seco River, Málaga, Spain), waste disposals (e, Hodgsons Creek, Victoria, Australia), and urban channels (f, Daró River, La Bisbal de l'Empordà, Girona, Spain).

Photos: Courtesy T. Datry (a and c), N. Bonada (b, d, and f), and A. Boulton (e).

The growing scientific understanding of IRES and their geomorphological, hydrological, biogeochemical, and ecological contributions to river networks and associated terrestrial landscapes (Chapters in Sections 2–4) has led to the legislative recognition in some countries of IRES as important waterways in their own right (Chapter 5.3) that are worthy of protection and conservation (Acuña et al., 2014). Most flowing-water legislation was originally developed for perennially flowing streams and rivers and is sometimes ill-suited to address the inherent variability of flow regimes, especially intermittence. Effective water governance for IRES needs a shift in public perceptions and expectations, greater appreciation of the importance of natural hydrological variability, and adoption of "systems thinking" to view IRES as key components interconnected with perennial surface waters, groundwaters, the atmosphere and surrounding land (Chapter 5.3). With these changes comes a need for novel biomonitoring approaches (e.g., Sheldon et al., 2010; Prat et al., 2014) that incorporate natural variability, and the effects of intermittence on assessment of water quality and appropriate biological indicators and reference conditions so that legislative controls can be enforced and implemented.

THEME 13

With increasing recognition of the values of IRES, conservation and restoration strategies modeled on those used in perennially flowing streams and rivers are being adopted but are likely to need refinements to explicitly protect or restore flow intermittence while adopting a catchment-level perspective.

Worldwide, IRES have been degraded by human activities such as water extraction, channel modification (e.g., sediment mining when dry), and wholesale clearance of catchment and riparian vegetation at alarming rates (Chapter 5.1). Given the limited appreciation of the ecosystem services of IRES (Chapter 5.2) and the slow pace of legislation to recognize and protect them (Chapter 5.3), there have been far fewer efforts to conserve specific IRES or restore damaged ones compared with perennial streams and rivers. To be successful, restoration of IRES should be planned and implemented within the context of catchments and landscapes rather than the more common site- or reach-level perspective (Bond and Lake, 2003; Beechie et al., 2010). Strategies similar to those used for restoring perennial rivers are likely to be useful (e.g., restoration of riparian zone vegetation, addition of instream habitat, and instigation of environmental flows) but these strategies must explicitly account for the functional importance of flow intermittence (Chapter 5.4). Therefore, it will be necessary to ensure periods when flow ceases and/or the channel is allowed to dry naturally, and refuges must be maintained and protected from water extraction, agricultural pollution, and other impacts. There also needs to be a detailed understanding of the stream and its catchment, including the physical and biological processes that operated historically, that operate now, and that are necessary to support the desirable state (Beechie et al., 2010) and can be achieved within the current and future context of rapidly changing environmental conditions. Suitable indicators of restoration success must be selected so that appropriate monitoring can be done, recognizing that recovery rates in IRES may differ from those in nearby perennial streams and rivers because of the effects of intermittent flow. The few examples of restoration of IRES illustrate the complexities of recovering functional ecosystems when periodic flow cessation and drying slow down revegetation or alter expected recovery trajectories (e.g., Bond and Lake, 2005).

THEME 14

IRES are complex social-ecological systems that require approaches such as Strategic Adaptive Management to adequately integrate the social, political, economic, and environmental complexities and focus on the benefits of colearning and coproduction of knowledge by all stakeholders.

IRES are seldom managed as social-ecological systems yet many have substantial cultural significance (Steward et al., 2012) and are frequently crucial sources of water and numerous other ecosystem services (Chapter 5.2). Although not valued as highly as perennial waters, they are still acknowledged as part of the social landscape (Armstrong et al., 2012) and constitute significant parts of many river networks across the globe. Consequently, IRES can be considered as complex social-ecological systems deserving Strategic Adaptive Management (SAM), involving transparent articulation of a desired state guided by a shared vision and supported by explicit objectives (Chapter 5.5). Within a context bounded by collective values and social, technological, economic, environmental, and political factors, SAM relies on agreement about key attributes (e.g., endangered fishes, riparian vegetation) and this focuses goals and subsequent monitoring programs to track the effectiveness of various management strategies (Kingsford and Biggs, 2012). Frequent evaluation that is focused on maximizing colearning and coproduction of knowledge by all stakeholders helps optimize the process (Chapter 5.5) and preserves the social dimension, acknowledging that different users have different values for a particular IRES.

6.3 REMAINING GAPS IN OUR KNOWLEDGE AND UNDERSTANDING OF THE ECOLOGY AND MANAGEMENT OF IRES

Not surprisingly, there are still many gaps in our knowledge and understanding of IRES. The preceding 20 chapters identify these gaps and propose promising directions for future research as well as some innovative solutions to help structure these endeavors. We summarize some of these gaps and potential solutions according to three broad areas: methodological, geographical and disciplinary, and conceptual gaps and opportunities.

METHODOLOGICAL GAPS AND OPPORTUNITIES

Almost every chapter acknowledges a variety of methodological constraints. One frequent constraint is the limited availability of inexpensive monitoring equipment that would enable adequate replication to encompass the large variability inherent in IRES. Another arises from scale-dependent limitations that include uncertainty about whether measurements made at, for example, the stream reach scale could be validly extrapolated to higher or lower scales. On the positive side, there is optimism that new technologies such as geochronological techniques (e.g., radiocarbon, luminescence, and cosmogenic radionuclide dating for tracking geomorphological processes, Tooth, 2012) and 16S rRNA sequencing for microbial studies (e.g., Zeglin, 2015) applied to IRES hold considerable promise for filling many of the current knowledge gaps. Although these new methods may not address all of the scale-dependent problems, they will add badly needed data from which meta-analyses and other synthetic studies might elucidate where mismatches of spatiotemporal scales are especially serious.

Many sampling techniques have been developed for use in perennial streams and rivers, and their applicability to different types of IRES must be carefully assessed, along with appropriate analyses of sensitivity and uncertainty. The same applies to the burgeoning field of modeling. For example, almost all of the currently used hydrological models are poor at simulating zero-flows (Chapter 2.2) and often need modification (e.g., Ivkovic et al., 2014) or even new models for successful application to simulate intermittence in IRES. In addition, fundamental hydrological data for IRES are often lacking because river gaging stations are preferentially placed on perennial streams and rivers (Chapter 1). This lack of data hampers assessment of the adequacy of models to simulate multiyear hydrographs in many IRES. Novel approaches may be needed such as citizen-science programs that record river flow states (Turner and Richter, 2011) or application of recent technologies such as multiple-probe data loggers and airborne imagery (Fig. 6.4, Chapters 2.2 and 2.3). Some of these techniques may also be useful for tracking the spread of invasive plants in IRES or monitoring recolonization pathways of invertebrates and fishes when flow resumes and of terrestrial and semiaquatic biota recolonization of the channel during drying.

These methodological constraints have several important implications for successful management of IRES. Firstly, the resulting lack of data (especially integrated hydrological, biogeochemical, and ecological information) may tempt managers to evade consideration of IRES or to attempt to use data from nearby perennial streams and rivers as a surrogate. Secondly, even where data are collected, the records are frequently incomplete. For example, using standard procedures to sample benthic aquatic macroinvertebrates as biomonitors (commonly done in perennial streams and rivers, Bonada et al., 2006) may not be feasible in IRES when the stream is dry. Even when aquatic macroinvertebrate samples can be collected, disentangling the likely ecological responses to anthropogenic inputs from those to natural intermittence (e.g., Chapters 3.1 and 3.2) is problematic. Habitat fragmentation caused by intermittence may also affect recolonization processes, confounding detection of anthropogenic impacts. Consequently, most biomonitoring approaches and tools developed in perennial systems are not applicable in IRES and must be adapted (Sheldon, 2005, Chapter 5.1). Alternatively, the use of terrestrial or hyporheic invertebrates as alternative indicators of biological condition during dry or zero-flow periods is being explored for IRES (e.g., Steward et al., 2012; Leigh et al., 2013) but results will still require cautious interpretation if they are to reliably indicate anthropogenic impacts during different hydrological phases. Thirdly, the inherent high variability of IRES means that greater numbers of samples and larger datasets are needed to adequately summarize physicochemical or biological condition, especially when pools are dwindling or soon after flow resumes. Methods that enable rapid and multipoint sampling of reliable indicators of ecosystem condition of IRES are badly needed to improve the information used to measure the success of restoration activities (Chapter 5.4) or SAM (Chapter 5.5). This presupposes an understanding of what constitutes "healthy" ecosystems in different IRES to define appropriate reference or target conditions, also a substantial knowledge gap for many parts of the world (see later).

GEOGRAPHICAL AND DISCIPLINARY GAPS AND OPPORTUNITIES

Although IRES are ubiquitous and occur on every continent (Chapter 1), research on these ecosystems is extremely uneven. This is evident when comparing a sample of the number of readily accessible English-language papers published on IRES from each country (Fig. 6.5). Much of what we know about IRES appears to be derived from work in just a handful of countries such as the United States, Australia, Spain, Portugal, and France (Fig. 6.5). Meanwhile, there are many regions where IRES are prevalent and yet it appears that little research has been published, leaving vast gaps in our regional knowledge and severely constraining global comparisons and syntheses. Furthermore, the types of IRES that have been most studied appear to be those that occur in Mediterranean and temperate climates (where flows and intermittence are relatively predictable and seasonal) or the larger IRES (e.g., Australian dryland rivers)

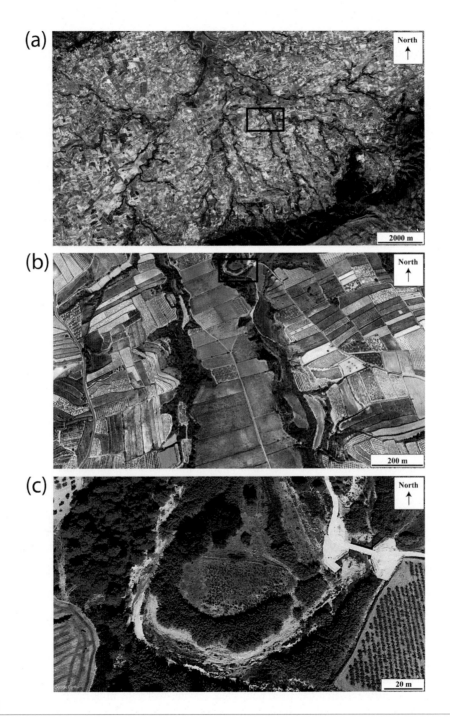

FIG. 6.4

Recent advances in technology such as airborne imagery at different scales (a–c, extracted from Google Earth Pro and showing the inland region of Alicante, Spain) and remote sensing are greatly enhancing our understanding of the relationships of IRES to the broader landscape and river networks worldwide.

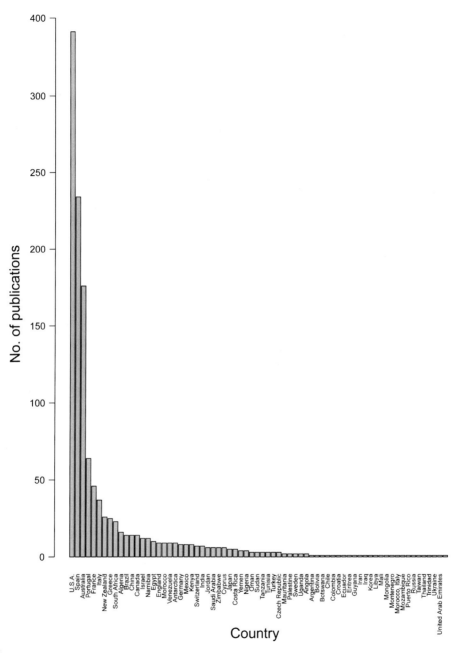

FIG. 6.5

Some 70% of the readily accessible English-language publications published in the field of IRES since 1900 (details on the terms used for the Web of Science search in Leigh et al., 2016a) describe IRES from only five countries; there are many parts of the world where IRES abound and yet very few papers have been published about these ecosystems.

in semiarid and arid climates. Far less is known about the geomorphology, hydrology, physicochemistry, and ecology of episodic and ephemeral streams and rivers with short-lived and often unpredictable flows punctuated by long dry periods. This uneven distribution and depth of knowledge may impede effective management and conservation of ephemeral streams (Boulton, 2014).

There is also a geographic imbalance in the amount of discipline-specific knowledge about IRES, and this is why we have covered geographical and disciplinary topics together. For example, geomorphologists have mostly focused on IRES in semiarid and arid regions (Chapter 2.1) whereas the majority of published research on biogeochemical responses to intermittence has been done in Mediterranean IRES (Chapters 3.2 and 4.1). Ecologically, IRES in tropical, alpine, karstic, and arctic regions are poorly known compared to IRES in temperate, Mediterranean, and semiarid regions (Fig. 6.6; Chapters 4.2–4.6).

FIG. 6.6

Karstic (a), tropical (b), forested headwater (c), and alpine (d) IRES are less well documented than Mediterranean (e) or temperate IRES (f).

Photos: Courtesy T. Datry (a–c, f), A. Foulquier (d), and N. Bonada (e).

The same geographic imbalance applies to various aspects of legislation, conservation, and management of IRES. In South Africa and Australia, for instance, legislative recognition of IRES is more advanced than in many other regions (Chapter 5.3), reflected in the extent to which SAM has been adopted in these systems (Chapter 5.5).

The net effect of the geographic/disciplinary imbalance in our understanding and knowledge of different types of IRES around the world is difficult to assess and perhaps has not been evident until doing the broad synthesis of multiple topics and disciplines required for this book. However, the imbalance is likely to heavily influence our current perceptions of the relative roles of different processes in IRES, especially given the different disciplinary emphases in different streams. Is it valid, for example, to extrapolate conclusions about biogeochemical processes in Mediterranean streams to interpret data on organic matter cycling in a dryland IRES where flow regimes and biota may be quite different? Until there is a more even geographic coverage of research across multiple related disciplines and different types of IRES, it is perhaps premature to propose reliable generalizations about the archetypic IRES, if one even exists. Consequently, our 14 themes described previously must be considered as tentative, and many of them could be treated as hypotheses to test in different climatic and regional settings as part of a multidisciplinary research program.

Happily, there are some substantial steps being taken to remedy the geographic and disciplinary gaps in our knowledge of IRES. One of these, arising from the IRBAS program described earlier, is "The 1000 Intermittent River project" (http://1000_intermittent_rivers_project.irstea.fr, Datry et al., 2016). This international initiative seeks to address both issues simultaneously, using discipline-specific approaches to experimentally explore the same ecosystem process (or processes) in as many IRES from as broad a geographic coverage as possible. For example, in the first year of the project's existence, the database included 210 IRES from 28 countries (Fig. 6.7). There is also a field experiment using consistent protocols to sample benthic organic matter during consecutive hydrological phases and from

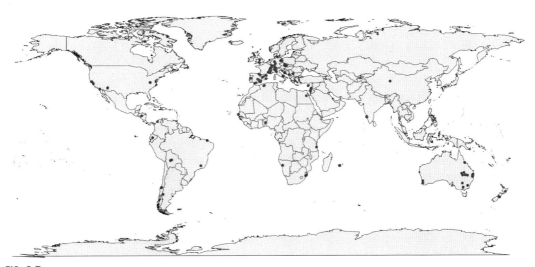

FIG. 6.7

Map showing the 210 IRES from 28 countries sampled during the first year of the "The 1000 Intermittent River project" (http://1000_intermittent_rivers_project.irstea.fr).

over 100 IRES in multiple countries that has been initiated to generate comparable data across a wide geographic array of IRES. Similar global field experiments in perennial rivers have yielded valuable insights into ecological processes such as organic matter decomposition (e.g., Boyero et al., 2016).

Large-scale efforts to translate current knowledge from different disciplines into tools for protecting and restoring IRES are also emerging. In Europe, the Life TRivers project (http://www.lifetrivers.eu/en) aims to produce tools that enable Mediterranean IRES to be incorporated into the Water Framework Directive, the legislation framework for rivers in Europe. In 2016, this initiative was extended to other types of IRES as a H2020 COST (European Cooperation in Science and Technology) Action involving 200 researchers and 40 managers from 30 countries in Europe and beyond. This synthesis effort (Science and Management of Intermittent Rivers and Ephemeral Streams, SMIRES, http://www.smires.eu/) combines research in hydrology, ecology, biogeochemistry, ecohydrology, social sciences, and environmental economics to address geographical and disciplinary gaps in IRES research, connect national and international scientific efforts on IRES, and synthesize fragmented knowledge and help translate it into useful tools for water managers and policy makers. These efforts may spawn similar initiatives in regions such as North America, South Africa, and Australia.

CONCEPTUAL GAPS AND OPPORTUNITIES

Most of the current concepts in river geomorphology, hydrology, water chemistry, ecology, and management have been developed and refined in perennially flowing rivers with only a few exceptions (e.g., the Telescoping Ecosystem Model, Fisher et al., 1998). Therefore, one of the biggest conceptual gaps is the extent to which contemporary concepts and their associated models apply to different types of IRES (Datry et al., 2014a,b). This is especially true when reaches with intermittent or ephemeral flow occur either upstream, downstream, or both of otherwise perennially flowing river networks. Are there gradual transitions or sharp discontinuities in, for example, material transport or ecological processes where flow changes from perennial to intermittent (or *vice versa*) that invalidate predictions by current conceptual models? How does intermittence and disruption of hydrological connectivity in one or more spatial dimensions (Chapter 2.3) affect riverine concepts with a central tenet of down-gradient flow? Almost every chapter in this book covered aspects of IRES that might defy the "conventional wisdom" portrayed by concepts developed in perennially flowing rivers.

However, several prominent characteristics of IRES lend themselves to testing or extending current conceptual frameworks, and have been explored to varying degrees. One example is the characteristic disturbance of flow intermittence, which may vary widely in predictability, severity, rate of onset, intensity, and timing in different IRES. Such disturbance, often alternating with sporadic floods, is central to disturbance theory and the allied concept of patch dynamics which is evident in IRES from geomorphological (Chapter 2.1) and hydrological processes (Chapters 2.2 and 2.3) through to effects on metasystems thinking (Chapter 4.9), evolution, and biogeographical aspects (Chapter 4.10) in IRES. Disturbance by intermittence and flooding has promoted development of adaptive functional traits across much of the aquatic, semiaquatic, and terrestrial biota in IRES to cope with associated habitat loss or fragmentation (Chapters 4.1–4.6 and 4.8–4.10). To what extent do flow cessation, drying, and rewetting represent disturbances (Lake, 2000) for these different groups of IRES biota and are the disturbances comparable in impact? Do these processes play major roles in defining the species habitat template for different groups of biota in IRES (Townsend and Hildrew, 1994)? Do more relevant conceptual frameworks for IRES lie in recent advances in metasystem or eco-evolutionary theory (Chapters 4.8–4.10)?

Other prominent characteristics of IRES were emphasized by Larned et al. (2010) in their paper proposing three models to address how biodiversity and biogeochemical processes in IRES are organized in space and time. These three models center on the dynamic habitat mosaic of wet and dry phases, the aquatic-terrestrial transitions and the pulsed biogeochemical dynamics of IRES (Larned et al., 2010). Thus far, these three models remain largely untested nor has a unified model been proposed and explored that explicitly incorporates the characteristic traits of IRES and their full spectrum of flow regimes.

A related conceptual gap, and one highly relevant to managing the impacts of human activities on IRES, is the extent to which anthropogenic IRES resemble natural ones in their biota, processes, and ecosystem services. Many human activities create or exacerbate flow intermittence (Chapters 5.1 and 5.3), producing "novel" ecosystems (*sensu* Hobbs et al., 2014) with "novel" biotic assemblages and perhaps ecosystem processes. Do these novel IRES function in the same way as natural IRES, sharing similar trophic structure (Chapters 4.1–4.7) and functional traits (Chapters 4.8–4.10)? To what extent do they yield the same ecosystem services (Chapter 5.2)? If functionally equivalent, are they worthy of the same legislative protection as natural IRES (Chapter 5.3) and will they follow similar recovery trajectories to natural IRES during restoration (Chapter 5.4)? Until we have reliable and valid conceptual models applicable to all types of IRES, the adoption of management strategies such as SAM is hampered because of their reliance on these models to predict the likely outcomes of various management strategies (Chapter 5.5).

Opportunities abound to test existing conceptual models in IRES and, if found wanting, refine the models or even develop new ones. These models should acknowledge that IRES are not aberrant or "inferior" ecosystems but are ecosystems whose flows are highly variable and characteristically intermittent. The most useful conceptual models will be those that span geomorphological, hydrological, biogeochemical, and ecological processes; apply across the broad spectrum of IRES worldwide; and have predictive power that satisfies theoretical and applied needs.

6.4 CONCLUSIONS

We fervently hope the collation of themes presented in this chapter together with the brief glimpse at some of the remaining gaps in our knowledge whets readers' intellectual appetites. Many of the themes are partly speculative and require validation. Only a few of the main gaps and opportunities have been identified here; the other chapters in this book describe many more. Perhaps no other type of coupled aquatic-terrestrial ecosystem has so many exciting opportunities to test current theories developed in perennial streams and rivers, to refine current theories, or to create new ones that are more relevant.

Even more important is that many natural IRES face serious human impacts, aggravated by a lack of public and political awareness of the significance of these ecosystems and their potential ecosystem services. Meanwhile, novel ecosystems with increased or new-found intermittence are becoming more abundant in many parts of the world where global climate change and intensifying human demands for fresh water reduce available supplies of this essential resource. We hope this book is a useful introduction to IRES for researchers and water resource managers as well as readers fortunate enough to live in catchments drained by IRES. Most of all, we dedicate it to all people who simply appreciate the changeable nature of rivers and their catchments, value their beauty, and realize their intrinsic worth—this book and this chapter are for you.

ACKNOWLEDGMENTS

We are grateful to Cath Leigh for the bibliographic synthesis underpinning Fig. 6.5 and Roland Corti for preparing Fig. 6.7. We also thank all the authors of the 20 chapters from which our conclusions have been drawn, and we acknowledge these authors' enthusiasm, commitment, and insights, along with those of all the referees, that have resulted in this book and culminated in this chapter. This book resulted from discussions and literature research conducted as part of the IRBAS (Intermittent River Biodiversity Analysis and Synthesis) working group supported by CESAB (Center for Synthesis and Analysis of Biodiversity), funded jointly by FRB (French Foundation for Research on Biodiversity), and ONEMA (French National Agency for Water and Aquatic Environments). We thank the many people who discussed ideas for this book while at CESAB or other IRBAS gatherings, and Cath Leigh and Rachel Stubbington for valuable comments on an earlier draft of this chapter.

REFERENCES

Acuña, V., Datry, T., Marshall, J., Barceló, D., Dahm, C.N., Ginebreda, A., et al., 2014. Why should we care about temporary waterways? Science 343, 1080–1082.

Arce, M.I., Sánchez-Montoya, M.M., Vidal-Abarca, M.R., Suárez, M.L., Gómez, R., 2014. Implications of flow intermittency on sediment nitrogen availability and processing rates in a Mediterranean headwater stream. Aquat. Sci. 76, 173–186.

Armstrong, A., Stedman, R.C., Bishop, J.A., Sullivan, P.J., 2012. What's a stream without water? Disproportionality in headwater regions impacting water quality. Environ. Manag. 50, 849–860.

Beechie, T.J., Sear, D., Olden, J.D., Pess, G.R., Buffington, J.M., Moir, H., et al., 2010. Process-based principles for restoring river ecosystems. Bioscience 60, 209–222.

Bogan, M.T., Boersma, K.S., Lytle, D.A., 2015. Resistance and resilience of invertebrate communities to seasonal and supraseasonal drought in arid-land headwater streams. Freshw. Biol. 60, 2547–2558.

Bonada, N., Rieradevall, M., Prat, N., Resh, V.H., 2006. Benthic macroinvertebrate assemblages and macrohabitat connectivity in Mediterrranean-climate streams of northern California. J. N. Am. Benthol. Soc. 25, 32–43.

Bond, N.R., Lake, P.S., 2003. Local habitat restoration in streams: constraints on the effectiveness of restoration for stream biota. Ecol. Manag. Restor. 4, 193–198.

Bond, N.R., Lake, P.S., 2005. Ecological restoration and large-scale ecological disturbance: the effects of drought on the response by fish to a habitat restoration experiment. Restor. Ecol. 13, 39–48.

Boulton, A.J., 2014. Conservation of ephemeral streams and their ecosystem services: what are we missing? Aquat. Conserv. Mar. Freshwat. Ecosyst. 24, 733–738.

Boyero, L., Pearson, R.G., Hui, C., Gessner, M.O., Pérez, J., Alexandrou, M.A., et al., 2016. Biotic and abiotic variables influencing plant litter breakdown in streams: a global study. Proc. R. Soc. B 283, 20152664.

Brock, M.A., Nielsen, D.L., Shiel, R.J., Green, J.D., Langley, J.D., 2003. Drought and aquatic community resilience: the role of eggs and seeds in sediments of temporary wetlands. Freshw. Biol. 48, 1207–1218.

Bruton, M.N., 1979. The survival of habitat desiccation by air-breathing clariid catfishes. Environ. Biol. Fish 4, 273–280.

Chester, E.T., Miller, A.D., Valenzuela, I., Wickson, S.J., Robson, B.J., 2015. Drought survival strategies, dispersal potential and persistence of invertebrate species in an intermittent stream landscape. Freshw. Biol. 60, 2066–2083.

Closs, G.P., Lake, P.S., 1994. Spatial and temporal variation in the structure of an intermittent-stream food web. Ecol. Monogr. 75, 2–21.

Corti, R., Datry, T., 2016. Terrestrial and aquatic invertebrates in the riverbed of an intermittent river: parallels and contrasts in community organisation. Freshw. Biol. 61, 1308–1320.

Costigan, K.H., Daniels, M.D., Dodds, W.K., 2015. Fundamental spatial and temporal disconnections in the hydrology of an intermittent prairie headwater network. J. Hydrol. 522, 305–316.

Datry, T., Arscott, D.B., Sabater, S., 2011. Recent perspectives on temporary river ecology. Aquat. Sci. 73, 453–457.

Datry, T., Larned, S.T., Tockner, K., 2014a. Intermittent rivers: a challenge for freshwater ecology. Bioscience 64, 229–235.

Datry, T., Larned, S.T., Fritz, K.M., Bogan, M.T., Wood, P.J., Meyer, E.I., et al., 2014b. Broad-scale patterns of invertebrate richness and community composition in temporary rivers: effects of flow intermittence. Ecography 37, 94–104.

Datry, T., Corti, R., Foulquier, R., Von Schiller, D., Tockner, K., 2016. One for all, all for one: a global river research network. EOS Earth Space Sci. News 97, 13–15.

De Groot, R.S., Wilson, M.A., Boumans, R.M., 2002. A typology for the classification, description and valuation of ecosystem functions, goods and services. Ecol. Econ. 41, 393–408.

Döll, P., Schmied, H.M., 2012. How is the impact of climate change on river flow regimes related to the impact on mean annual runoff? A global-scale analysis. Environ. Res. Lett. 7, 014037.

Ellery, W.N., McCarthy, T.S., Smith, N.D., 2003. Vegetation, hydrology, and sedimentation patterns on the major distributary system of the Okavango Fan, Botswana. Wetlands 23, 357–375.

Fisher, S.G., Grimm, N.B., Martí, E., Holmes, R.M., Jones, J.B., 1998. Material spiraling in stream corridors: a telescoping ecosystem model. Ecosystems 1, 19–34.

Fritz, K.M., Hagenbuch, E., D'Amico, E., Reif, M., Wigington, P.J., Leibowitz, S.G., et al., 2013. Comparing the extent and permanence of headwater streams from two field surveys to values from hydrographic databases and maps. J. Am. Water Resour. Assoc. 49, 867–882.

Hobbs, R.J., Higgs, E., Hall, C.M., Bridgewater, P., Chapin, F.S., Ellis, E.C., et al., 2014. Managing the whole landscape: historical, hybrid, and novel ecosystems. Front. Ecol. Environ. 12, 557–564.

Ivkovic, K.M., Croke, B.F.W., Kelly, R.A., 2014. Overcoming the challenges of using a rainfall–runoff model to estimate the impacts of groundwater extraction on low flows in an ephemeral stream. Hydrol. Res. 45, 58–72.

Kennard, M.J., Pusey, B.J., Olden, J.D., Mackay, S.J., Stein, J.L., Marsh, N., 2010. Classification of natural flow regimes in Australia to support environmental flow management. Freshw. Biol. 55, 171–193.

Kerezsy, A., Balcombe, S., Tischler, M., Arthington, A., 2013. Fish movement strategies in an ephemeral river in the Simpson Desert, Australia. Austral Ecol. 38, 798–808.

Kingsford, R.T., Biggs, H.C., 2012. Strategic Adaptive Management Guidelines for Effective Conservation of Freshwater Ecosystems in and Around Protected Areas of the World. IUCN WCPA Freshwater Taskforce, Australian Wetlands and Rivers Centre, Sydney.

Lake, P.S., 2000. Disturbance, patchiness and diversity in streams. J. N. Am. Benthol. Soc. 19, 573–592.

Larned, S.T., Datry, T., Arscott, D.B., Tockner, K., 2010. Emerging concepts in temporary-river ecology. Freshw. Biol. 55, 717–738.

Leff, L., Van Gray, J.B., Martí, E., Merbt, S.N., Romaní, A.M., 2016. Aquatic biofilms and biogeochemical processes. In: Romaní, A.M., Guasch, H., Balaguer, M.D. (Eds.), Aquatic Biofilms: Ecology, Water Quality and Wastewater Treatment. Caister Academic Press, Poole, pp. 89–108.

Leigh, C., Stubbington, R., Sheldon, F., Boulton, A.J., 2013. Hyporheic invertebrates as bioindicators of ecological health in temporary rivers: a meta-analysis. Ecol. Indic. 32, 62–73.

Leigh, C., Boulton, A.J., Courtwright, J.L., Fritz, K., May, C.L., Walker, R.H., et al., 2016a. Ecological research and management of intermittent rivers: an historical review and future directions. Freshw. Biol. 61, 1181–1199.

Leigh, C., Bonada, N., Boulton, A.J., Hugueny, B., Larned, S.T., Vander Vorste, R., et al., 2016b. Invertebrate assemblage responses and the dual roles of resistance and resilience to drying in intermittent rivers. Aquat. Sci. 78, 291–301.

McClain, M.E., Boyer, E.W., Dent, C.L., Gergel, S.E., Grimm, N.B., Groffman, P.M., et al., 2003. Biogeochemical hot spots and hot moments at the interface of terrestrial and aquatic ecosystems. Ecosystems 6, 301–312.

McHugh, P., Thompson, R.M., Greig, H.S., Warburton, H.J., McIntosh, A.R., 2015. Habitat size influences food web structure in drying streams. Ecography 38, 700–712.

Prat, N., Gallart, F., Von Schiller, D., Polesello, S., García-Roger, E.M., Latron, J., et al., 2014. The MIRAGE toolbox: an integrated assessment tool for temporary streams. River Res. Appl. 30, 1318–1334.

Romaní, A.M., Sabater, S., 1997. Metabolism recovery of a stromatolitic biofilm after drought in a Mediterranean stream. Arch. Hydrobiol. 140, 261–271.

Sabo, J.L., Finlay, J.C., Kennedy, T., Post, D.M., 2010. The role of discharge variation in scaling of drainage area and food chain length in rivers. Science 330, 965–967.

Sheldon, F., 2005. Incorporating natural variability into the assessment of ecological health in Australian dryland rivers. Hydrobiologia 552, 45–56.

Sheldon, F., Bunn, S.E., Hughes, J.M., Arthington, A.H., Balcombe, S.R., Fellows, C.S., 2010. Ecological roles and threats to aquatic refugia in arid landscapes: dryland river waterholes. Mar. Freshw. Res. 61, 885–895.

Snelder, T.H., Datry, T., Lamouroux, N., Larned, S.T., Sauquet, E., Pella, H., et al., 2013. Regionalization of patterns of flow intermittence from gauging station records. Hydrol. Earth Syst. Sci. 17, 2685–2699.

Steward, A.L., 2012. When the River Runs Dry: The Ecology of Dry River Beds. (Ph.D. thesis) Griffith University, Brisbane.

Steward, A.L., von Schiller, D., Tockner, K., Marshall, J.C., Bunn, S.E., 2012. When the river runs dry: human and ecological values of dry rivers. Front. Ecol. Environ. 10, 202–209.

Stubbington, R., Datry, T., 2013. The macroinvertebrate seedbank promotes community persistence in temporary rivers across climate zones. Freshw. Biol. 58, 1202–1220.

Timoner, X., Acuña, V., Frampton, L., Pollard, P., Sabater, S., Bunn, S.E., 2014a. Biofilm functional responses to the rehydration of a dry intermittent stream. Hydrobiologia 727, 185–195.

Timoner, X., Buchaca, T., Acuña, V., Sabater, S., 2014b. Photosynthetic pigment changes and adaptations in biofilms in response to flow intermittency. Aquat. Sci. 76, 565–578.

Tooth, S., 2012. Arid geomorphology: changing perspectives on timescales of change. Prog. Phys. Geogr. 36, 262–284.

Tooth, S., Nanson, G.C., 2011. Distinctiveness and diversity of arid zone river systems. In: Thomas, D.S.G. (Ed.), Arid Zone Geomorphology: Process, Form and Change in Drylands. third ed. Wiley-Blackwell, Chichester, pp. 269–300.

Townsend, C.R., Hildrew, A.G., 1994. Species traits in relation to a habitat templet for river systems. Freshw. Biol. 31, 265–275.

Turner, D.S., Richter, H.E., 2011. Wet/dry mapping: using citizen scientists to monitor the extent of perennial surface flow in dryland regions. Environ. Manag. 47, 497–505.

Von Schiller, D., Acuña, V., Graeber, D., Martí, E., Ribot, M., Sabater, S., et al., 2011. Contraction, fragmentation and expansion dynamics determine nutrient availability in a Mediterranean forest stream. Aquat. Sci. 73, 485–497.

Ward, J.V., 1989. The four-dimensional nature of lotic ecosystems. J. N. Am. Benthol. Soc. 8, 2–8.

Whitney, J.E., Gido, K.B., Martin, E.C., Hase, K.J., 2015. The first to go and the last to leave: colonization and extinction dynamics of common and rare fishes in intermittent prairie streams. Freshw. Biol. 61, 1321–1334.

Zeglin, L.H., 2015. Stream microbial diversity responds to environmental changes: review and synthesis of existing research. Front. Microbiol. 6, 454.

Index

Printed in the United States
By Bookmasters